Nucleation – a Transition State to the Directed Assembly of Materials

Leeds Beckett University, United Kingdom

30 March – 1 April 2015

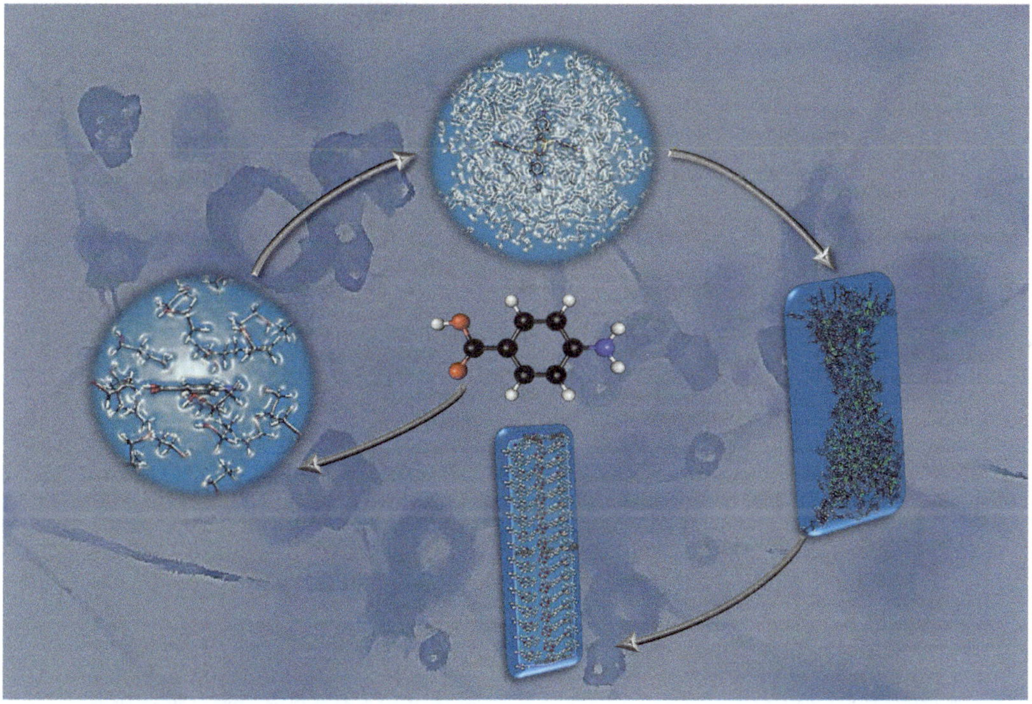

FARADAY DISCUSSIONS

Volume 179, 2015

ROYAL SOCIETY OF CHEMISTRY

The Faraday Division of the Royal Society of Chemistry, previously the Faraday Society, founded in 1903 to promote the study of sciences lying between Chemistry, Physics and Biology.

Editorial staff

Faraday Discussions (Print ISSN 1359-6640, Electronic ISSN 1364-5498) is published 9 times a year by the Royal Society of Chemistry, Thomas Graham House, Science Park, Milton Road, Cambridge, UK CB4 0WF.

Volume 179 ISBN-13: 978-1-78262-179-9

2015 annual subscription price: print+electronic £877, US $1635; electronic only £833, US $1553. Customers in Canada will be subject to a surcharge to cover GST. Customers in the EU subscribing to the electronic version only will be charged VAT. All orders, with cheques made payable to the Royal Society of Chemistry, should be sent to RSC Order Department, Royal Society of Chemistry, Thomas Graham House, Science Park, Milton Road, Cambridge, CB4 0WF, UK.
Tel +44 (0) 1223 432398; E-mail www.orders@rsc.org

If you take an institutional subscription to any RSC journal you are entitled to free, site-wide web access to that journal. You can arrange access via Internet Protocol (IP) address at www.rsc.org/ip. Customers should make payments by cheque in sterling payable on a UK clearing bank or in US dollars payable on a US clearing bank.

Printed in the UK

Faraday Discussions are unique international discussion meetings that focus on rapidly developing areas of chemistry and its interfaces with other scientific disciplines.

Scientific committee volume 179

Faraday standing committe on conferences

Nucleation – a Transition State to the Directed Assembly of Materials

Faraday Discussions

www.rsc.org/faraday_d

A General Discussion on Nucleation – a Transition State to the Directed Assembly of Materials was held in Leeds, UK on the 30th March, 1st and 2nd of April 2015.

CONTENTS

ISSN 1359-6640; ISBN 978-1-78262-179-9

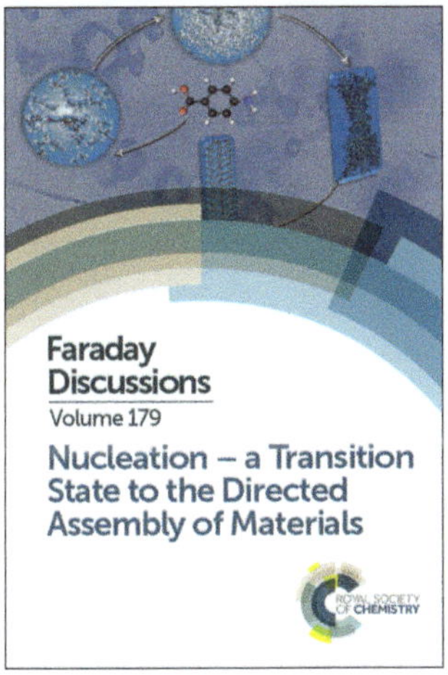

Cover
The process of PABA crystallization *via* the stages of solvated molecule, molecular clusters in solution, pre-nucleation clusters and crystal structure. Image courtesy of Dr Dimitrios Toroz, University of Leeds, UK.

INTRODUCTORY LECTURE

9 **Crystal nucleation from solutions – transition states, rate determining steps and complexity**
Roger J. Davey, Kevin R. Back and Rachel A. Sullivan

PAPERS AND DISCUSSIONS

27 **Recent advances in the understanding of two-step nucleation of protein crystals**
Maria A. Vorontsova, Dominique Maes and Peter G. Vekilov

41 **On the question of two-step nucleation in protein crystallization**
Andrea Sauter, Felix Roosen-Runge, Fajun Zhang, Gudrun Lotze, Artem Feoktystov, Robert M. J. Jacobs and Frank Schreiber

59 **High-resolution insights into the early stages of silver nucleation and growth**
Cornelia M. Völkle, Denis Gebauer and Helmut Cölfen

79 **Towards an understanding of the nucleation of alpha-*para* amino benzoic acid from ethanolic solutions: a multi-scale approach**
D. Toroz, I. Rosbottom, T. D. Turner, D. M. C. Corzo, R. B. Hammond, X. Lai and K. J. Roberts

115 **New *in situ* solid-state NMR techniques for probing the evolution of crystallization processes: pre-nucleation, nucleation and growth**
Colan E. Hughes, P. Andrew Williams, Victoria L. Keast, Vasileios G. Charalampopoulos, Gregory R. Edwards-Gau and Kenneth D. M. Harris

141 **Effect of mixing, concentration and temperature on the formation of mesostructured solutions and their role in the nucleation of DL-valine crystals**
Anna Jawor-Baczynska, Barry D. Moore and Jan Sefcik

155 **Molecular self-assembly and clustering in nucleation processes: general discussion**

199 **Measuring induction times and crystal nucleation rates**
Clément Brandel and Joop H. ter Horst

215 **Bioinspired magnetite formation from a disordered ferrihydrite-derived precursor**
Archan Dey, Jos J. M. Lenders and Nico A. J. M. Sommerdijk

227 **Chiral habit selection on nanostructured epitaxial quartz films**
Adrián Carretero-Genevrier, Martí Gich, Laura Picas, Clément Sanchez and Juan Rodriguez-Carvajal

235 **On a role of liquid intermediates in nucleation of gold sulfide nanoparticles in aqueous media**
Maxim Likhatski, Anton Karacharov, Alexander Kondrasenko and Yuri Mikhlin

247 **Time and space resolved methods: general discussion**

269 **Self-association of organic solutes in solution: a NEXAFS study of aqueous imidazole**
M. J. Thomason, C. R. Seabourne, B. M. Sattelle, G. A. Hembury, J. S. Stevens, A. J. Scott, E. F. Aziz and S. L. M. Schroeder

291 **Urea homogeneous nucleation mechanism is solvent dependent**
Matteo Salvalaglio, Marco Mazzotti and Michele Parrinello

309 **Influence of solvent on crystal nucleation of risperidone**
Donal Mealey, Jacek Zeglinski, Dikshitkumar Khamar and Åke C. Rasmuson

329 **A design equation for low dosage additives that accelerate nucleation**
Geoffrey G. Poon, Stefan Seritan and Baron Peters

343 **The many roles of mellitic acid during barium sulfate crystallization**
Matthew P. Boon, Sandra Freeman, Mark I. Ogden, Allan Oliveira, William R. Richmond, Brian W. Skelton and Franca Jones

359 **Modelling the stochastic behaviour of primary nucleation**
Giovanni Maria Maggioni and Marco Mazzotti

383 **Solvent and additive interactions as determinants in the nucleation pathway: general discussion**

421 **Do solid-to-solid polymorphic transitions in DL-norleucine proceed through nucleation?**
Joost A. van den Ende, Mireille M. H. Smets, Daniël T. de Jong, Sander J. T. Brugman, Bernd Ensing, Paul T. Tinnemans, Hugo Meekes and Herma M. Cuppen

437 **Zeolite synthesis in hydrated silicate ionic liquids**
Leen van Tendeloo, Mohamed Haouas, Johan A. Martens, C. E. A. Kirschhock, Eric Breynaert and Francis Taulelle

451 **Amorphous calcium phosphate phase-mediated crystal nucleation kinetics and pathway**
Shuqin Jiang, Haihua Pan, Yan Chen, Xurong Xu and Ruikang Tang

463 **Nucleation rate analysis of methane hydrate from molecular dynamics simulations**
Daisuke Yuhara, Brian C. Barnes, Donguk Suh, Brandon C. Knott, Gregg T. Beckham, Kenji Yasuoka, David T. Wu and Amadeu K. Sum

475 **Formation of new polymorphs without any nucleation step. Desolvation of the rimonabant monohydrate: directional crystallisation concomitant to smooth dehydration**
Baptiste Fours, Yohann Cartigny, Samuel Petit and Gérard Coquerel

489 **Localizing and inducing primary nucleation**
Zoubida Hammadi, Romain Grossier, Shuheng Zhang, Aziza Ikni, Nadine Candoni, Roger Morin and Stéphane Veesler

503 **Nucleation in complex multi-component and multi-phase systems: general discussion**

CONCLUDING REMARKS

543 **Concluding remarks**
Allan S. Myerson

ADDITIONAL INFORMATION

549 **Poster titles**

555 **List of participants**

Faraday Discussions

PAPER

Crystal nucleation from solutions – transition states, rate determining steps and complexity†

Roger J. Davey,* Kevin R. Back‡ and Rachel A. Sullivan§

Received 15th April 2015, Accepted 1st May 2015
DOI: 10.1039/c5fd00037h

This introductory paper offers a contemporary view of crystal nucleation. We begin with a molecular interpretation of the transition state and then revisit the use of classical nucleation theory as a means of obtaining molecular scale information from kinetic data. Traditional physical organic chemistry has always utilised the combination of kinetics and thermodynamics in order to gain insight over reaction pathways. Here we demonstrate for the cases of sucrose and *p*-aminobenzoic acid how solution chemistry, crystallography and kinetics come together to provide self-consistent pictures of the molecular scale processes occurring during nucleation. In this and a number of other systems desolvation of specific functional groups is highlighted as the rate determining step. Finally we move on to discuss the question of complexity, both from a phase and molecular perspective.

1. Introduction

In an obscure publication by Dunning and Shipman,[1] also available as ESI† published as part of the proceedings of the Tenth International Conference of Agricultural Industries held in Madrid in 1954, the authors considered many key features of a nucleation process. For example, they dealt with the issue of the transition state, offered a simple yet wonderfully insightful derivation of Classical Nucleation Theory (CNT) and described an experimental technique whereby they measured the nucleation rates of sucrose from aqueous solution over a range of temperatures. Using this combination of data and theory they concluded that desolvation was the rate determining step and that due to the complexity of its size and shape the sucrose molecule encountered an entropic barrier to its nucleation. Our current introductory paper pays homage to Dunning and

School of Chemical Engineering and Analytical Science, University of Manchester, Oxford Road, Manchester M13 9PL, UK. E-mail: roger.davey@manchester.ac.uk

† Electronic supplementary information (ESI) available: Copy of ref. 1. See DOI: 10.1039/c5fd00037h

‡ Current address: Pfizer Global R&D, Pharmaceutical Science, Ramsgate Road, Sandwich, Kent, UK CT13 9NJ.

§ Current address: AstraZeneca Ltd. Macclesfield, Cheshire, UK SK10 2NX.

Shipman's work and uses their themes – transition state; nucleation theory; experimental methods; rate determining steps and complexity – as a basis for the structure of this paper.

Of course Dunning and Shipman did not produce their work in a vacuum. They relied on the results of a century of study of crystallisation processes. In this world Friedrich Wilhelm Ostwald had been 'king', establishing many of the key aspects of crystallisation that we take for granted: the importance of the phase diagram in defining crystallisation conditions, the role of supersaturation in the control of nucleation, the concept of the metastable zone and of course his Rule of Stages relating to the appearance of crystal forms in systems with multiple solid phases. This was largely summarised in his 1897 paper[2] and it brought to an end a period during which crystallisation had been viewed as a largely uncontrollable and irreproducible process. However, it was not until 1926 [3] that Max Volmer developed a kinetic theory describing the relationship between supersaturation, interfacial tension and the rate of nucleation, initially for drops but later adapted in 1934 [4] by Kaischew and Stranski to cover the case of crystal nucleation. Overall this kinetic theory describes rather well the experimental observations of Ostwald and remains a major achievement in the field.

Whether or not these two scientists, Volmer and Ostwald ever met to discuss the questions in hand is unknown. Ostwald was born in 1853 and was Volmer's senior by some 32 years. By the time Volmer began his career in Berlin in 1916, Ostwald had been retired from Leipzig University for ten years. On chronological grounds a meeting seems unlikely. A search of the internet offered no further comment on this possibility but rather surprisingly did reveal a PDF (English translation) of the 1939 book[5] written by Volmer and entitled 'Kinetics of Phase Formation'. Although this book is frequently referenced we had never seen it before – it is illuminating. Reading Volmer's historical introduction seems to confirm that the two had never met. Of Ostwald he notes '*Ostwald's importance to the field rests upon the order and clarity he introduced into the chaotic multiplicity and confusion of existing data*'. On the question of the Rule of Stages he is critical, claiming that Ostwald ignored exceptions to the Rule revealing his prejudice in its favour. However it is his comment on Ostwald's response to the work of Gibbs that seems to confirm that the two never met. On metastability Volmer comments that '*the only valuable theoretical work remains completely unknown. It is to be found in the second part of Gibbs' great treatise 'On the equilibrium of heterogeneous substances' and is conveniently accessible in W. Ostwald's well-known German translation 'Thermodynamische Studien'. It is strange that Ostwald who was especially interested in the subject and who must have thoroughly penetrated the meaning of the work in the course of translating it, should have failed to recognise its importance. In his textbook of general chemistry which contains a brief summary of Gibbs' works, he merely says 'Passing over a few sections of more abstract subject matter concerned with the possibility of formation of a new phase within a homogeneous liquid… etc.*". Thus Volmer offers Ostwald a metaphorical 'slap on the wrist' but ironically of course he (presumably) relied on Ostwald's German translation of Gibbs work to enable him to develop what has now become CNT.

In recent years interest has been aroused by the possibility that CNT needs to be modified following the experimental discovery of pre-nucleation clusters, mesostructured solutions and liquid phase precursors to crystal nucleation. We do not discuss these issues but recognise that they are well covered in this discussion by the papers of Vorontsova *et al.* (DOI: 10.1039/c4fd00217b), Sauter *et al.* (DOI: 10.1039/c4fd00225c), Völke *et al.* (DOI: 10.1039/c4fd00269e) and Jawor-Baczynska *et al.* (DOI: 10.1039/c4fd00262h).

In what follows we will deal firstly with the contrasting physics and chemistry approaches to the transition state. Secondly we will say something about the relationship of this approach to CNT. As part of this we look briefly at experimental methods used and reprise Dunning and Shipman's conclusions in their interpretation of sucrose nucleation. One significant issue discussed is the extent to which solution and solid state chemistry can help us to formulate a molecular picture of the nucleation transition state and for the case of *p*-aminobenzoic acid we see how kinetic and transition state ideas come together to provide a self-consistent interpretation of data. We then move into the arena of nucleation in complex systems – complex in the sense of both molecular structure and phase behaviour.

2. The transition state

Dunning and Shipman's free energy co-ordinate diagram is reproduced here as Fig. 1. As physical chemists we are familiar with this simple form of the energy-reaction landscape, separating, as it does, reactants and products *via* a transition state situated at the highest point of the curve. In the case of nucleation this represents the passage from a metastable (supersaturated) solution over an activation barrier to the final state in which crystals are present. Dunning and Shipman equate the transition state to a cluster of molecules which CNT identifies as the critical-sized cluster (or embryo). They write '*However, there will be a certain sized embryo, the solubility of which is just equal to the concentration of the supersaturated solution. Embryos smaller than this critical size tend to redisperse – and those larger tend to grow. This critical size is the bottleneck of the process and corresponds to the activated state in ordinary kinetics*'. Versions of Fig. 1 have been used by other authors in attempting to describe the nucleation process – in their 1999 paper on concomitant polymorphism Bernstein *et al.*[6] used a similar figure to describe crystallisation in polymorphic systems and in 2002 Desiraju[7] used a version in his 'Cryptic Crystallography' commentary in Nature Materials to distinguish kinetic and thermodynamic products in crystallisation. In both these papers the transition state was again defined as the critical cluster of molecules. This notion of the transition state as a supramolecular assembly having the

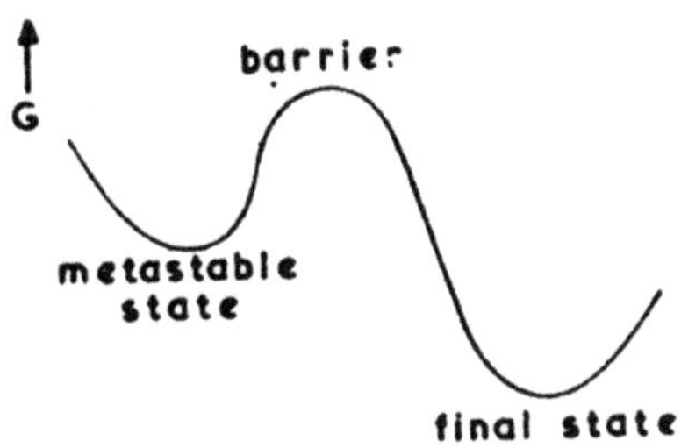

Fig. 1 The reaction co-cordinate – free energy landscape as visualised by Dunning and Shipman.[1]

packing of the final crystal structure derives from Volmer's physics approach and while successful in terms of deriving a kinetic theory it tells us nothing of the essential chemistry of the transition. In reaction chemistry, where molecules are built from covalent bonds, the idea of the transition state has led to an understanding of reaction rates based on molecular properties. The approach of the chemist is nicely summarised in Kirby's 1994 review:[8] '*Transition states are by far the most difficult of the relevant species which are experimentally accessible. Formally they can only be studied kinetically, because their only experimental manifestation is their effect on the rate of the reaction. A transition state, which lies by definition between the starting materials and products for a particular step of a reaction, is supposed to be closer in structure, because closer in free energy, to the higher energy species of the two... Crystal structures give us information specifically about ground states, averaged over a period of hours or even days; whereas transition states for bond making or breaking, as defined by transition state theory, do not exist for any significant length of time*'.

Hence the critical nucleus as a transition state poses a conundrum – while it does indeed represent a maximum in the free energy-size co-ordinate, it is also a reflection of the ground state packing of the product crystals. If on the other hand we consider the formation of non-covalent interaction to be nucleation's equivalent of covalent bond formation in reaction chemistry then for nucleation in solution it would surely seem reasonable to define the solvent as the leaving group in this reaction that sees two solute molecules come together either to form a dimer or to enlarge an existing cluster. A transition state viewed in this way will be closer in structure to the supersaturated state and might be approximated by the following intellectual process. Starting from knowledge of the first solvation shell of the solute in solution, one solvent molecule is removed and replaced with a partially solvated solute molecule so as to create an intermolecular contact that is present in the ground state crystal structure. This may be thought of conceptually as the first of many such steps leading to a critically-sized cluster. However, it leads our thinking away from concerns over the size, packing and surface free energy of a cluster and forces us to focus on the essential chemistry that must be performed before a cluster can emerge. It answers questions such as – which might be the most difficult solvent interactions to break? Can I choose a solvent that changes the chemistry and hence the transition state? Can I direct the formation of the solute–solute interaction that I want and hence control crystal structure? We have considered such questions before but never in such a formal way.[9,10] Here we consider three examples. In each case the ground state crystal structure is known from crystallography while the solution chemistry essential for this process comes from a number of different sources and is, where possible, supported by known structures of relevant solvated crystal forms.

2.1. Trimesic acid crystallising from DMSO

The case of trimesic acid (TMA) crystallising from DMSO has been discussed previously.[11] In this system there are two known solvates, the TMA·3DMSO trisolvate (XASFAA,[11]) which appears as a metastable phase with a lifetime of about 1 hour and the ultimate product of such a crystallisation, the monosolvate TMA·DMSO (TMADMS01,[12]). In the context of a transition state both of these structures are essentially ground states but we will consider the more stable monosolvate to be the true ground state in this system. In the trisolvate the DMSO is disordered and there are no acid–acid hydrogen bonds between TMA molecules. Indeed, in terms of FTIR the carbonyl environment in this trisolvate is identical to that in the solution from which it crystallises.[11] For this reason we take the crystal structure of the trisolvate as an approximate model for the supersaturated solution and this gives the TMA with its first solvation sphere as seen on the left hand side of Fig. 2. To transform this arrangement into the monosolvate by the addition of a second TMA molecule requires desolvation of one acid group and the creation of a new –C=O···H–O– hydrogen-bonded contact with an incoming TMA molecule. This creates the image of the transition state as seen in the centre of Fig. 2. Repetition of this process can then lead to a cluster having the packing of the ground state, monosolvate structure.

2.2. Hexamethylenetetramine crystallising from aqueous solutions

Hexamethylenetetramine (HMT) crystallises in a body centred cubic lattice from aqueous solutions (HXMTAM07,[13,14]). This is shown as the ground state on the right hand side of Fig. 3. There are no known polymorphs but a hexahydrate exists at temperatures below 13.5 °C (HXMTHH,[15]). The co-ordination of HMT in aqueous solutions is known rather precisely[16] from neutron scattering studies in the form of radial distribution functions. Thus in Fig. 3 we show on the left hand side the solution co-ordination *via* the spatial distribution of both HMT around HMT and the water oxygen (Ow) around HMT. It is clear that closest to a central HMT are water molecules, four of which are H-bonded to nitrogens and 4 interacting with the faces of the HMT cage. Beyond these eight that comprise the solvation shell are 4 HMT molecules poised ready to form –C–H···N– contacts as in the crystal structure. Fig. 3 then utilises the similarity in structure between the hexahydrate and the solution phase solvation shell to create a central image in which a leaving water molecule allows the formation of the strongest HMT–HMT contact yielding the transition state. It appears that in this case the solution co-ordination of HMT favours the creation of the ground state packing, a view supported by previously measured nucleation kinetics which show significant

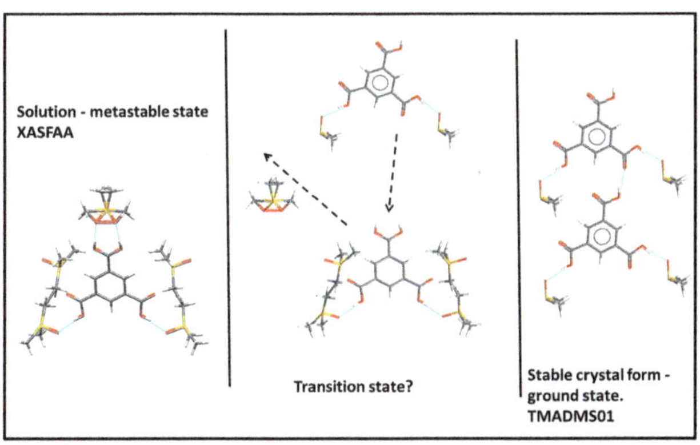

Fig. 2 The creation of a potential transition state (centre) from a combination of the solution co-ordination (left) and the ground state crystal structure (right). The case of trimesic acid and dimethylsulphoxide.

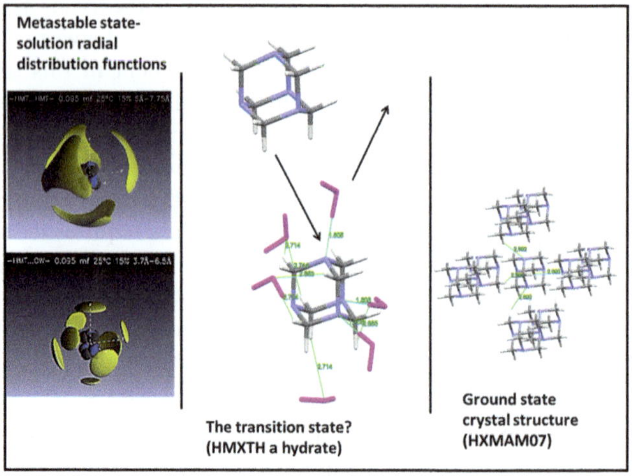

Fig. 3 The creation of a potential transition state (centre) from a combination of the solution co-ordination (left) and the ground state crystal structure (right). The case of HMT and water.

rates even at extremely low supersaturations with almost no thermodynamic dead zone.[14]

2.3. 5-Fluorouracil crystallising from water and nitromethane

In the case of 5-fluorouracil two polymorphs are known. Form I crystallises from water while Form II is prepared from dry nitromethane. Evidently even small amounts of water are sufficient to switch the outcome from Form II to I.[17] In order to understand this subtle effect Hamad *et al.*[18] performed molecular dynamics studies which linked the solution co-ordination of 5-fluorouracil to the polymorphic outcome. We have used these data as the basis for Fig. 4. This takes the

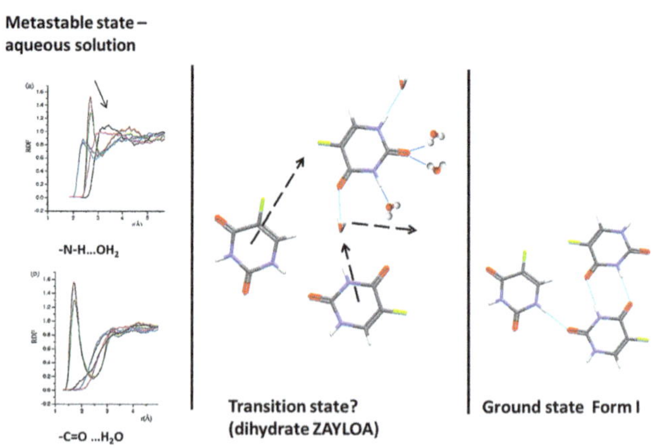

Fig. 4 The creation of a potential transition state (centre) from a combination of the solution co-ordination (left (taken from ref. 18 and reproduced with kind permission of the American Chemical Society)) and the ground state crystal structure (right). The case of 5-fluorouracil and water.

radial distribution functions computed in water together with the water positions observed in ZAYLOA (a hydrated co-crystal of 5-fluorouracil and theophylline,[19]) and the ground state Form I to create a transition state. In this case, following the reasoning of Hamad *et al.*[18] it appears that the transition is driven by the combination of the hydrophobic –F···F– interaction between neighbouring molecules and the difficulty of removal of water from the carbonyl and amine groups. This latter prevents the formation of the doubly H-bonded ribbon seen in Form II – this can only be formed by changing the nature of the transition state *via* the solvent. Thus, the use of dry nitromethane, which cannot form strong H-bonded interactions with the carbonyl groups of 5-fluorouracil, enables amide dimers to form the extended H-bonded ribbons of Form II.

In these three examples we have utilised the combination of solution and solid state chemistry in order to create an impression of a potential transition state. This appears to be a promising approach in that it directs our thinking away from questions of clusters and packing (which we can never resolve) towards the essential chemistry of the processes. It is evident that the NMR and NEXAFS techniques as reported in papers by Hughes *et al.* (DOI: 10.1039/c4fd00215f) and Thomason *et al.* (DOI: 10.1039/c5fd00005j) may have much to contribute to this topic. In the next section we take these ideas a step further by combining such structural insights with kinetic data.

3. Kinetic data and rate determining steps

Dunning and Shipman employ a direct approach to the measurement of nucleation rate. Using a specially constructed counting cell, solutions were subjected to different supersaturations and temperatures and the number of crystals appearing with time counted manually. These data yielded directly the nucleation rate. In more recent times workers have sought to perform measurements that reflect more directly the stochastic nature of nucleation. Turnbull[20] was probably the first to use liquid droplets to observe nucleation of metals and the 1978 paper of Keller *et al.*[21] describes very well how such a technique may be modified using emulsions to provide nucleation rates. Essentially many, many drops are kept under identical conditions and the probability that a drop contains a crystal is measured. From such data the nucleation rate may be obtained. More recently the use of emulsions has been superseded by high and medium throughput techniques using well plates,[22] microfluidic devices,[23] (see also in this discussion the paper by Hammadi *et al.* DOI: 10.1039/c4fd00274a) and induction time measurements,[24] (see also in this discussion the Brandel and ter Horst paper DOI: 10.1039/c4fd00230j). Whatever the methodology, the aim is to obtain *isothermal* nucleation rate *versus* supersaturation data to confront with an appropriate kinetic model. For nucleation the only model available is CNT[25] which gives a relationship between nucleation rate, J and supersaturation ratio, S of the form:

$$J = A_0 S \ln S^2 \exp\left(-\frac{B}{\ln S^2}\right) \tag{1}$$

The pre-exponential factor, A_0, is related to: z, the Zeldovich factor; f^*, the molecular attachment frequency of building units to a growing nucleus; and C_0, the concentration of nucleation sites. Both z and f^* are supersaturation dependant and using Kashchiev's evaluation of z[26] and realising that f^* is a linear function of supersaturation this leads to:

$$\ln \frac{J}{S \ln^2 S} = \ln A_0 - \frac{B}{\ln^2 S} \tag{2}$$

A plot of $\ln(J/S \ln^2 S)$ against $\ln^{-2} S$ should be linear. The slope yields the thermodynamic factor, B, from which the effective interfacial tension can be derived while the intercept gives $\ln A_0$. The product f^*C_0 can then be determined from the relationship:

$$f^*C_0 = \frac{A_0}{z} S \ln^2 S = \sqrt{12\pi B} A_0 S \tag{3}$$

Dunning and Shipman use their measured data to obtain the temperature dependence of A_0 from which they estimated the activation energy of the kinetic process to be 67 kJ mol^{-1}. This was attributed to the transport of a sucrose molecule from solution onto the nucleus and concomitant removal of four water molecules (*i.e.* 67 kJ mol^{-1} $\sim$ 4 hydrogen bonds). Such a desolvation process might be expected to be reflected in the aqueous solubility and it turns out that there is in fact a good linear correlation (not shown) between $\log A_0$ and the free energies of solvation $RT \ln X_s$ calculated using measured values of X_s, the aqueous mole fraction solubility of sucrose. Combining these data and interpretation with the known crystal structure of sucrose (SUCROS08,[27]) suggests that the process shown in Fig. 5 is the rate determining step in sucrose nucleation. Here the necessity to remove a total of 4 water molecules in order for a molecule to join a cluster along its *b*-axis is evident, matching rather well Dunning and Shipman's suggestion.

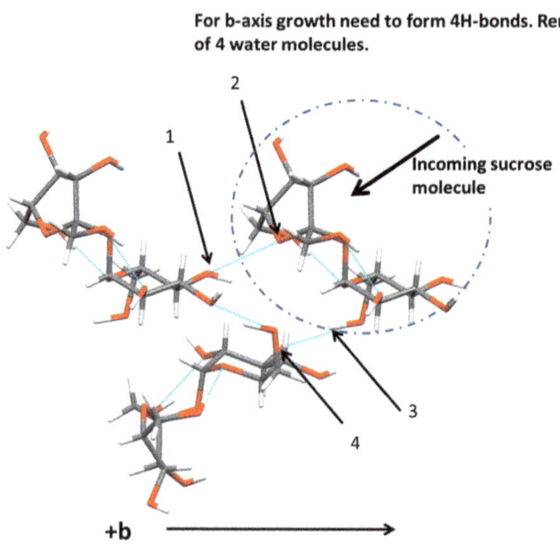

Fig. 5 The removal of 4 water molecules – the rate determining step in sucrose nucleation. The arrows and numbers highlight the 4 groups likely to be solvated by water molecules.

3.1. The nucleation of *p*-aminobenzoic acid (*p*ABA)

Overall given access to reliable and comparable kinetic data it seems that if we follow the kinetic rigour of Dunning and Shipman, together with additional insight into solution chemistry, it should be possible to define the rate determining step and to visualise a potential transition state which is consistent with both structural and kinetic factors. In this section we reprise and extend the recent report of Sullivan *et al.*[28] concerning the solvent dependent nucleation of the α polymorph of *p*ABA. This is a structure based on the carboxylic acid dimer (AMBNAC07,[28]). The induction time method was used[24] to measure nucleation rates at 20 °C in 3 solvents, 2-propanol, ethylacetate and acetonitrile and following eqn (1)–(3) the supersaturation dependence of f^*C_0/M was extracted. The authors made one major assumption that for data measured by the same researcher, in the same laboratory and using the same equipment C_0, the number of (heterogeneous) nucleation sites, is essentially fixed so that relative values of f^*C_0/M reflect the solvent effect on f^*, the attachment frequency of a molecule to a cluster. The results showed that this frequency was lowest in 2-propanol and highest in acetonitrile and these data are shown in Fig. 6 along with associated macroscopic crystal growth rate measurements. The latter are important since they provide an effective means of independently checking the relative magnitudes of the derived attachment kinetics since growth of a crystal is a macroscopic reflection of the growth of clusters. α-*p*ABA crystals grow as *b*-axis needles with their width corresponding to the *a*-axis.[29] The growth rate data measured on single seed crystals at 20 °C in a temperature controlled growth cell (Toroz *et al.*, DOI: 10.1039/c4fd00275j) are shown for both growth directions (*a* and *b*). It is clear that the relative order of growth rates with solvent follows the same pattern as the nucleation data with growth slowest from 2-propanol and fastest from acetonitrile. The growth curves show significant dead zones at low supersaturations suggesting surface nucleation controlled growth and their size show a similar solvent dependence. Taken together these data provide clear evidence that the kinetic factors in nucleation are controlled by the solvent choice. The available solution chemistry sheds further light on this. FTIR studies of the carbonyl group

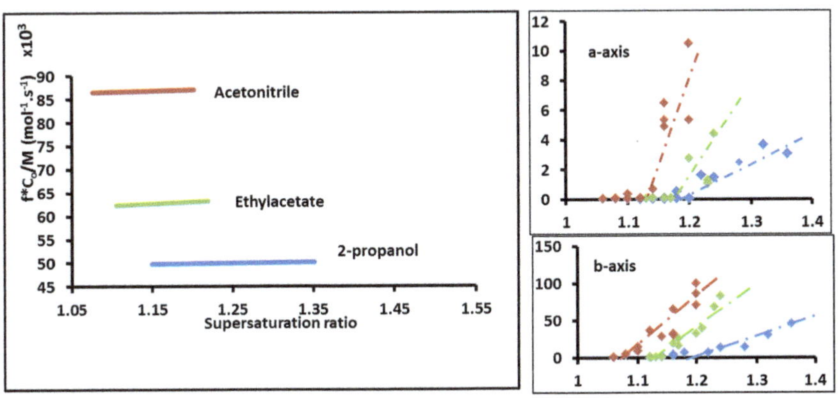

Fig. 6 Nucleation attachment frequencies (left) crystal growth rates (right, vertical axes) both as functions of supersaturation in the three solvents, 2-propanol (blue), ethylacetate (green) and acetonitrile (red). Growth rates are μm per minute.

in ethanol solution show two bands (1685 and 1715 cm^{-1}) having relative intensities which do not change with concentration in the range 0.056–0.9 M. In agreement with available neutron scattering data for benzoic acid in methanol,[30] this confirms that there are no carboxylic acid dimers in such solutions. Equivalent FTIR studies in acetonitrile show (1680 and 1710 cm^{-1}; 0.0012–0.38 M) that with increasing concentration the relative intensity of the 1710 : 1680 bands decreases from 15 to 1.4. This is a clear indication that the lower of the bands relates to a solute–solute interaction while the higher band is due to solvation. In ethylacetate interpretation of the spectra are hampered by overlap with the solvent. Related calculations of the solvation free energies (using a thermodynamic integration technique[28]) of both monomer and dimer in the three solvents confirmed the experimental result showing that in 2-propanol the solvated acid group was more stable than the dimer; in ethylacetate the two species seem finely balanced and in acetonitrile solvated dimers were slightly more stable than monomers. These results taken together suggest desolvation as the rate determining step in the growth of clusters. Plots of the relative values of f^* in the three solvents show a poor correlation (Fig. 7, left) with overall solvation as reflected in the relative solubilities ($\ln X_s$) but a good correlation (Fig. 7, right) with the calculated free energies for specific desolvation of the acid group. Hence we conclude that desolvation of the acid group and formation of acid dimers is the slowest, rate determining step in the nucleation pathway. This is certainly consistent with the relative growth rates of the slowest growing *a* direction since it is in the direction that dimers are created at the growing surfaces.[29]

Taking this information together and using the previously measured radial distribution functions for benzoic acid in methanol as a model of the solution we arrive at Fig. 8, a potential transition state for nucleation of *p*ABA from 2-propanol.

As a corollary to this insight, in their original work Sullivan *et al.*[28] argued that if indeed this was the transition state then the activation barrier for attachment could be lowered by use of a solvent in which the solute was pre-assembled (dimerised). Taking benzoic acid in toluene (*p*ABA is insoluble in non-polar solvents) they showed (FTIR) that these solutions contained preformed acid dimers and that the value of f^*C_0/M was almost two orders of magnitude higher than in *p*ABA in 2-propanol. This result showed, perhaps for the first time, that directed self-assembly is indeed possible through solvent selection.

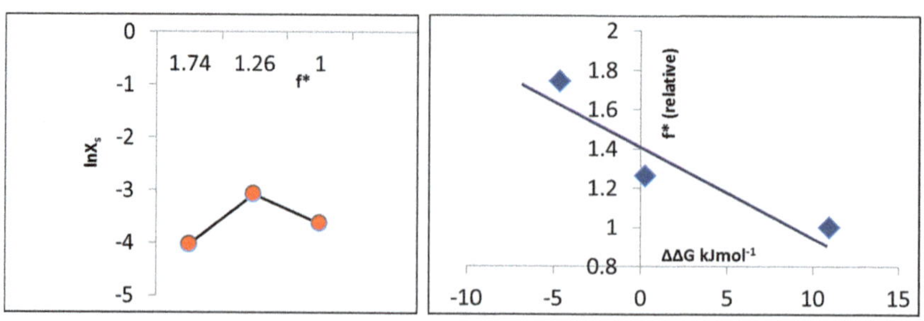

Fig. 7 Correlations of relative values of f^* for *p*ABA in different solvents with solvation parameters. Left with solubility and right with the solvation free energy of the carboxylic acid dimer.

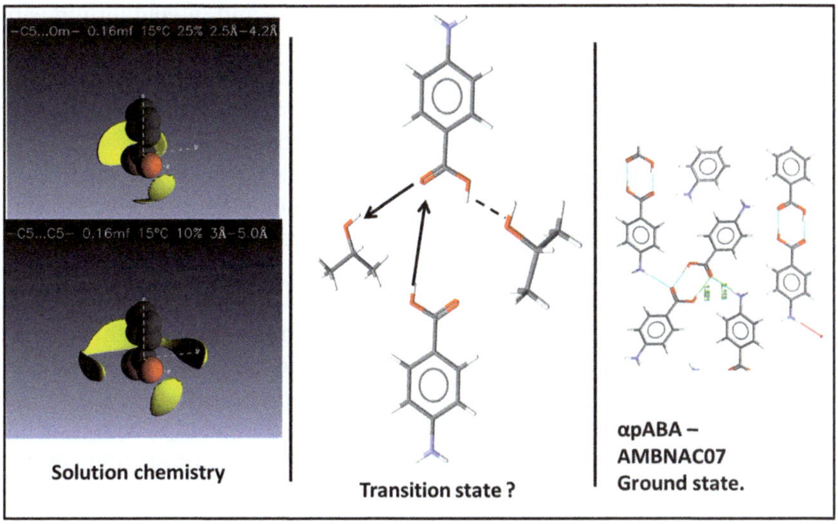

Fig. 8 A potential transition state for *p*ABA nucleating from 2-propanol (centre). Solution co-ordination, left, and ground state α polymorph, right.

A similar importance of desolvation as the rate determining step in nucleation has further been suggested by recent studies of salicylic acid[31] and riseridone (Mealy *et al.*, DOI: 10.1039/c4fd00223g). In both cases although the rate of nucleation has been related to the chosen solvent it does not correlate with overall solubility but rather with strong solvent binding at a specific site on the solute molecule. In the case of salicylic acid this implicates desolvation of the carboxylic acid group as the rate determining step while for risperidone it is the carbonyl group.

4. Complexity

There are a number of possible sources of complexity in crystal nucleation. Dunning and Shipman were most concerned with entropic effects. We discuss these here in the context of molecular conformation but mention first complexity originating from phase behaviour.

4.1. Phase complexity

Clearly polymorphism would be an obvious example of complex nucleation behaviour in which the pathway can follow more than one route.[32] We shall not discuss this issue further here, rather we concern ourselves with nucleation in ternary systems since these are of significant interest from the perspective of chiral separation, co-crystal discovery and molecular salt selection. We note that the famous Etter rules[33] for order of hydrogen bond formation are derived partly from experiments performed in ternary systems and yet we are not aware of any related, detailed nucleation studies. Three reports are worthy of mention, all involving cooling crystallisation. Of particular interest is the outcome of crystallisation experiments performed within the 3 phase region(s) of the ternary diagram where two solids phases are of equal stability and in equilibrium with solution having a eutectic (invariant) composition. Firstly, the case of mandelic acid is recalled which, in the ternary system $R/S/H_2O$, can crystallise either as pure enantiomers or as a stable (or metastable polymorph) racemic compound.[34] Secondly we mention carbamazepine/saccharin/methanol which crystallises as pure components together with a stable and metastable polymorph of the 1 : 1 co-crystal,[35] and thirdly the case of *p*-toluenesulfonamide (TSA)/triphenylphosphineoxide (Ph_3PO)/acetonitrile which crystallises as pure components together with 1 : 1 and 3 : 2 co-crystals[36] is discussed.

For mandelic acid[34] ternary compositions were selected to give solutions at 40 °C which on cooling to 25 °C allowed crystallisation within the three phase region (*S*/*RS*/solution). In this case the enantiomeric excess of the crystallised solids was measured as a function of time and the initially nucleated product was found to be rich in pure *S* suggesting that nucleation of *S* and (*R*,*S*) crystals is not concomitant but rather the first phase to be nucleated is the pure enantiomer, not the racemic compound. Only at longer equilibration times does the enantiomeric excess fall to the expected value.

In the carbamazepine–saccharin case the outcome of crystallisations[35] performed at constant solvent loadings but varying solute ratios was reported. In the two phase regions as expected only the pure forms appear. In three phase regions however, as the ternary diagram is traversed from carbamazepine-rich to saccharin-rich then close to the boundaries with the two phase regions only the pure co-former nucleates. In moving towards the co-crystal only region the appearance of the stable and metastable co-crystal polymorphs appears to depend on composition. Thus in moving across the ternary diagram at fixed solvent level we find that, although the phases all have the same stability in the three phase regions, kinetics favours the closest pure form. This behaviour is essentially mirrored in the (TSA)/(Ph_3PO)/acetonitrile system,[36] which is complicated further by the existence of both 3 : 2 and 1 : 1 co-crystals. In traversing the pure TSA/3 : 2 co-crystal region from left to right, the proportion of co-crystal in the product increases. In the region where Ph_3PO and the 1 : 1 co-crystal are stable it seems to be possible to nucleate mixtures which include the metastable 3 : 2 co-crystal along with pure Ph_3PO and the 1 : 1 co-crystal. Overall it appears that in these complex systems nucleation continues to occur stepwise with kinetic processes driving the outcomes so that even in regions where two phases have equal stability adjustment of composition can yield non-equilibrium mixtures.

4.2. Molecular complexity

Dunning and Shipman foresaw the possibility that because sucrose is a '*large and complicated molecule that it might have difficulty 'dovetailing' into the growing cluster. This might give rise to unexpected entropy effects*'. Of course entropic effects will arise for a number of reasons[37] associated with enhanced molecular mobility in the liquid phase. Most recently this issue has been considered specifically in the context of conformational flexibility which is seen as a potential constraint on the ease of crystallisation of large molecules.[38,39] Here we discuss one specific case of the molecule shown in Fig. 9. This is a pharmaceutical intermediate (3*S*)-oxolan-3-yl *N*-[(2*S*,3*R*)-3-hydroxy-4-[*N*-(2-methylpropyl)(4-nitrobenzene)sulfonamido]-1-phenylbutan-2-yl]carbamate; ethanol, which will be abbreviated to ONSC-

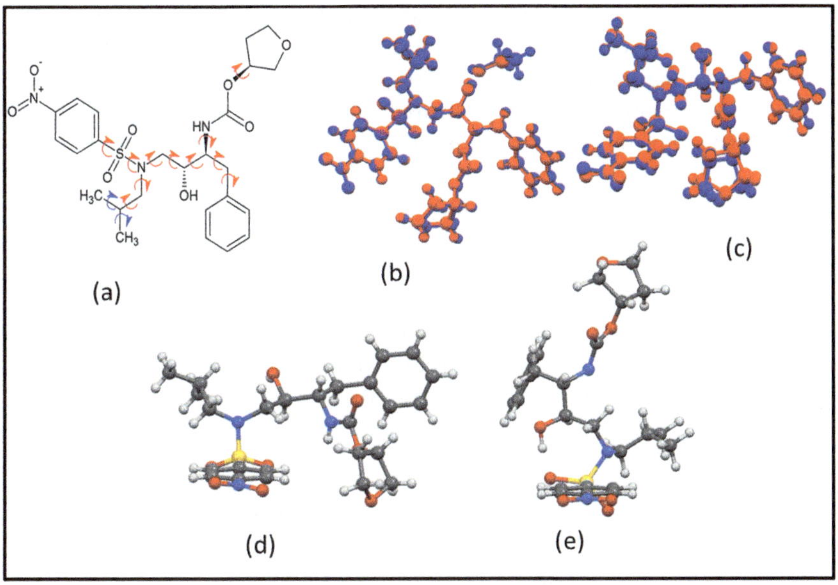

Fig. 9 (a) The molecular structure of ONSC with rotatable bonds highlighted in red. (b) An overlay of ONSC ethanolate (blue) and methanolate (red) conformations. (c) An overlay of ONSC ethanolate (blue) and hemihydrate (red) conformations. (d) ONSC-E crystal structure conformation. (e) Lowest energy structure from a Monte Carlo search.

E. In terms of complexity this molecule has at least 11 rotatable bonds and three chiral centres. If the general assumption is made that each rotatable bond has 3 low energy configurations and that they alter independently, then the theoretical maximum number of conformers for this molecule is $3^{11} = 177,147$. Obviously its potential energy surface is not accessible in the normal way by considering each bond in turn. As an alternative means of exploring conformational space a Monte Carlo search method was used together with the MM+ force field to generate a list of possible low energy conformations.[40] These were further optimised using *ab initio* calculations in Gaussian with each geometry being optimised at the HF/STO-3G level, followed by a single point energy calculation carried out at the B3LYP/6-31+G(d) level. This provided a list of 2611 conformations which were further reduced by the removal of duplicates to give a final list of 1850. From the lowest energy conformation only one other conformation was found to be within 2.5 kJ mol^{-1} and hence, at least in vacuum, the population will be highly biased towards these two conformers, neither of which match that in the crystal structures discussed below.

In terms of crystallisation ONSC is readily crystallised from solvents. In this study, acetone, ethylacetate, ethanol, methanol, 1-butanol, diethylether, 1-hexanol, acetonitrile, chloroform/ethanol, and ethylacetate/ethanol were all tested. Crystals obtained were needle-like with typical widths of 10 μm. Overall it proved extremely difficult to grow crystals wider than this. In cases where suitable crystals for XRD were recovered the crystal structure was always found to be solvated. Thus for example from ethanol/chloroform an ethanol solvate was obtained; from methanol a methanolate and from acetonitrile a hemihydrate.

The ethanolate and methanolate are isostructural, being channel solvates. The conformation of ONSC-E in the three crystal forms is compared in Fig. 9 with that of the lowest energy gas phase conformation discussed above. It is noted that in the crystal the –OH group of ONSC-E hydrogen bonds to the solvent whilst in the low energy vacuum conformer it forms an intra-molecular H-bond with the $-SO_2$ group.

Detailed crystallisation and solubility studies in the ternary system ONSC-E/ethylacetate/ethanol allowed the phase diagram in Fig. 10 to be determined at 25 °C. This confirmed the lack of a non-solvated form and defined the composition of stability of both the hemihydrate and the ethanolate. The temperature dependence and metastable zone width (MSZW) of the ethanolate in 3.7 : 1 ratio (v/v) of ethylacetate to ethanol solvent is also shown in Fig. 10. It is evident from these data that the stability domain of the ethanolate is very large and that its

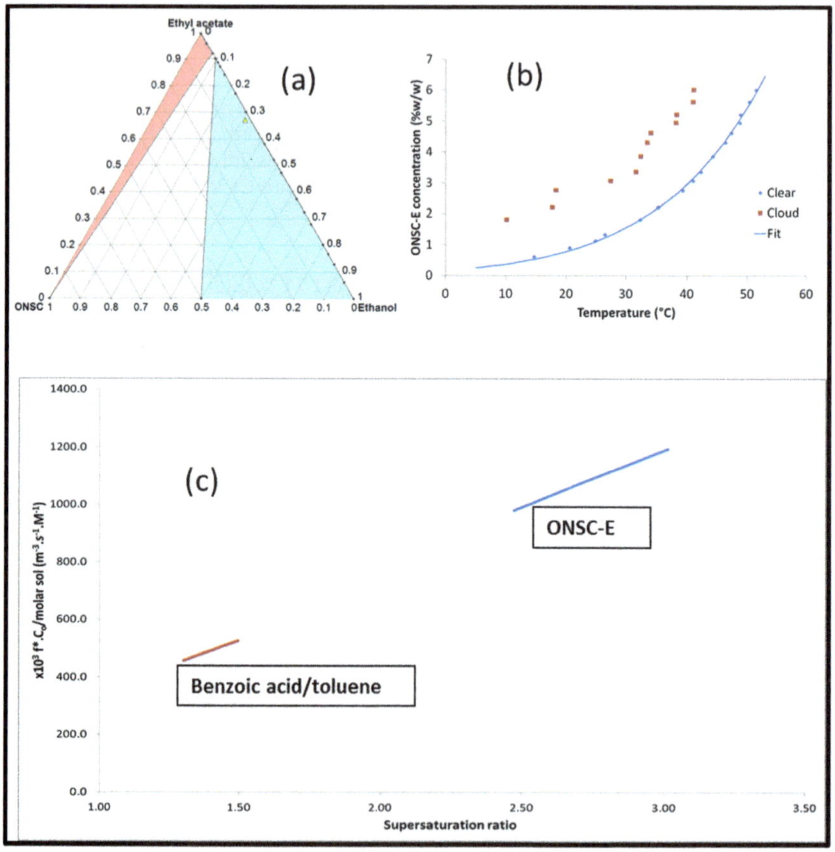

Fig. 10 (a) Phase diagram for the ternary system ONSC–ethylacetate–ethanol at 25 °C (mole fraction). The blue section highlights the region where the ethanol solvate is the stable form crystallised, and the red section where the hemihydrate is obtained. The yellow triangle indicates typical crystallisation conditions for the ethanolate in the system. (b) Solubility and MSZW of ONSC-E in the 3.7 : 1 (v/v) ethylacetate to ethanol system measured on the Crystal16. (c) The attachment frequency of ONSC-E nucleating from ethanol/ethylacetate solutions. Benzoic acid data is included for comparison.

crystallisation behaviour as evidenced by the MSZW data is highly regular and in no way abnormal. An attempt was made to measure the growth rate of a crystal in its needle direction (*a*-axis). A growth rate of 0.3 $\mu m\ min^{-1}$ was obtained at a supersaturation ratio of 3.46. It is noted that this is significantly slower than, for example, the *b*-axis growth rate of α-glycine, which grows at 61 $\mu m\ min^{-1}$ (ref. 41) at the lower supersaturation ratio of 1.58.

The nucleation kinetics of the ethanolate were determined at 15 °C using the induction time method[23] and the Crystal16 at supersaturation ratios (x/x_{sat}) of 2.47, 2.55, 2.68, 2.78, 2.94 and 3.01. The fitted growth times form these data were between 3000 and 6000 s. These are relatively long compared with benzoic acid (12 s) and this is consistent with the observation of unusually low growth rates of single crystals. Again using eqn (1)–(3) it was possible to deduce the value of f^*C_0/M (though correlations were poorer than in the *p*ABA case above due to the volatility of the solvent giving rise to crowning effects during measurement). As seen in Fig. 10 these appear to be of the same order of magnitude as benzoic acid. The effective interfacial tension was found to be 2 $mJ\ m^{-2}$, typical for such a low solubility system. Thus, despite the apparent number of degrees of freedom of this molecule its crystallisation behaviour, as a solvate, shows little out of the ordinary – its metastable zone is well characterised and its nucleation behaviour appears in line with benzoic acid and *p*-aminobenzoic acid. There is one major exception to this in that while its nucleation characteristics are normal, its growth rate appears to be exceptionally low. Whether or not this is in some way linked to conformation is yet to be established. What is clear, however, is that much of this regularity of behaviour arises from the role played by the solvent in moderating the conformation of the molecule. The lowest energy conformer found by computation does not appear in a crystal structure and no crystals of any non-solvated form were ever seen. In the context of this paper it might be concluded that the transition state for nucleation involves such favourable solvent binding that the result is always a solvate. It would appear that such is the stability of this solute–solvent dimer that conformation is, surprisingly, not an issue during nucleation; it only becomes a problem during growth.

5. Conclusions

Here we have suggested a way of combining data from different sources to give insight into the chemistry of nucleation – the processes which must happen in order for a molecule to move from its solvent environment into a cluster. Re-examination of the sucrose data of Dunning and Shipman reminds us of the role that carefully measured *isothermal* kinetic data can play in developing an understanding of nucleation processes. We have shown how all the structural and kinetic factors can come into play in the case of *p*ABA. This leads ultimately to the demonstration of the key determining role played by desolvation of specific functional groups and of the use of solvent selection as a way of achieving pre-assembly and enhanced nucleation attachment frequencies. Complexity comes in many forms and we have dealt here with two types that feel most relevant in today's climate – phase complexity and molecular conformations. Clearly little is really known about competitive nucleation in ternary systems where both pure phases and compound solids can form. Given the commercial importance of molecular salts and co-crystals this may be an area worthy of pursuit. As far as molecular conformation is concerned this is high on the agenda as an issue relating to the development of drugs with ever increasing molecular weights. However, there have been few studies in which problems of nucleation or crystal growth of relevant molecules have been reported and in the example of ONSC-E described here it seems that despite its 11 torsions it finds a way to crystallise using the solvent as a 'conformational' auxiliary.

Reflecting on Dunning and Shipman's 1954 study of nucleation in aqueous sucrose solutions it seems to us that their work showed remarkable foresight in dealing with many issues – the transition state, rate determining steps, molecular complexity – which remain highly relevant sixty years later. At the same time we are reminded of the strong tradition of physical chemistry in which kinetic measurements play a critical role in understanding the nature of transformative pathways, be they related to the creation of covalent chemical bonds or of packing interactions in crystalline products. Elucidation of the rate determining step can only be achieved through the combined use of kinetic and structural data. Today with the access to new, high throughput experimental possibilities for measuring nucleation rates, routine probes for exploring solution chemistry and the ever expanding crystal structure database (CSD), progress in unravelling the secrets of nucleation seems inevitable.

Finally, in praise of nucleation we offer a concluding Haiku.[42] A snapshot in 17 syllables:

From chaotic mess
the magical transition
yields forms evolving.

Acknowledgements

RJD and RAS acknowledge support from EPSRC and many helpful discussions with the Leeds/Manchester Critical Mass Project Team. KRB acknowledges support from GSK, SCI and EPSRC. RJD would like to express sincere thanks to Profs A. J Kirby, C. A. Hunter, S. N. Black, S. Woutersen, J. ter Horst, Dr G. Sadiq for essential discussions, to the CCDC who hosted a visit to Cambridge during which much of this paper was prepared and Trish for her unconditional support.

References

1 W. J. Dunning and A. J. Shipman, *Proc. Agric. Industries 10th International Conference*, Madrid 1954, pp. 1448–1456.
2 W. Ostwald, *Z. Phys. Chem.*, 1897, **22**, 289–330.
3 M. Volmer and A. Weber, *Z. Phys. Chem.*, 1926, **119**, 277–301.
4 R. Kaischew and I. N. Stranski, *Z. Phys. Chem., Abt. A*, 1934, **170**, 295–305.
5 M. Volmer, *Kinetik der Phasenbildung* in the series *Chemical Reaction*, ed. K. F. Bonhoeffer, Steinkopff, Dresden and Leipzig, 1939, vol. 4.
6 J. Bernstein, R. J. Davey and J. O. Henck, *Angew. Chem., Int. Ed.*, 1999, **38**, 3440–3461.
7 G. Desiraju, *Nat. Mater.*, 2002, **1**, 77–79.
8 A. J. Kirby, *Adv. Phys. Org. Chem.*, 1994, **29**, 87–183.

9 W. I. Cross, N. Blagden, R. J. Davey, R. G. Pritchard, M. A. Neumann, R. J. Roberts and R. C. Rowe, *Cryst. Growth Des.*, 2003, **3**, 151–158.
10 N. Blagden, W. I. Cross, R. J. Davey, M. Broderick, R. G. Pritchard, R. J. Roberts and R. C. Rowe, *Phys. Chem. Chem. Phys.*, 2001, **3**, 3819–3825.
11 R. J. Davey, M. Brychczynska, G. Sadiq, G. Dent and R. G. Pritchard, *CrystEngComm*, 2013, **15**, 856–859.
12 S. Bernès, G. Hernández, R. Portillo and R. Gutiérrez, *Acta Crystallogr., Sect. E: Struct. Rep. Online*, 2008, **64**, o1366.
13 M. Terpstra, B. M. Craven and R. F. Stewart, *Acta Crystallogr., Sect. A: Found. Crystallogr.*, 1993, **49**, 685–692.
14 J. R. Bourne and R. J. Davey, *J. Cryst. Growth*, 1976, **36**, 278–286.
15 T. C. W. Mak, *J. Chem. Phys.*, 1965, **43**, 2799–2805.
16 R. C. Burton, E. S. Ferrari, R. J. Davey, J. L. Finney and D. T. Bowron, *J. Phys. Chem. B*, 2009, **113**, 5967–5977.
17 A. T. Hulme, S. L. Price and D. A. Tocher, *J. Am. Chem. Soc.*, 2005, **127**, 1116–1117.
18 S. Hamad, C. Moon, C. R. A. Catlow, A. T. Hulme and S. L. Price, *J. Phys. Chem. B*, 2006, **110**, 3323–3329.
19 S. Zaitu, Y. Miwa and T. Taga, *Acta Crystallogr., Sect. C: Cryst. Struct. Commun.*, 1995, **51**, 1857–1859.
20 D. Turnbull, *J. Chem. Phys.*, 1952, **20**, 411–424.
21 D. M. Keller, R. E. Massey and O. E. Hillman Jr, *Can. J. Chem.*, 1978, **56**, 831–838.
22 O. Galkin and P. G. Vekhilov, *J. Phys. Chem. B*, 1999, **103**, 10965–10971.
23 D. Rossi, A. Gavriilidis, S. Kuhn, M. A. Candel, A. G. Jones, C. Price and L. Mazzei, *Cryst. Growth Des.*, 2015, **15**, 1784–1791.
24 S. Jiang and J. H. ter Horst, *Cryst. Growth Des.*, 2011, **11**, 256–261.
25 R. J. Davey, S. M. L. Schroeder and J. H. ter Horst, *Angew. Chem., Int. Ed.*, 2013, **52**, 2166–2179.
26 D. Kashchiev, *Nucleation: Basic Theory with Applications*, Butterworth-Heinemann, Oxford, 2000, p. 194, eqn 13.36.
27 R. C. Hynes and Y. Le Page, *J. Appl. Crystallogr.*, 1991, **24**, 352–354.
28 R. A. Sullivan, R. J. Davey, G. Sadiq, G. Dent, K. R. Back, J. H. ter Horst, D. Toroz and R. B. Hammond, *Cryst. Growth Des.*, 2014, **14**, 2689–2696.
29 R. A. Sullivan and R. J. Davey, *CrystEngComm*, 2015, **17**, 1015–1023.
30 R. C. Burton, E. S. Ferrari, R. J. Davey, J. L. Finney and D. T. Bowron, *J. Phys. Chem. B*, 2010, **114**, 8807–8816.
31 D. Khamar, J. Zeglinski, D. Mealey and Å. C. Rasmuson, *J. Am. Chem. Soc.*, 2014, **136**, 11664–11673.
32 J. F. B. Black, R. J. Davey, R. J. Gowers and A. Yeoh, *CrystEngComm*, 2015, DOI: 10.1039/c5ce00353a.
33 M. C. Etter, *Acc. Chem. Res.*, 1990, **23**, 120–126.
34 R. K. Mughal, R. J. Davey and S. N. Black, *Cryst. Growth Des.*, 2007, **7**, 225–228.
35 M. A. Oliveira, M. L. Peterson and R. J. Davey, *Cryst. Growth Des.*, 2011, **11**, 449–457.
36 D. M. Croker, R. J. Davey, A. C. Rasmuson and C. C. Seaton, *Cryst. Growth Des.*, 2013, **13**, 3754–3762.
37 S. N. Black, *Proc. R. Soc. A*, 2007, **463**, 2799–2811.
38 A. J. Cruz-Cabeza and J. Bernstein, *Chem. Rev.*, 2014, **114**, 2170–2191.

39 W. Du, A. J. Cruz-Cabeza, S. Woutersen, R. J. Davey and Q. Yin, *Chem. Sci.*, 2015, **6**, 3515–3524.
40 *HyperCube 8.0*, HyperChem 8.0, 2010, http://www.hyper.com.
41 R. Dowling, R. J. Davey, R. A. Curtis, G. Han, S. K. Poornachary, P. S. Chow and R. B. H. Tan, *Chem. Commun.*, 2010, **46**, 5924–5926.
42 P. A. Hunter, Private Communications, 1994–2015.

PAPER

Recent advances in the understanding of two-step nucleation of protein crystals

Maria A. Vorontsova,[a] Dominique Maes[b] and Peter G. Vekilov*[ac]

Received 10th November 2014, Accepted 2nd January 2015
DOI: 10.1039/c4fd00217b

The two-step mechanism of nucleation of crystals in solutions posits that the formation of crystal nuclei occurs within structures of extended lifetimes, in which the nucleating solute is at high concentration. The validity of this mechanism has been demonstrated for proteins, small-molecule organic and inorganic materials, colloids, and polymers. Due to large molecule sizes, proteins are an ideal system to study the details of this nucleation pathway, in particular the formation mechanisms of the nucleation precursors and the associated physico-chemical rules. The precursors of protein crystal nuclei are protein-rich clusters of sizes ~100 nm that contain 10 000–100 000 molecules and occupy less than 10^{-3} of the total solution volume. Here we demonstrate, using oblique illumination microscopy, the liquid nature of the clusters of the protein lysozyme and reveal their inhomogeneous structure. We test a hypothesis put forth by theory that clusters primarily consist of transient protein oligomers. For this, we explore how varying the strength of the Coulomb interaction affects the cluster characteristics. We find that the cluster's size is insensitive to variations of pH and ionic strength. In contrast, the addition of urea, a chaotropic agent that leads to protein unfolding, strongly decreases the cluster size. Shear stress, a known protein denaturant, induced by bubbling of the solutions with an inert gas, elicits a similar response. These observations support partial protein unfolding, followed by dimerization, as the mechanism of cluster formation. The amide hydrogen–deuterium exchange, monitored by nuclear magnetic resonance, highlights that lysozyme conformational flexibility is a condition for the formation of the protein-rich clusters and facilitates the nucleation of protein crystals.

Introduction

Protein structures are essential for understanding the fundamental processes in biology, are pivotal to the success of rational drug design, and enable numerous other laboratory and industry applications. Hence, structural genomics/

[a]Department of Chemical and Biomolecular Engineering, University of Houston, Houston, Texas, 77204-4004, USA. E-mail: vekilov@uh.edu
[b]Structural Biology Brussels, Vrije Universiteit Brussel, Pleinlaan 2, B-1050 Brussel, Belgium
[c]Department of Chemistry, University of Houston, Houston, Texas, 77204-5003, USA

proteomics projects around the world have invested considerable funds and effort in their determination. The most widely used method of protein structure determination is X-ray crystallography[1] and obtaining high quality crystals is a crucial prerequisite for progress in this field.[2–5] The difficulty of obtaining crystals of such complex molecules limits the success of these projects and brings the need for novel approaches in protein crystallization, based on understanding of the mechanisms and its constituent processes.[6–8]

Besides structural biology, the formation of protein crystals and similar ordered arrays of folded proteins are of interest for several fields of science and technology. They underlie several human pathological conditions. An example is the crystallization of hemoglobin C and the polymerization of hemoglobin S that cause, respectively, the CC and sickle cell diseases.[9–12] Another area which relies on protein crystals is pharmacy: crystal dissolution at a controlled rate is used to achieve sustained release of medications, such as insulin, interferon-α, or the human growth hormone.[13–17] Work is underway on the crystallization of other therapeutically-active proteins, *e.g.*, antibodies for foreign proteins, which can be dispensed as a microcrystalline preparation. If the administered dose consists of a few equidimensional crystallites, steady rates of medication release can be maintained for longer periods than for doses comprised of many smaller crystallites.

In all of these areas, control of crystallisation hinges on the ability to achieve and control nucleation, the first step of crystallization. Nucleation determines the main properties of the crystal population, including the crystal polymorph, the number of crystals, and their size and size distribution. The nucleation outcome favored in classical crystallography is to have a population consisting of one large crystal of a stable and robust polymorph, or, failing that, of several well-separated crystals of similar sizes and of a single polymorph. In the novel method of femtosecond X-ray protein nanocrystallography, which relies on crystals as small as 200 nm,[18] the need to grow the crystals after they have nucleated is nearly eliminated and nucleation emerges as the sole process to be controlled.

A novel outlook on protein nucleation in solution suggests that the nucleation of crystals is preceded by the formation of clusters of mesoscopic size composed of dense protein liquid.[19–23] Crystal nuclei then form inside the clusters.[24–26] Evidence of the action of this two-step mechanism has been provided for protein crystals[27,28] and sickle cell anemia fibers,[29] amyloid fibrils,[30,31] crystals of small-molecule organics,[25,32–34] colloids,[35–37] biominerals,[38–40] polymers,[41] and other substances. Evidence from multiple studies has indicated that the two-step mechanism is broadly applicable to protein crystallization.[27]

The likely precursor to crystal nuclei are protein-rich clusters. They have been observed in solutions of several proteins: lysozyme,[42,43] hemoglobin A and S,[44] glucose isomerase,[45] lumazine synthase,[46,47] insulin. The crucial role of the protein-rich clusters as locations of protein crystal nucleation has been supported by experimental evidence with hemoglobin S,[21,29,48] glucose isomerase,[45] and lysozyme;[27,49] crystal nuclei forming inside similar mesoscopic clusters have been directly observed for colloids,[36] which are larger and move slower than most molecules, and an ingeniously chosen organic system.[33] The size of the protein clusters varies from several tens to several hundreds of nanometers and their total volume fraction remains less than 10^{-3}. Their extended lifetimes indicate that the clusters are not concentration fluctuations. It was shown that these clusters are present even in homogeneous regions of the protein phase diagram[42,43] and their properties, average size and total population volume, do not undergo any singularity upon crossing the liquidus.[20,21] It was demonstrated that the cluster population volume responds to varying protein concentration in exact correspondence with Boltzmann predictions based on the free energy cost of regions of concentration as high as that in the clusters.[23,43,48] This observation indicates that the clusters are in equilibrium with the solution and excludes the possibility that the observed clusters represent aggregates of irreversibly denatured protein. On the other hand, the cluster size is independent of protein concentration. The decoupled behaviors of the cluster size and population volume are in stark contrast with the classical phase transition trends and highlight the unusual nature of the clusters. Experimental data and theoretical analyses that address the mechanisms determining the cluster size concluded that the size is determined by the dynamics of formation, diffusion and decay of protein oligomers of a limited lifetime. Hence, such oligomers are required for the clusters' existence.[43] The hypothesized oligomers are not expected to be bound by crystal contacts and, hence, are not crystal embryos. We hypothesize that their existence enables the clusters, which, in turn, facilitate crystal nucleation by lowering the free surface energy that limits the growth of the nuclei.

Here we address several open questions related to the oligomer mechanism of cluster formation: about the roles of electrostatic and water-structuring interactions, the liquid nature of the clusters, first suggested in previous work,[19–21] the uniformity of the cluster structure throughout their volume, and the significance of partial protein unfolding for cluster formation. We find that the classical interplay between electrostatic and hydrophobic interactions is significantly shifted to the latter ones so that the hydrophobic interactions become the dominant force for the cluster assembly. We employ a novel method of oblique illumination scattering microscopy and demonstrate the liquid nature of the clusters and suggest that they have an inhomogeneous structure with an inside core of high viscosity. We show that partial protein unfolding and hydrophobic stabilization appear to be the major factors contributing to oligomerization and cluster formation. The results presented here suggest that fine-tuning of water-structuring interactions promises to be a powerful tool to control the clusters' population and, in view of the cluster role in protein crystal nucleation, enhance or suppress the nucleation process.

For these tests we chose the protein lysozyme because of its high isoelectric point, pI = 11.35,[50] which results in a considerable molecular net positive charge in the range of tested pH (from $+15\bar{e}$ at pH 3.8 to $+8\bar{e}$ at pH 7.8). This high charge amplifies the significance of electrostatic forces in cluster formation. Furthermore, previous studies with lysozyme have revealed that lysozyme forms well detectable clusters in the homogeneous region of the lysozyme phase diagram[43] and this allows the study of cluster characteristics and behaviors before crystal nucleation occurs.

We used the second osmotic virial coefficient B_2 to characterize the response of lysozyme in the solution to modified electrostatic and water-structuring interactions. B_2 is an azimuthally and spatially averaged characteristic of the pairwise interaction potential, and revealed the attenuation of intermolecular repulsion with increasing ionic strength. We used static light scattering to determine B_2.[43] The response of the cluster population was characterized in terms of average cluster size R_2 and cluster population volume, expressed as a fraction of the total solution volume ϕ_2; R_2 and ϕ_2 were determined by dynamic light scattering.[20,21,23,42,43] We employed Brownian and oblique illumination microscopies to characterize the cluster population.

Materials and methods

Solution preparation

We used chicken egg white lysozyme (Affymetrix); KCl (Fisher); $(NH_4)_2SO_4$ and HCl, CH_3COOH (Fisher Scientific); NaCl and KOH (Mallinckrodt Chemicals); D_2O (Sigma-Aldrich); HEPES (Fisher and Calbiochem); we observed no difference between solutions of HEPES from the two companies.

Unless specified otherwise, all experiments were carried out in 20 mM HEPES buffer (*N*-2-hydroxyethylpiperazine-*N*′-2-ethanesulfonate) at pH 7.8 (adjusted with KOH). Since the pK_a of the respective acid is 7.5, the ionic strength of the buffer is ~0.667× of the total HEPES concentration. We determined the lysozyme concentration with a Beckman Coulter Du 800 Spectrophotometer using extinction coefficient $\varepsilon_{280\text{ nm}} = 2.64$ ml mg^{-1} cm^{-1}.[51] Stock solutions of ~150 mg ml^{-1} lysozyme in HEPES buffer were dialyzed over two days (44–50 hours). After dialysis, the solutions were diluted to a desired concentration and, if necessary, the ionic strength was adjusted by adding salts: NaCl, KCl or $(NH_4)_2SO_4$. All experiments were carried out at 22 °C. Prior to all measurements, the solutions were filtered through 0.22 μm PES syringe filters (Lightlabs).

Dynamic and static light scattering (DLS and SLS)

The DLS data were collected by ALV light scattering device equipped with He–Ne laser ($\lambda = 632.8$ nm, 35 mW) and ALV-5000/EPP Multiple tau Digital Correlator (ALV-Gmbh, Langen, Germany). The autocorrelation functions were acquired at 90° for 60 s. Data analysis was performed as in ref. 44 The viscosity of protein solutions used to evaluate clusters' hydrodynamic radius R_2 was determined independently as described in ref. 42 using OptiLink carboxylate-modified polystyrene microparticles ($d = 0.424$ μm). The static light scattering data on monomer interactions were collected at 90°. If the correlation function taken in parallel indicated the presence of clusters, we subtracted from the total scattered intensity the scattered intensity due to the clusters. The second virial coefficient B_2 was determined from Debye plots. For the refractive index increment of the solutions we used the value $dn/dc = 0.199 \pm 0.003$ ml g^{-1} obtained using a Brookhaven differential refractometer calibrated with KCl aqueous solutions at 25 °C, $\lambda = 620$ nm.

Brownian and oblique illumination microscopies

We employed Nanosight LM10-HS microscope (Nanosight Ltd) equipped with green laser with $\lambda = 532$ nm operating at temperature $T = 22 \pm 0.1$ °C to examine the Brownian motion of individual clusters in the tested solutions. The raw data of this method are movies of point spread functions of clusters undergoing Brownian motion. The rate of movie acquisition depends on the camera characteristics and settings, in our experiments it was ~25 fps. We determined the clusters' diffusivity by using a custom designed software package which analyzes individual cluster trajectories. The cluster radius R was calculated using the Stokes–Einstein equation accounting for previously determined viscosity. The number of clusters on each frame gave an estimate of the cluster concentration.

Oblique illumination microscopy (OIM) uses the same setup as Brownian microscopy with a more powerful objective lens and a faster, sensitive camera (S-PRI, AOS Technologies AG). OIM records the point spread functions of individual clusters, whose diameter ($2R_2 \approx 150$ nm) is comparable to the diffraction limit ($\lambda/2 \approx 250$ nm). We used this method to monitor the time resolved behavior of individual clusters.

Results and discussion

The role of Coulomb interactions in cluster formation

Coulomb interactions are a core of biophysics. They determine protein 3D structure,[52–57] substrate binding,[58,59] transitions between activated and deactivated states,[60,61] signal transmission,[62] and others. Proteins' structures were optimized by evolutionary design to prevent undesirable aggregation resulting from electrostatic interactions.[63,64] Hence, Coulomb forces were tested for their role in the formation of the mesoscopic protein rich clusters.

The strength of the intermolecular Coulomb forces depends on the charge of protein monomers and the Debye screening from free ions in a solvent, *i.e.* the ionic strength I of a solution. We probe the influence of electrostatics on our system by varying the ionic strength from 3 to 313 mM covering the physiological range of 100–200 mM. Low I values were obtained by varying the concentration of the HEPES buffer. For higher I's we used NaCl, KCl and $(NH_4)_2SO_4$. Dimensionless values of this parameter ranged from $B_2 = 114$ at the lowest probed $I = 3$ mM to $B_2 = -20$ at $I = 313$ mM, attaining the value of non-interacting hard spheres $B_2 = 4$ at an ionic strength of 50–60 mM.

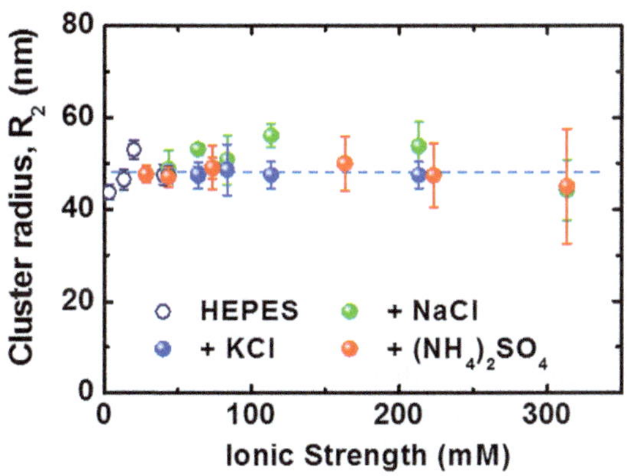

Fig. 1 The role of electrostatics in cluster formation: dependence of cluster radius R_2 on ionic strength. R_2 were evaluated from the cluster diffusion times, determined from the DLS correlation functions. The viscosity used in these evaluations was measured independently. The ionic strength of solutions was induced in four different ways: by changing concentration of HEPES buffer, and by adding salts KCl, NaCl, and $(NH_4)_2SO_4$.

We display the response of the cluster size to modified electrostatic forces in Fig. 1. We observe that cluster radius R_2 is independent of the ionic strength of the solution and on the salt identity used to increase I. The cluster volume fraction φ_2 (data not shown) decreased five-fold as the ionic strength increased to ~150 mM and stayed constant at higher I values.

We modified the monomers charge by changing the solution pH from 3.8 to 7.8 and performed identical tests. The results (data not shown) in terms of the second virial coefficient B_2 and diffusivity D reveal that intermolecular interactions are significantly affected by electrostatic forces. Similar to the behavior in response to increasing ionic strength, cluster volume fraction shows an apparent increase as the protein charge decreases at higher pH values, while the cluster size does not change within the experimental error.

The decoupled behaviors of R_2 and ϕ_2 at increasing I and pH indicate that these two characteristics of the cluster population are controlled by distinct mechanisms. This observation agrees with the oligomer mechanism of cluster formation by which R_2 is determined by the kinetics of decay of the oligomers accumulated in the clusters,[23] while ϕ_2 reflects the high free energy cost of bringing together positively charged molecules.[23,43] The lack of correlation between the cluster size R_2 and the solution ionic strength and pH indicates that cluster formation is not governed by Coulomb interactions. These observations exclude a colloid aggregation scenario based on the balance of van der Waals attraction and Coulomb repulsion[65,66] as the basis of the mesoscopic protein rich clusters.[66] Furthermore, these observations indicate that a mechanism relying on an electrostatically bound oligomer could not underlie the mesoscopic clusters in lysozyme solutions.

With many other studied proteins, clusters are observed at ionic strengths higher than 100 mM,[19,23,29,43,67,68] at which the Debye length is shorter than the molecular size and, hence, the lifetimes of electrostatically-bound oligomers would be insignificant. Hence, Coulomb-regulated colloid clustering and Coulomb-mediated oligomerization can be excluded as formation mechanisms of the mesoscopic clusters in solutions of these proteins.

Hydrophobic interactions in protein clusters' formation

Water structuring and hydrophobic interactions are often a dominant force in protein aggregation.[69–72] These forces are short-range and may play an important role in the stabilization of the protein-rich clusters, in which the protein concentration is ~450 mg ml^{-1} and the intermolecular separations are shorter than 1 nm.[23]

Urea is a chaotropic agent, which accumulates at the protein peptide backbone and breaks the water structure.[73] This mechanism underlies urea's action as a universal protein denaturant that causes complete molecule unfolding at concentrations 7–8 M.[74–76] We employed urea concentrations up to 3 M, significantly below the levels when complete protein unfolding occurs. The increase of the second virial coefficient B_2 from 38 to 52 upon increasing of urea concentration to 2 M, shown in Fig. 2a, suggests that it causes an increase of intermolecular repulsion. This urea-induced increase of repulsion is consistent with breaking water structures around the protein solvent-exposed backbone segments and weakening the hydrophobic attraction at low protein concentrations.

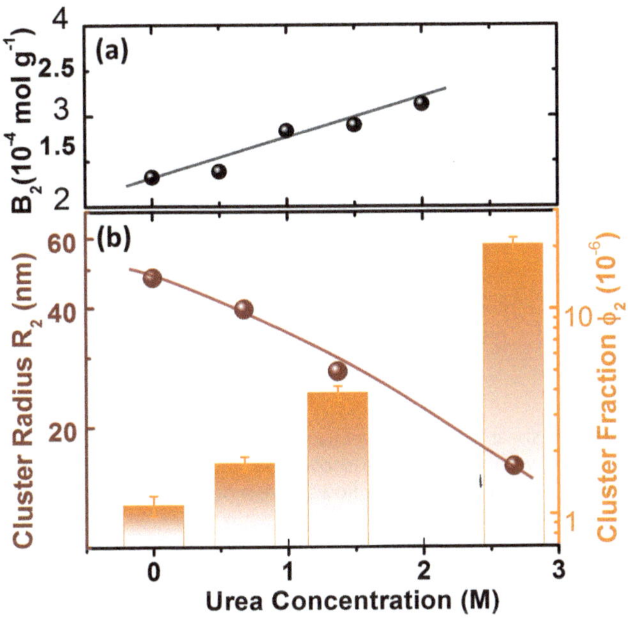

Fig. 2 The role of hydrophobic interactions in cluster formation. (a) Variation of the pairwise intermolecular interactions in lysozyme solutions, characterized in terms of the second virial coefficient B_2 (left ordinate, closed symbols) as a function of urea concentration. (b) The response of the cluster radius R_2 (left ordinate, spheres) and volume fraction ϕ_2 (right ordinate, bars) to urea concentration.

The cluster radius R_2 decreases three-fold at urea concentration 2.7 M while the clusters volume fraction decreases ten-fold. The decoupled behavior of these characteristics excludes the possibility of lysozyme denaturation and aggregation caused by urea and strongly supports the hypothesis of protein oligomerization contributing to cluster formation. Urea accumulates around solvent exposed segments of a protein backbone[77] shortening the lifetime of dynamic protein complexes. Following the theoretical model of cluster formation,[43] this leads to a smaller cluster radius R_2. At high protein concentration inside the clusters partial molecule unfolding caused by urea results in enhancement of attractive hydrophobic interactions. This stabilizes the concentrated protein phase and increases the cluster volume fraction ϕ_2.

Bubbling of a lysozyme solution with an inert gas is known to decrease the protein activity by causing molecules to unfold at the hydrophobic solvent–gas interface[78] or because of the induced shear stress.[79] We performed experiments in which lysozyme solutions were bubbled with nitrogen at different flow rates ranging from 0.2 to 0.5 L min^{-1} over the course of 2.5 hours. We did not register a change of pairwise intermolecular interactions, indicating that the majority of protein monomers was not affected by bubbling.

We show the evolution of cluster radius R_2 in Fig. 3. The reference solution with no bubbling was analyzed to verify the steadiness of cluster radius throughout the experiment. We noticed the increase of R_2 with increase of the gas

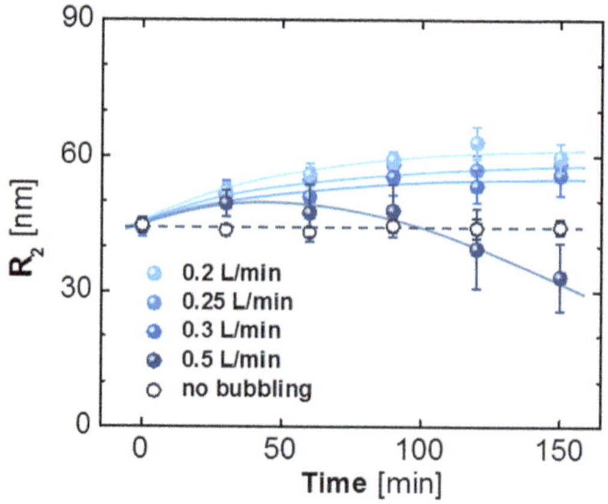

Fig. 3 The effects of shear flow on cluster formation. The response of the cluster radius R_2 to solution bubbling with N_2. The data shows that moderate bubbling rates (0.2–0.3 L min^{-1}) increase R_2; at 0.5 L min^{-1} shear flows have a negative effect on cluster formation and R_2 decreases with bubbling time.

flow rate from 0 to 0.3 L min^{-1} and increase of the bubbling time. This observation corresponds to a proposed cluster formation mechanism due to protein complex formation. Solution bubbling at moderate rates leads to partial protein unfolding, stabilization of protein oligomers, and increase of cluster radius. The decrease of cluster radius at the highest probed flow rate 0.5 L min^{-1} can result from the increase of shear rates in solutions which are not the subject of this discussion.

Molecular level view of oligomerization

The lysozyme folded chain, illustrated in Fig. 4a, contains two stabilized regions, called domains, linked by a relatively short and flexible sequence, called the hinge.[80] Domain swapping, schematically illustrated in Fig. 4b, is a mechanism of oligomerization observed for numerous proteins.[81] If the attraction between the domains weakens, a molecule can partially unfold, exposing its hydrophobic interfaces between the domains. The new conformation of unfolded monomers is energetically unstable. If two (or more) partially unfolded monomers come to a close proximity, a domain from one molecule may bind to a reciprocal domain from the other partially unfolded molecule. As a result, a dimer (or higher order oligomer) may form. Each step in this process is reversible. The free energy of the oligomer is lower than for its constituent unfolded monomers but higher than for their native conformation.

Even though there are no reports of domain swapped lysozyme oligomers, theoretical investigations reveal an internal molecular vibrational motion,[82,83] which occurs between the two domains and involves the displacement of the hinge region. This process is crucial for lysozyme activity and may lead to partial molecule unfolding. Nuclear magnetic resonance (NMR) studies of lysozyme have indicated different hydrogen–deuterium exchange rates in native *vs.* denatured states of the molecule[84] in D_2O-based solutions. These studies reveal that some hydrogens are more protected than others and the protection strictly depends on

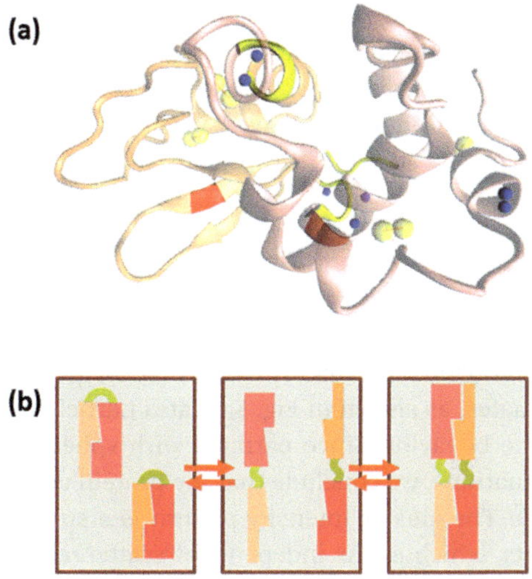

Fig. 4 Formation of protein oligomers. (a) Protein lysozyme structure (RCSB PDB: 2VB1). Two domains of the molecule are shown in pink and orange. The active center is highlighted in red, and the hinge in yellow. Yellow spheres represent the di-sulfide bridges. Blue spheres indicate the amide hydrogens of the amino acid residues, revealed by NMR, potentially involved in oligomer formation. (b) Schematic representation of the domain swapping mechanism.

the molecular conformation. We use this concept to test the involvement of partial protein unfolding in oligomerization leading to cluster formation.

Two protein solutions of high and low clusters volume fractions were prepared using D_2O-based HEPES buffer, 20 mM, pH 7.8. The control sample was at concentration 25 mg ml^{-1} at which the cluster volume fraction is low ($<10^{-7}$). To achieve a higher volume of the cluster population, we prepared a second similar solution with a protein concentration of 200 mg ml^{-1}.

In both solutions, hydrogen–deuterium exchange was allowed to proceed for 1 and 72 hours before it was stopped by quenching to pH 3.8 with a solution of acetic acid in D_2O.[85] Quenching diluted the concentration of the protein 8-fold. In the control solution, the concentration was brought back to 25 mg ml^{-1} by centrifugation. The protein concentration in the tested solution was equal to that and it was not centrifuged.

The amino acid residues with unexchanged hydrogens yield signal in ^{1}H–^{15}N HSQC 2D spectra, based on the natural abundance of ^{15}N in the protein. The results on the rates of amide hydrogen exchange can be summarized as following: all non-amide and most amide hydrogens exchanged rapidly at both concentrations; several residues equally preserved their amide hydrogens in both control and cluster containing solutions; no residue exhibited faster exchange rate in the control solution; several residues showed faster exchange rate in cluster-containing solution; these are F34, N39, K97, Q121, A122 at 1 hour, and K33, I98, V92 at 72 hour. We highlight these amide hydrogens in Fig. 4a. The faster exchange in the high concentration solution is counter-intuitive for a homogenously distributed protein and suggests that the protein in the clusters is partially unfolded; tests to eliminate the possibility that high concentration and not clusters contribute to the faster exchange are planned. Importantly, Fig. 4a shows that the H-atoms exposed in the high concentration solution are near the hinges between the two lysozyme structural domains.[80] These results support the notion that partial unfolding of the domains followed by dimerization, as schematically depicted in Fig. 4b, may be a part of this protein's cluster mechanism.

The liquid nature of protein-rich clusters

We employ oblique illumination microscopy (OIM) to monitor the dynamics of individual clusters.

Observations of individual clusters revealed that the point spread function of many clusters fluctuates, as shown in Fig. 5a. Latex particles of a similar size did not exhibit the same behavior. Since particles with spherical symmetry do not show intensity fluctuations, we conclude that the protein clusters have a complex asymmetrical shape. The cluster intensity pattern is a sum of the point spread functions of its parts, which act as independent scatterers. The intensity fluctuations of the point spread functions indicate that the clusters are characterized by a complex non-spherical shape which may also be fluctuating. We determined the characteristic time of the fluctuations, $\sim$60 ms. This time is much faster than the cluster rotation time, suggesting that the fluctuations reflect variations in cluster shape, *i.e.*, the clusters are liquid. The shape variations are likely due to Brownian collisions with solvent molecules.

In another experiment we introduced fluorescent particles in a protein solution and placed a filter between the cuvette and the objective lens which obstructed the scattered light and only allowed fluorescent intensity to register. The majority of fluorescent particles diffused independently of each other. Several groups of particles moved collectively. In Fig. 5b we show three particles which exhibited this behavior. The separation between them was $\sim$2 μm indicating that the particles were not bound to one another. After removing the fluorescence filter, we observed at their location a fluctuating intensity pattern, typical of a cluster. We conclude that the particles were trapped inside the cluster, confirming the liquid state of the clusters. We analyzed the evolution of the distances between the three fluorescent particles in the cluster and found that they do not change during the time of observation. This shows that the particles are

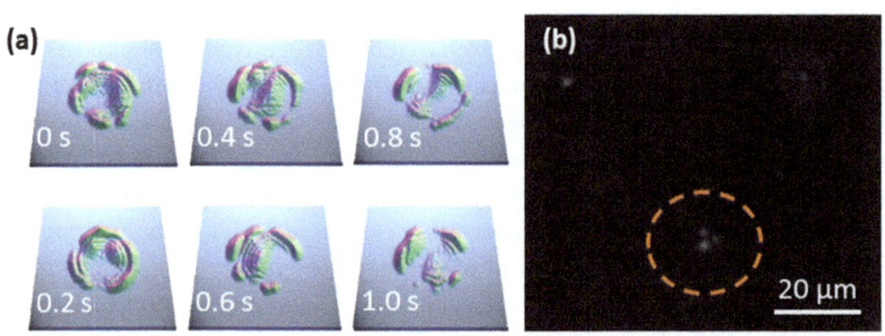

Fig. 5 The liquid cluster's nature. (a) Intensity snapshots of a protein cluster revealed by oblique illumination microscopy at times indicated in the images. (b) Fluorescent microspheres observed through a bandpass filter, which allows the transmittance of only fluorescent light. A dashed circle indicates three particles entrapped inside a cluster.

embedded in a medium of high viscosity. Lysozyme solutions at high concentrations tend to exhibit gelation[24,86] arresting the dynamics of the solution and increasing the viscosity. Thus, the observations in Fig. 5b suggest the presence of a viscous core inside the clusters.

Summary and conclusions

This investigation of the characteristics of the proteins-rich clusters in lysozyme solutions revealed several novel behaviors. We found that the Coulomb interactions do not significantly affect cluster populations: the cluster radius R_2 remains constant over the whole range of tested ionic strengths and pH, while the cluster volume fraction φ_2 slightly decreases. Experiments in which urea was added to the solutions revealed the importance of hydrophobic interactions for cluster formation. Bubbling experiments suggested the importance of partial protein unfolding in the clustering mechanism. Nuclear magnetic resonance tests located the regions of the lysozyme molecule that are potentially involved in partial unfolding during cluster formation, the interdomain hinges, and suggest that domain-swapped dimers may be the species that underlies cluster formation. Experiments with oblique illumination microscopy revealed the liquid nature of protein clusters and, for the first time, suggest an inhomogeneous structure, with a viscous core, of clusters.

Acknowledgements

This work was supported by NSF (grant MCB-1244568), NASA (NNX14AD68G and NNX14AE79G) and PRODEX (ESA-2004-070).

References

1 K. Wütrich, *Nat. Struct. Biol.*, 1998, **5**, 492–495.
2 M. F. Perutz and F. S. Mathews, *J. Mol. Biol.*, 1966, **21**, 199–202.
3 S. R. Simon, W. H. Konigsberg, W. Bolton and M. F. Perutz, *J. Mol. Biol.*, 1967, **28**, 451–454.
4 A. McPherson, *J. Cryst. Growth*, 1991, **110**, 1–10.
5 Z. Derewenda, *Structure*, 2004, **12**, 529–535.
6 S.-H. Kim, D. H. Shin, J. Liu, V. Oganesyan, S. Chen, Q. S. Xu, J.-S. Kim, D. Das, U. Schulze-Gahmen, S. R. Holbrook, E. L. Holbrook, B. A. Martinez, N. Oganesyan, A. DeGiovanni, Y. Lou, M. Henriquez, C. Huang, J. Jancarik, R. Pufan, n.-G. Choi, J.-M. Chandonia, J. Hou, B. Gold, H. Yokota, S. E. Brenner, P. D. Adams and R. Kim, *J. Struct. Funct. Genomics*, 2005, **6**, 63–70.
7 R. C. Stevens and I. A. Wilson, *Science*, 2001, **293**, 519–520.
8 S. K. Burley, S. C. Almo, J. B. Bonanno, M. Capel, M. R. Chance, T. Gaasterland, D. Lin, A. Sali, F. W. Studier and S. Swaminathan, *Nat. Genet.*, 1999, **23**, 151–157.
9 S. Charache, C. L. Conley, D. F. Waugh, R. J. Ugoretz and J. R. Spurrell, *J. Clin. Invest.*, 1967, **46**, 1795–1811.
10 R. E. Hirsch, C. Raventos-Suarez, J. A. Olson and R. L. Nagel, *Blood*, 1985, **66**, 775–777.

11 W. A. Eaton and J. Hofrichter, in *Advances in Protein Chemistry*, ed. C. B. Anfinsen, J. T. Edsal, F. M. Richards and D. S. Eisenberg, Academic Press, San Diego, 1990, vol. 40, pp. 63–279.
12 P. Vekilov, *Br. J. Haematol.*, 2007, **139**, 173–184.
13 J. Brange, *Galenics of Insulin*, Springer, Berlin, 1987.
14 M. L. Long, J. B. Bishop, T. L. Nagabhushan, P. Reichert, G. D. Smith and L. J. DeLucas, *J. Cryst. Growth*, 1996, **168**, 233–243.
15 S. Matsuda, T. Senda, S. Itoh, G. Kawano, H. Mizuno and Y. Mitsui, *J. Biol. Chem.*, 1989, **264**, 13381–13382.
16 S. Peseta, J. A. Langer, K. C. Zoon and C. E. Samuel, in *Annual Review of Biochemistry*, ed. C. C. Richardson, P. D. Boyer, I. B. Dawid and A. Meister, Annual Reviews, Palo Alto, 1989, vol. 56, pp. 727–778.
17 P. Reichert, C. McNemar, N. Nagabhushan, T. L. Nagabhushan, S. Tindal and A. Hruza, *Metal-interferon-alpha crystals, US Pat.*, 5,441,734, 1995.
18 H. N. Chapman, P. Fromme, A. Barty, T. A. White, R. A. Kirian, A. Aquila, M. S. Hunter, J. Schulz, D. P. DePonte, U. Weierstall, R. B. Doak, F. R. N. C. Maia, A. V. Martin, I. Schlichting, L. Lomb, N. Coppola, R. L. Shoeman, S. W. Epp, R. Hartmann, D. Rolles, A. Rudenko, L. Foucar, N. Kimmel, G. Weidenspointner, P. Holl, M. Liang, M. Barthelmess, C. Caleman, S. Boutet, M. J. Bogan, J. Krzywinski, C. Bostedt, S. Bajt, L. Gumprecht, B. Rudek, B. Erk, C. Schmidt, A. Homke, C. Reich, D. Pietschner, L. Struder, G. Hauser, H. Gorke, J. Ullrich, S. Herrmann, G. Schaller, F. Schopper, H. Soltau, K.-U. Kuhnel, M. Messerschmidt, J. D. Bozek, S. P. Hau-Riege, M. Frank, C. Y. Hampton, R. G. Sierra, D. Starodub, G. J. Williams, J. Hajdu, N. Timneanu, M. M. Seibert, J. Andreasson, A. Rocker, O. Jonsson, M. Svenda, S. Stern, K. Nass, R. Andritschke, C.-D. Schroter, F. Krasniqi, M. Bott, K. E. Schmidt, X. Wang, I. Grotjohann, J. M. Holton, T. R. M. Barends, R. Neutze, S. Marchesini, R. Fromme, S. Schorb, D. Rupp, M. Adolph, T. Gorkhover, I. Andersson, H. Hirsemann, G. Potdevin, H. Graafsma, B. Nilsson and J. C. H. Spence, *Nature*, 2011, **470**, 73–77.
19 O. Gliko, N. Neumaier, W. Pan, I. Haase, M. Fischer, A. Bacher, S. Weinkauf and P. G. Vekilov, *J. Am. Chem. Soc.*, 2005, **127**, 3433–3438.
20 O. Gliko, W. Pan, P. Katsonis, N. Neumaier, O. Galkin, S. Weinkauf and P. G. Vekilov, *J. Phys. Chem. B*, 2007, **111**, 3106–3114.
21 W. Pan, O. Galkin, L. Filobelo, R. L. Nagel and P. G. Vekilov, *Biophys. J.*, 2007, **92**, 267–277.
22 P. G. Vekilov, W. Pan, O. Gliko, P. Katsonis and O. Galkin, in *Lecture Notes in Physics, vol. 752: Aspects of Physical Biology: Biological Water, Protein Solutions, Transport and Replication*, ed. G. Franzese and M. Rubi, Springer, Heidelberg, 2008, pp. 65–95.
23 W. Pan, P. G. Vekilov and V. Lubchenko, *J. Phys. Chem. B*, 2010, **114**, 7620–7630.
24 O. Galkin and P. G. Vekilov, *Proc. Natl. Acad. Sci. U. S. A.*, 2000, **97**, 6277–6281.
25 B. Garetz, J. Matic and A. Myerson, *Phys. Rev. Lett.*, 2002, **89**, 175501.
26 P. G. Vekilov, *Cryst. Growth Des.*, 2004, **4**, 671–685.
27 P. G. Vekilov, *Cryst. Growth Des.*, 2010, **10**, 5007–5019.
28 Y. G. Kuznetsov, A. J. Malkin and A. McPherson, *J. Cryst. Growth*, 2001, **232**, 30–39.

29 O. Galkin, W. Pan, L. Filobelo, R. E. Hirsch, R. L. Nagel and P. G. Vekilov, *Biophys. J.*, 2007, **93**, 902–913.
30 A. Lomakin, D. S. Chung, G. B. Benedek, D. A. Kirschner and D. B. Teplow, *Proc. Natl. Acad. Sci. U. S. A.*, 1996, **93**, 1125–1129.
31 R. Krishnan and S. L. Lindquist, *Nature*, 2005, **435**, 765–772.
32 J. E. Aber, S. Arnold and B. A. Garetz, *Phys. Rev. Lett.*, 2005, **94**, 145503.
33 K. Harano, T. Homma, Y. Niimi, M. Koshino, K. Suenaga, L. Leibler and E. Nakamura, *Nat. Mater.*, 2012, **11**, 877–881.
34 D. Erdemir, A. Y. Lee and A. S. Myerson, *Acc. Chem. Res.*, 2009, **42**, 621–629.
35 M. E. Leunissen, C. G. Christova, A.-P. Hynninen, C. P. Royall, A. I. Campbell, A. Imhof, M. Dijkstra, R. van Roij and A. van Blaaderen, *Nature*, 2005, **437**, 235–240.
36 J. R. Savage and A. D. Dinsmore, *Phys. Rev. Lett.*, 2009, **102**, 198302.
37 T. H. Zhang and X. Y. Liu, *J. Phys. Chem. B*, 2007, **111**, 14001–14005.
38 E. M. Pouget, P. H. H. Bomans, J. A. C. M. Goos, P. M. Frederik, G. de With and N. A. J. M. Sommerdijk, *Science*, 2009, **323**, 1455–1458.
39 D. Gebauer, A. Volkel and H. Colfen, *Science*, 2008, **322**, 1819–1822.
40 L. B. Gower, *Chem. Rev.*, 2008, **108**, 4551–4627.
41 J. F. Wang, M. Muller and Z. G. Wang, *J. Chem. Phys.*, 2009, **130**, 154902.
42 Y. Li, V. Lubchenko and P. G. Vekilov, *Rev. Sci. Instrum.*, 2011, **82**, 053106.
43 Y. Li, V. Lubchenko, M. A. Vorontsova, L. Filobelo and P. G. Vekilov, *J. Phys. Chem. B*, 2012, **116**, 10657–10664.
44 W. Pan, O. Galkin, L. Filobelo, R. L. Nagel and P. G. Vekilov, *Biophys. J.*, 2007, **92**, 267–277.
45 M. Sleutel and A. E. S. Van Driessche, *Proc. Natl. Acad. Sci. U. S. A.*, 2014, **111**, E546–E553.
46 O. Gliko, N. Neumaier, W. Pan, I. Haase, M. Fischer, A. Bacher, S. Weinkauf and P. G. Vekilov, *J. Am. Chem. Soc.*, 2005, **127**, 3433–3438.
47 O. Gliko, W. Pan, P. Katsonis, N. Neumaier, O. Galkin, S. Weinkauf and P. G. Vekilov, *J. Phys. Chem. B*, 2007, **111**, 3106–3114.
48 V. Uzunova, W. Pan, V. Lubchenko and P. G. Vekilov, *Faraday Discuss.*, 2012, **159**, 87–104.
49 W. Pan, A. B. Kolomeisky and P. G. Vekilov, *J. Chem. Phys.*, 2005, **122**, 174905.
50 L. R. Wetter and H. F. Deutsch, *J. Biol. Chem.*, 1951, **192**, 237–242.
51 K. C. Aune and C. Tanford, *Biochemistry*, 1969, **8**, 4579–4585.
52 Y. Goto, L. J. Calciano and A. L. Fink, *Proc. Natl. Acad. Sci. U. S. A.*, 1990, **87**, 573–577.
53 R. Fairman, K. R. Shoemaker, E. J. York, J. M. Stewart and R. L. Baldwin, *Proteins: Struct., Funct., Bioinf.*, 1989, **5**, 1–7.
54 W. G. J. Hol, L. M. Halie and C. Sander, *Nature*, 1981, **294**, 532–536.
55 R. Fairman, K. R. Shoemaker, E. J. York, J. M. Stewart and R. L. Baldwin, *Biophys. Chem.*, 1990, **37**, 107–119.
56 L. Xiao and B. Honig, *J. Mol. Biol.*, 1999, **289**, 1435–1444.
57 J. M. Thornton, *Nature*, 1982, **295**, 13–14.
58 R. H. Meltzer, E. Thompson, K. V. Soman, X. Z. Song, J. O. Ebalunode, T. G. Wensel, J. M. Briggs and S. E. Pedersen, *Biophys. J.*, 2006, **91**, 1302–1314.
59 C. F. Anderson and M. T. Record, *Annu. Rev. Phys. Chem.*, 1982, **33**, 191–222.
60 E. Ozkirimli, S. S. Yadav, W. T. Miller and C. B. Post, *Protein Sci.*, 2008, **17**, 1871–1880.

61 A. M. Dean and D. E. Koshland Jr, *Science*, 1990, **249**, 1044–1046.
62 L. Hemsath, R. Dvorsky, D. Fiegen, M. F. Carlier and M. R. Ahmadian, *Mol. Cell*, 2005, **20**, 313–324.
63 J. P. K. Doye, A. A. Louis and M. Vendruscolo, *Phys. Biol.*, 2004, **1**, p9–p13.
64 J. P. K. Doye and W. C. K. Poon, *Curr. Opin. Colloid Interface Sci.*, 2006, **11**, 40–46.
65 F. Sciortino, S. Mossa, E. Zaccarelli and P. Tartaglia, *Phys. Rev. Lett.*, 2004, **93**, 055701.
66 S. B. Hutchens and Z.-G. Wang, *J. Chem. Phys.*, 2007, **127**, 084912.
67 B. G. Mohr, C. M. Dobson, S. C. Garman and M. Muthukumar, *J. Chem. Phys.*, 2013, **139**, 121914.
68 M. Sleutel and A. E. Van Driessche, *Proc. Natl. Acad. Sci. U. S. A.*, 2014, **111**, E546–E553.
69 R. D. Hills Jr and C. L. Brooks iii, *J. Mol. Biol.*, 2007, **368**, 894–901.
70 G. Favrin, A. Irbäck and S. Mohanty, *Biophys. J.*, 2004, **87**, 3657–3664.
71 B. Raman, E. Chatani, M. Kihara, T. Ban, M. Sakai, K. Hasegawa, H. Naiki, M. Rao Ch and Y. Goto, *Biochemistry*, 2005, **44**, 1288–1299.
72 L. A. Munishkina, J. Henriques, V. N. Uversky and A. L. Fink, *Biochemistry*, 2004, **43**, 3289–3300.
73 M. Auton, L. M. F. Holthauzen and D. W. Bolen, *Proc. Natl. Acad. Sci. U. S. A.*, 2007, **104**, 15317–15322.
74 A. Das and C. Mukhopadhyay, *J. Phys. Chem. B*, 2008, **112**, 7903–7908.
75 A. Caballero-Herrera, K. Nordstrand, K. D. Berndt and L. Nilsson, *Biophys. J.*, 2005, **89**, 842–857.
76 L. Hua, R. Zhou, D. Thirumalai and B. J. Berne, *Proc. Natl. Acad. Sci. U. S. A.*, 2008, **105**, 16928–16933.
77 M. Auton, L. M. F. Holthauzen and D. W. Bolen, *Proc. Natl. Acad. Sci. U. S. A.*, 2007, **104**, 15317–15322.
78 M. Caussette, A. Gaunand, H. Planche, S. Colombié, P. Monsan and B. Lindet, *Enzyme Microb. Technol.*, 1999, **24**, 412–418.
79 I. B. Bekard, P. Asimakis, J. Bertolini and D. E. Dunstan, *Biopolymers*, 2011, **95**, 733–745.
80 J. A. McCammon, B. R. Gelin, M. Karplus and P. G. Wolynes, *Nature*, 1976, **262**, 325–326.
81 M. P. Schlunegger, M. J. Bennett and D. Eisenberg, *Adv. Protein Chem.*, 1997, **50**, 61–122.
82 J. A. McCammon, B. R. Gelin, M. Karplus and P. G. Wolynes, *Nature*, 1976, **262**, 325–326.
83 A. De Simone, R. W. Montalvao, C. M. Dobson and M. Vendruscolo, *Biochemistry*, 2013, **52**, 6480–6486.
84 S. E. Radford, M. Buck, K. D. Topping, C. M. Dobson and P. A. Evans, *Proteins: Struct., Funct., Bioinf.*, 1992, **14**, 237–248.
85 Y. Bai, J. S. Milne, L. Mayne and S. W. Englander, *Proteins: Struct., Funct., Bioinf.*, 1993, **17**, 75–86.
86 M. Muschol and F. Rosenberger, *J. Chem. Phys.*, 1997, **107**, 1953–1962.

Faraday Discussions

PAPER

On the question of two-step nucleation in protein crystallization†

Andrea Sauter,[a] Felix Roosen-Runge,[b] Fajun Zhang,*[a] Gudrun Lotze,[c] Artem Feoktystov,[d] Robert M. J. Jacobs[e] and Frank Schreiber[a]

Received 18th November 2014, Accepted 7th January 2015
DOI: 10.1039/c4fd00225c

We report a real-time study on protein crystallization in the presence of multivalent salts using small angle X-ray scattering (SAXS) and optical microscopy, focusing particularly on the nucleation mechanism as well as on the role of the metastable intermediate phase (MIP). Using bovine beta-lactoglobulin as a model system in the presence of the divalent salt $CdCl_2$, we have monitored the early stage of crystallization kinetics which demonstrates a two-step nucleation mechanism: protein aggregates form a MIP, which is followed by the nucleation of crystals within the MIP. Here we focus on characterizing and tuning the structure of the MIP using salt and the related effects on the two-step nucleation kinetics. The results suggest that increasing the salt concentration near the transition zone *pseudo-c*** enhances the energy barrier for both MIPs and crystal nucleation, leading to slow growth. The structural evolution of the MIP and its effect on subsequent nucleation is discussed based on the growth kinetics. The observed kinetics can be well described, using a rate-equation model based on a clear physical two-step picture. This real-time study not only provides evidence for a two-step nucleation process for protein crystallization, but also elucidates the role and the structural signature of the MIPs in the nonclassical process of protein crystallization.

1. Introduction

Studies of the early stage of nucleation in various systems have revealed new features which cannot be explained using the classical nucleation theory.[1–6] A large body of experimental results supported by theory and simulations suggest

[a] *Institut für Angewandte Physik, Universität Tübingen, Auf der Morgenstelle 10, 72076 Tübingen, Germany. E-mail: fajun.zhang@uni-tuebingen.de; Fax: +49 7071 29 5110; Tel: +49 7071 29 78670*
[b] *Institut Laue-Langevin, 6, rue Jules Horowitz, 38042 Grenoble Cedex 9, France*
[c] *European Synchrotron Radiation Facility, 71 avenue des Martyrs, 38043 Grenoble Cedex 9, France*
[d] *Jülich Centre for Neutron Science JCNS, Forschungszentrum Jülich GmbH, Outstation at MLZ, Lichtenbergstrasse 1, 85747 Garching, Germany*
[e] *Department of Chemistry, Chemistry Research Laboratory, University of Oxford, Mansfield Road, Oxford, OX1 3TA, UK*

† Electronic supplementary information (ESI) available. See DOI: 10.1039/c4fd00225c

that a metastable intermediate phase (MIP) exists before the final crystal structure is formed,[7–21] *i.e.* the solutes in the supersaturated solution form, in a first step, either small clusters or a macroscopic dense liquid phase. In the second step, the nucleation occurs within the MIPs. This two-step nucleation mechanism was originally proposed by ten Wolde and Frenkel for crystallization of a colloidal system with short-range attraction and near the critical point of the metastable liquid–liquid coexistence line.[14] The two-step nucleation mechanism can be considered as an example of Ostwald's step rule in the microscopic world.[22] Since then, this concept has been widely used under various conditions.[1–6,17–20,23]

While the two-step mechanism seems plausible for certain experiments, direct observation of such a process is not easy. Recently, direct visualization of the crystallization kinetics and pathways of nucleation in colloidal crystallization became possible. This provided detailed information on the MIP and the transition in colloidal suspensions. Colloidal systems exhibit similar phase behavior to atomic and molecular systems, and their large particle sizes enable visualization on a single-particle level. Using this technique, Tan *et al.* studied the liquid–solid phase transition and observed the formation of a metastable precursor under their experimental conditions, regardless of the final state and the interaction potential.[24] Peng *et al.* studied the kinetics of a solid–solid phase transition using single-particle resolution video microscopy. They observed that the transition between two different solid states occurs *via* a two-step diffusive nucleation pathway involving liquid nuclei.[25] This pathway is favored in comparison to one-step nucleation, because the energy of the solid–liquid interface is lower than that between the solid phases.

While these experimental observations on colloidal systems demonstrate that two-step nucleation follows Ostwald's step rule for simple liquids,[16,19,26,27] the application of this concept to other systems, in particular protein crystallization, is still challenging. The dimensions of proteins on the nanometer scale limit the applicability of optical methods for the study of MIP formation. Because of the larger size and slow dynamics of colloids, the microstructural arrangement of colloidal particles typically relaxes in a time scale of seconds, which leads to various non-equilibrium phenomena in these systems. Moreover, the interaction potentials in these colloidal systems are isotropic, whereas the effective protein–protein interactions are often non-isotropic, involving hydrophobic and electrostatic patches as well as ion-bridges. The quantitative description remains poorly understood.[28–30]

For these systems, in which direct visualization is not possible, another method has to be developed to characterize the MIP and the nucleation and growth kinetics. Here, we argue that two-step nucleation can be distinguished from the classical one-step process by following the overall crystallization kinetics. When a MIP exists, particular care should be paid to distinguishing the sequential and parallel pathways. The sequential pathway corresponds to real two-step nucleation, in which the crystals nucleate from the MIP. The parallel pathway consists of the parallel formation of the MIP and crystals, in two one-step nucleation events from the liquid. The different pathways are illustrated in Fig. 1.

For a one-step nucleation and growth mechanism with or without MIPs, nucleation and growth are mainly determined by the supersaturated initial solution. The overall growth kinetics is expected to show an increase first and then to saturate later. In contrast, for a two-step mechanism, nucleation may

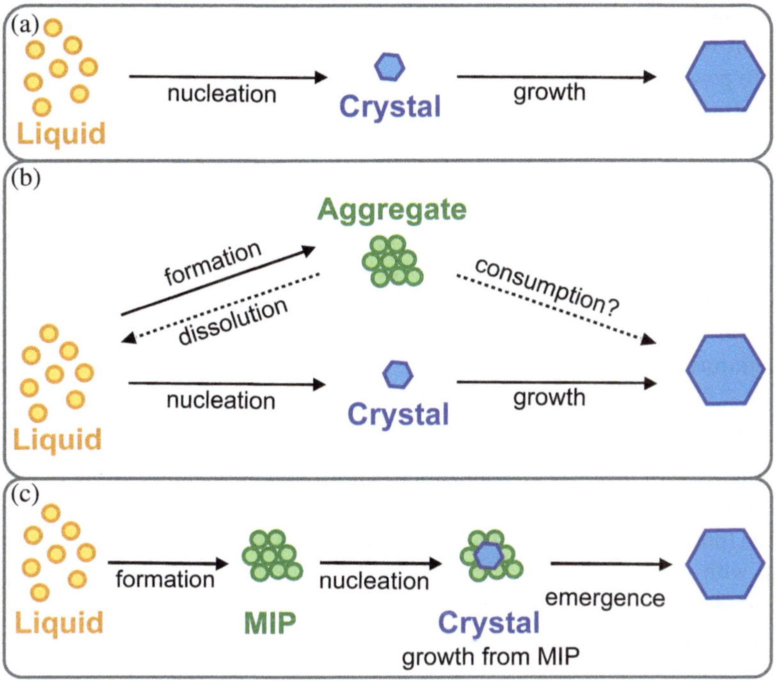

Fig. 1 Illustrations of different pathways for crystal nucleation and growth. (a) Classical one-step pathway. (b) Parallel process of aggregate formation and crystal nucleation. (c) Multi-step process with a precursor (MIP) for nucleation.

occur in both the MIPs and the initial supersaturated solution, and the nucleation rates are different due to the different interfacial energies. Furthermore, the subsequent crystal growth rates are different too. Therefore, a characteristic two-step growth kinetics is expected, dominated by the MIP and the initial solution, respectively. More details will be discussed based on the rate equation model in this paper. Indeed, in our recent study using time-resolved SAXS and optical microscopy, we identified the kinetic signature of two-step nucleation in the crystallization of the protein β-lactoglobulin (BLG) in the presence of $CdCl_2$.[31] The first step of kinetics is dominated by the two-step nucleation and growth within the MIP, *i.e.* protein aggregates form as a MIP, followed by the nucleation of crystals within the MIP. During this period, the number of crystals increases with time, but the growth rate is slow due to the low mobility of surrounding aggregates. The second step of kinetics is dominated by the growth in the dilute phase: the consumption of the MIP leads to the exposure of crystals to free molecules in the dilute phase, which accelerates the crystal growth. In this work, we focus on the effect of the structural property of the MIP on the nucleation and growth of the two-step nucleation mechanism.

2. Experimental

2.1 Materials and sample preparation

The globular protein β-lactoglobulin (BLG) from bovine milk (product no. L3908), $CdCl_2$ (202908) was purchased from SIGMA-ALDRICH. For sample preparation, appropriate amounts of salt stock solution, millipore water and protein stock solution were mixed. Stock solutions were prepared by dissolving the salt or protein powder in deionized (18.2 MΩ) and degassed millipore water. The protein concentration of stock solutions was determined by UV absorption measurements using an extinction coefficient of 0.96 l g^{-1} cm^{-1}, at a wavelength of 278 nm.[32] All samples in this work were prepared without additional buffer, since buffers can affect the phase behavior of proteins and the solubility of salts. The pH of the solutions was monitored using a Seven Easy pH instrument from Mettler Toledo. The pH values for all experimental conditions were above the pI = 5.2 of BLG.[33] Therefore, cation binding was the main driving force of charge inversion instead of pH.[33,34,41] All experiments were performed at a room temperature of 293 ± 1 K.

2.2 Optical microscopy

Time-dependent protein crystallization was followed by transmission optical microscopy using an AXIOSCOPE.A1 from ZEISS. The protein stock solution was filtered (pore size 100 μm) in advance. Samples were prepared using a micro-batch setup with two hydrophobically coated glass slides sealed by silicone (sample thickness ≈ 250–300 μm). Images were taken by an integrated AXIOCAM ICC5 camera.

2.3 Small-angle X-ray and neutron scattering (SAXS and SANS)

Small angle X-ray scattering (SAXS) measurements were performed at the ESRF, Grenoble, France at the beamline ID02. Different energies of 16 047 eV and 12 460 eV were used for two different beamtimes, and the sample-to-detector distance was 2 m with an accessed q-range of 0.06 to 4.3 nm^{-1} or 0.04 to 3.9 nm^{-1}, respectively. *Ex situ* measurements were performed using a flow capillary cell. For real-time measurements, samples were measured using quartz capillaries in a vertical capillary holder; these were quickly loaded and transferred to the sample station after sample preparation. Measurements started about 2–3 min after mixing and took place every couple of minutes during the entire crystallization process. The beam position in the sample was shifted after each exposure (duration 0.05 s) to avoid radiation damage. For further details on the beamline, calibration and data collection, see ref. 35.

SANS measurements were carried out at KWS1 at FRMII, Munich, Germany. The applied sample-to-detector distances were 1.5 and 8 m, which covers a q-range from 0.04 nm^{-1} to 3.1 nm^{-1} at a wavelength of 7 Å ($\Delta\lambda/\lambda = 10\%$). Protein-salt solutions in D_2O were filled in rectangular quartz cells with a path-length of 2 mm. The beam size on the sample was 6 mm × 12 mm. Plexiglas was used as a secondary standard to calibrate the absolute scattering intensity. The data correction and absolute intensity calibration were obtained using QtiKWS software.[36]

3. Results

3.1 Experimental phase diagram of BLG with $CdCl_2$ or $ZnCl_2$

We first describe the experimental phase diagram of our system which provides the basis for the following kinetic studies on protein crystallization. Our studies of globular proteins in solutions containing multivalent metal ions have revealed complex phase behavior including reentrant condensation (RC), metastable liquid–liquid phase separation (LLPS) and crystallization.[34,37–43] A similar experimental phase diagram, like the one shown in Fig. 2, has been observed for several proteins in solution in the presence of trivalent metal ions. The physical mechanism of this RC behavior is due to the effective charge inversion of proteins and a cation-mediated attraction, presumably by intermolecular bridges of multivalent cations.[38,44] With an isoelectric point below 7, the proteins used in our work are acidic at a neutral pH. At low salt concentrations, proteins carry negative net charges, and the electrostatic repulsion stabilizes the solution. When trivalent metal ions are added into the solution, the binding of metal ions to the carboxyl groups on the protein surface reduces the effective net charge. Above a certain salt concentration c^*, electrostatic repulsion is not strong enough to balance the attractive potential, and samples phase separate and become turbid ("regime II"). The interesting observation for this system is that with a further increase in salt concentration, the continuous binding of metal ions to the protein surface leads to a charge inversion, which again establishes the long-range electrostatic repulsion. Therefore above a second boundary (experimentally rather broad), c^{**}, the solutions become completely clear again. The charge inversion and the effective attraction mediated *via* multivalent metal ions have been further investigated by experiments, simulations and theoretical studies.[41,42,44–46]

Previous studies on β-lactoglobulin (BLG) systems with divalent salts $ZnCl_2$ and $CdCl_2$ showed a similar experimental phase behavior (Fig. 2).[31,47] For these divalent salts, the samples above a certain salt concentration become gradually less turbid, but not completely clear again, and this transition zone is denoted as *pseudo-c^{**}*. Both boundaries induced by $CdCl_2$ and $ZnCl_2$ are remarkably similar as shown in Fig. 2. In comparison to the trivalent salt YCl_3, both transitions are

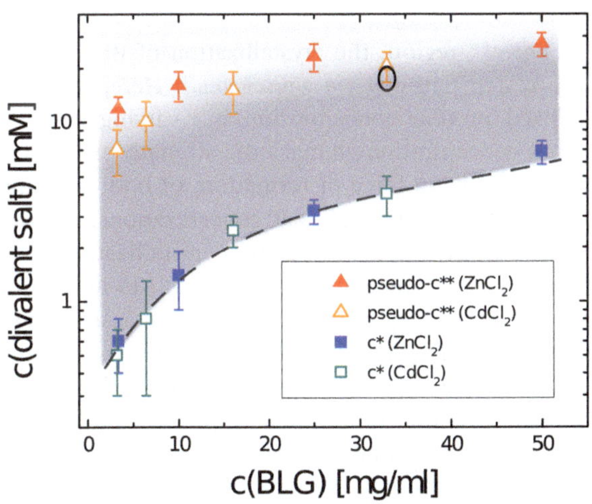

Fig. 2 Experimental phase diagram of BLG with $ZnCl_2$ and $CdCl_2$.[31,47] Both salts share a similar c^*, below and above which the protein solutions are clear and turbid, respectively. With a further increase in salt concentration, samples become gradually less turbid, but not completely clear again, and this transition zone is denoted as *pseudo-c^{**}*. The black ellipse marks the conditions mainly used in this work (33 mg ml^{-1} BLG with 15–20 mM $CdCl_2$).

shifted towards higher salt concentrations.[47] Although the reentrant effect is not complete, a charge inversion with increasing divalent salt concentration has been observed in both cases (ESI, Fig. S1†). Note that the phase behavior shown in Fig. 2 was only observed for BLG, not for bovine or human serum albumin (data not shown), which suggests a more specific interaction between these divalent ions and BLG.[47]

We emphasize that the observed protein condensation is not caused by a change in the protein structure induced by $CdCl_2$, $ZnCl_2$ or other multivalent salts; this was demonstrated in previous work using Fourier transform infrared (FTIR) and circular dichroism spectroscopy for a broad protein and salt concentration range.[31,48] Both techniques indicate no significant change in the secondary structure of the protein. Moreover, the successful growth of high-quality crystals and fine structural analysis confirm that the proteins are still in their native state.[38,47]

3.2 Crystallization followed by optical microscopy

Crystallization of BLG in the presence of $CdCl_2$ and $ZnCl_2$ was observed within a wide range of the above shown phase diagram (Fig. 2).[31,47] Slightly below c^*, no indications of a multi-step process were observed: crystals nucleate directly from a solution and grow larger without any visible aggregation or LLPS. In contrast, in the transition zone of *pseudo*-c^{**}, protein aggregates quickly form after sample preparation and the solution observed by optical microscopy becomes more turbid. In the case of $ZnCl_2$, the aggregates further relax into a liquid-like network before crystallization starts.[47] During crystallization, the turbid solution becomes gradually clearer, indicating that the aggregates are consumed. If the aggregates are removed by centrifugation, the number of crystals or nucleation rate can be significantly reduced, as has also been shown in other protein systems.[20] The two different growth pathways have been observed for the same protein, BLG, with a trivalent salt, YCl_3.[38,47]

Within the *pseudo*-c^{**} region, the crystallization of BLG in the presence of $CdCl_2$ follows a two-step pathway *via* aggregates.[31] Here, we study a series of samples with a fixed protein concentration but varying salt concentrations. Variation of the salt concentration changes the strength of the attractive potential, and therefore the driving force of formation, of both MIP and nucleation. Crossing *pseudo*-c^{**} from low to high salt concentrations, the driving force is reduced. Fig. 3 shows a time series of images of crystallization, for samples with 33 mg ml^{-1} BLG and $CdCl_2$ concentrations of 17, 18, 19 and 20 mM. Below 17 mM, the sample solution is deep in regime II, where massive amorphous aggregates form and precipitate quickly from the solution due to the strong attraction. Although crystallization can still be observed, only a small fraction of the amorphous aggregates can turn into the crystalline phase. Above 20 mM, solutions become nearly clear, and only small aggregates are visible under a microscope. Crystallization in these solutions becomes extremely slow, or they do not crystallize at all.

From the images shown in Fig. 3, one can see that with 17 mM salt, large aggregates are still formed, but the nucleation rates are also high. We emphasize that nearly all crystals are associated with the network. Hardly any crystals form in the dilute phase. In contrast to the high nucleation rate, the crystal growth period is short. After about 2 h, visible change can no longer be observed, and the

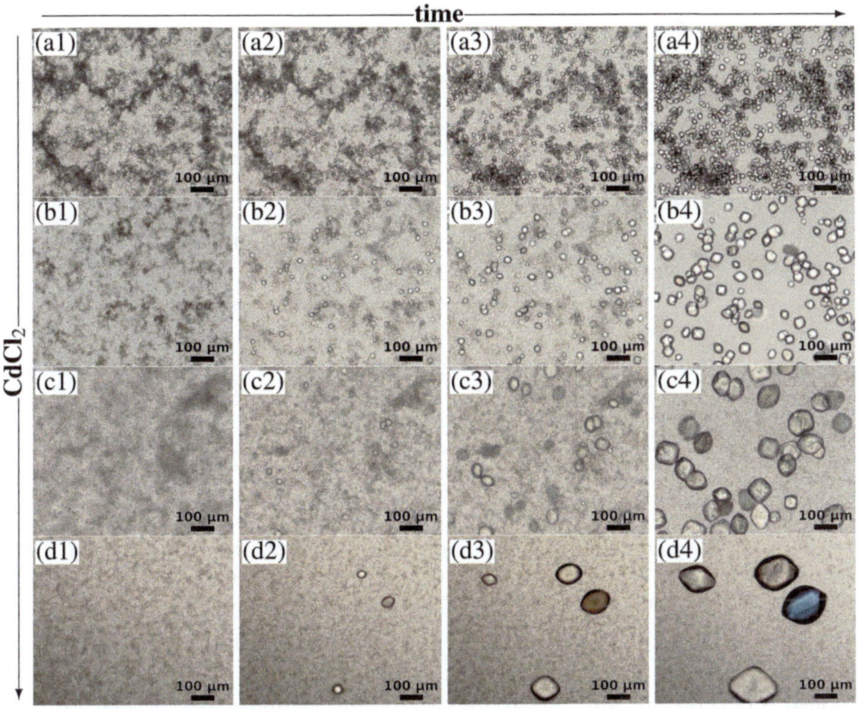

Fig. 3 Image series of protein crystallization for samples with a BLG concentration of 33 mg ml^{-1} and $CdCl_2$ at different times: (a) 17 mM, (b) 18 mM, (c) 19 mM and (d) 20 mM (increasingly less turbid). (a1) 0, (a2) 50 min (a3) 75 min (a4) 100 min after preparation. (b1) 0 (b2) 75 min (b3) 100 min (b4) 3 h. (c1) 0, (c2) 100 min, (c3) 3 h, (c4) 6 h. (d1) 0, (d2) 3 h, (d3) 6 h and (d4) 10 h. Videos of the crystallization process in (a) and (d) can be found in the ESI.†

resulting system contains a large number of small crystals, while most of the network of amorphous aggregates has turned into the crystalline phase. When the salt concentration is increased by just 1 mM, to 18 mM, the overall phenomenological picture changes dramatically. Large aggregates are still visible, but are not well connected to each other. The number of crystals is reduced, but the final crystals are bigger. When the salt concentration is further increased, protein aggregates become smaller and the number of crystals is further reduced, but the crystal size is larger. In the end, the MIP is consumed by crystal growth and the solutions become clear. The average number of crystals, normalized for an area of 1 mm^2, as a function of time is plotted in Fig. 4 for three conditions. The number of crystals initially increases with time, then saturates. The nucleation rates in the early stage are obtained from the slope of a linear fit; they are 1.44 ± 0.08, 0.32 ± 0.03 and 0.08 ± 0.01 min^{-1} for 18 mM, 19 mM and 20 mM $CdCl_2$, respectively. This decrease in the nucleation rate is expected, as the driving force is reduced with increasing salt concentration.

3.3 Structure of MIP revealed by SAXS and SANS

Due to the limited resolution of optical microscopy, SAXS and SANS were used to study the structure and the role of the MIP on the crystallization process. Firstly,

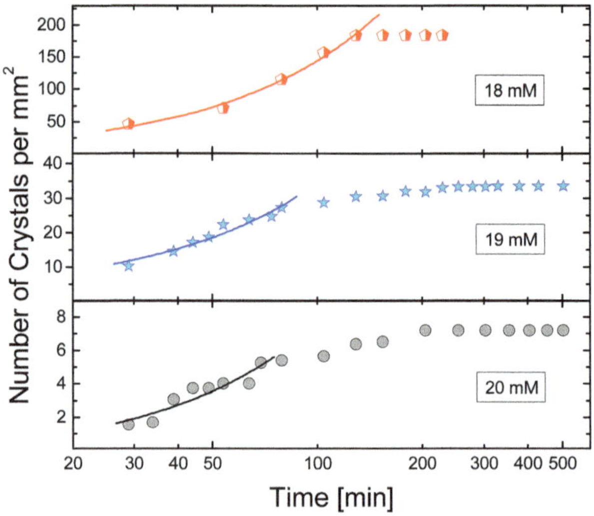

Fig. 4 Number of crystals as a function of time in a normalized area from Fig. 3 for samples with 18 mM, 19 mM and 20 mM $CdCl_2$. For the 19 and 20 mM plots, three series of picture sets were analyzed and averaged for better statistics.

typical SAXS results are shown in Fig. 5a for samples with a low protein concentration of 6.5 mg ml^{-1} and $CdCl_2$ concentrations covering all three regimes. In regime I, the scattering curve (with 0.5 mM salt) is dominated by the form factor of the BLG dimer, which is in agreement with the literature that BLG occurs predominantly as a dimer at room temperature and a pH between 3.5 and 7.5.[49] In regime II, with 1 mM salt, the high q part is still dominated by the form factor, but in the low q region, the increase in intensity indicates the formation of protein aggregates. In the third regime (12–90 mM, not all data are shown), a new feature forms shortly after preparation at $q \approx 0.7$ nm^{-1} and a sharp peak occurs at 2.2 nm^{-1}, as indicated by the arrows in this figure. A previous study of BLG in solution in the presence of YCl_3 has shown that this maximum corresponds to the monomer–monomer (M–M) correlation, due to the bridging effect.[50] Here, in the presence of $CdCl_2$, the peak is sharper. A possible explanation is the formation of a highly ordered fiber-like structure, which gives such a diffraction peak corresponding to the axial translation of the subunit (BLG monomer). Similar diffraction peaks have been observed in the solution scattering of F-actin.[51]

Fig. 5b presents SAXS data for samples under similar conditions as for the real-time optical microscopy study in Fig. 3. The SAXS profiles were collected during the crystallization process at a time at which the broad peak at $q \approx 0.7$ nm^{-1} has already developed, but Bragg peaks have not yet appeared. With 15 mM salt, the sample is located deep in regime II. While large aggregates are visible using optical microscopy (Fig. 3a), the SAXS profile shows a rather weak shoulder at $q \approx$ 0.7 nm^{-1}, indicating less order within the large aggregates. Increasing the salt concentration increases the ordering inside of the aggregates, as the broad peak becomes more pronounced. The data set with 17 mM salt was collected at a different beamtime with higher resolution, and also shows the monomer–monomer correlation peak at q around 2.2 nm^{-1}.

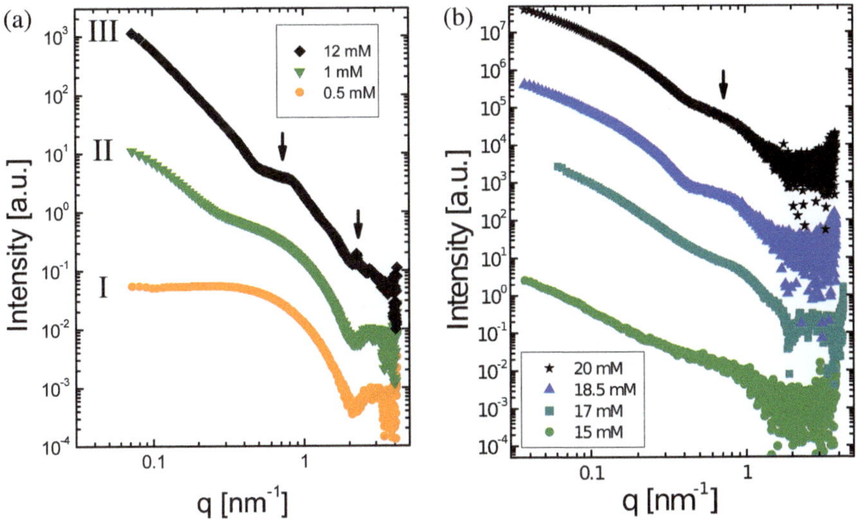

Fig. 5 (a) SAXS measurements at a protein concentration of 6.5 mg ml^{-1}, with $CdCl_2$ concentrations covering all three regimes of RC (marked by I, II, III). (b) Selected SAXS profiles at the time when the broad peak is well developed for samples in regime III with 33 mg ml^{-1} BLG and 15, 17, 18.5 and 20 mM $CdCl_2$, respectively. Data are shifted upward for clarity.

We further measured a similar set of samples using SANS. In this case, the samples were prepared in D_2O instead of H_2O for better contrast. Neutron scattering causes no radiation damage to the proteins in solution, and most importantly, SANS measurements apply a large beam size (6 mm × 12 mm), which covers a much larger volume containing a reasonable number of crystals. This is in contrast to SAXS measurements with a beam size of 0.3 × 0.15 mm, which contains a limited number of crystals in the case of the crystallization in Fig. 3c and d. The measurements were carried out for freshly prepared samples, 2 h and 6 h after preparation, respectively. The results shown in Fig. 6 suggest that the samples behave very similarly as to how they do in H_2O. First of all, in all cases, the fresh samples show the broad peak at $q \approx 0.7$ nm^{-1} which has been identified as the characteristic structural feature of the MIP. However, due to the low time resolution, the details of the development of this peak cannot be obtained. Nevertheless, for the sample with 20 mM salt, it is clear that this broad peak develops and becomes more pronounced after 2 h. In addition to this broad peak, the monomer–monomer correlation peak at q around 2 nm^{-1} is pronounced in all cases, which is in good agreement with the SAXS measurements. The slight shift of the monomer–monomer peak to a low q value in SANS is due to the hydration effect.[52] Secondly, after a certain time, smeared Bragg peaks appear for all samples. At the same time, the broad peak (MIP) reduces in its intensity or completely disappears. Although the low resolution of SANS at the high-q region smears the Bragg peaks, their positions are consistent with the SAXS measurements.

Both SAXS and SANS measurements reveal a similar structural feature of the MIP, *i.e.* the local ordering within the large protein aggregates, characterized by a

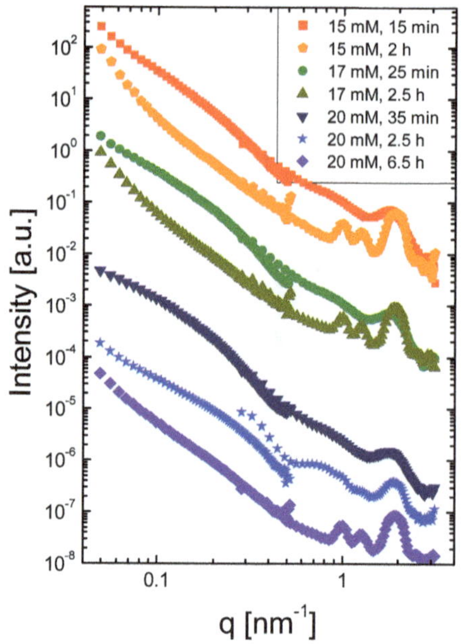

Fig. 6 BLG 33 mg ml^{-1} with 15, 17 and 20 mM $CdCl_2$ in D_2O measured by SANS. Data are shifted upward for clarity. The mismatch of the SANS curves (especially visible for 20 mM after 2.5 h) was caused by measurements at different detector positions, together with the fast kinetics occurring in the sample.

broad peak at $q \approx 0.7$ nm^{-1} and the monomer–monomer correlation peak at q around 2 nm^{-1}. As discussed in the following section, the broad peak is closely related to nucleation and crystal growth. It thus becomes the structural signature of the MIP.

3.4 Crystallization kinetics followed by real-time SAXS

To extract information on the underlying crystallization process, we employed real-time SAXS measurements of the crystallization kinetics with high time and structural resolution. Fig. 7 shows examples of time-resolved SAXS measurements for 33 mg ml^{-1} BLG with 17 (a) and 20 mM $CdCl_2$ (b) in 3D surface illustration and 2D projection; additional data for samples with 15 and 18.5 mM $CdCl_2$ are shown in the ESI, Fig. S2.† The bottom 2D projections are created by dividing all curves by the first one, and therefore visualize the ongoing changes in the system with time. Selected $I(q,t)/I(q,t=0)$ curves are further presented in Fig. 7c and d. The SAXS curves of both samples feature a strong increase at a low q which hardly changes with time; these indicate the presence of large aggregates, consistent with the observation by optical microscopy. With increasing time, a broad peak located at $q \approx 0.7$ nm^{-1} develops, which has been assigned to the nucleation precursors (MIP).[31] Once Bragg peaks appear, the most pronounced ones at 1.01 and 1.27 nm^{-1} overlap with the broad peak. The intensity of the Bragg peaks increases with time, while the broad peak shrinks. In the $I(q,t)/I(q,t=0)$ plots (Fig. 7c and d), it is clear that the broad peak appears before the Bragg peaks and becomes stronger with time. At the end of crystallization, the broad peak shrinks and eventually disappears (Fig. 7a and c).

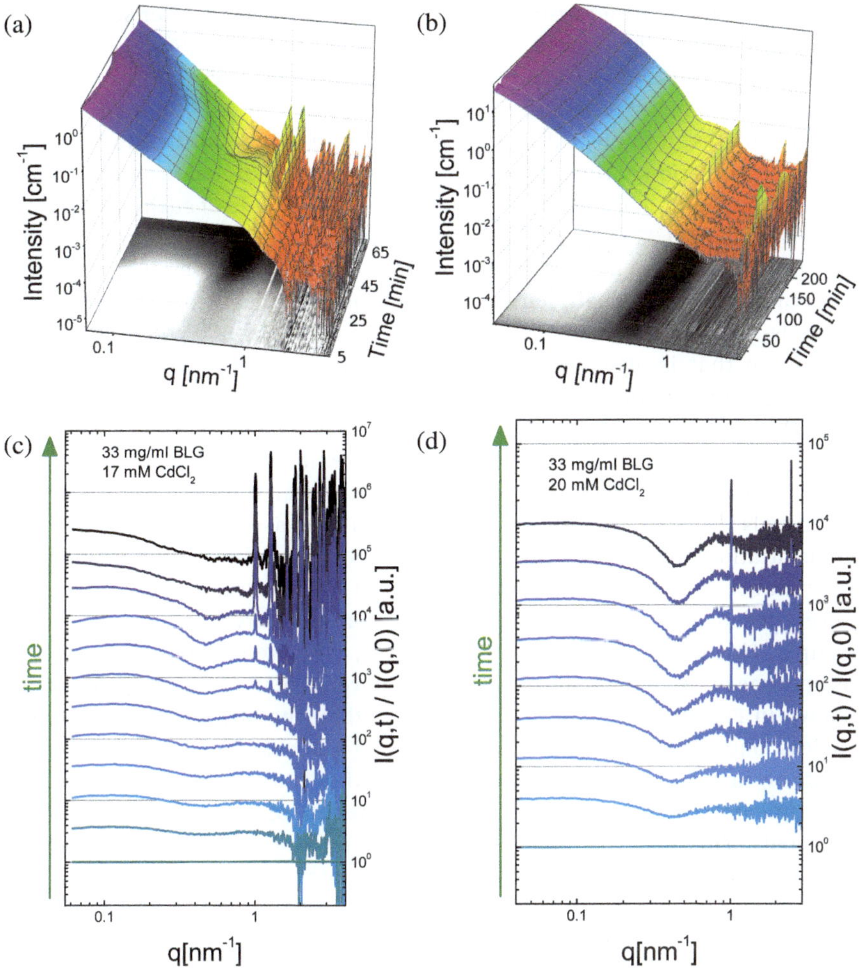

Fig. 7 Protein crystallization followed by real-time SAXS measurements. Samples contain 33 mg ml^{-1} BLG with 17 mM $CdCl_2$ (a & c) and 20 mM $CdCl_2$ (b & d), located in the transition zone of *pseudo-c*** (*i.e.* at the upper end of regime II). (a & b) Three-dimensional plots of $I(q,t)$. The bottom projections are created from $I(q,t)/I(q,t=0)$ and thus monitor the evolution of the curves with time. (c & d) Selected $I(q,t)/I(q,t=0)$ curves for the evolution of Bragg peaks and the amorphous peak. Data are shifted upward by a constant factor for better visibility.

From the optical microscopy experiments (Fig. 3) one observes that the MIP forms before crystallization starts, and is consumed during crystal growth. From real-time SAXS measurements, one can see that the typical broad peak for the MIP follows the same development: it appears first and develops, and once crystallization starts it reduces in intensity and eventually disappears. Based on these observations, we propose using the relative change of the area of this peak (representative of the MIP) and the two Bragg peaks which overlap with it, to quantify the relationship between the MIP and the crystalline phase as a function of time. At this point, we use the concept of crystallinity from semi-crystalline polymer systems for further data analysis.[53] After subtraction of the intensity at the minimum, the broad peak in the $I(q,t)/I(q,t=0)$ curves was fitted by a scaled Gaussian function and the Bragg peaks by two further (sharp) Gaussians.[31] The crystallization kinetics can be followed by the enveloped area of the broad region, A_{interm}, and the area of the Bragg peaks, A_{Bragg}, as a function of time. This method is further illustrated in an animation that can be found in the ESI†. Fig. 8a displays an example of such an analysis, for a sample with 33 mg ml^{-1} BLG and 17 mM $CdCl_2$. The development of the MIP (A_{interm}) shows a maximum at around 40 min, and the overall crystallinity (A_{Bragg}) has a plateau between 40 and 60 min, and then grows faster. Interestingly, the overall growth rate, *i.e.* the first derivative of A_{Bragg} on time, gives a maximum that is also located around 40 min, indicating that the overall crystal growth rate in the early stage strongly depends on the development of the MIP.

We have performed real-time SAXS measurements on all four salt conditions followed by optical microscopy. However, as seen from Fig. 3, the number of crystals decreases and the size of crystals increases with increasing salt concentration. This makes the real-time SAXS measurements challenging, as the number of crystals within the illuminated volume drops significantly. We have tried to compensate for this by measuring more positions from the sample. This is partially successful. However, the time resolution is reduced, as only one or two out of ten spots show the development of the Bragg peaks. As shown in Fig. 8b, less data for Bragg peaks than the MIP are shown. Nevertheless, one can still recognize the interesting kinetics: first the MIP develops relatively fast, and it then becomes saturated after 120 min. Within the current experimental time scale, only a minor fraction of crystalline phase was detected. The experimental observations on the kinetics, particularly including the non-monotonous crystallization rate (red dashed line in Fig. 8a), agree well with a simple model, which will be discussed in the following section.

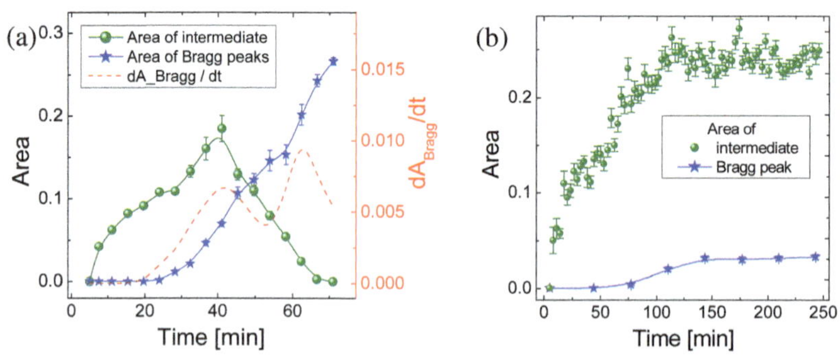

Fig. 8 Time-dependence of the area of the broad peak (A_{interm}) and the Bragg peaks (A_{Bragg}). Green spheres (left axis) show $A_{\text{interm}}(t)$, the integral of the broad Gaussian function connected to the intermediate. Blue stars (left axis) show $A_{\text{Bragg}}(t)$, the integral of the two Bragg peaks in this area. (a) BLG 33 mg ml^{-1} with 17 mM $CdCl_2$. For clarity and better statistics, at each time point three data points of $A_{\text{interm}}(t)$ and $A_{\text{Bragg}}(t)$ were merged into one. The dashed red line corresponds to the first derivative of the area of Bragg peaks as a function of time (right axis). (b) BLG 33 mg ml^{-1} with 20 mM $CdCl_2$.

4. Modeling with rate equations

In order to also compare the observed kinetic features more quantitatively to possible crystallization scenarios, we employed rate equation models. As essential variables of the modeling, L, I, C_I and C_L denote the mass fractions of free monomers in the liquid, in the intermediate, crystals in the intermediate and crystals in the liquid, respectively. The three paradigmatic cases for the crystallization process shown in Fig. 1 are presented as follows:

Classical nucleation $L \rightarrow C_L$

The classical nucleation theory contains a one-step nucleation process of the critical nucleus from the homogeneous solution. After nucleation, crystallites grow larger from the solution. In terms of modeling, $\Delta_{nL} = k_n L$ and $\Delta_{gL} = k_{gL} L C_L$ represent the nucleation and the growth term with rates k_n and k_{gL}. Using these, the process can be easily modeled by the following set of rate equations:

$$\partial_t L = -\Delta_{nL} - \Delta_{gL} \quad (1)$$

$$\partial_t C_L = \Delta_{nL} + \Delta_{gL}$$

Parallel process $I \leftrightarrow L \rightarrow C_L$

In this nonclassical process, we assume that in addition to the one-step crystallization process, a reversible intermediate is formed in the solution, competing with crystallization for the free monomers. In addition to the nucleation and growth term from the classical one-step nucleation, we include the formation of the intermediate with the term $\Delta_I = k_I(L - I)$. The corresponding set of rate equations reads

$$\partial_t L = -\Delta_I - \Delta_{nL} - \Delta_{gL} \quad (2)$$

$$\partial_t I = \Delta_I$$

$$\partial_t C_L = \Delta_{nL} + \Delta_{gL}$$

Two-step nucleation process $L \rightarrow I \rightarrow C_I \rightarrow C_L$

For true two-step nucleation, the formation of an intermediate is followed by the nucleation of crystallites within the intermediate. After the subsequent growth of the crystallites has consumed the intermediate, crystals emerge into the remaining liquid and grow further. Table 1 describes the steps in more detail and lists the rate terms. Using these, the rate equations read

$$\partial_t L = -\Delta_I - \Delta_{gL} \quad (3)$$

$$\partial_t I = \Delta_I - \Delta_{nI} - \Delta_{gI}$$

$$\partial_t C_I = \Delta_{nI} - \Delta_e + \Delta_{gI}$$

$$\partial_t C_L = \Delta_e + \Delta_{gL}$$

The results of the three models are shown in Fig. 9 with the model parameters from another study[31] (see also figure caption). For the classical nucleation (Fig. 9a), nucleation and growth of crystals speed up, until the liquid is halfway consumed, and saturate until complete consumption of the liquid. In the parallel process (Fig. 9b), the intermediate competes with crystallization and slows down the crystal nucleation and growth considerably. It is important to notice, however, that the crystallization rate is homogeneously increasing until saturation. By contrast, the two-step process (Fig. 9c) shows the formation of a plateau in the crystal mass fraction. After increasing at small times, the crystallization rate drops again considerably, whereas it increases monotonically for both one-step cases until saturation (Fig. 9d). This particular feature of the two-step process is caused by the nucleation and slow growth in the intermediate, while the crystals grow faster once emerged into solution. Thus, the occurrence of the plateau indicates the presence of a multi-step nucleation process.

The large number of model parameters does not allow for a reliable extraction of nucleation rates *via* model fits to the kinetic analysis of the SAXS data. We emphasize that this is not a problem of the model or the data quality, but the complex pathway, which involves many coupled processes. As an example, the amount of crystal nucleation within the MIP depends not only on the rate k_n, but also on the amount of MIP. The latter is mainly determined by the supersaturation represented by L_0, at least for fast formation of the MIP compared to crystal

Table 1 Summary of the proposed two-step process. From the intermediate formed firstly from the liquid, crystals nucleate and grow until they consume the intermediate. Subsequently, the crystals emerge into the liquid and grow from the free monomers. $(x)_+$ stands for x for positive x and zero otherwise. L, I, C_I and C_L denote the mass fractions of free monomers in the liquid, in the intermediate, crystals in the intermediate and crystals in the liquid, respectively

Processes	Rates	Parameters
1. Formation of intermediate until the remaining solution is stable	$\Delta_I = k_I(L - L_0)_+$	L_0 mass fraction of stable solution k_I formation rate
2. Crystal nucleation within the intermediate	$\Delta_{nI} = k_n I$	k_n nucleation rate
3. Crystal growth within the intermediate	$\Delta_{gI} = k_{gI} I C_I$	k_{gI} growth rate in intermediate
4. Crystals emerge into the liquid once intermediate is consumed	$\Delta_e = k_e C_I(\alpha_I C_I - I)_+$	α_I critical ratio of intermediate and crystals k_e emergence rate
5. Crystal growth in the liquid	$\Delta_{gL} = k_{gL} L C_L$	k_{gL} growth rate in liquid

nucleation. Based only on fitting the kinetic data from SAXS, an decrease in L_0 will consequentially cause a decrease in k_n. While additional information *e.g.* on the supersaturation are therefore required for a real quantitative fit, the qualitative signatures of the two-step process, such as the plateau in the crystal fraction, can still be used to provide evidence for the scenario of a two-step nucleation process.

Nevertheless, this simple model can reproduce the experimental crystallization kinetics at different conditions. This can be achieved either by varying the rate parameters or by choosing the amount of MIP which approximately approaches $L - L_0 = 0.2$. This value can be determined experimentally by following the protein concentration in the supernatant over time using UV-visible

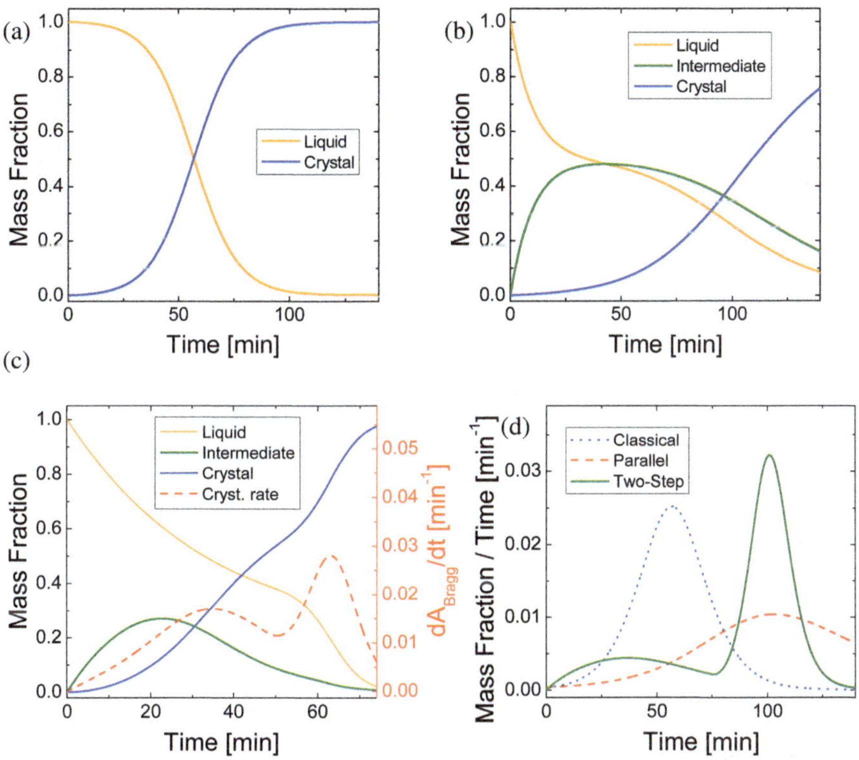

Fig. 9 Comparison between different crystallization scenarios: (a) classical nucleation from a homogeneous liquid and subsequent growth of the crystallites. (b) Parallel processes, representing the formation of an intermediate and independent nucleation from the liquid with subsequent growth of the crystallites. (c) Two-step nucleation process involving the formation of an intermediate from which the crystals nucleate (see Table 1 for further details). The following parameter values were used to show the good qualitative agreement of the model with the data set in Fig. 8a: $k_e = 0.15$ min^{-1}, $L_0 = 0.2$, $\alpha_I = 0.2$, $k_{gL} = 0.6$ min^{-1}, $k_I = 0.03$ min^{-1}, $k_n = 0.02$ min^{-1}, $k_{gI} = 0.2$ min^{-1}. An additional model plot reproducing the data set in Fig. 8b can be found in the ESI, Fig. S3.† (d) Comparison of the crystallization rates d*C*/d*t*. While one-step and parallel nucleation processes show a monotonous speed-up until saturation, the two-step process can have a non-monotonous signature with two maxima. The model parameters were chosen for all three models to be the same: $k_I = 0.05$ min^{-1}, $k_n = 0.02$ min^{-1}, $k_{gL} = 1.0$ min^{-1}, $k_{gL} = 0.2$ min^{-1}, $k_e = 1.0$ min^{-1}, $L_0 = 0.7$, $\alpha_I = 0.2$.

spectroscopy. A tentative experiment for a sample with 20 mM $CdCl_2$ leads to a value of 0.2. Further experiments are needed to refine this parameter.

5. Conclusions and outlook

We have investigated the two-step nucleation process of protein crystallization in solutions by following the overall crystallization kinetics, using real-time optical microscopy and SAXS. The experimental results, together with a rate equation model, provide evidence of two-step nucleation in the early stage of crystallization. The BLG-salt ($CdCl_2$) solutions were chosen at the transition zone of *pseudo-c***, where small aggregates form after sample preparation. These protein aggregates serve as the metastable intermediate phase (MIP) during crystallization.

SAXS and SANS reveal that the MIP shows a certain local ordering instead of random aggregates, as monitored by a broad shoulder at intermediate $q \approx 0.7$ nm^{-1}, and a monomer–monomer correlation peak at q around 2 nm^{-1}. Real-time SAXS results show that the crystallization kinetics is proportional to the development of the MIP in the early stage of crystallization, *i.e.* there appears a local maximum in the crystallization rate at the maximum quantity of the intermediate. In the late stage of crystallization, a plateau develops due to the transition from nucleation controlled in the early stage, to growth controlled after the consumption of MIP. This transition in the overall crystallization kinetics is a typical feature for two-step nucleation in the early stage. These experimentally observed kinetics can be reproduced using a rate equation model.

For further real-time measurements, we note that the smaller beam size and scattering volume of SAXS can be compensated by using SANS. The combination of real-time SAXS and SANS could provide more systematic information of the crystallization kinetics.

Acknowledgements

A. Sauter acknowledges a fellowship of the Landesgraduiertenförderung. The authors thank T. Stehle (IFIB, Universität Tübingen) for sharing lab resources. This work was supported by the DFG.

References

1 L. C. Jacobson, W. Hujo and V. Molinero, *J. Am. Chem. Soc.*, 2010, **132**, 11806–11811.
2 H. Cölfen and M. Antonietti, *Mesocrystals and nonclassical crystallization*, John Wiley & Sons, West Sussex, 2008.
3 D. Gebauer, A. Völkel and H. Cölfen, *Science*, 2008, **322**, 1819–1822.
4 D. Gebauer, M. Kellermeier, J. D. Gale, L. Bergström and H. Cölfen, *Chem. Soc. Rev.*, 2014, **43**, 2348–2371.
5 A. F. Wallace, L. O. Hedges, A. Fernandez-Martinez, P. Raiteri, J. D. Gale, G. A. Waychunas, S. Whitelam, J. F. Banfield and J. J. De Yoreo, *Science*, 2013, **341**, 885–889.
6 R. P. Sear, *Int. Mater. Rev.*, 2012, **57**, 328–356.

7 O. Gliko, N. Neumaier, W. Pan, I. Haase, M. Fischer, A. Bacher, S. Weinkauf and P. G. Vekilov, *J. Am. Chem. Soc.*, 2005, **127**, 3433–3438.
8 O. Gliko, W. Pan, P. Katsonis, N. Neumaier, O. Galkin, S. Weinkauf and P. G. Vekilov, *J. Phys. Chem. B*, 2007, **111**, 3106–3114.
9 W. Pan, O. Galkin, L. Filobelo, R. L. Nagel and P. G. Vekilov, *Biophys. J.*, 2007, **92**, 267–277.
10 W. Pan, P. G. Vekilov and V. Lubchenko, *J. Phys. Chem. B*, 2010, **114**, 7620–7630.
11 Q. Chen, P. G. Vekilov, R. L. Nagel and R. E. Hirsch, *Biophys. J.*, 2004, **86**, 1702–1712.
12 Y. Liu, X. Wang and C. B. Ching, *Cryst. Growth Des.*, 2010, **10**, 548–558.
13 T. Schilling, H. J. Schöpe, M. Oettel, G. Opletal and I. Snook, *Phys. Rev. Lett.*, 2010, **105**, 025701.
14 P. R. T. Wolde and D. Frenkel, *Science*, 1997, **277**, 1975–1978.
15 S. B. Hutchens and Z.-G. Wang, *J. Chem. Phys.*, 2007, **127**, 084912.
16 G. I. Tóth, T. Pusztai, G. Tegze, G. Tóth and L. Gránásy, *Phys. Rev. Lett.*, 2011, **107**, 175702.
17 H. J. Schöpe, G. Bryant and W. van Megen, *Phys. Rev. Lett.*, 2006, **96**, 175701.
18 T. H. Zhang and X. Y. Liu, *J. Am. Chem. Soc.*, 2007, **129**, 13520–13526.
19 J. R. Savage and A. D. Dinsmore, *Phys. Rev. Lett.*, 2009, **102**, 198302.
20 M. Sleutel and A. E. S. Van Driessche, *Proc. Natl. Acad. Sci. U. S. A.*, 2014, **111**, E546–E553.
21 A. Bijelic, C. Molitor, S. G. Mauracher, R. Al-Oweini, U. Kortz and A. Rompel, *ChemBioChem*, 2015, **16**, 233–241.
22 P. Rein ten Wolde and D. Frenkel, *Phys. Chem. Chem. Phys.*, 1999, **1**, 2191–2196.
23 P. G. Vekilov, *Cryst. Growth Des.*, 2004, **4**, 671–685.
24 P. Tan, N. Xu and L. Xu, *Nat. Phys.*, 2014, **10**, 73–79.
25 Y. Peng, F. Wang, Z. Wang, A. M. Alsayed, Z. Zhang, A. G. Yodh and Y. Han, *Nat. Mater.*, 2015, **14**, 101–108.
26 T. Kawasaki and H. Tanaka, *Proc. Natl. Acad. Sci. U. S. A.*, 2010, **107**, 14036–14041.
27 J. F. Lutsko and G. Nicolis, *Phys. Rev. Lett.*, 2006, **96**, 046102.
28 R. Piazza, *Curr. Opin. Colloid Interface Sci.*, 2004, **8**, 515–522.
29 R. P. Sear, *Curr. Opin. Colloid Interface Sci.*, 2006, **11**, 35–39.
30 F. Zhang, M. W. A. Skoda, R. M. J. Jacobs, R. A. Martin, C. M. Martin and F. Schreiber, *J. Phys. Chem. B*, 2007, **111**, 251–259.
31 A. Sauter, F. Roosen-Runge, F. Zhang, G. Lotze, R. M. J. Jacobs and F. Schreiber, *J. Am. Chem. Soc.*, 2015, **137**, 1485–1491.
32 H. A. Sober, *CRC Handbook of Biochemistry: Selected data for molecular biology*, CRC, 1970.
33 U. M. Elofsson, M. A. Paulsson and T. Arnebrant, *Langmuir*, 1997, **13**, 1695–1700.
34 F. Zhang, S. Weggler, M. J. Ziller, L. Ianeselli, B. S. Heck, A. Hildebrandt, O. Kohlbacher, M. W. A. Skoda, R. M. J. Jacobs and F. Schreiber, *Proteins: Struct., Funct., Bioinf.*, 2010, **78**, 3450–3457.
35 T. Narayanan, Synchrotron small-angle X-ray scattering, in *Soft Matter Characterization*, ed. R. Borsali and R. Pecora, Springer, 2008, p. 899ff.
36 V. Pipich, QtiKWS: *user-friendly program for reduction, visualization, analysis and fit of SA(N)S data*, http://www.qtikws.de, 2012.

37 F. Zhang, M. W. A. Skoda, R. M. J. Jacobs, S. Zorn, R. A. Martin, C. M. Martin, G. F. Clark, S. Weggler, A. Hildebrandt, O. Kohlbacher and F. Schreiber, *Phys. Rev. Lett.*, 2008, **101**, 148101.
38 F. Zhang, G. Zocher, A. Sauter, T. Stehle and F. Schreiber, *J. Appl. Crystallogr.*, 2011, **44**, 755–762.
39 F. Zhang, R. Roth, M. Wolf, F. Roosen-Runge, M. W. A. Skoda, R. M. J. Jacobs, M. Sztucki and F. Schreiber, *Soft Matter*, 2012, **8**, 1313–1316.
40 F. Zhang, F. Roosen-Runge, A. Sauter, M. Wolf, R. M. J. Jacobs and F. Schreiber, *Pure Appl. Chem.*, 2014, **86**, 191–202.
41 F. Roosen-Runge, B. S. Heck, F. Zhang, O. Kohlbacher and F. Schreiber, *J. Phys. Chem. B*, 2013, **117**, 5777–5787.
42 E. Jordan, F. Roosen-Runge, S. Leibfarth, F. Zhang, M. Sztucki, A. Hildebrandt, O. Kohlbacher and F. Schreiber, *J. Phys. Chem. B*, 2014, **118**, 11365–11374.
43 D. Soraruf, F. Roosen-Runge, M. Grimaldo, F. Zanini, R. Schweins, T. Seydel, F. Zhang, R. Roth, M. Oettel and F. Schreiber, *Soft Matter*, 2014, **10**, 894–902.
44 F. Roosen-Runge, F. Zhang, F. Schreiber and R. Roth, *Sci. Rep.*, 2014, **4**, 7016.
45 A. Kubíčková, T. Křížek, P. Coufal, M. Vazdar, E. Wernersson, J. Heyda and P. Jungwirth, *Phys. Rev. Lett.*, 2012, **108**, 186101.
46 R. Akiyama and R. Sakata, *J. Phys. Soc. Jpn.*, 2011, **80**, 123602.
47 A. Sauter, M. Oelker, G. Zocher, F. Zhang, T. Stehle and F. Schreiber, *Cryst. Growth Des.*, 2014, **14**, 6357–6366.
48 L. Ianeselli, F. Zhang, M. W. A. Skoda, R. M. J. Jacobs, R. A. Martin, S. Callow, S. Prévost and F. Schreiber, *J. Phys. Chem. B*, 2010, **114**, 3776–3783.
49 M. Verheul, J. S. Pedersen, S. P. F. M. Roefs and K. G. de Kruif, *Biopolymers*, 1999, **49**, 11–20.
50 F. Zhang, F. Roosen-Runge, A. Sauter, R. Roth, M. W. A. Skoda, R. Jacobs, M. Sztucki and F. Schreiber, *Faraday Discuss.*, 2012, **159**, 313–325.
51 T. Sato, T. Shimozawa, T. Fukasawa, M. Ohtaki, K. Aramaki, K. Wakabayashi and S. Ishiwata, *Biophysics*, 2010, **6**, 1–11.
52 F. Zhang, F. Roosen-Runge, M. W. A. Skoda, R. M. J. Jacobs, M. Wolf, P. Callow, H. Frielinghaus, V. Pipich, S. Prevost and F. Schreiber, *Phys. Chem. Chem. Phys.*, 2012, **14**, 2483–2493.
53 G. Strobl, in *The physics of polymers*, Springer, Berlin, Heidelberg, New York, 3rd edn, 2007, ch. 5.

Faraday Discussions

PAPER

High-resolution insights into the early stages of silver nucleation and growth†

Cornelia M. Völkle, Denis Gebauer* and Helmut Cölfen*

Received 17th December 2014, Accepted 12th January 2015
DOI: 10.1039/c4fd00269e

Nucleation and growth of silver nanoparticles has already been investigated with various experimental and computational tools. However, owing to inherent problems associated with the analytical characterization of nucleation processes, there is a general lack of experimental data regarding the earliest precursors and smallest Ag(0) clusters. Here, we address this problem by the application of Synthetic Boundary Crystallization Ultracentrifugation, utilizing a multiwavelength detector for the first time, complemented by a specialized titration assay. These techniques shed new light on silver nanoparticle precursors existing in the pre-nucleation regime, and the initially nucleated ensemble of nanoclusters. For the first time, we present experimental data of UV-Vis spectra for fractionated silver clusters. These allow for unsurpassed insights into the sequence of nucleation and early growth species as well as their optical properties.

Introduction

Nucleation is the fundamental event in many phase transitions, and thus is of great importance for various reactions and processes.[1] Thanks to its great relevance, nucleation has been intensely studied for more than 100 years. The first quantitative concept to describe nucleation phenomena, Classical Nucleation Theory (CNT), was based upon the ideas of Gibbs,[2] and was eventually finalized by the pioneering work of Becker and Döring.[3] This theory has been modified over the years,[4–7] and alternative approaches have also been introduced, which do not rely upon any assumptions about the homogeneity of nuclei, and do not require the division of their free energy into a bulk and a surface contribution.[8] Later, the so-called two-step mechanism was introduced to explain the considerable deviations between experimentally determined nucleation rates and CNT predictions, especially close to the critical point.[9–11] Recently, the view on nucleation has been further extended by the so-called pre-nucleation cluster pathway.[12] In this case, thermodynamically stable (with respect to monomers in solution) pre-nucleation

Department of Chemistry, Physical Chemistry, University of Konstanz, Universitätsstr. 10, D-78457 Konstanz, Germany. E-mail: denis.gebauer@uni-konstanz.de; helmut.coelfen@uni-konstanz.de

† Electronic supplementary information (ESI) available: Movie showing the sedimentation in the AUC, which is presented as snapshots in Fig. 1. See DOI: 10.1039/c4fd00269e

clusters serve as fundamental solute precursors to phase-separated entities *via* a nanoscopic liquid–liquid separation event.

All of these different pathways require experimental verification, and detailed molecular insights are needed to expand and refine the models—or even refute their applicability. This is, by principle, very difficult to obtain for nucleation processes, since the relevant species are small, highly dynamic, and also transient. A further complication arises from the fact that nucleation processes can be very fast, owing to the high levels of supersaturation that are, for example, crucial to achieve narrow particle size distributions,[13] and thus challenge existing analytical methods.

Currently, one of the best-suited techniques to time-resolved studies of nucleation, as well as early particle growth processes, is X-ray scattering. With synchrotron light sources, a time resolution of 200 ms can be achieved.[14] More recently, reaction times down to 100 ms could be studied even on a standard laboratory instrument, by taking advantage of a continuous flow experiment.[15] With a stopped-flow device and synchrotron radiation, species can be observed in time intervals as short as 10 ms after mixing.[16] The same time resolution was achieved in a reaction tube and free reactant jet, coupled to a rapid mixing device; this was achieved by observing the reaction products with SAXS and TEM, after shooting the grids through the free jet and rapidly cooling in liquid nitrogen.[17] Here, the formation of liquid droplets was observed during the very early stages, strongly suggesting that CNT is not applicable in this case. With a free jet set-up coupled to laminar mixing, the time resolution at a synchrotron beamline could be increased by an impressive factor of 100, allowing the observation of reaction times as short as 75–100 μs.[18,19] This time resolution seems to be the current limit of SAXS or WAXS experimentation at synchrotron beamlines. While this is certainly sufficient to observe early growth species, and perhaps also nucleation phenomena for many systems, SAXS gives only average particle size and shape information, which can be complemented by crystal structure information *via* simultaneous WAXS investigation. Nevertheless, it is certainly desirable to observe a sequence of growth species from pre-nucleation species and intermediates to larger nanoparticles, and to determine their composition and properties. While a free reactant jet is in principle highly suitable for such analyses – since it allows different analytical methods to be applied along the jet and thus the capture of species formed at different times even if they require long integration times – it is not always possible to capture the nucleation and early growth species, due to the current time resolution limit of approximately 75–100 μs.[18,19]

To circumvent this problem, an analytical ultracentrifuge (AUC) method was developed, called "Synthetic Boundary Crystallization Ultracentrifugation".[20,21] The idea behind this approach is to bring two reactants together in a so-called synthetic boundary cell. Speeding up the ultracentrifuge rotor leads to the formation of a sharp reaction boundary. If the overlaid reactant is added in tiny amounts, the chemical reaction that leads to the nucleation and growth of particles will occur only for a short time. Following this, particle growth of nucleated species is no longer possible, because they leave the reaction zone and become subject to fractionation, and characterization, in the AUC.[20,21] AUC was shown to provide very high particle size resolution in the Ångström range, even for complex mixtures and sizes down to below 1 nm, with ultimate statistical relevance, as every particle of the ensemble is detected.[22,23] Moreover, it proved to be one of the very few reliable techniques to quantitatively characterize pre-nucleation clusters,[24] since electron microscopy is limited by insufficient statistical sample sizes, leading to fundamental issues when it comes to the statistically significant detection of very small and rare species in solution.[25–27] Mass spectrometry, on the other hand, has the inherent problem that it is not clear *per se* if the detected species truly reflect the solution composition, since some species might be problematic to transfer into the gas phase, or might be even generated artificially upon ionization.[25]

The advantage of Synthetic Boundary Crystallization Ultracentrifugation is that the technique works especially well for fast reacting systems, where the chemical reaction is finished within the first seconds of the experiment; this therefore provides well-defined initial conditions for the nucleation of the ensemble of sedimenting particles and the startpoint of sedimentation. However, the advantage of high resolution analysis could not be utilized at the time when the method was first introduced, due to the lack of suitable methods for the evaluation of the experiments.[20,21] Nowadays, powerful evaluation programs like SEDFIT[28] and ULTRASCAN[29] are available, and provided that the reaction is fast, a static sample (*i.e.* the particles do not grow upon fractionation) will be investigated, which can be evaluated with algorithms developed for band sedimentation experiments.[30] These yield the sedimentation coefficient distribution – even corrected for diffusion broadening – that can be converted into the corresponding particle size distribution.

In addition, a UV-Vis multiwavelength detector has been developed within the so-called open AUC project,[31] which allows for the measurement of UV-Vis spectra at each point in the AUC cell, and thus combines the excellent size resolution of AUC with spectral information for each of the detected species.[32–34] Such a detector is especially promising for the investigation of the nucleation and early growth species of semiconductors and metal nanoparticles, because of their size-dependent optical and electronic properties.[35,36] This would allow for particle size analyses with Ångström resolution, yielding the UV-Vis absorption spectra for each identified species, potentially even starting in the pre-nucleation regime.

The nucleation of silver nanoparticles has been investigated with a series of different techniques.[37] The reduction of silver salts by sodium borohydride and tetraoctylammonium acetate was first studied utilizing UV-Vis spectrometry by Rothenberg *et al.*[38] Henglein *et al.* studied the γ-irradiation induced formation of colloidal silver from $AgClO_4$ solutions, where citrate was used as a stabilizer.[39] They postulated two growth mechanisms that depend on the citrate concentration: the condensation of small silver clusters (type-I), and the reduction of Ag^+ on the particles *via* radical-to-particle electron transfer (type-II). When the concentration of citrate was too low, coalescence of the formed silver nanoparticles occurred. Harada and Katagiri studied the photo-reduction of $AgClO_4$ solutions in the presence of poly(*N*-vinyl-2-pyrrolidone) (PVP) utilizing SAXS.[40] They found that the rate of nucleation strongly depends on the initial metal concentration, which is at least qualitatively consistent with CNT, whereas the early growth processes can be subdivided into three distinct stages; reduction–nucleation, Ostwald ripening, and particle coalescence. The nucleation process yielded small particles of $\sim$2.5 nm average radius, whereas the ripening and coalescence processes led to particles of $\sim$11.5 nm average radius. All of this is, in principle, consistent with the Finke–Watzky mechanism;[41] this stipulates that the first step in metal nanocluster formation *via* reduction with hydrogen is slow continuous nucleation, and the second step is autocatalytic surface growth, which is not controlled by diffusion. This mechanism provides several key predictions, such as a certain predominance of magic-number size nanoclusters, or the use of face-selective capping agents as a way to block the autocatalytic surface growth and, thereby, to provide a means to develop tailor-made nanoclusters. Richards *et al.* utilized a combination of TEM and UV-Vis techniques to study the growth of Ag-nanoparticles from a $[(PPh_3)_2Ag(O_2CC_{13}H_{27})]$ precursor.[42] They found that classical nucleation and growth, aggregative nucleation and growth, and Ostwald ripening occurred in consecutive experimental stages, whereby growth was dominated by the aggregative regime. Takesue *et al.* utilized time resolved SAXS experimentation (0.18 ms time resolution) to elucidate the aggregation-based formation of silver nanoparticles from silver nitrate solutions, and corroborated the existence of three distinct stages.[43] Their analyses showed that the silver nanoparticles were essentially formed from precursors that related to a peak diameter of ~0.7 nm, which corresponds to the size of the Ag_{13} cluster. The study suggests that this magic cluster is most likely the elementary growth unit which agglomerates to form silver nanoparticles. Woehl *et al.* studied the electron-beam induced nucleation of silver nanoparticles from silver nitrate solutions by means of liquid-cell TEM, whereby the electron beam current could be used to control the growth mechanism.[44] Low electron beam dosage led to reaction-limited growth (yielding faceted nanocrystals), but high electron beam dosage resulted in diffusion-limited growth that yielded spherical crystals. Quantitative assessments of nucleation and growth rates suggested that the results could be explained by classical models. A gradual colour change from clear to yellow in silver nitrate solutions with sodium acrylate in the presence of NaOH was observed by Nishimura *et al.*, who utilized UV-Vis and XAFS spectroscopy to study the process of Ag-nanoparticle formation in detail.[45] They found that the nucleation rate dramatically increased with an increase in NaOH concentration, and developed an alternative reaction channel *via* Ag^+ species (such as $Ag(OH)_x$). These can serve as intermediates in the reduction, increasing the formation rate of the nanoparticles in the presence of NaOH. Investigating the formation of silver nanoparticles from silver perchlorate, with and without PVP, by means of time-resolved *in situ* SAXS, Polte *et al.* found a four-stage mechanism.[15] Without PVP, within the first 100 ms, particles of ~1 nm size almost instantaneously formed, followed by a rapid increase in their number. The particles then grew to an average size of 4.6 nm, whereas the particle number decreased. Again, this is consistent with a growth mechanism *via* aggregation of nanoscale entities. With PVP, the precursor particles of ~1 nm in size could be stabilized significantly before coalescence occurred.

All of the above shows that there is a general lack of high-resolution data, especially for the formation of the first tiny nanoparticle – or rather nanocluster – precursors, which appear to be fundamental growth units for subsequent Ag nanoparticle growth.[43] The motivation of the present study is to apply, for the first time, a multiwavelength UV-Vis detection in Synthetic Boundary Ultracentrifugation experiments, and obtain unsurpassed information about the nucleation and early particle growth stages of silver nanoparticles. This model system has been chosen because it is literature known (see above), and allows for the combination of AUC experiments with titration experiments, which have proven to be very successful for the characterization of the early stages of $CaCO_3$ precipitation.[24]

The titration assay is based upon the slow mixing of precursor solutions, and the continuous generation of supersaturation, which eventually leads to precipitation. This process can be followed *in situ* by parallel pH titration and ion potential measurements, and can be further complemented by redox potential and conductivity measurements.[46] The titration assay thus provides quantitative information on free and bound precursor species, which can be used to characterize pre- and post-nucleation solute associates using suitable thermodynamic models of speciation.[24] Moreover, the experimentation can be used to identify, categorize and quantify the multiple effects of additives during precipitation, and can thus provide novel insights into the mechanisms of additive-mediated crystallization control.[47,48]

Materials and methods

Analytical ultracentrifugation

For the AUC measurements, an UV-Vis multiwavelength AUC has been used. The setup has been described in the literature.[32–34] The rotor speed was 60 000 rpm. All experiments were performed at 25 °C. In this work, we applied a 12 mm charcoal filled Epon double-sector synthetic boundary cell of the Vinograd type. This cell utilizes the centrifugal force to form a reaction boundary by layering $NaBH_4$ (ρ = 0.9974 g mL^{-1}) in a reservoir onto a more dense $AgNO_3$/CTAB solution (ρ = 1.1044 g mL^{-1}) in one sector of a double-sector centrepiece *via* a thin capillary. The two solutions build a sharp boundary, at which the reaction takes place. Although the diffusion rate of the ions is very high, the fast reaction to an insoluble Ag cluster/nanoparticle[43] prevents extensive diffusion broadening of the boundary.

The synthesis of the silver nanoparticles is roughly based on the recipe of Pal *et al.*[49] 6 μL 10 mM $NaBH_4$ were layered onto 290 μL 0.1 mM $AgNO_3$ and 0.05 M CTAB solution in a 1 : 1 mixture of H_2O and D_2O. $AgNO_3$ was purchased from Roth, $NaBH_4$ from Merck and cetyltrimethylammoniumbromid (CTAB) from Acros. All chemicals were used as received.

The data were evaluated using the SEDFIT software Vers. 14.1 and the models Analytical Zone Centrifugation ls-g*(s) called g(s) in the following for simplicity, and Analytical Zone Centrifugation c(s). The partial specific volume of the sample was taken as 0.0953 mL g^{-1}.

Titration assay

The principal set-up is described in detail elsewhere.[46] Here, we utilized an Ag-redox electrode (Metrohm, no. 6.0430.100), and a double junction pH probe (Metrohm, no. 6.0269.100), where the outer reference electrolyte was 3 M KNO_3. Solutions for the titration were based on a recipe by Doty *et al.*[50] In brief, an aqueous 2.5 mM $AgNO_3$ and 2.5 mM trisodium citrate hydrate solution was dosed into 50 mL freshly prepared 10 mM $NaBH_4$ solution at a rate of 0.1 mL min^{-1}. The pH of the $NaBH_4$ solution was initially around ~9.8, was then set to pH 9.73, and was subsequently held constant by automatic counter-titration with 0.25 M HNO_3. The experimental duration was 1500 s.

For the measurements of UV-Vis spectra, a 600 μL sample was drawn after 148 s, 370 s, 570 s, 750 s, 940 s, 1130 s, 1320 s and 1500 s of silver addition, and was directly measured each time on a Varian Cary 50 UV-Vis spectrometer, utilizing the software Varian UV Scan Version 3.99(339). In another set of experiments, 600 μL samples were drawn after 100 s, 160 s, 190 s, 220 s, 250 s, 280 s, 310 s, 340 s, 370 s, 400 s, 500 s, 600 s, 800 s, 1000 s, 1200 s and 1400 s of silver addition, and were measured accordingly, but after the titration assay was finished (ripening for *ca.* 10–15 minutes for each sample).

$AgNO_3$ was purchased from Roth, $NaBH_4$ from Merck and trisodium citratehydrate from Sigma-Aldrich. All chemicals were used as received.

Results and discussion

Analytical ultracentrifugation

Overlaying of a $NaBH_4$ solution onto the $AgNO_3$ solution upon speeding up the ultracentrifuge rotor leads to an immediate reaction and the formation of Ag nanoparticles. Due to the fast reaction, particles can be generated and analysed at the very early stages, which is immediately obvious from their very slow sedimentation, even at the AUC's maximum speed of 60 000 rpm (corresponding to a centrifugal field strength of 280 000*g*). The raw data show that the sedimentation is so slow that the peak in the sedimentation profile spreads out completely, thanks to diffusion broadening, before the sample has reached the bottom of the AUC cell (Fig. 1 and Movie SI 1†).

Following the sedimentation profile peak intensity evolution with time (Fig. 2), the intensity increases during the first 6 scans (red scans in Fig. 2). This indicates

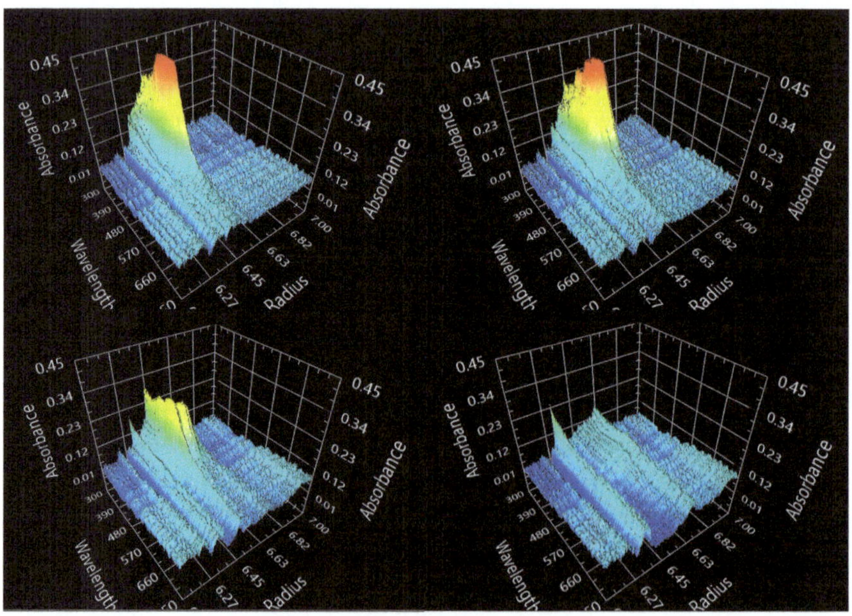

Fig. 1 Experimental scans taken at different times of sedimentation (top left 13.7 min, top right 38.8 min, bottom left 80.9 min, bottom right 169.1 min). See also Movie SI 1.†

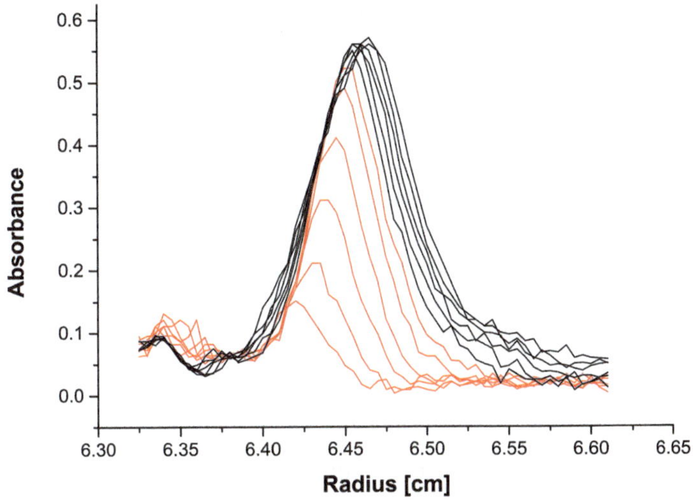

Fig. 2 Experimental radial absorption scans at 450 nm after the $NaBH_4$ solution was overlaid over the $AgNO_3$ solution. A slight movement of the reaction boundary towards the cell bottom can be observed as a result of the *ca.* 2-times faster BH_4^- diffusion as compared to Ag^+.[51,52]

that the amount of reducing agent was still slightly too high, so it was only used up completely after 617 seconds, due to the need for ion diffusion for further reaction, despite the very fast reduction in the ms time range.[43] Therefore, these scans were excluded from further evaluation of the sedimentation coefficient distribution, as this requires static conditions for the sample, which were only reached after 617 seconds. Although the relevant ion diffusion coefficients are 1.6 $\times 10^{-10}$ $m^2 s^{-1}$ (Ag^+)[51] and 8.4 $\times 10^{-10}$ $m^2 s^{-1}$ (BH_4^-),[52] which would correspond

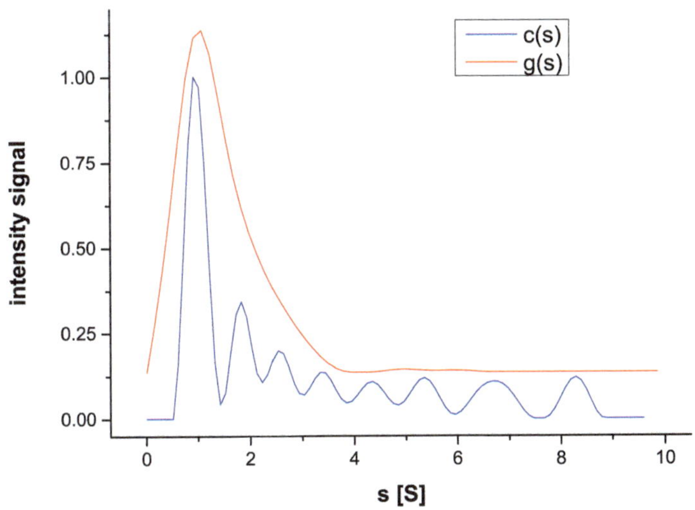

Fig. 3 Sedimentation coefficient distribution *g*(*s*) and the diffusion broadening-corrected distribution *c*(*s*), for silver nanoparticles formed in a CTAB solution.

to diffusion ranges of *ca.* 0.75 mm resp. 1.8 mm after 617 s, the reaction boundary still does not spread significantly, as a result of the opposite diffusion direction of the reacting ions and the immediate reaction.

The sedimentation coefficient distribution $g(s)$ confirms the slow sedimentation of the nanoparticles, showing a maximum of 0.92 S (Fig. 3). After correcting the distribution for peak broadening due to diffusion, much finer details of the sedimentation coefficient distribution can be observed, indicating the presence of eight species in the mixture. Since the model-free calculated $g(s)$ envelopes the diffusion-corrected $c(s)$, the diffusion-corrected distribution is relevant.

The sedimentation coefficient s can be converted to the particle diameter d using a modified form of the Svedberg equation valid for hard spheres:

$$d = \sqrt{\frac{18\eta s}{\rho_P - \rho_S}} \quad (1)$$

with solvent viscosity η and density ρ, with index P indicating the quantities for the sedimenting particle and index S for the solvent.

The problem with this conversion is that the particle density is typically unknown – especially for very small particles, or clusters, where the stabilizer shell contributes significantly to the particle density. Therefore, the density has to be estimated by applying a core–shell model, with the core having the bulk silver density of 10.49 g mL^{-1} and the shell having the density of CTAB (0.50 g mL^{-1}). This approach yields particle-size dependent densities.[53] For these calculations, the thickness of the shell is very important – much more than the density of the surface layer. This is beneficial, since the density of a surface layer consisting of solvent and surfactant is not precisely known.[53]

We have performed a calculation of the particle size distribution, assuming a CTAB double layer with a thickness of the stretched CTAB molecule of 1.89 nm (Table 1). It becomes immediately obvious that very small species were detected in the performed AUC experiment. The very small particle size further confirms that the reaction leading to $Ag^{(0)}$ formation was definitely finished, and that the species, which were detected, no longer grew during the sedimentation experiment, at least after 617 seconds. Otherwise, much larger particle sizes would have been detected, since time-resolved synchrotron XRD has revealed nanoparticle sizes >5 nm already after 1 ms after reactant mixing.[43]

Table 1 Particle diameters of silver cores according to a Ag core/CTAB shell model, and for pure (uncoated) silver nanoparticles

Sedimentation coefficient (s)	Particle Ag-core size with CTAB layer (nm)	Particle size for neat silver particles (nm)
0.9	1.38	0.40
1.8	1.50	0.56
2.5	1.58	0.66
3.3	1.66	0.76
4.4	1.76	0.86
5.4	1.86	0.96
6.7	1.96	1.06
8.3	2.08	1.18

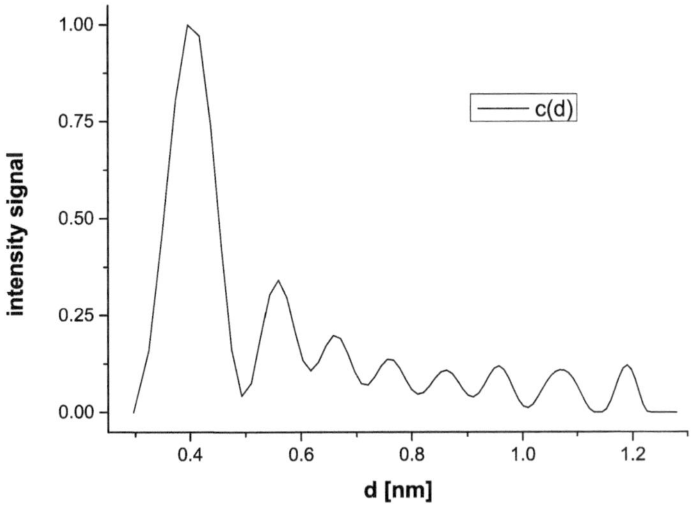

Fig. 4 Particle size distribution of silver nanoparticles formed in a CTAB solution calculated with the bulk density of silver. The presented particle sizes are those of the silver core.

However, it appears unrealistic that a continuous CTAB layer coats nanoparticles as small as 1.5–5 nm. Therefore, this calculation is certainly a limiting case, that does not reflect reality particularly well. The other extreme case is to assume that no CTAB stabilizer adsorbs on the very early-formed particles. In this case, the density of silver can be used directly as the particles' density. This particle size distribution is shown in Fig. 4. The real particle density will then be somewhere in between these two extremes, which gives us the possibility of determining a realistic particle size range for each of the 8 detected species.

The range of particle sizes determined for the eight detected species between the two particle density extremes is rather large in terms of relative deviations (Table 1). However, when considering the absolute deviations resulting from the two limiting cases of realistic densities, it is obvious that the resolution between the detected species is in the Ångström range.

It is certainly realistic to assume that the stabilizer shell is, unlike the CTAB bilayer, very small, and therefore the particle core sizes should be closer to that for the uncoated silver than to the CTAB coated nanoparticles. Certainly, hydration layers will play a role, but adding a monolayer of water to the nanoparticles leaves the calculated particle size virtually unchanged (data not shown). Making this assumption, we can preliminarily allocate literature known species to the species detected in the AUC experiment.

It has been suggested that Ag as well as Au clusters grow as icosahedra, representing thermodynamically preferable states.[54] Therefore, the first fully completed silver icosahedron is the Ag_{13} cluster of 0.7 nm diameter (diameter Ag(0) 0.288 nm; Ag(I) 0.252 nm).[37] The second full-shell silver icosahedron, Ag_{55}, is 1.2 nm in diameter.[37] Unfortunately, although we can detect the particle sizes with Ångström resolution, the unknown particle density still creates a certain level of uncertainty, so that the direct identification of distinct species is not yet directly possible with the data in Table 1.

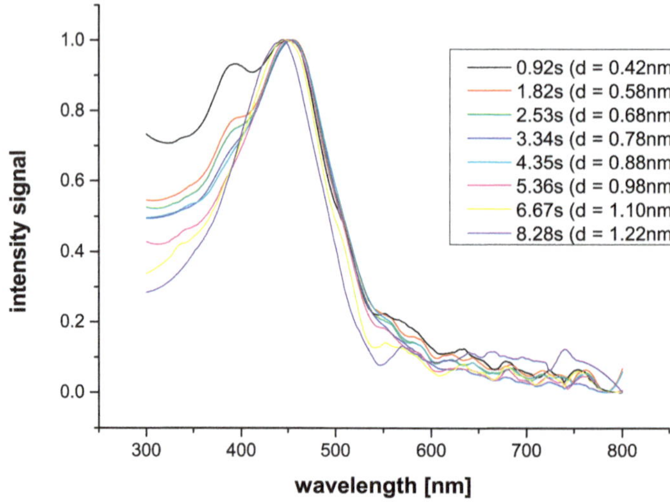

Fig. 5 Overlay of UV-Vis scans calculated according to eqn (2) for silver nanoparticles for all species. The particle size is given including a water monolayer.

However, due to the unique possibilities of the multiwavelength detector, which can often already provide important information on samples directly from the experimental raw data,[55,56] we can try to allocate the UV-Vis spectra of the eight detected species in Fig. 4. Currently, SEDFIT[28,30] allows for the evaluation of synthetic boundary experiments but does not yet allow for evaluation of multiwavelength data, and so far ULTRASCAN[57] only has commonly available routines for multiwavelength evaluation of boundary sedimentation runs. This currently hinders us in directly allocating a spectrum to a certain nanoparticle size in the diffusion corrected particle size distribution $c(d)$.

However, we can rearrange the equation for the calculation of sedimentation coefficient s from the boundary position r at a certain time:

$$r = r_{\mathrm{m}} e^{s\omega^2 t} \tag{2}$$

with radius r (index m refers to the meniscus), sedimentation coefficient s, angular velocity of the rotor ω, and time t when the scan was recorded.

This allows for the allocation of UV-Vis spectra taken at a location r corresponding to a scan time t for a given sedimentation coefficient of one of the eight species in Fig. 3. It is therefore possible to obtain one UV-Vis spectrum for each scan. Since a number of scans were taken, several UV-Vis spectra can be averaged, improving the data quality. The overlay of these scans is shown in Fig. 5 for all eight species. We notice two absorption maxima for the smallest species at 400 and 450 nm. The maximum at 400 nm decreases for the next three larger species until it vanishes completely, while the 450 nm peak slightly but continuously blue-shifts from 451 nm to 443 nm with increasing particle size.

It must be noted that these spectra are still affected by neighbouring species in the sedimenting boundary, because a diffusion correction of the radial and spectral range is not yet possible for synthetic boundary experiments. However, since the sedimentation transport prevails over diffusion transport with time, due to the $t^{1/2}$ dependence of mean displacement by diffusion, averaging the scans – especially those at later experimental times – sharpens out the spectra for the species identified in the particle size distribution.

Titration assay

In order to complement the AUC analyses, we have investigated the early stages of Ag-nanoparticle formation, utilizing a titration assay. In brief, 2.5 mM aqueous $AgNO_3$ solution was dosed at a constant rate of 0.1 mL min^{-1} into 50 mL aqueous 10 mM $NaBH_4$, where 2.5 mM citrate was present as a capping agent in the silver solution. A direct conceptual transfer of the AUC experiment was unfortunately not possible, owing to the rather low solubility of CTAB at room temperature. This did not matter that much in the AUC experiments; the low solubility of CTAB is a problem in the titration assay, as these experiments take a while, and so the CTAB precipitates. The AUC experiments are faster and require a 25 times smaller $AgNO_3$ concentration than the titrations. Therefore, in the AUC experiments, CTAB typically does not precipitate.

In order to avoid precipitation of the CTAB capping agent in the titration experiments, and any possible interference with the early stages of Ag precipitation, we utilized citrate instead (note that citrate does not act as a reduction agent in the case of silver).[39]

The pH of the sodium borohydride solution was kept constant at pH 9.73 ± 0.05 using 0.25 M HNO_3 in an automatic counter-titration. On one hand, this is

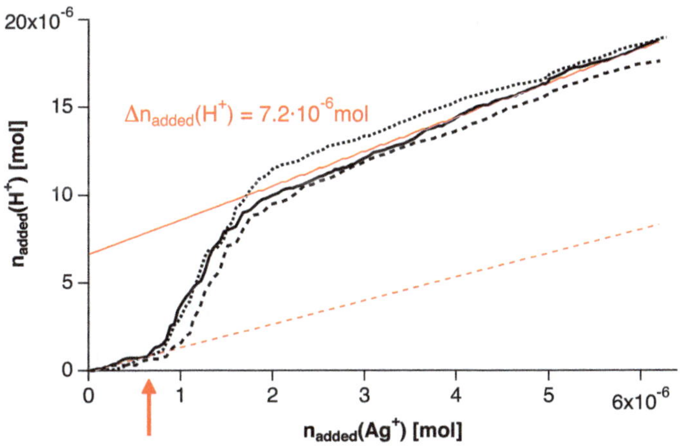

Fig. 6 Development of the amount of H^+ required to maintain a constant pH of 9.73 in automatic counter-titration using HNO_3, upon continuous addition of Ag^+ to 50 mL 10 mM $NaBH_4$ solution. Three independent repetitions are shown (continuous, dashed, and dotted black lines). The background titration (red line representing a linear fit of the average development after the addition of 2×10^{-6} mol Ag^+) is due to the continuous generation of HBO_2 in the borohydride solution, which is formally independent of silver nucleation. The dashed red line represents a linear fit of the average development until ~7 $\times 10^{-7}$ mol Ag^+ have been added. The different slopes of the red lines indicate that the rate of $NaBH_4$ decomposition decreases with time. The red arrow indicates the nucleation event of $Ag^{(0)}$ clusters, whereas the value of Δn_{added} (red text) gives the offset of the background titrations at this point. For explanation see text.

necessary, owing to the continuous decomposition of borohydride, $NaBH_4 + 4H_2O \rightarrow NaB(OH)_4 + 4H_2\uparrow$, which produces hydroxide ions according to $B(OH)_4^- \rightleftharpoons B(OH)_3 + OH^-$.

However, on the other hand, there is a sudden increase in the HNO_3 titration upon the continuous addition of the silver solution (Fig. 6, red arrow), while maintaining a constant pH of 9.73 ± 0.05. The curves are well reproducible, and show a distinct increase in the titration rate when 7–8 × 10^{-7} mol Ag^+ have been added, indicating a sudden release of base in the system (the pH offset before and after this event is <0.02 pH units). Considering the chemistry of silver(I) reduction by borohydride, $2Ag^+ + 2BH_4^- \rightarrow 2Ag(0) + B_2H_6 + H_2\uparrow$ and $B_2H_6 + 6H_2O \rightarrow 2B(OH)_3 + 6H_2\uparrow$, and given the weak Lewis acidity of boric acid, this observation cannot be explained by a sudden reduction of Ag^+ at this point (red arrow, Fig. 6). Quite the contrary; reduction of Ag^+ by BH_4^- is expected to lead to a minor decrease in pH, due to the generation of boric acid.

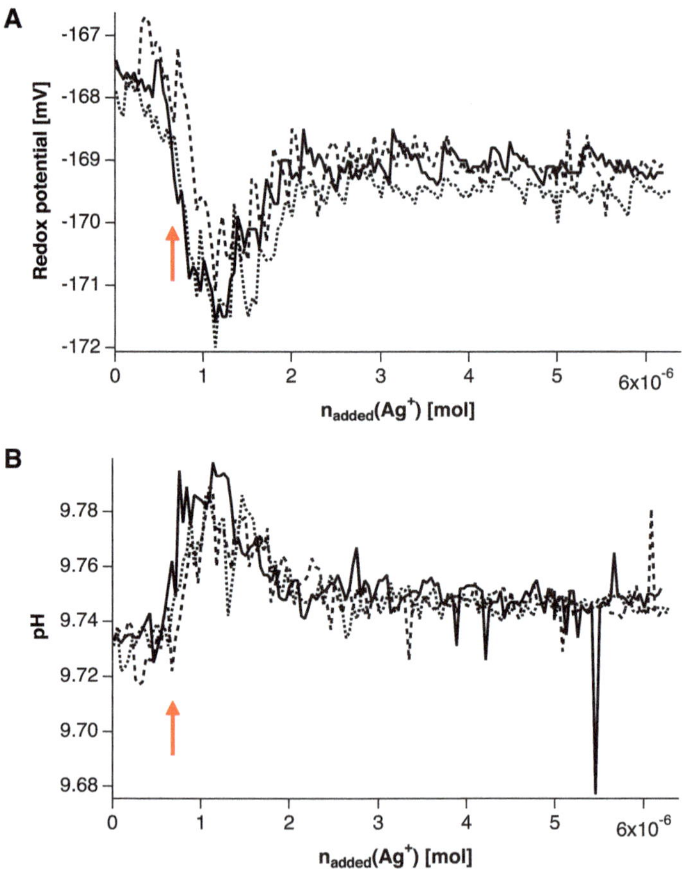

Fig. 7 Redox potential (A) and pH (B) upon continuous addition of Ag^+ to 50 mL 10 mM sodium borohydride solution. Three independent repetitions are shown, (continuous, dashed, and dotted black lines, according to Fig. 6). The pH constancy is maintained within ±0.05 pH units (B), whereas the changes in redox potential are due to the concurrent minute pH changes within experimental accuracy. Red arrows indicate the nucleation point as in Fig. 6. For explanation see text.

Considering the signal of the redox electrode (Fig. 7A), we cannot detect any significant changes that would not directly correlate with the minute changes in pH (Fig. 7B). This suggests that Ag(I) reduction cannot be detected within experimental accuracy throughout the experiment. Furthermore, any decrease in pH due to Ag(I) reduction by BH_4^- is likely to be concealed by the continuous decomposition of $NaBH_4$ (see above). Consequently, the redox reactions alone cannot explain the distinct release of base after the addition of *ca.* 7–8 × 10^{-7} mol Ag^+ to 50 mL 10 mM $NaBH_4$ solution (red arrows in Fig. 6 and 7).

UV-Vis spectra drawn from samples from the titration assay at different times are shown in Fig. 8, which highlight that a distinct change occurs upon crossing this characteristic point. Before the obvious sudden release of base (which occurs at an added amount of 0.62 μmol Ag^+, black spectra in Fig. 8), there is no shoulder at the onset of the water absorption band at ~250 nm, and the peak at ~394 nm is absent. Directly after the characteristic event (at an added amount of 1.5 μmol

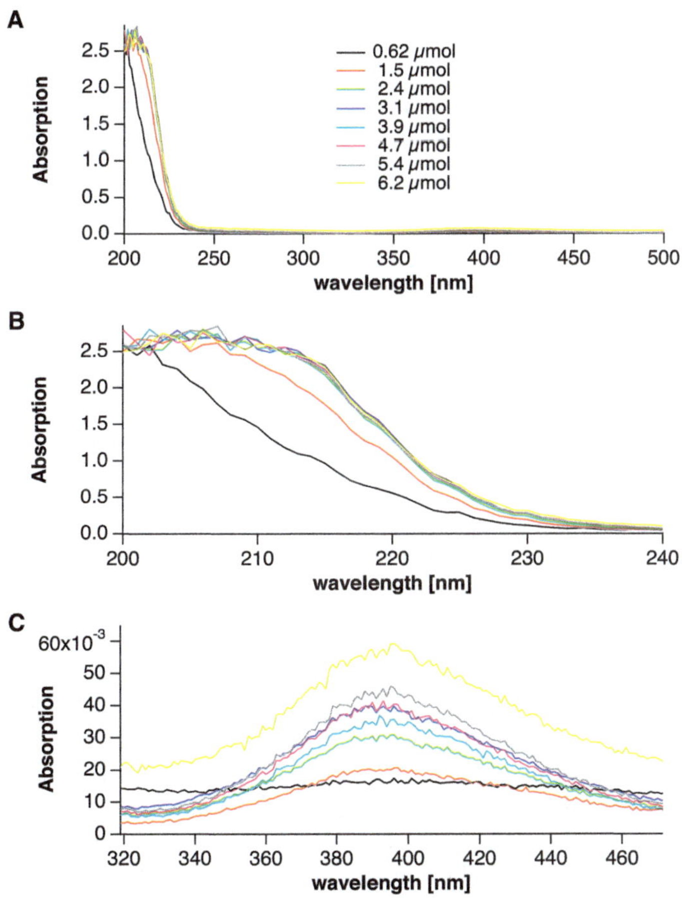

Fig. 8 UV-Vis spectra directly measured after solutions were drawn at different times from the titration assay. The respective amounts of substance quoted in the legend (A, overview) correspond to the added amounts of Ag^+ at the time of sampling. The spectra were collected directly after the samples were drawn, and ripening effects do occur (see Fig. 9). B and C show magnified spectral regions. For explanation see text.

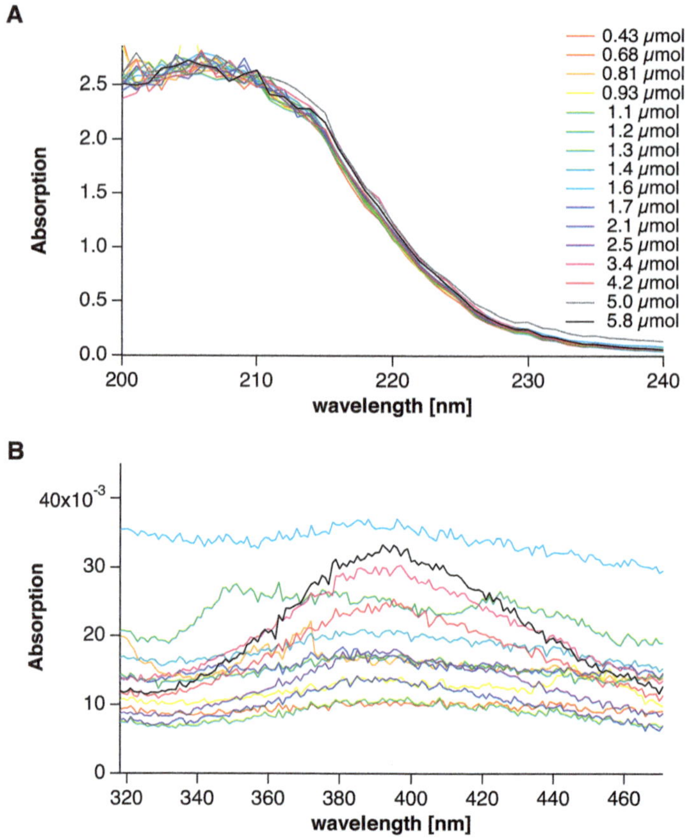

Fig. 9 UV-Vis spectra of solutions drawn at different times from the titration assay and allowed to ripen for *ca.* 10–15 minutes prior to measurement. The respective amounts of substance quoted in the legend correspond to the added amounts of Ag^+ at the time of sampling. A and B correspond to the spectral regions shown in Fig. 8B and C, respectively.

Ag^+, red spectra in Fig. 8), a shoulder develops at around 210 nm, and the peak at ~395 nm steadily increases upon further addition of silver, indicative of the generation of more and more silver nanoparticles. Altogether, we can conclude that the characteristic event of base release (red arrows in Fig. 6 and 7) corresponds to the nucleation event of metallic silver nanoparticles. It remains unclear whether or not there is a sudden onset of the redox reaction, but our data clearly shows a sudden release of base.

When the solutions are allowed to ripen for *ca.* 25 minutes, the spectra of the samples drawn at different times throughout the experiment no longer differ with regard to the absorption at ~250 nm (Fig. 9A). This suggests that all states are critically metastable, with induction times on the order of minutes. However, the continuous increase in the intensity of the absorption peak at ~395 nm shows that more and more silver nanoparticles are formed when an increasing amount of silver has been added (Fig. 9B). Within experimental accuracy, we cannot detect any shift in this absorption band during the experimental duration.

In any case, the titration data provide evidence that there is a distinct pre-nucleation regime, and show that the transformation of pre-nucleation precursors towards nucleated silver(0) clusters/nanoparticles leads to a distinct generation of free base. In our experiments, this occurs at an overall silver concentration of ~20 μM, whereas basic species released from an amount of 0.7–0.8 μmol silver ions – and/or atoms – require neutralization by *ca.* 7 μmol H^+ (Fig. 6, red arrow). This may be explained by the formation of alternative Ag^+ species within the silver precursor solution, as well as upon addition of the silver solution to the borohydride solution. These may be $Ag(OH)_x$ as suggested by Nishimura *et al.*[45] Thus, significant amounts of hydroxide ions may be released from nucleation precursors upon the nucleation of silver(0) clusters and nanoparticles. However, the release of bound citrate molecules upon nucleation cannot be excluded, and likely contributes to the increased counter-titration rate. Thus, determination of the average composition of Ag-pre-nucleation species requires further experiments, *e.g.* employing silver ion-specific electrodes, and varying the capping agent concentration. In addition, future AUC experiments may provide insight into the sizes of these species, and whether they may or may not relate to highly dynamic pre-nucleation clusters, which are observed in an increasing number of systems.[12]

Conclusions

The titration assay shows that the nucleation of metallic silver nanoparticles/clusters is preceded by a distinct pre-nucleation stage, which involves silver species bound to OH^- (ref. 45) and most likely citrate ions. The redox state of silver species bound within these precursors remains currently unknown, however. It is a matter of speculation as to whether the structural form of these pre-nucleation species relates to highly dynamic chain-like assemblies, as suggested for calcium carbonate and phosphate pre-nucleation clusters.[58] In analogy to the pre-nucleation cluster pathway,[12] internal condensation of such precursors could underlie the event of phase separation, and be accompanied by a sudden release of distinct amounts of base, most likely OH^-, and citrate. This event occurs at minor overall silver concentrations in $NaBH_4$, showing that the nucleation of metallic silver is virtually instantaneous in the Synthetic Boundary Crystallization Ultracentrifugation experiment discussed herein. Thus, the titration assay provides new insights, especially into the very early (pre-)nucleation stages of silver nanoparticle/cluster formation. Future experiments will be designed to characterize these precursors, with respect to composition and size, in more detail.

The multiwavelength AUC experimentation allows the identification of different spectra for 8 distinct species formed upon virtually instantaneous nucleation in the reactive boundary zone (Fig. 3). However, the differences between the spectra are not very pronounced, since they have not (yet) been corrected for the diffusion of neighbouring species. Diffusion correction is at present only possible for the radius domain, yielding $c(s)$.[28] Thus, the hydrodynamic information has Ångström resolution, but not yet the corresponding spectra; this is a future task for the open AUC project.[31]

Extraction of spectra from the multiwavelength scans (Fig. 5) shows that the nanoparticle/cluster species exhibit UV-Vis absorption maxima at ~400 nm and

~450 nm. This shows that the AUC experimentation can be utilized to extract UV-Vis spectra of defined Ag nanoclusters. This allows quantitative investigations into the effects of ligands and additives on size-dependent UV-Vis spectra, which could be complemented by *ab initio* modelling and simulation approaches in the future, potentially yielding novel insights into quantum effects.

The detected absorption maxima are typical for silver nanoclusters and nanoparticles. However, it must be noted that the interface of the Ag nanoparticles, the surrounding phase, and the type and quantity of the capping agent have profound effects on absorption maxima, as small changes in the electronic properties of the surrounding medium lead to strong peak shifts.[59] Hence, the UV-Vis spectra from the titration assay (citrate-capped) cannot be directly compared to those from the multiwavelength AUC experimentation (CTAB-capped). A 400 nm absorption peak was reported for Ag_5 and Ag_9 mixtures in water;[60] although the capping agent differs, this may be the early species also observed in the AUC and titration experiments. This comparison, albeit with differing capping agents, suggests the occurrence of Ag_5 and Ag_9 clusters in both titrations and AUC experimentation. Regarding the evaluation of the derived sedimentation coefficients with respect to particle sizes, this in turn suggests that no CTAB is adsorbed on the very small Ag clusters. Instead, calculation of the particle size, taking into account the density effect of a single water hydration layer, appears most realistic for the very small species detected. Assuming spherical particles and cubic dense sphere packing allows the calculation of the corresponding atom numbers for each of the species (Table 2). The absence of any CTAB layer on small Ag clusters may be crucial to explain aggregative growth regimes.

The derived atom numbers are very small, underpinning the idea that the earliest species formed upon nucleation have been detected. The smallest and most abundant species appears to be a single atom of Ag(0), or linear nanoclusters consisting of Ag_3, which may not be unambiguously differentiated given the uncertainty of particle densities. Larger clusters appear to form from this basic unit, whereas we cannot detect Ag_{13}-mediated growth, or a preferable cluster of this size assuming spherical particles. However, the thermodynamically preferable icosahedral Ag_{13} (0.7 nm) and Ag_{55} (1.2 nm)[37] could be also allocated to the 2.5 *s* resp. 8.3 *s* species.

Table 2 Sedimentation coefficients, particle core sizes of an Ag cluster/particle with a one molecule thick water layer, and the corresponding atom numbers assuming spherical particles and cubic most dense packing

Sedimentation coefficient (*s*)	Particle core size with water layer (nm)	Atoms
0.9	0.42	1–2
1.8	0.58	4
2.5	0.68	7
3.3	0.78	11
4.4	0.88	15
5.4	0.98	21
6.7	1.10	30
8.3	1.22	41

However, the formation of silver nanoclusters is clearly kinetically controlled during the initial stages investigated here, which does not necessarily lead to the formation of thermodynamically preferable states.[61] Sizes seem to be defined from AUC but so far, no distinct correlation to a particle structure/size is possible due to the unknown particle density.

Within the notions of CNT, the very small sizes of silver clusters suggest that the size of the critical nucleus is already close to $n = 1$ at very low silver concentrations, and the barrier to phase separation observed in the titration experiments cannot be immediately rationalized. However, the occurrence of alternative silver pre-nucleation precursors could give rise to barriers within the notion of a pre-nucleation cluster pathway towards silver nanoparticles. Thus, silver nucleation in aqueous solution may be another example for phase separation *via* pre-nucleation clusters.

Acknowledgements

DG is a Research Fellow of the Zukunftskolleg of the University of Konstanz, and is supported by the Fonds der Chemischen Industrie. We acknowledge funding within the NSF-DFG Materials World Network (GE 2278/6-1; CO 194/12-1).

Notes and references

1 D. Kashchiev, *Nucleation: Basic Theory with Applications*, Butterworth-Heinemann, Oxford, 2000.
2 J. W. Gibbs, *Trans. Conn. Acad. Arts Sci.*, 1876, **3**, 108–248.
3 R. Becker and W. Döring, *Ann. Phys.*, 1935, **24**, 719–752.
4 J. Frenkel, *J. Chem. Phys.*, 1939, **7**, 538–547.
5 J. B. Zeldovich, *Acta Physicochim. URSS*, 1943, **18**, 1–22.
6 A. Dillmann and G. E. A. Meier, *Chem. Phys. Lett.*, 1989, **160**, 71–74.
7 I. J. Ford, A. Laaksonen and M. Kulmala, *J. Chem. Phys.*, 1993, **99**, 764–765.
8 J. W. Cahn and J. E. Hilliard, *J. Chem. Phys.*, 1959, **31**, 688–699.
9 P. R. tenWolde and D. Frenkel, *Science*, 1997, **277**, 1975–1978.
10 P. G. Vekilov, *Cryst. Growth Des.*, 2004, **4**, 671–685.
11 P. G. Vekilov, *Nanoscale*, 2010, **2**, 2346–2357.
12 D. Gebauer, M. Kellermeier, J. D. Gale, L. Bergström and H. Cölfen, *Chem. Soc. Rev.*, 2014, **43**, 2348–2371.
13 V. K. Lamer and R. H. Dinegar, *J. Am. Chem. Soc.*, 1950, **72**, 4847–4854.
14 T. Narayanan, O. Diat and P. Bosecke, *Nucl. Instrum. Methods Phys. Res., Sect. A*, 2001, **467**, 1005–1009.
15 J. Polte, R. Erler, A. F. Thunemann, S. Sokolov, T. T. Ahner, K. Rademann, F. Emmerling and R. Kraehnert, *ACS Nano*, 2010, **4**, 1076–1082.
16 J. Bolze, D. Pontoni, M. Ballauff, T. Narayanan and H. Cölfen, *J. Colloid Interface Sci.*, 2004, **277**, 84–94.
17 H. Haberkorn, D. Franke, T. Frechen, W. Goesele and J. Rieger, *J. Colloid Interface Sci.*, 2003, **259**, 112–126.
18 B. Marmiroli, G. Grenci, F. Cacho-Nerin, B. Sartori, E. Ferrari, P. Laggner, L. Businaro and H. Amenitsch, *Lab Chip*, 2009, **9**, 2063–2069.
19 B. Marmiroli, G. Grenci, F. Cacho-Nerin, B. Sartori, P. Laggner, L. Businaro and H. Amenitsch, *Nucl. Instrum. Methods Phys. Res., Sect. B*, 2010, **268**, 329–333.

20 L. Börger and H. Cölfen, *Prog. Colloid Polym. Sci.*, 1999, **113**, 23–28.
21 L. Börger, H. Cölfen and M. Antonietti, *Colloids Surf., A*, 2000, **163**, 29–38.
22 H. Cölfen and T. Pauck, *Colloid Polym. Sci.*, 1997, **275**, 175–180.
23 H. Cölfen, H. Schnablegger, A. Fischer, F. C. Jentoft, G. Weinberg and R. Schlogl, *Langmuir*, 2002, **18**, 3500–3509.
24 D. Gebauer, A. Völkel and H. Cölfen, *Science*, 2008, **322**, 1819–1822.
25 M. Kellermeier, R. Rosenberg, A. Moise, U. Anders, M. Przybylski and H. Cölfen, *Faraday Discuss.*, 2012, **159**, 23–45.
26 E. M. Pouget, P. H. H. Bomans, J. Goos, P. M. Frederik, G. de With and N. Sommerdijk, *Science*, 2009, **323**, 1455–1458.
27 H. Lange, *Part. Part. Syst. Charact.*, 1995, **12**, 148–157.
28 P. Schuck, *Biophys. J.*, 2000, **78**, 1606–1619.
29 B. Demeler, T. L. Nguyen, G. E. Gorbet, V. Schirf, E. H. Brookes, P. Mulvaney, A. O. El-Ballouli, J. Pan, O. M. Bakr, A. K. Demeler, B. I. H. Uribe, N. Bhattarai and R. L. Whetten, *Anal. Chem.*, 2014, **86**, 7688–7695.
30 P. Schuck, http://www.analyticalultracentrifugation.com/default.htm, 2014.
31 H. Cölfen, T. M. Laue, W. Wohlleben, K. Schilling, E. Karabudak, B. W. Langhorst, E. Brookes, B. Dubbs, D. Zollars, M. Rocco and B. Demeler, *Eur. Biophys. J.*, 2010, **39**, 347–359.
32 S. K. Bhattacharyya, P. Maciejewska, L. Börger, M. Stadler, A. M. Gülsün, H. B. Cicek and H. Cölfen, *Prog. Colloid Polym. Sci.*, 2006, **131**, 9–22.
33 H. M. Strauss, E. Karabudak, S. Bhattacharyya, A. Kretzschmar, W. Wohlleben and H. Cölfen, *Colloid Polym. Sci.*, 2008, **286**, 121–128.
34 J. Walter, K. Lohr, E. Karabudak, W. Reis, J. Mikhael, W. Peukert, W. Wohlleben and H. Cölfen, *ACS Nano*, 2014, **8**, 8871–8886.
35 Y. Wang and N. Herron, *J. Phys. Chem.*, 1991, **95**, 525–532.
36 W. P. Halperin, *Rev. Mod. Phys.*, 1986, **58**, 533–606.
37 N. T. K. Thanh, N. Maclean and S. Mahiddine, *Chem. Rev.*, 2014, **114**, 7610–7630.
38 J. Wang, H. F. M. Boelens, M. B. Thathagar and G. Rothenberg, *ChemPhysChem*, 2004, **5**, 93–98.
39 A. Henglein and M. Giersig, *J. Phys. Chem. B*, 1999, **103**, 9533–9539.
40 M. Harada and E. Katagiri, *Langmuir*, 2010, **26**, 17896–17905.
41 M. A. Watzky and R. G. Finke, *J. Am. Chem. Soc.*, 1997, **119**, 10382–10400.
42 V. N. Richards, N. P. Rath and W. E. Buhro, *Chem. Mater.*, 2010, **22**, 3556–3567.
43 M. Takesue, T. Tomura, M. Yamada, K. Hata, S. Kuwamoto and T. Yonezawa, *J. Am. Chem. Soc.*, 2011, **133**, 14164–14167.
44 T. J. Woehl, J. E. Evans, L. Arslan, W. D. Ristenpart and N. D. Browning, *ACS Nano*, 2012, **6**, 8599–8610.
45 S. Nishimura, D. Mott, A. Takagaki, S. Maenosono and K. Ebitani, *Phys. Chem. Chem. Phys.*, 2011, **13**, 9335–9343.
46 M. Kellermeier, H. Cölfen and D. Gebauer, in *Research Methods in Biomineralization Science*, ed. J. J. D. Yoreo, 2013, vol. 532, pp. 45–69.
47 D. Gebauer, H. Cölfen, A. Verch and M. Antonietti, *Adv. Mater.*, 2009, **21**, 435–439.
48 A. Verch, D. Gebauer, M. Antonietti and H. Cölfen, *Phys. Chem. Chem. Phys.*, 2011, **13**, 16811–16820.
49 T. Pal, T. K. Sau and N. R. Jana, *Langmuir*, 1997, **13**, 1481–1485.

50 R. C. Doty, T. R. Tshikhudo, M. Brust and D. G. Fernig, *Chem. Mater.*, 2005, **17**, 4630–4635.
51 R. R. Johnston and M. Spiro, *J. Phys. Chem.*, 1967, **71**, 3784–3790.
52 H. P. Mo and T. C. Pochapsky, *J. Phys. Chem. B*, 1997, **101**, 4485–4486.
53 J. A. Jamison, K. M. Krueger, C. T. Yavuz, J. T. Mayo, D. LeCrone, J. J. Redden and V. L. Colvin, *ACS Nano*, 2008, **2**, 311–319.
54 G. Schmid, *Chem. Soc. Rev.*, 2008, **37**, 1909–1930.
55 E. Karabudak, C. Backes, F. Hauke, C. D. Schmidt, H. Cölfen, A. Hirsch and W. Wohlleben, *ChemPhysChem*, 2010, **11**, 3224–3227.
56 E. Karabudak, W. Wohlleben and H. Cölfen, *Eur. Biophys. J.*, 2010, **39**, 397–403.
57 B. Demeler, http://www.ultrascan.uthscsa.edu/, 2014.
58 R. Demichelis, P. Raiteri, J. D. Gale, D. Quigley and D. Gebauer, *Nat. Commun.*, 2011, **2**, 590.
59 M. Farrag, M. Thamer, M. Tschurl, T. Burgi and U. Heiz, *J. Phys. Chem. C*, 2012, **116**, 8034–8043.
60 A. Ledo-Suarez, J. Rivas, C. F. Rodriguez-Abreu, M. J. Rodriguez, E. Pastor, A. Hernandez-Creus, S. B. Oseroff and M. A. Lopez-Quintela, *Angew. Chem., Int. Ed.*, 2007, **46**, 8823–8827.
61 W. W. Laxson and R. G. Finke, *J. Am. Chem. Soc.*, 2014, **136**, 17601–17615.

Faraday Discussions

PAPER

Towards an understanding of the nucleation of alpha-*para* amino benzoic acid from ethanolic solutions: a multi-scale approach

D. Toroz,* I. Rosbottom, T. D. Turner, D. M. C. Corzo, R. B. Hammond, X. Lai and K. J. Roberts

Received 22nd December 2014, Accepted 7th January 2015

DOI: 10.1039/c4fd00275j

The molecular assembly and subsequent nucleation of *para*-amino benzoic acid (PABA) from ethanolic solutions is probed using a multi-scale and multi-technique approach. This is applied by examining and interrelating information regarding the molecular, solution-state, cluster, solid-state and surface structures to understand why the alpha form of PABA is crystallised in preference to its low temperature beta form. Calculations suggest that conformational changes within the solute molecule play little or no role in directing the nucleation of either the alpha or beta crystal forms. Combined *ab initio* and molecular dynamics calculations of the stability of small clusters in solution suggests that the hydrogen-bonded carboxylic acid dimers, present in the alpha structure, are the most stable in solution and play a major role in the self-assembly and polymorphic expression of the alpha form in ethanol in preference to the beta form. These calculations are in good agreement with X-ray small-angle scattering analysis which reveals the presence of PABA clusters in ethanol which are consistent with the size and shape of a carboxylic acid dimer. SAXS studies also reveal the presence of larger cluster structures in a size range 10–40 nm which appear to grow, perhaps reflecting a change in the balance between monomers and dimers within the solution during the nucleation process. The results of crystallisation-kinetics experiments indicate an instantaneous nucleation mechanism where the number of instantaneously nucleated crystallites is calculated to be 1360–660 nuclei per ml and the subsequent growth is found to be only rate limited by diffusion of the growth unit to the crystallite surface. A linear dependence of growth rate with respect to supersaturation is observed for the (0 1 −1) capping face, which is associated with strong $\pi-\pi$ stacking interactions. This is consistent with a solid-on-solid mechanism associated with surface roughened growth and concomitant poor lattice-perfection. Conversely, the side (1 0 −1) surface has a growth mechanism consistent with a 2D nucleation birth and spread mechanism.

Institute of Particle Science and Engineering and Institute of Process, Research and Development, School of Chemical and Process Engineering, University of Leeds, Leeds, LS2 9JT, UK

Hence, these mechanisms result in very fast growth along the *b*-axis and the needle-like morphology that is observed for alpha-PABA.

1 Introduction

Research into the molecular-scale mechanism that underpins the directed assembly of crystalline materials from the solution phase into a solid-state structure through a crystallisation process is currently of significant interest.[1,2] The crystallisation of organic crystals is important to the pharmaceutical and fine chemicals industries in terms of their application as crystalline ingredients or formulation components as used for a variety of materials and products such as drugs, pigments, agrochemicals and confectionery. Crystallisation under moderate supersaturation conditions is generally understood to comprise an initial 3D nucleation stage followed by the faceting of the nucleated cluster and its subsequent development to macroscopic dimensions through the 2D surface-mediated growth on the crystal's habit planes. Both the nucleation and growth stages are important with respect to their role in directing many of the key physico-chemical properties (see Fig. 1) of a crystal and/or the same when formulated into a product. Such attributes can be important in terms of a material's or product's practical utility and subsequent performance.[3–6] Examination of Fig. 1 thus reveals the importance of being able to understand and potentially control the initial stages of the crystallisation process and, in particular, the inter-relationship between the structural nature of the initial nucleated clusters and the kinetics associated with their formation and the physical and chemical properties of the crystalline materials produced.[7,8]

In general, the process of nucleation is postulated to occur from the initial formation of molecular clusters to which further molecules attach to the cluster until it exceeds a critical size and a particle nucleates[9,10] and subsequently further develops through growth. There is a rich body of literature in the elucidation of thermodynamic data concerning the nucleation process, and this was covered in a recent review by Davey *et al.*[11] However, elucidation of the kinetic and structural aspects of the nucleation process has proved more challenging. Recent studies have demonstrated how advancements in experimental methods, at the one millilitre scale, have resulted in further development of theory describing the kinetic aspects and solution structure of the nucleation process utilising statistical methods.[12–14] As an alternative, increases in computing power have vastly improved the sophistication and physical relevance of simulations of solution

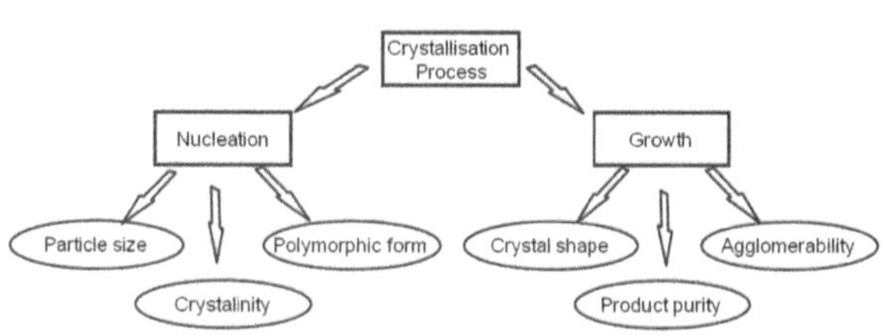

Fig. 1 Schematic showing the role played by crystallisation fundamental parameters (nucleation and growth) on directing the physical properties of solid forms.

structure, self-assembly and pre nucleation clustering.[15–18] This has, in turn, further increased the information about aspects of the nucleation process that are difficult, or even impossible, to measure experimentally. Reflecting these developments, a combination of advanced experimental and simulation techniques have been utilised to probe the nucleation process for the alpha form of a model compound *para*-amino benzoic acid (PABA) in ethanol (EtOH).

PABA contains the kind of hydrogen bonding (H-bonding) functional groups that are typical of those present in many pharmaceutical compounds and is known to crystallise in two well-defined polymorphs, alpha and beta (Fig. 2). A more recent study of the crystallisation of PABA from aqueous solutions containing selenous acid has identified a third polymorph with an orthorhombic structure.[19] However, since only the alpha and beta polymorphs have been observed to form in EtOH, considerations of polymorphic competition through solute clustering will only encompass the alpha and beta forms.

The alpha and beta forms share an enantiotropic relationship, albeit interestingly even crystallisation experiments attempted below the transition temperature often yield the alpha form. Experimental crystallisation of the beta polymorph reported in the literature is often accompanied by concomitant crystallisation of the alpha form.[20] Slurry experiments for PABA in EtOH above and below 13.8 °C (the transition temperature determined) by Glennon *et al.*[21] have revealed a reversible phase change, with alpha to beta conversion below 13.8 °C and beta to alpha conversion above 13.8 °C.

Despite the study from Glennon *et al.*[21] appearing to confirm the thermodynamic stability of the two forms as a function of temperature in EtOH, there has still to be any reported spontaneous crystallisation of the beta form from EtOH under any conditions. Hence, here we attempt to rationalise this preference for nucleation of the alpha form by attempting to understand in more detail the molecular, solution-state, pre-nucleation structure, nucleation and growth behaviour of alpha-PABA. The molecular level data consist of an understanding of how conformational changes of individual solute molecules influence the nucleation behaviour. Further molecular modelling is utilised to assess the relative stability of putative, molecular building blocks (dimers, tetramers and octamers) in EtOH and, complementing these calculations, *in situ* small angle X-ray scattering (SAXS) experiments of saturated EtOH solutions attempt to reveal

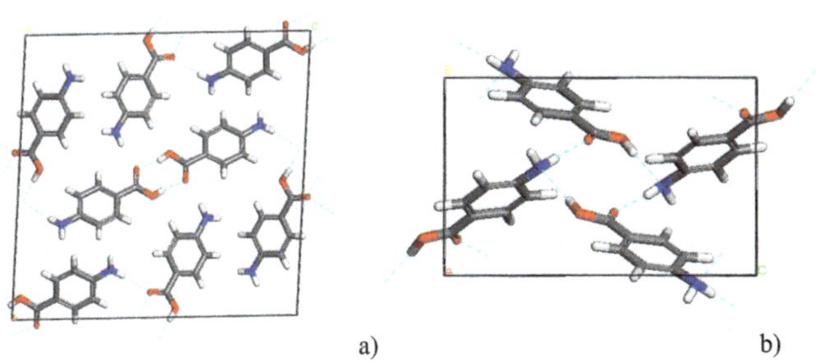

Fig. 2 (a) Unit cell of alpha-PABA with two molecules in the asymmetric unit and eight molecules in the unit cell; packing is comprised of carboxylic acid H-bonding dimers and NH⋯O H-bonds. (b) Beta-PABA unit cell showing 4-membered H-bonding ring containing NH⋯O and OH⋯N interactions.

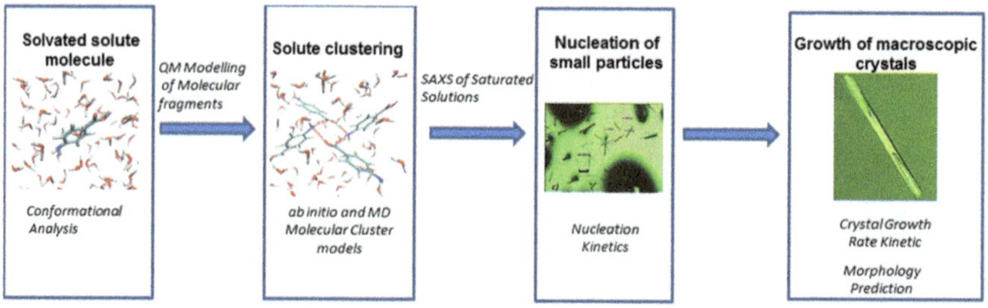

Scheme 1 Themes of research topics and techniques followed for the examination of the crystalline form of *p*-amino benzoic acid. A combination of modelling and experimental techniques are used to probe each stage of the nucleation and growth process.

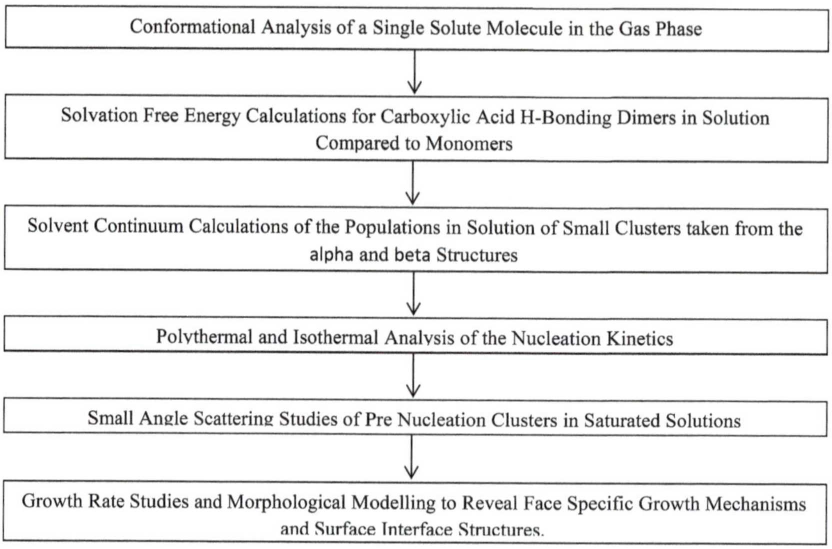

Scheme 2 Flow chart of computation and experimental studies combined to reveal more information on the nucleation of alpha-PABA.

solution and pre nucleation cluster structures. Finally, the nucleation and growth kinetics are investigated and the intermolecular interactions that influence the solute clustering and nucleation are compared to the important intermolecular interactions that control the growth of the final macroscopic particle. These processes are summarised in Schemes 1 and 2.

Hence this study attempts to rationalise how each stage of the crystallisation process influences not only the final macroscopic crystal observed, but also how each stage impacts on the subsequent stages of the crystallisation process.

2 Materials and methods

2.1. Materials

Crystallographic information for all the published structures of PABA is shown in Table 1.

Table 1 Crystallographic details of structures submitted to the CSD

Ref code	Polymorph	a (Å)	b (Å)	c (Å)	Beta (°)	Space group	Reference
AMBNAC 01	Alpha	18.55	3.86	18.64	93.56	$P2_1/n$	Lai, 1965 (ref. 22)
AMBNAC 06	Alpha	18.57	3.84	18.63	93.56	$P2_1/n$	Athimoolam, 2007 (ref. 23)
AMBNAC 04	Beta	6.27	8.58	12.36	100.13	$P2_1/n$	Gracin, 2005 (ref. 24)

All experimental work in this paper was carried out with alpha-PABA and 99.9% absolute ethanol both purchased from Sigma Aldrich. The AMBNAC01 and AMBNAC06 structures for the alpha form differ mainly in the geometry of NH_2 group, the AMBNAC01 being slightly pyramidal and the AMBNAC06 being planar. Interestingly the beta-PABA crystal structure is also pyramidal. In this study the computational experiments were performed on the AMBNAC04 and 06 structures, and in recognition of the structures with different NH_2 geometry, the conformational space of this functional group was also explored.

2.2. Computational methods

2.2.1. Molecular conformational analysis. Conformational analysis of *para*-amino benzoic acid was performed using a hierarchical selection scheme based on electronic structure theory methods using an in-house computer code.[25–27] The initial conformers created were optimized at the Hartree–Fock level of theory using a 3-21G* basis set (HF/3-21G*) to identify the lowest energy conformers. The resulting minima obtained from the HF calculations were optimized at the B3LYP/6-31G* level of theory. Next, the lowest lying conformers resulting from B3LYP were optimized at the MP2/6-31G* level of theory. The initial starting geometry of the PABA molecule was taken from one of the molecules described in the asymmetric unit of the alpha form (AMBNAC 04). The flexible torsions are shown in Fig. 3 and the optimal step sizes were determined for the hydroxyl group (180° for A) and more flexible bonds (30° for B and C).

The hierarchical selection method using these step sizes generated 288 conformers. All the geometry optimizations were performed using the Gaussian 09 package.[28]

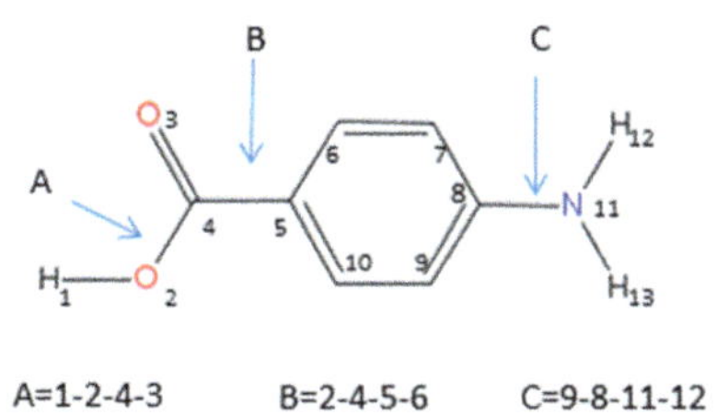

Fig. 3 Atomic labelling and definition of the dihedral angles considered in the hierarchical selection scheme.

A potential energy curve was computed for the transition between a pyramidal and a planar amino group in an isolated molecule. The transition energy was evaluated by performing single point energy calculations along the dihedral angle that defines the pyramidal amino group in a range -140 to $180°$. The method used to scan this dihedral angle was the B3LYP functional using the 6-31G* basis set.

2.2.2. Molecular building block methodology. To select dimers as putative building blocks for the self-assembly of the crystal structures of PABA, the strengths of the pairwise intermolecular interactions within both crystal structures were evaluated using HABIT.[29] The operational details of HABIT are reported extensively in previous publications.[30,31] The calculations employ the "atom–atom approximation" which states that the energy between two molecules is equal to the energy between all of the constituent atom–atom pairs. The interactions are calculated using an atomistic force field, in this case the Momany force field was employed[32] which is well parameterised for organic molecules containing carboxylic acid and amine functionality.

Selected pairwise interactions were then evaluated using the COSMO-RS[7] approach. This approach utilises two steps. Firstly, the optimisation whereby the molecule or cluster is optimised using a quantum mechanical calculation within a conducting solvent dielectric continuum.[33] The QM optimisation utilised an M06/6-31G* level of theory as this includes a dispersion term that is important for the accuracy of these calculations. Secondly, this calculation is combined with a fast statistical thermodynamics calculation which allows prediction of the propensity of the selected molecular conformers or clusters in solution.[34–37] In this study, this approach was used for the dimers, tetramers and octamers.

2.2.3. Calculation of solvation free energies for molecular fragments. The solvation free energies of PABA fragments were examined with the use of molecular dynamics simulations (MD) utilising the thermodynamic integration (TI) technique. All the molecular dynamics simulations were calculated with Gromacs.[38] The thermodynamic integration technique was applied in ethanol from its gas phase at 278 K and 1 atm. The final state with $\lambda_{\mathrm{coupl}} = 1$ is defined as one PABA molecule fully solvated. The initial state $\lambda_{\mathrm{coupl}} = 0$ is defined as turning off the electrostatic and van der Waals interactions between solute and solvent molecules. Nine intermediate windows were used from the initial state to the final state with an interval 0.1. For each window, the system was equilibrated for 100 ps and then sampled for 500 ps. The topology files for the clusters were created with Antechamber[39] using Ambertools and the atom types defined from the General Amber Force Field (GAFF).[40] For the electrostatic potential RESP charges were used based on the MP2/aug-ccpvtz level of theory derived from the Antechamber within Ambertools.

2.2.4. Morphological prediction and surface interface modelling. Pairwise intermolecular energy calculations can be summed within the crystal lattice and used to calculate the lattice energy (E_{cr}) with the interactions then partitioned with respect their position in a growth slice of thickness d_{hkl} into slice energy (E_{sl}) and attachment energy (E_{att})[41] terms according to eqn (1):

$$E_{\mathrm{cr}} = E_{\mathrm{sl}} + E_{\mathrm{att}} \tag{1}$$

where E_{sl} is defined as the energy released upon formation of a growth slice and E_{att} is defined as the energy per molecule released when a slice crystallizes on a face. In this approach it is expected that the slowest growing faces will be morphologically important at the crystal surface and thus the relative growth rate of an individual face is taken as being proportional to the calculated attachment energy[42] for the crystal surface. The attachment energies were calculated on the morphologically important faces as predicted by the BFDH rule using MORANG.[43] The relative attachment energies of each face were expressed as centre-to-face distances and then used to create a Wulff plot to represent the external growth morphology using Shape.[44]

2.3. Poly-thermal and isothermal turbidimetric measurements

Poly-thermal crystallisation experiments were carried out at the 1.5 ml scale with 300 rpm micro magnetic-bar stirring using an Avantium Crystal 16 [45] unit. Turbidimetric detectors were used to monitor the crystallisation and dissolution temperatures, T_C and T_{dis} respectively.

Solutions prepared at concentrations of 170, 180, 190 and 200 g kg^{-1} were heated and cooled in a pre-programmed cycle between 5 °C and 50 °C at rates of heating and cooling of 0.1, 0.3, 0.5, 0.7 and 1.0 °C min^{-1}. Each heating/cooling experiment was repeated five times at each concentration to obtain average values of T_{dis} and T_C, where T_{dis} is determined when the transmission value is 100% and T_C is determined when the transmission % decreases below 90%.

Following a methodology presented elsewhere,[46,47,48] plots were obtained for the relative critical undercooling (u_c) values against cooling rate (q) in ln–ln coordinates. The slope of the best linear fit of these data indicates the nature of the nucleation mechanism through the application of the "rule of three" *i.e.* instantaneous (IN) (slope < 3) or progressive (PN) (slope > 3). The dimensionless parameter u_c can be calculated from eqn (2)

$$u_c = \frac{\Delta T_c}{T_e} \tag{2}$$

Likewise the value of the critical undercooling, ΔT_c, can be calculated from eqn (3)

$$\Delta T_c = T_e - T_C \tag{3}$$

where T_e is the equilibrium solubility temperature.

According to the KBHR approach,[46,47] to proceed further with the analysis, equations 10 to 16 in ref. 48 could be used in the case of PN or alternatively equations 19 to 24 in the case of IN. The derivation of these expressions is subjected to compliance with the inequalities presented in eqn (4)

$$u_c < 0.1,\ au_c < 1 \tag{4}$$

Here a is the dimensionless, molecular latent-heat of crystallisation defined by eqn (5)

$$a = \frac{\lambda}{k\ T_e} \tag{5}$$

where k is the Boltzmann constant and λ is the molecular latent-heat of crystallisation.

The value of q_0, a parameter related to the dependence of u_c on q in the case of IN, can be related to the number of crystallites C_0 at the point of nucleation t_0 through eqn (6),

$$q_0 = \left[\frac{k_v\ C_0}{(n+1)^d\ \alpha_{det}}\right]^{\frac{1}{md}} (a)^n\ K_G\ T_e \tag{6}$$

where C_0 is the concentration of crystallites at the detection point, α_{det} is the relative volume of crystallites at the detection point, and n and m are growth exponents for the growth mechanism of the crystallites; with $n = 1$ associated with diffusion of solute across a crystal/solution interface, and $n = 2$ indicating the presence of screw dislocations within the crystallite. The parameter m can take a value between 0.5 and 1, with $m = 0.5$ indicating undisturbed diffusion of the solute to the crystal surface and $m = 1$ indicating growth through diffusion of solute through a stagnant layer around the crystal.[49] The value of d is the dimensionality of the crystal growth, *i.e.* 1 for needle-like crystallites. K_G is the overall growth rate of the crystal and k_v is the crystallite growth shape factor *e.g.* $2A_0$ for needle-like crystals (where A_0 is the fixed needle cross-sectional area).

The value of q_0 can be obtained through a Nyvlt type relationship, eqn (7), by plotting the q *vs.* u_c in ln–ln coordinates.

$$\ln q = \ln q_0 + (n + 1)\ln u_c \tag{7}$$

For the isothermal analysis of alpha-PABA crystallising from ethanol, solutions with concentrations of 180 g kg^{-1} and 200 g kg^{-1} were prepared. The measurements were carried out at supersaturations (S) ranging from 1.03–1.10 and 1.07–1.15 respectively (Table 8). The values of S were calculated using the recorded MSZW data from section 3.4.1 which gave a MSZW of 5–9 °C in the concentration range studied.

The solutions were then heated to 10 °C above the dissolution temperature and held for one hour to ensure complete dissolution of the solute. The solution was then subjected to rapid cooling at 5 °C min^{-1} to the desired holding temperature within the MSZW. The measured induction time, τ, was taken as the difference from the start time of cooling to the time of the measured transmittance decrease indicating crystallisation. Experiments at each supersaturation were repeated 8 times at the relevant concentration to provide an average induction time measurement.

The induction time data (τ as a function of S) were analysed using classical nucleation theory[50] according to eqn (8)

$$\ln\left\{\tau\left(S(S-1)^{md}\right)^{\frac{1}{md}}\right\} = \ln k_{md} + \frac{B}{(1+md)(kT\ \ln(S))^2}. \tag{8}$$

The value of d is taken as 1 for a needle-like crystallite, γ_{eff} is the interfacial tension, and B is a term related to the molecular volume v_0 and the interfacial tension. Through this relationship a plot of $\ln\left\{\tau\left(S(S-1)^{md}\right)^{\frac{1}{md}}\right\}$ *vs.* $\frac{1}{T^3\ (\ln S)^2}$ will yield a straight line with the slope equal to $\frac{16\pi\ v_0 2\gamma_{eff}{}^3}{3(1+md)k^3}$ whereby γ_{eff} can be calculated. If a spherical critical nucleus is assumed, the radius r^* and number of molecules i^*, within the critical nucleus can be calculated using eqn (9) and (10),

$$r^* = \frac{2\gamma_{\text{eff}} v_0}{kT \ln S} \tag{9}$$

$$i^* = \frac{4\pi (r^*)^3}{3 v_0} \tag{10}$$

2.4. Small angle X-ray scattering studies

The scattered intensity $I_{(q)}$ can be evaluated by the general eqn (11), where φ is the particulate number density, $P_{(q)}$ is the form factor which describes the electron-density distribution of the particle and $S_{(q)}$ is the structure factor which details the distribution of the scatterers in the solution.[51]

$$I_{(q)} = \varphi P_{(q)} S_{(q)} \tag{11}$$

The unified fit method in the program Irena[52] was used to model the Guinier and power law regime observed within the recorded data. The Unified model for a structural level is as follows;

$$I_{(q)} \approx G \exp\left(-q^2 {R_g}^2/3\right) + B\left\{\left[\operatorname{erf}\left(\frac{qR_g}{6^{\frac{1}{2}}}\right)\right]^3 / q\right\}^P \tag{12}$$

Hence the size (R_g) and fractal dimensionality (P) can be extracted for various structurally-related scattering levels. Where $P < 3$ indicates presence of mass fractals, $3 < P < 4$ indicates surface fractals dimensionality and $P > 4$ indicates diffuse interfaces.[53,54] A detailed derivation of the unified fit method and its application to nucleation of an organic material from solution can be found in the literature.[55–57]

Fitting of simple geometrical form factors to scattering data was carried out using Scatter.[58] The form factors, $P_{(q)}$, can be calculated from the scattering amplitudes $F_{(q)}$ from hypergeometric functions of the form;

$$P_{(q)} = {F_{(q)}}^2 = \left\langle {F_1}^2\left(\frac{d+2}{2} - \frac{q^2R^2}{4}\right)\right\rangle \tag{13}$$

where R is the cross-sectional radius of a sphere or cylinder and d is the dimensionality of the 3D shape; $d = 3$ for spheres, $d = 2$ for cylinders. Form factors of cylinders and spheres take into account size distributions of the cross-sectional radii only, and the distribution type used in this instance was a Schulz–Zimm distribution.[59] These form factors were fitted separately to the experimental data using the simplex and Marquadt algorithms whereby the primary beam intensity, $I_{(0)}$, the sphere radius and cylinder length were free parameters in the fitting process, and a homogenous core was assumed for the particle geometry.

SAXS measurements were taken at the SAXS-1 beam line at the Brazilian Synchrotron Light Laboratory, Campinas, Brazil.[60] The liquid sample cell of 2 mm thickness was temperature controlled. The diameter of the 8.3 keV primary X-ray beam was 1.5 mm. The X-ray beam chamber, including the collimator, sample cell, and up to the detector, was fully vacuumed to 10^{-2} mbar. The Pilatus 300 K detector for SAXS pattern collection had a dimension of 84 × 107 mm with a pixel size of 172 micron. This set up allowed a q space range of 0.128–5.4 nm^{-1}. Data collection times were 29 s with 1 s for detector binning. The SAXS pattern intensities collected were then corrected for primary beam intensity variation, transmission, and a background from a clean sample of ethanol was subtracted.

Alpha-PABA solutions in ethanol with the concentration of 247 g kg^{-1} saturated at 50 °C were prepared. The solutions were pre-heated and kept at 55 °C while being transferred into the sample cell in the beam line. The cell temperature was monitored using a thermocouple inserted into the cell next to the sample environment. During a poly-thermal cooling cycle from 55–30 °C at a rate of 0.1 °C min^{-1}, SAXS patterns were collected.

2.5. Crystal growth rate measurements

The crystal growth rates of alpha-PABA in EtOH were measured in an analogous approach to that used in a recent study.[61] In this, a 0.8 ml parallelepiped UV cuvette cell was filled with a supersaturated solution and a seed crystal. The cuvette cell was placed inside a glass windowed growth cell sat on an IMT-2 inverted microscope. Water was flowed round the cuvette cell to control the temperature, first at 40 °C to slightly dissolve the seed and remove surface imperfections, then the temperature was crashed to 20 °C where crystal growth was observed. Images were captured at 1 or 2 minute intervals using the Infinity

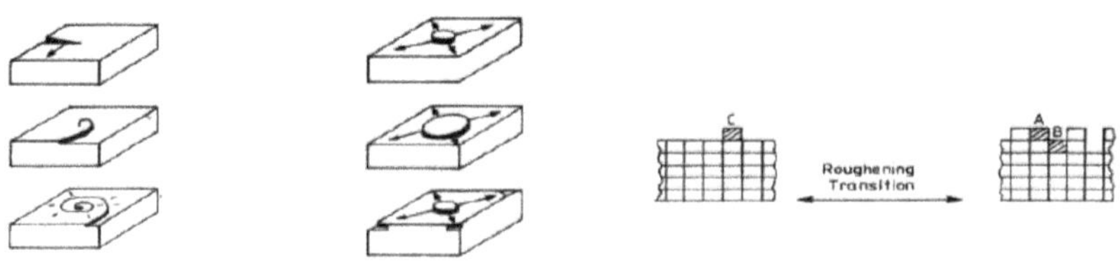

BCF: The incorporation of growth units onto the stepped surface provided by protrusion of dislocations leads to the formation of a growth spiral over the crystal surface creating a permanent source of growth steps at the crystal surface.

$$R_{hkl} = A\sigma \tanh\left(\frac{B}{\sigma}\right) \quad (14)$$

B & S: In the absence of steps surfaces develop through the nucleation (birth) and growth (spread) a monolayer. After nucleation, further molecules can absorb and integrate into the existing monolayer thus enabling it to spread over the surface followed, in turn, by further 2D nucleation events when the surface layer has fully spread over the surface.

$$R_{hkl} = A_1 \sigma^{5/6} \exp\left(\frac{5}{6}\right) \quad (15)$$

RIG: At high supersaturation, the growth interface undergoes surface roughening providing through this abundant sites for surface integration with a lot more step and kink sites thus resulting in a much higher growth rate.

$$R_{hkl} = A\sigma \quad (16)$$

Fig. 4 Diagrams and short description of three major growth mechanisms that can be elucidated from relating growth rate to supersaturation. The BCF growth *via* spirals, B & S growth *via* 2D nucleation and the surface and RIG *via* a roughening polynucleation mechanism. Eqn (14) represents BCF, eqn (15) B & S and eqn (16) RIG.

Analyse software, which was also used to measure the length and width of the imaged crystals, which defined the growth rates of the (0 1 −1) and (1 0 −1) surfaces respectively. These growth rates were measured over a supersaturation range (σ) of 0.03–0.20. In addition, interfacial angles between the side and capping faces were measured in ImageJ and compared to calculated interfacial angles from Morang to aid in identifying the capping faces.

The growth rate *vs* supersaturation measurements were fitted to three different equations that represent crystal growth mechanisms, Burton Carbrera and Frank (BCF),[62] Birth and Spread (B & S),[63,64] and Rough Interfacial Growth (RIG)[65] shown in Fig. 4.

The measured face specific growth rates ($R_{\rm hkl}$) as a function of supersaturation (σ) were fitted to eqn (14), (15) and (16) using origin pro, and the best fit in terms of R^2 value was deemed to represent the crystal growth mechanism for this surface.

3 Results and discussion

3.1. Conformational analysis of an isolated PABA molecule and the associated transition energy profile for the amino group

Fig. 5a shows the most stable conformers located for the PABA molecule in the gas phase. The first conformer (1) alpha has a planar amino group, whereas the second conformer (2) has a pyramidalized amino group. The third conformer (3) is characterized by a 180° rotation of the hydroxyl group of the carboxylic acid moiety.

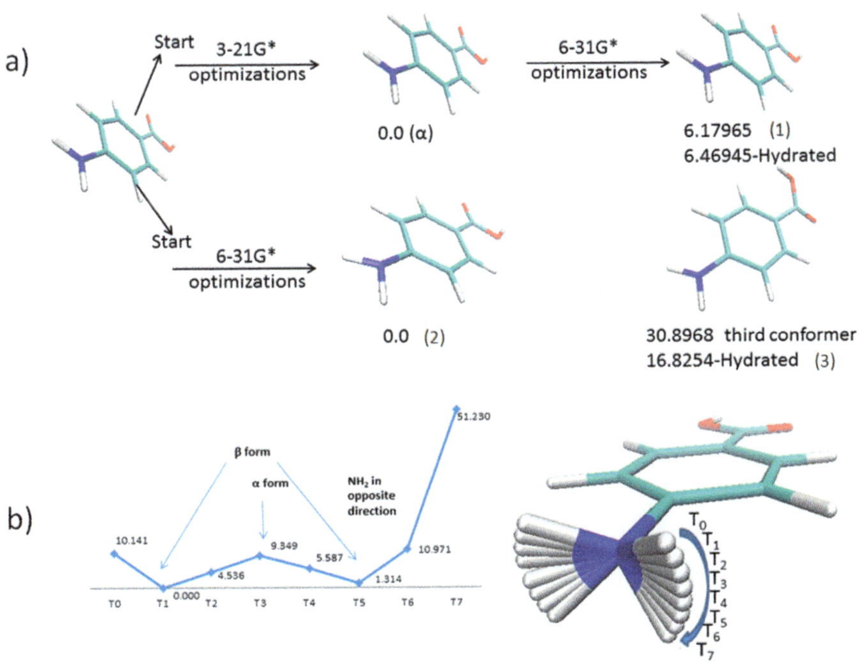

Fig. 5 (a) The three conformers of PABA ((1), (2), and (3)) found from the two selection runs. Relative energies in kJ mol^{-1} (above). (b) Transition energy profile of the dihedral that defines the amino group in PABA (a).

The outcomes of the conformational analysis were found to depend on the size of the basis set employed in the preliminary geometry optimisation step as indicated in Fig. 5a. When a small basis set was used the amine group remains planar (1) even following subsequent optimization with a larger basis set. However, the global energy-minimum (2) is located directly when a larger basis set is used for the preliminary geometry optimisation step. The energy difference in the gas phase between beta conformer (2), which is the most stable one, and the alpha conformer (1) is 6.19 kJ mol^{-1}. The third conformer found is less stable than the other two conformers by 30.89 kJ mol^{-1} compared to beta conformer (2).

To investigate the apparent stability of the planar amine function, a potential energy curve was computed for the transition from a planar to a pyramidal amine group. As can be seen from Fig. 5b there is an energy barrier of about 9.35 kJ mol^{-1} between the planar alpha and beta pyramidal arrangements of the amine moiety. This energy barrier is probably the reason that the smaller basis set optimization stabilizes the planar amine group. The calculation also revealed from the force constants calculated for the normal modes that the conformation with a planar amine group represents a local energy minimum.

The conformational analysis shows that the conformations found in the crystal structures are very similar to the lowest energy conformers found in the gas phase. This suggests that the molecular conformation plays a limited role in the nucleation process. This behaviour is typical of inflexible molecules as was shown by the study of aspirin by Ouvard and Price.[66] Conversely, it has been observed in previous studies that flexible molecules undergo significant conformational change from gas phase to crystal structure.[5,6,8,67] These results therefore prompted a more rigorous investigation into the role of the intermolecular interactions on the nucleation of the alpha form.

3.2. Solvation free energies calculated for the carboxylic acid dimer of PABA

To calculate the solvation free energies of the H-bonding carboxylic acid dimers two harmonic potentials were applied to retain the H-bonding between the molecules during the thermodynamic integration technique, as shown in Fig. 6.

Table 2 shows that two molecules forming a carboxylic acid dimer are preferred over two non-interacting molecules in ethanol with a free energy difference of about 5 kJ mol^{-1} at 293 K, which suggests it is likely that carboxylic acid dimers form in solution prior to nucleation. Recently, similar computations for PABA predicted that carboxylic acid dimer formation is more favoured in acetonitrile compared to ethyl acetate and iso-propyl alcohol and these predictions were found to be in good agreement with experimental nucleation data,[51]

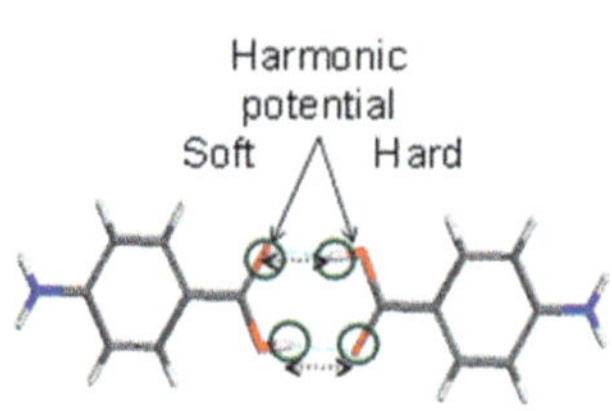

Fig. 6 Application of harmonic potentials to retain the dimerization of the two PABA molecules.

Table 2 Calculated solvation free energies of the carboxylic acid dimer and two molecules of PABA not forming carboxylic acid dimers in ethanol at 293 K and 1 atm. Energies in kJ mol^{-1}

Carboxylic acid dimer	Two PABA molecules not forming –COOH dimers	$\Delta\Delta G$
−84.07 ± 1.46	−79.19 ± 1.49	−4.88

Table 3 Strongest intermolecular interactions from the alpha-PABA structure. Interactions are visualised and labelled on the packing diagram Fig. 7

Bond	Multiplicity	Distance (Å)	Total (kcal mol^{-1})
Aα	1	8.07	−7.10
Bα	1	8.03	−6.94
Cα	2	3.84	−2.68
Dα	2	3.84	−2.58
Eα	1	7.91	−1.90
Fα	1	7.82	−1.28

Table 4 Strongest intermolecular interactions from the beta-PABA structure. Interactions are visualised and labelled on Fig. 8

Bond	Multiplicity	Distance (Å)	Intermolecular energy (kcal mol^{-1})
Aβ	1	4.17	−2.57
Bβ	2	8.11	−2.45
Cβ	2	5.73	−2.39
Dβ	2	6.74	−1.46

highlighting the importance of this molecular building block in other solvents. The predicted stability of the carboxylic acid dimer in EtOH prompted a detailed comparison of the stabilities the carboxylic acid H-bonded dimer motif when compared with other molecular building blocks extracted from the crystal structures of the alpha and beta polymorphs.

3.3. Dielectric solvent continuum analysis of molecular building blocks in ethanolic solutions

3.3.1. Stability of dimers in ethanol. The strength of pairwise, intermolecular interactions in the alpha and beta crystalline structures are presented in Tables 3 and 4:

Fig. 7 and Table 3 show the strongest interactions in the alpha structure are the OH···O carboxylic acid H-bonding dimers. Interestingly, the second strongest interaction is the head-to-head π–π stacking interaction. Fig. 8 and Table 4 show the head-to-tail offset π–π stacking interaction is found to be the strongest in the

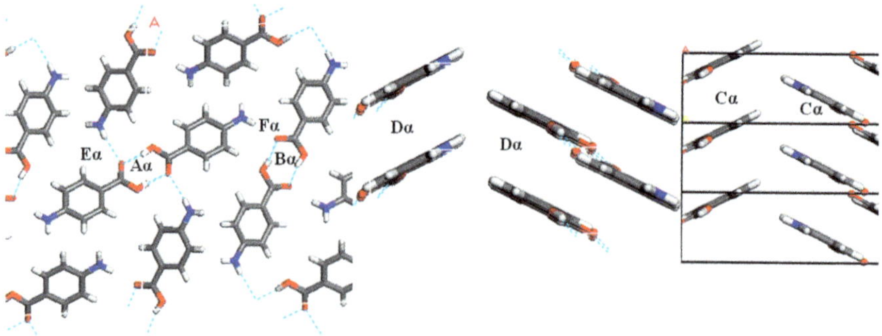

Fig. 7 Strongest H-bonding intermolecular interactions of alpha-PABA viewed down the *b*-axis (left) and strongest intermolecular π–π stacking interactions down the *a*-axis (right).

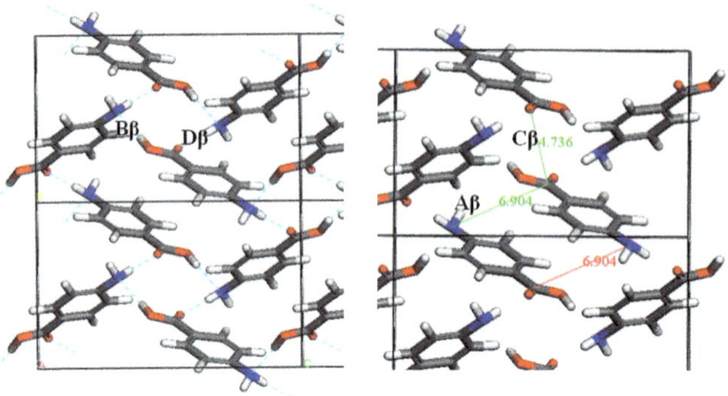

Fig. 8 Strongest intermolecular interactions of beta-PABA. H-bonding ring interactions shown (left) and π–π stacking interactions shown (right).

beta structure, albeit with interaction energies very similar to that of the OH···N and NH···O interactions that make up the 4-membered H-bonding ring structure. These data were used to select ten dimers to analyse using the solvent dielectric continuum approach, and these dimers were then used to select tetramers and octamers present in both structures for further analysis.

The ten strongest dimers selected from the crystal structures alpha and beta are shown in Fig. 9.

Fig. 10 suggests the cluster containing the carboxylic acid dimer is the most stable in ethanol (DA1). The energies of the non-hydrogen bonded dimers (DA4, DB5 and DB6) showed that the most stable non-H-bonding dimer is the head-to-tail stacking motif (DB5). Relative to the H-bonding dimers, the head-to-tail π–π stacking cluster has comparable stability to the DA2 (NH···O) and DB4 (OH···N) motifs.

Fig. 10b shows the dominance in propensity of the carboxylic acid H-bonding dimers in ethanol above any of the other strong pairwise intermolecular interactions identified from the crystal structures of alpha and beta.

3.3.2. Stability of tetramers and octamers in ethanol. Fig. 10c shows that the H-bonding tetramers are the most stable in ethanol. As was observed for the dimers, the cluster which contains two carboxylic acid dimers (TA1) is found to be the most energetically stable cluster.

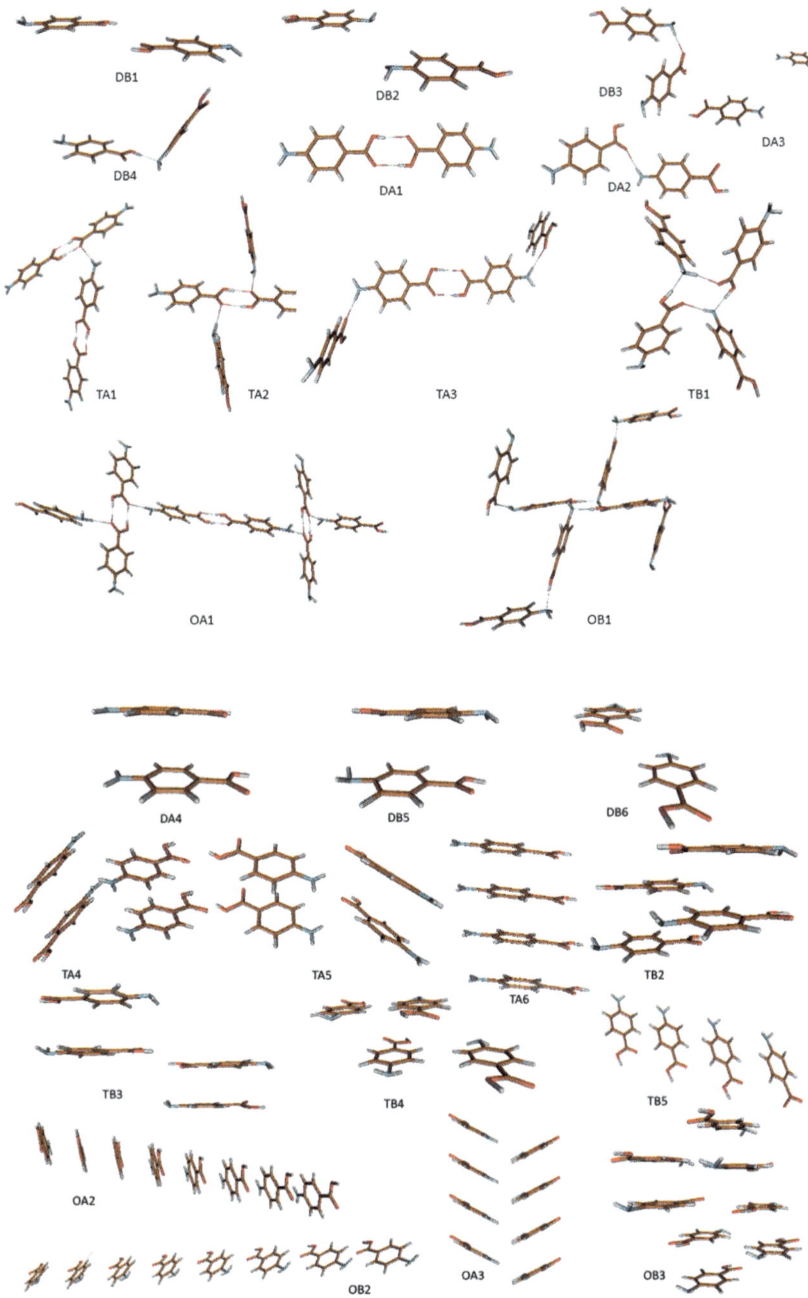

Fig. 9 Fragments depicted from the crystal structure of alpha and beta polymorphs of PABA. Hydrogen bonded fragments: dimers (DB1, DB2, DB3, DB4, DA1, DA2, DA3), tetramers (TA1, TA2, TA3, TB1), octamers (OA1, OB1). $\pi-\pi$ stacking fragments: dimers (DA4, DA5, DB6), tetramers (TA4, TA5, TA6, TB2, TB3, TB4, TB5), octamers (OA2, OA3, OB2, OB3).

The large difference in energy between the H-bonding and non-H-bonding tetramers was reflected in the solution propensities calculated. The TA1 structure appears to dominate due to its composition of two H-bonding carboxylic acid

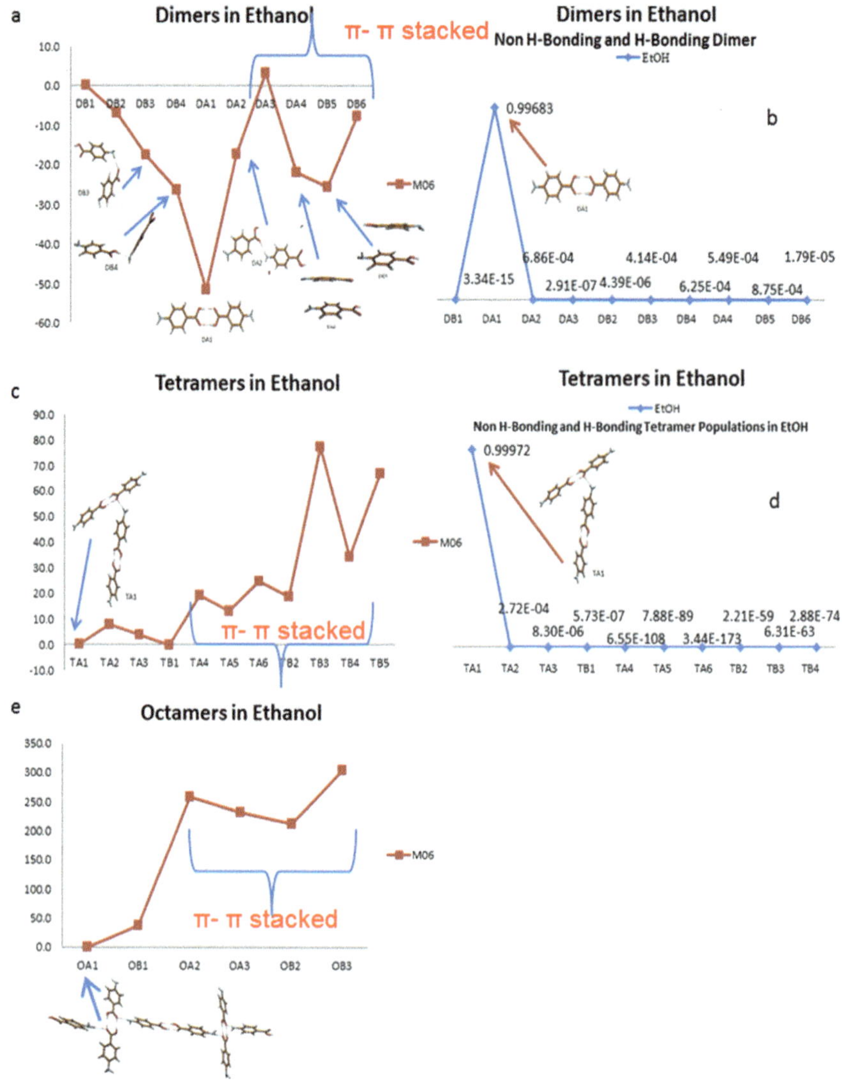

Fig. 10 (a) Relative energies (in kJ mol^{-1}) of the ten dimers from alpha and beta. (b) Normalised weight factor (population) of ten most important H-bonding and non-H-bonding dimers. (c) Relative energies (in kJ mol^{-1}) of the tetramers from alpha and beta. (d) Normalised weight factor of ten most important H-bonding and non-H-bonding tetramers from alpha and beta. (e) Relative energies of the octamers built from the dimers and tetramers. M06/6-31g*(red) in ethanol.

dimers. This particular motif was the most stable over the tetramer containing one carboxylic acid dimer and two NH···O hydrogen bonds, suggesting that the NH···O interactions are secondary to the OH···O interactions when considering the overall packing of the structure.

Fig. 10e reveals that for the two octamers selected, the cluster that facilitates three hydrogen bonded interactions (OA1) was found to be the most stable in ethanol, whilst the π–π stacked dimers were found to be much less stable than H-bonded ones in a range of 200 to 400 kJ mol^{-1}.

Table 5 Average dissolution and crystallisation temperatures for ethanolic alpha-PABA solutions with calculated standard deviations and critical undercooling from eqn (3)

170 g kg^{-1} rate (°C min^{-1})	T_{dis} (°C)	T_C (°C)	ΔT_c
0.1	32.05 ± 0.78	22.40 ± 1.13	6.91
0.3	32.18 ± 0.87	16.10 ± 1.57	13.21
0.5	33.14 ± 1.10	7.85 ± 2.62	21.46
0.7	35.63 ± 1.10	5.40 ± 0.14	23.91
1	41.90 ± 3.42	−0.67 ± 0.32	29.98
	$\mathbf{T_e = 29.31}$		

The calculated stability of dimers, tetramers and octamers in solution suggest that the formation of the carboxylic acid H-bonded dimers in solution could be the first step in the self-assembly of alpha-PABA and the reason why, under conditions of spontaneous crystallisation, the alpha form dominates in EtOH.

3.4. PABA nucleation kinetics from ethanolic solutions

3.4.1. Poly-thermal analysis. The measured dissolution and crystallisation temperatures were analysed using the KBHR methodology. The calculated ΔT_c

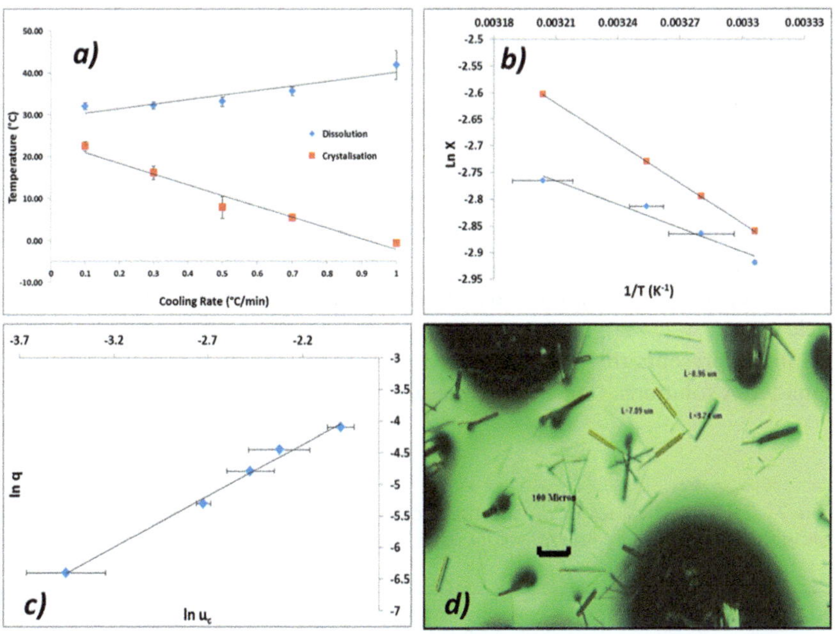

Fig. 11 (a) Plot of T_{dis} and T_C as a function of cooling rate recorded for a solution concentration of 170 g kg^{-1} (b) van't Hoff plot showing the solubility from crystal 16 measurements of alpha-PABA in ethanol (diamonds) and the ideal solubility of alpha-PABA in ethanol (squares) (c) plot of q vs. u_C in ln–ln coordinates of alpha-PABA ethanolic solution at 170 g kg^{-1} (d) optical micrograph of alpha-PABA crystals collected after a crystallisation at a concentration of 170 g kg^{-1}.

Table 6 Calculated values of critical undercooling, u_c as a function of concentration and cooling rate

Conc. (g kg^{-1})	T_e (°K)	Slope of ln q vs. ln u_c	u_c	au_c	$u_c < 0.1$, $au_c < 1$ (yes/no)	R^2 of linear fit to ln q vs. ln u_c
170	302.46	1.63	0.083	0.68	Yes	0.99
180	304.85	1.76	0.066	0.54	Yes	0.96
190	307.32	1.71	0.052	0.41	Yes	0.85
200	312.13	1.51	0.042	0.34	Yes	0.86

values and corresponding standard deviations for the case of 170 g kg^{-1} solution concentration are given in Table 5.

It can be seen from Fig. 11a that the crystallisation temperature is highly dependent on the cooling rate, indicating that the crystallisation is kinetically controlled. This behaviour can also be understood by analysing the size of the MSZW which exceeds 40 °C at higher cooling rates; this indicates that solutions of ethanolic alpha-PABA can be significantly undercooled for long periods of time, particularly at low concentrations.

Fig. 11b shows a van't Hoff plot of the measured solubility curve from measured values of T_{dis} and that this is below the ideal solubility curve, indicative of solute–solute interactions being favoured in solution. This observation correlates well with the results of the solvation free energy calculations showing that a solvated carboxylic acid dimer is more stable than two isolated, solvated molecules.

Fig. 11c shows a plot of critical undercooling *vs.* cooling rate in ln–ln coordinates for the concentration 170 g kg^{-1}. Table 6 highlights the results of this 'rule of three' analysis for all concentrations of ethanolic alpha-PABA solutions, listing the regression values for the linear fits and the inequalities test for the data. The results show that for all concentrations the slope of these plots is <3 identifying the mechanism of nucleation for ethanolic alpha-PABA solutions in the studied concentration range to be instantaneous, whereby all nuclei are formed at the same time, hence the solution will contain a fixed number of crystallites following nucleation.

Table 7 presents the data for q_0 and n for the 4 concentrations of ethanolic alpha-PABA solutions. Considering that n can only take a value between 1 and 2,

Table 7 Calculated values of q_0 and the growth exponent, n as a function of concentration for alpha-PABA in ethanol solutions from the slope and intercept of the linear fit to q vs. u_c in ln–ln coordinates

Conc (g kg^{-1})	a	ln (q_0)	$n+1$	q_0 (K s^{-1})	n	C_0 (m^{-3})	Nuclei/ 1 ml	ΔT_c range/°C
170	8.17	0.77	1.63	2.16	0.63	1.36×10^9	1364	6.91–29.98
180	8.08	0.07	1.76	1.07	0.76	6.81×10^8	681	7.13–21.30
190	7.94	0.20	1.71	1.23	0.71	7.85×10^8	785	5.09–16.57
200	7.93	0.05	1.51	1.05	0.51	6.63×10^8	663	3.81–13.43

Table 8 Measured induction times (τ) and supersaturation ratios for ethanolic alpha-PABA solutions. 180 g kg^{-1} and 200 g kg^{-1} solution concentrations were cooled to within the metastable zone at the holding temperatures stated

200 g kg^{-1} holding temperature (°C)	T (°K)	Average τ (s)	Equilibrium conc (g kg^{-1})	S (supersaturation ratio)
30	303	3746	174.40	1.15
32	305	4780	180.29	1.11
33	306	5850	183.30	1.09
34	307	7700	186.37	1.07

the calculated values of n are more likely to be 1 rather than 2. The cases where n is equal to 1 are indicative of a system whereby crystallite growth is rate-limited by the diffusion of solute molecules towards the growing crystallites.

The parameter q_0 is related to C_0 through eqn (6) and with the assumption of the following parameters; where $d = 1$, $m = 1$, $\alpha_{det} = 1 \times 10^{-6}$, $K_G = 1 \times 10^{-8}$ m s^{-1}, $k_v = 1.28 \times 10^{-10}$. The value of d was assumed for a needle-like crystallite; m was assumed to be 1 for growth mediated by diffusion of the solute through a stagnant layer around the crystal. K_G is the overall growth rate of the crystal and was assumed from a combination of *in situ* growth studies of alpha-PABA in ethanol solutions (section 3.6). k_v was calculated from $k_v = 2A_0$, where A_0 is the fixed cross-sectional area in the direction of the long axis of the needle, which is nearly square shaped. The widths of alpha-PABA crystals were measured using optical microscopy after poly-thermal cooling crystallisation experiments; an example is shown in Fig. 11d. The measured values of the width were found to be in the range 6–10 μm, a middle value of this range, 8 μm, was chosen as a good approximate value and used to calculate A_0.

The values calculated for C_0 at the various solution concentrations are presented in Table 7. The values calculated for ethanolic alpha-PABA solutions were found to show good agreement with previous studies of a similar organic molecule exhibiting instantaneous nucleation, aspirin. The values are in the range of 6.6×10^8–1.3×10^9 m^{-3} for PABA and 1×10^7–1.28×10^{10} m^{-3} for aspirin.[68]

The poly-thermal data presented highlight how alpha-PABA nucleates and subsequently crystallises from ethanolic solutions, where nuclei seem to be stable to large values of under cooling as was shown by observation of very large MSZWs. The crystallisation thus appears to be kinetically dependent indicated by the large variation of T_C as a function of cooling rate. These results are consistent with the fact that steeper T_C (q) lines, such as the ones obtained in this case, would deliver u_c (q) lines with lower slopes such as in the case of IN, in which all the nuclei are instantaneously formed only after a sufficient supersaturation level is reached. This fact might indicate that although the van't Hoff analysis shows that solute–solute interactions are preferable in solution, the chemistry related to the process of de-solvation, to further give way to the formation of clusters' building units, would still be somewhat rate limiting in the nucleation process.

3.4.2. Isothermal analysis. Induction time, τ, data were collected as a function of supersaturation, S, for a 200 g kg^{-1} concentration of ethanolic PABA solution, and are presented in Table 8.

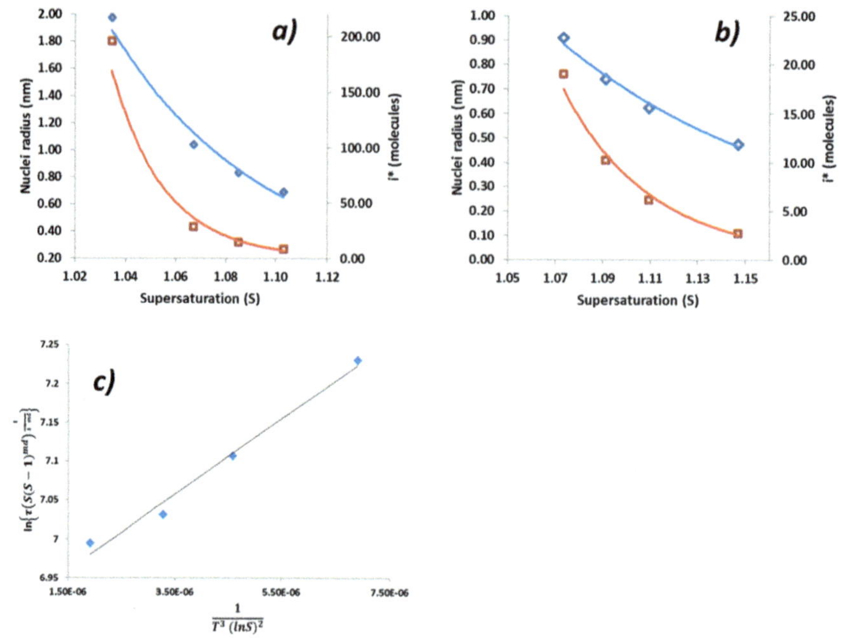

Fig. 12 (a) r^* (blue) and i^* (red) *vs.* supersaturation for at 180 g kg^{-1} concentration, (b) r^* (blue) and i^* (red) *vs.* supersaturation for at 200 g kg^{-1} concentration (c) plot of the experimental induction time data to obtain Sl_{exp} for a solution concentration of 200 g kg^{-1}.

Interfacial tensions, γ_{eff}, were calculated for each concentration using eqn (8), and the calculated slope (Sl_{exp}) from the linear relationship which is equal to $\frac{16\pi\ \nu_0 2\ \gamma_{eff}{}^3}{3(1+md)k^3}$. The linear relationship for a concentration of 200 g kg^{-1} is shown in Fig. 12c, where the fit to the data is represented by $y = 48\,641x + 6.8876$.

The range of calculated values of r^* and i^*, together with the slope of the linear fit and the resulting value of γ_{eff} for the two concentrations of ethanolic alpha-PABA solutions are presented in Table 9. The calculated values of i^* for the 180 g kg^{-1} solution shows increasing values from 8–195 molecules as a function of decreasing supersaturation, ranging from 1.03–1.10, and for the 200 g kg^{-1} solution the values are lower at 3–19 molecules, which is expected for a higher supersaturation. These values are represented graphically in Fig. 12a and b for a solution concentration of 180 and 200 g kg^{-1} respectively.

Table 9 Calculated values of r^* and i^* from the extrapolated values of γ_{eff} for ethanolic α-PABA solutions at the concentrations 180 and 200 g kg^{-1}

Conc g kg^{-1}	Sl_{exp}	R^2	γ_{eff} (mJ m^{-2})	r^* (nm)	i^* (molecules)
180	52 935	0.94	0.85	0.69–1.98	8–195
200	48 641	0.98	1.31	0.48–0.91	3–19

Sullivan *et al.* have recently reported calculated values of γ_{eff} for alpha-PABA in acetonitrile, 2-propanol and ethyl acetate to be in the range of 1.33–2.44, however the higher values in this case were calculated for a much larger range of supersaturations when compared to the range used in these experiments. The calculated interfacial tension and r^* values at 200 g kg^{-1} shows good agreement with the calculated value of PABA in acetonitrile by Sullivan *et al.*,[12] where the supersaturation range is very similar in both studies; as such this provides some confidence in the data presented here. The low values of the calculated interfacial tension for the 180 g kg^{-1} solution are likely due to the low supersaturation range studied and hence the range of r^* values is large; 8–195.

The critical cluster sizes calculated, 0.48–1.98 nm, seem relatively small considering the observed stability of PABA in ethanol from the poly-thermal analysis. To further probe cluster size information during nucleation, small angle X-ray scattering experiments were conducted to reveal this size information during a cooling crystallisation of alpha-PABA in ethanol.

3.5. Small angle X-ray scattering studies of alpha-PABA in ethanolic solutions during cooling

Fig. 13a shows a three dimensional plot of the fully corrected data from the 0.1 °C min^{-1} cooling crystallisation experiment, plotted as intensity against momentum transfer, q, against time in minutes. The data which reflect the PABA system during the cooling process as the solution enters the metastable zone and subsequently crystallises reveal two regions of interest. The first region is the low q region centred around 0.15 nm^{-1} which exhibits intensity and gradient increases during the cooling process, indicating increasing numbers and size of scatterers. The second region is the high q region around 3–4 nm^{-1} which exhibits increases in slope and also intensity decreases throughout the cooling process (Fig. 13b), suggesting the scatterers are increasing in size but decreasing in number.

The high q region was found to be well represented in q space and exhibited a Guinier region followed by a power law regime. This latter area is consistent with scattering from the monomer of PABA or small aggregations of 1–2 nm in size, such as a dimer of PABA. The low q region was not found to be very well differentiated in the data due to the intensity cut off by the beam-stop, which limited the low q region beyond 0.1 nm^{-1}. The scattering in this region can be expected to be due to larger aggregates in the region of 10–40 nm, however a complete analysis of this region was not found to be possible due to the low number of data points and as a result the Guinier region at low q was not observable.

Fig. 13c shows the intensity variation for three regions of the q range; low (0.15 nm^{-1}), medium (2 nm^{-1}) and high (4 nm^{-1}) values of q. The intensity from the low q region exhibits an increase at around 60 min, this corresponds to the time the solution temperature entered the meta-stable zone. This could indicate the formation of the large molecular aggregations or clusters seen at low q. The data are consistent with the size and numbers of these clusters growing as evidenced by the intensity increase throughout the cooling process until around 230 min. The intensity at this latter stage of the cooling process was then found to decrease, which would be consistent with these larger clusters having grown beyond the range of q space observable by the detector. The intensity was found to

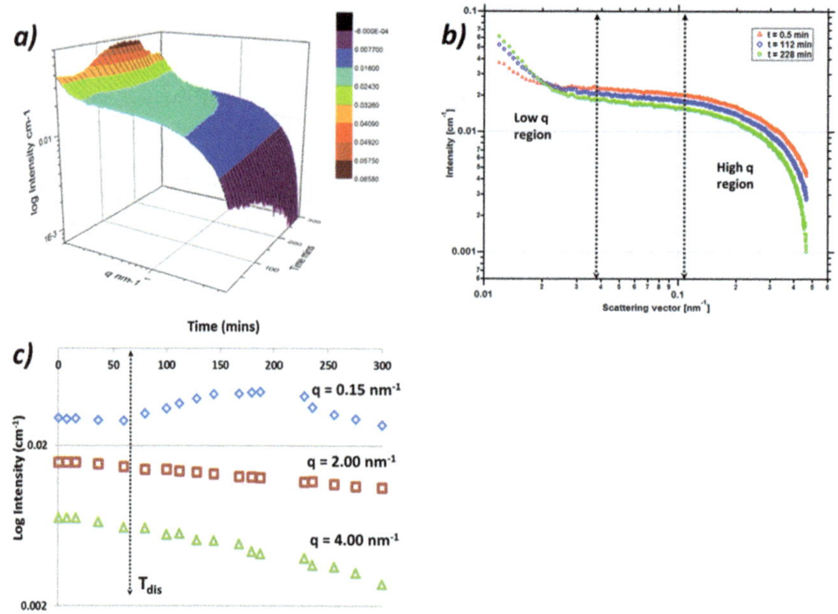

Fig. 13 (a) 3D log–log plot of absolute $I_{(q)}$ *vs.* q as a function of time during a poly-thermal cooling crystallisation at 0.1 °C min^{-1}, (b) log–log $I_{(q)}$ *vs.* q at the start of the cooling profile, 0.5 min, the middle of the cooling profile, 112 min and at I_{max} for the low q region, 228 min (c) (q) progression in log coordinates throughout the cooling process for three regions of q space; 0.15 nm^{-1}, 2.00 nm^{-1} and 4 nm^{-1}.

change at 2 and 4 nm^{-1} showing a general decrease throughout the cooling process which is likely due to the molecules of PABA being consumed by the nucleation process and consequently by the subsequent crystallisation process, whereby molecules of PABA would be expected to transfer to the larger clusters observed in the low q region.

The data collected were analysed using the unified fit model, which allowed various related structural levels of the system to be analysed with respect to the radius of gyration (R_g) and to the power law value (P) to give a calculated fit to the

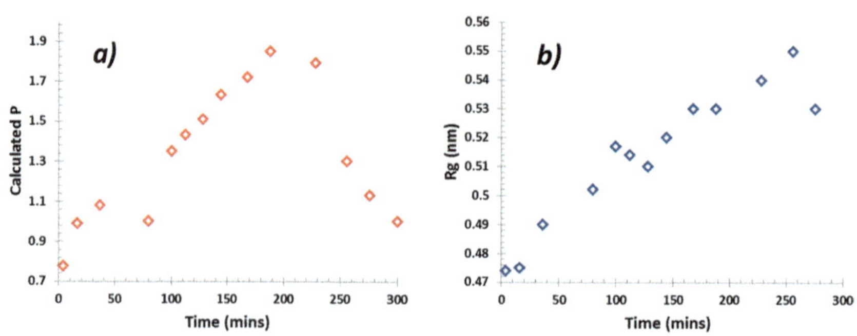

Fig. 14 (a) Progression of the power law value (P) at low q values and (b) R_g (Å) progression at high q during the cooling crystallisation.

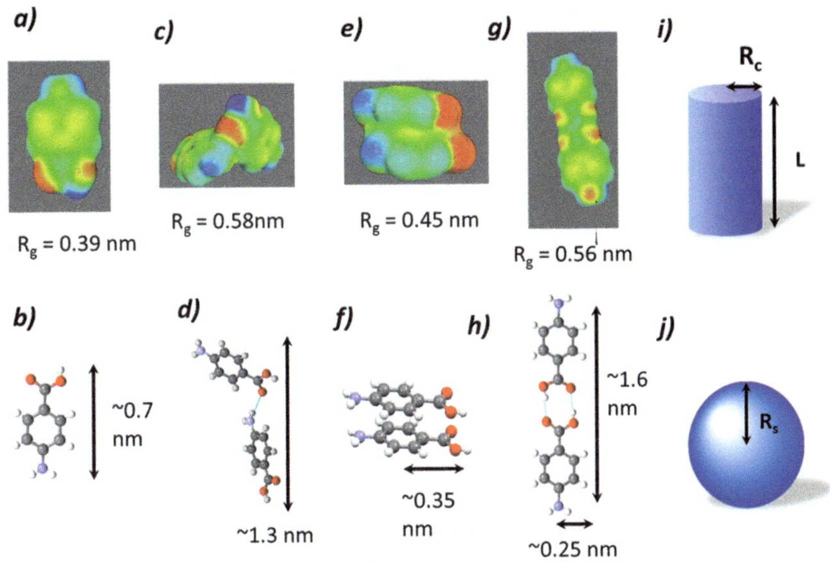

Fig. 15 (a), (c), (e) and (g) represent surface charge density distributions of the single molecule, NH···O dimer, stacked dimer and carboxylic acid H-bonding dimer of PABA respectively, indicating the calculated R_g values in Crysol. (b), (d), (f) and (h) are the ball and stick models of the single molecule, NH···O dimer, stacked dimer and carboxylic acid H-bonding dimer respectively indicating the molecular dimensions, (i) and (j) are the geometric form factors for a cylinder and sphere indicating the parameters used to model the scattering intensity $I_{(q)}$.

data. The R_g values of the monomer NH···O dimer, stacked dimer and carboxylic acid H-bonding dimer structure for alpha-PABA calculated in Crysol[69] were found to be 0.39 nm, 0.58 nm, 0.45 nm and 0.56 nm respectively. The structural models and the calculated R_g values are shown in Fig. 15a, c, e and g. The R_g progression, measured from the unified fit analysis of the experimental high q region, shows an increase from the initial value of 0.47 nm, to its maximum at 0.55 nm, prior to crystallisation (Fig. 14b). This seems to be consistent with a model of molecular aggregation whereby a distribution of monomers and dimers of PABA exist in solution prior to crystallisation. This distribution seems to shift towards a dimer structure as the supersaturation increases but the data obtained could not reveal the exact structure of this dimer. However, comparison of the calculated R_g values for the stacked dimer, NH···O dimer and the carboxylic acid H-bonding dimer with the measured values from the unified fit analysis shows the maximum observed R_g exceeds the calculated R_g for the stacked dimer. This suggests the molecular aggregations observed at this stage of the cooling process do not exhibit this stacking motif but may suggest formation of the NH···O or carboxylic acid H-bonding dimer. The calculated P values for the low q region are shown in Fig. 14a which shows a progression from 0.8 to a maximum of 1.9 as a function of time, after which the maximum in the power law value was found to rapidly drop, consistent with the intensity drop-off of the low q region. A decay of intensity of q^{-1} or q^{-2}, taking a fractal approach to the system, would indicate the presence of a mass fractal dimension. The increase of the P value from 1–2 indicates increased

Table 10 Results from the program Scatter for fitting of cylindrical and spherical form factors, $P_{(q)}$, to the low q region of the experimental scattering curves as a function of time. Fitted parameters were $I_{(0)}$, L and R_s. χ^2 values are also provided to highlight the goodness-of-fit to the data

Time (min)	Cylinder $I_{(0)}$	Cylinder L (nm)	Cylinder χ^2	Sphere $I_{(0)}$	Sphere R_s (nm)	Sphere χ^2
0.5	2.11×10^{-2}	1.15	1.11	2.20×10^{-2}	0.64	0.62
36	2.09×10^{-2}	1.14	1.17	2.15×10^{-2}	0.64	0.68
80	1.96×10^{-2}	1.18	2.02	2.04×10^{-2}	0.65	0.99
128	1.90×10^{-2}	1.30	2.05	1.95×10^{-2}	0.66	1.01
168	1.79×10^{-2}	1.30	1.50	1.88×10^{-2}	0.68	1.66
228	1.71×10^{-2}	1.45	2.61	1.72×10^{-2}	0.69	1.43
256	1.65×10^{-2}	1.48	2.65	1.67×10^{-2}	0.70	2.07
300	1.56×10^{-2}	1.53	2.45	1.55×10^{-2}	0.68	2.00

ordering of the cluster structures in the low q region. This may represent the presence of disordered aggregations of molecules in the size range of 10–40 nm which increase in size and also become more ordered as the solution supersaturates. The dimer structures identified from the high q region are likely transferred to the growing larger aggregates of the low q region during this process.

To further investigate the proposed dimer cluster structure observed during the unified fit analysis, fitting of the experimental data was carried out using simple geometric form factors, $P_{(q)}$, of a cylinder and a sphere. This was rationalised by analysing the surface charge density distributions, calculated using the conducting solvent, di-electric continuum approach,[32] in ethanol for a NH···O dimer, stacked dimer and the carboxylic acid H bonding dimer, shown in Fig. 15c, e and g respectively. A simple comparison can be made for an carboxylic acid H bonding dimer or NH···O represented by a cylindrical form factor, with the stacked dimer represented by a spherical form factor indicated in Fig. 15i and j.

Taking both of these form factors into account and fitting to the high q region good fits to the data were achieved, as can be seen from the χ^2 values in Table 10. As expected the value of $I_{(0)}$ in both cases falls as time increases; this represents a decrease in the number of scatterers as the small monomer/dimer units are consumed by the larger growing structures, ~10–40 nm, seen in the low q region.

The spherical radii of the fitted form factor increases from 0.64 nm to a maximum value of 0.70 nm during the cooling process and this value appears too large when compared to a maximum radii of 0.35 nm for the stacked dimer. This overestimation of the calculated value of R_s maybe due to the assumption of a spherical form factor for the stacked dimer rather than a more elongated structure as seen in Fig. 15f. The cylindrical form factor length increases from 1.15 nm to 1.53 nm as shown in Table 10, this is an intermediate value between the monomer length, 0.7 nm and the carboxylic acid H-bonding dimer length, 1.6 nm. The maximum calculated cylinder length of 1.53 nm is larger than the NH···O dimer dimensions, and this could indicate that the carboxylic acid H-bonding dimer is the structure favoured due to better agreement to the molecular dimensions.

This observation could indicate that the carboxylic acid H-bonding dimer is favoured prior to nucleation and is the important building block in the

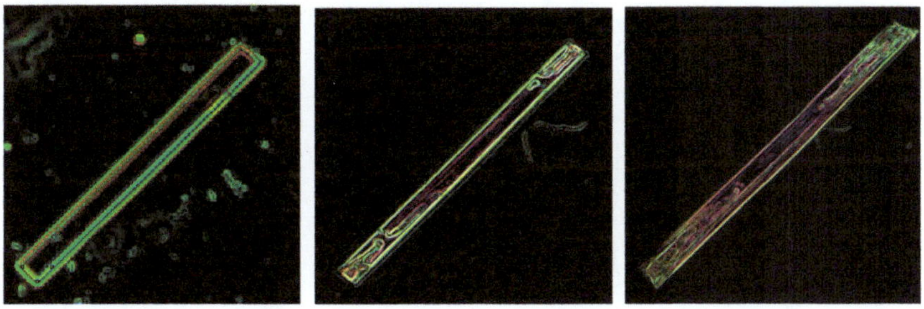

Fig. 16 (a) Alpha-PABA crystal grown at $S = 1.03$ EtOH after 10 min of growth. (b) Alpha-PABA crystal grown at $S = 1.11$ EtOH after 2 min of growth. (c) Alpha-PABA crystal grown at $S = 1.11$ EtOH after 6 min of growth. Pictures are an area of 0.75×0.75 mm.

self-assembly of alpha-PABA. This is supported by data in section 3.3 that calculates the strongest pairwise interaction in the alpha-PABA crystal structure is the carboxylic acid H-bonding dimer. In addition, the carboxylic acid H-bonding dimer is also predicted to be the most favoured interaction in solution from calculations in ethanol from section 3.3, which provides further support of this interpretation.

3.6. Growth rate measurements and morphological and surface modelling

Fig. 16 shows the morphology of alpha-PABA crystals grown in EtOH at various super saturations. The BFDH list suggests that the (0 1 ¬1), (1 ¬1 0) and (1 1 ¬1) surfaces are morphologically important at the capping ends of the needle and the attachment energy calculations predict similar relative growth rates for each face. Therefore the interfacial angle between the (1 0 ¬1) surface and the capping face was measured to identify the correct Miller index for that face.

Fig. 16a reveals that for low supersaturations the interfacial angle between the side (1 0 −1) surface and the capping face stays constant at around 90° throughout the observed growth, which is consistent with the interfacial angle between the (0 1 −1) and (1 0 −1) faces. However, increasing the supersaturation was found to create roughening and unstable growth resulting in poor lattice perfection observed at the capping faces, and the interfacial angles are observed to change as the crystal grows. Fig. 16b shows the interfacial angle to be around 90° after two minutes of growth at $\sigma = 0.11$, however Fig. 16c shows after 6 min of growth the angle has changed to around 70°, which does not match any of the calculated interfacial angles between the (1 0 −1) and likely capping faces. The attachment energy morphology predicts the capping face to be the (1 −1 0) face as the attachment energies of the (1 −1 0) and (0 1 −1) surfaces are very similar; considering the interfacial angles measured, it appears the original face that facets is the (0 1 −1) surface and the roughening changes the interfacial angle, therefore the (0 1 −1) faces are assumed to be the capping face.

Fig. 17b shows the fit of growth rate along the long axis of the needle (0 1 −1) as a function of supersaturation corresponds to an RIG mechanism even at low supersaturations. Conversely, fitting of the width growth rate, *i.e.* normal to the (1 0 −1) surface as a function of supersaturation, corresponds to a B & S growth

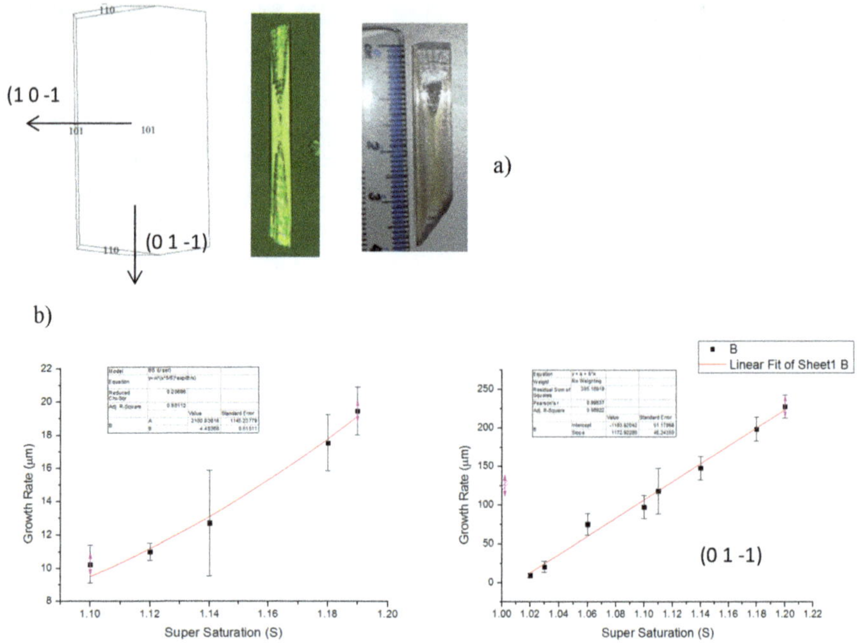

Fig. 17 (a) Attachment energy morphological prediction of the alpha-PABA structure (left), an optical micrograph of alpha-PABA grown from EtOH (0.3 × 1.2 mm) (middle) and a large alpha-PABA crystal grown from slow solvent evaporation from EtOH:NMe mix. (b) Alpha-PABA length (0 1 −1) and width (1 0 −1) growth rates measured as a function of supersaturation. (0 1 −1) surface fits a rough interfacial growth mechanism even at low supersaturations while the (1 0 −1) surface fits birth and spread above $\sigma = 0.1$.

mechanism above $\sigma = 0.1$. Almost no growth is observed below $\sigma = 0.1$ which suggests the presence of a dead zone below $\sigma = 0.1$ for the (1 0 −1) face. Comparing between the predicted and observed growth morphologies shown in Fig. 17a (left and right respectively) reveals a more needle-like morphology is the observed crystal habit in comparison to that which is predicted. This difference perhaps reflects the difference in growth rates and mechanism pertaining to the (1 0 −1) and (0 1 −1) faces.

The interactions which contribute to the attachment energy of the (0 1 −1) and (1 0 −1) surface were identified in Fig. 18 and 19 along with Tables 11 and 12 to understand the pairwise interactions that contribute to the growth.

Fig. 18 and Table 11 reveal the growth of the (1 0 −1) surface is dominated by H-bonding interactions as the COOH and NH_2 functional groups are directed along the growth normal of this surface. Considering growth from ethanol, binding of the ethanol molecules could provide some competition for the PABA molecules at this surface, as the OH group of the ethanol molecules would be expected to H-bond to the exposed C=O groups. Hence the de-solvation at this surface could be slower than at other surfaces and slow down the growth rate of this surface.

Fig. 19a shows the space fill model of the (0 1 −1) surface and how the PABA molecules stack in a wave-like pattern, therefore this forms the π–π interactions

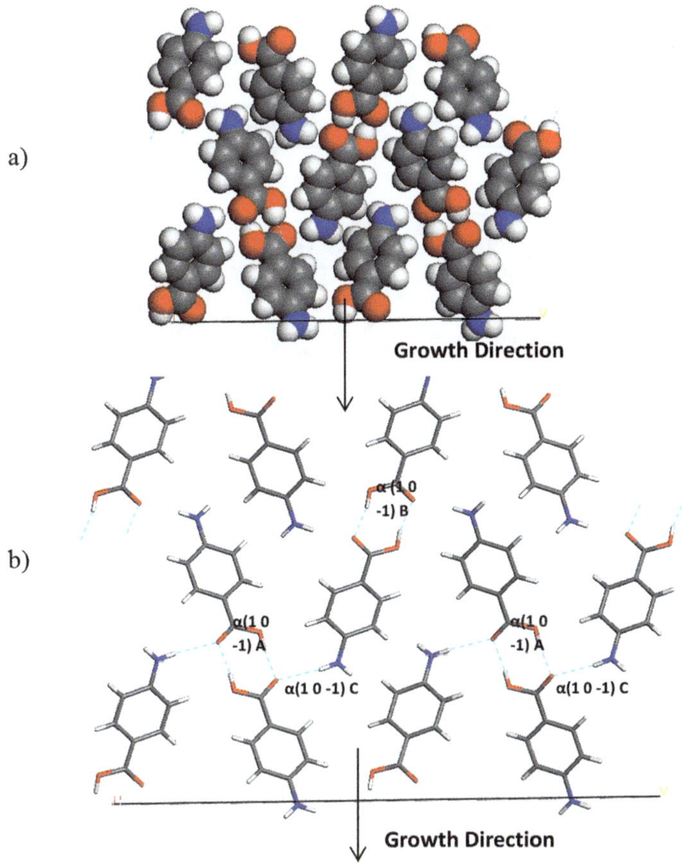

Fig. 18 (a) Space fill and (b) stick model of the alpha-PABA (1 0 −1) surface. The stick model has strongest intermolecular interactions exposed at this surface contributing to growth marked.

that facilitate the fast growth along the long axis of the needle. The H-bonding carboxylic acid dimers also contribute to the attachment energy of this surface, but the direction of the intermolecular interaction actually is almost perpendicular to the growth normal, compared to the π–π stacking interactions which are almost parallel to the growth normal. The H-bonding functional groups are orientated away from the surface at the (0 1 −1) surface which suggests they are less likely to interact with the polar ethanol molecules, which could mean that the PABA molecules face less competition to bind to this surface and incorporate into the lattice. The extra competition created by the EtOH binding at the (1 0 −1) surface could lead to an increase in the amount of PABA molecules attempting to bind at the capping (0 1 −1) surface leading to an increase in supersaturation around these faces. This appears to result in the rough interfacial growth mechanism and disorder at this surface which is demonstrated by the appearance of the sector zoning effects.

Interestingly, though the supersaturation *vs.* growth rate measurements of the (0 1 −1) fit with a roughening growth mechanism, the images of the crystals do

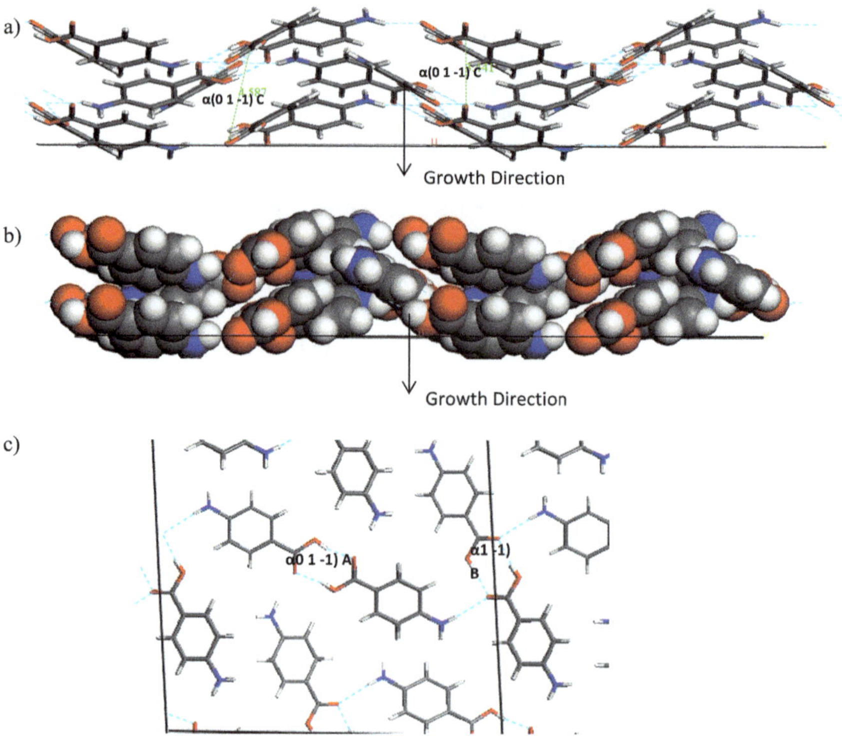

Fig. 19 (a) Space fill and stick model of the alpha-PABA (0 1 −1) surface. (b) The stick model has strongest intermolecular interactions exposed at this surface contributing to growth marked. (c) Top view of the alpha-PABA (0 1 −1) surface growing towards the plane of the page. H-bonding interactions are marked on the surface.

Table 11 Strongest interactions contributing to the growth of the PABA (1 0 −1) surface predicted by HABIT 98

Bond	Multiplicity	Distance (Å)	Intermolecular energy (kcal mol^{-1})
Alpha (1 0 −1) A	1	8.07	−7.10
Alpha (1 0 −1) B	1	8.03	−6.94
Alpha (1 0 −1) C	2	7.91	−1.90

Table 12 Strongest interactions contributing to the growth of the PABA (0 1 −1) surface

Bond	Multiplicity	Distance (Å)	Intermolecular energy (kcal mol^{-1})
Alpha (0 1 −1) A	1	8.07	−7.10
Alpha (0 1 −1) B	1	8.03	−6.94
Alpha (0 1 −1) C	2	3.84	−2.68

not exhibit obvious roughening and the growth facets in this direction appear to be relatively smooth. Fig. 19a shows the molecules attach by π–π stacking interactions which would be expected to exhibit strong molecular recognition when molecules diffuse from solution to the surface, resulting in the extremely fast growth observed at this surface. It is possible that even though the surface is growing by a rough polynucleation mechanism, the strong molecular recognition allows the molecules to attach in an ordered manner and create a smoother facet than normally seen from growth by the RIG mechanism.

It would seem that even though the solution structure and self-assembly of the alpha form appears to be facilitated by H-bonding interactions, the growth of the alpha form is dominated by π–π stacking interactions along the long axis of the needle. It has been suggested in a recent study by Lovette and Doherty[70] that growth spirals would contribute to the growth of the capping faces of alpha-PABA at low supersaturations. However, in this work, the growth rate *vs.* supersaturation curves suggests, in contrast, that the growth is *via* a polynucleation roughening mechanism even at low supersaturations. This conclusion also correlates with the analysis of the polythermal data which proposes that the early stages of growth would be dominated by diffusion through solution rather than interface kinetics mediated by the presence of growth spirals. This would suggest that a degree of interfacial roughening along the long axis of the needle occurs in all conditions in EtOH.

4 Conclusions

Using a multi scale approach, the nucleation of PABA has been examined, and the key outcomes are summarised in Table 13.

This study reveals that the dominant nucleation of the alpha form of PABA over the beta form in ethanolic solutions is most likely due to the stability and favourable formation of H-bonding carboxylic acid dimers in solution. The lack of significant change between the gas phase stable conformers and conformers in the crystal structure suggests that the intermolecular interactions, rather than any conformational change, play the most significant role in crystallisation of the material.

Ab initio continuum solvent calculations of the stability of small clusters (up to 8 molecules), taken from the crystal structures of the alpha and beta polymorphs, were consistent with clusters containing the H-bonding carboxylic acid motif always being more stable than those that do not contain this motif. Combined with MD solvation free energy calculations that suggest formation of the H-bonding carboxylic acid dimer is more favoured than two individual solvated molecules in solution, these calculations suggest that this dimer formation is an important building block in the self-assembly of the alpha-PABA crystal structure.

Analysis of experimentally obtained solubility data was found to be consistent with the propensity for solute–solute interactions in good agreement with the solvation free energy calculations. The large variations of T_C as a function of cooling rate observed from the poly-thermal nucleation analysis suggest that the crystallisation of alpha-PABA is strongly kinetically dependent. This might to some extent explain the interesting observation that the alpha-form is observed to nucleate below the transition temperature. These results also predict nucleation of PABA to be instantaneous and the number of crystallites formed to be small,

Table 13 Summary of the key outcomes of the techniques used to probe the nucleation of PABA. The section containing the detail of the key outcomes also highlighted.

Techniques		Subsection	Outcomes
Modelling	Conformation analysis	3.1	– Three distinct conformers found – One with beta crystal structure conformation (the most stable) – One with the alpha crystal structure conformation
Modelling	Solvation free energies	3.2	– Two molecules forming a carboxylic acid dimer preferred over two non-interacting molecules
Modelling	Solid state crystal chemistry dominating intermolecular interactions	3.3	– Alpha packing dominated carboxylic acid H-bonds and head-head π–π stacking – Beta packing dominated by 4-membered NH…O and OH…N H-bonding ring and head-tail offset stacking interactions
Modelling	Cluster modelling-populations	3.3	– Carboxylic acid dimer most stable in ethanol – Tetramers and octamers containing two and four carboxylic acid dimers found the most stable in solution
Modelling	Morphology and growth interfaces	3.6	– Prediction of a lathe like morphology – Side (1 0 −1) surface grows *via* attachment of hydrogen bonds – Capping (0 1−1) surface grows *via* attachment of π–π stacking interactions.
Experimental	Solubility analysis	3.4.1	– Less than ideal solubility with solute–solute interactions preferred
Experimental	Polythermal nucleation analysis	3.4.1	– Nucleation is rate-limiting step prior to crystallisation – Instantaneous nucleation mechanism – Crystallite number range at nucleation point: 860–1957
Experimental	Isothermal nucleation analysis	3.4.2	– Critical cluster sizes 0.48–1.98 nm

Table 13 *(Contd.)*

Techniques		Subsection	Outcomes
			– Critical cluster size larger for higher solution concentration
Experimental	SAXS	3.5	– High q region reveals a model consistent with the formation of a carboxylic acid dimer as supersaturation increases – Low q region consistent with larger cluster structures formed during undercooling of solution which increase in size and fractal dimensionality with time
Experimental	Growth rates	3.6	– (0 1 −1) growth rates 45–230 μm min^{-1} over $\sigma = 0.03$–0.20 – (1 0 −1) growth rates 10–20 μm min^{-1} over $\sigma = 0.1$–0.2. Dead zone below $\sigma = 0.1$ for this face
Experimental	Crystal growth mechanisms	3.6	– (1 0 −1) surface grows *via* 2D nucleation birth and spread mechanism – (0 1 −1) surface grows *via* a poly nucleation rough interfacial growth mechanism

indicating that the nuclei sizes may be large. The isothermal analysis calculates relatively small cluster sizes, 0.5–2 nm, less than might be anticipated compared to the low number of instantaneously nucleated crystallites from the poly-thermal analysis.

SAXS results from supersaturated solutions in the high q range suggest there is a distribution of monomers and dimers in solution, and that during the cooling process the distribution shifts towards an increase in dimers prior to a nucleation event, correlating with both the QM and MD solvation modelling and the analysis of the solubility data. Modelling of the R_g values for a π–π stacked dimer and the H-bonding carboxylic acid dimer show that the experimental R_g values prior to nucleation are more consistent with the H-bonding carboxylic acid dimer, correlating well with calculations of the stability of this motif in solution. Analysis of the low q range reveal larger structures consistent with a possible size range of 10–40 nm which increase in size as the solution supersaturates and the number of monomers/dimer structures in solution decreases. The increase in the value of the fractal dimension (P) from 1–2 during the super-cooling of the solution suggests increased ordering of the pre-nucleation structures as the solution supersaturates.

Though it appears that the H-bonding carboxylic acid dimers play a key role in the nucleation of the alpha-form, morphological prediction and surface interface modelling suggest that the growth is dominated by π–π stacking interactions. Growth rate measurements indicate that the alpha-form grows by a poly-nucleation roughening mechanism along the long axis of the needle even at low supersaturations, even though the optical microscopy observations appear to show a stable, planar facet. This roughening mechanism suggests that the majority of the growth is controlled by diffusion of molecules to the surface, and the attachment of molecules *via* the π–π stacking motif probably results in a solid–solid integration mechanism at the (0 1 −1) surface, suggesting that any growth spirals present at this surface are not influencing the growth. This correlates with poly-thermal measurements of the growth exponent (n) that suggest the growth is rate-limited by diffusion of solute across the crystal solution interface, rather than by the presence of growth spirals.

List of symbols

A_0	Fixed needle cross sectional area (m^2)
a	Dimensionless molecular latent heat of crystallisation
B	Dimensionless thermodynamic parameter
C_e	Equilibrium solution concentration (m^{-3})
C_0	Concentration of instantaneously nucleated crystallites (m^{-3})
d	Dimensionality of crystallite growth
E_{cr}	Lattice energy
E_{att}	Attachment energy
E_{sl}	Slice energy
i^*	Number of molecules in critical nucleus
k	Boltzmann constant (J K^{-1})
K_G	Growth rate constant $m^{(1/m)}$ s^{-1}
m, n	Crystallite growth exponents
P	Power law value
q	Cooling rate (K s^{-1})
q (SAXS)	Forward scattering vector (nm^{-1})
q_0	Parameter in the $u_{c(q)}$ dependence for both PN and IN (K s^{-1})
r^*	Critical nucleus radius (m)
R_g	Radius of gyration (nm)
S	Supersaturation ratio
σ	Absolute supersaturation
T_0	Temperature at which crystallites are instantaneously nucleated (K)
T_{dis}	Dissolution temperature (K)
T_C	Crystallisation temperature (K)
T_e	Equilibrium dissolution temperature (K)
ΔT_c	Critical undercooling for crystallisation (K)
u_c	Relative critical undercooling for crystallisation
v_0	Volume of solute molecule in crystal (m^3)
alpha_{det}	Detectable fraction of crystallised volume
γ_{eff}	Effective interfacial tension of crystal nucleus in 3D HEN (J m^{-2})
λ	Molecular latent heat of crystallisation (J)
Sl_{exp}	Slope of induction time data plotted using linear relationship of eqn (8)

List of abbreviations

RIG	Rough interfacial growth
B & S	Birth and spread model
BCF	Burton–Carbera–Frank model
BFDH	Bravais–Friedel–Donnay–Harker
MD	Molecular dynamics
QM	Quantum mechanics
TI	Thermodynamic integration
PABA	*para*-Amino benzoic acid
EtOH	Ethanol
KBHR	Kashchiev–Borissova–Hammond–Roberts approach
MSZW	Metastable zone width
GAFF	Generalised Amber force field
RESP	Restrained electrostatic potential
vdW	van der Waals
λ_{coupl}	Coupling parameter indicates the level of change that has taken place between two states

Acknowledgements

The authors gratefully acknowledge the UK's EPSRC for the funding of this research through a joint collaborative Critical Mass project between the Universities of Leeds and Manchester (grant references EP/IO14446/1 and EP/IO13563/1) which forms part of the doctoral studies of two of us (IR and TDT). We also gratefully acknowledge the Critical Mass team at the University of Manchester lead by Professor's Davey and Schroeder for stimulating discussions that inspired and contributed to the research and ideas presented in this paper. One of us (DMCC) gratefully acknowledges Infinium UK for the funding of a PhD studentship. We are most grateful to Professor Andreas Klamt (COSMOlogic and University of Regensbeurg) for his valuable advice on the implementation and suitability of the COSMO-RS method with clusters of PABA. We would like to acknowledge and thank the Brazilian Synchrotron Light Laboratory Campinas for allocation of scattering beam-time, also to the beam line scientist, Dr Mateus Cardoso and his team Dr Jean Marie Polli and Mr Tiago Kalile for their efforts and helpful discussion. We also thank Dr Tomasz Stawski for very helpful and stimulating discussions regarding SAXS analysis.

References

1 R. Breslow and M. V. Tirrel, *Beyond the Molecular Frontier: Challenges for Chemistry and Chemical Engineering*, National Research Council of the National Academies, 2003, ISBN 2003100913.

2 EPSRC Grand Challenge Network in Directed Assembly, Web http://beyondthemolecule.org.uk, accessed 12 March 2014.

3 R. B. Hammond, K. Pencheva and K. J. Roberts, *J. Phys. Chem. B*, 2005, **109**, 19950.

4 R. B. Hammond, K. Pencheva and K. J. Roberts, *Faraday Discuss.*, 2007, **136**, 87.

5 R. B. Hammond, K. Pencheva, K. J. Roberts and T. Auffret, *J. Pharm. Sci.*, 2007, **96**, 1967.

6 R. B. Hammond, K. Pencheva and K. J. Roberts, *CrystEngComm*, 2012, **14**, 1069.

7 R. B. Hammond, C. Ma, K. J. Roberts, P. Y. Ghi and R. K. Harris, *J. Phys. Chem.*, 2003, **107**, 11820.

8 R. B. Hammond, R. S. Hashim, C. Y. Ma and K. J. Roberts, *J. Pharm. Sci.*, 2006, **95**, 2361.

9 M. Volmer and A. Weber, *Zeitschrift Fur Physikalische Chemie–Stochiometrie Und Verwandtschaftslehre*, 1926, vol. 119, p. 277.

10 D. Kashchiev, *Nucleation: basic theory with applications*, Butterworth-Heinemann, Oxford, 2000.

11 R. Davey, S. L. M. Schroeder and J. H. ter Horst, *Angew. Chem., Int. Ed.*, 2013, **52**, 2166.

12 R. A. Sullivan, R. J. Davey, G. Sadiq, G. Dent, K. R. Back, J. H. ter Horst, D. Toroz and R. B. Hammond, *Cryst. Growth Des.*, 2014, **14**, 2689.

13 D. M. C. Corzo, A. Borissova, R. B. Hammond, D. Kashchiev, K. J. Roberts, K. Lewtas and I. More, *CrystEngComm*, 2014, **16**, 974.

14 S. Jiang and J. H. ter Horst, *Cryst. Growth Des.*, 2011, **11**, 256–261.

15 D. Gebaur, M. Kellermeier, J. D. Gale, L. Bergstrom and H. Coelfen, *Chem. Soc. Rev.*, 2014, **43**, 2348.

16 A. Gavezotti and G. Filippini, *Chem. Commun.*, 1998, 287.

17 S. Hamad, C. Moon, R. A. Catlow, A. T. Hulme and S. L. Price, *J. Phys. Chem.*, 2006, **110**, 3323.

18 A. Mattei and T. Li, *Cryst. Growth Des.*, 2014, **14**, 2709.

19 R. Benali-Cherif, R. Takouachet, E. E. Bendeif and N. Benali-Cherif, *Acta Crystallogr., Sect. C: Struct. Chem.*, 2014, **3**, 323.

20 S. Gracin and Å. C. Rasmuson, *Cryst. Growth Des.*, 2004, **4**, 1013.

21 H. Hao, M. Barrett, Y. Hu, W. Su, S. Ferguson, B. Wood and B. Glennon, *Org. Process Res. Dev.*, 2012, **16**, 35.

22 T. F. Lai and R. E. Marsh, *Acta Crystallogr.*, 1967, **22**, 885.

23 S. Athimoolan and S. Natarajan, *Acta Crystallogr., Sect. C: Cryst. Struct. Commun.*, 2007, **63**, O514.

24 S. Gracin and A. Fisher, *Acta Crystallogr., Sect. E: Struct. Rep. Online*, 2005, **61**, O1242.

25 D. Toroz and T. Van Mourik, *Mol. Phys.*, 2006, **104**, 559.

26 D. Toroz and T. Van Mourik, *Mol. Phys.*, 2007, **105**, 209.

27 D. Toroz and T. Van Mourik, *Phys. Chem. Chem. Phys.*, 2010, **12**, 3463.

28 M. J. Frisch, G. W. Trucks, H. B. Schlegel, G. E. Scuseria, M. A. Robb, J. R. Cheeseman, G. Scalmani, V. Barone, B. Mennucci, G. A. Petersson, H. Nakatsuji, M. Caricato, X. Li, H. P. Hratchian, A. F. Izmaylov, J. Bloino, G. Zheng, J. L. Sonnenberg, M. Hada, M. Ehara, K. Toyota, R. Fukuda, J. Hasegawa, M. Ishida, T. Nakajima, Y. Honda, O. Kitao, H. Nakai, T. Vreven, J. A. Montgomery Jr, J. E. Peralta, F. Ogliaro, M. Bearpark, J. J. Heyd, E. Brothers, K. N. Kudin, V. N. Staroverov, R. Kobayashi, J. Normand, K. Raghavachari, A. Rendell, J. C. Burant, S. S. Iyengar, J. Tomasi, M. Cossi, N. Rega, J. M. Millam, M. Klene, J. E. Knox, J. B. Cross, V. Bakken, C. Adamo, J. Jaramillo, R. Gomperts, R. E. Stratmann, O. Yazyev, A. J. Austin, R. Cammi, C. Pomelli, J. W. Ochterski, R. L. Martin,

K. Morokuma, V. G. Zakrzewski, G. A. Voth, P. Salvador, J. J. Dannenberg, S. Dapprich, A. D. Daniels, Ö. Farkas, J. B. Foresman, J. V. Ortiz, J. Cioslowski, and D. J. Fox, *Gaussian 09*, Gaussian, Inc., Wallingford CT, 2009.
29 G. Clydesdale, R. Docherty and K. J. Roberts, *Comput. Phys. Commun.*, 1991, **64**, 311.
30 R. Docherty, G. Clydesdale, K. J. Roberts and P. Bennema, *J. Phys. D: Appl. Phys.*, 1991, **24**, 89.
31 R. Docherty and K. J. Roberts, *J. Cryst. Growth*, 1988, **88**, 159.
32 F. A. Momany, L. M. Carruthers, *et al.*, *J. Phys. Chem.*, 1974, **78**, 1595.
33 A. Klamt and G. Schuurmann, *J. Chem. Soc., Perkin Trans. 2*, 1993, 799.
34 F. Eckert and A. Klamt, *AIChE J.*, 2002, **48**, 369.
35 A. Klamt, F. Eckert, *et al.*, *J. Comput. Chem.*, 2002, **23**, 275.
36 A. Klamt, *Fluid Phase Equilib.*, 2003, **206**, 223.
37 J. Palomar, V. R. Ferro, *et al.*, *Ind. Eng. Chem. Res.*, 2007, **46**, 6041.
38 B. Hess, C. Kutzner, D. van der Spoel and E. Lindahl, *J. Chem. Theory Comput.*, 2008, **4**, 435.
39 J. Wang, W. Wang, P. A. Kollman and D. A. Case, *J. Mol. Graphics Modell.*, 2006, **25**, 247260.
40 J. Wang, R. M. Wolf, J. W. Caldwell, P. A. Kollman and D. A. Case, *J. Comput. Chem.*, 2004, **25**, 1157.
41 Z. Berkovitch-Yellin, *Am. Chem. Sci. J.*, 1985, **107**, 8239–8253.
42 P. Hartmann and W. G. Perdok, *Acta Crystallogr.*, 1955, **8**, 29–52.
43 R. Docherty, K. J. Roberts and E. Dowty, *Comput. Phys. Commun.*, 1988, **51**, 423.
44 E. Dowty, *Am. Mineral.*, 1980, **65**, 465.
45 Crystal 16 details, http://www.crystallizationsystems.com/en/crystal16/.
46 D. Kashchiev, A. Borissova, R. B. Hammond and K. J. Roberts, *J. Cryst. Growth*, 2010, **312**, 698.
47 D. Kashchiev, A. Borissova, R. B. Hammond and K. J. Roberts, *J. Phys. Chem. B*, 2010, **114**, 5441.
48 D. M. C. Corzo, A. Borissova, R. B. Hammond, D. Kashchiev, K. J. Roberts, K. Lewtas and I. More, *CrystEngComm*, 2014, **16**, 974.
49 D. Kashchiev and A. Firoozabadi, *J. Cryst. Growth*, 2003, **250**, 499.
50 D. Kashchiev, *Nucleation: basic theory with applications*, Butterworth-Heinemann, Oxford, 2000.
51 O. Kratky, in *Small Angle X-ray Scattering*, ed. O. Glatter and O. Kratky, Academic Press, London, 1982.
52 J. Ilavsky and P. R. Jemian, *J. Appl. Crystallogr.*, 2009, **42**, 347.
53 P. W. Schmidt, *J. Appl. Crystallogr.*, 1991, **24**, 414.
54 H. Boukari, J. S. Lin and M. T. Harris, *J. Colloid Interface Sci.*, 1997, **194**, 311.
55 G. J. Beaucage, *J. Appl. Crystallogr.*, 1995, **28**, 717.
56 G. J. Beaucage, *J. Appl. Crystallogr.*, 1996, **29**, 134.
57 C. U. Segre, S. Chattopadhyay, D. Erdemir, J. M. B. Evans, J. Ilavsky, H. Amenitsch and A. S. Myerson, *Cryst. Growth Des.*, 2005, **5**(no. 2), 523.
58 S. Förster, L. Apostol and W. Bras, *J. Appl. Crystallogr.*, 2010, **43**(no. 3), 639.
59 S. Förster and C. Burger, *Macromolecules*, 1998, **31**, 879.
60 Brazilian synchrotron details, http://lnls.cnpem.br/.
61 T. T. H. Nguyen, R. B. Hammond, K. J. Roberts, I. Marziano and G. Nichols, *CrystEngComm*, 2014, **16**, 4568.
62 W. K. Burton, N. Cabrera and F. Frank, *Philos. Trans. R. Soc., A*, 1951, 299–358.

63 M. Ohara and R. C. Reid, in *Modeling Crystal Growth Rates from Solution*, Prentice-Hall, 1973, pp. 225.

64 J. van der Eerden, P. Bennema and T. Cherepanova, in *Progress in Crystal Growth and Characterisaton*, 1973, vol. 1, pp. 219–254.

65 P. Bennema, *Morphology of Crystals*, Terra Scientific, Tokyo, 1987.

66 C. Ouvard and S. L. Price, *Cryst. Growth Des.*, 2004, **4**, 1119.

67 K. J. Roberts, R. Docherty, P. Bennema and L. A. Jetten, *J. Phys. D: Appl. Phys.*, 1993, **26**, 7.

68 K. Pencheva, Ph.D. thesis, University of Leeds, England, 2006.

69 D. I. Svergun, C. Barberato and M. H. J. Koch, *J. Appl. Crystallogr.*, 1995, **28**, 768.

70 M. A. Lovette and M. F. Doherty, *Cryst. Growth Des.*, 2013, **13**, 3341.

Faraday Discussions

PAPER

New *in situ* solid-state NMR techniques for probing the evolution of crystallization processes: pre-nucleation, nucleation and growth

Colan E. Hughes, P. Andrew Williams, Victoria L. Keast, Vasileios G. Charalampopoulos, Gregory R. Edwards-Gau and Kenneth D. M. Harris*

Received 10th November 2014, Accepted 25th November 2014
DOI: 10.1039/c4fd00215f

The application of *in situ* techniques for investigating crystallization processes promises to yield significant new insights into fundamental aspects of crystallization science. With this motivation, we recently developed a new *in situ* solid-state NMR technique that exploits the ability of NMR to selectively detect the solid phase in heterogeneous solid–liquid systems (of the type that exist during crystallization from solution), with the liquid phase "invisible" to the measurement. As a consequence, the technique allows the first solid particles produced during crystallization to be observed and identified, and allows the evolution of different solid phases (*e.g.*, polymorphs) present during the crystallization process to be monitored as a function of time. This *in situ* solid-state NMR strategy has been demonstrated to be a powerful approach for establishing the sequence of solid phases produced during crystallization and for the discovery of new polymorphs. The most recent advance of the *in situ* NMR methodology has been the development of a strategy (named "CLASSIC NMR") that allows both solid-state NMR and liquid-state NMR spectra to be measured (essentially simultaneously) during the crystallization process, yielding information on the complementary changes that occur in both the solid and liquid phases as a function of time. In this article, we present new results that highlight the application of our *in situ* NMR techniques to successfully unravel different aspects of crystallization processes, focusing on: (i) the application of a CLASSIC NMR approach to monitor competitive inclusion processes in solid urea inclusion compounds, (ii) exploiting liquid-state NMR to gain insights into co-crystal formation between benzoic acid and pentafluorobenzoic acid, and (iii) applications of *in situ* solid-state NMR for the discovery of new solid forms of trimethylphosphine oxide and L-phenylalanine. Finally, the article discusses a number of important fundamental issues relating to practical aspects, the interpretation of results and the future scope of these techniques, including: (i) an assessment of the smallest size of solid particle that can be detected in

School of Chemistry, Cardiff University, Park Place, Cardiff CF10 3AT, Wales, UK. E-mail: HarrisKDM@cardiff.ac.uk

in situ solid-state NMR studies of crystallization, (ii) an appraisal of whether the rapid sample spinning required by the NMR measurement technique may actually influence or perturb the crystallization behaviour, and (iii) a discussion of factors that influence the sensitivity and time-resolution of *in situ* solid-state NMR experiments.

1. Introduction

Crystal nucleation and growth processes[1,2] are encountered in many different scientific fields and are crucially important in many aspects of chemical, pharmaceutical and biological sciences. Progress to improve our fundamental understanding of crystallization processes has important practical applications, including the development of new ways to exert control over the polymorphic form of crystals produced in industrial applications. Deepening our fundamental physico-chemical understanding of crystallization relies significantly on the development and application of new experimental strategies, particularly those that allow direct *in situ* monitoring of the process. In general, crystallization processes are governed by kinetic factors and meta-stable polymorphs are often produced rather than the thermodynamically stable polymorph (in this context, polymorphs[3–10] are defined as crystalline materials that have identical chemical composition but different crystal structures). Furthermore, crystallization processes often evolve through a sequence of different solid forms before reaching the final crystallization product and details of the time-evolution of the process can depend critically on the exact conditions of the crystallization experiment. In order to optimize and control crystallization in such situations, it is essential to understand the sequence of events involved in the evolution of the solid form, rather than simply characterizing the final crystalline phase collected at the end of the process. In this regard, experimental strategies that allow direct *in situ* monitoring of crystallization processes are clearly essential. A wide variety of experimental techniques have been used for *in situ* studies of crystallization,[11] including scattering techniques (*e.g.*, X-ray diffraction, small-angle X-ray scattering, neutron diffraction and small-angle neutron scattering), spectroscopic methods (*e.g.*, infrared, Raman and X-ray absorption spectroscopies) and microscopy (*e.g.*, atomic force microscopy). However, until the present time, solid-state NMR spectroscopy has not been used extensively in this regard.

Although solid-state NMR is a powerful and versatile technique for investigating structural and dynamic properties of solids, adapting this technique for *in situ* studies of chemical processes is associated with specific technical challenges. These challenges include the fact that high-resolution solid-state NMR spectra are usually recorded under conditions of rapid sample spinning (at frequencies typically around 10 000 revolutions per second) and the fact that the system under investigation must be located in a sealed rotor within a confined and relatively inaccessible space inside the magnet of the solid-state NMR spectrometer. For these reasons, the development of *in situ* solid-state NMR techniques has typically lagged behind other *in situ* experimental methods.

Recently, we developed a new *in situ* solid-state NMR technique[12,13] for monitoring the evolution of the solid phase as a function of time during crystallization from solution. This technique has been shown to yield new insights into several issues relating to crystallization processes, particularly the time-evolution of different polymorphs (or other solid forms) and the transformations between polymorphs that occur during the crystallization process. Very recently,[14] we proposed and demonstrated an extension of the *in situ* NMR strategy involving combined liquid-state and solid-state NMR measurements. This new implementation of the strategy, termed "CLASSIC NMR" (Combined Liquid- And Solid-State *In situ* Crystallization NMR), allows the evolution of both the solid phase and the liquid phase to be probed as a function of time during the same crystallization experiment, yielding significantly deeper insights on the behaviour of the entire crystallization system.

After giving an overview of the foundations of these *in situ* solid-state NMR techniques, including illustrative examples of their applications, we focus on presenting new results that highlight the capability of *in situ* solid-state NMR to deepen our understanding of crystallization processes. In addition, a critical discussion of several fundamental aspects of the methodology is presented. Issues covered include an appraisal of the possible effects of rapid sample rotation (an intrinsic requirement for recording high-resolution solid-state NMR data) and sample confinement on the pathways and outcome of crystallization processes, and an assessment of the extent to which knowledge gained from *in situ* NMR studies may be exploited in the optimization of bulk crystallization experiments.

2. Essential background to high-resolution solid-state NMR

A basic premise underlying the development of solid-state NMR for *in situ* studies of crystallization processes is the well-established fact that high-resolution solid-state NMR can be utilized for identification of different polymorphic forms of solids.[4,9,15] In the case of organic solids, for example, the high-resolution solid-state ^{13}C NMR spectrum contains, in principle, one peak for each crystallographically distinguishable carbon atom in the crystal structure (although, in practice, the actual number of observed peaks may be less than this number as a consequence of accidental peak overlap). The peak positions (*i.e.*, the isotropic chemical shifts) in the high-resolution solid-state ^{13}C NMR spectrum depend both on the environment of the ^{13}C nucleus within the molecule and also on the local environment surrounding the ^{13}C nucleus within the crystal structure. As a consequence, for a given ^{13}C site within an organic molecule, the peak positions in high-resolution solid-state ^{13}C NMR spectra can typically differ within a range of about $\pm$5 ppm between different polymorphs (or other solid forms) due to the dependence of the isotropic chemical shift on the crystal structure. Of course, larger differences in peak positions may arise between polymorphs that contain significantly different conformations or different tautomeric forms of the molecule. In this section, we give a very brief introduction to the experimental methods for measuring high-resolution solid-state NMR spectra of powder samples of organic solids.

In the case of high-resolution solid-state ^{13}C NMR of organic solids with natural isotopic abundances, the primary contributions to line-broadening arise from chemical shift anisotropy and from direct ^{13}C–^{1}H dipole–dipole interactions. The line-broadening is reduced (resulting in high-resolution ^{13}C NMR spectra) by a combination of rapid magic-angle sample spinning (MAS),[16] typically at spinning frequencies of around 10 kHz, and high-power ^{1}H decoupling.

Another experimental technique that is commonly applied in the measurement of high-resolution solid-state ^{13}C NMR spectra of organic materials is ^{1}H → ^{13}C cross polarization (CP).[9,17–19] In this technique, rather than directly exciting the ^{13}C nuclei followed by detection of the ^{13}C NMR signal, the procedure is to excite the ^{1}H nuclei, followed by transfer of magnetization from the ^{1}H nuclei to the ^{13}C nuclei (*i.e.*, the "cross-polarization" process) and detection of the resulting ^{13}C NMR signal. In the measurement of high-resolution solid-state ^{13}C NMR spectra of powder samples, the major benefit of using ^{1}H → ^{13}C CP is that it gives a significant enhancement of signal intensity compared to normal direct-excitation ^{13}C NMR measurements.

In the context of applying high-resolution solid-state ^{13}C NMR for *in situ* studies of crystallization from solution, the use of ^{1}H → ^{13}C CP has the critical additional benefit of allowing selective detection of the ^{13}C NMR signal only from the solid component of the heterogeneous solid–liquid system that exists during the crystallization process (see Section 3 for more details). An important aspect of the ^{1}H → ^{13}C CP technique is that the recycle delay in the NMR experiment (the time between successive applications of the pulse sequence) is dictated by ^{1}H spin–lattice relaxation (and not by ^{13}C spin–lattice relaxation). This feature is advantageous because, for organic materials, ^{1}H spin–lattice relaxation times (T_1) are typically much shorter than ^{13}C spin–lattice relaxation times, allowing more rapid repetition of the pulse sequence (in general, the recycle delay is taken as $5 \times T_1$), hence reducing the time required to record the solid-state ^{13}C NMR spectrum. For direct excitation ^{13}C NMR measurements, on the other hand, the recycle delay is dictated by the ^{13}C spin–lattice relaxation and, therefore, typically requires longer recycle delays than for ^{1}H → ^{13}C CP measurements.

More details of these techniques and other technical aspects of solid-state NMR can be found elsewhere.[18,19]

3. Experimental aspects of *in situ* solid-state NMR studies of crystallization processes

In our *in situ* solid-state NMR strategy for monitoring crystallization processes, a homogeneous (undersaturated) solution is initially prepared inside the NMR rotor at elevated temperature (T_H) and crystallization is then induced by decreasing the temperature rapidly to a specific target temperature (T_L) at which the solution is supersaturated (Fig. 1). Crystallization is thermodynamically favoured at the target temperature and the time-dependence of the crystallization process is monitored by repeatedly recording high-resolution solid-state NMR spectra as a function of time.

Clearly, the time-resolution of the *in situ* monitoring of the crystallization process depends on the time required to record an individual NMR spectrum of adequate quality to identify and distinguish the different solid forms present during the evolution of the system. Thus, optimization of the sensitivity of the measurement is also important, allowing solid-state NMR spectra of adequate quality to be recorded in the shortest possible time. More discussion of sensitivity

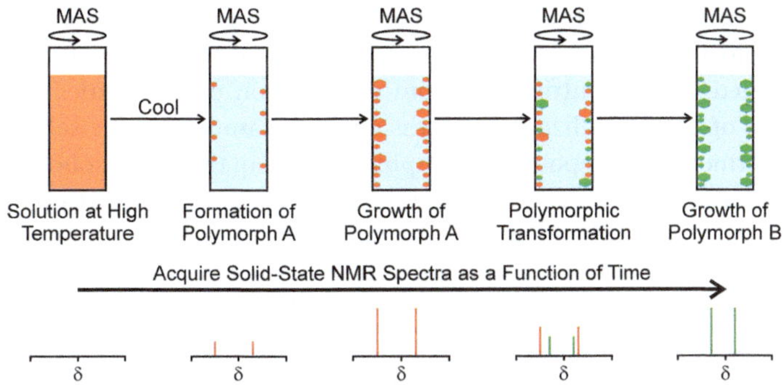

Fig. 1 Schematic of the strategy for *in situ* solid-state NMR studies of crystallization processes, illustrated by the case of a system in which the crystallization process initially produces a meta-stable polymorph A (red) followed by a polymorphic transformation to produce a more stable polymorph B (green). The corresponding changes in the measured solid-state NMR spectrum as a function of time are shown at the bottom.

and time-resolution in *in situ* NMR studies of crystallization is given in Section 6.4.

As discussed in Section 2, sufficiently good spectral resolution is also required in order to identify and assign the solid phases present at different stages of the crystallization process. Clearly, it is desirable to be able to detect and identify the first solid particles produced in the crystallization process, although at this stage the amount of solid present in the crystallization system is generally very low. Maximizing the sensitivity of the NMR measurement is thus also very important for enhancing the prospects of observing very early stages of the process.

Prior to our development of the *in situ* NMR techniques described above, the prospect of using solid-state NMR for *in situ* studies of crystallization from solution was limited by the difficulty of securely sealing a fluid phase inside an NMR rotor such that MAS could be carried out at several kHz without problems arising from leakage of the liquid from the rotor. Fortunately, advances in rotor technology (see Fig. 12 in Section 6) that allow solutions to be securely sealed inside NMR rotors for MAS experiments have taken place in recent years and this technical development paved the way for the types of experiment described here.

An important feature of the *in situ* solid-state NMR strategy is that it exploits the opportunity afforded by NMR of complete selectivity in detecting only the solid component in the crystallization system, such that the dissolved solute and solvent are undetected in the measurement. In the case of organic materials, such discrimination between solid and solution phases is achieved by recording ^{13}C NMR spectra under conditions of ^{1}H $\rightarrow$ ^{13}C cross polarization (CP). As molecules in the solid phase and the solution phase generally have substantially different dynamic behaviour, measurements under normal conditions for ^{1}H $\rightarrow$ ^{13}C CP give rise to a signal only from the solid phase. Thus, even if only a small fraction of the solute has crystallized (*e.g.*, in the early stages of crystallization), only the solid particles contribute to the measured NMR spectrum, while the dissolved solute molecules (present in much greater amount in the early stages of crystallization) are "invisible" to the measurement.

In contrast, for *in situ* studies of crystallization processes based on X-ray or neutron scattering techniques, scattering occurs both from the crystalline solid particles (giving rise to Bragg diffraction peaks) and from the solution phase (giving rise to a broad background scattering). As a result, the scattering data may be dominated by the contribution from the solution phase, particularly in the early stages of the crystallization process when the amount of the solid phase is low. Furthermore, it is important to emphasize that, in the *in situ* solid-state NMR experiment, essentially the entire crystallization system (*i.e.*, the sample inside the NMR rotor) is studied. In the case of *in situ* X-ray or neutron scattering experiments, on the other hand, a finely focused incident beam is used, which typically interrogates only a fraction of the sample inside the *in situ* cell. Finally, we note that, unlike the situation for diffraction-based techniques, solid-state NMR has no requirement that the sample must be crystalline. Thus, high-resolution solid-state NMR spectra can be measured both for crystalline and amorphous solids, and therefore *in situ* solid-state NMR studies of crystallization processes provide the opportunity to observe both crystalline and amorphous solid forms that may be produced at different stages of the process.

4. Illustrative examples of applications of *in situ* solid-state NMR techniques to study crystallization processes

4.1 Monitoring polymorphic evolution during crystallization

Glycine ($H_3^+NCH_2CO_2^-$) is widely adopted as a model system for studies of crystallization and polymorphism.[20–38] Under ambient conditions, three polymorphs of glycine (denoted α, β and γ) are known,[20–24] with relative stabilities γ > α > β.[28,39] The three polymorphs are readily distinguished by high-resolution solid-state ^{13}C NMR as the isotropic ^{13}C chemical shift for the carboxylate group is distinctly different in the three polymorphs (176.5, 175.5 and 174.5 ppm for the α, β and γ polymorphs respectively).[30] The consensus in the literature is that crystallization from water at neutral pH produces the meta-stable α polymorph. However, an early publication suggested[22] that crystallization from deuterated water may promote the formation of the γ polymorph, although systematic studies of this isotope effect were reported only recently,[34,36] in which it was demonstrated *inter alia* that deuteration (even at levels as low as 1%) does significantly increase the probability of obtaining the γ polymorph. Crystallization of glycine from methanol/water is reported[32] to promote the formation of the β polymorph.

To explore the effect of deuteration, we carried out separate *in situ* solid-state ^{13}C NMR studies[12] of crystallization of glycine from water with natural isotopic abundances (H_2O; Fig. 2a) and from deuterated water (D_2O; Fig. 2b) (in the latter case, the total level of deuteration of all exchangeable hydrogen sites in the system was 86%). In crystallization from H_2O, we simply observe the formation and growth of the α polymorph as a function of time, with no other polymorphs observed throughout the duration (13 h) of the experiment. In crystallization from D_2O, the α polymorph is again observed as the first solid form, suggesting that the same nucleation pathway is followed in both H_2O and D_2O. However, about 1.5 h after commencing the experiment, a peak characteristic of the γ polymorph appears in the ^{13}C NMR spectrum. The amount of the γ polymorph then increases rapidly while the amount of the α polymorph diminishes at the same rate,

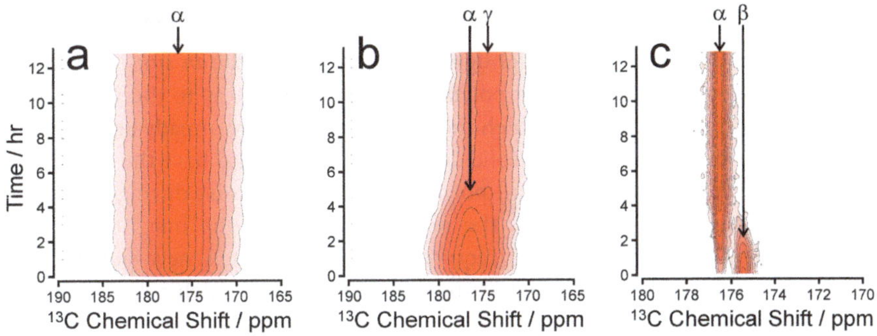

Fig. 2 *In situ* $^1H \rightarrow {}^{13}C$ CP NMR spectra (showing only the carboxylate region) recorded as a function of time for glycine crystallizing from (a) H_2O, (b) D_2O and (c) H_2O/methanol. The known peak positions for the α, β and γ polymorphs are indicated. Note that the sample of glycine used for the experiments in (a) and (b) was ^{13}C labelled in both carbon sites, whereas the sample of glycine used for the experiment in (c) was ^{13}C labelled only in the carboxylate site. As a consequence, the ^{13}C linewidth is significantly narrower in (c). In each case, intensity contour intervals are defined on a logarithmic scale.

consistent with a solution-mediated polymorphic transformation. For each of the two isotopomeric systems, the final polymorph obtained at the end of the *in situ* solid-state NMR study is consistent with the preferred polymorphic outcome in conventional laboratory crystallization experiments carried out under the same conditions and over the same total period of time.

To investigate the growth and stability of the β polymorph, crystallization was initiated by adding methanol to an aqueous solution of glycine[32] in an NMR rotor just prior to insertion into the NMR spectrometer for the *in situ* solid-state ^{13}C NMR study (Fig. 2c).[13] In the first spectrum recorded, the solid is identified as a virtually pure sample of the β polymorph (with a very small amount of the α polymorph also present). The amount of the β polymorph then decreases progressively, transforming into the α polymorph. Clearly, the *in situ* solid-state ^{13}C NMR study allows the timescale of this polymorphic transformation to be established and indicates that a viable strategy for isolating the β polymorph would be to stop the crystallization experiment within a few minutes of triggering the crystallization process.

Another example of polymorphic evolution has been revealed by *in situ* solid-state ^{13}C NMR studies of crystallization of *m*-aminobenzoic acid (*m*-ABA) from methanol. Five polymorphs of *m*-ABA are known[40,41] and the crystal structures of four polymorphs (forms II, III, IV and V) have been determined. In forms I, III and IV, the *m*-ABA molecules exist as zwitterions whereas, in forms II and V, the *m*-ABA molecules are non-zwitterionic. Each polymorph is uniquely distinguished by its high-resolution solid-state ^{13}C NMR spectrum.[14] In laboratory crystallization experiments, the outcome of crystallization of *m*-ABA from methanol is unpredictable, yielding either form I, form III or mixtures of forms I and III. Our *in situ* solid-state ^{13}C NMR study[42] led to a full understanding of this situation, as the results demonstrate clearly that form I is the initial crystallization product, followed by a polymorphic transformation (within a few hours of the formation of form I) to produce form III. Thus, depending on the time at which crystals are collected in laboratory crystallization experiments, the product may be either pure form I, a mixture of forms I and III, or pure form III.

4.2 Discovery of new (unknown) solid forms by *in situ* solid-state NMR studies

The examples discussed so far highlight the capability of the *in situ* solid-state NMR strategy to observe transient polymorphs on the pathway to the final crystallization product, in cases for which these polymorphs were already known and well characterized. Clearly, *in situ* solid-state NMR studies also have the potential to reveal new (previously unknown) polymorphs that may exist during crystallization processes.[43] In such cases, the results provide unique insights on the specific crystallization conditions required to produce each new polymorph observed, including the appropriate "time-window" during which a specific new polymorph is present.

To illustrate the discovery of new solid forms, we consider an *in situ* solid-state ^{13}C NMR study of co-crystal formation involving urea and 1,10-dihydroxydecane. Co-crystals of even-chain α,ω-dihydroxyalkanes [$HO(CH_2)_nOH$ with $n = 6$, 8, 10, 12, 14, 16] and urea (in 1 : 2 molar ratio) have been shown[44] to exhibit well-defined structure types characterized by two features: (i) double-stranded hydrogen-bonded urea ribbons in which the two strands are either parallel or antiparallel, and (ii) an angle between the axis of the α,ω-dihydroxyalkane and the positive direction of the urea strand which is either acute or obtuse. Specifically, the antiparallel/acute structure is found for $n = 6$, the parallel/acute structure is found for $n = 8$ and the antiparallel/obtuse structure is found for $n = 10$–16. Surprisingly, in laboratory crystallization experiments using a wide range of different conditions, no polymorphism has been observed for any member of this family. However, in the *in situ* solid-state ^{13}C NMR study of the crystallization of 1,10-dihydroxydecane and ^{13}C-labelled urea from methanol, the ^{13}C chemical shift for the initial crystallization product is clearly different from that of the known co-crystal phase (Fig. 3). After about 40 min, a new peak characteristic of the known co-crystal phase appears and thereafter grows very rapidly. At the end of the experiment (11.4 h), only the known co-crystal phase is present (confirmed by *ex situ* powder X-ray diffraction). Comparison of the ^{13}C chemical shift for urea in the new solid form present in the early stages of the crystallization process with those for urea in other members of the family of α,ω-dihydroxyalkane-(urea)$_2$ co-crystals leads to the assignment that the new transient solid form may be a 1,10-dihydroxydecane-(urea)$_2$ co-crystal with the parallel/acute structure type.

4.3 CLASSIC NMR

As discussed in Section 3, the *in situ* solid-state NMR methodology has recently been extended to exploit the fact that a solid-state NMR spectrometer can be used to study both the liquid phase and the solid phase in a heterogeneous solid–liquid system, simply by changing the pulse sequence used to record the NMR data. Specifically, by alternating between two different pulse sequences in an *in situ* NMR study of crystallization, alternate solid-state NMR and liquid-state NMR spectra are recorded, yielding essentially simultaneous information on the time-evolution of both the solid phase and the liquid phase during the crystallization process (Fig. 4). This strategy (called "CLASSIC NMR") reveals the complementary changes that occur in the solid phase and in the liquid phase as a function of time during crystallization. The time-dependence of the amount and polymorphic identity of the solid phase is established from the solid-state NMR spectra, while changes in the solution-state speciation and modes of molecular aggregation in

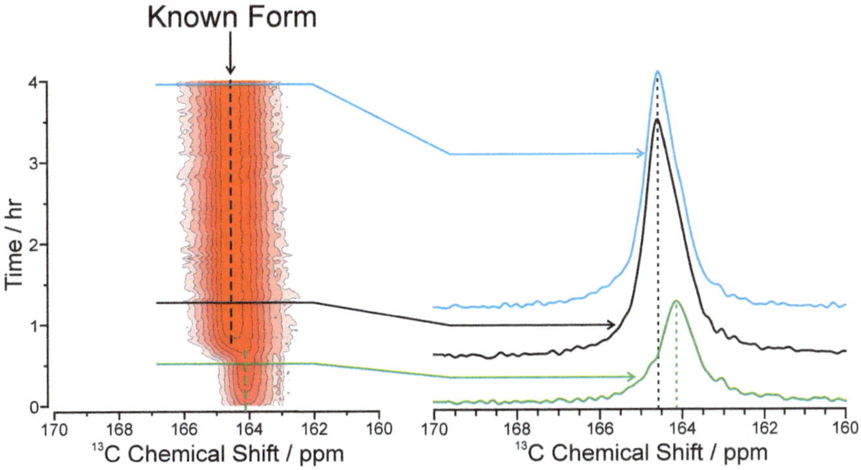

Fig. 3 *In situ* solid-state ^{13}C NMR spectra (showing the region of the spectrum containing the isotropic peak for urea) recorded as a function of time during crystallization of 1,10-dihydroxydecane and ^{13}C-labelled urea from methanol. Intensity contour intervals are defined on a logarithmic scale.

solution are monitored from the time-evolution of the liquid-state NMR spectrum. Although critical nucleation entities are generally too small to be observed (see Section 6.1), understanding the conditions under which nucleation takes place potentially allows improved models for nucleation to be proposed.

The CLASSIC NMR strategy has been demonstrated[14] in a ^{13}C NMR study of crystallization of *m*-aminobenzoic acid (*m*-ABA) from dimethyl sulfoxide (DMSO). Selective detection of the solid phase was achieved using ^{1}H → ^{13}C CP with high-power ^{1}H decoupling, while selective detection of the liquid phase was achieved using direct ^{13}C excitation with no ^{1}H decoupling and a relatively short recycle delay (the absence of ^{1}H decoupling and the short recycle delay ensure that no significant signal is detected from the solid phase).

The solid-state component of the CLASSIC ^{13}C NMR data showed that, about 2 h after the start of the measurements, form I of *m*-ABA appeared, with the amount then increasing during the next 6 h. No other polymorphs were observed during the crystallization process. The liquid-state component of the CLASSIC ^{13}C NMR data also did not change for the first 2 h of the experiment (*i.e.*, the period

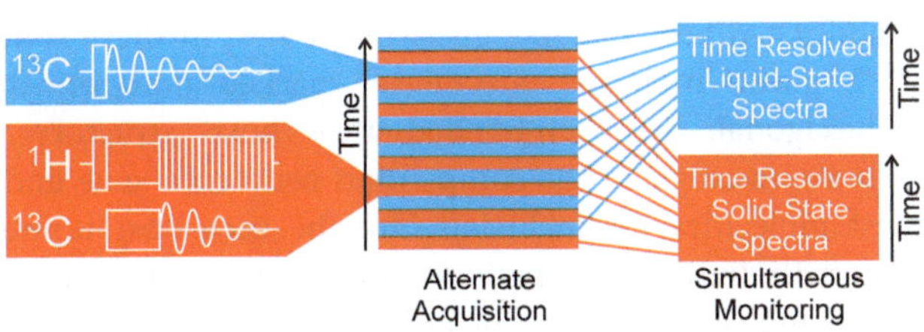

Fig. 4 Schematic of the experimental procedure for CLASSIC NMR, which allows the evolution of both the solid phase and the liquid phase in a crystallization experiment to be investigated separately and simultaneously.

before crystallization commenced), after which significant changes were observed. The integrated intensity of the peaks for *m*-ABA in the liquid-state NMR spectra decreased during the next 6 h, reflecting the fact that the solution phase becomes progressively more dilute as crystallization proceeds.

Significantly, systematic changes are observed in the liquid-state ^{13}C chemical shifts for *m*-ABA and the solvent DMSO as a function of time during the crystallization process (some peaks shift to more positive chemical shifts and other peaks shift to more negative chemical shifts). At the start of the measurement, the system is a supersaturated solution. After crystallization begins, the supersaturation decreases with time and by the end of the crystallization process, the system is an equilibrium saturated solution. It is known from independent studies that, in an equilibrium saturated solution of *m*-ABA in DMSO, the *m*-ABA molecules exist in the non-zwitterionic form. From this knowledge, and by rationalizing the change in the ^{13}C chemical shift for each ^{13}C site in *m*-ABA observed as a function of time during the CLASSIC NMR experiment, significant insights can be gained on the nature of the supersaturated solution that exists at the start of the crystallization experiment. Specifically, the observed changes in liquid-state ^{13}C chemical shifts are consistent with the supersaturated solution of *m*-ABA in DMSO at the start of the crystallization experiment having (a) a higher proportion of zwitterionic *m*-ABA molecules and/or (b) a higher proportion of non-zwitterionic *m*-ABA molecules present in hydrogen-bonded aggregates, relative to a saturated solution of *m*-ABA in DMSO. Both scenarios (a) and (b) represent an increased degree of protonation of the NH_2 group of *m*-ABA and are plausible solution-state precursors to the $O^- \cdots H–N^+$ hydrogen bonds that exist between *m*-ABA zwitterions in the crystal structure of form I. Although we cannot distinguish whether situation (a) or situation (b) is predominant, the CLASSIC NMR results nevertheless give clear insights into the nature of the speciation and interactions that exist in the supersaturated pre-nucleation solution of *m*-ABA in DMSO prior to crystallization, relative to those in the saturated solution at the end of the crystallization process.

5. New results from *in situ* NMR studies of crystallization processes

5.1 Monitoring competitive inclusion processes in solid inclusion compounds by a CLASSIC NMR approach

Recently, we have exploited the opportunity to use *in situ* NMR to explore the crystallization of conventional urea inclusion compounds with a variety of long chain alkane and α,ω-dibromoalkane guest molecules. In these solid inclusion compounds,[45–49] "guest" molecules are densely packed along one-dimensional tunnels in a hexagonal, hydrogen-bonded urea "host" structure (Fig. 5). In particular, we have investigated the competition between alkane and α,ω-dibromoalkane guest molecules for inclusion into the urea host tunnel structure during crystal growth. In these competitive crystallization experiments, two types of potential guest molecule are present in the solution phase and compete for inclusion within the urea host structure.

We report a combined liquid-state and solid-state *in situ* NMR study of crystallization from a solution comprising urea, 1,8-dibromooctane and

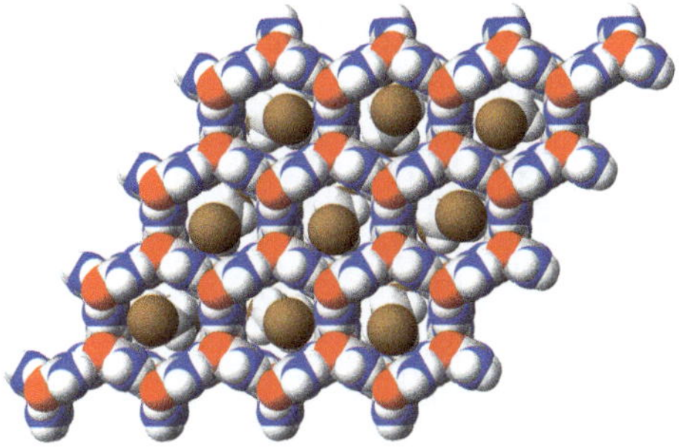

Fig. 5 Crystal structure of a conventional urea inclusion compound, viewed along the direction of the one-dimensional tunnels in the urea host structure.

tetradecane (approximate molar ratios 10 : 1 : 2, respectively) in methanol. The solution was prepared at 50 °C and then cooled to 20 °C at a constant rate over 11 h. Direct-excitation ^{13}C NMR spectra were recorded (with ^{1}H decoupling) throughout the cooling process and for a further 6 h after reaching 20 °C. In contrast to most organic solids, the guest molecules in urea inclusion compounds undergo rapid reorientational motions within the urea host tunnel structure. As a consequence, direct-excitation ^{13}C NMR spectra recorded under conditions normally used to record liquid-state ^{13}C NMR spectra detect a signal from the guest molecules in the solid urea inclusion compounds. Thus, the direct-excitation ^{13}C NMR spectra recorded in the *in situ* study of crystallization of urea inclusion compounds contain signals from the guest molecules both in the solid urea inclusion compounds and in the liquid phases present. On the other hand, $^{1}H \rightarrow {}^{13}C$ CP measurements give much weaker signals for the guest molecules in the urea inclusion compounds. Consequently, only direct-excitation ^{13}C NMR spectra were recorded in the present case (Fig. 6).

The peak assignments in the *in situ* ^{13}C NMR spectra (see Fig. 6) are made by reference to samples of urea inclusion compounds and solutions prepared independently of the *in situ* ^{13}C NMR study. At the start of the *in situ* study (at 50 °C), both 1,8-dibromooctane and tetradecane are present (together with urea) dissolved in methanol and peaks are observed in the ^{13}C NMR spectrum due to these components. However, an additional set of peaks is present, arising from a separate liquid phase of pure tetradecane (clearly, under the conditions of this experiment, the tetradecane was only partially dissolved in the methanol solution). As the system is cooled, the amount of the liquid phase of tetradecane diminishes and new peaks emerge concomitantly in positions characteristic of tetradecane guest molecules in the urea host structure. Significantly, no peaks are present in the positions expected for 1,8-dibromooctane guest molecules in the urea host structure, clearly indicating the preferential inclusion of tetradecane over 1,8-dibromooctane in the formation of the urea inclusion compound. Indeed, it is well established[50] that inclusion of longer guest molecules over shorter guest molecules is favoured within a homologous family, although the nature of end-group substituents on the guest molecules may also influence the preferential uptake of guest molecules within the urea tunnel structure.

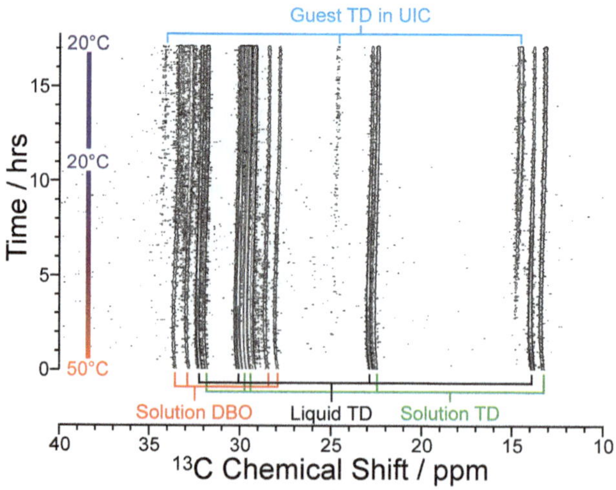

Fig. 6 *In situ* ^{13}C NMR spectra recorded using the direct-excitation ^{13}C pulse sequence during crystallization of a urea inclusion compound (UIC) from a solution containing urea, 1,8-dibromooctane (DBO) and tetradecane (TD) in methanol. Peaks for the DBO and TD molecules in different phases are labelled. In addition to the solution phase and the solid (urea inclusion compound) phase, a pure liquid tetradecane phase is also present. From these results, there is no evidence for uptake of DBO guest molecules in the urea inclusion compound in this experiment.

Fig. 7 shows the relative amounts of tetradecane in the three different phases as a function of time during the crystallization process (determined by integration of the peak for the methyl group in each phase). The amount of tetradecane in the solution phase remains essentially constant throughout the experiment, whereas the amount of the pure liquid tetradecane phase decreases and the amount of tetradecane guest molecules in the solid urea inclusion compound increases. From these results, it is clear that the formation of the urea inclusion compound containing tetradecane guest molecules occurs within the solution phase. Moreover, while tetradecane is consumed from the

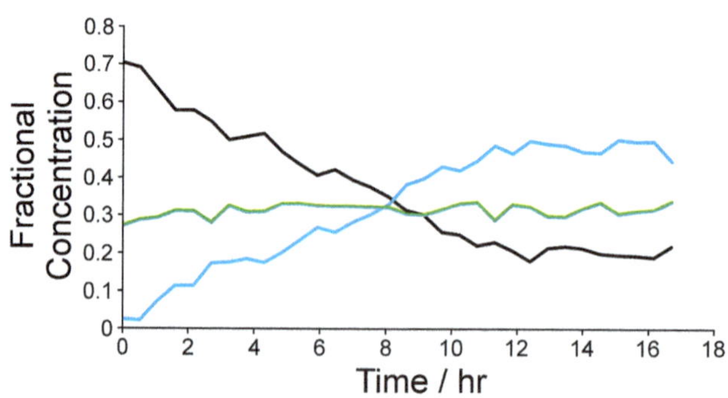

Fig. 7 Time-dependence of the relative amounts of tetradecane molecules in the three distinct phases present in the crystallization system (guest molecules in the urea inclusion compound, blue; solution phase, green; pure liquid tetradecane, black) determined from the intensities of the methyl peaks in the *in situ* ^{13}C NMR spectra.

solution phase by this process, it is evident from Fig. 7 that the partitioning of tetradecane between the solution phase and the pure liquid tetradecane phase adjusts in order to maintain a constant concentration within the solution phase, leading to a diminution in the amount of the pure liquid tetradecane phase present as a function of time.

5.2 Exploiting liquid-state NMR to gain insights on crystallization processes

In some cases, long $T_1(^1H)$ relaxation times for organic solids are such that the recycle delays required to record solid-state ^{13}C NMR spectra using the $^1H \rightarrow {}^{13}C$ CP technique are very long. In such cases, the time to record an individual solid-state ^{13}C NMR spectrum within the context of an *in situ* NMR study of crystallization would be prohibitively long to allow acceptable time-resolution in monitoring the process. In such cases, however, liquid-state ^{13}C NMR spectra can still be recorded quickly (as ^{13}C spin–lattice relaxation times for the liquid phase are significantly shorter) and may still yield significant insights into the processes occurring. This situation is encountered for benzoic acid and its derivatives.

Following an initial *in situ* liquid-state ^{13}C NMR study of crystallization of benzoic acid,[51] we report here studies of crystallization from a solution containing equimolar amounts of benzoic acid (BA) and pentafluorobenzoic acid (PFBA) in d_2-dichloromethane. BA and PFBA are known to form a 1 : 1 co-crystal[52] in which the two components exist exclusively as hetero-molecular BA···PFBA pairs, linked by the well-known double hydrogen-bonded carboxylic acid dimer motif. A relevant question is whether all possible solution-state pairings BA···BA, BA···PFBA and PFBA···PFBA exist in the solution state prior to crystallization and whether there is any change in the nature of these interactions as crystallization proceeds.

Ex situ studies indicate that the known co-crystal phase is the only crystallization product from this solvent. However, very distinct behaviour is observed in our *in situ* liquid-state NMR studies of the crystallization process. Fig. 8 shows the results from separate *in situ* ^1H NMR and *in situ* ^{19}F NMR studies of crystallization from a solution comprising BA and PFBA (in 1 : 1 molar ratio) in d_2-dichloromethane (with time resolution of 28 s for the ^1H NMR study and 25 s for the ^{19}F NMR study).[53] In both sets of data, a sudden change in chemical shifts is observed (at 38 min in the ^1H NMR data and at 10 min in the ^{19}F NMR data), corresponding to very fast co-crystal formation. The existence of a lag time before crystallization occurs and variability in the length of this lag time in different experiments is typical for this system and explains the difference in the time at which the sudden change in chemical shifts occurs in the *in situ* ^1H NMR and ^{19}F NMR experiments.

In the *in situ* ^1H NMR spectra, the largest change in ^1H chemical shift occurs for the carboxylic acid group, which shifts by *ca.* −0.4 ppm. The fact that only one ^1H resonance is observed for the carboxylic acid groups in the system indicates that rapid proton exchange must occur between BA and PFBA molecules, as well as between different interaction environments of these molecules (*e.g.*, the BA···BA, BA···PFBA and PFBA···PFBA hydrogen-bonded pairings). The ^1H chemical shifts for the phenyl ring of BA move in the opposite direction by *ca.* +0.1 ppm. The change in the ^1H chemical shift for the carboxylic acid groups

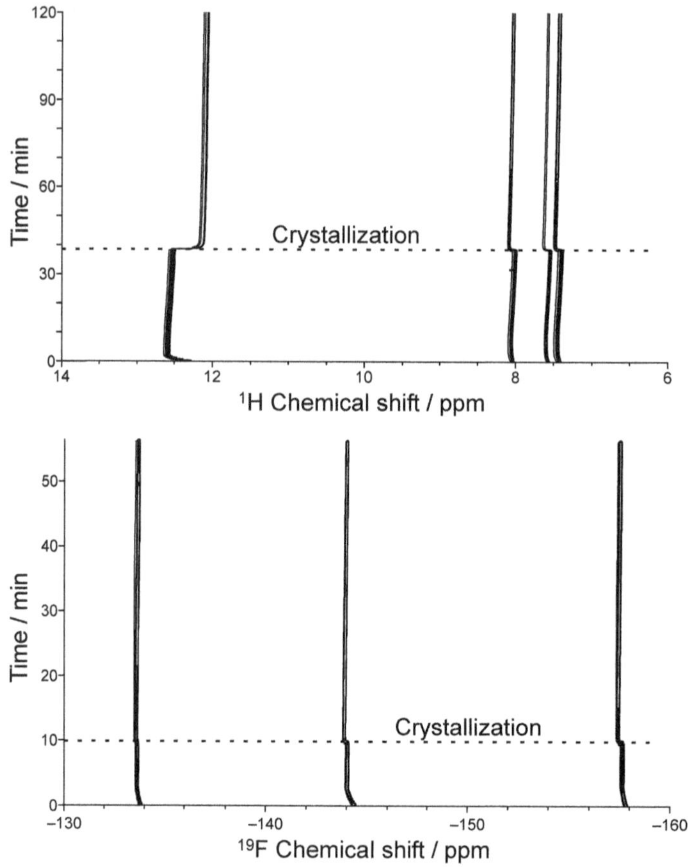

Fig. 8 *In situ* liquid-state ^{1}H NMR (top) and ^{19}F NMR (bottom) spectra recorded (in separate experiments) as a function of time for co-crystallization of benzoic acid and pentafluorobenzoic acid from d_2-dichloromethane.

suggests that, in the supersaturated solution that exists prior to crystallization, the proportion of molecules in solution that exist in hydrogen-bonded dimers is higher than that in the more dilute solution that exists at the end of the crystallization process (it is well established that the ^{1}H chemical shift of carboxylic acids is more positive for a hydrogen-bonded dimer than a non-hydrogen bonded monomer). In the *in situ* liquid-state ^{19}F NMR study, the three ^{19}F chemical shifts all move in the same direction by between +0.06 ppm and +0.23 ppm. This observation suggests that the aryl rings of both the BA and PFBA molecules in the solution state experience similar changes as a consequence of the rapid decrease in solution-state concentration that accompanies crystallization, probably indicative of a change in the extent of π-stacking of the aromatic rings of both BA and PFBA molecules.

5.3 Polymorph discovery

An example illustrating the discovery of a hitherto unknown solid phase concerns our *in situ* solid-state ^{31}P NMR study of crystallization of trimethylphosphine oxide (TMPO) from toluene. Only one crystal structure has been reported for TMPO, with one molecule in the asymmetric unit (and hence one ^{31}P NMR environment). Correspondingly, the solid-state ^{31}P NMR spectrum of a powder sample of TMPO has only one isotropic peak and a single peak in each spinning sideband. In the *in situ* ^{31}P NMR study of crystallization from toluene, even the first spectrum recorded (see Fig. 9) reveals that a substantial amount of solid was present at the earliest stages of the crystallization process. Interestingly, this spectrum has two isotropic peaks, while each spinning sideband is a single peak. One isotropic peak corresponds to the known phase of TMPO, while the new isotropic peak (indicated in Fig. 9) must represent a new solid form of TMPO. The observation that the ^{31}P NMR spectrum of the new solid form has an isotropic peak with no spinning sidebands under the ^{1}H → ^{31}P CPMAS NMR measurement conditions is intriguing and suggests that the TMPO molecules are in a dynamic regime that is sufficiently slow to allow a signal to be generated in ^{1}H → ^{31}P CP, but sufficiently rapid to average the ^{31}P chemical shift anisotropy such that spinning sidebands are eliminated. We propose that the additional isotropic peak observed in the crystallization of TMPO may arise from an amorphous phase or perhaps a rotator phase solid (we note that the shape of the TMPO molecule is almost isotropic, given the similar steric character of the oxygen and methyl substituents, which is a common feature in rotator phase materials).

5.4 A new solid form of L-phenylalanine

Recently, L-phenylalanine (L-Phe) has been shown to display abundant polymorphism, with four polymorphs reported so far. The first form to be structurally characterized,[54] now denoted form I of L-Phe, has recently been re-determined[55] (including correction of an earlier reported "re-determination"[56]). In the past two years, three new polymorphs have been reported, one discovered by our group, denoted form II,[57] followed by form III[58] and form IV.[55] Our work[57] also revealed the existence of monohydrate and hemihydrate phases of L-Phe.

In situ solid-state ^{13}C NMR studies have been carried out on the crystallization of L-Phe from ethanol/water. Under these conditions, L-Phe has been reported[59] to produce a solid form different to the only polymorph (form I) that was known at that time. The results of our *in situ* ^{13}C NMR study are shown in Fig. 10, together with solid-state ^{13}C NMR spectra recorded for powder samples of the monohydrate and hemihydrate of L-Phe and form I of anhydrous L-Phe. Although the *in situ* ^{13}C NMR data were recorded at high magnetic field (20 T corresponding to a ^{13}C Larmor frequency of 213.8 MHz) over 21 h, the signal-to-

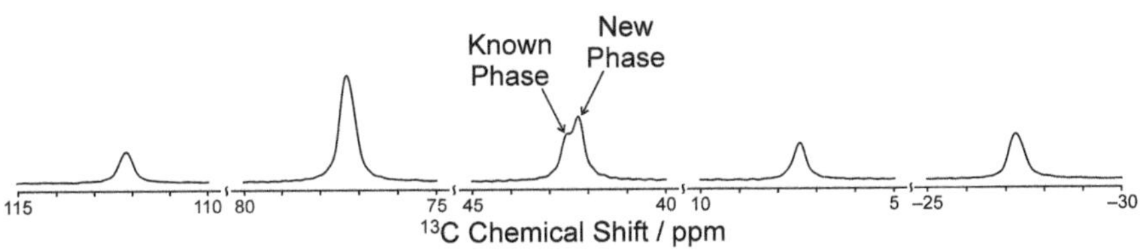

Fig. 9 The first ^{1}H → ^{31}P CP NMR spectrum recorded during the *in situ* ^{31}P NMR study of crystallization of TMPO from toluene, showing the isotropic peaks (in the range 42 to 43 ppm) and several spinning sidebands.

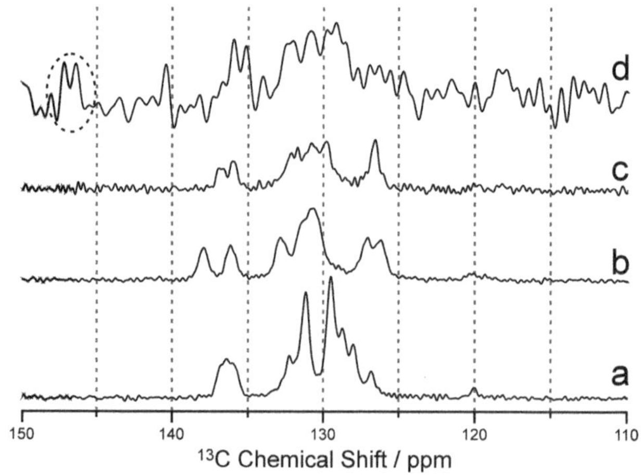

Fig. 10 High-resolution solid-state ^{13}C NMR spectra (recorded using $^{1}H \rightarrow {}^{13}C$ CP) for: (a) polymorph I of anhydrous L-Phe, (b) the hemihydrate of L-Phe, (c) the monohydrate of L-Phe, and (d) the solid-phase formed in our *in situ* solid-state ^{13}C NMR study of the crystallization of L-Phe from ethanol/water (the peaks around 147 ppm discussed in the text are highlighted).

noise ratio is rather low as a result of slow ^{1}H spin–lattice relaxation (requiring a long recycle delay in the NMR measurement) and a low concentration of the crystallization solution (resulting in only a small amount of the solid phase being formed). In spite of the poor quality of the *in situ* ^{13}C NMR data, however, new peaks (particularly around 147 ppm; see Fig. 10) are clearly identified that are not present in the ^{13}C NMR spectra of other solid forms of L-Phe shown in Fig. 10, specifically form I of L-Phe and the hemihydrate and monohydrate phases (solid-state ^{13}C NMR spectra of the other polymorphs of L-Phe have not been measured as form II can only be prepared by rigorous drying of the hemihydrate and we have not attempted to prepare forms III and IV). However, periodic DFT calculations using the CASTEP code (Accelrys, San Diego, CA)[60] (which allows NMR spectra to be calculated from known crystal structures[61,62]) indicate that the ^{13}C NMR spectra of forms II and III do not have an isotropic peak with ^{13}C chemical shift around 147 ppm. DFT could not be applied to form IV as the crystal structure is disordered.

At this stage, the new solid form observed in the *in situ* solid-state NMR study is not yet identified, although the three primary possibilities are: (i) form IV, (ii) a new polymorph of L-Phe, or (iii) a new solvate of L-Phe. A new solvate phase is perhaps the most likely, although no ethanol solvates of L-Phe have been reported. Nevertheless, we note that the hemihydrate and monohydrate phases are channel hydrates with closely related structures (in which the water molecules are located in channels created by the packing of L-Phe molecules), with form II of anhydrous L-Phe essentially representing the "empty" channel structure.[57] It is certainly plausible to suggest that other types of solvent molecules, such as ethanol, could be accommodated within this L-Phe channel structure. Attempts to isolate the new solid form (to allow crystal structure determination and other analysis) have so far been unsuccessful.

6. A critical appraisal of practical aspects, interpretation of results and the scope and limitations of *in situ* NMR studies of crystallization processes

In this section, we discuss a number of issues that are relevant to the practical implementation of the *in situ* solid-state NMR methodology, as well as considering important factors that serve to define the scope and limitations of the technique. In this discussion, we address some key questions that are often raised about the *in situ* solid-state NMR technique, including: (i) what is the smallest size of solid particle that can be detected, and (ii) is the crystallization process influenced by the application of the rapid sample rotation required to record high-resolution solid-state NMR spectra?

6.1 What is the smallest size of solid particle that can be detected in *in situ* solid-state NMR studies of crystallization?

A question frequently asked in connection with our technique relates to the smallest size of solid particle that can be observed in the *in situ* cross-polarization (CP) NMR measurements, with implications on the earliest stage of the crystallization process that can actually be detected. Unfortunately, establishing this limit experimentally is very challenging. One experimental approach could be to carry out systematic studies of a set of samples each consisting of solid particles with a known, narrow range of sizes in liquid suspension. However, we have not been able to obtain a suitable set of samples of this nature. Consequently, we estimate the limiting size based on theoretical considerations.

As emphasized above, a key feature of our *in situ* solid-state NMR strategy is that selectivity in measuring only the NMR spectrum of the solid phase may be achieved by the use of a CP pulse sequence. The key limitation for obtaining a signal by CP is the rate of tumbling of the solid particle, since this motion "scrambles" the orientation of the dipolar couplings responsible for the polarization transfer. The tumbling motion is characterized by a correlation time, τ_c, which is related to the volume (V) of the particle, the viscosity (η) of the solvent and temperature (T), through the Stokes–Einstein–Debye relation[63]

$$\tau_c = \frac{6V\eta}{k_B T l(l+1)}, \tag{1}$$

where l is the order of the spherical harmonic for the interaction ($l = 2$ for dipolar coupling). Nowacka *et al.*[64] established that, for a $^1H \rightarrow {}^{13}C$ CP experiment to give a signal under similar conditions to those used in our experiments, the limiting value of τ_c is *ca.* 10 μs. Using eqn (1) and taking the bulk viscosity of water ($\sim10^{-3}$ Pa s), the limiting volume is estimated to be $V \approx 4 \times 10^7$ Å^3, corresponding to a sphere with radius of around 210 Å. To put this value in context, particles of glycine of this size would contain around 5×10^5 molecules. Although a number of approximations and assumptions are inherent in this derivation, it nevertheless offers a reasonable estimate of the smallest size of solid particle that could be detected in an *in situ* solid-state NMR study of crystallization. Clearly, the estimate could be refined in any particular case by taking into account known features of the experimental conditions, such as knowledge of the viscosity of the solution.

Although the size of the critical nucleation entity in crystallization from solution may vary significantly between different crystallization systems and for a given crystallization system under different experimental conditions, it is anticipated that the critical clusters involved in the nucleation of most organic materials would comprise significantly fewer than 5×10^5 molecules. Thus, the smallest solid particles that are observable under the conditions of *in situ* solid-state NMR studies employing the $^{1}H \rightarrow {}^{13}C$ CP technique are likely to represent post-nucleation stages of the crystallization process. Nevertheless, as illustrated in the case of crystallization of glycine from methanol/water in Section 4.1, identification of the polymorphic form of the earliest solid particles detected in the crystallization process can nevertheless provide important clues about the probable nucleation pathway, which cannot be obtained from *ex situ* identification of the final crystallization product collected at the end of the crystallization process.

6.2 Does the NMR measurement technique itself influence the crystallization process?

Another important question is whether the rapid sample rotation required to record high-resolution solid-state NMR spectra may actually affect the pathway and/or the final outcome of the crystallization process. More broadly, it is important to consider the effects of all the physical conditions that the crystallization system is subjected to in our experiments, including (i) the pressure exerted on the sample due to rapid spinning, (ii) the nature of the rotor materials in contact with the sample and (iii) the containment of the sample under the mother liquor throughout the experiment.

With regard to (i), the pressure distribution within a sample subjected to spinning in a cylindrical rotor can be calculated if the density distribution is known. Specifically, the pressure imposed by sample spinning is given by

$$p(r) = \omega^2 \int_{r_0}^{r} \rho(r')r' \, dr', \tag{2}$$

where ω is the spinning frequency in rad s^{-1} and $\rho(r)$ is the density of the sample at distance r from the rotation axis of the rotor (see Fig. 11). The term r_0 represents the radius of any cylindrical void present at the centre of the rotor (and co-axial with the rotation axis), which arises if the sample does not completely fill the rotor. If the density is assumed to be constant throughout the sample, eqn (2) simplifies to

$$p(r) = \frac{1}{2}\omega^2 (r^2 - {r_0}^2)\rho \tag{3}$$

Although the assumption of uniform density may be justifiable in some cases (*e.g.*, for samples with low compressibility), a more realistic form of eqn (3) would take into account the distribution of density of the sample inside the rotor (*i.e.*, $\rho(r)$ as in eqn (1)). In the case of a homogeneous solution inside the rotor prior to crystallization, the solution must respond to the pressure gradient (lowest pressure at the centre of the rotor and greatest pressure at the walls) by redistribution

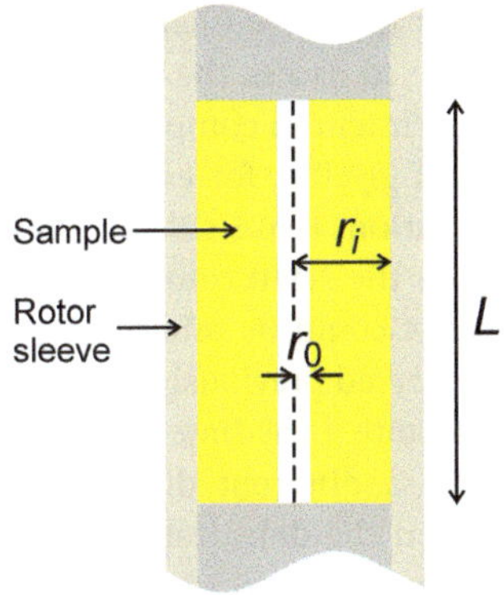

Fig. 11 Schematic of a sample inside an NMR rotor undergoing magic-angle spinning, including definitions of terms used in eqn (2)–(5).

of mass towards the outer part of the rotor to create a density gradient across the sample (again, with lowest density at the centre of the rotor and greatest density at the walls). Such a density gradient may have important implications for the ensuing crystallization process, as it may result in a non-uniform distribution of concentration across the solution. Furthermore, given the distribution of pressure within the spinning sample (and the fact that solubility is a function of pressure), the solubility may also be non-uniform across the sample. Clearly, the confluence of these issues creates significantly more complex crystallization conditions than those that exist in a conventional laboratory crystallization experiment, and a more detailed quantitative understanding of the effects of sample rotation is merited. Intuitively, however, we anticipate that the inhomogeneities introduced into the crystallization system as a consequence of the sample spinning probably represent relatively minor non-uniformities which, in the majority of cases, will not significantly affect the behaviour of the crystallization process.

Returning to the question of pressure, we note that the pressure at the walls of the rotor is independent of the density distribution through the sample and is given by

$$p(r_\mathrm{i}) = \frac{1}{2}\omega^2\left(r_\mathrm{i}^{\,2} - r_0^{\,2}\right)\rho_\mathrm{av}, \tag{4}$$

where r_i is the internal radius of the rotor and ρ_av is the average density of the sample, equal to the static density in the absence of compression. Under these circumstances, eqn (4) becomes

$$p(r_\mathrm{i}) = \frac{m\omega^2}{2\pi L} \tag{5}$$

where m is the total mass of the sample and L is the length of the sample cavity in the rotor.

Using eqn (4), for a rotor with internal radius $r_\mathrm{i} = 1.5$ mm containing water and spinning at MAS frequency $\omega = 12$ kHz with a cylindrical void $r_0 = 0.5$ mm at the centre of the rotor, the pressure at the rotor walls is $p(r_\mathrm{i}) = 57$ atm. This value rises to $p(r_i) = 64$ atm for a full rotor (*i.e.*, $r_0 = 0$) at the same MAS frequency. Armed with these values, it is possible to assess the effect of spinning on the crystals formed in the *in situ* NMR experiment in comparison to high pressure studies on powder samples. In the case of glycine, for example, the pressures observed to induce polymorphic transformations (≥0.5 GPa or 5000 atm)[65–69] are significantly higher than those generated in the NMR rotor subjected to MAS. Nevertheless, polymorphism in other systems could be affected by pressures of the order of those generated by MAS, as observed previously in *in situ* solid-state NMR studies of dehydration of sodium acetate trihydrate.[70] In this case, the dehydration process was found to lead to a different distribution of the polymorphs of anhydrous sodium acetate when the dehydration was carried out under conditions of rapid sample spinning compared to dehydration of a static (non-spinning) sample.

Sample rotors for MAS NMR experiments are composed of a sleeve and one or two sealing caps (depending on whether the sleeve is open at one end or both ends). Schematics of two different designs of sealing system are shown in Fig. 12. The sleeves are usually made from zirconia, whilst the caps may be composed of a variety of materials including Kel-F® and Teflon® with some designs using silicone or fluorosilicone rubber O-rings. Clearly, these materials interact with the crystallization solution and may potentially play a role in the nucleation process.[33]

Finally, another key distinction between *in situ* NMR experiments and *ex situ* studies of crystallization is that the *in situ* NMR technique observes the crystallization product directly within the crystallization solution without removal from the mother liquor. Potentially, this feature could lead to differences in the solid form observed in the *in situ* NMR study compared to the form observed in laboratory crystallization experiments in which the crystallization product is characterized *ex situ* (*e.g.*, by powder X-ray diffraction or solid-state NMR) following removal from the crystallization solution. Clearly, it is well known that certain solid forms may only be stable when submerged in the mother liquor and

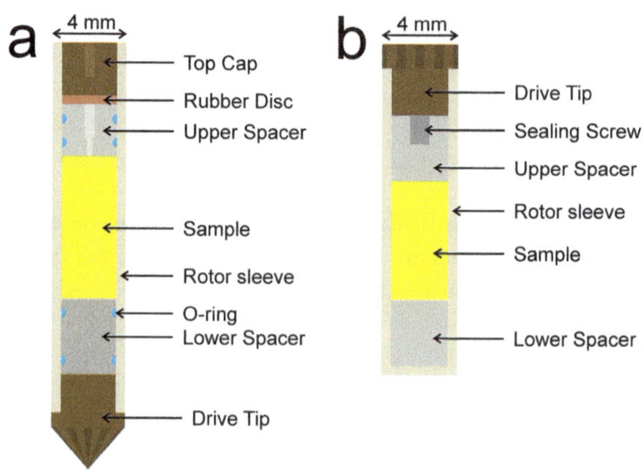

Fig. 12 Examples of designs that have overcome the challenges of sealing liquid-containing samples inside NMR rotors for use in rapid MAS experiments, as required for *in situ* solid-state NMR studies of crystallization from solution. The rotor designs shown represent schematics of sealing systems used by Varian (left) and Bruker (right).

transform rapidly to a more stable form on removal from the crystallization solution.

It is clear from the issues discussed in this section that certain aspects of the instrumentation and experimental methodology required to carry out *in situ* solid-state NMR studies of crystallization from solution have the potential to influence the pathway and outcome of the crystallization process, including (perhaps most importantly) the effect of the pressure generated in the solution by rapid sample spinning and the consequent effect on the density distribution within the solution. However, in our experience so far, we have not yet observed any systems for which the final solid form(s) extracted from the NMR rotor following an *in situ* crystallization study does not match the final solid form(s) resulting from crystallization in the laboratory under the same conditions and carried out over a comparable period of time.

6.3 Some practical aspects

An important practical aspect of our *in situ* NMR studies of crystallization processes is accurate temperature calibration. This issue is particularly crucial with regard to ensuring reliable transferability of experimental information derived from normal laboratory experiments (*e.g.*, information on solubilities of materials and crystallization conditions required to produce specific polymorphs, *etc.*) into the context of *in situ* NMR studies. In our experience, the temperature control equipment on NMR spectrometers is often inaccurate for samples undergoing MAS, largely due to the heating effect of the rapid sample spinning but also due to a lack of adequate calibration of the equipment. Consequently, we rely on calibrated values of temperature which incorporate the heating effect of MAS at the specific MAS frequency used. The discrepancy between the temperature indicated by the temperature control equipment and the actual temperature can be as large as 30 °C. We use calibrations based on measurements on lead nitrate[71] and methanol.[72–74] A calibration using glycol is also available for higher temperatures than those covered by the methanol calibration.[72]

As discussed in Section 3, a technical advance that was pivotal to the successful development of the *in situ* solid-state NMR technique was the availability of NMR rotors that can be used for magic-angle spinning of liquid containing samples, without leakage of liquid. Depending upon the type of NMR rotor used, various sealing systems are available (see Fig. 12). An important practical consideration for these rotors is whether the solvent to be used in a crystallization experiment is compatible with the material used to manufacture the seals on the rotor. While water is compatible with all such materials, certain organic solvents may soften the rotor materials, leading to leakage of liquid particularly at the elevated temperature required to achieve dissolution.

6.4 Sensitivity and time-resolution

As mentioned in Section 3, achieving sufficiently good time resolution in *in situ* solid-state NMR studies is an essential requirement for observing all the phases involved in the crystallization process, including short-lived solid forms that have only a transient existence. Clearly, the time-resolution of the *in situ* study is dictated by the time required to record an individual NMR spectrum with adequate signal-to-noise ratio to allow the solid forms present to be identified. A number of factors are important in this regard, including: (i) the NMR receptivity of the selected nucleus, (ii) the applied magnetic field strength, (iii) the NMR relaxation properties and (iv) the total amount of solid phase that can be produced in the crystallization experiment.

First, the NMR receptivity of a selected nucleus depends on fundamental properties of the nucleus and on the natural isotopic abundance. The only opportunity for the experimenter to increase the receptivity is to increase the isotopic abundance by using an isotopically labelled sample. Of the nuclei discussed here, the natural isotopic abundances are ^{1}H (99.99%), ^{13}C (1.05%), ^{19}F (100%) and ^{31}P (100%), so the prospect of isotopic labelling is relevant only in the case of ^{13}C NMR. Clearly, ^{13}C labelling is highly desirable for *in situ* ^{13}C NMR studies, although it is not always essential and may not be feasible for synthetic reasons. However, to illustrate the advantages of ^{13}C labelling, the *in situ* solid-state ^{13}C NMR study of crystallization of 1,10-dihydroxydecane/urea (using ^{13}C-labelled urea) discussed in Section 4.2 had a time-resolution of 2.7 min (ref. 43) whereas the solid-state component of the CLASSIC ^{13}C NMR study of crystallization of *m*-ABA from DMSO (using natural isotopic abundances) discussed in Section 4.3 had a time-resolution of 38.4 min.[14]

Second, in order to achieve maximum sensitivity in NMR measurements and hence to achieve the best possible time-resolution for *in situ* studies, the experiments should be carried out using an NMR spectrometer with the highest available applied magnetic field. For this reason, most of our recent research on the application of the techniques described here has been carried out at the UK National High-Field (850 MHz) Solid-State NMR Facility.

Third, as discussed in Section 2, the main sample-dependent factor controlling the time required to record an NMR spectrum is the recycle delay (the time delay that must be left between successive applications of the NMR pulse sequence) which depends on the spin–lattice relaxation time (T_1) of one type of nucleus present in the material. In the case of solid-state ^{13}C NMR spectra recorded using $^{1}\mathrm{H} \rightarrow {}^{13}\mathrm{C}$ CP, the recycle delay is dictated by the ^{1}H spin–lattice relaxation time.[75] In the case of organic solids, the presence of a mobile group (*e.g.*, methyl or ammonium) in the molecule usually ensures that $T_1(^{1}\mathrm{H})$ is less than 1 s (for molecules of the size we have typically investigated). However, in the absence of a mobile group, $T_1(^{1}\mathrm{H})$ is generally much longer. Clearly, solid-state ^{13}C NMR spectra can be recorded much more rapidly for materials with short $T_1(^{1}\mathrm{H})$, which is conducive for good time-resolution in *in situ* solid-state ^{13}C NMR studies of crystallization of organic materials.

Finally, the total amount of the solid phase that can be produced in the crystallization experiment clearly depends on the concentration of the solute and the differential of solubility as a function of temperature, recalling that, in our *in situ* NMR strategy, crystallization is induced by decrease of temperature. Thus, in our experimental protocol, the solute must be completely soluble at the high temperature (T_{H}) at which the homogeneous solution is prepared and should be insoluble at low temperature (T_{L}) at which the *in situ* NMR measurements are carried out. Thus, if the molality of the initial solution prepared at T_{H} is denoted m_{o}, and using S to denote solubility, we require that $S(T_{\mathrm{H}}) > m_{\mathrm{o}}$ and $S(T_{\mathrm{L}}) < m_{\mathrm{o}}$. Clearly, at some temperature (T_{C}) between T_{H} and T_{L}, the system must satisfy $S(T_{\mathrm{C}}) = m_{\mathrm{o}}$ and crystallization becomes thermodynamically favoured for $T \leq T_{\mathrm{C}}$. Clearly, to ensure that a significant amount of solid phase is produced during the crystallization experiment, the difference in solubilities $S(T_C) - S(T_L)$ should be as large as possible. Hence, for a given crystallization system, the three factors under the control of the experimenter (*i.e.*, the molality m_o and temperature T_H at the start of the experiment and the target crystallization temperature T_L) should be chosen such that T_H is as high as possible, T_L is as low as possible and m_o should be only slightly lower than $S(T_H)$. Under these circumstances, T_C lies close to T_H and the difference in solubilities $S(T_C) - S(T_L)$ is maximized (for a given value of T_L).

For a given magnetic field strength, we have found that the "worst case scenario" occurs when the absolute change in solubility, $S(T_C) - S(T_L)$, is small, the isotopic abundance of the observed nucleus is low, and $T_1(^1H)$ is long (*e.g.*, longer than 30 s). Under such conditions, even at high applied magnetic field, acquisition of a solid-state NMR signal on a timescale useful for studying crystallization processes is probably impossible. However, if just one of the three factors is favourable, we can reasonably expect to be able to observe the crystallization process with adequate time-resolution, especially when measurements are carried out at high applied magnetic field.

7. Concluding remarks

We have shown, both in our previous publications[12–14,42,43,51] and in the new results discussed in this paper, that *in situ* solid-state NMR studies offer a variety of new opportunities for studying different aspects of crystallization processes. These opportunities include observing polymorphic evolution, discovering new polymorphs and other solid phases, developing strategies for preparing unstable polymorphs and identifying changes in the solution state that occur during the crystallization process. In each case, *in situ* NMR yields information that cannot be obtained from *ex situ* studies or, in some cases, from other *in situ* experimental techniques. In particular, the recently developed CLASSIC NMR experiment extends significantly the scope and capability of *in situ* monitoring of crystallization processes as it is unique in providing simultaneous and complementary information on the time-evolution of both the liquid and solid phases during crystallization. We forecast with confidence that further developments of the methodology for *in situ* NMR studies of crystallization, together with the application of these techniques to a wide range of different types of systems, will significantly advance our fundamental understanding of crystallization processes in the years to come.

Acknowledgements

We are grateful to EPSRC (studentship funding to VKL and GREG) and Agilent (studentship funding to VKL) for financial support, and to the Action «Supporting Postdoctoral Researchers» of the Operational Program "Education and Lifelong Learning" (Action's Beneficiary: General Secretariat for Research and Technology of Greece), co-financed by the European Social Fund and the Greek State, for a postdoctoral fellowship to VGC (Project Code: PE4 1236). We are grateful to the UK 850 MHz Solid-State NMR Facility for the award of significant amounts of time for *in situ* solid-state NMR studies of crystallization. This facility is funded by EPSRC, BBSRC and the University of Warwick, including part funding through

Birmingham Science City Advanced Materials Projects 1 and 2 supported by Advantage West Midlands and the European Regional Development Fund. We particularly appreciate the assistance of the Facility Manager, Dr Dinu Iuga.

References

1 B. Kahr and J. M. McBride, *Angew. Chem., Int. Ed.*, 1992, **31**, 1–26.
2 Crystal Growth and Nucleation, *Faraday Discuss.*, 2006, **136**.
3 J. D. Dunitz, *Pure Appl. Chem.*, 1991, **63**, 177–185.
4 J. Bernstein, IUCr Monographs on Crystallography, *Polymorphism in Molecular Crystals*, Oxford University Press, Oxford, 2002.
5 R. J. Davey, *Chem. Commun.*, 2003, 1463–1467.
6 J. Bernstein, *Chem. Commun.*, 2005, 5007–5012.
7 S. Chen, H. Xi and L. Yu, *J. Am. Chem. Soc.*, 2005, **127**, 17439–17444.
8 S. Ahn, F. Guo, B. M. Kariuki and K. D. M. Harris, *J. Am. Chem. Soc.*, 2006, **128**, 8441–8452.
9 R. K. Harris, *Analyst*, 2006, **131**, 351–373.
10 S. L. Price, *Acc. Chem. Res.*, 2008, **42**, 117–126.
11 N. Pienack and W. Bensch, *Angew. Chem., Int. Ed.*, 2011, **50**, 2014–2034.
12 C. E. Hughes and K. D. M. Harris, *J. Phys. Chem. A*, 2008, **112**, 6808–6810.
13 C. E. Hughes and K. D. M. Harris, *Chem. Commun.*, 2010, **46**, 4982–4984.
14 C. E. Hughes, P. A. Williams and K. D. M. Harris, *Angew. Chem., Int. Ed.*, 2014, **53**, 8939–8943.
15 K. D. M. Harris and J. M. Thomas, *J. Solid State Chem.*, 1991, **94**, 197–205.
16 E. R. Andrew, A. Bradbury and R. G. Eades, *Nature*, 1959, **1958**, 1802–1803.
17 S. R. Hartmann and E. L. Hahn, *Phys. Rev.*, 1962, **128**, 2042–2053.
18 M. J. Duer, *Introduction to Solid-State NMR Spectroscopy*, Blackwell Publishing Ltd., Oxford, 2004.
19 D. C. Apperley, R. K. Harris and P. Hodgkinson, *Solid State NMR: Basic Principles & Practice*, Momentum Press LLC, New York, 2012.
20 G. Albrecht and R. B. Corey, *J. Am. Chem. Soc.*, 1939, **61**, 1087–1103.
21 Y. Iitaka, *Acta Crystallogr.*, 1960, **13**, 35–45.
22 Y. Iitaka, *Acta Crystallogr.*, 1961, **14**, 1–10.
23 P.-G. Jönsson and Å. Kvick, *Acta Crystallogr., Sect. B: Struct. Crystallogr. Cryst. Chem.*, 1972, **28**, 1827–1833.
24 Å. Kvick, W. M. Canning, T. F. Koetzle and G. J. B. Williams, *Acta Crystallogr., Sect. B: Struct. Crystallogr. Cryst. Chem.*, 1980, **36**, 115–120.
25 T. N. Drebushchak, E. V. Boldyreva and E. S. Shutova, *Acta Crystallogr., Sect. E: Struct. Rep. Online*, 2002, **58**, o634–o636.
26 V. A. Drebushchak, E. V. Boldyreva, T. N. Drebushchak and E. S. Shutova, *J. Cryst. Growth*, 2002, **241**, 266–268.
27 B. A. Garetz, J. Matic and A. S. Myerson, *Phys. Rev. Lett.*, 2002, **89**, 175501.
28 E. V. Boldyreva, V. A. Drebushchak, T. N. Drebushchak, I. E. Paukov, Y. A. Kovalevskaya and E. S. Shutova, *J. Therm. Anal. Calorim.*, 2003, **73**, 409–418.
29 E. S. Ferrari, R. J. Davey, W. I. Cross, A. L. Gillon and C. S. Towler, *Cryst. Growth Des.*, 2002, **3**, 53–60.
30 R. E. Taylor, *Concepts Magn. Reson., Part A*, 2004, **22A**, 79–89.

31 C. S. Towler, R. J. Davey, R. W. Lancaster and C. J. Price, *J. Am. Chem. Soc.*, 2004, **126**, 13347–13353.

32 I. Weissbuch, V. Y. Torbeev, L. Leiserowitz and M. Lahav, *Angew. Chem., Int. Ed.*, 2005, **44**, 3226–3229.

33 M. Xu and K. D. M. Harris, *J. Phys. Chem. B*, 2007, **111**, 8705–8707.

34 C. E. Hughes, S. Hamad, K. D. M. Harris, C. R. A. Catlow and P. C. Griffiths, *Faraday Discuss.*, 2007, **136**, 71–89.

35 S. K. Poornachary, P. S. Chow and R. B. H. Tan, *Cryst. Growth Des.*, 2007, **8**, 179–185.

36 C. E. Hughes and K. D. M. Harris, *New J. Chem.*, 2009, **33**, 713–716.

37 C. Chen, O. Cook, C. E. Nicholson and S. J. Cooper, *Cryst. Growth Des.*, 2011, **11**, 2228–2237.

38 G. J. Han, S. Thirunahari, P. S. Chow and R. B. H. Tan, *CrystEngComm*, 2013, **15**, 1218–1224.

39 G. L. Perlovich, L. K. Hansen and A. Bauer-Brandl, *J. Therm. Anal. Calorim.*, 2001, **66**, 699–715.

40 A. Théorêt, *Spectrochim. Acta, Part A*, 1971, **27**, 11–18.

41 P. A. Williams, C. E. Hughes, G. K. Lim, B. M. Kariuki and K. D. M. Harris, *Cryst. Growth Des.*, 2012, **12**, 3104–3113.

42 K. D. M. Harris, C. E. Hughes and P. A. Williams, *Solid State Nucl. Magn. Reson.*, 2015, **65**, 107–113.

43 C. E. Hughes, P. A. Williams, T. R. Peskett and K. D. M. Harris, *J. Phys. Chem. Lett.*, 2012, **3**, 3176–3181.

44 J. Martí-Rujas, B. M. Kariuki, C. E. Hughes, A. Morte-Ródenas, F. Guo, Z. Glavcheva-Laleva, K. Taştemür, L.-L. Ooi, L. Yeo and K. D. M. Harris, *New J. Chem.*, 2011, **35**, 1515–1521.

45 K. D. M. Harris, *J. Solid State Chem.*, 1993, **106**, 83–98.

46 K. D. M. Harris, *J. Mol. Struct.*, 1996, **374**, 241–250.

47 K. D. M. Harris, *Chem. Soc. Rev.*, 1997, **26**, 279–289.

48 F. Guillaume, *J. Chim. Phys. Phys.-Chim. Biol.*, 1999, **96**, 1295–1315.

49 K. D. M. Harris, *Supramol. Chem.*, 2007, **19**, 47–53.

50 K. D. M. Harris, P. E. Jupp and S.-O. Lee, *J. Chem. Phys.*, 1999, **111**, 9784–9790.

51 C. E. Hughes, P. A. Williams and K. D. M. Harris, in preparation.

52 D. Albesa-Jové, B. M. Kariuki, S. J. Kitchin, L. Grice, E. Y. Cheung and K. D. M. Harris, *ChemPhysChem*, 2004, **5**, 414–418.

53 Owing to the proximity of the ^{1}H and ^{19}F Larmor frequencies, hardware modifications are required to convert between ^{1}H and ^{19}F NMR measurements, so it is not feasible to record both ^{1}H and ^{19}F NMR spectra in the same crystallization experiment. As a consequence, ^{1}H and ^{19}F NMR spectra were measured for two separate crystallization experiments on the same sample.

54 I. Weissbuch, F. Frolow, L. Addadi, M. Lahav and L. Leiserowitz, *J. Am. Chem. Soc.*, 1990, **112**, 7718–7724.

55 F. S. Ihlefeldt, F. B. Pettersen, A. von Bonin, M. Zawadzka and C. H. Görbitz, *Angew. Chem., Int. Ed.*, 2014, **53**, 13600–13604.

56 M. D. King, T. N. Blanton and T. M. Korter, *Phys. Chem. Chem. Phys.*, 2012, **14**, 1113–1116.

57 P. A. Williams, C. E. Hughes, A. B. M. Buanz, S. Gaisford and K. D. M. Harris, *J. Phys. Chem. C*, 2013, **117**, 12136–12145.

58 E. Mossou, S. C. M. Teixeira, E. P. Mitchell, S. A. Mason, L. Adler-Abramovich, E. Gazite and V. T. Forsyth, *Acta Crystallogr., Sect. C: Cryst. Struct. Commun.*, 2014, **70**, 326–331.
59 M. H. Frey, J. A. Di Verdi and S. J. Opella, *J. Am. Chem. Soc.*, 1985, **107**, 7311–7315.
60 S. J. Clark, M. D. Segall, C. J. Pickard, P. J. Hasnip, M. J. Probert, K. Refson and M. C. Payne, *Z. Kristallogr.*, 2005, **220**, 567–570.
61 A. L. Webber, L. Emsley, R. M. Claramunt and S. P. Brown, *J. Phys. Chem. A*, 2010, **114**, 10435–10442.
62 R. K. Harris, P. Hodgkinson, C. J. Pickard, J. R. Yates and V. Zorin, *Magn. Reson. Chem.*, 2007, **45**, S174–S186.
63 P. Debye, *Polar Molecules*, The Chemical Catalog Company, New York, 1929.
64 A. Nowacka, P. C. Mohr, J. Norrman, R. W. Martin and D. Topgaard, *Langmuir*, 2010, **26**, 16848–16856.
65 E. V. Boldyreva, S. N. Ivashevskaya, H. Sowa, H. Ahsbahs and H.-P. Weber, *Dokl. Phys. Chem.*, 2004, **396**, 111–114.
66 A. Dawson, D. R. Allan, S. A. Belmonte, S. J. Clark, W. I. F. David, P. A. McGregor, S. Parsons, C. R. Pulham and L. Sawyer, *Cryst. Growth Des.*, 2005, **5**, 1415–1427.
67 E. V. Boldyreva, S. N. Ivashevskaya, H. Sowa, H. Ahsbahs and H.-P. Weber, *Z. Kristallogr.*, 2005, **220**, 50–57.
68 S. V. Goryainov, E. N. Kolesnik and E. V. Boldyreva, *Phys. B*, 2005, **357**, 340–347.
69 S. V. Goryainov, E. V. Boldyreva and E. N. Kolesnik, *Chem. Phys. Lett.*, 2006, **419**, 496–500.
70 M. Xu and K. D. M. Harris, *J. Am. Chem. Soc.*, 2005, **127**, 10832–10833.
71 A. Bielecki and D. P. Burum, *J. Magn. Reson., Ser. A*, 1995, **116**, 215–220.
72 A. L. Van Geet, *Anal. Chem.*, 1968, **40**, 2227–2229.
73 A. L. Van Geet, *Anal. Chem.*, 1970, **42**, 679–680.
74 A. E. Aliev and K. D. M. Harris, *Magn. Reson. Chem.*, 1994, **32**, 366–369.
75 F. Bloch, *Phys. Rev.*, 1946, **70**, 460–474.

Faraday Discussions

PAPER

Effect of mixing, concentration and temperature on the formation of mesostructured solutions and their role in the nucleation of DL-valine crystals

Anna Jawor-Baczynska,[a] Barry D. Moore[b] and Jan Sefcik*[a]

Received 12th December 2014, Accepted 29th January 2015
DOI: 10.1039/c4fd00262h

We report investigations on the formation of mesostructured solutions in DL-valine–water–2-propanol mixtures, and the crystallization of DL-valine from these solutions. Mesostructured liquid phases, similar to those previously observed in aqueous solutions of glycine and DL-alanine, were observed using Dynamic Light Scattering and Brownian microscopy, in both undersaturated and supersaturated solutions below a certain transition temperature. Careful experimentation was used to demonstrate that the optically clear mesostructured liquid phase, comprising colloidal mesoscale clusters dispersed within bulk solution, is thermodynamically stable and present in equilibrium with the solid phase at saturation conditions. Solutions prepared by slow cooling contained mesoscale clusters with a narrow size distribution and a mean hydrodynamic diameter of around 200 nm. Solutions of identical composition prepared by rapid isothermal mixing of valine aqueous solutions with 2-propanol contained mesoscale clusters which were significantly larger than those observed in slowly cooled solutions. The presence of larger mesoscale clusters was found to correspond to faster nucleation. Observed induction times were strongly dependent on the rapid initial mixing step, although solutions were left undisturbed afterwards and the induction times observed were up to two orders of magnitude longer than the initial mixing period. We propose that mesoscale clusters above a certain critical size are likely to be the location of productive nucleation events.

Introduction

Crystallization is widely used for particle formation, purification and separation in the chemical and pharmaceutical industries. Phase separation of a solute from

[a]EPSRC Centre for Innovative Manufacturing in Continuous Manufacturing and Crystallization, Department of Chemical and Process Engineering, University of Strathclyde, 75 Montrose Street, Glasgow G1 1XJ, UK. E-mail: jan.sefcik@strath.ac.uk; Tel: +44 (0)141 548 2410

[b]WestCHEM, Department of Pure and Applied Chemistry, University of Strathclyde, 295 Cathedral Street, Glasgow G1 1XL, UK. E-mail: b.d.moore@strath.ac.uk; Tel: +44 (0)141 548 2301

supersaturated solution to form a crystalline phase involves two main steps: nucleation, the birth of a stable crystal nucleus, and crystal growth, the addition of material onto an existing crystal lattice. In the last twenty years, significant efforts have been made to better understand the mechanism of the nucleation process, and a wide variety of techniques have been used to probe the relationship between molecular self-organization in solutions and melts, the formation of nuclei and the resultant crystal polymorphs. Molecular self-organization due to solute–solute and solute–solvent interactions have been examined in undersaturated, saturated and supersaturated crystallizing solutions using X-ray Small-Angle Scattering (SAXS),[1–3] Wide-Angle X-ray Diffraction (WAXD),[1] Fourier Transform Infrared Spectroscopy (FTIR),[4] Nuclear Magnetic Resonance (NMR),[5] Small Angle Neutron Scattering (SANS),[6] Empirical Potential Structure Refinement (EPSR)[7] and molecular modelling.[8,9] A common finding from these studies has been that solutes associate to form clusters in supersaturated solutions.[10–14] However, identification of a direct relationship between the solute molecular clusters and the subsequently formed crystal has proved challenging.[15–18] Two recently published reviews by Davey *et al.* and Gebauer *et al.*[19,20] discuss various molecular self-assembly pathways in nucleating solutions, including liquid–liquid macroscopic phase separation as well as the formation of pre-nucleation clusters in a wide range of organic and inorganic systems. A number of experimental and theoretical studies have pointed towards the formation of liquid-like clusters as the first step towards crystal nucleation.[21–25] Recent studies of the crystal nucleation of various small molecules highlighted non-classical mechanisms, including the formation of pre-nucleation clusters or colloidal scale intermediates on the crystal nucleation pathway.[20,26]

Despite numerous experimental and theoretical studies of non-classical nucleation mechanisms, of both small and large molecules, many questions remain open regarding the kinetics and thermodynamics of the formation of liquid-like mesospecies or pre-nucleation clusters, and what the mechanisms of solid phase formation are in such clusters. The nature of mesoscale clusters is not well understood and it is often not clear if they present a separate phase from the surrounding solution or not, if they are in equilibrium with the surrounding solution or not, or if there is an energetic barrier to their formation or not. Better understanding the link between the pre-nucleation clusters, intermediate structures in solutions (under-and supersaturated) and molecular organisation in the final crystalline product will provide the crucial insight required for better control of the final particle attributes; for example, purity, morphology, polymorphism, size and size distribution.

Metastable microphases with liquid-like properties have been observed in concentrated protein solutions[11,21,23] but have also been reported in small molecule systems.[2] Macroscopic liquid–liquid phase separation during a crystallization process has been also observed in certain classes of macromolecular systems, such as proteins[28–31] and polymers.[32,33] Bonnett *et al.* have studied liquid–liquid phase separation during the cooling-induced crystallization of an organic substance from solvent mixtures.[27] Formation of quasi-emulsions (consisting of apparently stable droplets with diameters of 100–1000 nm) during antisolvent precipitation of small organic compounds was reported by Wang *et al.*[34] The optically clear quasi-emulsion can increase induction time, stabilizing crystallizing solution for days, and can have an impact on polymorph formation.[35]

Recent studies on the effects of fluid shear on the kinetics of primary nucleation pointed towards the possible role of colloidal scale clusters in crystal nucleation in butylparaben solutions in ethanol[36] and glycine solutions in water.[25]

Recently there have been a number of reports describing the formation of mesostructured crystals and their transformation to single crystals *via* non-classical pathways. Mesocrystals have been reported to be formed with both inorganic salts and organic compounds, usually but not exclusively in the presence of stabilising polymers.[37–42] An example of a system where mesocrystals were formed without polymers is that of the amino-acid DL-alanine in aqueous solution. Our previous study of undersaturated and supersaturated aqueous solutions of DL-alanine revealed that mesoscale clusters with a size range from 100 to 300 nm exist well below solid–liquid equilibrium concentration (saturation limit). It has been concluded that the mesoscale clusters are not a separate phase, and the system has been described as a thermodynamically stable mesostructured liquid containing solute-rich domains dispersed within bulk solution.[43]

We have previously reported the presence of submicron-sized mesospecies in supersaturated aqueous glycine solutions, and their likely role in the nucleation of glycine crystals.[44] We have also previously studied the crystallization of DL-valine from 2-propanol–water solutions using Small Angle Static Light Scattering and turbidimetry measurements, which indicated the presence of mesoscale clusters in optically clear supersaturated solutions.[45] Here we further investigated the crystallization of DL-valine from 2-propanol–water solutions, mesoscale clusters in these solutions and their role in the nucleation of DL-valine crystals. We investigated these using either slow cooling or rapid initial mixing of valine aqueous solutions with 2-propanol, followed by isothermal crystallization.

Experimental

Materials

All chemicals and solvents were of laboratory reagent grade and were used without further purification: 2-propanol (VWR International Ltd.), DL-valine, puriss. ≥ 99.0%, NT (Fluka). Deionised water was supplied from an in-house Millipore Water System.

Dynamic light scattering

The mean hydrodynamic diameter of the clusters was determined with an HPPS from Malvern Instruments, HPP5001, equipped with temperature control, using photon correlation spectroscopy in backscattering mode, at a scattering angle of 173° and a laser light wavelength of 632 nm. Autocorrelation function decays were fitted using a cumulant method. The preparation of samples for DLS had to be conducted with the utmost care, because scattering artefacts can be introduced while the species of interest can be inadvertently removed. It was particularly important to check the compatibility of the solvents and mixtures with all contacted materials, and to ensure that filtration was carried out rigorously on all starting solutions and solvents but not on the crystallizing samples.

Brownian microscopy/nanoparticle tracking analysis

Cluster size distributions were determined with a Nanosight LM10 instrument with a temperature control unit. Nanosight NTA2.1 software was used to analyze videos and calculate the size and concentration of nanodroplets. The camera setting of the instrument was set using the 'Autosettings' option on the software to allow the software to optimize the shutter and gain settings. The sample was introduced into the viewing unit, and an image of the particles' scattering of the laser light was captured by a CCD camera attached to a microscope. A video of the sample was recorded and processed, with each observed individual particle 'tracked' by the nanoparticle tracking analysis software. The diffusion coefficient from the mean squared displacement of the particle tracked was calculated, and substituted into the Stokes–Einstein equation to obtain the particle hydrodynamic diameter.

Solubility measurements

The solubility of DL-valine in water–2-propanol solvent mixture (1 : 1, v/v) was determined in a temperature range from 25 °C to 55 °C using HPLC. Samples were prepared at fixed temperature in an incubator by mixing an excess of crystalline DL-valine with the solvent mixture (water–2-propanol, 1 : 1, v/v). The samples were left to equilibrate in the incubator whilst being continually stirred using the magnetic mixer plate for 24 hours, at the set temperature being investigated. The supernatant solution was separated by filtration using a glass syringe and disposable 0.1 μm PTFE syringe filters. 2 ml of the filtered sample was then added to another vial and diluted with 2 ml of a pre-filtered 1 : 1 2-propanol–water mixture. The filtration step was also carried out in the incubator. All equipment used in this step had been left overnight in the incubator, to ensure that there was no change in temperature during filtration which could cause some valine to crystallize out. This process was repeated for samples prepared at four different temperatures: 55 °C, 40 °C, 30 °C and 25 °C. This gave an indication of the solubility across a wide range of temperatures. The reversed-phase HPLC technique was used for the determination of DL-valine concentration. Prepared samples were injected into a Water2697 Separations Module with a 2487 Dual Wavelength detector using a Water Symmetry C18 3.8 × 150 mm column. The flow rate within the column was 1 ml min^{-1} and the run time for each sample was 5 minutes. The injection volume was 100 μl, and between each injection the needle was washed with pure deionised water. Each experiment was repeated at least three times and an average solubility was calculated using data from each experiment, as shown in Table 1.

Formation of supersaturated solutions

Supersaturated solutions were formed either by rapidly mixing equivolume aliquots of 2-propanol with aqueous valine solution at 25 °C (antisolvent crystallization), or by cooling a solution prepared at 50 °C (cooling crystallization).

Antisolvent crystallization

A saturated 2-propanol solution was prepared by intense mixing of 2-propanol and excess valine (2–3 g solute in 1–2 l of solvent) for at least 6 hours, in a glass

Table 1 Solubility of DL-valine in water–2-propanol solvent mixture (1 : 1, v/v) at different temperatures

Temperature (°C)	Solubility (mg $ml_{solvent}^{-1}$)	Error (+/−)
25	14.87	0.05
30	17.3	0.33
40	23.5	1.02
55	38.6	1.56

bottle. The solutions were filtered by a 0.1 μm polytetrafluoroethylene (PTFE) syringe filter using a glass syringe and stored in a glass bottle. Before using these solutions, filtration was applied again to remove any newly formed crystals that may have arisen from laboratory temperature changes over time.

Aqueous valine solutions were prepared by the dissolution of a predetermined weighed amount of valine in distilled water, filtered with a 0.2 μm polypropylene (PP) syringe filter using a glass syringe, and stored in a glass bottle. Fresh aqueous solutions were prepared at least once a week in order to avoid bacterial contamination.

All batch antisolvent crystallization experiments were prepared at 25 °C in small glass vials (diameter 25 mm), with a volume of 5 ml, and were mixed for 60, 90 or 180 seconds using a magnetic mixer with a stirring speed of 975 rpm, (magnetic stirrer: 12 mm × 4.5 mm; octagonal). Appropriate volumes of aqueous solution were added to 2-propanol with a pipette; then after mixing, samples were transferred with a pipette into a quartz cuvette for DLS measurements.

Cooling crystallization

A supersaturated solution of DL-valine was prepared by the dissolution of a predetermined weighed amount of solid valine in distilled water that had been filtered with a 0.2 μm polypropylene (PP) syringe filter using a glass syringe. The aqueous sample was kept in a small glass vial with a screw-on cap, at 50 °C, for a few hours. The valine aqueous solution was then mixed with an equal volume of pre-heated 2-propanol that had been filtered with a 0.1 μm polytetrafluoroethylene (PTFE) syringe filter using a glass syringe. Mixtures were typically left for 24 hours at 50 °C prior to cooling, to expel any bubbles. All syringes, filters and cuvettes were preheated in an incubator at 50 °C to avoid premature cooling of the solution. DLS sample cell holders were preheated to 50 °C before the solution was inserted. Data collection was started immediately and the temperature ramped down from 50 °C to 25–15 °C.

Results and discussion

Supersaturated and undersaturated solutions prepared by cooling

Supersaturated solutions were prepared by mixing hot filtered aqueous valine and 2-propanol at 50 °C, and slowly cooling it to ambient temperature. At 50 °C only small molecular clusters of up to 1–2 nm were observed, but below about 35 °C colloidal-scale species with a mean hydrodynamic diameter of about 200 nm were found to spontaneously form, accompanied by small molecular clusters (see Fig. 1a). The colloidal (mesoscale) clusters formed in cooled solutions had a relatively narrow size distribution (100–300 nm) and upon storage over many hours there was no change in their average size or size distribution.

After reheating solutions back to above 35 °C, the mesoscale clusters rapidly disappeared again. As shown in Fig. 1b, during both the cooling and heating cycles the sizes of the mesoscale clusters were similar to each other at any particular temperature. The presence of these species cannot be explained by scattering from solid impurities or bubbles. Heating of a solution typically helps to remove soluble gases, and subsequent cooling would not lead to nucleation of bubbles, as the solubility of gases in liquids is expected to increase with decreasing temperature.

Near saturated solutions in contact with solid phase

When valine solutions were cooled to the desired temperature where they were then left undisturbed under isothermal quiescent conditions, crystallization was very slow. Solutions remained transparent for many hours until a small number of large valine crystals eventually appeared at the bottom of the cell. Even in solutions equilibrated for more than 12 hours, with valine crystals present, the mesoscale clusters persisted. The reversible appearance of the mesoscale clusters in undersaturated solutions at a specific temperature, and their coexistence with valine crystals in saturated and supersaturated solutions, indicates that an equilibrium phase state has been reached in these solutions.

In another experiment, most of the liquid supernatant in equilibrium with grown crystals was carefully removed with a pipette, without disturbing the crystals, and was replaced with a slightly undersaturated filtered (using a 0.1 μm PTFE syringe filter) valine solution (14.2 mg ml^{-1}). DLS data showed a weak signal from mesoscale clusters that still remained, so the supernatant was again removed and carefully replaced with the filtered undersaturated solution. This treatment lowered the concentration of mesoscale clusters to below the detectable limit (Fig. 2a).

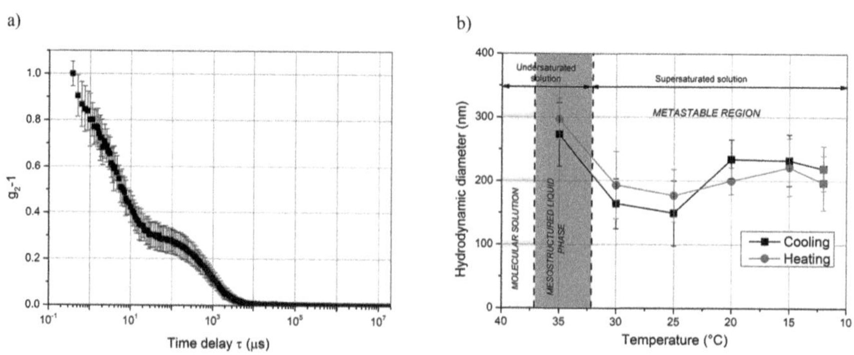

Fig. 1 (a) Dynamic light scattering autocorrelation functions $g_2(\tau) - 1$ measured at 25 °C, (b) average hydrodynamic diameter of mesoscale clusters, cooling–heating experiment, $C_{val} = 19$ mg ml^{-1}, 2-propanol–water, 1 : 1, v/v.

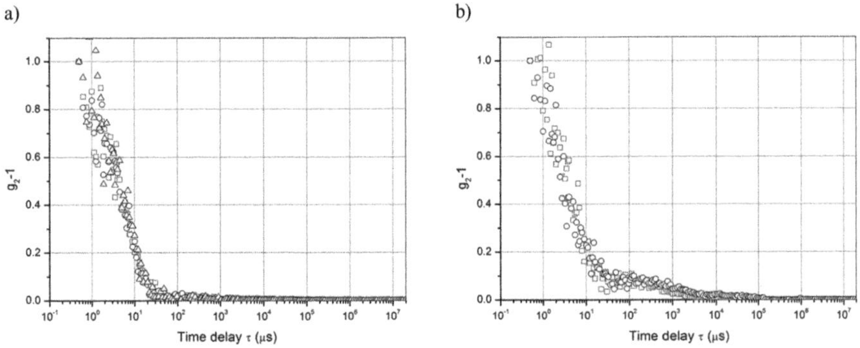

Fig. 2 Dynamic light scattering autocorrelation functions $g_2(\tau) - 1$, (a) filtered (0.1 μm PTFE) 14.2 mg ml^{-1} DL-valine–2-propanol–water solution with valine crystals at 25 °C, (b) 14.2 mg ml^{-1} DL-valine–2-propanol–water filtered solution with valine crystals at 35 °C. Different signs correspond to repetition of the same experiment.

Crystals were still clearly present, bound to the bottom of the cell out of the optical path (the laser beam was about 1 cm above crystals), and so the system was left undisturbed at 25 °C for 24 hours to equilibrate. After this time, DLS data showed a very weak fluctuating signal consistent with the presence of a small population of mesoscale clusters. The undisturbed cell was then gradually heated in the DLS cell, whereupon stronger constant signals started to be observed. On reaching 35 °C the shape of the DLS signal was clearly very similar to that obtained on cooling, but was still less intense (Fig. 2b).

The appearance of mesoscale clusters upon gentle heating is consistent with the source of the mesoscale clusters being the valine crystals at the base of the cell. Diffusion of mesoscale clusters away from the surface of the crystals would be expected to be quite slow in an isothermal stationary cell. However, heating provides a convection current that can serve to transport mesoscale clusters into the beam path. Support for this transport limitation was obtained by gentle rocking of the DLS cell to homogenise the mixture without dislodging the crystals. Immediately a much stronger signal was observed for the mesoscale clusters, and this could still be observed after a further one week of equilibration (Fig. 3).

Our observations as described above are consistent with the formation of mesostructured liquid phases, comprised of mesoscale clusters dispersed within a bulk valine solution. These mesostructured liquid phases appear to be thermodynamically stable over a range of valine concentrations and temperatures. Similar phase behaviour has been previously observed in aqueous solutions of other amino acids, glycine and DL-alanine,[43] and the role of such structures in the nucleation mechanism of glycine crystals was investigated.[44] The question addressed further here will be the role of the mesoscale clusters in valine crystal nucleation and how the presence of such structures will affect the antisolvent crystallization process of valine. Before considering that, let us investigate mesoscale clusters in solutions prepared by isothermal mixing of aqueous valine with 2-propanol.

Supersaturated and undersaturated solutions prepared by isothermal mixing

Solutions were prepared by mixing an aqueous valine solution with an equal volume of 2-propanol in a small glass vial, using a magnetic stirring bar, which

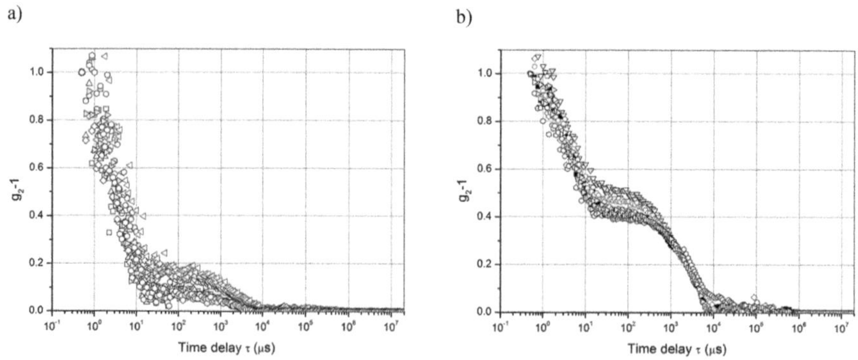

Fig. 3 Dynamic light scattering autocorrelation functions $g_2(\tau) - 1$, of 14.2 mg ml^{-1} DL-valine–2-propanol–water filtered solution with valine crystals, after 1 week keeping at 25 °C, (a) 35 °C, (b) 35 °C after gentle rocking, different signs corresponds to repetition of the same experiment.

was then transferred to a DLS cell that had been preheated to the same temperature. DLS measurements were then obtained during the induction time period, until interference from larger scattering species (*e.g.*, sedimenting crystals) became too intense. Immediately following mixing, small molecular clusters as well as colloidal scale mesoscale clusters could be detected in the optically clear solutions, as in solutions prepared by slow cooling. Control experiments showed that colloidal scale scattering species were not present in filtered solutions prior to mixing. Similarly, they could not be detected when mixing aqueous valine and propanol to form dilute solutions at very low supersaturation ($S = 0.2$). The species could not therefore be ascribed to solid impurities or (the formation of) (nano)bubbles, and were therefore attributable to the presence of valine-rich mesoscale clusters. Investigations were carried out on mixtures prepared over a range of valine concentrations, from undersaturated to supersaturated solutions.

As the valine concentration was increased, the mean size of the mesoscale clusters in supersaturated solutions varied, depending upon the method that was used to prepare them. In undersaturated solutions mesoscale clusters were of similar average size, regardless of whether they were prepared by cooling or isothermal mixing, although uncertainties in their size estimates were larger because of the fluctuating signals. However, in supersaturated solutions prepared by isothermal mixing, the mean cluster size was significantly larger than in those prepared by slow cooling (see Fig. 4 and 5). The reasons for this variation in size distribution with the method of solution preparation are not certain, but it seems reasonable to assume that larger mesoscale clusters arise from the coalescence of smaller ones in metastable supersaturated solutions subject to an intense period of initial mixing.

Effect of initial mixing period on induction time

Solutions were prepared by mixing an aqueous valine solution with an equal volume of 2-propanol in a small glass vial, using a magnetic stirring bar. To determine the mixing efficiency (*i.e.* how quickly full supersaturation could be achieved upon mixing the aqueous valine solution with propanol), a competitive

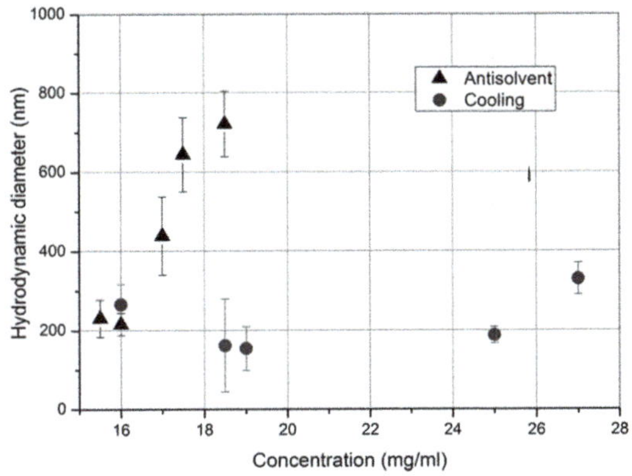

Fig. 4 Average hydrodynamic diameter of mesoscale clusters as a function of valine concentration and preparation method: cooling and antisolvent (rapid isothermal mixing, mixing time 90 s), 2-propanol–water, 1 : 1, v/v.

reaction procedure was carried out using the same solvent ratios, vials and mixers.[46] These experiments showed that a microscopically homogenous mixture of aqueous solution and propanol was formed in a few seconds. Thus, the observed enhancement of crystallization rates upon application of agitation well beyond a few seconds are not attributable to the rate at which supersaturation has been reached.

The agitation was then continued for pre-determined periods to give total agitation times of between 60 and 180 seconds. Typically, 2.5 ml of an aqueous valine solution of known concentration was admixed with 2.5 ml of 2-propanol in a 20 ml vial to give a solution of specified supersaturation. The time before the appearance of the first crystals (induction time) was found to vary as a function of the period and method of mixing, as well as the supersaturation. The induction time was determined by visual inspection, based on the appearance of the first crystal in the solution after or during the initial mixing step. Induction times as a function of valine supersaturation for different initial mixing periods are shown in Fig. 6. It was observed that the initial mixing period had a large impact on the

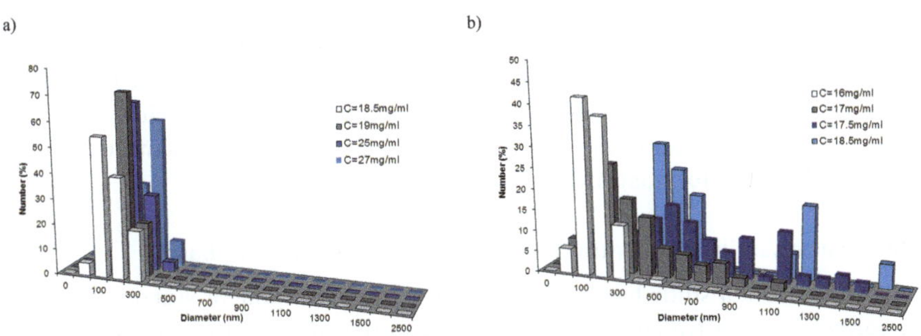

Fig. 5 Size distribution of mesoscale clusters determined by Brownian microscopy as a function of valine concentration in 2-propanol–water, 1 : 1, v/v, at 25 °C: (a) solutions prepared by cooling, (b) isothermal mixing, mixing time 90 s.

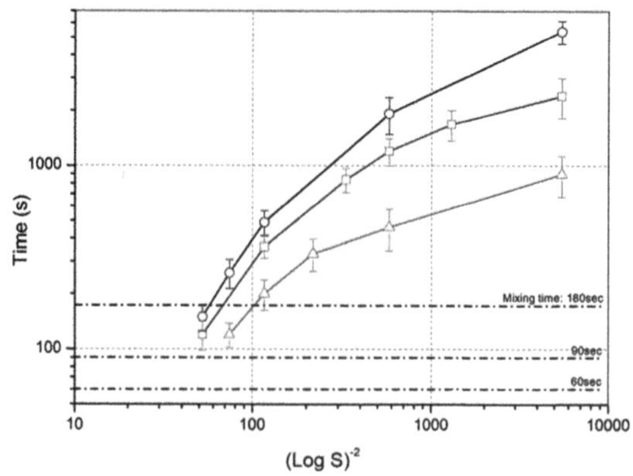

Fig. 6 Mean induction times as a function of initial supersaturation; solutions prepared by rapid isothermal mixing, mixing times: 60 s (circles), 90 s (squares), 180 s (triangles), 2-propanol–water, 1 : 1, v/v. Each experiment was repeated at least 20 times.

time to appearance of the first crystals. This effect was retained when the air–solvent interface was eliminated and magnetic stirring was applied within a completely full vial.

In all these experiments, the actual formation of the supersaturated solution was very rapid, since it required only seconds to achieve complete mixing of the two fully miscible solvents. Variations in the induction times were therefore influenced by subsequent agitation carried out over periods at least 10–20 times greater than this. Such behaviour does not seem to fit with a classical nucleation mechanism applied to the bulk molecular solution because, once a supersaturated solution is produced, the process of molecular association to form critical nuclei is not expected to be affected by the period or intensity of mixing, unless the critical nuclei or their precursors are at colloidal length scale. This is because a size on the order of 100s nm is required for the Peclet number, which relates characteristic times of convection and diffusion, to reach an order of unity. Therefore, while molecular clusters on the order of 1 nm would not be expected to be influenced by shear rates typically encountered in laboratory conditions, mesoscale clusters would be. Furthermore, when the initial mixing conditions were kept identical but the supersaturation varied, a standard plot of log induction time *versus* log reciprocal of supersaturation squared was strongly deviating from what would be expected from the classical nucleation theory; see Fig. 6. In order for the classical nucleation theory to become applicable here, one may need to understand composition and internal dynamics of mesoscale clusters, as it appears that supersaturation values based on bulk solution concentrations are not directly useful for quantifying the driving force for nucleation.

Molecular crystallization pathways in mesostructured liquid phases

This experimental study has revealed that the three-component system of valine dissolved in a 1 : 1 (v/v) water–2-propanol mixture exhibits a thermodynamically stable mesostructured liquid phase. It should also be noted that the presence of two solvents means that as additional valine is added or as the system is cooled, the compositions of mesoscale clusters can adjust in order to minimise the overall free energy of the system. Thus, the distribution of valine, water and 2-propanol between the mesoscale clusters and surrounding solution may be fixed at any particular composition and temperature, but continuously variable as the total composition or temperature is changed. In this system, the mesostructured liquid is expected to be the most thermodynamically stable phase from the point at which monomers or small molecular cluster assemble to mesoscale clusters (critical mesoscale cluster concentration)[43] up to a composition where the crystalline phase becomes stable (solubility). At higher concentrations the mesostructured liquid phase will continue to be the most kinetically accessible phase, but since it is supersaturated it will be metastable with respect to the solid phase.

In supersaturated solutions prepared by cooling, mesoscale clusters were relatively small and in a narrow size range of 100–300 nm. Even at relatively high supersaturations, all valine solutions prepared by cooling remained optically clear for many hours when left undisturbed. Few large crystals were typically observed after 1–2 days. However, even gentle swirling or pipetting of these solutions resulted in the rapid onset of nucleation. Although secondary nucleation was considered, it is not expected that a gentle flow without any stirrer or impeller would induce massive secondary nucleation in solutions where no visible crystals could be observed. On the other hand, when solutions were prepared by isothermal rapid initial mixing followed by quiescent isothermal crystallization, induction times were 1–2 orders of magnitude shorter than in solutions of the identical compositions prepared by cooling. We have also seen that the application of rapid initial mixing results in a much larger size of mesoscale clusters (Fig. 4 and 5).

We hypothesize that once mesoscale clusters reach a critical size, a much more rapid nucleation process can be accessed, and crystallization follows quickly. Thus, the greater the number of large mesoscale clusters formed, the faster the overall rate of crystal formation will be. It can be speculated that large mesoscale clusters may form by shear-induced coalescence during the initial mixing period. Alternatively, adsorption of mesoscale clusters onto the vial surface may well

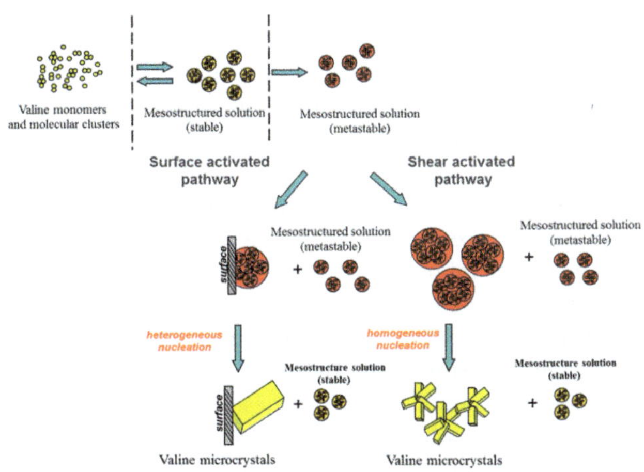

Fig. 7 Proposed pathways for the crystallization of amino acid (DL-valine) involving mesoscale clusters: surface and shear-activated pathways.

provide an alternate, albeit slower, coalescence pathway. Surface-assisted coalescence would then trigger crystal nucleation in a similar way. Consistent with this hypothesis, it was often observed that valine crystals formed from cooled solutions were bound to the side of the vial. Fig. 7 provides a pictorial summary of the proposed nucleation crystallization pathways.

Borrowing from the classical nucleation theory and building on what we proposed previously[44] smaller mesoscale clusters may not contain sufficient amount of valine to be able to form stable nuclei able to survive and grow in the surrounding bulk solution. This could certainly be the case if as anticipated the mesoscale clusters were to contain a significant volume fraction of solvent. The critical mesoscale cluster size for productive nucleation would then correspond to the cluster volume that contains just sufficient amount of valine to be able to form a viable growing crystal. In this model it can be imagined that mesoscale clusters of all sizes intermittently sample the crystalline phase and most commonly they simply reform as the nascent crystal is not viable in the bulk solution. Thus the probability of a nucleation event occurring within small supersaturated mesoscale clusters, such as those formed on cooling, will be very low. Only when a critical cluster volume has been reached will the probability of productive nucleation become high enough for it to occur at a significant rate.

Conclusions

In this work we have experimentally studied antisolvent and cooling crystallization of valine from water–2-propanol mixtures. DLS data showed that valine solutions contain two kinds of clusters present in the optically clear solution: molecular clusters (diameters of up to 1–2 nm) and solute-rich mesoscale clusters (diameters of hundreds nanometers). Within cooled solutions mesoscale clusters were around 200 nm in diameter and remained unchanged in size and monodisperse over periods of many hours. Samples prepared by isothermal rapid initial mixing contained larger mesoscale clusters, and a significant effect of the initial mixing step used to form supersaturated valine solutions on the mesoscale cluster size was observed. Solutions prepared by rapid isothermal mixing crystallized at least two orders of magnitude faster than solutions of identical compositions prepared by cooling. Careful experimentation showed that mesoscale clusters dispersed within bulk valine solution together form a thermodynamically stable mesostructured liquid phase, that at the crystal solubility limit exists in equilibrium with valine crystals.

That the mesostructured liquid is a genuine phase of matter can be demonstrated by the appearance of stable mesoscale clusters on mixing the filtered solvent with an excess of valine crystals. The mesoscale clusters were found to arise directly from the dissolution of solid, showing that the solid and mesostructured liquid phases are present together in thermodynamic equilibrium. Although they are clearly visible using DLS, it is important to recognise that the total amount of valine contained within the mesoscale clusters is very small, and careful experimental design was required to obtain these data. Large mesoscale clusters obtained by shear- or surface-assisted coalescence are proposed to be the likely location of productive nucleation events leading to the formation of valine crystals.

Notes and references

1 R. J. Davey, W. Liu, M. J. Quayle and G. J. T. Tiddy, *Cryst. Growth Des.*, 2002, **2**, 269–272.

2 S. Chattopadhyay, D. Erdemir, J. M. B. Evans, J. Ilavsky, H. Amenitsch, C. U. Segre and A. S. Myerson, *Cryst. Growth Des.*, 2005, **5**, 523–527.

3 D. Erdemir, S. Chattopadhyay, L. Guo, J. Ilavsky, H. Amenitsch, C. U. Segre and A. S. Myerson, *Phys. Rev. Lett.*, 2007, **99**, 115702.

4 S. Parveen, R. J. Davey, G. Dent and R. G. Pritchard, *Chem. Commun.*, 2005, 1531–1533.

5 R. A. Chiarella, A. L. Gillon, R. C. Burton, R. J. Davey, G. Sadiq, A. Auffret, M. Cioffi and C. A. Hunter, *Faraday Discuss.*, 2007, 136.

6 T. J. Sorensen, P. C. Sontum, J. Samseth, G. Thorsen and D. Malthe-Sorenssen, *Chem. Eng. Technol.*, 2003, **26**, 307–312.

7 R. C. Burton, E. S. Ferrari, R. J. Davey, J. Hopwood, M. J. Quayle, J. L. Finney and D. T. Bowron, *Cryst. Growth Des.*, 2008, **8**, 1559–1565.

8 S. Hamad, C. E. Hughes, C. R. A. Catlow and K. D. M. Harris, *J. Phys. Chem. B*, 2008, **112**, 7280–7288.

9 R. J. Davey, K. Allen, N. Blagden, W. I. Cross, H. F. Lieberman, M. J. Quayle, R. Righini, L. Seton and G. J. T. Tiddy, *CrystEngComm*, 2002, 257–264.

10 J. F. Lutsko and G. Nicolis, *Phys. Rev. Lett.*, 2006, **96**, 046102.

11 P. G. Vekilov, *Interdisciplinary Transport Phenomena*, 2009, vol. 1161, pp. 377–386.

12 J. Sefcik and A. V. McCormick, *Chem. Eng. Sci.*, 1999, **54**, 3513–3519.

13 A. S. Myerson and P. Y. Lo, *J. Cryst. Growth*, 1990, **99**, 1048–1052.

14 J. W. Mullin and C. L. Leci, *Philos. Mag.*, 1969, **19**, 1075–1077.

15 S. Parveen, R. J. Davey, G. Dent and R. G. Pritchard, *Chem. Commun.*, 2005, 1531–1533.

16 R. J. Davey, G. Dent, R. K. Mughal and S. Parveen, *Cryst. Growth Des.*, 2006, **6**, 1788–1796.

17 R. A. Chiarella, A. L. Gillon, R. C. Burton, R. J. Davey, G. Sadiq, A. Auffret, M. Cioffi and C. A. Hunter, *Faraday Discuss.*, 2007, **136**, 179–193.

18 S. Hamad, C. E. Hughes, C. R. A. Catlow and K. D. M. Harris, *J. Phys. Chem. B*, 2008, **112**, 7280–7288.

19 R. J. Davey, S. L. M. Schroeder and J. H. ter Horst, *Angew. Chem., Int. Ed.*, 2013, **52**, 2–16.

20 D. Gebauer, M. Kellermeier, J. D. Gale, L. Bergstrom and H. Colfen, *Chem. Soc. Rev.*, 2014, **43**, 2348–2371.

21 P. G. Vekilov, *Cryst. Growth Des.*, 2004, **4**, 671–685.

22 D. Erdemir, A. Y. Lee and A. S. Myerson, *Acc. Chem. Res.*, 2009, **42**, 621–629.

23 G. Gliko, W. Pan, P. Katsonic, N. Neumaier, O. Galein, S. Weinkauf and P. G. Vekilov, *J. Phys. Chem. B*, 2007, **11**, 3106–3114.

24 M. B. Sweatman, R. Fartaria and L. Lue, *J. Chem. Phys.*, 2014, **140**, 124508.

25 C. Forsyth, P. A. Mulheran, C. Forsyth, M. D. Haw, I. S. Burns and J. Sefcik, *Cryst. Growth Des.*, 2015, **15**, 94–102.

26 D. Erdemir, A. Y. Lee and A. S. Myerson, *Acc. Chem. Res.*, 2009, **42**, 621–629.

27 P. E. Bonnett, K. J. Carpenter, S. Dawson and R. J. Davey, *Chem. Commun.*, 2003, **6**, 698–699.

28 P. Rein den Wolde and D. Frenkel, *Science*, 1997, **277**, 1975–1978.
29 V. G. Taratuta, A. Holschbach, G. M. Thurston, D. Blankschtein and G. B. Benedek, *J. Phys. Chem.*, 1990, **94**, 2140–2144.
30 D. F. Rosenbaum and C. F. Zukoski, *J. Cryst. Growth*, 1996, **169**, 752–758.
31 N. Asherie, N. Lomakin and G. B. Benedek, *Phys. Rev. Lett.*, 1996, **77**, 4832–4835.
32 H. L. Chen, J. C. Hwang, J. M. Yang and R. C. Wang, *Polymer*, 1998, **39**, 6983–6989.
33 Y. S. Soh, J. H. Kim and C. C. Gryte, *Polymer*, 1995, **36**, 3711–3717.
34 X. Wang, J. M. Gillian and D. J. Kirwan, *Cryst. Growth Des.*, 2006, **6**, 2214–2227.
35 X. Wang and D. J. Kirwan, *Cryst. Growth Des.*, 2006, **6**, 2228–2240.
36 J. Liu and Å. C. Rasmuson, *Cryst. Growth Des.*, 2013, **13**, 4385–4394.
37 A. W. Xu, Y. Ma and H. Cölfen, *J. Mater. Chem.*, 2007, **17**, 415–449.
38 M. Niederberger and H. Cölfen, *Phys. Chem. Chem. Phys.*, 2006, **8**, 3271–3287.
39 H. Cölfen and S. Mann, *Angew. Chem.*, 2003, **42**, 2350–2365.
40 D. D. Medina and Y. Mastai, *Cryst. Growth Des.*, 2008, **8**, 3646–3651.
41 D. Schwahn, Y. Ma and H. Cölfen, *J. Phys. Chem. C*, 2007, **111**, 3224–3227.
42 Y. Ma, H. Cölfen and M. Antonietti, *J. Phys. Chem. B*, 2006, **110**, 10822–10828.
43 A. Jawor-Baczynska, B. D. Moore, H. S. Lee, A. V. McCormick and J. Sefcik, *Faraday Discuss.*, 2013, **167**, 425–440.
44 A. Jawor-Baczynska, J. Sefcik and B. D. Moore, *Cryst. Growth Des.*, 2013, **13**, 470–478.
45 M. Variny, S. Alvarez de Miguel, B. D. Moore and J. Sefcik, *J. Dispersion Sci. Technol.*, 2008, **29**, 617–620.
46 A. Brown, PhD thesis, University of Strathclyde, 2012.

DISCUSSIONS

Molecular self-assembly and clustering in nucleation processes: general discussion

Sarah Price, Stéphane Veesler, Haihua Pan, Ken Lewtas, Mireille Smets, Bart Rimez, Allan Myerson, Colan Hughes, Alan Hare, Fajun Zhang, Hugo Meekes, Marco Mazzotti, Ian Rosbottom, Dikshitkumar Khamar, Joost van den Ende, Laszlo Fabian, Simon Black, Francis Taulelle, Martí Gich, Peter Vekilov, Dimitrios Toroz, Celso Aparecido Bertran, Jan Sefcik, Sven Schroeder, Samuel Booth, Ake Rasmuson, Eric Breynaert, Elena Simone, Robert Hammond, Richard Sear, Jim de Yoreo, Roger Davey, Jamshed Anwar, Radoljub Ristic, Diana Milena Camacho Corzo, Kevin Roberts, Kenneth Harris, Helmut Colfen and Thomas Turner

DOI: 10.1039/c5fd90036k

Peter Vekilov opened a general discussion of the introductory lecture by Roger Davey: I think we need to distinguish between two kinetic barriers that determine the rate of crystal nucleation. The first one, ΔG^*, is due to the creation of a surface between a newly created crystal nucleus and is treated by the classical nucleation theory. The second barrier, $\Delta G^{\neq}$, is due to the creation of chemical bonds between molecules that join the crystal and those already in the crystal. Typically, this barrier is interpreted as the high energy point during the stretching of the bonds between a solute molecule and the associated solvent molecules; it is treated by the transition-state theory. Since bonds are created in all events of molecular attachment to pre-critical or post-critical clusters, starting from the first crystal dimer, the transition state barrier superimposes on the free energy barrier to create a free energy relief that accounts for both barriers and look like the one depicted in Fig. 1. It seems that title of our conference focuses on the classical nucleation barrier when it refers to nucleation as a transition state to materials assembly. On the other hand, you seem to focus on the barrier for creation of crystal bonds. Is my interpretation correct?

Roger Davey responded: Yes you are correct. This is a very elegant way to put it. Thank you.

Jim De Yoreo commented: Many of the papers cited in the talk are not ones with which I am familiar and many of the papers that come to mind in the context of the talk were not cited. I believe this is because there is a division between the

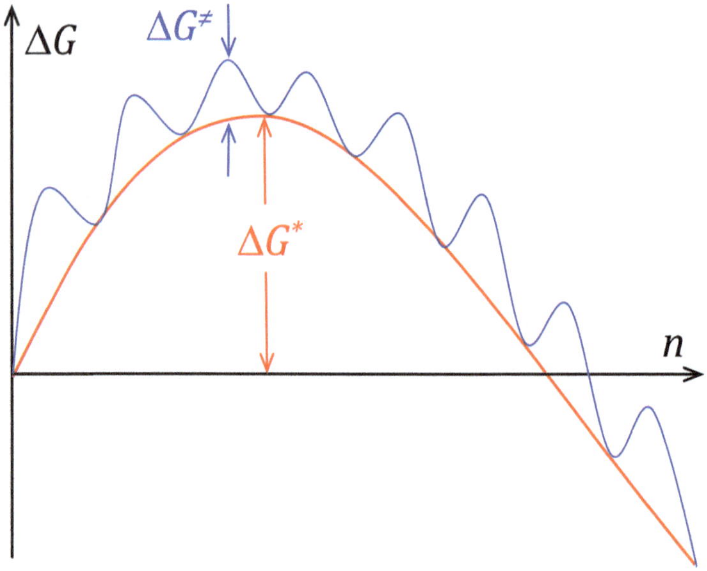

Fig. 1 Free energy relief composed of the transition state barrier and the free energy barrier.

organic and inorganic crystallization communities. I hope this meeting will serve as a transition state between the two.

Jim De Yoreo continued: The description of the transition state you presented is somewhat static; it contrasts with my picture of a cluster that is a dynamical structure with molecules attaching and detaching in different ways and changing its bonds with neighbors in the cluster so that there is no single type of bound state that defines the transition, simply because the energy landscape is complex. How do you account for the complexity in your picture of a transition state during nucleation?

Roger Davey replied: This really goes to the heart of what I was trying to suggest. Of course you are right, the reality in nucleation, just as in reaction chemistry, is that the situation is dynamic – molecules are on the move, colliding and diffusing. Some of these collisions will lead to successful desolvation and stepwise formation of clusters and in this sense indeed there is no single state that defines the transition. However this does not stop us from thinking about what these steps might be and which solvent–solute interactions must be replaced by solute–solute interactions in the process. To do this requires not only that we use our knowledge of crystal packing but also that we build up our molecular level understanding of solvation. The static picture that results is thus a fantasy but its power is to make us think about what must be happening.

Francis Taulelle asked: During crystallization desolvation occurs and lattice energy formation takes over. In case of coordinated solvent to crystallizing entities, desolvation takes place. The solvent may participate and also the solvate crystal. How would you disentangle the two phenomena to evaluate their contribution to the free energy of crystal formation?

Roger Davey responded: So as I understand the question you are saying that there can be two sorts of solvent molecule – those that must be removed to allow crystallisation and those that remain co-ordinated to the solute and end up in a solvated crystal structure. Understanding why solvates form remains of course a challenge but must be related to both the nature of solvent binding and the ability of a molecule to yield a close packed structure.

Allan Myerson remarked: If we look at the nucleation free energy diagram in polymorphic systems we often represent a kinetic but less stable phase has having a lower barrier. In addition we often make the general observation at a fixed supersaturation the less stable polymorphs often nucleate first even though they will have a lower supersaturation driving force (due to their higher solubility). What has not been addressed is the molecular conformation and symmetry of the nucleating form related to the most stable form. Since a set of molecules with known crystal structures and stabilities are marketed as pharmaceutical molecules, this might be a interesting study to conduct.

Ken Lewtas opened a general discussion of the paper by Peter Vekilov: I would like to ask about your work using oblique illuminated microscopy using crossed polarisers, This technique is new to me and it looks very interesting. I will look up the details later but can you tell me: (i) if there are any practical limitations; (ii) can you use it on any material (I am particularly interested in non-aqueous media); (iii) can you also use phase-contrast or differential interference contrast in this technique and; (iv) I assume that the minimum sizes that can be observed are limited to the normal restrictions of optical microscopy, is this correct?

Peter Vekilov answered: Oblique illumination microscopy is useful to identify and characterize heterogeneous objects in a fluid. The objects can be solid or another liquid. A significant advantage of this technique is that it provides images of individual droplets or aggregates, in contrast to dynamic light scattering, where the signal is collectively generated by a population of new phase domains. In our implementation, the method requires about 300 microliters of solution that is pumped in the chamber; this may be a too large volume for some systems. Other limitations are that the difference in refractive index between the solution and the studied objects needs to be significant; if the solution contains large monomers, they may obstruct the signal form the aggregates. If the aggregates occupy a high volume fraction, the solution layer can be made thinner.

There should not be a problem with non-aqueous media; you can implement it on any microscope and combine with other advanced techniques: phase contract, DIC, or fluorescence microscopy.

Since the technique detects scattered light, object smaller than the diffraction limit can be detected. In fact, the clusters that we observe are often about 150 nm in diameter. The images are of the point spread functions of the observed objects; if their concentration is so high that the distance between the objects in the plane of the image becomes comparable to the diffraction limit, the individual aggregates will not be resolved.

Jim De Yoreo asked: The mechanism you propose seems very specific to lysozyme or other proteins that undergo oligomerization to create clusters in which nucleation by monomers then occurs. Do you think it should be a general process for all proteins or do you think it's specific to lysozyme and some other limited set of protein systems?

Peter Vekilov answered: The mechanism of formation of the mesoscopic protein-rich clusters that we have proposed[1–3] appears to be the only way to explain the existence for clusters of size 100 nm and larger that contain 10 000–100 000 protein molecules. This mechanism relies on the formation of transient protein oligomers, whose lifetime and diffusivity determine the cluster size. This part of the mechanism is common for all proteins. Our investigations of this mechanism with lysozyme have indicated that the oligomers that lead to formation of clusters are domain-swapped dimers. It is likely that this mechanism will be valid for other single-chain proteins that do not form different (Coulomb, van der Waals, *etc.*) oligomers. For proteins that consist of multiple chains, such as haemoglobin, which contains four chains, or insulin, with six, or lumazine synthase, with 24, it is possible that the oligomers that underlie cluster formation are mis-assembled protein molecules with a different number of chains or with improperly embedded chains.

1. W. Pan, P. G. Vekilov and V. Lubchenko, *J. Phys. Chem. B*, 2010, **114**, 7620–7630.
2. H. Y. Chan, V. Lankevich, P. G. Vekilov and V. Lubchenko, *Biophys. J.*, 2012, **102**, 1934–1943.
3. Y. Li, V. Lubchenko, M. A. Vorontsova, L. Filobelo and P. G. Vekilov, *J. Phys. Chem. B*, 2012, **116**, 10657–10664.

Kevin Roberts asked: The research space underpinning the concept of two-step nucleation is a very interesting and a topical one particularly in terms of its inter-relationship to the process of oiling out whereby liquid–liquid separation takes place in highly supersaturated solutions. Do you think that the two processes can be regarded as analogous in terms of the physico-chemical processes involved or are they quite distinctly different?

Often it has been found that systems that do oil out can fail to crystallise or perhaps crystallise very slowly.[1] In our previous studies of citric acid the formation quite large clusters, *i.e.* in the micron size range, have been observed prior to crystallisation.[2] Similar effects have been seen monosodium glutamate.[3] In both cases the phase separation effect precedes the crystallisation onset in terms of their respective induction times. What is not clear is the structural nature, *i.e.* molecular organisation, density *etc*, of the dispersed phase clusters with respect to that of the continuous phase. There are many questions to ponder: which of the two phases is the denser? Does one of the phases scavenge the impurities that perhaps are impeding the nucleation process? Where does the crystallisation process initiate, *i.e.* in the cluster, at the cluster/solution interface or within the bulk of the continuous phase?

1. J. Lu, Y.-P. Li, J. Wang, Z. Li, S. Rohani and C. B. Ching, Study on the Oiling-out and Crystallization for the Purification of Idebenone, *Org. Process Res. Dev.*, 2012, **16**(3), 442–446.
2. H C Groen and K J Roberts, Nucleation, growth and pseudo-polymorphic behaviour of citric acid as monitored *in situ* by attenuated total reflection Fourier transform infrared spectroscopy, *J. Phys. Chem. B*, 2001, **105**, 10723–10730.
3. Heidi Grön, On-line Monitoring and Control of Supersaturation in Batch Crystallisers for Organic Fine Chemical Products using ATR-FTIR Spectroscopy, PhD thesis, Heriot-Watt University, 2001.

Peter Vekilov answered: I will try to address these issues with examples from our studies of protein crystal nucleation and draw bridges to the nucleation pathways of small molecules. We find that with proteins, the macroscopic dense liquid phase is distinct from the protein-rich clusters that serve as nucleation precursors. The difference is fundamental: the macroscopic dense liquid is a property of the protein solution,[1,2] its equilibrium behaviors are fully described by the solution thermodynamics,[3] the evolution of its structure (droplets that nucleate, grow and coalesce to form macroscopic domains) fully complies with predictions of classical new phase theories, such as Lifshitz-Slyozov.[4] This is in contrast with the behaviors of the protein-rich clusters, for which only the equilibrium characteristics (the fraction of the total protein captured in the clusters) are regulated by the solution thermodynamics.[5,6] The cluster phase consists of relatively monodisperse domains of mesoscopic sizes that never reach macroscopic dimensions.[7–11] This highly unusual latter observation suggests that the clusters consist of transient protein complexes (sometimes called oligomers or dimers), in which the mechanism of complexation depends on the properties of the protein.[10]

How these insights transfer to small-molecule (organic and ionic) solutions is an open question. Recent simulations of calcium carbonate solution suggest that metastable liquids are possible in such solutions.[12] Whether the nucleation precursors observed and surmised in calcite nucleation[13,14] are droplets of this metastable dense liquid or clusters of an entirely different phase is an open question. Regarding the question of nucleation and oiling-out, in protein solutions, the macroscopic dense liquid may, in some cases, support nucleation,[15,16] while in other cases, completely suppress it.[17] The nucleation suppression in dense liquids has been attributed to high viscosity that ensues after gelation.[17] This latter situation is a clear analogue to oiling out. Note however, that the suppression of nucleation in macroscopic dense liquids is not a universal behavior and some macroscopic dense liquids may enhance crystal nucleation. I do not think we have clear answers to the other questions raised by Professor Roberts: does the dense liquid sequester the impurities that impede nucleation; does nucleation occur on the cluster/solution interface or in the cluster bulk. They are fundamental questions that need to be addressed in future efforts.

1. J. A. Thomson, P. Schurtenberger, G. M. Thurston and G. B. Benedek, *Proc. Natl. Acad. Sci. USA*, 1987, **84**, 7079–7083.
2. A. Lomakin, N. Asherie and G. B. Benedek, *J. Chem. Phys.*, 1996, **104**, 1646–1656.
3. D. N. Petsev, X. Wu, O. Galkin and P. G. Vekilov, *J. Phys. Chem. B*, 2003, **107**, 3921–3926.
4. M. Shah, O. Galkin and P. G. Vekilov, *J. Chem. Phys.*, 2004, **121**, 7505–7512.
5. Y. Li, V. Lubchenko, M. A. Vorontsova, L. Filobelo and P. G. Vekilov, *J. Phys. Chem. B*, 2012, **116**, 10657–10664.
6. V. Uzunova, W. Pan, V. Lubchenko and P. G. Vekilov, *Faraday Disc.*, 2012, **159**, 87–104.
7. O. Gliko, N. Neumaier, W. Pan, I. Haase, M. Fischer, A. Bacher, S. Weinkauf and P. G. Vekilov, *J. Amer. Chem. Soc.*, 2005, **127**, 3433–3438.
8. O. Gliko, W. Pan, P. Katsonis, N. Neumaier, O. Galkin, S. Weinkauf and P. G. Vekilov, *J. Phys. Chem. B*, 2007, **111**, 3106–3114.
9. W. Pan, O. Galkin, L. Filobelo, R. L. Nagel and P. G. Vekilov, *Biophys. J.*, 2007, **92**, 267–277.
10. W. Pan, P. G. Vekilov and V. Lubchenko, *J. Phys. Chem. B*, 2010, **114**, 7620–7630.
11. Y. Li, V. Lubchenko, M. A. Vorontsova, L. Filobelo and P. G. Vekilov, *J. Phys. Chem. B*, 2012, **116**, 10657–10664.
12. A. F. Wallace, L. O. Hedges, A. Fernandez-Martinez, P. Raiteri, J. D. Gale, G. A. Waychunas, S. Whitelam, J. F. Banfield and J. J. De Yoreo, *Science*, 2013, **341**, 885–889.
13. D. Gebauer, A. Volkel and H. Colfen, *Science*, 2008, **322**, 1819–1822.

14. E. M. Pouget, P. H. H. Bomans, J. A. C. M. Goos, P. M. Frederik, G. de With and N. A. J. M. Sommerdijk, *Science*, 2009, **323**, 1455–1458.
15. O. Galkin, K. Chen, R. L. Nagel, R. E. Hirsch and P. G. Vekilov, *Proc. Natl. Acad. Sci. USA*, 2002, **99**, 8479–8483.
16. D. Vivares, E. Kaler and A. Lenhoff, *Acta Crystallogr. Sect. D: Biol. Crystallogr.*, 2005, **61**, 819–825.
17. P. G. Vekilov, *Cryst. Growth Des.*, 2004, **4**, 671–685.

Radoljub Ristic asked: In your experiment, strictly speaking, after L–L separation the denser liquid droplet has three different "phases". One of them is L–L interface, another one is the liquid–glass cover interface and the third one is the bulk of the precursor. My question is, can we determine experimentally and unambiguously the origin of the nucleation of an observed crystalline particle(s)? Which of the three phases is likely to be most creative in terms of nucleation and why?

Peter Vekilov answered: In several cases nucleation of crystals in dense liquid droplets has been directly observed: several examples of colloids,[1,2] an organic system,[3] sickle cell haemoglobin,[4] calcite[5] and others. In many other cases, the evidence of the action of two-step nucleation is indirect, but still compelling: protein crystals,[6,7] calcite,[8] polymers and others. The second question, as I understand it, emphasizes the difference between the bulk and the interface of the precursor clusters. This is a valid and important issue. In the cases where the direct nucleation was observed, the precursor clusters were small and consisted of several tens of molecules or particles. This makes differentiation between their bulk and their interface moot. Clearly, more work is needed to address this question.

1. M. E. Leunissen, C. G. Christova, A-P. Hynninen, C. P. Royall, A. I. Campbell, A. Imhof, M. Dijkstra, R. van Roij and A. van Blaaderen, *Nature*, 2005, **437**, 235–240.
2. J. R. Savage and A. D. Dinsmore, *Phys. Rev. Lett.*, 2009, **102**, 198302.
3. K. Harano, T. Homma, Y. Niimi, M. Koshino, K. Suenaga, L. Leibler and E. Nakamura, *Nat. Mater.*, 2012, **11**, 877–881.
4. O. Galkin, K. Chen, R. L. Nagel, R. E. Hirsch and P. G. Vekilov, *Proc. Natl. Acad. Sci. USA*, 2002, **99**, 8479–8483.
5. E. M. Pouget, P. H. H. Bomans, J. A. C. M. Goos, P. M. Frederik, G. de With and N. A. J. M. Sommerdijk, *Science*, 2009, **323**, 1455–1458.
6. M. Sleutel and A. E. van Driessche, *Proc. Natl. Acad. Sci. USA*, 2014, **111**, E546–553.
7. P. G. Vekilov, *Crys. Growth Des.*, 2010, **10**, 5007–5019.
8. D. Gebauer, A. Völkel, H. Cölfen, Stable Prenucleation Calcium Carbonate Clusters, *Science*, 2008, **322**, 1819–1822.

Fajun Zhang asked: In your paper, you mentioned that the clusters have a concentration of up to 450 mg ml^{-1}, that is about 33% in volume fraction. Based on established phase diagram of lysozyme, this volume fraction is most likely in the arrested state. If this is true, certainly they cannot act as a precursor for protein crystallization. However, as the clusters are about 100 nm in diameter, as we know that on the nanoscale, a significant fraction of molecules is at the interface with a higher flexibility than those in the bulk, so, it seems still possible for them to trigger nucleation; can you comment?

Peter Vekilov responded: Indeed solutions of many proteins at concentrations above 150 mg ml^{-1} form gels of high viscosity in which most dynamics is slowed or arrested. On the other hand, there are numerous examples of solutions of other proteins and small molecules (haemoglobin is one of them[1]) that are fluids with moderate viscosity even at very high concentrations. Thus arrested dynamics in the clusters is not an inevitable necessity.

I fully agree with your second point. The theoretical model of the protein-rich clusters[2] predicts a periphery in which the properties transition from those in the cluster core to those in the solution. Thus, even if the matter in the cluster core is fully gelled, the one in the periphery may still be fluid and may be the likely location of crystal nucleation. In support of this conjecture, we note that an earlier paper on the nucleation of protein crystals assisted by macroscopic dense liquid noted that crystals are found predominantly on the surface of the dense phase domains.[3]

1. O. Galkin, K. Chen, R. L. Nagel, R. E. Hirsch and P. G. Vekilov, *Proc. Natl. Acad. Sci. USA*, 2002, **99**, 8479–8483.
2. W. Pan, P. G. Vekilov and V. Lubchenko, *J. Phys. Chem. B*, 2010, **114**, 7620–7630.
3. Y. G. Kuznetsov, A. J. Malkin and A. McPherson, *J. Cryst. Growth*, 2001, **232**, 30–39.

Stéphane Veesler opened a general discussion of the paper by Fajun Zhang: I do not agree that the fact that the growth rate goes through a maximum is evidence that crystals are growing inside the aggregates. It can be due to the growth of crystals in presence of a LLPS (liquid–liquid phase separation). There is a paper from E. Vlieg *et al.*[1] showing different growth rates when crystals are growing in the presence and in the absence of a dense phase due to a LLPS. Therefore, your kinetic measurements can be explained by the growth of the crystals at the expense of the dense phase followed by the growth in homogeneous solution when all the dense phase has disappeared.

1. M. C. R. Heijna, W. J. P. van Enckevort, and E. Vlieg, *Phys. Rev. E: Stat., Nonlinear, Soft Matter Phys.*, 2007, **76**, 011604.

Hugo Meekes added: If we have an intermediate phase, stable or metastable, like the aggregates you observe, one could still have aggregation resulting in this phase simultaneously with the final crystalline phase forming somewhere else in the sample as the concentration in the sample will not be homogeneous. So I do not agree that your two slopes in parallel up to a maximum, seemingly at the same time, are a proof for the nucleation of the final phase directly from the intermediate phase. Do you agree?

Fajun Zhang replied: Both questions in fact describe a scenario similar to the parallel growth pathway as we discussed in Fig.1 of our paper. This means that crystal nucleation occurs independent of the LLPS aggregates and the crystal would grow at the expense of the dense phase or aggregates.

It is worth noting that the coincidence of the maxima of the crystallization rate and of the amount of intermediate does not prove the nucleation directly from the intermediate. However, the coincidence is consistent with a two-step picture and, in combination with the plateau in the crystallization rate, hard to explain by parallel scenarios without *ad hoc* modifications. As discussed in our paper and also a previous publication,[1] if a parallel scenario were the case, the initial nucleation rate would only depend on the condition of the dilute protein solution and the dense phase or the aggregates only serve as a reservoir. If we assume that the crystal growth can consume all the available material, then the overall growth kinetics would feature two stages with different slopes. However, this cannot explain our experimental data where a plateau is reached before a further fast growth regime. Our experimentally determined growth kinetics suggests a different picture of crystallization. In the early stage, the nucleation rate (and overall crystal growth) depends on the dense phase or aggregates; in the second stage, when the dense phase or aggregates are consumed, or as described in our paper, when the crystal growth front is exposed to the dilute phase, then the crystal growth kinetics depends mainly on the dilute phase. This scenario can explain the growth kinetics presented in our paper.

In addition, we have performed another experiment to demonstrate the role of the aggregates on nucleation (see Fig. 2). For the same sample solutions, we removed the aggregates by centrifugation quickly after sample preparation. The resulting clear solution can still crystallize, but much slower and with significantly less crystals. This experiment demonstrates that the crystal nucleation is closely related to the aggregates.

The number of nucleation events also increases in the beginning and reaches a maximum at the maximum amount of the amorphous phase. In addition, the faster growth in a later step leads to a plateau in the crystallization rate. This cannot be easily explained by the depletion of the surrounding of the crystals as it was described in the paper by Heijna *et al.*[2] The crystal fraction should rather be a curve starting with one slower slope and then turning up to a faster slope. Furthermore, even if the surrounding medium changed from LLPS (which was never observed in the system BLG+$CdCl_2$) to homogeneous, it would do so because of consumption of molecules due to crystal growth – in particular, the homogeneous phase would have a much lower mean concentration than the initial LLPS medium. This is in contrast to the case studied in the paper of Heijna *et al.*[2] (with a similar average concentration), and the significant concentration difference would most likely slow down the crystallization much faster than the slight increase of the rates from LLPS to homogeneous reported by Heijna *et al.*

In conclusion, we remain convinced that our model is appropriate, but we are not in a position to rule out all alternative explanations.

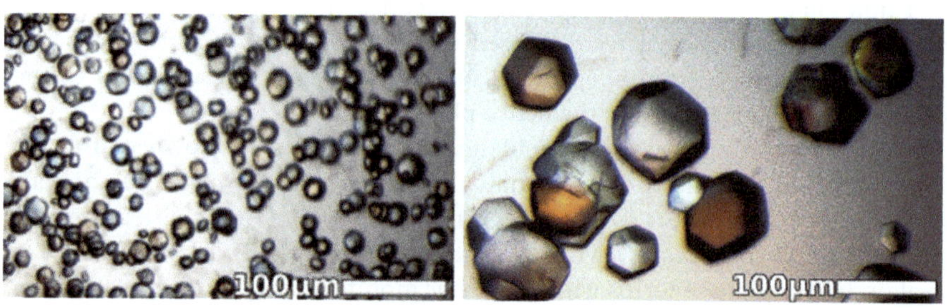

Fig. 2 Crystallization of 3.3 mg ml^{-1} BLG with 7 mM $CdCl_2$ at 20 °C. Left: without centrifuging about 2.5 h. Right: with centrifuging after ~20 h. Both pictures show the state where the crystals do not grow anymore to a large extent.

1. A. Sauter, F. Roosen-Runge, F. Zhang, G. Lotze, R. M. J. Jacobs, F. Schreiber, Real-Time Observation of Nonclassical Protein Crystallization Kinetics, *J. Am. Chem. Soc.*, 2015, **137**, 1485–1491.
2. M. C. R. Heijna, W. J. P. van Enckevort, and E. Vlieg, *Phys. Rev. E: Stat., Nonlinear, Soft Matter Phys.*, 2007, **76**, 011604.

Ian Rosbottom asked: Since the aggregates only appear at high supersaturations, it seems like a kinetic amorphous phase that then transforms quicker than nucleation of the stable crystal happens. At low supersaturation the thermodynamically stable crystal appears, do you think the aggregates are an absolutely necessary precursor to the crystal or are they kinetic product of high supersaturation?

Fajun Zhang replied: We have not claimed that aggregates are a necessary precursor, but we have argued and demonstrated that they occur as one precursor under the specific conditions. From the real-time SAXS data, we observe the evolution of a (not yet crystalline) structural feature within the aggregates which is then followed by crystal nucleation. From our real-time SAXS and optical microscope results, we consider these aggregates as "precursor" for crystal nucleation. The thermodynamic nature of the intermediate is open for future investigation. While it is for sure metastable with respect to crystallization and thus only is observed for kinetic reasons, the exact relation to a liquid–liquid phase separation has not been fully established yet. In particular, for high supersaturation, the formation of the intermediate may also occur *via* a partially arrested spinodal decomposition (which results in a transient gel state) without a nucleation step. This corresponds to a special case of the "two-step" nucleation mechanism.

Hugo Meekes returned to the paper by Peter Vekilov: If first you have a kinetic phase followed by a phase transition to the final phase, what is the difference between this two step mechanism and the classic Ostwald rule of steps, as Ostwald could not make a difference between an amorphous phase and a crystalline phase?

Peter Vekilov responded: Ostwald[1] noted that sometimes the crystals of a polymorph with higher solubility, *i.e.*, with lower crystallization free energy, form earlier than crystals of a polymorph with lower solubility, *i.e.*, with higher crystallization free energy. Thus, a less stable phase forms faster than a more stable one. This was an empirical observation and its pretense of generality not only to phase transitions but to all chemical processes was poorly justified. When we think of the Ostwald rule of stages, we often mean the formulation of Stranski and Totomanov,[2] who demonstrated that a less stable phase may have a lower nucleation barrier and form first. Neither the original nor the Stranski–Totomanov formulation of the Ostwald rule of stages assume any participation of the earlier appearing phase in the nucleation of a subsequent phase. In contrast, the two-step mechanism states that clusters form first and serve as preferred locations for the formation of crystal nuclei, *i.e.*, crystal nucleation inside the clusters is much faster than outside them.

1. W. Ostwald, *Z. Phys. Chem.*, 1897, **22**, 289–330.
2. I. Stranski and D. Totomanov, *Z. Phys. Chem.*, 1933, **A163**, 399–408.

Roger Davey commented: I am confused about how your results demonstrate that the liquid phase is actually part of a two-step nucleation process. How do we know that this process is not akin to a polymorphic transition in which the liquid dissolves at the expense of the crystalline form?

Peter Vekilov replied: In several cases nucleation of crystals in dense liquid droplets has been directly observed: several examples of colloids,[1,2] an organic system,[3] sickle cell hemoglobin,[4] calcite[5] and others. In many other cases, the evidence of the action of two-step nucleation is indirect, but still compelling: protein crystals,[6,7] calcite,[8] polymers and others. For instance, in ref. 6–8, three powerful arguments in favor of the two step mechanism are presented: 1. The dependences of the homogeneous nucleation rate on concentration and temperature are broken and non-monotonic, in stark contrast with the prediction of the classical nucleation theory assuming direct nucleation, but are in agreement with a two-step model; 2. The value of the pre-exponential factor of the nucleation rate expression is 10 orders of magnitude lower than the prediction of the classical theory assuming direct nucleation – the pre-exponential factor is a linear function of concentration and it is very hard to image anything that could bring it so low short of a dramatically different nucleation mechanism. On the other hand, the low pre-exponential factor is natural consequence of crystal nucleation occurring in clusters that occupy about 10^{-6} of the solution volume and have viscosity around 100 cP; 3. The rates of nucleation strictly scale with the volume occupied by the dense liquid clusters.

Note that the clusters observed in protein solutions capture a very low fraction of the protein, *ca.* 10^{-5}, so that their dissolution has a minor effect on the nucleation and growth of crystals that might nucleate independently of them. This is contrast to polymorphic transitions, in which the first formed polymorph captures a major part of the solute.

1. M. E. Leunissen, C. G. Christova, A-P. Hynninen, C. P. Royall, A. I. Campbell, A. Imhof, M. Dijkstra, R. van Roij and A. van Blaaderen, *Nature*, 2005, **437**, 235–240.
2. J. R. Savage and A. D. Dinsmore, *Phys. Rev. Lett.*, 2009, **102**, 198302.
3. K. Harano, T. Homma, Y. Niimi, M. Koshino, K. Suenaga, L. Leibler and E. Nakamura, *Nat. Mater.*, 2012, **11**, 877–881.
4. O. Galkin, K. Chen, R. L. Nagel, R. E. Hirsch and P. G. Vekilov, *Proc. Natl. Acad. Sci. USA*, 2002, **99**, 8479–8483.
5. E. M. Pouget, P. H. H. Bomans, J. A. C. M. Goos, P. M. Frederik, G. de With and N. A. J. M. Sommerdijk, *Science*, 2009, **323**, 1455–1458.
6. M. Sleutel and A. E. van Driessche, *Proc. Natl. Acad. Sci. USA*, 2014, **111**, E546–553.
7. P. G. Vekilov, *Crys. Growth Des.*, 2010, **10**, 5007–5019.
8. D. Gebauer, A. Völkel, H. Cölfen, Stable Prenucleation Calcium Carbonate Clusters, *Science*, 2008, **322**, 1819–1822.

Jan Sefcik said: It is often assumed that dense liquid clusters and similar phenomena present in two-step nucleation scenarios represent separate phases because the clusters are so large. What is the evidence that these are separate phases and not just giant clusters in the mesostructure liquid? How could one prove this? For example, we have provided experimental evidence for equilibrium mesostructured clusters in aqueous solutions of glycine and DL-alanine[1,2] and water-isopropanol solutions of DL-valine (Jawor-Baczynska *et al.*, DOI: 10.1039/c4fd00262h) and we believe that these systems comprise single liquid (mesostructured) phases, as the clusters observed seem not to be able to exist as macroscopic phases. It is important to know the number of phases and whether these are separate phases when using the Gibbs phase rule.

1. A. Jawor-Baczynska, J. Sefcik and B. D. Moore, *Cryst. Growth Des.*, 2013, **13**, 470–478.
2. A. Jawor-Baczynska, B. D. Moore, H. S. Lee, A. V. McCormick and J. Sefcik, *Faraday Discuss.*, 2013, **167**, 425–440.

Peter Vekilov replied: The Gibbs definition of a phase has no constraints on the size of the phase domains and explicitly specifies that a phase may be contained in multiple domains.[1–3] Thus, insisting that a domain is of size greater than the diffraction limit so that it can be detectable by optical microscopy seems unnecessary. In the spirit of Gibbs, I would argue that a phase needs to have clearly recognizable bulk, distinct from the interface. Thus, clusters smaller than 10 molecules, in which all constituents are in contact with the solution, are clearly not a new phase. On the other hand, the protein-rich clusters contain about 100 000 molecules. A spherical cluster that contains 100 000 molecules will have about 60 molecules along a diameter. Since in solutions of moderate ionic strength (0.1 M and higher) the range of interactions is 1 nm and shorter, only the first two or three molecules closest to the cluster surface will be affected by the proximity of the dilute solution and may be considered to belong to an interfacial layer. The other 55 or so will have bulk properties. Since these are the majority of the protein molecules in the cluster, we view the clusters as a phase.

Clusters in different systems may vary in their properties. One needs to consider the presence of molecules whose properties are not affected by the proximity to the cluster surface to decide whether clusters of interest can be viewed as an individual phase or not. The question of whether clusters are a separate phase or not also depends on the questions that we are asking about them. Some questions can be answered without assumptions on that matter; others cannot.

1. J. W. Gibbs, *The Scientific Papers of J. Willard Gibbs, Vol. 1: Thermodynamics*, Oxbow Press, Woodbridge, Connecticut, USA, 1993.
2. J. W. Gibbs, *Trans. Connect. Acad. Sci.*, 1878, **3**, 343–524.
3. J. W. Gibbs, *Trans. Connect. Acad. Sci.*, 1876, **3**, 108–248.

Richard Sear asked: If your clusters of protein molecules are a well-defined thermodynamic phase (liquid or metastable polymorph) they should have a well defined solubility (at given temperature, salt concentration *etc*). Do you think that is correct? If so have you measured this solubility?

Peter Vekilov responded: The clusters do not have a unique solubility, *i.e.*, a solution concentration that would lead to cluster growth if exceeded and to cluster dissolution if not reached. Because of the chemical reaction

$$n \text{ monomer} \leftrightarrows \text{oligomer}$$

clusters form even at low solute concentrations and the amount of protein captured in the clusters is an exponential function of the solution concentration.[1,2]

In detail, solubility is defined as the concentration or activity of a solution in equilibrium with another phase. If two phases are in equilibrium, the pressure and temperature and the chemical potentials of all components in them are equal. If a single component transitions between phases in equilibrium or several components transition independently of each other, then we can define unique solubilities, as illustrated below for the case of a single solute that transitions between a solution and a dense liquid, or a solution and crystal.

For the chemical potential of the solute, component 2, in a dense liquid–dilute solution equilibrium, we get

$$\mu_2^{\text{dense}} = \mu_2^{\text{dilute}} \text{ and}$$

$$\mu_2^{\text{dense}} = \mu_{2,0}^{\text{dense}} + RT\ln\gamma_{2,\text{dense}}C_{2,\text{dense}} \text{ and}$$

$$\mu_2^{\text{dilute}} = \mu_{2,0}^{\text{dilute}} + RT\ln\gamma_{2,\text{dilute}}C_{2,\text{dilute}}.$$

Since, in most cases, the dense and dilute phases only differ in solute concentration,

$$\mu_{2,0}^{\text{dense}} = \mu_{2,0}^{\text{dilute}},$$

we get for the concentrations of the two phases in equilibrium

$$\ln\gamma_{2,\text{dense},e}C_{2,\text{dense},e} = \ln\gamma_{2,\text{dilute},e}C_{2,\text{dilute},e};$$

details about the dense liquid–dilute solution are available in ref. 3. At each temperature and pressure the concentrations of the two phases in equilibrium are uniquely defined. $C_{2,\text{dilute},e}$ may be considered to be the solubility of the dense phase.

In a crystal-solution equilibrium, we get

$$\mu_2^{\text{crystal}} = \mu_2^{\text{solution}} \text{ and}$$

$$\mu_2^{\text{crystal}} = \mu_{2,0}^{\text{crystal}} \text{ and } \mu_2^{\text{solution}} = \mu_{2,0}^{\text{solution}} + RT\ln\gamma_2C_2,$$

where $\mu_{2,0}^{\text{crystal}}$ and $\mu_{2,0}^{\text{solution}}$ are the standard values of the respective chemical potentials. Since they both do not depend on the concentration C_2, we get

$$\ln\gamma_{2,e}C_{2,e} = (\mu_{2,0}^{\text{crystal}} - \mu_{2,0}^{\text{solution}})/RT,$$

i.e., an equilibrium activity $\gamma_{2,e}C_{2,e}$ that is only a function of temperature and pressure.

The protein-rich clusters are in equilibrium with the solution,[1,2] *i.e.*,

$$\mu_2^{\text{cluster}} = \mu_2^{\text{solution}} \text{ and}$$

$$\mu_{2,0}^{\text{cluster}} + RT\ln\gamma_{2,\text{cluster}}C_{2,\text{cluster}} = \mu_{2,0}^{\text{solution}} + RT\ln\gamma_{2,\text{solution}}C_{2,\text{solution}}.$$

However, in this case, $\mu_{2,0}^{\text{cluster}}$ is not constant, but depends on the composition of the clusters that likely includes protein oligomers, and the exchange of the oligomers with the solution and their conversion to solute monomers. Hence, a $C_{2,\text{solution}}$ that only depends on temperature and pressure and would represent the cluster solubility cannot be defined.

1. Y. Li, V. Lubchenko, M. A. Vorontsova, L. Filobelo and P. G. Vekilov, *J. Phys. Chem. B*, 2012, **116**, 10657–10664.
2. V. Uzunova, W. Pan, V. Lubchenko and P. G. Vekilov, *Faraday Disc.*, 2012, **159**, 87–104.
3. D. N. Petsev, X. Wu, O. Galkin and P. G. Vekilov, *J. Phys. Chem. B*, 2003, **107**, 3921–3926.

Allan Myerson opened a general discussion on the theme of molecular self-assembly and clustering in nucleation processes: If you do experiments and use simple molecules such as inorganic salts which have very narrow metastable zones they will be easy to nucleate and have short induction times. If you go to simple organic molecules, at similar supersaturations, the metastable zone becomes wider and the nucleation induction times are longer.

If we look at a simple two-step nucleation mechanism we can say that the second step which is arrangement into the correct lattice structure is fast for simple molecules, which generally have more symmetry in their structures, and longer for more complex molecules, which have less symmetric crystal structures.

This increase in nucleation induction time and metastable zone width with molecular complexity is something which many people have observed over the years.

Roger Davey replied: I think this is a point well taken. I would make one further comment however – do these observations really require a new theory for their interpretation? The rate reductions you talk about could quite easily be accounted for in CNT through the impact of molecular size and complexity on the attachment frequency. What we really need are more detailed kinetic data so that we can properly confront our theories rather than making assumptions. In this case an ancillary question might be "do the growth rates follow the same trends?" – if they do then perhaps this is indeed more about cluster growth than cluster structure.

Peter Vekilov commented: A very intriguing question is under what conditions one may expect two-step nucleation to turn into direct nucleation? Is there a parameter that can be varied to select either direct or two-step nucleation? The selection of the nucleation mechanism is governed by the rate associated with all the possible mechanisms, we see the action of the fastest mechanism. This is a direct corollary of the principle of fastest increase of the entropy of the universe. It seems that in many cases the precursor clusters are in equilibrium with the solution, thus, the chemical potential of the solute in the clusters is equal to that in the solution and the supersaturation that drives crystal nucleation is equal for both pathways. The rate of two-step nucleation is determined by the volume occupied the precursor clusters; the viscosity of the liquid in them; and the surface free energy between the liquid in the clusters and the crystal.[1,2] The rate of direct nucleation is determined by the solution viscosity, which is lower than in the clusters, and the surface free energy between the solution and the crystal nucleus, which is significantly higher than the one in the clusters.[3] Note the nucleation rate is a linear function of the solution viscosity and exponential cubic of the surface free energy; this discrepancy ensures the selection of two-step nucleation in many cases. Owing to the higher surface free energy, increasing supersaturation lowers the barrier for direct nucleation more strongly than the barrier for two-step nucleation. This effect is counteracted by the increase of the cluster volume at higher solution concentration and supersaturation.[4,5] Still, direct nucleation appears more likely at very high supersaturations.

1. W. Pan, A. B. Kolomeisky and P. G. Vekilov, *J. Chem. Phys.*, 2005, **122**, 174905.
2. P. G. Vekilov, *Cryst. Growth Des.*, 2010, **10**, 5007–5019.
3. D. Kashchiev, *Nucleation. Basic theory with applications*, Butterworth, Heinemann, Oxford, 2000.
4. W. Pan, P. G. Vekilov and V. Lubchenko, *J. Phys. Chem. B*, 2010, **114**, 7620–7630.
5. V. Uzunova, W. Pan, V. Lubchenko and P. G. Vekilov, *Faraday Discuss.*, 2012, **159**, 87–104.

Fajun Zhang replied: This is a very interesting and important question! Several works in the literature have demonstrated that specific interactions (patchy interaction) may be important for the formation of the intermediate during protein crystallization. Here we want to mention the work by S. Whitelam,[1,2] who used a lattice model of anisotropic particle systems to study the influence of the interplay between nonspecific and specific interactions on the nucleation pathways. Their simulation results indicate that for a given specific interaction, the strong nonspecific interactions lead to liquid-like precursors for crystal nucleation. If the nonspecific interaction is weak, crystallization can proceed *via* ordered nuclei. Based on the simulation results, they propose that one can in principle control the crystallization pathway by altering independently the nonspecific and specific interactions.

Our system may indeed provide a good experimental model system to address this prediction,[3,4] in connection with a consistent theoretical description using patchy particles.[5] A study on the protein BLG crystallized in the presence of YCl_3 and $ZnCl_2$ has demonstrated that the yttrium and zinc ions bridge the neighbouring proteins and create new contacts in the lattice.[3,4] Therefore, the salt concentration provides a control parameter for the specific interactions.[5] The nonspecific interaction on the other hand can be tuned by temperature. Indeed, we observed experimentally that at low salt concentration, the crystal growth follows the classical picture. Increasing the salt concentration to a certain region, the protein solutions first become turbid, forming metastable protein clusters or dense liquid droplets depending on protein concentration, later crystals appear. Crystal growth consumes the intermediates and leaves only crystals in the end.[3,4]

1. S. Whitelam, Control of Pathways and Yields of Protein Crystallization through the Interplay of Nonspecific and Specific Attractions, *Phys. Rev. Lett.*, 2010, **105**(8), 088102.
2. T. K. Haxton, S. Whitelam, Design rules for the self-assembly of a protein crystal, *Soft Matter*, 2012, **8**(13), 3558–3562.
3. F. Zhang, G. Zocher, A. Sauter, T. Stehle, F. Schreiber, Novel approach to controlled protein crystallization through ligandation of yttrium cations, *J. Appl. Crystallogr.*, 2011, **44**, 755–762.
4. A. Sauter, M. Oelker, G. Zocher, F. Zhang, T. Stehle, F. Schreiber, Nonclassical Pathways of Protein Crystallization in the Presence of Multivalent Metal Ions, *Cryst. Growth Des.*, 2014, **14**(12), 6357–6366.
5. F. Roosen-Runge, F. Zhang, F.; Schreiber, R. Roth, Ion-activated attractive patches as a mechanism for controlled protein interactions, *Sci. Rep.*, 2014, **4**, 7016.

Ake Rasmuson remarked: With reference to the comment from Peter Vekilov: is it possible for the same compound to have a classical nucleation mechanism at high supersaturation when the critical nucleous is small and a two-step mechanism at low supersaturation when the critical nucleous is large or *vice versa*?

Peter Vekilov responded: I would say yes, this seems like a realistic scenario.

Jamshed Anwar presented some results on the role of clusters in secondary nucleation: I would like take the opportunity to present our recent results.[1] Secondary nucleation is nucleation that occurs in the presence of crystalline solute material. It forms the basis for most industrial crystallization processes where crystal seeds are employed to promote bulk crystallization. It is particularly important in continuous crystallization, where a certain amount of the product crystal is always present to induce nucleation in the fresh feed.

What actually happens in secondary nucleation has eluded us for some time now. Mechanical attrition of crystal (large) seeds is known to induce nucle.[1] However, more subtle effects such as movement of a crystal in a moderately supersaturated solution or the movement of the solution past a stationary crystal surface can also produces copious nuclei,[2] and how this occurs has remained unexplained.

We have employed molecular dynamics (MD) simulations, wherein the trajectories of interacting molecules are simulated. The solute and solvent molecules were represented by simple, single-particle models based on the Lennard-Jones (LJ) interaction. Such simple models are appropriate since the nucleation processes of interest are generic, being observed in a wide class of materials. An implication is that the findings are also generic and should be applicable to a wide class of materials.

A series of large-scale MD simulations were carried out in which a crystal slab representing a seed crystal, was immersed in a set of solutions of increasing supersaturation (see Fig. 3).

Three distinct regimes were observed. At low supersaturations, there was little clustering of the solute molecules and individual solute molecules attached themselves to the seed surface resulting in crystal growth as expected. At very high supersaturations, the solute molecules very quickly formed clusters which then nucleated. This too is expected as the chemical potential is sufficient to induce spontaneous nucleation in the bulk solution.

The interesting results come from the intermediate supersaturation. Here we get formation of solute clusters which tend to remain stable but do not nucleate spontaneously. When some of these clusters make contact with the crystal seed surface, they immediately nucleate. Interestingly, any clusters that make contact with the newly formed nuclei also nucleate.

Therefore, the nuclei breeding journey in secondary nucleation involves the molecules in solution forming clusters which, on attaching to the crystal seed surface, form new nanocrystals that are loosely attached to the seed surface. The solvent can shear the weakly-tethered new crystals from the seed crystal surface allowing the surfaces to be further available for a repeat process and the new crystallites to go on to act as seed crystals themselves.

Focusing on clustering, we observe a spectrum of behaviour as a function of supersaturation. At low supersaturations, there appears to be no clustering and

Fig. 3 Snapshot from a molecular dynamics simulation trajectory (135 ns into the simulation) showing the nucleation of a solute cluster on coming in contact with the crystal seed surface and subsequent nucleation of another cluster on coming in contact with the previously nucleated cluster. The solvent molecules are not shown.

molecules are largely present as individual entities dispersed in the solvent. As the supersaturation is increased, the solutes molecules show an increasing tendency to cluster and reveal exchange of solute molecules with the bulk solution. At higher supersaturations still, the molecules form clusters which aggregate and form larger structures.

1. J. Anwar, S. Khan, L. Lindfors, *Angew. Chem., Int. Ed.*, DOI: 10.1002/anie.201501216, in press.
2. Y. Cui, A. S. Myerson, *Cryst. Growth Des.*, 2014, **14**, 5152–5157.
3. (a) M. A. Larson, *Chem. Eng. Commun.*, 1981, **12**, 161–169; (b) H. E. C. Powers, *Nature*, 1960, **188**, 289–291.

Roger Davey commented: I think that this is such nice work. I can still remember conferences in the 1980s when chemical engineers (Larson, Estrin, DeJong) talked about nucleation induced by shear and really we had little idea how this worked or if was even possible. These simulations confirm all that these ideas were in fact correct.

Peter Vekilov remarked: Your simulations show that secondary nuclei are created according to the following scenario: amorphous clusters form in the growth medium; the clusters, by growth and diffusion, reach the surface of a crystal; the influence of the periodically ordered field of the crystal induces fast crystallization of the cluster; since the cluster is still only weakly attached to the crystal, it breaks off creating an additional crystal nucleus. There are numerous experimental observations of amorphous clusters that attain crystal order upon contact with a crystal, two examples are given in ref. 1 and 2. However, one of your results appears very intriguing: the abundance of clusters in the solution. What assumptions in your model lead to this clustering? How general is this presence of clusters? What are the typical cluster sizes and do they vary with solution conditions? What fraction of your solute is entrapped in the clusters? What experimental systems should be expected to exhibit similar clustering?

1. O. Gliko, N. Neumaier, W. Pan, I. Haase, M. Fischer, A. Bacher, S. Weinkauf and P. G. Vekilov, *J. Am. Chem. Soc.*, 2005, **127**, 3433–3438.
2. D. Li, M. H. Nielsen, J. R. I. Lee, C. Frandsen, J. F. Banfield and J. J. De Yoreo, *Science*, 2012, **336**, 1014–1018.

Jamshed Anwar responded: Thank you for alerting me to the experimental observations of amorphous clusters that attain crystallinity upon contact with a crystal.

The molecules in the simulations are single-particle models based on the Lennard-Jones interaction. The MD simulations are carried out at constant temperature and pressure conditions and the molecules are free to sample their phase space under those conditions. There are no other assumptions. The size of our molecules is taken to be 0.47 nm. Given that scaling, the cluster size range varies from a few nm to about 20 nm.

The clustering behaviour depends on the supersaturation. At low supersaturations the molecules remain largely independent and in the presence of the crystal seed give rise to crystal growth. The tendency to cluster increases at higher supersaturations. The models are generic and hence the results should be applicable to a wide spectrum of materials including proteins.

As to the extent of solvent incorporation in the clusters, this depends on the strength of the interaction between the solute and solvent. We are currently looking into this.

Celso Aparecido Bertran opened a general discussion of the paper by Helmut Cölfen: In the experiments with Analytical Ultracentrifuge Method (AUC), was the size distribution of the Ag^0 particles observed in Fig 3 and Table 1 of your paper a consequence of the presence of the micelles of CTAB surfactant, that was added to the initial solution of Ag^+ ions? Could the micelles act like as "nanoreactors" to form the 8 different Ag-core size aggregates observed?

Helmut Cölfen responded: We do not think that the very small species, which we report here, have CTAB bound at all because of the molecular dimensions. CTAB will certainly form micelles in the aqueous environment we work in, but the micelle core will then be hydrophobic. This means that the silver salts and the reductant will not enter the micelle core and therefore the micelles will not act as a nanoreactor. If CTAB is bound to the Ag clusters/particles in water, it should form a double layer with dimensions exceeding the size of the smallest Ag cores. This would create such bending as well as steric pressure that a CTAB double layer is not seen to stabilize the smallest nanoparticles.

Samuel Booth communicated: You mention that in your titration study, due to the time required and concentration, relative to the AUC experiment, that you replaced CTAB with citrate. Would it be possible to perform this reaction at a higher temperature to avoid CTAB precipitation? Also is there a reason why CTAB was chosen in preference to citrate in your AUC experiment?

Helmut Cölfen communicated in reply: In the AUC experiment, CTAB was chosen as a stabilizer as it worked well with gold nanoparticles (these results have not been published). AUC experiments were also performed with citrate as the stabilizer, but the obtained particles seemed to be a lot larger than with CTAB, so CTAB seemed to be a better stabilizer.

Regarding the setup, it is possible to perform titration experiments at higher temperatures, and the solubility of CTAB increases with the temperature, so in principle this should be possible. But there would probably arise problems from the temperature dependent decomposition of sodium borohydride.

Alan Hare communicated: At first sight it may seem natural to propose regular icosahedral geometry for a M_{13} cluster, because $C + V = 13$, where the respective number of closed and external vertices are $C = 1$ and $V = 2(1 + 5)$. This is not the only geometric possibility, however. (For example, a tetrahedral arrangement $V = 4(3 + \delta)$ could exist, in which $\delta_i = \{[0, 1], twice\}_{i=1}^{4}$. In the special case of all $\delta_i = 0$, again $V = 12$. Incidentally: like the cube, though unlike the regular icosahedron, the Biliński rhombic dodecahedron, comprising four golden rhombohedra, is a 3d space-filler.)

It seems that there may exist clusters $M_{n(r)}$ where $r = \{i + 2\}_{i=1}^{4}$, and the corresponding Fibonacci numbers are $F(r) = \{2, 3, 5, 8\}$. In this case, many reported $Ag_{n(r)}$ clusters appear to fit an equation of the kind,

$$|G_1(r)\rangle = \Phi_{F(r)+1}\,|G_1(1)\rangle + \Phi_{F(r)}\,|G_2(1)\rangle;$$

where strangely, the $\Phi_{F(r)}$ themselves are the Fibonacci numbers $\{\Phi_2 = 1, \Phi_3 = 2, \Phi_5 = 5, \Phi_8 = 21\}$. In this series there are no missing numbers to explain: both $|G_1(1)\rangle$ and $|G_2(1)\rangle$ would represent Ag_1; and n_r would equal $\Phi_{F(r)} + \Phi_{F(r)+1}$, or the series $\{n_2 = 3, n_3 = 5, n_5 = 13, n_8 = 55\}$. This system resembles the pair of simultaneous equations describing "Penrose in 3d", itself a special case $n = 10 - i$ $(i = 1,2)$ of the enlargement

$$|G_{Vi}(x + 1)\rangle = \Phi_{n+1}|G_{V1}(x)\rangle + \Phi_n|G_{V2}(x)\rangle,$$

in which each polyhedron is self-similar and enlarged by the linear factor, $\tau^3 =$ 2+sqrt(5); so that the fractal dimension $d = 3\kappa \ln\tau / \ln 2$, suggesting asymmetric Cantor dust in 3d, of capacity dimension $\kappa_{capacity} = 3$. This peculiar characterisation raises the possibility of a quasi-crystal which may be "nowhere dense". A large entropic contribution might favour it. My conjecture is that neither polyhedron of a pair $\{|G_{V1}(x)\rangle, |G_{V2}(x)\rangle\}$ in which both are nowhere dense may be space-filled contiguously to every edge by similar polyhedra $|G_{V1}(x - 1)\rangle$ or $|G_{V2}(x - 1)\rangle$ in any combination ($x \geq 2$). I am looking then to see if we can test for the physical existence of Cantor dust.

It would be interesting to know what atom numbers might be derived if we were to assume golden rhombohedra instead of spherical particles, with face-sharing or perhaps edge- or vertex-sharing instead of cubic packing.

Helmut Cölfen communicated in reply: In the ultracentrifugation experiments, we obtain the hydrodynamic size of the equivalent sphere. This means that we cannot distinguish between different similar-sized polyhedra from the hydrodynamic data. Therefore we used the most dense sphere packing for our calculation of the atom numbers for the detected species knowing that they represent an unrealistic dense packing case. Nevertheless, the atom numbers already allow a determination of the size range of the early nucleation species. However, if a structure would be suggested by modelling, we could calculate the hydrodynamic size and compare it with the data we obtained. If it fits to a species, we can say that this detected species might be the modelled one but there may be many other equivalent possibilities for fitting structures too. However, since we can obtain UV/Vis spectra for the detected species, we might have a unique identifier at hand to work out which early nucleation species are detected.

Alan Hare communicated: Can the sphere be regarded as the circumsphere of a candidate polyhedron?

For example, if the M_{13} cluster were to be ascribed the $C + V$ vertices of a regular icosahedron ($V = 12$), then its edge length $a = R/\text{sqrt}((5+\text{sqrt}(5))/8)$, where R is the circumradius.

Helmut Cölfen communicated in reply: The hydrodynamic equivalent sphere is not exactly the circumsphere of a polyhedron. It is smaller since it represents the size of the sphere, which hydrodynamically behaves the same as the polyhedron. For polyhedra, it will be a good assumption to use the sphere with the same volume as the polyhedra.

Alan Hare said: I see. Thank you, Professor.

So the distance between nearest neighbour atoms in the candidate icosahedral cluster would then depend on the hydrodynamic radius:

$$5(3 + \text{sqrt}(5))a^3/12 = 4\pi R^3/3, \text{ or } a = 2R(2\pi/5(3 + \text{sqrt}(5)))^{1/3}$$

Helmut Cölfen communicated in reply: Yes, correct.

Colan Hughes opened a discussion of the papers by Peter Vekilov, Fajun Zhang and Helmut Cölfen: All three studies in this section introduce deuterium at some point in the form of D_2O. We have shown in our study of glycine that this can have an influence on the product of a crystallization. Has this possibility been considered?

Helmut Cölfen answered: We did not study the influence of H_2O *vs.* D_2O.

Fajun Zhang answered: For protein crystallization in the presence of di- or trivalent salts, changing the solvent to D_2O can be important. However, for the current study of BLG crystallization with $CdCl_2$, we have presented the SANS data in Fig. 6 for which the samples were prepared in D_2O. The results indicate no influence on protein crystallization in this case.

Peter Vekilov replied: Prior to our NMR characterization of solution that contain clusters and those that do not, we tested cluster formation and evolution over 80 h in D_2O. We found that the cluster sizes, volume fractions, concentrations and the evolutions of these there variables are identical to those in H_2O.[1]

1. Y. Li, V. Lubchenko, M. A. Vorontsova, L. Filobelo and P. G. Vekilov, *J. Phys. Chem. B*, 2012, **116**, 10657–10664.

Roger Davey asked: Is there the possibility that by interchange of techniques each could learn from the other? Could the techniques be complimentary in a positive way?

Helmut Cölfen answered: Yes, the techniques can be complimentary in a positive way. Analytical Ultracentrifugation is a well-established technique for protein characterization. It can detect protein aggregation as well as work out equilibrium constants and stoichiometries in case of reversible interactions. This would be very useful information to understand the very early stages of the two-step nucleation process found for protein systems. If the particle/droplet size in this process would be known by an independent technique like light scattering, it would be possible to work out the sample density. This would enable you to distinguish a liquid from an amorphous or crystalline sample. So, there is much common interest calling for cooperations within the community.

Fajun Zhang added: Yes, certainly, the combination of ultra-centrifugation, dynamic light scattering (DLS), optical microscopy and real-time SAXS can provide deeper understanding on the early stage of nucleation.

It is important to note the difference in their application conditions. Optical microscopy has a relatively low spatial resolution. While it is not easy to distinguish the exact nucleation sites on the intermediate, it provides direct and solid information on the number and the shape and size of the crystals with time. SAXS on the other hand provides the internal structure changes before nucleation, which is crucial to establish the relationship between the intermediate and the crystal nucleation. DLS is a very convenient technique to monitor the protein clusters as presented by Prof. Vekilov (Vorontsova, *et al.*, DOI: 10.1039/c4fd00217b). In our work, we have also employed DLS to study the cluster formation in protein solutions in the presence of multivalent metal ions.[1] Importantly, in DLS studies, the clusters normally have a low volume fraction to avoid the problem of multiple-scattering.

1. D. Soraruf, F. Roosen-Runge, M. Grimaldo, F. Zanini, R. Schweins, T. Seydel, F. Zhang, R. Roth, M. Oettel, F. Schreiber, F. Protein cluster formation in aqueous solution in the presence of multivalent metal ions – a light scattering study, *Soft Matter*, 2014, **10**(6), 894–902.

Peter Vekilov responded: Absolutely

Eric Breynaert directed a question to Fajun Zhang: Fig. 5 in your manuscript demonstrates that all stages in the nucleation process exhibit distinct SAXS patterns. This indicates that the evolution of the scattering intensity for the different species that can be observed in *in situ* SAXS measurements should allow a demonstration of whether this nucleation is a real two step process, or if the aggregates initially formed redissolve in favor of crystals nucleated outside the aggregates. In case of a real two-step nucleation and crystallisation within the aggregates, at some point Bragg reflections will start to develop without changes to the size, shape or concentration of the aggregates.

Fajun Zhang answered: There may be a bit of a misunderstanding. The SAXS curves shown in Fig. 5 of the paper do not correspond to different stages of crystallization in time. In fact, they are the static SAXS data for samples with fixed protein concentration and different salt concentrations and therefore different positions in the phase diagram (Fig. 2 of the paper).

Fig. 7 of our paper shows SAXS profiles measured in real-time during the crystallization process. We remark that the structural feature characterizing the intermediate is indeed conserved during the period where crystal peaks appear. The concentration of the intermediate changes (see our Fig. 8), since the time scales of crystal nucleation and formation of the intermediate are not separated by several order of magnitudes.

Martí Gich said: Was the video you showed in your presentation recorded in the same conditions as those of the SAXS experiment? I ask this because the scattering experiment in combination with the rate equations modeling is very convincing to support the existence of a two-step nucleation which assumes a crystal nucleation inside the aggregates, but in the video the aggregates are progressively less visible around the growing crystal suggesting a parallel process of aggregate consumption. Did you form very small crystals which couldn't be seen in the video?

Fajun Zhang replied: The presented SAXS analysis was performed for a sample with 33 mg ml^{-1} BLG and 17 mM $CdCl_2$. The shown microscopy video was of a sample with the same protein concentration but a slightly higher salt concentration of 19 mM $CdCl_2$. As shown in our Fig. 3, a higher salt concentration results in fewer more homogeneously distributed aggregates and fewer, larger crystals. The observation by optical microscopy is not sufficient to determine the role of the aggregates which is why we apply SAXS (which naturally is sensitive to smaller crystals, too).

Ian Rosbottom returned to the paper by Peter Vekilov: Where do you think the nucleation is occurring in the cluster? Is it in the centre or closer to the interface? Are there any significant trends as to where the crystals nucleate with respect to the clusters, *i.e.* do the crystals appear where the clusters are most dense or where the clusters appear first?

Peter Vekilov responded: The theoretical model of the protein-rich clusters[1] predicts a periphery in which the properties transition from those in the cluster core to those in the solution. Thus, even if the matter in the cluster core is fully gelled, the one in the periphery may still be fluid and may be the likely location of crystal nucleation. In support of this conjecture, we note that an earlier paper on the nucleation of protein crystals assisted by macroscopic dense liquid noted that crystals are found predominantly on the surface of the dense phase domains.[2]

1. W. Pan, P. G. Vekilov and V. Lubchenko, *J. Phys. Chem. B*, 2010, **114**, 7620–7630.
2. Y. G. Kuznetsov, A. J. Malkin and A. McPherson, *J. Cryst. Growth*, 2001, **232**, 30–39.

Bart Rimez remarked: You detect in your lysozyme folding mechanism that was revealed by NMR (Fig 4 in your paper) that you find a sort of dimer configuration. I was wondering if that sort of structure could be related with an actual template for crystal growth.

Peter Vekilov responded: The putative configuration of the dimer that leads to cluster formation in lysozyme solutions is not related to the structure of crystals that are composed on monomers. We think that the presence of domain-swapped dimers is necessary for the formation of clusters. However, the clusters also contain a significant fraction of monomers. Since the clusters are in equilibrium with the solution,[1,2] the chemical potentials of both monomers and dimers in the clusters is equal to those in the solution. Since the difference of the chemical potential of the monomers between the solution and crystals provides the driving force for crystallization, the presence of dimers in the clusters does not lower the supersaturation. On the other hand, the crowded environment of the clusters lowers the surface energy of the crystal nuclei and this leads to faster nucleation of crystals composed on monomers.

1. W. Pan, P. G. Vekilov and V. Lubchenko, *J. Phys. Chem. B*, 2010, **114**, 7620–7630.
2. V. Uzunova, W. Pan, V. Lubchenko and P. G. Vekilov, *Faraday Discuss.*, 2012, **159**, 87–104.

Jim de Yoreo remarked: I think I heard you describe what sounds like one-step nucleation; it still occurs *via* monomers, but in a special heterogeneous environment created by disordered clusters of dimers. That is, the monomers do not come together into a disordered form and then nucleate into an ordered crystal, rather monomers within disordered clusters of dimers – that are in equilibrium with the monomers outside the clusters – get together to form an ordered crystal nucleus. Is that correct?

Peter Vekilov answered: After the first step, the formation of the clusters has occurred, there is only one more step left, the formation of crystal nuclei inside it. So yes, the scenario highlighted by you is mostly correct, with one revision: the clusters contain both monomers and dimers and it is the monomers that assemble into crystal nuclei. You seem to imply that a two-step mechanism can only be really two-step if all molecules that partake in step 1 engage in step 2. This seems an artificial requirement.

Dimitrios Toroz asked: From the molecular modelling point of view, it would be interesting to calculate the free energy of dimerization for the proteins examined in this paper and monitor from the differences of the PMF depths the various orientations which are directly related to the shape of the proteins. My question is to what extent these calculations have been performed so far and if there was any agreement with the measured experimental data?

Peter Vekilov answered: We did exactly as you suggest for properly folded lysozyme molecules.[1] We accounted for coulomb, van der Waals, ion bridges, and protonation–deprotonation equilibria. We found that such molecules may form dimers at specific orientations. However, the binding energy of the strongest of the possible dimers was much too low to explain the cluster sizes observed in lysozyme solution (according to the existing model of cluster formation,[2,3] the cluster size $R_{cl} = (\tau_{dimer}D_{dimer})^{1/2}$, τ_{dimer} depends on the dimer bond). This is one of the main arguments in favor of partial protein unfolding as a perquisite for cluster formation with this protein.

1. Ho Y. Chan, V. Lankevich, Peter G. Vekilov and V. Lubchenko, *Biophys. J.*, 2012, **102**, 1934–1943.
2. W. Pan, P. G. Vekilov and V. Lubchenko, *J. Phys. Chem. B*, 2010, **114**, 7620–7630.
3. Y. Li, V. Lubchenko, M. A. Vorontsova, L. Filobelo and P. G. Vekilov, *J. Phys. Chem. B*, 2012, **116**, 10657–10664.

Diana Milena Camacho Corzo presented some slides on the determination of nucleation mechanisms and kinetics from analysis of polythermal data: A recently developed methodology was used to assess nucleation for α-PABA (Toroz *et al.*, DOI: 10.1039/c4fd00275j) from polythermal data. As this methodology is of general application to crystallisation from solutions I thought it would be of interest to present some more detail about this approach.

The methodology makes use of the KBHR (Kashchiev–Borissova–Hammond–Roberts) model, which can be applied broadly to crystallisation from solutions. This is so because the KBHR model was derived starting from the KJMA (Kolmogorov–Johnson–Mehl–Avrami) formula, which is a general expression that follows the increase of the relative volume of crystals (α) in a solution with time.

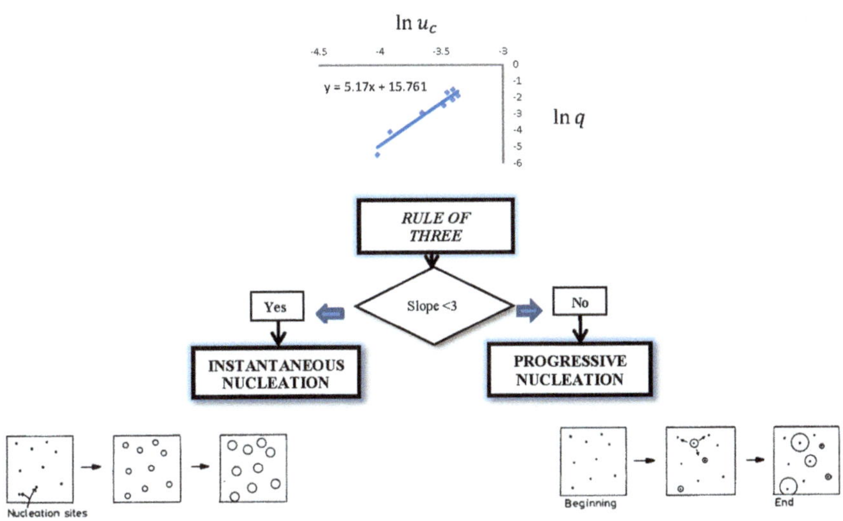

Fig. 4 Assessment of nucleation mechanism from polythermal data using the KBHR model.

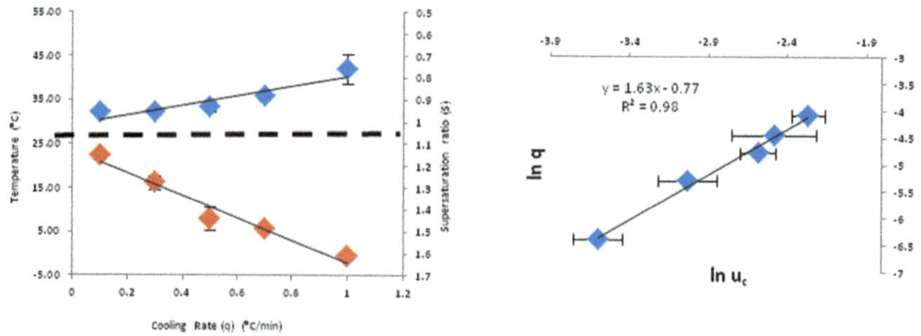

Fig. 5 Example of application KBHR approach to the analysis of polythermal data collected for α-PABA crystallising from ethanol Con 170 g Kg^{-1}.

Specifically the analysis of polythermal data through the KBHR model, allows differentiation between the cases of progressive nucleation (PN) and instantaneous nucleation (IN), from the slope of a line dependence of relative critical undercooling (μ_c) for crystallisation on the cooling rate (q) (Fig. 4). The relative critical undercooling can be obtained from an expression that uses the values of crystallisation temperature T_c, obtained experimentally as a function of q, and the solubility data, *i.e.* $\mu_c = (T_e - T_c) \times T_e^{-1}$.

IN is the case when all crystals are nucleated practically at the initial moment $t = 0$ and then only grow irreversibly. PN refers to the case when the crystals are continuously nucleated during the process. Once the mechanism is established, the use of a set of expressions related to each case[1-3] will allow obtaining of nucleation parameters such as interfacial tension (γ) and in turn nucleation rates for PN and the concentration of crystals (C_0) at the nucleation point for IN. The differentiation between the two mechanisms is relevant as it can give an

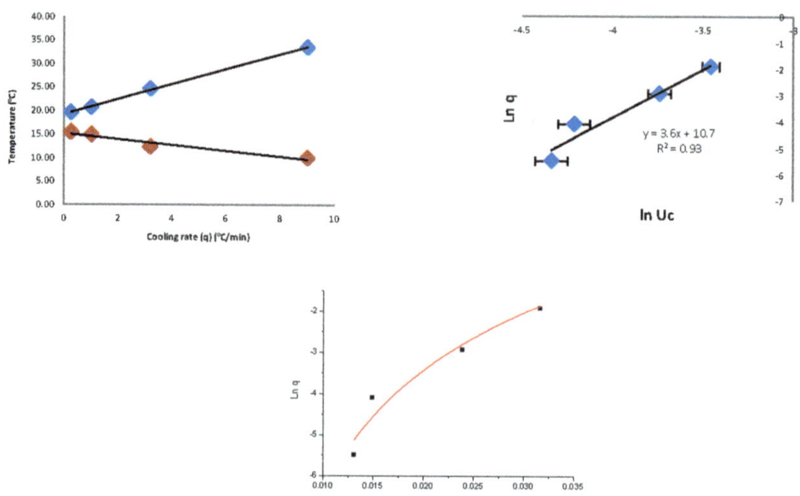

Fig. 6 Example of application KBHR approach to the analysis of polythermal data collected for methyl stearate crystallising from kerosene Con (250 g l^{-1}).[3]

indication as to how easily nucleation can occur as a function of crystallisation environment.

The application of this methodology to α-PABA crystallising from ethanol solutions showed that crystals are formed by means of IN as inferred from collected experimental polythermal data (Fig. 5). The concentration of PABA nuclei C_0 at the nucleation point, for different solution concentrations, was found to be in the order of $6.6 \times 10^8 - 1.3 \times 10^9$ m^{-3} (Toroz *et al.*, DOI: 10.1039/c4fd00275j).

Comparatively, an example of the PN case can be observed for methyl stearate crystallising from kerosene Fig. 6.

In this case the application of the methodology allowed obtaining of the interfacial tensions, critical radius (r^*) and nucleation rates (J) in the range of 1.21–1.91 mJ m^2,0.7–0.9 nm and $5.1 \times 10^{16} - 7.9 \times 10^{16}$ nuclei ml^{-1} s^{-1} respectively.

More examples can be found in recent publications for the crystallisation of: potassium sulphate in aqueous solution,[1] rubidium, sodium dodecyl sulphate in aqueous solution and aspirin in ethanolic aqueous solution[2]and for methyl stearate crystallising from kerosene.[3] The later reference presents the derivation of an expression allowing determination of a system nucleation mechanism from previous Nyvlt-type data analysis and shows that there is a good agreement between nucleation parameters obtained by the isothermal method and those obtained by means the polythermal method.

1. D. Kashchiev, A. Borissova, R. B. Hammond, K. J. Roberts, Effect of cooling rate on the critical undercooling for crystallization, *J. Cryst. Growth*, 2010, **312** 698–704.
2. D. Kashchiev, A. Borissova, R. B. Hammond, K. J. Roberts, Dependence of the Critical Undercooling for Crystallization on the Cooling Rate, *J Phys. Chem. B*, 2010, **114** 5441–5446.
3. D. Camacho, A. Borissova, R. Hammond, D. Kashchiev, K. Roberts, K. Lewtas, I. More, Nucleation mechanism and kinetics from the analysis of polythermal crystallisation data: methyl stearate from kerosene solutions, *CrystEngComm*, 2014, **16** 974–991.

Sarah Price remarked: Are the results you've described for PABA typical and how widely could the technique be used for other molecular systems?

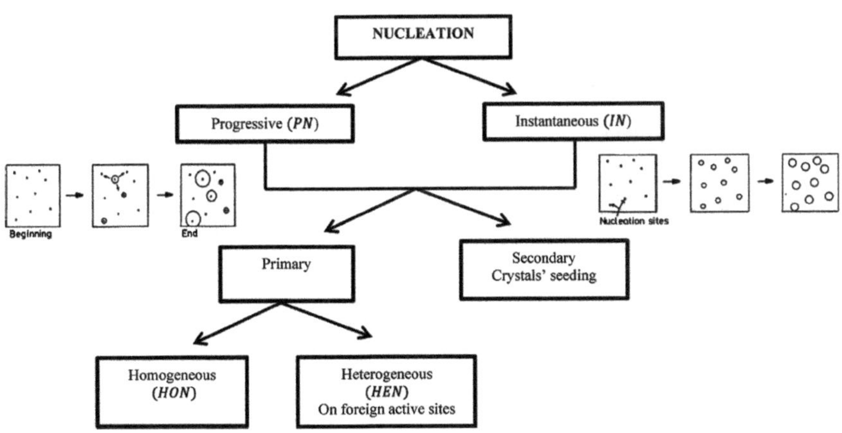

Scheme 1

Diana Milena Camacho Corzo replied: The methodology that was used to assess nucleation for α-PABA can be applied to any other system as it is a methodology that makes use of the KBHR (Kashchiev–Borissova–Hammond–Roberts) model, which can be applied broadly to crystallisation from solutions. This is so because the KBHR model was derived starting from the KJMA (Kolmogorov–Johnson–Mehl–Avrami) formula, which is a general expression that follows the increase of the relative volume of crystals in a solution with time.

The concept underpinning the derivation of these expressions is that the volume of crystals V_c results from nucleation of materials points at a rate $J(t)$ (m^{-3} s^{-1}) which then only expand irreversibly in the radial direction with growth rate $G(t)$.[1] Thus, the model applies to systems whose phase transformation occurs under conditions allowing the formation of statistically many nuclei. In this case, the poly-nuclear mechanism can manifest by either of two ways known as instantaneous (IN) or progressive (PN) nucleation. IN is the case when all crystals are nucleated practically at the initial moment $t = 0$ and then only grow irreversibly. PN refers to the case when the crystals are continuously nucleated during the process.

Specifically the KBHR model, which is used in this methodology, allows establishing which one of these mechanisms is ruling the nucleation process from polythermal experimental data and once the mechanism is established, to obtain nucleation parameters such as interfacial tension and in turn nucleation rates for the case of PN and the concentration of crystals at the nucleation point in the case of IN. A schematic diagram highlighting the various mechanisms important in nucleation processes is presented in Scheme 1.

The application of this methodology to α-PABA crystallising from ethanol solutions showed that crystals are formed by means of IN as inferred from collected experimental polythermal data (Fig 11c of Toroz *et al.*, DOI: 10.1039/c4fd00275j). According to this methodology, a plot of relative critical undercooling μ_c (eqn 2 in the paper) *vs* cooling rate (q) in ln–ln coordinates will deliver the nucleation mechanism from the slope of the line that best fits these data. In this case a slope <3 indicates IN. Furthermore, the concentration of PABA nuclei C_0 at the nucleation point was found to be in the order of $6.6 \times 10^8 - 1.3 \times 10^9$ m^{-3}. Examples of the application of this methodology to different solutions systems, in which both cases of IN and PN are analysed, can be found in recent publications for the crystallisation of: potassium sulphate in aqueous solution,[2] rubidium, sodium dodecyl sulphate in aqueous solution and aspirin in ethanolic aqueous solution[3] and for methyl stearate crystallising from kerosene.[4]

1. D. Kashchiev, *Nucleation basic theory with applications*, Butterworth Heinemann, Oxford ; Boston, 2000.
2. D. Kashchiev, A. Borissova, R. B. Hammond, K. J. Roberts, Effect of cooling rate on the critical undercooling for crystallization, *J. Cryst. Growth*, 2010, **312**, 698–704.
3. D. Kashchiev, A. Borissova, R. B. Hammond, K. J. Roberts, Dependence of the Critical Undercooling for Crystallization on the Cooling Rate, *J Phys. Chem. B*, 2010, **114**, 5441–5446.
4. D. Camacho, A. Borissova, R. Hammond, D. Kashchiev, K. Roberts, K. Lewtas, I. More, Nucleation mechanism and kinetics from the analysis of polythermal crystallisation data: methyl stearate from kerosene solutions, *CrystEngComm*, 2014, **16**, 974–991.

Sarah Price opened a general discussion of the paper by Dimitrios Toroz: The crystallisation of *p*-amino benzoic acid (PABA) is clearly complicated: in 1965

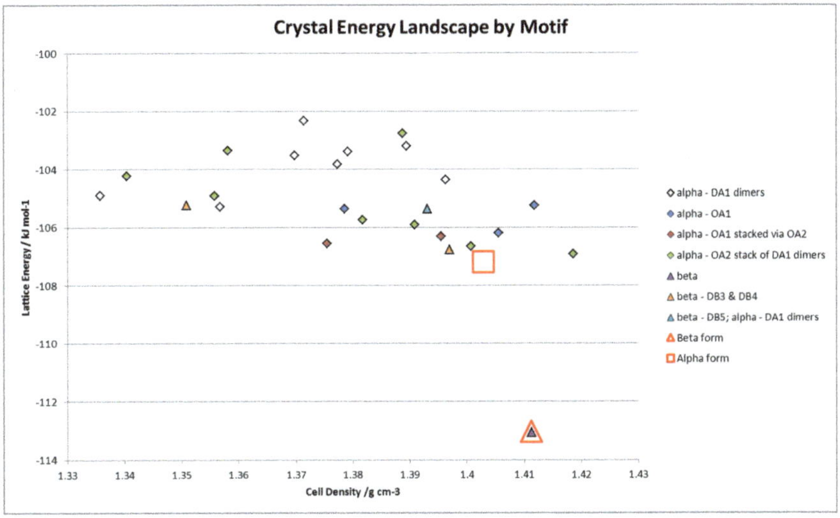

Fig. 7 Energy landscape of computer-generated $Z' = 1$ crystal structures for PABA, with each lattice energy minimum denoted by a symbol showing which of the motifs of Fig 9 of the paper by Toroz *et al.*, are found in the structure. The corresponding minima for the α and β forms are shown by open red symbols: the α form was not found in the search as it is $Z' = 2$.

Kilean *et al.*[1] reported finding three modifications, all polysynthetic twins with huge unit cells, and an inability to find a previously reported form with rhombic plates. Polymorph V, as recently reported by Benali-Cherif et al.,[2] $Pna2_1$, $Z' = 2$, has a common structure for a double layer of molecules with the α form, differing only in the direction of molecules in the next layer. We performed a "Crystal Structure Prediction" study on PABA in 2009 [3] and the relative energies and densities of the 25 lowest energy $Z' = 1$ structures are shown in Fig. 7, coded by the packing motifs identified by Toroz *et al.* (DOI: 10.1039/c4fd00275j).

There are clearly a large number of crystal structures that are only very slightly less stable than the $Z' = 2$ structure of the α form, and have at least the carboxylic acid dimer in common, and often larger clusters of molecules. There are clearly a number of crystal packings of PABA which have a lot in common with α (form V observed and others computer generated) and are very close in energy. To what extent would these alternative structures be equally well accounted for by the results and analysis in your paper? Do these alternative crystal structures show that disorder/growth mistakes in the growth of the crystal are likely? To what extent are the α crystals being formed in your experiments structurally pure? Could these alternative packings of the carboxylic acid dimer account for the growth along the needle axis?

1. R. C. G. Kilean, *et al.*, *Acta Crystallogr.*, 1965, **19**, 482.
2. R. Benali-Cherif, R. Takouachet, E.-E. Bendeif and N. Benali-Cherif, *Acta Crystallogr., Sect. C: Struct. Chem.*, 2014, **70**, 323–325.
3. P. G. Karamertzanis, A. V. Kazantsev, N. Issa, G. W. A Welch, C. S. Adjiman, C. C. Pantelides and S. L. Price, Can the Formation of Pharmaceutical Co-Crystals Be Computationally Predicted? II. Crystal Structure Prediction. *J. Chem. Theory Comput.*, 2009, **5**, 1432–1448.

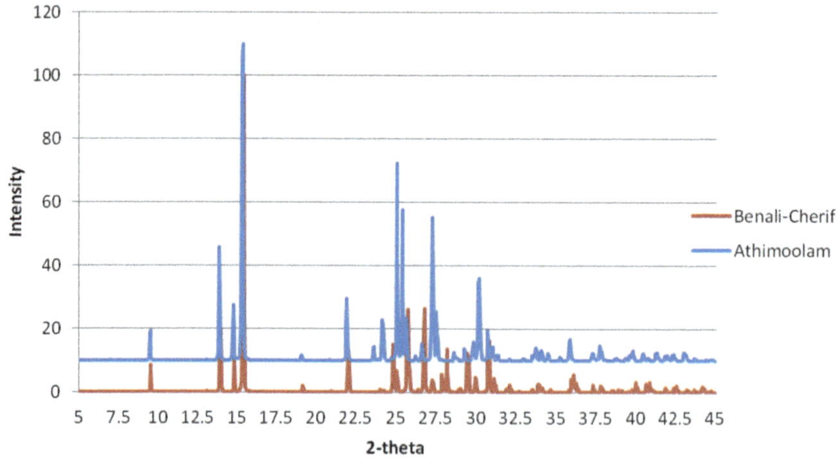

Fig. 8 Powder X-ray diffraction patterns simulated from structures published by Benali-Cherif (red) and Athimoolam (blue).

Dimitrios Toroz responded: The structural variation of the alpha form is indeed an interesting point. With regards to the crystal structure, we did initially have doubts over the structure ourselves within the project, and re-solved the structure twice, each time getting the $Z' = 2$ crystal structure reported first by Lai and later by Athimoolam. Indeed, we even had samples examined by solid state NMR from which the observations are consistent with two molecules in the asymmetric unit rather than one due to the amount of carbon resonances solved ortho to the acid group. Nevertheless, the solid state NMR does only find one carboxylic acid peak, but this is perhaps not surprising due to the very similar state of the two molecules in the asymmetric unit.

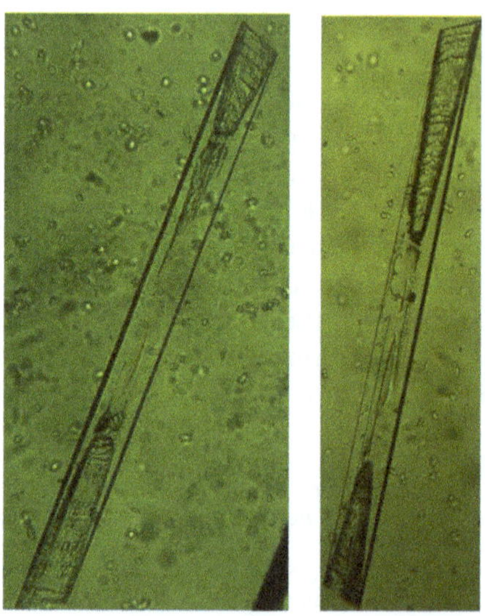

Fig. 9 α-PABA crystals grown from ethanol showing variations in the size of the solvent inclusions at the capping faces of the sector zones.[2]

Regarding the third structure most recently published by Benali-Cherif *et al.*,[1] the calculated powder patterns of both structures are very similar (Fig. 8). It would be very easy to mistake between the powder patterns of these structures, and it is obviously not feasible to do solid state NMR on each sample we have crystallised throughout the project.

Though at a glance the XRD patterns look similar for the low order reflections, the high order reflections show enough variation to discriminate between the structures. Hence, we feel that when we have characterised the crystals we have produced, we are fairly confident the PXRD patterns that have been analysed are representative of the α-PABA structure with $Z' = 2$.

The roughening along the long axis of the needle is an interesting characteristic of α-PABA. The solvent inclusions that occur along this long axis can be observed to often not be symmetrical, *i.e.* it appears that one end of the crystal can sometimes grow faster than the other (Fig. 9).

The significantly different observable sizes of the inclusion sector zoning at the capping faces is quite common for these crystals, leading us to believe there was potential that the crystal was polar. We therefore attempted to solve the crystal in polar super groups of $P2_1/n$, but did not manage to get a sensible refinement. It would be interesting to attempt to solve a full single crystal X-ray diffraction data set within the $Pna2_1$ space group and compare the quality of fit to the $P2_1/n$ fit.

Though we believe the bulk crystal structure is correct, there is a curiosity over whether the sector zoning effects observed at the end of the crystal could cause any observable changes to the local surface structure compared to the bulk. We believe that the sector zoning is caused by the π–π stacking interactions having strong intermolecular recognition to the molecules in solution, and the potential that molecules lie flat at the surface of the crystal, and therefore have to do little work to rearrange and adsorb into the lattice, in a similar manner to the study of benzoic acid[3] needle like crystals that contain π–π stacking interactions.

Polarisation dependent near edge X-ray absorption fine structure (NEXAFS) spectroscopy can provide information on the local intra- and inter-molecular structure at the surface of crystals, specifically the orientation of molecules within a small local environment. This technique was used to probe the local geometry of large single crystals by aiming the beam at the centre of the crystal (Fig. 10), and

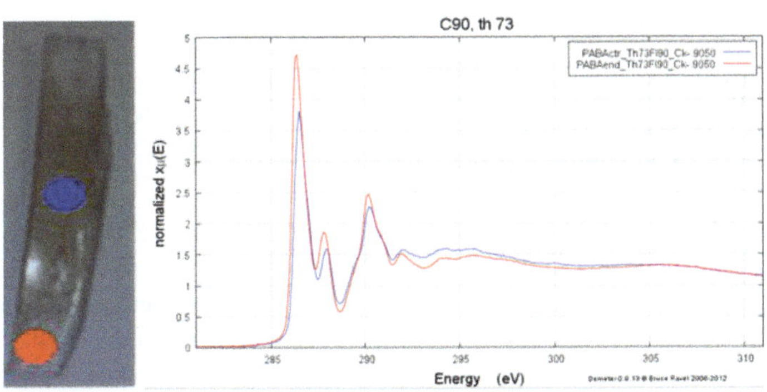

Fig. 10 Carbon K-edge NEXAFS spectra for a large single crystal.[2] Centre (blue) and end (red) of the crystal.

then at the end sector zone of the crystal, and comparing the spectra. If the crystal structure is truly periodic then the two spectra can be expected to be almost identical.

Fig. 10 shows the significant increase in the intensities of resonances associated with the π orbitals and a slight decrease in the resonances associated with the σ orbitals in the spectra measured at the end of the crystal and the centre respectively. In polarisation dependent NEXAFS, when the beam polarisation is parallel to a molecular orbital, a strong resonance is expected. Since the direction of the π orbitals and σ orbitals are almost perpendicular to each other, the changes in intensity of the resonances suggest a slight change in orientation of the molecules. This could support the hypothesis that the speed of growth along the long axis of the needle means that the sector zones associated with the capping faces have poor lattice perfection and disorder of the molecules, which would not be obvious from bulk XRD measurements

Overall we believe that the needles we have crystallised are consistent with α-PABA with $Z' = 2$. However, we concede that the needles often have poor lattice perfection along the long axis of the needle, and that could result in subtle surface structural changes, *i.e.* relaxation in crystal packing. We do not believe this results in a different crystal structure. We do not, however, have any definitive evidence for the crystal structure being different the from the $P2_1/n$ α-PABA structure that we have based this work on.

1. R. Benali-Cherif, R. Takouachet, E.-E. Bendeif and N. Benali-Cherif, *Acta Crystallogr., Sect. C: Struct. Chem.*, 2014, **70**, 323–325.
2. Ian Rosbottom, unpublished results, University of Leeds, 2015.
3. P. McArdle, Y. Hu, A. Lyons and R. Dark, *CrystEngComm*, 2010, **12**, 3119–3125.

Kenneth Harris asked: In the polymorph prediction calculations that you have described for *p*-aminobenzoic acid (PABA), only the non-zwitterionic form was considered. Did you also carry out calculations using the zwitterionic form of PABA?

It is well known that, for *o*-aminobenzoic acid[1] (OABA) and *m*-aminobenzoic acid[2] (MABA) some of the known polymorphs contain the zwitterionic form whereas others contain the non-zwitterionic form. In contrast, all the polymorphs of PABA[3] for which crystal structures have been reported so far contain only the non-zwitterionic form of the molecule. Is there any fundamental reason that PABA may have a lesser propensity for crystallizing in the zwitterionic form in comparison to OABA and MABA?

1. W. H. Ojala and M. C. Etter, *J. Am. Chem. Soc.*, 1992, **114**, 10288–10293.
2. P. A. Williams, C. E. Hughes, G. K. Lim, B. M. Kariuki and K. D. M. Harris, *Cryst. Growth Des.*, 2012, **12**, 3104–3113.
3. S. Gracin and Å. C. Rasmuson, *Cryst. Growth Des.*, 2004, **4**, 1013–1023.

Sarah Price responded: No, we did not carry out any calculations on the zwitterionic form, as in our 2009 study[1] we were very aware that the method used was unlikely to be adequate for zwitterions. Since then, the method has been used for predicting new polymorphs of creatine,[2] by carefully selecting which groups of structures have to be refined using periodic, dispersion-corrected, electronic structure calculations. This takes into account the errors in the model used to generate the structures caused by the neglect of the different polarisation within

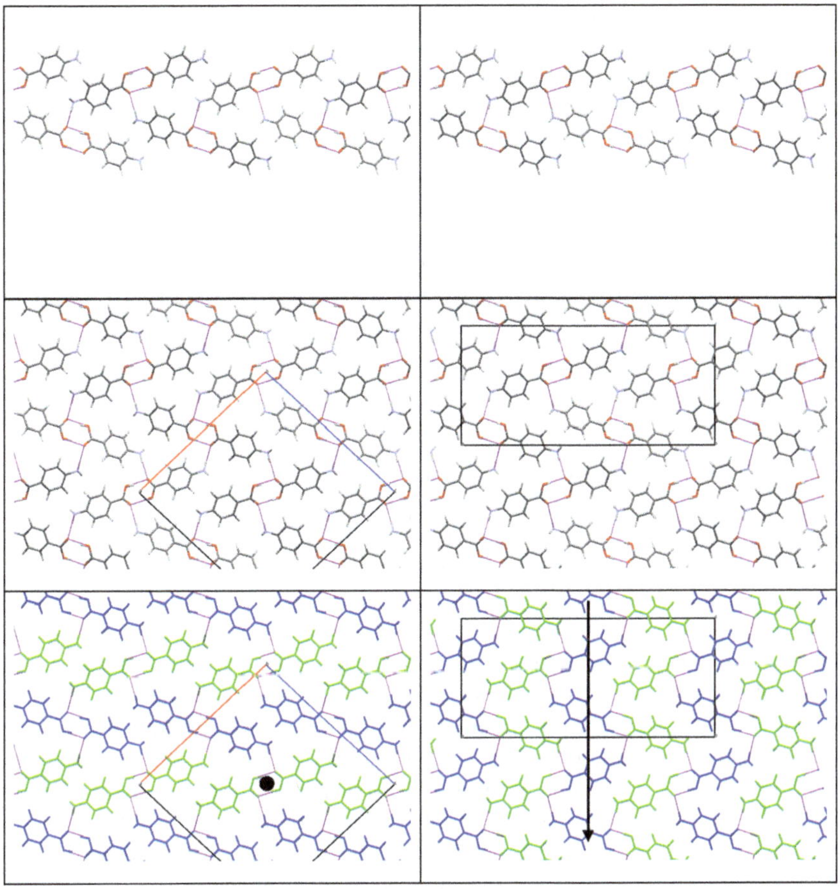

Fig. 11 Left: AMBNAC01 ($P2_1/c$, $Z' = 2$) . Right: AMBNAC09 ($Pna2_1$, $Z' = 2$). Top: Common double layer of the two structures. Middle: Additional layer, with the molecules at different tilts with respect to the direction of view in the two structures. Bottom: Molecules coloured by symmetry. The two symmetry-independent molecules in AMBNAC01 are related by a screw axis in AMBNAC09 (shown by black arrow), and the two symmetry-independent molecules in AMBNAC09 are related by an inversion centre in AMNBAN01 (shown by black circle). Perpendicular to the direction of view, both structures have mirror planes.

and between the zwitterionic creatine molecules in different conformations. A "Crystal Structure Prediction" study of OABA and MABA would be a good test of whether current lattice energy evaluation methods are capable of correctly balancing the energy differences between zwitterionic and non-zwitterionic polymorphs.

1. P. G. Karamertzanis, A. V. Kazantsev, N. Issa, G. W. A. Welch, C. S. Adjiman, C. C. Pantelides and S. L. Price, Can the Formation of Pharmaceutical Co-Crystals Be Computationally Predicted? II. Crystal Structure Prediction, *J. Chem. Theory Comput.*, 2009, **5**, 1432–1448.
2. D. E. Braun, M. Orlova and U. J. Griesser, Creatine: Polymorphs Predicted and Found, *Cryst. Growth Des.*, 2014, **14**, 4895–4900.

Roger Davey commented: You mentioned a new polymorph of *p*-aminobenzoic acid which I suppose is the one published recently from a French group? As I recall it is a non-centric structure. Would it be possible that it could in fact be a twinned version of the alpha structure?

Sarah Price answered: From the space group type of argument, they are unlikely to be a simple twin. However, as shown in Fig. 11, the two structures are very similar, with a common double layer and a difference in tilt of the molecules between these layers. The authors claim that the structure is a polymorph.[1] Hence this appears to be experimental evidence for PABA being able to pack in other structures closely related to α.

1. R. Benali-Cherif, R. Takouachet, E.-E. Bendeif and N. Benali-Cherif, *Acta Crystallogr., Sect. C: Struct. Chem.*, 2014, **70**, 323–325.

Ian Rosbottom commented: With regard to the non-centro-symmetric polymorph of PABA crystallised in 2014, it does indeed resemble the alpha structure. Whether or not there is any competition between this and the alpha polymorph when the crystal is packing is not something we had particularly considered, but we have regularly checked the polymorph that we have crystallised when we obtained needles using PXRD, and it was always alpha PABA. The rough interfacial growth (RIG) mechanism is interesting, and I believe that this is due to the favourable π–π stacking interactions along the long axis of the needle. The RIG mechanism is observed from ethanolic solutions, a polar protic solvent. The dominating π–π interactions along the long axis of the needle are primarily hydrophobic and would not be expected to interact as strongly with solvent compared to the side $(1\ 0\ -1)$ surface, where the growth is dominated by OH$\cdots$O H-bonds through exposed carboxylic acid groups at the surface. It would be expected that the EtOH molecules could also bind strongly to these groups and hence slow down growth, and possibly force the majority of the PABA molecules to the end of the needle where they could interact more uninhibited through the π orbital ring interactions. In addition, as is shown in the paper, the SAXS results suggest that there is a distribution of monomers and carboxylic acid H-bonding dimers present in the supersaturated solution prior to nucleation. If this is true, then incorporation of the dimers would be especially favoured along the long axis of the needle. I feel that this combination of unhindered π–π interactions and possibly increased supersaturation at the tip of the needle encourages the RIG growth mechanism.

Colan Hughes addressed Ian Rosbottom and Thomas Turner: In the discussion on PABA, you said that you're always looking for nice crystals. When you don't get nice crystals, have you examined the samples you did get by powder X-ray diffraction to see which forms have been produced?

Ian Rosbottom answered: Often when we get nice crystals it is by solvent evaporation in a dish, and even then there are much less nice crystals in the dish alongside the nice single crystals. Since I added NMe to my ethanol solutions to lower the aspect ratio and improve the perfection of the crystals, I considered I could have made the solvate and crushed up the less perfect crystals for powder XRD and it was alpha PABA.

Thomas Turner added: A number of alpha PABA samples have been tested using powder X-ray diffraction under a range of crystallisation conditions over the course of this work, in particular samples which have turned a yellow colour have been compared to standard alpha PABA. The data has been fit using Rietveld refinement to compare a fit to the crystallographic structure and no significant differences were observed, however, although unlikely, this does not rule out the possibility that a very small amount of another polymorph is formed below the detectable level of an XRD measurement.

Simon Black addressed Dimitrios Toroz and Roger Davey: Can you see the PABA dimers using spectroscopy?

Roger Davey responded: Yes, I believe so. I discussed this in my introductory lecture. I would point out however that we have only looked at compositions up to saturation. The paper by Toroz *et al.* (DOI: 10.1039/c4fd00275j) takes us into the supersaturated domain.

Dimitrios Toroz answered: Recent work by Sullivan *et al.*[1] presented results of infrared spectroscopy data of PABA in acetonitrile, ethyl acetate and iso-propyl alcohol, using dilution studies to try to observe peak shifts or peak splitting related to breaking of the dimer in solution. The results were not conclusive but seem to indicate that there are no dimers formed at the concentrations studied. However it should be noted these experiments were not carried out in ethanol and were not probing the supersaturated state as in the SAXS experiments carried out on PABA. Planned work to study the saturated state using *in situ* FTIR during a cooling crystallisation will hopefully reveal the structure of the saturated solution in ethanol.

1. R. A. Sullivan, *et al.*, *Cryst. Growth Des.*, 2014, **14**, 2689.

Laszlo Fabian commented: One thing you looked at is the radius of gyration. Are you getting out single numbers or can you get in some vein a distribution? Based on our molecular dynamics simulations, I would expect a wider distribution of sizes. I wouldn't expect a static picture, but a much more dynamic one, with molecules associating and dissociating continuously. This applies both to cluster formation and to forming and breaking individual hydrogen bonds. Can you see any indication in your experiments of such dynamic behaviour?

Dimitrios Toroz answered: I agree that almost certainly a distribution of sizes for clusters and aggregations of molecules would be expected in such a dynamic process, however molecular dynamics simulations will give $\sim$ ns snapshot of the structural picture, unfortunately in reality the experimental data in this case were recorded over thirty seconds. This means that the scattering patterns we record are a spatially and temporally averaged picture of the system over the time frame of data collection, so the values extrapolated in the radius of gyration are also a thirty seconds average number. We do highlight that in the case of the monomer and dimer structures a population between the two species is likely and that the population shifts towards dimers on cooling.

Fajun Zhang said: I have a comment on the interpretation of the SAXS data in your paper. Due to the limited q range, the data in the high q region is already truncated. For better understanding of the shape, size and its distribution of the nanoparticles, higher q data are needed.

Dimitrios Toroz answered: This is indeed the case. Due to the restricted, low q space covered, the data at the high q region become truncated; this was addressed in the analysis by considering only the q space up to 3.5 nm^{-1} when calculating the radius of gyration in the unified fit analysis and also when fitting the form factors. The high q region would be very interesting to study in more detail and this will be done in the future with planned neutron scattering experiments which should provide a larger q range and improved scattering length density.

Jan Sefcik said: The SAXS data which are shown in Fig. 13 of your paper are very interesting. They seem to show that scattering from larger clusters (at least tens of nm) is present from the very beginning of the cooling process even in undersaturated solutions. Similar observations were previously made in aqueous solutions of glycine.[1,2] Would you be able to comment on their role as pre-nucleation clusters and how it relates to the critical nucleus size estimated elsewhere in your paper? You presented two ways to look at the nucleation process; how would you fit it all together?

1. S. Chattopadhyay, *et al.*, *Cryst. Growth Des.*, 2005, **5**, 523–527.
2. A. Jawor-Baczynska, J. Sefcik and B. D. Moore, *Cryst. Growth Des.*, 2013, **13**, 470–478.

Dimitrios Toroz responded: The clusters you are referring to do appear to precede the supersaturated state, however due to the lack of q space in this region it is not possible to calculate any size or distribution parameters. As such I would not like to comment on their role as "pre-nucleation clusters"; to do so would be presuming a number of structural parameters and aggregation pathways of the nucleation process. As a comparison to the glycine work in the two publications you quote, there is an interesting comparison to be made between the similarities in the data. However the two systems, PABA and glycine, are very different in terms of their solution chemistry. For example glycine is highly soluble in water and hence hydrophilic in nature and so it could be postulated that glycine produces a more than ideal solution in water. Comparing this to alpha PABA which is highly hydrophobic and even in ethanol (the solvent in which these measurements were produced) shows less than ideal behaviour as can be seen from the van't Hoff plot in Fig. 11b in the paper. In this sense then it would be expected that solute clustering could likely occur in alpha PABA. Regardless of this difference in solution chemistry there still remains an interesting similarity across these multiple studies using multiple techniques and these observations should be expanded with future work in this field. Further investigations into these larger structures are certainly needed and upcoming neutron scattering experiments on this system will hopefully reveal an enhanced understanding of the structural pathway which leads to their formation through experimental focus on the low q regime.

Radoljub Ristic remarked: I find this undertaking intriguing and inspiring. The ultimate goal is to rationalise why the alpha form of PABA is crystallised in preference to its low temperature beta form. For this purpose, you made use of a multi-scale and multi-technique approach. One of the techniques was small-angle X-ray scattering (SAXS).

There is a general consensus that the accurate assessment of SAXS data is still a rather difficult task. Subjective interpretation of data is still required to fully assess the data quality and to ensure that the conclusions that are drawn are not erroneous. While X-ray crystallography and NMR have quantitative standards by which data quality can be assessed, such as minimum d-spacing, chemical shift dispersion or peak line width, SAXS has no equivalents of such values.

In your experimental work, the recorded SAXS data are coming from a solution displaying multiple-size-scale and shape of the solution inhomogeneities (clusters). Therefore, the use of the unified approach, the combination of two-parameter Guinier-exponential and two-parameter power-law fits, is most appropriate at the current state of the art. However, both laws have limitations such as applicability to diluted systems, inability to provide structural information on a molecular level and last but not least, the difficulties related to interpretation of unpredictable surface-fractal-like power-law features over a limited range of diffraction vector q because of the overlap of various nano/micro features. Bearing in mind this and the fact that you are dealing with non-diluted systems, I would like to know how you coped with the above difficulties and to which extent your SAXS data interpretation of cluster size/shape are informative.

Thomas Turner responded: I agree that SAXS data analysis is not a trivial task and indeed neither is data collection for organic systems especially in the study of nucleation where a dynamic complex structural process can be expected. Therefore as part of this a number of people were involved in not only the data collection but also the data analysis and as such are acknowledged in the paper. I should add that you are correct in saying that the model is indeed quite simplistic by splitting the scattering curve into various sections related to populations of scatterers. However as there are a limited number of published works in this field, mainly due to the relative infancy of the experimental technology and analysis methodology which allows these experiments to be undertaken, a more complex approach is simply not available at this time. Work in relation to this project is currently being undertaken to try and link molecular dynamics simulations with recorded SAXS data reported in this paper in an attempt to provide a more detailed analytical approach through modelling techniques.

In terms of using SAXS and the concentration issues this is true and can be a problem particularly in protein crystallography where aggregation of particles is not wanted as structure determination relies on visualisation of the individual scattering envelope. However, in this case I should stress that this is a fundamental application and as such we do want to observe molecular aggregation because that is what is interesting in nucleation. The concentration used is a product of the poor scattering power of PABA in relation to the surrounding solvent (ethanol) and as such high concentrations are required in order to observe a scattering signal above the background; this is a reflection of the experimental limitations in the field at this time and unfortunately cannot be overcome (at present).

Regarding the Guinier and power law fit question, these laws have been applied in a number of publications,[1–3] and interestingly have also been applied to a number of polymer systems[4] which are not only concentrated but highly dense phases which would be expected to undergo aggregation and the Unified fit approach seems to describe these systems well. I do agree that no concrete conclusions should be drawn from the power law fitting (fractal approach) of the low q space in this paper due to the limited number of data points in this region and as such I speculate as to what this could be interpreted as in the paper rather than drawing sound conclusions. It is interesting that the power law fit does increase in order at the exact moment the cell temperature enters the metastable zone ~50 °C and keeps increasing as supersaturation increases suggesting these larger clusters of molecules not only increase in size but possibly in order too. This observation is further supported by the fact it is repeatable over different experimental conditions.

In terms of the inability to scrutinise molecular structure I completely agree and this is not what SAXS is used for; the size limitations of small angle scattering in terms of the diameter of a spherical scatterer are around 1 nm depending on the experimental set up, beyond this is the realm of diffraction which we are not considering in these experiments. The goal in this work was to try to observe scattering functions which relate to the size and shape of certain clusters of molecules which could be expected for this system. The SAXS data is supported in this publication by calculations of the most stable molecular synthons which could be present within the supersaturated state; additionally further efforts to understand the nucleation pathway of alpha PABA are planned to build on the results of this work.

1. G. Beaucage, *J. Appl. Crystallogr.*, 1996, **29**, 134.
2. H. Boukari, J. S. Lin, M. T. Harris, *J. Colloid Interface Sci.*, 1997, **194**, 311
3. G. Beaucage, H. K. Kammler and S. E. Pratsinis, *J. Appl. Crystallogr.*, 2004, **37**, 523.
4. G. Beaucage, *J. Appl. Crystallogr.*, 1995, **28**, 717.

Sarah Price remarked: This paper is putting together the results of a lot of different techniques to try to give an overall understanding of the nucleation of an amino benzoic acid. Thus the limitations in the amount of information that can be gathered from the SAXS experiments is complemented by other evidence, and the confidence in the interpretation is carefully described in the abstract.

Dimitrios Toroz responded: This is certainly true that SAXS data generally requires other experimental and modelling techniques to verify any conclusions made from the analysis. During this study synchrotron SAXS was used because this technique provides a useful tool to perceive structural details of a system during dynamic changes in a system, such as nucleation.

However, studying organic materials using SAXS can be particularly challenging due to contrast issues, as the scattering length densities are similar in ethanol, 7.5×10^{-6} Å^2, and in PABA, 12.3×10^{-6} Å^2. This similarity in "scattering power" means that recording useable data can be challenging due to the low signal from the solute. Progress has however been made more recently with faster and more sensitive detectors combined with more powerful radiation sources to allow organics to be studied in more detail; in particular a number of studies into the nucleation and solution structure of glycine have been carried out.[1] Although the SAXS data is limited in the sense that the whole structural pathway to the nucleation of alpha PABA is not revealed, the data gives some initial insights into larger cluster formation of organic materials and also the solution self-assembly in the saturated state. Unfortunately the mid q range data is poorly defined in this case and so structural information regarding medium sized clusters of the order of 5–20 nm, if they indeed are formed during nucleation, are not observed, possibly due to short lifetimes or because the structures are absent altogether. This interpretation of the experimental data will be supported by molecular dynamics simulations where the scattering pattern from simulations will be developed and compared to the experimental scattering pattern to get a better understanding of the overall q range data. The questions left by these experiments, namely what are the size and distribution of the larger clusters, will hopefully be addressed in upcoming experiments which will utilise neutron scattering to allow the lower q regions to be scrutinised and also give an enhanced scattering contrast between the solute and solvent.

1. S. Chattopadhyay, *et al.*, *Cryst. Growth Des.*, 2005, **5**, 523.

Kevin Roberts commented: I feel it might be helpful to the discussion to clarify the role of SAXS in these experiments on PABA nucleation. Critically, I don't think anyone is saying that SAXS is a technique with the capability to identify the polymorphic form of the nucleating molecular species. The information content in SAXS data is obviously much less and hence not comparable with (say) that of powder XRD data which would normally be used for the purposes of polymorphic characterisation.

In this work, by examining the scattering function as a function of q, SAXS is simply used to try and measure the size and size development of the nucleating clusters and, through modelling, offer up an opportunity to see what particle shape models would be consistent with the observed data. PABA in ethanolic solutions, due to the limited electron density contrast between solute and solution, is extremely challenging from a SAXS perspective and this needs to be borne in mind when examining the data.

Here we merely have data which tells you that in this range we can get some tantalising and reproducible information about the early crystal growth stages which, for the high q range, is consistent with an increase in the proportion of dimers in solution with respect to that of the monomers. The latter is significant mindful of the fact that the α-polymorphic form expected from the process studied is characterised by a strong carboxylic acid dimer packing motif. The low q range shows the development of larger clusters in the 40 nm size range which grow in concentration during the crystallisation process and hence are indicative of a developing crystallisation process.

Unfortunately, the mid q range is featureless as a function of q and thus provides little insight regarding the further development of the dimers into the larger structures observed. The reasons for this are not currently clear but may reflect a wide dispersity of cluster sizes and/or the transient lifetimes associated with scattering process within the mid q region of the data.

We are hopeful the planned SANS experiments will in the future provide a much improved scattering contrast and concomitantly a better insight to the molecular assembly process for PABA from the solution phase.

Elena Simone communicated: Did you try to measure the dimerization constant of PABA in ethanol? Or did you check in the literature? I am aware of some experimental data for OABA in several organic solvents using UV-Vis.[1]

1. L. Zapala, *et al.*, *Biophys. Chem.*, 2009, **140**, 91–98.

Dimitrios Toroz communicated in reply: The UV-Vis spectra of PABA in ethanol were not studied in this work and so we have not calculated the dimerization constant, however, this is a very interesting suggestion and one which should be considered in future work. As an alternative, simulations relating to this question were performed, in which the preference for dimers to form or not in solution was calculated using molecular dynamics simulations as well as in other solvents.[1] A similar preference is observed for acetonitrile whereas dimers are not preferred in IPA. The calculations showed that in ethanol dimers are preferred in solution in comparison to two isolated monomers.

1. R. A. Sullivan, *et al.*, *Cryst. Growth Des.*, 2014, **14**, 2689.

Mireille Smets opened a general discussion of the paper by Kenneth Harris: The difference in the ^{13}C NMR spectra of polymorphs with a reasonably similar structure is in the order of a few ppm. How well did your DFT calculations agree with the observed ^{13}C NMR spectra for the polymorphs of L-phenylalanine with a known crystal structure and was it really possible to exclude some structures for the unknown phase?

Kenneth Harris answered: With careful referencing of the computed values of the isotropic ^{13}C chemical shifts, the errors between the experimental ^{13}C NMR data and the computed ^{13}C NMR data for form I of L-phenylalanine are less than 1 ppm. With regard to the assignment of the solid-state ^{13}C NMR spectrum observed for a "new" solid phase during an *in situ* ^{13}C NMR study of crystallization, ideally we would prefer to compare the measured *in situ* ^{13}C NMR spectrum to the experimentally measured ^{13}C NMR spectra for all known solid forms of the molecule of interest. However, in the absence of experimental data for a particular known solid form, computed values of the isotropic ^{13}C chemical shifts for this known solid form could reasonably be used to rule out this form if the computed values of the isotropic ^{13}C chemical shifts differ by more than about 1 ppm from those observed in the experimental *in situ* ^{13}C NMR study.

Mireille Smets asked: What would you estimate is the error in your DFT calculations in ppm?

Kenneth Harris responded: From a comparison of experimental ^{13}C NMR data and computed ^{13}C NMR data for a number of materials, we would estimate that the error is generally within 1 ppm. The error can be significantly smaller in many cases, but generalization is not straightforward as the error typically varies between different ^{13}C environments in the crystal structure. As the differences in isotropic ^{13}C chemical shifts for a given ^{13}C environment in different polymorphs can typically be as large as ±5 ppm, the periodic DFT calculations are sufficiently accurate to allow polymorph assignments to be made with a reasonable degree of confidence. Clearly, as the methodology for periodic DFT calculations continues to improve, the degree of confidence in making such assignments should also be expected to improve in the future. My response has been focused on the absolute error between experimental ^{13}C NMR data and computed ^{13}C NMR data. Clearly, the relative errors in the computed ^{13}C NMR data for different polymorphs is also a relevant issue, which would merit a detailed systematic investigation.

Joost van den Ende asked: You state that the increase of the γ and the decrease of the α polymorph as observed in Fig. 2b of your paper is consistent with a solution-mediated polymorphic transition. Would it also be consistent with a solid–solid polymorphic transition?

Kenneth Harris answered: As observed in Fig. 2b, the transformation from the α polymorph to the γ polymorph of glycine during the *in situ* NMR study of crystallization from D_2O occurs over a period of a few hours. For crystalline samples of the α polymorph of glycine kept under rigorously dry conditions, the transformation from the α polymorph to the γ polymorph takes place over a substantially longer period of time, of the order of several days or weeks. On this basis, we may reasonably rule out that the transformation occurring during the *in situ* NMR study of crystallization is a direct solid–solid transition. Instead, we know from independent studies that crystalline samples of the α polymorph of glycine kept under humid atmospheres transform to the γ polymorph much more quickly than under rigorously dry conditions, which suggests that the interaction of water with the surfaces of the crystals of the α polymorph may promote the transformation to the γ polymorph. Similarly, the fact that the crystals of the α polymorph formed initially in crystallization from water can transform over a few hours to the γ polymorph may reasonably be attributed to the role of interactions of the solvent with the surfaces of the crystals of the α polymorph. On this basis, we assign the transformation as solvent mediated. Clearly, we also cannot rule out the possibility that secondary nucleation of the γ polymorph occurs after the initial nucleation and growth of the crystals of the α polymorph, which would constitute another plausible mechanism for the polymorphic transformation to occur during the crystallization process, given that the α polymorph is meta-stable relative to the thermodynamically stable γ polymorph.

Alan Hare asked: Thinking back to the "old-style classic" experiments performed by Akitt in Leeds some 40 years ago: I am wondering if there are aluminium-27 species you think might contribute to a vindication of your approach?

Kenneth Harris answered: There are certainly very good prospects for applying our *in situ* NMR strategy to investigate the formation of aluminium-containing materials, and there are many important materials of this type for which insights on crystallization pathways would provide valuable new information. We have already carried out *in situ* ^{27}Al NMR studies in a project relating to cement formation, and we will publish the results soon. Another relevant example is the formation of aluminophosphate materials, an area that we are also currently exploring.

Alan Hare said: Some systems exist that are already well-characterised as monoclinic (here I am thinking of gibbsite and bayerite, in particular); might these be useful? Your time factors remind me also of some electron spin resonances; those studied, for example, using the nitroxide radical in zeolite cavities. However, I can see that you would wish to avoid perturbation of the system by a strong field such as from the vanadyl ion.

Kenneth Harris answered: In principle, our techniques may be usefully applied to such systems, except in the case of the materials comprising radicals in zeolites that you mention. In general, NMR studies of radical-based materials are challenging as a consequence of severe line broadening.

Ken Lewtas asked: The new multiple pulse sequence technique is very interesting indeed. I work on many projects involving multiple component formulations which (usually) have crystalline or glassy polymers, small molecules of varying T_gs and, frequently, solvents to adjust the properties. Could I extrapolate your work to such systems and examine both the solid and solution part of such formulations at the same time?

Kenneth Harris responded: Yes, our *in situ* solid-state NMR strategy and the new CLASSIC NMR strategy would be very powerful approaches for exploring several aspects of the types of multi-component system that you describe, provided of course that the different components in these systems are amenable to giving NMR spectra of reasonable quality on a suitable timescale.

Haihua Pan opened a general discussion of the paper by Anna Jawor-Baczynska†: I would like to ask you about your measurement of the solubility; In your experiments, you used an excess amount of valine and then filtered using a 100 nm pore size. However, in your paper, you have proven that even in a slightly undersaturated solution there are particles with more than 200 nm in diameter. Could the filtration with 100 nm pore size lead to the loss of solute, and lead to the underestimate of solubility?

Jan Sefcik responded: Mesoscale clusters in DL-valine solutions contain only a very small fraction of the overall solute present in the liquid phase, similar to previous observations in glycine and DL-alanine solutions.[1,2] Therefore their removal by filtration does not significantly influence concentration of the solute in the bulk solution.

1. A. Jawor-Baczynska, J. Sefcik and B. D. Moore, *Cryst. Growth Des.*, 2013, **13**, 470–478.
2. A. Jawor-Baczynska, B. D. Moore, H. S. Lee, A. V. McCormick and J. Sefcik, *Faraday Discuss.*, 2013, **167**, 425–440.

† Anna Jawor-Baczynska's paper was presented by Jan Sefcik, University of Strathclyde, Glasgow, UK.

Richard Sear remarked: Are your mesoscale clusters liquid-like in the sense that molecules in them rapidly exchange with the surrounding solution?

Jan Sefcik replied: That is a very good question. I would expect so but that needs to be tested.

Richard Sear further asked: Would it be possible to apply the technique of FRAP (Fluorescence Recovery After Photobleaching) to study the exchange of molecules between the mesoscale clusters and the surrounding solution? FRAP is commonly used in biology to study turnover rates in clusters.[1] This would, I think, need a photobleachable dye that partitions into your mesoscale clusters, combined with a confocal microscope.

1. T. Schwarz-Romond, C. Merrifield, B. J. Nichols and M. Bienz, *J. Cell Sci.*, 2005, **118**, 5269.

Jan Sefcik responded: I think this is a very good suggestion and certainly worth trying. It would need to be checked though if a dye has any significant effect on formation or dynamics of clusters themselves.

Sven Schroeder said: Is there a possibility to combine light scattering with other spectroscopic techniques to obtain a deeper understanding of speciation and internal structure in the mesophases you are observing?

Jan Sefcik answered: It would be certainly possible and very informative to combine scattering studies of mesostructued liquid phases with suitable spectroscopic techniques to get information about speciation, internal structure and dynamics within mesoscale clusters. Since mesoscale clusters typically contain only very small fraction of the total solute present in the liquid phase, highly sensitive techniques would be most likely to be useful.

Robert Hammond asked: During the discussion it has been suggested that there could be a partial analogy between the mesoscale, liquid-like structures observed in DL-valine–water–2-propanol mixtures and the structures the association colloids form. In this context, have you considered using fluorescence probes to interrogate the different environments in the solution that are suggested by your experimental observations? Since the structures are persistent steady-state fluorescence measurements would probably suffice and might provide information about the polarity and micro-viscosity within the structures.

Jan Sefcik replied: Fluorescence measurements would be certainly worth trying. We have not made any systematic studies using fluorescence so far. Of course it would need to be checked whether fluorescence dyes have any significant effect on the formation or dynamics of clusters themselves.

Robert Hammond said: Do you observe any changes at the solution/air interface that accompany the formation of the mesoscale liquid-like structures in the bulk of the solution? For example is there any change in the surface tension?

Jan Sefcik responded: We have not investigated solution/air interface in these systems but it is certainly worth looking into.

Dikshitkumar Khamar remarked: In addition to the size of clusters, the number of clusters are equally important. Is it completely inappropriate to look at the number of clusters by DLS while comparing two samples especially when the samples are repetitively showing an order difference in the derived count rate? Or is this something that should only be looked at with a Brownian microscopic technique?

Jan Sefcik responded: Light scattering intensities at a given scattering angle depend on cluster properties such as size, structure and optical contrast (refractive index). For clusters with given properties the corresponding contribution to the scattered intensity is then proportional to the number of clusters present in a scattering volume. Therefore if all clusters were of same size, structure and composition, and were sufficiently dilute to neglect their mutual interactions, the scattering intensity (after subtracting background scattering from the bulk solution) would be proportional to the number concentration of clusters. So for monodisperse uniform systems this may be a reasonable approximation. On the other hand, even for Brownian microscopy, there is usually a distribution of intensities from individual clusters detected and only those with sufficient optical contrast will be picked up by image processing software used for data analysis.

Marco Mazzotti commented: I find the results presented in your Fig. 4, 5 and 6, about cluster size distributions under different preparation conditions and corresponding induction times very interesting, but somewhat under-exploited. I wonder if you could comment on how cluster size distributions and induction times can be correlated? Another important question is whether the induction times have been measured a number of times large enough to build representative statistics?

Jan Sefcik answered: It can be seen from data reported in Fig. 4, 5 and 6 that larger cluster sizes are correlated with shorter induction times as discussed in our paper. We agree that these data can be exploited further, for example in terms of analysing induction time distributions to estimate nucleation kinetic parameters and relating these to suitable moments of cluster size distributions. This will be further reported elsewhere. Induction times were measured at least 20 times for each experimental condition and the trends reported in Fig. 6 are indeed statistically significant.

Simon Black said: Are you saying that these supersaturated solutions remember whether they have been prepared by cooling or by anti-solvent addition?

Jan Sefcik answered: We can certainly say that these supersaturated solutions have history dependent properties, so that depending on how they were prepared there are different distributions of mesoscale clusters observed. I would not necessarily say that these solutions "remember" anything, they are just different.

Dikshitkumar Khamar followed on from the question by Simon Black: In our work with Brownian microscopy and DLS techniques in relation to the effect of solution history on nucleation, we see a clear effect of pre-treatment conditions (dissolution temperature, dissolution time, stirring *etc.*) on the cluster size and numbers. This eventually affects the induction time; we could not probe such changes with vibration spectroscopy even after using multivariate data analysis but DLS and Brownian microscopy produce reproducible results suggesting the effect of solution history on nucleation.

Jan Sefcik replied: As correlations between mesoscale clustering in supersaturated solutions and nucleation kinetics are being established, the next challenge will be to identify any relevant causal relationships and test hypotheses related to mechanisms through which clusters participate in nucleation processes.

Faraday Discussions

PAPER

Measuring induction times and crystal nucleation rates

Clément Brandel†[a] and Joop H. ter Horst*[b]

Received 22nd November 2014, Accepted 27th January 2015
DOI: 10.1039/c4fd00230j

A large variation is observed in induction times measured under equal conditions in 1 ml solutions. Ruling out experimental errors, this variation originates from the nucleation process. The induction time distribution is explained by the stochastic nature of nucleation if the number of nuclei formed is approaching 1 per vial. Accurate heterogeneous crystal nucleation rates were determined from the induction time distributions on a 1 ml scale for racemic diprophylline in two solvents. The difference in nucleation behaviour in the two solvents originates from the energy barrier for nucleation, which is much higher in the solvent in which induction times are much longer. In addition the pre-exponential factor for the crystal nucleation rate in both solvents is rather low compared to predictions using Classical Nucleation Theory. Unfortunately, concentration and surface characteristics of the effective heterogeneous particles are not known which clouds a further molecular interpretation.

1. Introduction

Crystal nucleation rate predictions still are unreliable, differing by orders of magnitude with the experimental ones.[1,2] At the same time, crystal nucleation rates are difficult to measure. This is due to the strongly non-linear behaviour making it necessary to accurately control supersaturation. Further, the nucleus, generally of a size of 1–1000 molecules, can as yet not be observed experimentally due to analytical limitations and the difficulty in the localization of the nucleus. Another complication is the occurrence of growth, agglomeration and secondary nucleation during the primary crystal nucleation measurement, which can distort the experimentally-obtained nucleation rates.

[a]*Normandie Université, Crystal Genesis Unit, SMS, Université de Rouen, EA3233, IMR 4114, F-76821 Mont Saint-Aignan Cedex, France*

[b]*EPSRC Centre for Innovative Manufacturing in Continuous Manufacturing and Crystallisation (CMAC), Strathclyde Institute of Pharmacy and Biomedical Sciences, University of Strathclyde, 161 Cathedral Street, Glasgow G4 0RE, UK. E-mail: Joop.terHorst@strath.ac.uk; Tel: +44 (0)141 548 2858*

† Current address: Particle Sciences, Devices & Eng – UK R&D Platform Technology & Science, GSK, Gunnels Wood Road, Stevenage, Hertfordshire, SG1 2NY, UK, Email: clement.m.brandel@gsk.com

Accurately known crystal nucleation rates as a function of supersaturation under well-defined conditions would enable a molecular level interpretation of the nucleation process.[3] Thorough nucleation rate measurements using the double pulse technique[4] for instance lead to the experimental identification of the two-step crystal nucleation mechanism.[5] Such molecular level interpretations will lead to the development of new crystal nucleation theories with improved predictive power.

Recently, a new measurement method was proposed that determines the crystal nucleation rate in stirred solutions using the variations in induction time measurements under equal conditions in small volumes.[6] The aim of this paper is to discuss the limitations and advantages of the new method. First, induction time measurements under equal conditions are discussed after which it is shown that a conventional analysis does not arrive at reasonable nucleation rates. It is then shown that in these measurements a single nucleus mechanism is occurring for which a model is derived. The model can be used to determine nucleation rates from the obtained induction time distributions. Nucleation rates of the compound diprophylline are analysed and discussed in the light of the Classical Nucleation Theory.

2. Experimental

Racemic diprophylline (Sigma-Aldrich, ≥99%), isonicotinamide (Sigma-Aldrich, ≥99%), 4-hydroxyacetophenone (Fluka Analytical, ≥98%), ethanol (Sigma-Aldrich, ≥99.8%), isopropylalcohol, dimethylformamide and ethyl acetate (J. T. Baker, ≥99.5%) were used as received. Saturation temperatures and induction times were measured using a 1.5 ml multiple reactor system (Crystal16, Avantium Amsterdam). It can hold 16 standard HPLC glass vials (1.8 ml) and measures the transmission of light through a sample in the vials. A temperature recalibration of the Crystal16 was performed to make sure the actual temperature within the vials at the constant temperature of the induction time measurement coincided with the set temperature of the device. This means that the difference in the constant temperature between vials was brought back to its minimum. The Erlenmeyer flask, volumetric flask, pipettes, funnels and spatula were cleaned with 95% ethanol and dried prior to use to minimize the introduction of particulates and dissolved impurities.

To reduce concentration differences between sample vials a 25 ml solution was prepared by dissolving the right amount of solid material in the solvent at elevated temperature. A heated bottle-top dispenser was used to dispense 1 ml of clear solution into each vial. A small magnetic stirrer was introduced in each vial and the vial caps were strongly fixed to avoid evaporation of the solvent. They were then placed in the Crystal16 setup.

A heat–hold–cool cycle was repeated at least 4 times with a controlled stirring speed of 700 rpm. Firstly, the sample was heated at a rate of 0.3 °C min^{-1} to 60 °C, which was well above the saturation temperature. Then, this temperature was maintained for 60 minutes to make sure that crystals were dissolved to obtain a clear solution. Then, the clear solution was cooled down to a constant temperature with a fast cooling rate of 5 °C min^{-1}. The moment the set temperature was reached was taken as time zero, after which the constant temperature was maintained for at least 5 hours. The induction time was taken as the difference between the time at which transmission of light started to decrease and time zero. The start and end time could be determined with an accuracy of 1 second. After 5 hours a subsequent heat–hold–cool cycle was started. In this way a total of at least 64 induction time measurements were obtained per supersaturation ratio value.

The vials were weighted before and after the experiments in order to check that evaporation of the solvent had not taken place. In the case the difference of weight was significant (≥0.05%) the corresponding induction times were not considered in the analysis.

Care was taken to measure model systems that were not prone to crowning behaviour where a circle of crystals would form just above the liquid level. This was also checked during the measurements.

3. Crystal nucleation rate measurement method

3.1. Induction times

The induction time t_i is defined as the time progress between the instance of constant supersaturation creation and the instance of detection of crystals.[7] Fig. 1 shows a typical induction time measurement on a 1 ml scale in a stirred supersaturated solution. The instance of reaching constant supersaturation was chosen to be the moment that the set temperature reached the desired constant temperature. The actual temperature in the vial will probably show a slight delay. Therefore, series of measurements in which induction times of close to zero are measured should be carefully treated.

The response of the transmission of light to the presence of crystals is quite sharp. However, the transmission of light detection does not detect the nucleus. It detects the moment that the suspension contains a certain volume fraction of crystals. Therefore, the moment that the transmission of light decreases is delayed in respect to the moment that the nucleus appears.

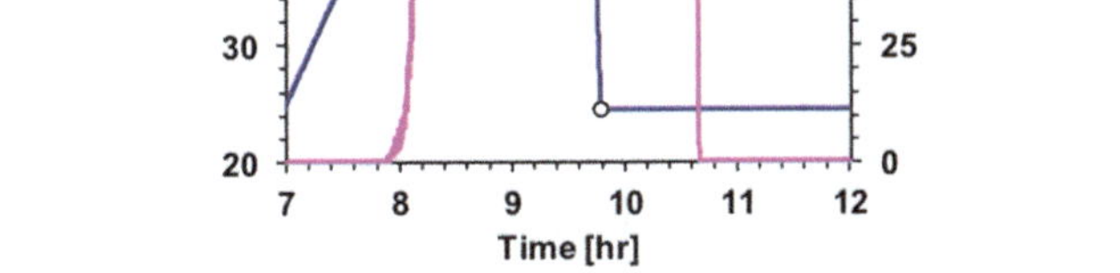

Fig. 1 An induction time t_i measurement using temperature to control supersaturation and light to detect crystals. The start of the t_i measurement, indicated by the circle on the temperature line, is the time of reaching the set temperature of 25 °C. The end of the measurement, indicated by the circle on the light transmission line, is the time at which crystals are detected by a decrease in the transmission of light through the sample. The measured induction time was roughly 2900 seconds.

Fig. 2 shows the induction times measured for two vials in 8 consecutive experiments under equal conditions. While the induction time can be measured under well-defined and well-controlled conditions, a large variation in the induction time values can be seen. The series of measurements show induction times respectively ranging from 156 to 3044 s and from 390 to 1673 s.

The induction time variations within a measurement series in a single vial, therefore, show large variations. Weighing the vials before and after the series of experiments indicated that no solvent evaporation took place. Solvent evaporation would lead to an increase in supersaturation during the series of measurements and thus to a decreasing induction time throughout the series of measurements. Also the measured clear point of the formed suspensions did not show changes and thus did not indicate a change in supersaturation. Also no indications of decomposition of solvent or model compound were obtained.

Both series of measurements seem to vary similarly. These variations are not the result of concentration differences between the vials or within the series. Buffer solutions were used for the vials making sure that the concentration difference between the vials was as small as possible. It seems that the variations are a result of the crystallization in the vials.

3.2. The conventional model for induction time analysis

One way of analysing induction times is to assume detection upon filling of a certain fraction of the volume by the crystalline phase. If it is assumed that the suspension is formed by only primary nucleation and growth, the induction time can be written as:[7]

$$t_i = \left(\frac{3\alpha}{\pi G^3 J}\right)^{1/4} \tag{1}$$

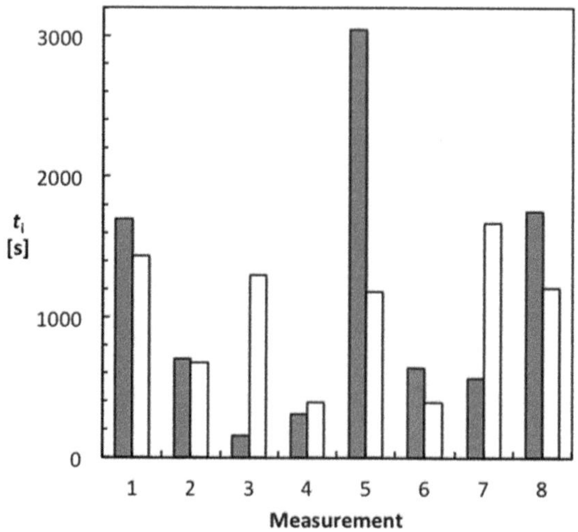

Fig. 2 Two series of 8 subsequently measured induction times under equal conditions in a single vial. The system is 4-hydroxy acetophenone in ethyl acetate at a supersaturation of $S = 1.3$.

where it was assumed that spherical particles are formed which grow with a linear growth rate G. The equation further contains the detectable crystal volume fraction α and the crystal nucleation rate J. The nucleation rate J can be determined from the induction time t_i by assuming a value for the growth rate G and volume fraction α.

$$J = \frac{3\alpha}{\pi G^3 t_i^{\,4}} \tag{2}$$

By assuming a value of $G = 10^{-7}$ m s^{-1} and of $\alpha = 0.1$ we arrive at crystal nucleation rates varying from 1 to 161.000 ml^{-1} s^{-1} (10^6 to 161×10^9 m^{-3} s^{-1}) for the induction times in Fig. 2. This is a variation over 5 orders of magnitude. These variations in nucleation rate cannot be explained by variations in conditions such as temperature and concentration since these were very small. The variations therefore would have to be the result of the crystallisation process itself. However, having these kinds of induction time variations for nucleation rates of more than 1 per vial per second seem not to be possible because of the large number of crystals that would be present in the vials, diminishing stochastic behaviour.

3.3. The single nucleus mechanism

Visualization of the stirred solution in 3 ml volumes showed the presence of a large single crystal before a crystal suspension was detected by the decrease in the transmission of light through the sample.[8] Fig. 3 shows a clear solution, the single parent crystal and the suspension after detection through a decrease in the transmission of light. The observation of the single crystal relied on the probability that on the small number of occasions that it passed the camera it also would be captured by the camera (1 picture per second). Probably therefore, not all visualization experiments resulted in the observation of a single crystal. However, this single nucleus mechanism was observed on a 3 ml scale in a large number of model systems, among which isonicotinamide in ethanol, butanol, nitrobenzene, nitromethane, paracetamol and succinic acid in water as well as 4-hydroxy acetophenone in ethyl acetate. Increasing the solution volume at constant supersaturation decreases the induction time variation because more nuclei form in the solution. Similarly, variations in metastable zone widths

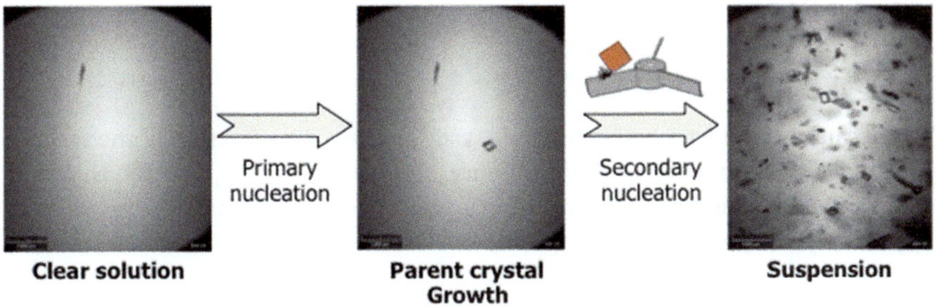

Fig. 3 The single nucleus mechanism where a single parent crystal nucleates in a clear solution. After growth of this parent crystal secondary nucleation occurs and a suspension is forming.

reduced in larger volumes.[9] While for paracetamol in water there were no variations anymore on a 1 liter scale, there was still significant variation for isonicotinamide in ethanol on the same scale.

If the single crystal is large enough it can cause secondary nucleation either through crystal–stirrer collisions (attrition) or some surface breeding mechanism. Then, only a single crystal is responsible for the generation of the entire final crystal population in the suspension. Under concomitant polymorphism conditions of isonicotinamide in solvent mixtures it was tested whether cooling crystallisation in 3 ml volumes resulted in pure polymorphic products or in polymorphic mixtures. A series of 10 stirred cooling crystallisation experiments lead to 7 of them crystallizing as pure form I and 3 as pure form II while no mixtures were observed.[10] Other systems showed similar behaviour.

These experiments indicate that a single nucleus mechanism is occurring. After supersaturation generation a single crystal nucleates and grows until it is large enough for secondary nucleation to occur. Then quickly a full suspension of crystals, all originating from the same single crystal, is formed. The crystallisation process in such small volumes then is split up into different periods in which the subsequent processes of crystallisation occur. First, a single crystal nucleation event occurs. Then, growth occurs up to a macroscopic size. Finally, secondary nucleation occurs due to the presence of this large single crystal in the stirred solution. Since the subprocesses of crystallisation are effectively separated, a simple model can be made to capture this process as well as the variations in it.

3.4. A model for the single nucleus mechanism

The variation of induction times can be explained if the formation of the suspension is the result of a single nucleation event. This event depends on the probability of the formation of a nucleus in a certain timeframe. The lower the nucleation rate, the lower the probability would be to form a nucleus in this timeframe. At constant supersaturation this can be captured through the Poisson distribution, which describes the probability P_m of the formation of m nuclei in a certain timeframe t.

$$P_m = \frac{N^m}{m!}\exp(-N) \tag{3}$$

where $N(t) = JVt_J$ is the average number of nuclei formed in the timeframe t_J with the rate J and within a volume V. The probability $P^*(t)$ that one or more nuclei are formed is then:

$$P^* = 1 - P_0 = 1 - \exp(-JVt_J) \tag{4}$$

We can only detect a suspension of crystals occupying a fraction of the volume which gives a time delay t_g, the difference between the time t_J and the time t of detection of crystals. If this growth time is equal for all detections, the probability $P(t)$ that crystals are detected is given by:

$$P(t) = 1 - P_0 = 1 - \exp(-JV(t - t_g)) \tag{5}$$

The above equation is a description of the probability $P(t)$ that crystals are detected in a supersaturated solution sample as a function of time. It depends on the nucleation rate J, the sample volume V and the growth time t_g. This equation is valid starting at time t_g. Below time t_g the probability $P(t)$ would be zero. For very large times t the probability would become 1 (Fig. 4).

3.5. Nucleation rate determination

The probability $P(t)$ is also experimentally accessible. It can be determined by measuring a large number of induction times under absolutely equal conditions as we did in the case of HAP in ethyl acetate shown in Fig. 2. The experimental probability $P(t)$ can then be determined from:

$$P(t) = \frac{M^{+}(t)}{M} \tag{6}$$

where M is the total number of induction time measurements and $M^{+}(t)$ is the number of measurements in which crystals are already detected at time t. Fig. 5 shows a typical induction time distribution using a series of 120 induction times divided over 8 vials. The smallest induction time measured was 156 s while the largest was 4100 s. After an initial time period within which no crystals were detected for all samples, the probability $P(t)$ quickly increases and levels off towards a probability of 1. This is exactly the curve shape that is captured in the model for the single nucleus mechanism proposed earlier.

As an example the cumulative probability distribution function is fitted to the experimentally determined probability distribution using a nucleation rate $J = 865 \pm 20\ \mathrm{m^{-3}\ s^{-1}}$ and growth time $t_g = 263 \pm 20$ s. Model and data can be fitted quite well, only for the higher values of $P(t)$ is a deviation between them apparent. It thus seems that the method can be used to determine crystal nucleation rates.

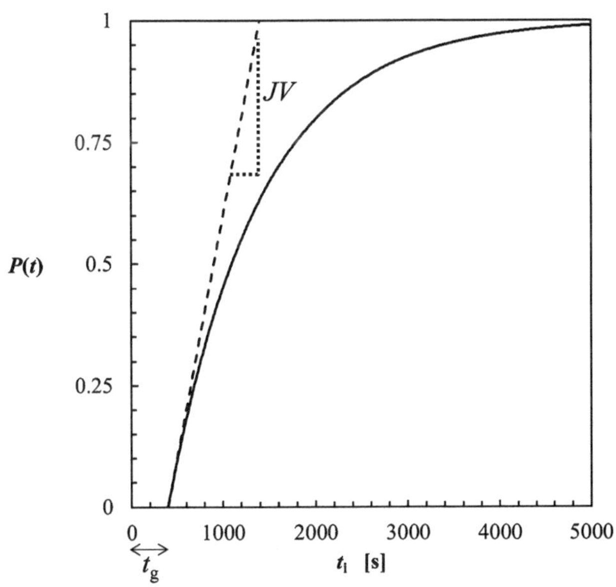

Fig. 4 The probability from eqn (5) as a function of time. The relation between equation and graph is indicated by t_g and JV. The graph was constructed with a growth time $t_g = 400$ s and product of volume V and nucleation rate J of $JV = 0.001\ \mathrm{s^{-1}}$.

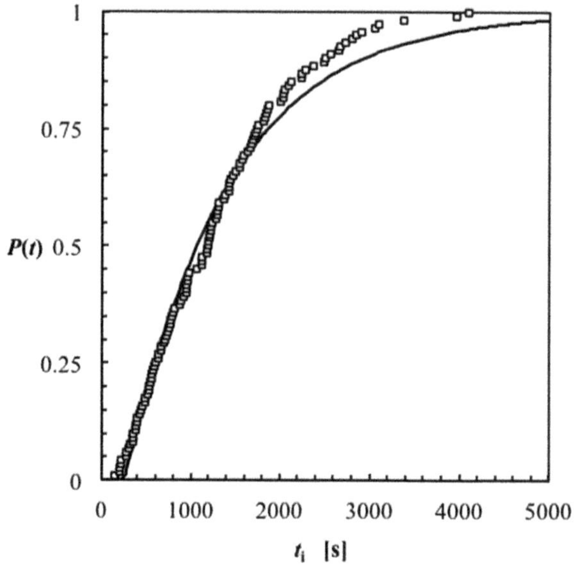

Fig. 5 The induction time distribution of 120 induction times measured for HAP in ethanol at a supersaturation $S = 1.3$ and temperature $T = 25$ °C. The distribution function (eqn (5)) was fitted using a nucleation rate $J = 865$ m^{-3} s^{-1} and a growth time $t_g = 263$ s.

3.6. Heterogeneous crystal nucleation

Usually it is assumed that crystal nucleation from solution occurs through heterogeneous nucleation: crystals form onto foreign particles such as dust particles or on interfaces such as that between glass walls, stirrer surface or air and the solution. To investigate the kind of heterogeneous nucleation occurring during the crystal nucleation rate measurements using the model system isonicotinamide in ethanol 3 types of experiments were done. Next to the base

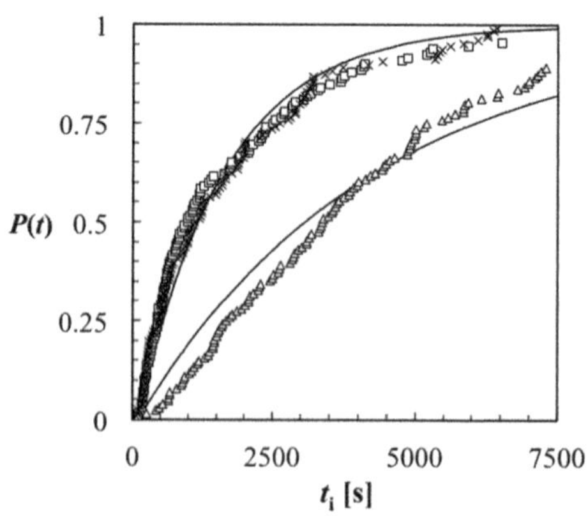

Fig. 6 The induction time probability distribution of isonicotinamide in ethanol at a supersasturation of $S = 1.44$. □: untreated vial and solution; ×: using silanized vials; △: using filtered solutions.

experiments the solution was filtered or the vials were silanized. These 3 series of induction time measurements are shown in Fig. 6.

While the silanization of the glass surfaces of the vials does not seem to have an effect on the distribution, the solution filtration has a marked effect. Longer induction times are needed to reach the same induction time probability: apparently the nucleation rate decreased. It can be seen that this is the nucleation rate and not the growth time because all probability curves intercept the time-axis at around the same point. It thus can be concluded that the crystal nucleation takes place on suspended heterogeneous particles in the solution that can be partially removed by solution filtration.[11] Unfortunately, the concentration and surface functionality of these heterogeneous particles are not known. This is actually generally true for the heterogeneous particles in crystallisation processes.

4. Nucleation rate analysis & interpretation

The theophylline derivate diprophylline (DPL, Fig. 7), used in the treatment of obstructive airway diseases such as bronchial asthma, is marketed as a racemic compound. Racemic DPL is a polymorphic compound[12] which can form as the marketed stable form RI and a metastable form RII. Supersaturated DPL solutions have difficulty crystallizing showing in a persistence of the metastable supersaturated solution: it took at least 3 hours before crystals were observed in a stirred solution in dimethylformamide (DMF) at 15 °C at a supersaturation ratio $S = 2.8$.[13] While prone to large metastable zone widths, polar solvents like dimethylformamide (DMF) result in DPL form RI. In the less polar solvent DPL the metastable zone is much smaller while form RII crystallizes. At 20 °C the solubility of RI in DMF is 7.0 w% while that of RII in IPA is 0.5 w%. During the gravimetric solubility measurements no transformation of RII to RI occurred in IPA. Crystallization experiments showed that only form RII formed in IPA. In DMF only form RI formed. To compare and analyse crystal nucleation rates for the different polymorphs we have measured induction time distributions of DPL from DMF and IPA.

4.1. Induction time distribution measurements

Fig. 8 shows a number of experimental induction time distributions for different supersaturation ratios of racemic DPL in IPA and DMF at 20 °C. It can be seen that the distributions become broader at lower supersaturations indicating lower nucleation rates. The point where the probability starts to increase from zero is larger for smaller supersaturation ratios, indicating an increase in the growth time t_g with decreasing supersaturation ratios.

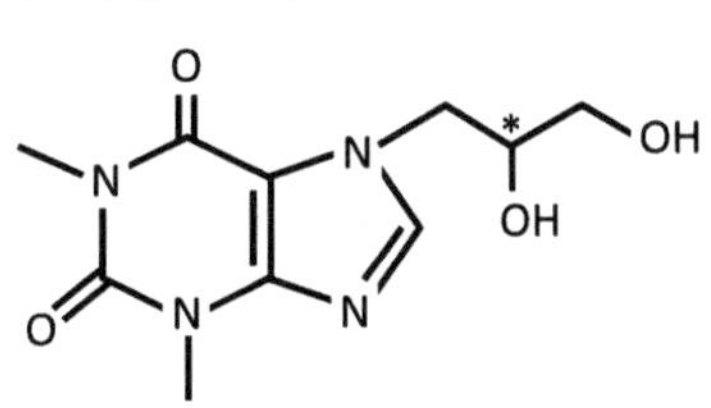

Fig. 7 The molecular structure of diprophylline.

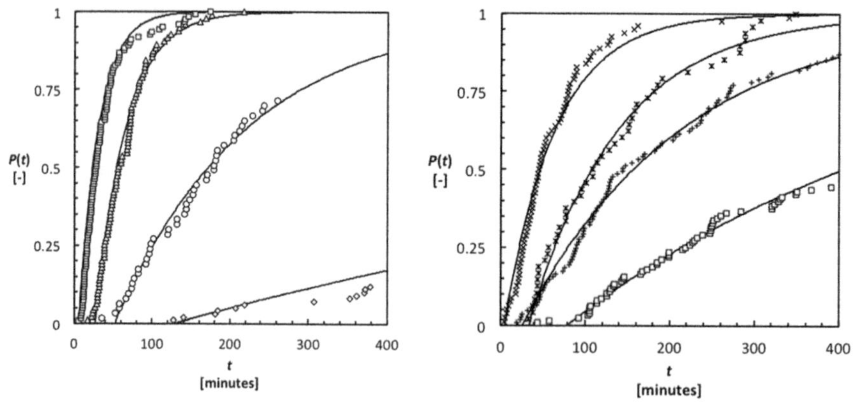

Fig. 8 The experimental induction time probability distributions of diprophylline (DPL) in IPA and DMF at different weight fraction-based supersaturation ratios S. Also the fits to eqn (5) are shown. Left: DPL in IPA: ◇ − $S = 1.5$; ○ − $S = 1.7$; △ − $S = 3.0$; □ − $S = 3.0$. Right: DPL in DMF: □ − $S = 3.0$; + − $S = 3.5$; * − $S = 4.0$; × − $S = 4.2$. Note that time is given in minutes rather than seconds.

A remarkable difference between the 2 graphs in Fig. 8 is the prevailing supersaturation ratio. While in IPA supersaturation ratios of $S = 1.5$ to 3 are responsible for the measured distributions, in DMF they are $S = 3$ and up. Looking at the measurements at $S = 3$ in both graphs, for DPL in IPA and DMF induction times from, respectively, 5.9 to 173 and 43 to 424 minutes are measured. All curves possess the typical distribution shape as in Fig. 4. Fitting eqn (5) to the experimental induction time distributions in Fig. 8 gives the values in Table 1 for the nucleation rate J and growth time t_g at different supersaturations. Apparently the nucleation of RII from IPA is much faster than the nucleation of RI (but also RII) from DMF. It is interesting to investigate whether this is due to the nucleation kinetics or the energy barrier for nucleation.

4.2. Nucleation rate factors

The obtained crystal nucleation rates at different supersaturations can now be analysed in terms of the Classical Nucleation Theory. The crystal nucleation rate equation from Classical Nucleation Theory is given by:[1,7]

$$J = AS \exp\left(-\frac{B}{\ln^2 S}\right) \tag{7}$$

although also other dependencies of the pre-exponential term on the supersaturation ratio S are known in literature.[7,14] The variables A and B are supersaturation ratio S independent variables and can be referred to as the kinetic factor or pre-exponential factor A and thermodynamic or exponential factor B. The equation can be rearranged:

$$\ln\frac{J}{S} = \ln A - \frac{B}{\ln^2 S} \tag{8}$$

Table 1 Crystal nucleation rates J and growth times t_g determined from the fit of eqn (5) to the experimental induction time probability distributions of DPL in IPA and DMF. The error in the fit is less than 10%

	DPL RII from IPA		DPL RI from DMF	
S [—]	J [m^{-3} s^{-1}]	t_g [min]	J [m^{-3} s^{-1}]	t_g [min]
1.5	11.6	128		
1.7	97.0	52		
2.0	429	27		
3.0	668	5.9	35.7	79
3.5			88.7	26
4.0			157	34
4.2			275	4.7

This shows that the experimental nucleation rate data can be plotted as $\ln(J/S)$ *versus* $\ln^{-2} S$ which would result in a straight line with intercept $\ln A$ and slope $-B$. This is done in Fig. 9 for the nucleation rates reported in Table 1 of DPL in IPA and DMF. It can be seen that the points are to some extent on a straight line. The fitting procedure results in a value of $A = 576$ m^{-3} s^{-1} and $B = 0.68$ for DPL from IPA and $A = 499$ m^{-3} s^{-1} and $B = 4.57$ for DPL from DMF (Table 2). While the thermodynamic parameter B is an order of magnitude larger for DPL in DMF compared to that in IPA, the kinetic parameter is similar in size.

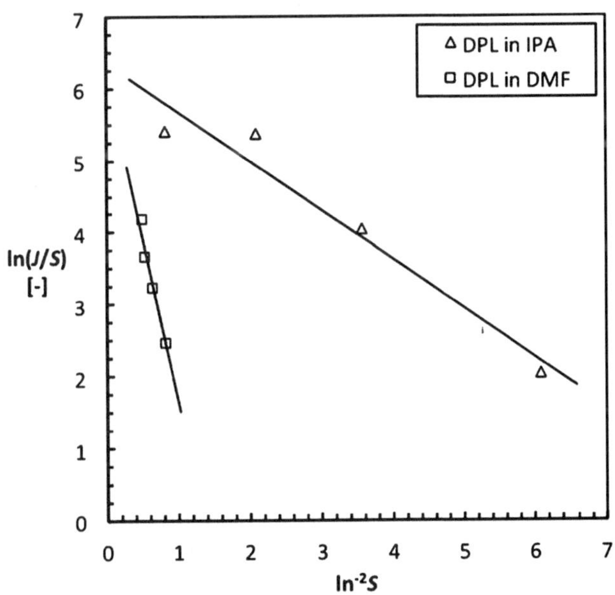

Fig. 9 The experimental crystal nucleation rates of DPL from IPA and DMF (Table 1) plotted following eqn (8) in order to determine the kinetic factor A and the thermodynamic factor B.

Table 2 Values for the pre-exponential kinetic factor A and the thermodynamic factor B determined for a number of different systems using the same crystal nucleation rate measurement method

System	A [m^{-3} s^{-1}]	B	Reference
DPL form RII in IPA	576 ± 235	0.68 ± 0.11	This work
DPL form RI in DMF	499 ± 221	4.57 ± 0.70	This work
m-Aminobenzoic acid in 50 w% water–ethanol	870×10^3	3.6	6
L-Histidine in water	36.3×10^3	1.1	6
Isonicotinamide in ethanol	5.86×10^3	0.323	11
p-Aminobenzoic acid in acetonitrile	36.3×10^3	0.016	14
p-Aminobenzoic acid in ethyl acetate	20.5×10^3	0.053	14
p-Aminobenzoic acid in 2-propanol	10.9×10^3	0.078	14
Benzoic acid in toluene	61.8×10^3	0.46	14

4.3. Interpretation of the thermodynamic factor B

The thermodynamic parameter B is much higher for DPL nucleation from DMF than from IPA. This indicates that at the same supersaturation ratio S the crystal nucleation work, which is the energy barrier for nucleation, is much larger for DPL nucleation from DMF. The nucleation work W^* can be calculated using $W^*/kT = B/\ln^2 S$. For a supersaturation ratio of $S = 3$ this gives values of $W^* = 0.6$ and $3.8kT$ for DPL nucleation from respectively IPA and DMF. These nucleation work values seem on the low side, especially for DPL nucleation from IPA with its nucleation work below $1kT$.

In Section 3.4 it was established that heterogeneous rather than homogeneous nucleation takes place for isonicotinamide in ethanol. Most probably this is also the case for DPL. For heterogeneous nucleation the thermodynamic factor B is related to the creation and reduction of surface areas with different interfacial energies. It is a balance between the 3 interfacial energies that play a role: the interfacial energy between nucleus and heterogeneous particle, nucleus and solution, and solution and heterogeneous particle. This creates an energy barrier (nucleation work) lower than that for homogeneous nucleation in absence of a heterogeneous particle in which only a single interfacial energy plays a role: that between the nucleus and the solution. Therefore, an effective interfacial energy γ_{ef} can be introduced that captures the balance between all interfacial energies involved.

The relation between heterogeneous nucleation work W^* and the thermodynamic parameter B for heterogeneous nucleation is given by:[1,7]

$$\frac{W^*}{kT} = \frac{4c^3 v^2 \gamma_{ef}}{27(kT)^3 \ln^2 S} = \frac{B}{\ln^2 S} \tag{9}$$

In which shape factor $c = (36\pi)^{1/3}$ for a spherical particle, temperature $T = 298$ K and molecular volume $v = 269 \times 10^{-30}$ m^3 can be assumed for DPL to calculate the effective interfacial energy $\gamma_{ef} = 3.34$ and 6.29 mJ m^{-2} from the thermodynamic parameters $B = 0.323$ and 4.48 of DPL in respectively IPA and DMF. The effective interfacial energy for DPL nucleation from DMF is much larger and therefore the nucleation work is higher and the nucleation proceeds much slower.

Crystal solubility is connected to the interfacial energy between crystal and solution.[15] The interfacial energy between DPL crystals and the solution can be calculated from the solubilities to be around $\gamma = 29.9$ and 15.6 mJ m^{-2} for DPL in IPA and DMF respectively. Since the solubility in IPA is much smaller than in DMF the interfacial energy is twice as large. Interestingly, while the interfacial energy γ is much larger for DPL in the solvent IPA compared to the solvent DMF, the effective interfacial energy γ_{ef} is much smaller in IPA.

The relation between the interfacial energy γ between crystal and solution and the effective interfacial energy γ_{ef} is given by the activity factor $\psi = \gamma_{ef}/\gamma$. Effectively, the interfacial energy between crystal and solution is substantially decreased by the presence of the heterogeneous particles. Table 3 shows the calculated values for DPL nucleation from IPA and DMF. The calculated value for the activity factor of $\psi = 0.11$ and 0.40 seem characteristic values for heterogeneous crystal nucleation. They also seem to indicate that the heterogeneous particles in IPA are more effective in reducing the energy barrier compared to those in DPL solutions in DMF. However, care should be taken with the method to determine interfacial energies from solubilities because it is based on a constant derived from supposedly homogeneous crystal nucleation rate measurements of ionic compounds.[16]

4.4. Interpretation of the kinetic factor *A*

The pre-exponential factor *A* found for DPL is roughly equal in both solvents. The theoretical value of the kinetic factor *A* is estimated to be 10^{15} to 10^{25} m^{-3} s^{-1}.[7,15] This is orders of magnitude larger than the experimental values for *A* of DPL nucleation. Other compounds also show low pre-exponential factors ranging from 10^3 to 10^6 m^{-3} s^{-1}. The experimental kinetic factor *A* is thus orders of magnitude smaller than theoretically is expected. In order to analyse this discrepancy the different parts that constitute the kinetic factor should be looked at. The pre-exponential term *AS* from eqn (8) describes the statistical process that a building unit attaches to the nucleus.[1] It is the product of the concentration of nucleation sites C_0, the attachment frequency f^* and the Zeldovich factor z.

$$AS = zf^*C_0 \tag{10}$$

The Zeldovich factor z expresses that clusters larger than the nucleus size still have a probability to decay rather than to grow out to macroscopic sizes.[1] It further corrects for the use of an equilibrium nucleus concentration rather than the actual nucleus concentration in the derivation of the crystal nucleation rate

Table 3 Interfacial energy γ calculated from the solubility, effective interfacial energy γ_{ef} from the nucleation rate measurements and activity factor $\psi = \gamma_{ef}/\gamma$ for DPL in IPA and DMF

System	γ [J m^{-2}]	γ_{ef} [J m^{-2}]	$\psi = \gamma_{ef}/\gamma$
DPL form RII in IPA	29.9	3.34	0.11
DPL form RI in DMF	15.6	6.29	0.40

equation. It does not seem likely that the Zeldovich factor is the origin of the discrepancy between theoretical and experimental values.

The attachment frequency f^* of building units to the nucleus determines how many nuclei of size n^* chance to supernuclei of size $(n^* + 1)$ per unit of time. During this process, building units must transfer from being dissolved and solvated to being partially desolvated and adsorbed on the nucleus surface. For an interfacial transfer controlled attachment frequency, the rate that building units become adsorbed can be given by:[1,7]

$$f^* = \lambda A^* D \frac{X}{d} \tag{11}$$

where A^* is the nucleus surface area, X is the concentration of building units in the solution and d reflects a distance representing the jump of the building unit from dissolved to partially adsorbed. The parameter D is some kind of diffusion coefficient that describes the transfer of a building unit from the nucleus neighbourhood to being partially adsorbed. The sticking coefficient λ reflects that there are building units in the nucleus neighbourhood not transferring to being adsorbed.

The attachment frequency thus depends on how the solute is present in the solution. The solute might have a solvent shell that has to be removed before it can be partially adsorbed onto the nucleus. There could also be different associates present in solution that differ in concentration X and diffusion coefficient D. Isonicotinamide for instance shows widely different self-association behaviour in different solvents resulting in varying polymorphic crystallisation outcomes.[17]

If the discrepancy between theoretical and experimental A comes from the attachment frequency f^*, the sticking coefficient λ, the concentration X or the diffusion coefficient D would be orders of magnitude smaller than theoretically predicted. Since this attachment process is difficult to access in experiments, the study of it with molecular simulations would enable a more accurate and sound description, leading to increased understanding and predictability of crystal nucleation rates.

The concentration of nucleation sites C_0 is related to the concentration of heterogeneous particles. If only 1 nucleus forms on a heterogeneous particle, the concentration C_0 coincides with that of the heterogeneous particles. Apart from their concentration, their surface functionality is important since it influences the interfacial energies involved in the heterogeneous nucleation work, the effective interfacial energy, the activity factor and thus the thermodynamic factor B. From the analysis of the factor B we found an activity factors of $\psi = 0.11$ and 0.40. Unfortunately, the concentration of particles onto which crystal nucleation occurs is not known. We can say that if the discrepancy between theoretical and experimental A comes from the concentration of nucleation sites, the actual concentration of nucleation sites is orders of magnitude smaller than assumed in the theoretical prediction. Anyway, the heterogeneous particles responsible for heterogeneous nucleation cannot be separately characterized. This makes it impossible to interpret the concentration of nucleation sites C_0 and with it, the kinetic factor A.

The way forward in crystal nucleation research lies in the interpretation of crystal nucleation rate data. However, a good interpretation cannot be done without knowing anything about the effective heterogeneous particles and the self-association and solvation processes in the solution. A first attempt was given for isonicotinamide of which the role of self-association on template crystallisation was recently studied.[18]

The nucleation rate equation from Classical Nucleation Theory is backed by a thorough and detailed physical description.[7] Apparently, this description does not significantly accurate capture the true molecular aspects of the process. We would first have to establish the defects in this description by Classical Nucleation Theory before developing new predictive theories.

5. Conclusions

The variation in induction times of experiments under equal conditions results from crystal nucleation behaviour. There seems to be a single nucleus mechanism in operation where a parent crystal is formed which, through secondary nucleation, is responsible for the formation of the suspension. The proposed crystal nucleation measurement method based on induction time probability distributions allows the determination of the nucleation rate as a function of supersaturation. It is established that heterogeneous nucleation on particles in the solution is occurring.

The crystal nucleation rates of DPL from IPA and DMF were measured as a function of the supersaturation. The thermodynamic factor B in the nucleation rate equation was determined to be much higher for DPL in DMF than for DPL in IPA. Therefore, the energy barrier for nucleation is much higher in DMF than in IPA. The analysis of the kinetic factor A of the nucleation rate equation from the Classical Nucleation Theory shows that the pre-exponential kinetic term is orders of magnitude lower than the theoretical value, possibly because either the attachment frequency or the concentration of heterogeneous particles is much lower than theoretically predicted. A thorough analysis of the deficit of Classical Nucleation Theory can be performed by using well-defined functionalized template particles in solutions having different kinds of self-association and solvation.

Acknowledgements

The authors would like to acknowledge Shanfeng Jiang, Somnath Kadam, Nuria Llort Gual and Samir Kulkarni for doing the measurements in Section 3 and discussing the data. They would further like to thank Gerard Coquerel, Samuel Petit, Roger Davey, Sven Schroeder and Hugo Meekes for interesting discussions on the topic. JtH thanks the Dutch funding agency STW and the companies Avantium, DSM, Mettler-Toledo and Syncom for funding part of the research described here. JtH thanks the EPSRC Centre for Innovative Manufacturing in Continuous Manufacturing and Crystallisation (http://www.cmac.ac.uk) for supporting this work (EPSRC funding under grant reference: EP/I033459/1).

References

1 S. Auer and D. Frenkel, *Nature*, 2001, **409**, 1020–1023.
2 N. M. Dixit, A. M. Kulkarni and C. F. Zukoski, *Colloids Surf., A*, 2001, **190**, 47–60.

3 R. J. Davey, S. L. M. Schroeder and J. H. ter Horst, Nucleation of organic crystals – a molecular perspective, *Angew. Chem.*, 2013, **52**, 2166–2179.

4 O. Galkin and P. G. Vekilov, *J. Phys. Chem. B*, 1999, **103**, 10965–10971.

5 P. G. Vekilov, *Cryst. Growth Des.*, 2010, **10**, 5007–5019.

6 S. Jiang and J. H. ter Horst, Crystal nucleation rates from probability distributions of induction times, *Cryst. Growth Des.*, 2011, **11**, 256–261.

7 D. Kashchiev, *Nucleation: Basic Theory with Applications*, Butterworth-Heinemann, Oxford, 2000.

8 S. S. Kadam, H. J. M. Kramer and J. H. ter Horst, Combination of a single primary nucleation event and secondary nucleation in crystallization processes, *Cryst. Growth Des.*, 2011, **11**, 1271–1277.

9 S. S. Kadam, S. A. Kulkarni, R. Coloma Ribera, A. I. Stankiewicz, J. H. ter Horst and H. J. M. Kramer, A new view on the metastable zone width during cooling crystallization, *Chem. Eng. Sci.*, 2012, **72**, 10–19.

10 S. A. Kulkarni, H. Meekes and J. H. ter Horst, Polymorphism control through a single nucleation event, *Cryst. Growth Des.*, 2014, **14**(3), 1493–1499.

11 S. A. Kulkarni, S. S. Kadam, H. Meekes, A. I. Stankiewicz and J. H. ter Horst, Crystal nucleation kinetics from induction times and metastable zone widths, *Cryst. Growth Des.*, 2013, **13**(6), 2435–2440.

12 U. J. Griesser, M. E. Auer and A. Burger, The polymorphic drug substances of the European pharmacopoeia, part 10: Diprophylline, *Sci. Pharm.*, 1999, **67**, 319–330.

13 C. Brandel, Y. Amharar, J. Rollinger, U. Griesser, Y. Cartigny, S. Petit and G. Coquerel, Impact of molecular flexibility on double polymorphism, solid solutions and chiral discrimination during crystallization of Diprophylline enantiomers, *Mol. Pharmaceutics*, 2013, **10**(10), 3850–3861.

14 R. A. Sullivan, R. J. Davey, G. Sadiq, G. Dent, K. R. Back, J. H. ter Horst, D. Toroz and R. B. Hammond, Revealing the roles of desolvation and molecular self-assembly in crystal nucleation from solution: Benzoic and *p*-Aminobenzoic acid, *Cryst. Growth Des.*, 2014, **14**, 2689–2696.

15 D. Kashchiev and G. M. Van Rosmalen, Review: nucleation in solutions revisited, *Cryst. Res. Technol.*, 2003, **38**(7–8), 555.

16 A. Mersmann, *J. Cryst. Growth*, 1990, **102**, 841.

17 S. A. Kulkarni, E. S. McGarrity, H. Meekes and J. H. ter Horst, Isonicotinamide self-association: the link between solvent and polymorph nucleation, *Chem. Commun.*, 2012, **48**, 4983–4985.

18 S. A. Kulkarni, C. C. Weber, A. S. Myerson and J. H. ter Horst, Self-association during heterogeneous nucleation onto well-defined templates, *Langmuir*, 2014, **30**(41), 12368–12375.

Faraday Discussions

PAPER

Bioinspired magnetite formation from a disordered ferrihydrite-derived precursor

Archan Dey,[a] Jos J. M. Lenders[ab] and Nico A. J. M. Sommerdijk*[ab]

Received 19th November 2014, Accepted 2nd January 2015
DOI: 10.1039/c4fd00227j

We show that by reacting ferrihydrite (FeH) with $Fe^{(II)}$ ions and subsequently increasing the pH, magnetite is formed through a multi-step nucleation process mediated by monodisperse FeH–$Fe^{(II)}$ primary particles. The interaction of these primary particles with a transient green rust phase leads to the formation of smaller secondary particles which form the feedstock for magnetite formation. Surprisingly, the presence of a polypeptide additive prevents the formation of green rust as an $Fe^{(II)}$-rich intermediate phase, and leads to the formation of amorphous aggregates of FeH–$Fe^{(II)}$ particles which subsequently transform into the final magnetite nanocrystals. The observation of multiple transitions and the involvement of disordered precursor phases in this bioinspired crystallization route is important for our understanding of the nucleation of magnetite in geological and biological environments, and may lead to new approaches in the sustainable synthesis of this technologically important mineral.

1 Introduction

Magnetite (Fe_3O_4) is a widespread ferrimagnetic mineral encountered in many biological and geological systems.[1] It also has a broad array of technological applications, *e.g.* in magnetic data storage materials, inks, ferrofluids and scavengers in water purification, but also in magnetic resonance imaging and controlled drug delivery.[2–4] The magnetic properties of Fe_3O_4 nanoparticles strongly depend on their size and morphology. Consequently, controlling the growth of these crystalline particles is essential to fine-tuning their properties for specific applications. As magnetotactic bacteria produce such iron oxide particles with precisely controlled shape and size under ambient conditions,[5] the synthesis of magnetite under biomimetic conditions, *i.e.* at ambient temperatures and in aqueous media, is now a field of increasing interest.[6] It is clear that only through a detailed understanding of the mechanisms underlying the nucleation and growth

[a]*Laboratory of Materials and Interface Chemistry and Soft Matter CryoTEM Research Unit, Department of Chemical Engineering and Chemistry, Eindhoven University of Technology, P.O. Box 513, 5600 MB Eindhoven, The Netherlands. E-mail: N.Sommerdijk@tue.nl*
[b]*Institute for Complex Molecular Systems, Eindhoven University of Technology, P.O. Box 513, 5600 MB Eindhoven, The Netherlands*

in biomimetic magnetite formation will the production of optimized magnetite nanoparticles eventually become feasible.

Magnetotactic bacteria synthesize 30–140 nm magnetite crystals with often elongated morphologies, which are both crystallographically and magnetically aligned in chains.[5] The formation of these nanoparticles takes place in magnetosome vesicles,[5,7] where their nucleation and growth is controlled through their interaction with a variety of biomolecules. Even though the mechanistic pathway involved has been a topic of debate for many years,[8–10] it was recently established that ferrihydrite is present as a precursor inside the magnetosome vesicle prior to the formation of magnetite.[11–13] Although many questions still remain concerning the exact pathway of magnetite biomineralization, it is clear that it concerns two important steps: (1) the local formation of significant quantities of ferrihydrite and (2) the conversion to magnetite through a reaction with $Fe^{(II)}$.

Following our initial report that magnetite can be formed *via* transient ferrihydrite-like primary particles,[14] we recently developed a bioinspired synthesis route, which allows the formation of ferrimagnetic magnetite nanocrystals from a ferrihydrite precursor.[15] This procedure involves the reaction of ferrihydrite with $Fe^{(II)}$ which is present in the solution, upon increasing the pH through the in-diffusion of ammonia vapor, and allows polypeptides to interact with the developing crystals to control their size, shape and dispersibility.[16]

Here we investigate the nucleation of magnetite from a ferrihydrite precursor, separating the addition of $Fe^{(II)}$ from the pH increase by ammonia in-diffusion. Using high-resolution cryo-TEM with selected-area electron diffraction, we reveal that the conversion of ferrihydrite to magnetite follows a multi-step process. This involves green rust as a transient phase, of which the formation and disintegration is mediated by two different populations of 1.5–2.0 nm and 1.0–1.5 nm particles, respectively. Strikingly, the formation of green rust is bypassed in the presence of a random polypeptide of glutamic acid, lysine and alanine monomers, such that magnetite formation now proceeds through an intermediate devoid of any long-range order, consisting of an assembly of these nanoscopic particles.

2 Results

Magnetite was synthesized using ferrihydrite (FeH) as a precursor, with the latter obtained from $FeCl_3 \cdot 6H_2O$ through a modified literature procedure.[17] The obtained gel-like precursor material was identified as 6-line FeH using powder X-ray diffraction (PXRD) after freeze-drying, and did not change its crystallinity or phase purity upon storage at 4 °C for a month (Fig. 1a and b). Cryogenic transmission electron microscopy (cryo-TEM) demonstrated that the gel-like FeH phase consisted of a highly hydrated network of nanometer-sized particles (Fig. 2).[14] For cryo-TEM, ~100 nm thin films of aqueous samples were rapidly vitrified in liquid ethane (−183 °C) to obtain amorphous ice layers, which were subsequently studied at temperatures low enough to avoid their recrystallization (<−140 °C).[18]

The transformation of ferrihydrite to magnetite was carried out by the addition of 0.55 equivalents of $FeCl_2 \cdot 4H_2O$ under a N_2 atmosphere, and the subsequent increase of the solution pH by NH_3 in-diffusion.[15] After 16 hours a black precipitate was isolated, which PXRD showed was phase-pure magnetite (Fig. 1c).

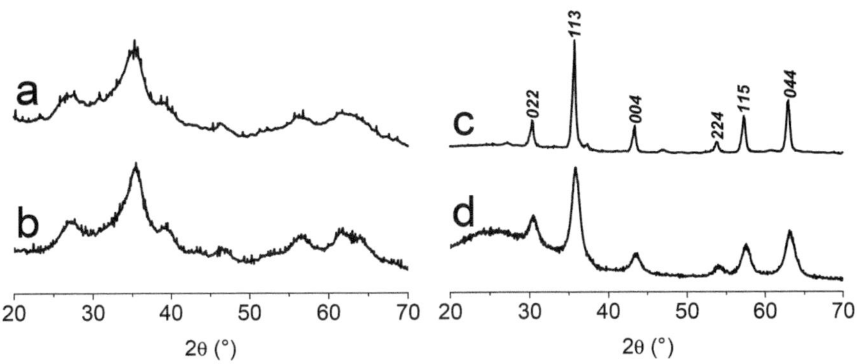

Fig. 1 Powder XRD patterns of freeze-dried samples. (a) Freshly prepared 6-line ferrihydrite (FeH) from $FeCl_3 \cdot 6H_2O$. (b) FeH stored at 4 °C for a month, showing no signs of altered crystallinity or phase purity. (c) Phase-pure magnetite obtained after 16 hours reaction time without additives. The 6 major reflections are indexed to common Fe_3O_4 crystal planes. (d) Phase-pure magnetite obtained after 16 hours reaction time in the presence of the additive **EKA**, showing broader peaks as compared to the control.

To obtain an insight into the mechanism of this conversion process, the development of the morphology and structure was monitored using cryo-TEM and low-dose selected-area electron diffraction (SAED), and the pH was monitored with a pH electrode. Upon addition of $Fe^{(II)}$ to the FeH dispersion at pH ~ 5, cryo-TEM analysis showed the formation of a more monodisperse population of particles, with diameters of 1.5–2.0 nm, within the initial FeH gel (Fig. 3). This was similar to the primary particles observed previously for the controlled coprecipitation of $Fe^{(II)}$ and $Fe^{(III)}$.[14] Additionally, in this case the primary particles did not show any long-range order in the SAED results (Fig. 3b, inset), but did show improved stability towards electron beam exposure, meaning that they could resist a higher electron dose before visible damage occurred. This observation

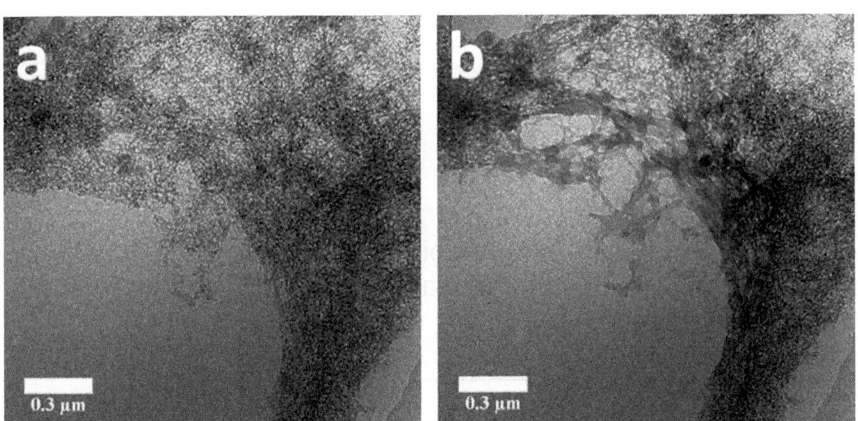

Fig. 2 Cryo-TEM images of the FeH precursor, showing its gel-like character and beam-sensitive nature; (a) and (b) are two subsequent images that were taken from the same area.[14] Reprinted with permission from the Nature Publishing Group, Copyright 2013.

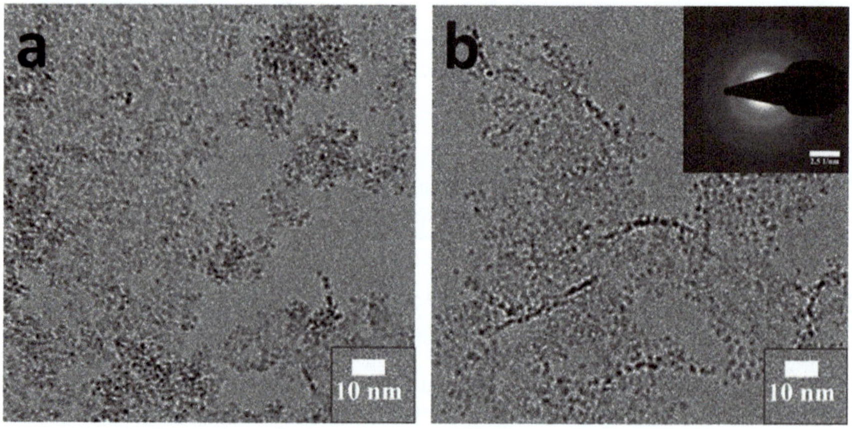

Fig. 3 Cryo-TEM images of FeH in the presence of $Fe^{(II)}$ at pH ~ 5; (a) and (b) are different areas in the same sample, showing the dehydrated FeH and 1.5–2.0 nm particles of FeH stabilized by $Fe^{(II)}$ ions. The inset shows a typical SAED pattern, confirming the absence of long-range order.

suggests that their formation was accompanied by a dehydration of the original FeH gel.

Upon exposure of the FeH–$Fe^{(II)}$ nanoparticle suspension to NH_3 vapor, the pH curve showed an increase with two local maxima (bumps), the first one at pH ~ 7.5 and the second one at pH ~ 8.7 (Fig. 4), suggesting a two-step conversion process involving the uptake of base. Cryo-TEM revealed that the first bump in the pH curve after ~1 hour corresponded to the formation of hexagonal crystals with diameters of 200–500 nm (Fig. 5a), which on the basis of low-dose SAED (inset) were tentatively identified as green rust, $Fe^{(II)}{}_3Fe^{(III)}(OH)_8Cl\cdot H_2O$.[19] A more detailed inspection of these hexagonal crystals showed cracks, suggestive of an ongoing dissolution process alongside their formation.

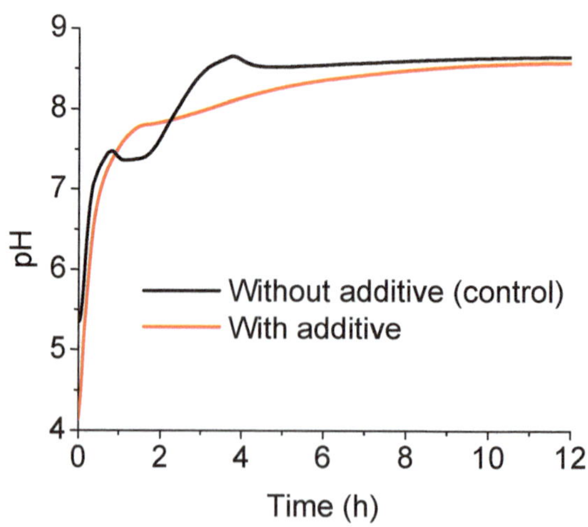

Fig. 4 pH curves through time for magnetite synthesis from FeH and $Fe^{(II)}$ by NH_3 in-diffusion in the absence (control, black) and presence (red) of the additive **EKA**.

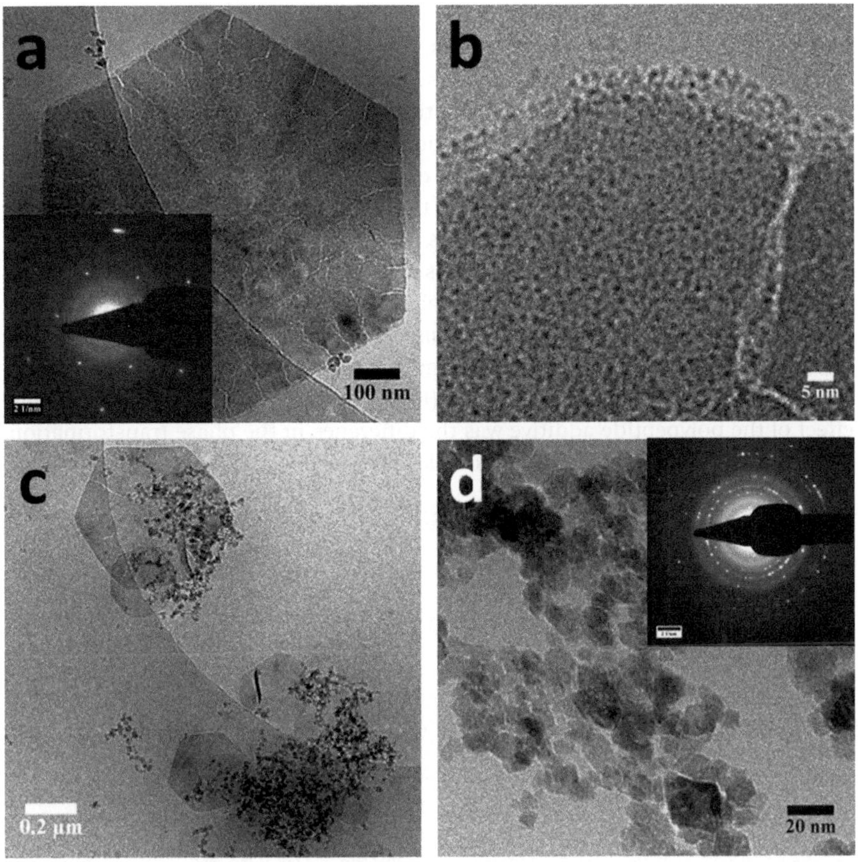

Fig. 5 Cryo-TEM images of the control experiment. (a) Typical example of a hexagonal crystal of green rust after ~1 hour reaction time. The inset shows the corresponding SAED pattern with 3.9 nm^{-1} spacings.[19] (b) Higher-magnification image of a corner of the green rust crystal in (a), revealing that it is covered with spherical nanoparticles, suggesting that the green rust is converted into these particles. (c) Co-localized formation of magnetite (product) near the green rust (intermediate) after ~4 hours reaction time. (d) The final product, consisting of polydisperse magnetite nanoparticles, rarely showing an octahedral morphology. The inset shows a typical SAED pattern, displaying the common reflections.

Interestingly, both the surfaces and the edges of these disintegrating crystals were covered with particles that from the obtained cryo-TEM images were determined to be 1.0–1.5 nm in size (Fig. 5b). Although these values are smaller than those quoted for the FeH–$Fe^{(II)}$ particles described above, it is important to stress that the Fresnel fringes associated with the defocus values used to visualize these particles do not allow us to make a definite distinction between the two observed populations. Unfortunately, their co-localization with the green rust crystals also prohibited their further detailed structural analysis with SAED. After the second bump in the pH curve (~4 hours), magnetite crystals started to form on the hexagonal platelets and eventually became the single product, at the expense of both the platelets and the nanoparticles present (Fig. 5c and d).

More information on these events was obtained when the reaction was performed at a 4-fold higher $Fe^{(II)}$ concentration. At this point the cryo-TEM images showed hexagonal platelets with highly corrugated surfaces covered with large numbers of nanometer-sized particles, of which the diameters were again measured as 1.0–1.5 nm (Fig. 6a). These images suggest that the particles are actually formed by the dissolution of the hexagonal crystals, rather than the crystals being formed through the accretion of these particles. Moreover, the cryo-TEM images also showed what seemed to be the next step in the crystallization process, *i.e.* the formation of magnetite through the accretion of the 1.0–1.5 nm particles (Fig. 6b–d), similar to what has been reported previously.[14]

When the mineralization reaction was carried out in the presence of a random copolymer of glutamic acid (E: ~40 mol%), lysine (K: ~30 mol%) and alanine (A: ~30 mol%) (**EKA**),[20] the PXRD analysis of the final magnetite crystals showed significant peak broadening (Fig. 1d), suggesting a decrease in particle size, which was indeed confirmed using cryo-TEM (Fig. 7b). However, the most remarkable effect of the polypeptide additive was the difference in the phase transformation sequence. The pH curve no longer displayed discrete local maxima (Fig. 4), but

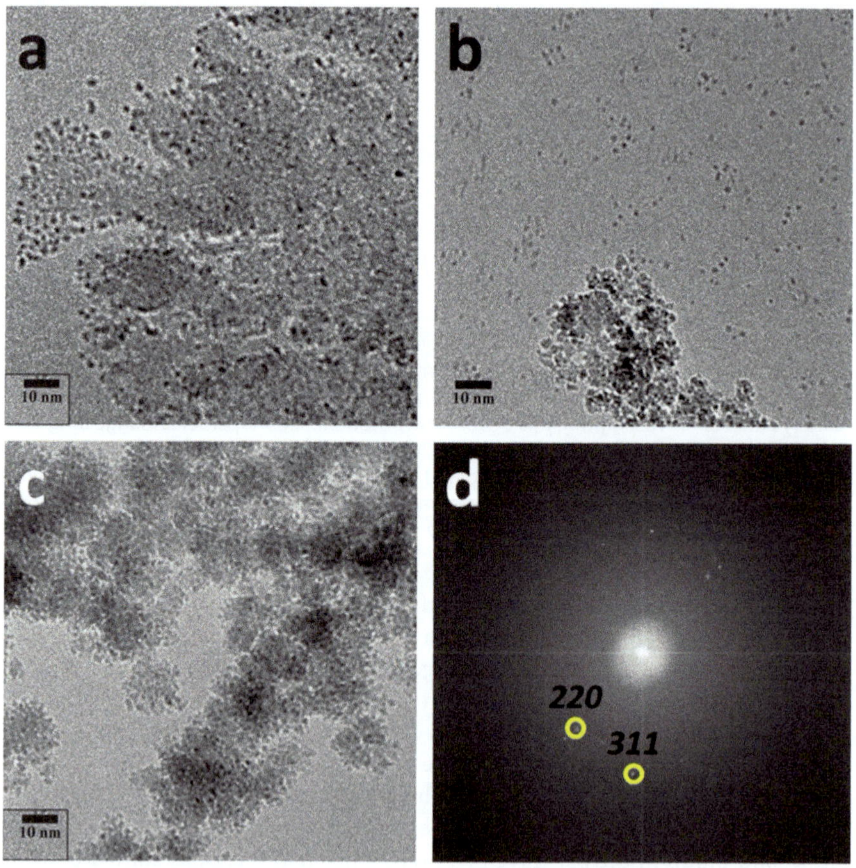

Fig. 6 Cryo-TEM images of the control experiment at a 4-fold higher Fe$^{(II)}$ concentration. (a) Formation of monodisperse 1.0–1.5 nm particles through green rust dissolution. (b) Formation of magnetite from these nanoparticles. (c) The product, consisting of magnetite crystals surface-decorated with nanoparticles. (d) FFT of (c), confirming the presence of magnetite crystal planes.

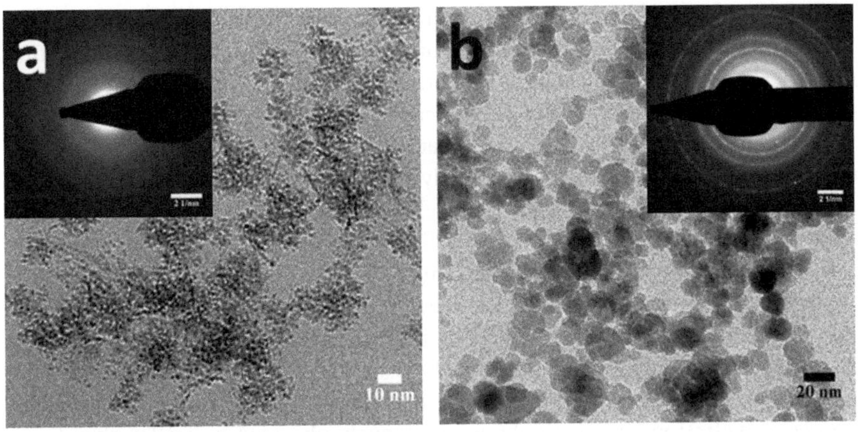

Fig. 7 Cryo-TEM images of the experiment in the presence of the additive **EKA**. (a) Assembly of the 1.5–2.0 nm particles into aggregates after ~1.5 hours reaction time. The inset shows a typical SAED pattern, confirming the absence of long-range order. (b) The final product after >12 hours reaction time, consisting of smaller and more uniform magnetite nanoparticles as compared to the control. The inset shows a typical SAED pattern, displaying the 6 major reflections.

showed a transition point after ~1.5 hours at pH ~ 7.8, followed by a gradual further increase up to pH ~ 8.5. Cryo-TEM showed that the transition point at pH ~ 7.8 marked the assembly of the 1.5–2.0 nm particles into aggregates, for which SAED indicated no long-range order (Fig. 7a). Although this transition point (a discontinuation of the pH increase during the continuous influx of NH_3) is indicative of a phase transition, only in a later stage (>12 hours) was the presence of crystalline magnetite confirmed using cryo-TEM with SAED (Fig. 7b).

3 Discussion

Here we show the formation of magnetite by the reaction of ferrihydrite (FeH) with $Fe^{(II)}$ under an influx of ammonia. Upon exposure of FeH to $Fe^{(II)}$ ions (0.55 equivalent), monodisperse FeH–$Fe^{(II)}$ nanoparticles with dimensions of 1.5–2.0 nm are formed, which we have previously termed primary particles. These primary particles show, in addition to a more monodisperse size distribution, an increased stability under electron beam exposure compared to the gel-like FeH phase they nucleate from. The observation of a monodisperse particle size suggests that the interaction of FeH with the ferrous and chloride ions results in FeH–$Fe^{(II)}$ nanoparticles with a specific composition that favors this well-defined size distribution. Meanwhile the improved stability against electron beam irradiation suggests that this surface reaction is accompanied by a dehydration of the original FeH nanoparticles.

Upon increasing the pH, we observed next a solid intermediate phase, consisting of single-crystalline hexagonal platelets, which we tentatively identified as green rust. Judging from the local minimum observed in the pH curve, these platelets are formed by a discrete nucleation event. The observation of large numbers of nanometer-sized particles attached to their surfaces – and of thinner areas on the crystal surfaces where discrete primary particles can no longer be observed – is suggestive of an accretion-based growth process in which the $Fe^{(III)}$-rich primary particles react with ferrous hydroxide species that become increasingly insoluble when the pH increases above 7.

However, the "fjord"-like cracks observed in the plate-like crystals suggests that during their growth they are already disintegrating. This process was even more prominently observed at higher $Fe^{(II)}$ concentrations (2.2 equivalent with respect to $Fe^{(III)}$). Images showing the disintegrating platelets also reveal their close association with apparently even smaller (1.0–1.5 nm) nanoparticles, which are located on what used to be the platelet surfaces as well as on the inside of the "fjords". We tentatively interpret these observations as showing the disintegration of the hexagonal plates due to a reaction with the excess primary particles, yielding smaller secondary particles with a higher $Fe^{(II)}$ content and a lower degree of hydration. Interestingly, we previously observed a similar shrinkage from ~2 nm to ~1 nm in size for primary particles attaching to the surface of magnetite.

The next step in the process is the formation of magnetite, which according to the second bump in the pH curve is again a discrete nucleation event. The cryo-TEM images show that magnetite nanocrystals form on the surfaces of the hexagonal platelets and grow through the accretion of similar nanometer-sized particles. Although is it difficult to differentiate between the sizes of the primary particles (estimated 1.5–2.0 nm) and the secondary particles (estimated 1.0–1.5 nm) from the current cryo-TEM images, we propose that it is the secondary particles, with an increased $Fe^{(II)}/Fe^{(III)}$ ratio, which are still adsorbed on the hexagonal platelets that form the feedstock for magnetite nucleation and growth.

Strikingly, the addition of a polypeptide composed of almost equal amounts of acidic (Glu), basic (Lys) and neutral (Ala) amino acids (**EKA**) let to quite a different crystallization mechanism, in which the formation of green rust was completely bypassed. Apart from a transition point at pH ~ 7.8 in the pH curve, which marked the aggregation of the FeH–$Fe^{(II)}$ primary particles, no further signs of a specific nucleation point were observed, despite the fact that magnetite with a less polydisperse size distribution than the control was isolated as the single product. Although acidic polypeptides have been shown to bind both $Fe^{(II)}$ and $Fe^{(III)}$,[16] is unlikely that **EKA** effectively competes with the formation of ferrihydrite above pH ~ 5. However, it is likely that **EKA** binds the far more soluble $Fe^{(II)}$, thereby effectively inhibiting the formation of the green rust at pH 7.5. Nevertheless, as the pH increases to ~7.8, it may no longer be able to prevent the reaction of the $Fe^{(II)}$ ions with the primary particles. We propose that this results in the formation of the secondary particles, and that these form aggregates which, one by one – possibly depending on the aggregate size – will transform into magnetite through the uptake of ferrous hydroxyl species. Given the fact that magnetite nanocrystals with a narrower size distribution were formed, the absence of a distinct nucleation event suggests that the nucleation of magnetite must occur from FeH–$Fe^{(II)}$ aggregates that are already close in dimension to the final crystals; also that the size of the latter is mostly determined by the nucleation event, rather than through a growth process. Although it seems very likely that the same aggregate-to-crystal formation also occurs in the absence of the additive, the observation of a clear nucleation point for magnetite suggests that the shape and transformation kinetics of the aggregates are modulated through their interaction with the green rust platelets.

4 Conclusion

Starting from ferrihydrite (FeH), we have demonstrated the formation of magnetite through a multi-step nucleation process, mediated by primary particles that are stabilized by their interaction with $Fe^{(II)}$ ions. We propose that through their interaction with a transient green rust phase, the primary particles transform into secondary particles with increased $Fe^{(II)}$ contents and reduced hydration levels, and that these form the feedstock for magnetite formation. Furthermore, it was demonstrated that the presence of an iron-binding polypeptide additive **EKA** prevents the formation of green rust as a $Fe^{(II)}$-rich intermediate phase, and that this leads to the formation of amorphous aggregates of FeH–$Fe^{(II)}$ particles which eventually transform into the final magnetite nanocrystals.

The observation of multiple transitions in the conversion of ferrihydrite to magnetite upon reaction with $Fe^{(II)}$ ions, and the involvement of disordered precursor phases, is important for our understanding of the nucleation of this important magnetic material in geological and biological environments. It may also lead to new approaches in the sustainable synthesis of this technologically important mineral.[21] Our study further shows the power of cryo-TEM in revealing the development of structure and morphology at the nanoscopic level.

5 Experimental

Synthesis of ferrihydrite and magnetite

A literature procedure was modified for the syntheses in this manuscript.[17] Ferrihydrite (FeH) was obtained by the addition of 200 mM NaOH solution at a rate of 1 mL min^{-1} to 50 mL of 40 mM $FeCl_3 \cdot 6H_2O$ solution, until a pH of 7.5 was reached. This reaction was carried out in a glass beaker under stirring, and the pH was monitored using a computer-controlled commercial titration setup (Metrohm). A stock solution was prepared from the obtained dispersion by dilution, and was stored at 4 °C for further use.

For magnetite synthesis, 25 mL of the diluted FeH dispersion (2.5 mM $Fe^{(III)}$) was degassed (10–15 min) with a N_2 flow, followed by the addition of 7.0 mg $FeCl_2 \cdot 4H_2O$ (1.4 mM $Fe^{(II)}$) and 650 µL $E_{40}K_{30}A_{30}$ (stock solution: 3.33 mg mL^{-1}, degassed with N_2). To increase the pH, a 6 vol% NH_3 solution was used inside a glove box, under a N_2 atmosphere and constant stirring, to keep a homogeneous pH gradient in the reaction mixture. During the in-diffusion of NH_3, the pH was monitored using a computer-controlled commercial titration setup (Metrohm).

Powder X-ray diffraction (PXRD)

PXRD measurements were performed on a Rigaku Geigerflex powder diffractometer with Bragg-Brentano geometry, using copper radiation at 40 kV and 30 mA and a graphite monochromator to eliminate Cu K_β radiation. Samples were prepared on cover glasses from dried ground powder. The PXRD patterns were acquired by step scans from $2\theta = 20°$ to $2\theta = 70°$ with step sizes of 0.01° to 0.02° and an appropriate dwell time.

(Cryogenic) transmission electron microscopy (Cryo-TEM)

For standard dry TEM, 200 mesh Cu grids with continuous carbon films (Agar Scientific) were used. Sample preparation involved dropping 3 μL aqueous dispersion onto a TEM grid, blotting using filter paper, and allowing the grid to dry in air. For cryo-TEM, 200 mesh Cu grids with Quantifoil R 2/2 holey carbon films (Quantifoil Micro Tools GmbH) were used. Sample preparation was performed using an automated vitrification robot (FEI Vitrobot Mark III) for plunging in liquid ethane.[22] All TEM grids were surface plasma treated for 40 seconds using a Cressington 208 carbon coater prior to use. TEM samples were studied on an FEI Technai 20 (type Sphera) operated at 200 kV, equipped with a LaB_6 filament and a 1k × 1k Gatan CCD camera, or on the TU/e cryoTITAN (FEI, http://www.cryotem.nl) operated at 300 kV, equipped with a field emission gun (FEG), a post-column Gatan Energy Filter (GIF) and a post-GIF 2k × 2k Gatan CCD camera. Gatan DigitalMicrograph (including DiffTools) and ImageJ were used for TEM image and SAED pattern analysis.

Acknowledgements

A.D. and N.A.J.M.S are supported by a VICI grant from the Dutch Science Foundation – Chemical Sciences (NWO–CW). The work of J.J.M.L. is supported by NanoNextNL, a micro and nanotechnology consortium of the government of The Netherlands and 130 partners.

References

1 R. M. Cornell and U. Schwertmann, *The Iron Oxides (Structure, Properties, Reactions, Occurences and Uses)*, Wiley-VCH, Weinheim, 2003.

2 C. T. Yavuz, J. T. Mayo, W. W. Yu, A. Prakash, J. C. Falkner, S. Yean, L. Cong, H. J. Shipley, A. Kan, M. Tomson, D. Natelson and V. L. Colvin, *Science*, 2006, **314**, 964–967.

3 S. Laurent, D. Forge, M. Port, A. Roch, C. Robic, L. Vander Elst and R. N. Muller, *Chem. Rev.*, 2008, **108**, 2064–2110.

4 A.-H. Lu, E. L. Salabas and F. Schüth, *Angew. Chem., Int. Ed.*, 2007, **46**, 1222–1244.

5 D. Faivre and D. Schüler, *Chem. Rev.*, 2008, **108**, 4875–4898.

6 T. Prozorov, D. A. Bazylinski, S. K. Mallapragada and R. Prozorov, *Mater. Sci. Eng., R*, 2013, **74**, 133–172.

7 R. Blakemore, *Science*, 1975, **190**, 377–379.

8 R. B. Frankel, G. C. Papaefthymiou, R. P. Blakemore and W. O'Brien, *Biochim. Biophys. Acta Mol. Cell Res.*, 1983, **763**, 147–159.

9 D. Faivre, L. H. Böttger, B. F. Matzanke and D. Schüler, *Angew. Chem., Int. Ed.*, 2007, **46**, 8495–8499.

10 S. Staniland, B. Ward, A. Harrison, G. van der Laan and N. Telling, *Proc. Natl. Acad. Sci. U. S. A.*, 2007, **104**, 19524–19528.

11 J. Baumgartner, G. Morin, N. Menguy, T. Perez Gonzalez, M. Widdrat, J. Cosmidis and D. Faivre, *Proc. Natl. Acad. Sci. U. S. A.*, 2013, **110**, 14883–14888.

12 M. L. Fdez-Gubieda, A. Muela, J. Alonso, A. García-Prieto, L. Olivi, R. Fernández-Pacheco and J. M. Barandiarán, *ACS Nano*, 2013, **7**, 3297–3305.

13 M. I. Siponen, P. Legrand, M. Widdrat, S. R. Jones, W.-J. Zhang, M. C. Y. Chang, D. Faivre, P. Arnoux and D. Pignol, *Nature*, 2013, **502**, 681–684.
14 J. Baumgartner, A. Dey, P. H. H. Bomans, C. Le Coadou, P. Fratzl, N. A. J. M. Sommerdijk and D. Faivre, *Nat. Mater.*, 2013, **12**, 310–314.
15 J. J. M. Lenders, C. L. Altan, P. H. H. Bomans, A. Arakaki, S. Bucak, G. de With and N. A. J. M. Sommerdijk, *Cryst. Growth Des.*, 2014, **14**, 5561–5568.
16 J. J. M. Lenders, H. R. Zope, A. Yamagishi, P. H. H. Bomans, A. Arakaki, A. Kros, G. de With and N. A. J. M. Sommerdijk, *Adv. Funct. Mater.*, 2015, **25**, 711–719.
17 H. Liu, M. Ma, M. Qin, L. Yang and Y. Wei, *J. Solid State Chem.*, 2010, **183**, 2045–2050.
18 H. Friedrich, P. M. Frederik, G. de With and N. A. J. M. Sommerdijk, *Angew. Chem., Int. Ed.*, 2010, **49**, 7850–7858.
19 I. R. McGill, B. McEnaney and D. C. Smith, *Nature*, 1976, **259**, 200–201.
20 V. Dmitrović, J. J. M. Lenders, H. R. Zope, G. de With, A. Kros and N. A. J. M. Sommerdijk, *Biomacromolecules*, 2014, **15**, 3687–3695.
21 C. L. Altan, J. J. M. Lenders, P. H. H. Bomans, G. de With, H. Friedrich, S. Bucak and N. A. J. M. Sommerdijk, *Chem Eur. J.*, 2015, **21**, 6150–6156.
22 M. R. Vos, P. H. H. Bomans, P. M. Frederik and N. A. J. M. Sommerdijk, *Ultramicroscopy*, 2008, **108**, 1478–1483.

Faraday Discussions

PAPER

Chiral habit selection on nanostructured epitaxial quartz films†

Adrián Carretero-Genevrier,‡*[a] Martí Gich,‡*[b] Laura Picas,[c] Clément Sanchez[def] and Juan Rodriguez-Carvajal*[g]

Received 15th December 2014, Accepted 14th January 2015

DOI: 10.1039/c4fd00266k

Understanding the crystallization of enantiomorphically pure systems can be relevant to diverse fields such as the study of the origins of life or the purification of racemates. Here we report on polycrystalline epitaxial thin films of quartz on Si substrates displaying two distinct types of chiral habits that never coexist in the same film. We combine Atomic Force Microscopy (AFM) analysis and computer-assisted crystallographic calculations to make a detailed study of these habits of quartz. By estimating the surface energies of the observed crystallites we argue that the films are enantiomorphically pure and we briefly outline a possible mechanism to explain the habit and chiral selection in this system.

Chiral objects, defined by the lack of mirror plane or center of symmetry, are ubiquitous in nature from molecules to galaxies.[1] The term chiral comes from the Greek, meaning "hand" as this is the prototypical example of chiral object: the mirror image of a right hand is not the same right hand but a left hand. The pairs of objects which, like our hands, are identical except for their opposite orientations are named enantiomorphs (enantiomers in the case of molecules), and

[a]*Institut des Nanotechnologies de Lyon (INL) CNRS-Ecole Centrale de Lyon, 36 avenue Guy de Collongue, 69134 Ecully, France. E-mail: adrien.carretero-genevrier@ec-lyon.fr*

[b]*Institut de Ciència de Materials de Barcelona, ICMAB, Bellaterra 08193, Catalonia, Spain. E-mail: mgich@icmab.es*

[c]*Centre de Biochimie Structurale (CBS), UMR 5048, Montpellier, 34090, France*

[d]*Collège de France, UMR 7574, Chimie de la Matière Condensée de Paris, Paris, F-75005, France*

[e]*Sorbonne Universités UPMC Univ Paris 06, UMR 7574, Chimie de la Matière Condensée de Paris, Paris, F-75005, France*

[f]*CNRS, UMR 7574, Chimie de la Matière Condensée de Paris, Paris, F-75005, France*

[g]*Institut Laue-Langevin, Diffraction Group, F-38042 Grenoble Cedex, France. E-mail: rodriguez-carvajal@ill.eu*

† Electronic supplementary information (ESI) available: Experimental protocols. AFM images supporting other examples performed on quartz films. Detailed information on the morphological calculations from AFM images using the crystallographic calculator program (CrysCalCon) existing in the CrysFML repository. Extended representation of the surface energies of h1 in the $P3_121$ and $P3_221$ space groups. See DOI: 10.1039/c4fd00266k

‡ These authors contributed equally to this work.

chiral molecules and crystals typically exist in two forms only distinguishable by their contrasting optical activity consisting in rotating the plane-polarized light by an equal amount but in opposite directions. For instance, two forms of quartz (α-SiO_2) are found in the lithosphere, levorotatory crystals (L-quartz, space group $P3_121$) containing right-handed helices of SiO_4 tetrahedra along their trigonal c-axis, and the dextrorotatory enantiomorph (D-quartz space group $P3_221$) displaying left-handed helices. Natural quartz occurrences are racemic, *i.e.* the ratio of right and left enantiomorphs is 50 : 50.[2] The occurrence of racemates is pervasive in the supramolecular and molecular scales, with the notable exception of biomolecules. Indeed, cells only produce levorotatory aminoacids and dextrorotatory sugars. The striking homochirality of living matter is still not understood, most likely because it is connected to the origins of life.[3] It is believed that homochirality might have originated from a mechanism yielding enantiomeric excesses further amplified to attain pure chirality[4] but this remains an open question. Among the different mechanisms to produce enantiomeric excesses,[5] the asymmetric adsorption of one enantiomer on quartz surfaces was demonstrated to account for a significant degree of selectivity[6] but it is compromised by the random handedness of natural quartz.

In this context, it is relevant to consider if a system of chirally pure crystals can emerge from an enantiomorphic mixture and to study the mechanisms governing such a process. Interestingly, Viedma has quite recently shown how a racemic population of crystals can achieve a complete chiral purity in a process made possible by the combination of an autocatalytic secondary nucleation and a dissolution–recrystallization.[7]

Here we report on the spontaneous selection of one among two distinct variants of the crystalline habit in thin polycrystalline quartz films epitaxially grown on silicon substrates. We argue that the surface energy of each crystalline habit variant is minimized by one of the two enantiomorphs and that the films are expected to be homochiral. Since both quartz enantiomorphs are expected to nucleate with the same probability on (100)-Si surfaces, our purpose is also discussing what mechanisms allow obtaining chirally pure film surfaces in this system.

We prepared dense polycrystalline epitaxial films of α-quartz on (100)-Si substrates by the thermal devitrification of mesoporous amorphous silica films infiltrated with strontium, barium, or calcium catalysts.[8,9] The amorphous silica films were dip-coated in a water–ethanolic solution of a Si-alcoxide containing CTAB surfactant that allowed us to obtain a hexagonal mesoporous structure with 5 nm pore size. Once the crystallization of this silica matrix occurs after thermal treatment at 900 °C, the porosity of the film collapses, yielding a dense quartz film.[8] X-ray diffraction (XRD) studies of the films revealed an epitaxial growth, with the (100) quartz planes parallel to the (100)-Si substrate and the [210] axis being the crystal growth direction (*i.e.* [100]*, in the reciprocal space).[8]

The main findings of the present work arise from the comprehensive analysis of the surfaces of different film samples by Atomic Force Microscopy (AFM). Fig. 1 presents AFM images of two different quartz films (panel a–d and e–h, respectively). In Fig. 1a and e we can see the topography of large areas from these films.

Detailed scans within the zones indicated by the dashed squares (Fig. 1b and f) reveal that both films are polycrystalline and present a dense arrangement of crystals with their upper faces displaying tetragonal irregular convex shapes

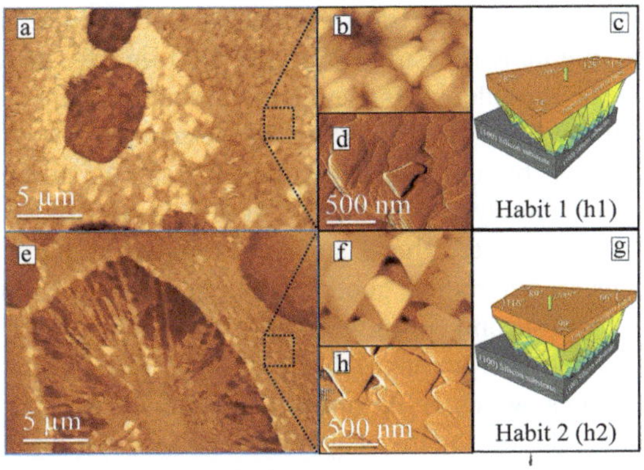

Fig. 1 AFM topographical and phase images of polycrystalline quartz films surface displaying both h1 and h2 habits (a, b, d and e, f, h, respectively). 3D schematic representation showing the angles (c and g, respectively) extracted from the AFM phase images for the h1 and h2 habits (d and h, respectively).

below 500 nm in size. The study of more than 10 films prepared by the same approach invariably lead us to observe in a given film only one (see Fig. S1†) of the two distinct crystalline habits sketched in Fig. 1c and g, hereafter labelled h1 and h2. Moreover, we found that roughly one half of the studied films presented h1 and the other half to h2. Considering the equiprobable appearance of both habits and that only one type of habit can be found in different zones of the same film, it is reasonable to assume that some mechanism of habit selection is at play during the crystallisation and growth of the films.

From images of the AFM phase such as the ones displayed in Fig. 1g and h we made detailed measurements of the inner angles defined by the edges of the top tetragonal faces of h1 and h2, which are reported in the sketches of Fig. 1c and g and in Table 1.

In order to get more insight into the crystalline habits of these quartz films we calculated the different habit solutions compatible with the experimental

Table 1 List of calculated possible lateral planes intercepting the observed top plane (100) in the AFM images at the edge [01−2] for both, h1 and h2 habits of quartz films

Experimental angle, h1	Theoretical angle, h1	Crystallographic directions, h1	Planes, h1
87°	86.959°	[0−31]∧[013]	(−131)
128°	126.977°	[0−31]∧[01−3]	(0−1−3)
71°	73.032°	[01−3]∧[010]	(0−3−1)
74°	73.032°	[013]∧[010]	(001)
Experimental angle, h2	**Theoretical angle, h2**	**Crystallographic directions, h2**	**Planes, h2**
90°	90°	[01−2]∧[010]	(001)
116°	118.644°	[010]∧[001]	(010)
89°	85.949°	[0−2−1]∧[001]	(012)
66°	65.407°	[01−2]∧[0−2−1]	(0−2−1)

observation of a (100) out-of-plane texture and a top face characterized by the four inner angles experimentally measured from the AFM images. This was achieved by writing a simple module in the general crystallographic calculator (CrysCalCon) based on the CrysFML library[10] which selects the lowest index pairs of edges $[uvw]_1$ and $[uvw]_2$ being both perpendicular to [100]* and forming angles close to those experimentally observed (see ESI Fig. S2–S5† for details).

These calculations indicated the existence of trapezoidal habit solutions presenting inner angles in remarkably good agreement with those obtained by AFM for both h1 and h2 (see second column in Table 1). Notably, the calculations also allowed us to obtain the crystallographic planes corresponding to the lateral facets of the crystallites, which could not be imaged by AFM, and its inclination with respect to the substrate plane. Fig. 2 summarizes the orientations of the edges (in red) and the indices of the lateral planes and their orientation with respect to the substrate plane (in blue).

An additional solution presenting low index lateral faces at 90° with respect to the substrate was found to be compatible with h1 and h2. This was however disregarded because we observed lateral faces at about 60° with respect to the substrate in a scanning electron microscopy study of a film cross section (see Fig. 3).

Note that the top faces of h1 and h2 present a definite handedness and closely resemble the specular images of one another. It is therefore natural to consider if these distinct habits correspond to quartz crystals of different chirality (*i.e.* crystallizing in the $P3_121$ and $P3_221$ space groups). However, a direct determination of the chirality of the quartz films is not trivial. In the present case the standard identification of enantiomorphs based on the sense of the rotation imparted to plane-polarized light fails because the small thickness of the films results in rotations ($\sim 10^{-4\circ}$) which cannot be reliably detected with conventional systems.

Different enantiomorphs can also be distinguished with dedicated techniques not readily available to us such as anomalous scattering[11] or three-beam X-ray diffraction[12] both requiring synchrotron radiation and crystals of the appropriate size, or circularly-polarized Raman spectroscopy which also needs a special setup.[13]

Nevertheless, we have exploited the relationships between crystal morphology and crystal structure to get more insight into the chirality of epitaxial quartz films on silicon. It is well known that the crystalline habit is composed by slow-growing faces characterized by a low number of dangling bonds per unit area and a

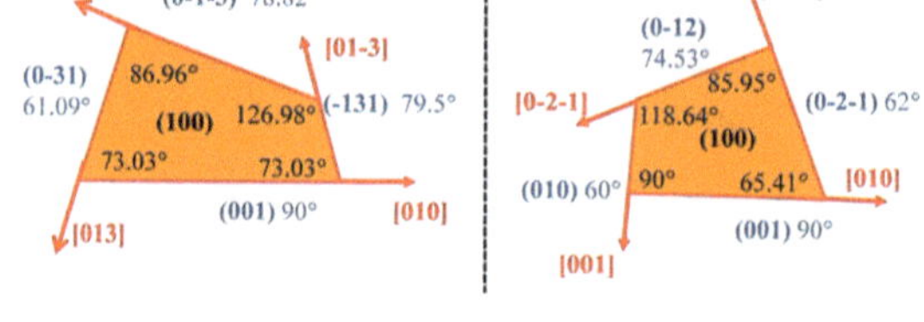

Fig. 2 Habit reconstruction from AFM images (in red, orientations of the edges and in blue, their orientation with respect to the substrate plane).

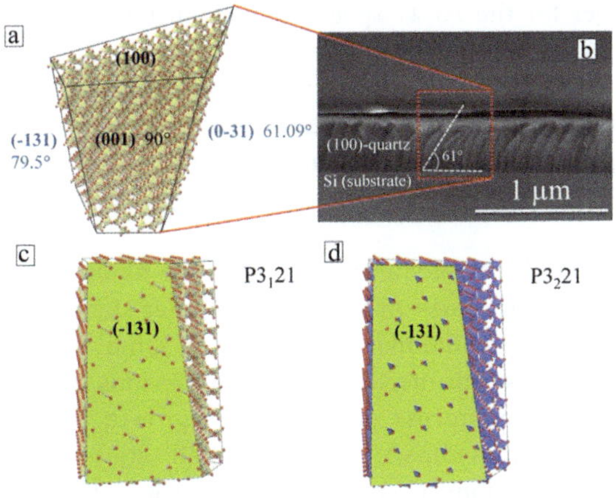

Fig. 3 Crystal morphology of quartz crystallites superimposed on its structural model (a). Cross-section FEG-SEM image showing the inclination of the lateral faces in quartz films (b). Representation of the (−131) surfaces in $P3_121$ or $P3_221$ space groups (c, d). In red, oxygen atoms contained within (−131) surfaces.

correspondingly low surface energy. We shall see that the surface energies of h1 and h2 habits are substantially different depending on whether we consider crystallization in $P3_121$ or $P3_221$ space groups and we believe that this is a powerful argument supporting the homochirality of the epitaxial quartz films.

We have made a comparative estimation of the surface energies of h1 and h2 in the $P3_121$ and $P3_221$ space groups by evaluating the number of dangling bonds per unit area for each of the surfaces of these habits using the VESTA[14] software (see Fig. 3a and 4).

Taking into account the strongly covalent character of the silicon–oxygen bonds we have considered complete SiO_4 tetrahedra and we have equated the number of dangling bonds to that of oxygen atoms contained within or above the crystal facets. Let us take the case of h1 to illustrate our point. Fig. 3c and d clearly show that the number of oxygen atoms in a (−131) surface with the same area is

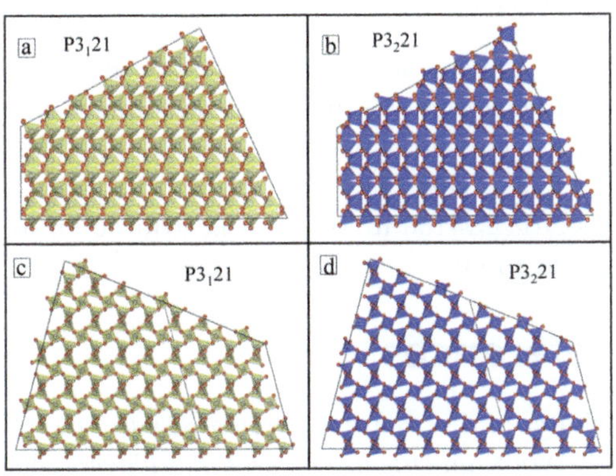

Fig. 4 (100) habit reconstruction from AFM images superimposed on their structural models in $P3_121$ or $P3_221$ space groups (crystal structures are represented by co-ordination polyhedra and gray and red balls represent Si and O atoms, respectively).

significantly larger for the $P3_121$ space group than for $P3_221$. In particular we found about 0.06 at. $Å^{-2}$ for $P3_121$ compared to about 0.04 at. $Å^{-2}$ for $P3_221$, implying that a h1 habit with a $P3_121$ structure would present (−131) faces with 50% higher surface energy. Similarly, the surface energies of (0−31) and (013) are about 20% and 7% lower for a $P3_221$ structure, while no differences are found in the case of (100), (−100) and (001) surfaces (see images of these surfaces in Fig. S6 and S7†).

The crystallization mechanism underpinning the habit and enantiomorphic selection illustrated in the present work is still under study and far from being understood. However we would like to comment on some aspects that we believe can be relevant to explain this mechanism. In our previous studies on this system we found evidence of a process of competitive growth occurring during the crystallization of the quartz films.[8,9] Also, the crystallization of amorphous SiO_2 is assisted by alkaline earth ions in a process occurring at high temperatures. In this scenario, we conjecture that the conditions might exist for a process of secondary nucleation to operate amplifying enantiomorphic fluctuations, analogous to the mechanism proposed to explain the emergence the enantiomorphic purity in the $NaClO_3$ model system.[7,15]

Conclusions

In this work we have presented a combined AFM study with computer-assisted crystallographic calculations showing the existence of two single trapezoidal habits of epitaxial quartz films on silicon. The estimated number of dangling bonds per unit area for each of the surfaces of these habits provides insight into the homochirality of epitaxial quartz films. Importantly, the number of oxygen atoms in the (−131) surface was significantly larger for the $P3_121$ space group than for $P3_221$. Therefore, our study evidences that these quartz surfaces are likely to be intrinsically chiral and thus, reveals an enantiomer-selective habit of epitaxial quartz films. We believe that this work might open new perspectives in the field of enantioselective surface chemistry and into the mechanisms governing crystallization of chirally pure systems.

Acknowledgements

ACG acknowledges the Collège de France foundation for financial support and IMPC for use FEG-SEM facilities. We thank David Montero and Simon Scheuring (U1006, Marseille) for technical support. Cellule Energie INSIS-CNRS awarded ACG with project PEPS (1D-RENOX). We thank the European Commission awarded PCIG09-GA-2011-294168 to MG, and Beatriu de Pinos and Foundation Pierre-Gilles de Gennes to LP.

Notes and references

1 B. Aryal, S. R. Acharya and W. Saurer, *Astrophys. Space Sci.*, 2007, **307**, 369–383.
2 K. Evgenii and T. Wolfram, *Origins Life Evol. Biospheres*, 2000, **30**, 431–434.
3 J. L. Bada, *Nature*, 1995, **374**, 594–595.
4 Y. Saito and H. Hyuga, *Rev. Mod. Phys.*, 2013, **85**, 603–621.
5 W. Bonner, *Origins Life Evol. Biospheres*, 1991, **21**, 59–111.

6 W. A. Bonner, P. r. Kavasman, F. S. Martin and J. J. Flores, *Science*, 1974, **186**, 143–144.
7 C. Viedma, *Phys. Rev. Lett.*, 2005, **94**, 065504.
8 A. Carretero-Genevrier, M. Gich, L. Picas, J. Gazquez, G. L. Drisko, C. Boissiere, D. Grosso, J. Rodriguez-Carvajal and C. Sanchez, *Science*, 2013, **340**, 827–831.
9 G. L. Drisko, A. Carretero-Genevrier, M. Gich, J. Gazquez, D. Ferrah, D. Grosso, C. Boissiere, J. Rodriguez-Carvajal and C. Sanchez, *Adv. Funct. Mater.*, 2014, **24**, 5494–5502.
10 J. Rodríguez-Carvajal and J. Gonzalez-Platas, CompComm Newsletter, 2003, **1**, 50, Freely available at http://www.iucr.org/resources/commissions/crystallographic-computing/newsletters, The source code of the CrysFML library can be found in the repository: http://forge.epn-campus.eu/projects/crysfml/repository. The program CrysCalCon (Crystallographic Calculator Console program) can also be found in the above repository in the directory: Program_Examples/Cryst_Calculator_console.
11 J. M. Bijvoet, A. F. Peerdeman and A. J. Vanbommel, *Nature*, 1951, **168**, 271–272.
12 K. Hummer and E. Weckert, *Acta Crystallogr., Sect. A: Found. Crystallogr.*, 1995, **51**, 431–438.
13 L. D. Barron, in *Chirality at the Nanoscale*, Wiley-VCH Verlag GmbH & Co. KGaA, 2009, pp. 1–27DOI: 10.1002/9783527625345.ch1.
14 K. Momma and F. Izumi, *J. Appl. Crystallogr.*, 2011, **44**, 1272–1276.
15 D. K. Kondepudi, R. J. Kaufman and N. Singh, *Science*, 1990, **250**, 975–976.

Faraday Discussions

PAPER

On a role of liquid intermediates in nucleation of gold sulfide nanoparticles in aqueous media

Maxim Likhatski,* Anton Karacharov, Alexander Kondrasenko and Yuri Mikhlin

Received 11th December 2014, Accepted 20th January 2015
DOI: 10.1039/c4fd00258j

Previously, we found a series of fluid nanoscale intermediates preceding nucleation of gold sulfide in the reaction between aqueous $HAuCl_4$ and sodium sulfide. Here, the effects of temperature, addition of an inert electrolyte and some other factors on characteristics of the "dense liquid" intermediates and the formation of solid nuclei were studied using UV-vis absorption spectroscopy, DLS, zeta-potential measurements, and SAXS. It was revealed, in particular, that the negatively-charged interfaces of the dense liquid species critically impede their fusion into large enough dense droplets in which nucleation can take place. The nucleation and following coagulation of poorly crystalline gold sulfide proceed not instantly but progressively as the dense liquid droplets arise, with the process being sharply accelerated by the injection of NaCl or temperature increase.

Introduction

Recently, we explored the reduction of aqueous gold chloride complexes with sulfide ions yielding gold, gold sulfide nanoparticles or their mixture, depending on the reactant ratio,[1–3] which are of vivid interest for geochemistry and mineral processing of gold, and for preparation of Au_2S nanoparticles and Au_2S–Au hybrid nanomaterials for plasmonics, surface enhanced Raman spectroscopy (SERS), and biomedical applications.[4–7] A series of long-lived pre-crystallization intermediates was found to arise in the reaction with the molar ratio $Na_2S/HAuCl_4$ higher than 2 at room temperature.[3] The "soft" droplet-like species from 10–50 nm to about 200 nm and more in the lateral size were directly imaged using *in situ* tapping-mode atomic force microscopy (AFM) (Fig. 1), in good agreement with time-resolved dynamic light scattering (DLS) and small-angle X-ray scattering (SAXS) data. The findings were rationalized in terms of accumulation of reduced $Au_mS_n^{-}\cdot pH_2O$ solutes (referred as pre-nucleation clusters after Gebauer and

Institute of Chemistry and Chemical Technology of the Siberian Branch of the Russian Academy of Sciences, Akademgorodok, 50/24, Krasnoyarsk, 660036, Russia. E-mail: lixmax@icct.ru

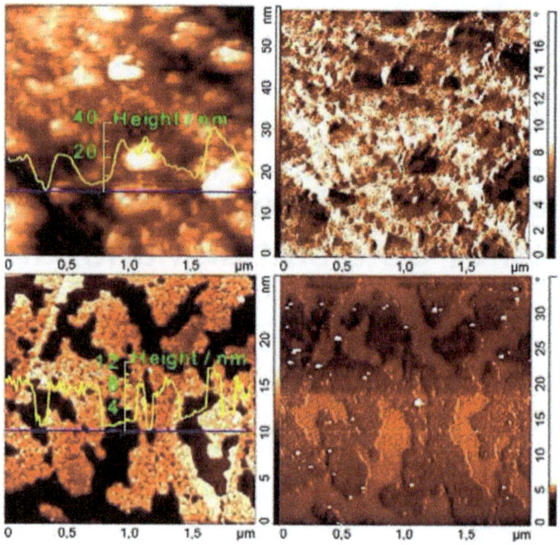

Fig. 1 Tapping-mode AFM height (left panels) and phase images of the products formed with $Na_2S/HAuCl_4$ ratio of 3 acquired *in situ* within about 2 h after mixing the reagents (upper panels) and *ex situ* of a solution drop dried on HOPG in air (adapted from ref. 3). One can see voluminous species in the solution producing a thin film upon drying mM $HAuCl_4$ (32.5 ml) in air and thermostated at the desired temperature.

Cölfen[8]) in dense liquid domains of various dimensions, including "liquid clusters" less than about 20 nm, globules about 50 nm and their submicrometer aggregates. Nucleation was suggested, in accordance with the two step nucleation mechanism,[9–12] to proceed within "dense droplets" large enough to accumulate the necessary amount of Au–S substance formed *via* coalescence of the above fluid species, which probably is a slow stage determining the rate of gold sulfide formation at room temperature. We believe that there are several sorts of dense liquid "droplets" accumulating the pre-nucleation species which arise, evolve with time, temperature, *etc.*, and serve as submicrometer "reactors" for nucleation in the case of the Au–S system. This results in a very long, up to few days, existence of the liquid intermediates, making possible their detection and characterization before abrupt sedimentation of poorly crystalline gold sulfide occurs. The phenomena are important for understanding both the fundamentals of nucleation and crystal growth in aqueous solutions and the specific chemistry of colloidal gold and gold sulfide in sulfide-containing media. Moreover, there are experimental evidences of an involvement of fluid intermediates in the well-known synthesis of Au NPs by citrate reduction of aqueous tetrachloroaurate,[13,14] formation of PtS and PdS nanoparticles, and some other media with effectively retarded chemical reaction or/and nucleation stages.

Here, we report the results of a study on the effect of temperature, in different electrolytes and some other factors on properties of the fluid intermediates formed in the aqueous $HAuCl_4$–Na_2S system with the molar ratio 1 : 3. A set of *in situ* experimental techniques including UV-vis absorption spectroscopy, DLS, zeta-potential measurements and SAXS were applied in an attempt to monitor the changes of dimensions and properties of dense liquid species and to detect the appearance of solid nuclei. The elevation of temperature to 40–70 °C and an addition of NaCl solution at room temperature were employed in order to initiate nucleation of gold sulfide and to relate that with evolution of the transient liquid species.

Experimental

In all the experiments performed deionized Milli-Q Pure water with no additional purification was used. A portion of 10 mM sodium sulfide solution (3 ml) was quickly added to agitated 0.3 mM $HAuCl_4$ and thermostated at the desired temperature, freshly prepared from 0.1 M $HAuCl_4$ stock solution (Krasnoyarsk Plant of Non-Ferrous Metals). The solution was agitated for one min, then agitation was switched off. All the chemicals were analytical grade and used as received. UV-vis absorption spectra were acquired in the wavelength range 200 nm to 1200 nm using an Evolution 300 spectrophotometer (Thermo Scientific) in a thermostated 1 cm quartz cell. Dynamic light scattering studies and zeta-potential measurements were performed with a Zetasizer Nano ZS device (Malvern Instruments) at scattering angle 173°. The intensity-weighted mean size is normally used in this study due to its high reliability, although the results of multimodal fitting also were taken into consideration.

SAXS measurements were performed at the Austrian SAXS beamline at the ELETTRA synchrotron, Trieste, Italy.[15] The time-resolved experiments were conducted with an aliquot of the solution immediately after mixing of the reactants loaded in a standard X-ray quartz capillary (inner diameter of 1.5 mm) sealed and temperature-controlled with the Peltier heating/cooling sample holder (Anton Paar). The SAXS data were acquired at a photon energy of 8 keV applying a 2D Pilatus 100 K detector system. The common procedures of radial averaging and normalization were employed to obtain 1D $I(q)$ *vs.* q curves, where $q = (4\pi/\lambda)\sin\theta$ is wave vector, $I(q)$ is scattering intensity, $\lambda = 0.154$ nm is the wavelength used, and 2θ is the scattering angle.

Results

UV-vis spectroscopy

Fig. 2 shows typical UV-vis absorption spectra of the yellow-brown reaction media with the initial composition of 0.3 mM $HAuCl_4$ + 1 mM Na_2S, which, after a few seconds of reaction, become smooth curves having neither the peak of tetrachloroaurate at 315 nm nor plasmon maxima of gold nanoparticles. The absorbance magnitudes at a wavelength of 420 nm and the plots of $(\alpha E_{ph})^{1/2}$ *vs.* E_{ph} (α is the absorption coefficient, E_{ph} is the discrete photon energy) were further used to estimate temporal changes in the spectra intensity and shape and in the band gap width E_g of the disordered matter produced, assuming an indirect electronic transition. The absorption increases in the starting period (about 30 min) and stays almost constant for as long as several days at room temperature. Addition of a strong electrolyte, 5 mM sodium chloride, that approximately doubled the ionic strength of this rather stable solution (after 2 h reaction), causes some increase in the optical absorption, which then decreases gradually over about 2 h and more rapidly after that. Except for the starting period, the E_g values vary in the same manner. The absorbance at an elevated temperature (70 °C) grows rapidly in the first minutes and then slowly changes till a sharp drop, while the E_g magnitude

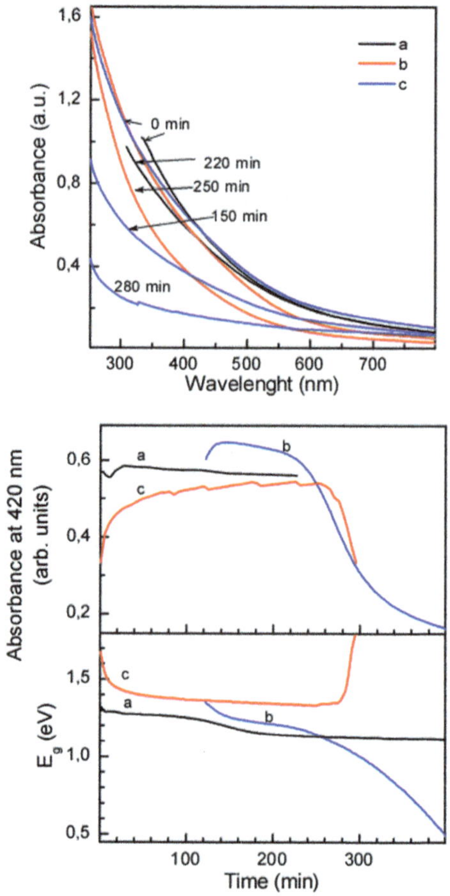

Fig. 2 Typical UV-vis absorption spectra and absorbance at 420 nm and the band gap width *vs.* reaction time for 0.3 mM $HAuCl_4$ + 1 mM Na_2S solution at (a) 21 °C, (b) after addition of 0.05 M NaCl to the solution reacted for 2 h, and (c) at 70 °C.

alters antipathetically. In general, the absorbance at enhanced temperatures is lower and the band gap is wider than at the room temperature. Both in the high temperature and the NaCl addition experiments, formation of a bulky black residue and solution bleaching related with the optical absorption decay were observed. Moreover, minor amounts of suspended solids could be seen by the naked eye in these systems well before the bulky sedimentation. The findings confirm that the absorbance is due to soluble (colloidal) products, whereas the pronounced reduction of the spectrum intensity is caused by coagulation of solid particles, which were found to be composed of poorly crystalline cuprite-type Au_2S somewhat enriched with sulfur but easily decomposed to sulfur-capped metal gold nanoparticles.[1–3] The value referred as the indirect band gap E_g that varies in the range 1.7 eV to 1.2 eV is thought to be determined by Au(I)–S bonding, decreasing with shortening the bond length as the nanometer liquid clusters coalesce into larger droplets and the matter within the droplets becomes denser and more ordered (see below). The steep E_g decline in the media with additional NaCl may be due to Au–Cl bonding in the solution with low residual concentration of sulfide ions. It should be mentioned that the process rates, especially at the elevated temperatures, notably depend on agitation, air access and other factors. In some cases, instability regions several min long were observed after sharp initial changes of the absorbance and E_g and just before the bulky precipitation at 60–70 °C.

Dynamic light scattering

Fig. 3 shows average hydrodynamic size (D_h) of scattering particles as a function of time at two temperatures and after the addition of sodium chloride. At room temperature, the scatterers grow to several hundred nm over several hours, in accordance with the dimensions of "soft" droplet-like entities directly observed in the solution using *in situ* AFM.[3] The addition of NaCl solution after 2 h reaction results in a seeming decrease of the particles and then their rapid growth up to a micrometer and bigger. Interestingly, the mean hydrodynamic size of the particles usually increases more slowly at 70 °C. The DLS data suggest a polydisperse composition of the systems under many conditions, particularly in the initial stages and before the precipitation. A multimodal fit of the correlation functions (the quantitative results are not quite reliable and are not shown in figures) shows that the main volume contribution is provided by particles as small as 10–20 nm at room temperature[3] and 3–10 nm at 70 °C in the conditional stability regions, although their scattering intensity is minor. The situation in the case where NaCl is added is even more complicated as the volumes of different size scatterers are often comparable.

Zeta-potential

Zeta-potentials measured in the systems (Fig. 4) are always negative, with the magnitude increasing from −10 mV to −40 ± 5 mV in the initial stage when rather fast growth of the species was observed. The potential values stay nearly constant in the quasi-stationary regions, and the one approaches zero as gold

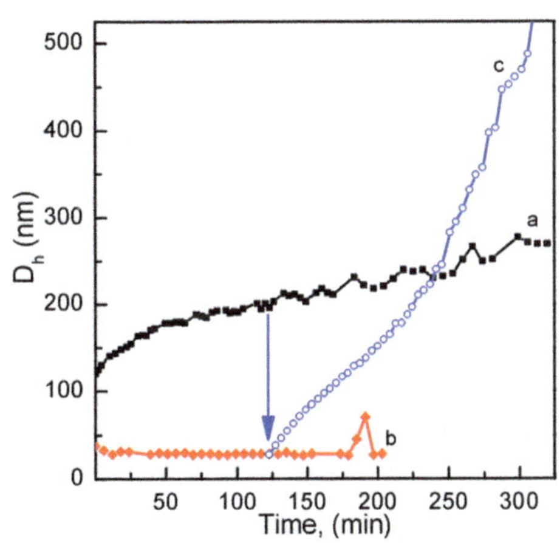

Fig. 3 Average hydrodynamic diameter of the species determined using DLS *vs.* reaction time for 0.1 mM $HAuCl_4$ + 0.3 mM Na_2S media at (a) 21 °C, (b) 70 °C and (c) after addition of 0.05 M NaCl to the solution reacted for 2 h at 21 °C.

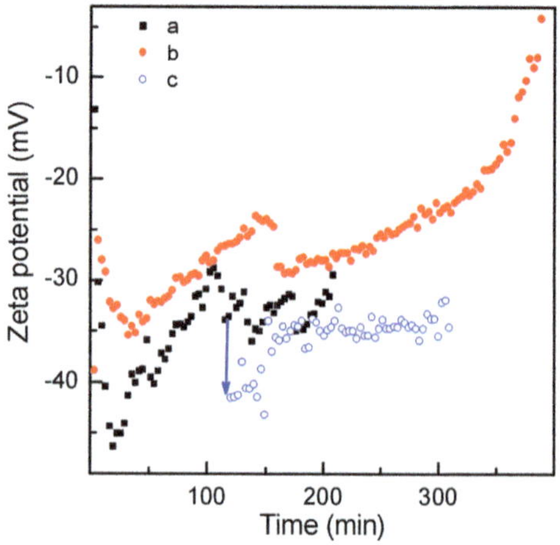

Fig. 4 Evolution of zeta-potentials with reaction time of the species formed in 0.3 mM $HAuCl_4$ + 1 mM Na_2S solution at (a) 21 °C, (b) 65 °C and (c) after addition of 0.05 M NaCl to the solution reacted for 2 h at 21 °C.

sulfide residue precipitates. One can conclude from the comparison of the data on the zeta-potential and the hydrodynamic diameter time variation that the entities are stabilized, first of all, due to the electrically-charged interfaces, with the negative charge seemingly produced by both hydrosulfide anions and chloride-ions. The injection of additional sodium chloride initially causes an increase in the negative zeta-potential of the species (Fig. 4c). In the next stage, the zeta-potential reduces owing to higher electrolyte concentration, promoting further aggregation. It is worth noting that the study of zeta-potential time evolution was complicated by the impact of the imposed electric field on properties of the hydrosols. To minimize side effects, the experiments were also conducted using fresh portions of the solution sampled at pre-determined time intervals; principally the same results were obtained.

SAXS

In contrast to DLS, SAXS patterns acquired at the wave vector q above 0.1 nm^{-1} are sensitive to nanoparticles smaller than 20–30 nm, allowing monitoring of their evolution and, in principle, detection of the formation of solid nuclei. The polydispersity of the scatterer size was taken into account by using the Schultz–Zimm statistical distribution of the scatterer core radii. As demonstrated above, the particles tend to form aggregates. Hence, a model taking into account some stickiness of the scatterers should be applied. For simplicity, the hard sphere potential combined with rectangular well attractive potential[16] was considered to model the inter-particles' interaction in decoupling approximation.[17] In the simplest case, the intensity of the scattered radiation is given by

$$I(q) = \text{scale}\langle F^2(q)\rangle_R \, S(q, R_{\text{HS}}, \lambda, \varepsilon, \varphi) + c_{\text{p}}q^{-n} + \text{bkg}$$

here q is the value of the scattering vector, bkg is the constant contribution to the scattering background

$$F(q) = (\rho_{\text{scatterer}} - \rho_{\text{media}})Vf(qR)$$

is the form factor amplitude of a scatterer having electron density $\rho_{\text{scatterer}}$, and ρ_{media} is the electron density of the media where the scatterers exist; $f(qR) = (3\sin qR - qR\cos qR)/q^3R^3$ is the form factor of a sphere with radius R, and V is the volume of a sphere of that radius. $\langle F^2(q)\rangle_R$ stands for averaging over scattering potentials of the scatterers of different radii R, with Schultz distribution being the weighting function. The averaging was performed by a numerical integration. $S(q, R_{\text{HS}}, \lambda, \varepsilon, \varphi)$ is the structure factor based on the sticky hard-sphere inter-scatterer interaction potential model.[16] Here, R_{HS} is the radius of the hard sphere, φ is the apparent volume fraction of the scatterers in the solution, with $\lambda - 1$ and ε representing the range and strength of the attraction. The radius of a scatterer R was assumed to be equal to that of the hard sphere. Because of the agglomeration of scatterers takes place, we have added a generalized power law term c_pq^{-n} to the fitting function. When it was required, the second fraction of the sticky hard-sphere scatterers was considered.

SAXS curves acquired at temperatures changing in a programmable manner are shown in Fig. 5 (top). The arrows shown on the plot of the mean diameter *vs.* time (bottom) mark the start of temperature elevations, with the rate of heating being 20 °C min^{-1}. In 5 min after the start of the reaction at room temperature, the scattering curves can be fitted using a single fraction of sticky hard spheres with diameter ~8.5 nm, and some growth was observed with time. The main transformations of the SAXS curves occur within about 1 h after the start of the reaction, with results of the fits showing the species growth from 8.5 nm to 13 nm. After 110 min of the reaction and the temperature of 70 °C being reached, the single fraction model function fits the experimental data unsatisfactorily, and, therefore, the second fraction of sticky hard spheres should be added to the model functions for $I(q)$. The initial mean diameter of the second fraction scatterers is of 1.32 ± 0.4 nm and quite rapidly increases with time. This can be rationalized in terms of nucleation and growth of solid nanoparticles, but not liquid species observed before this point. Then, rapid aggregation of solid nanoparticles results in partly suspended black bulk residue.

Almost immediately on adding 0.05 M NaCl solution to the system reacted at ambient temperature for 2 hour, comparatively tiny species of ~4 nm attributable to solid nanoparticles were revealed together with the ~15 nm species (liquid clusters) (Fig. 5, curves b). The volume of the nanoparticles estimated from the $I(0)$ magnitude continuously decreased with time, probably, due to their aggregation and sedimentation.

Discussion

The experimental data obtained in this and previous work[1,3] demonstrate that the reduction of tetrachloroaurate by excessive sodium sulfide in the concentrations well above the stability range of sulfide and chloride complexes of Au(I) produces fluid droplet-like entities whereas solid particles form in quantities below the detection limit of DLS, SAXS and other techniques. In addition, the ^{1}H NMR DOSY technique was applied in an attempt to specify water molecules related to the dense droplets. It was found that the average diffusion coefficient is slower by about 10% in the media containing dense liquid intermediates. However, the

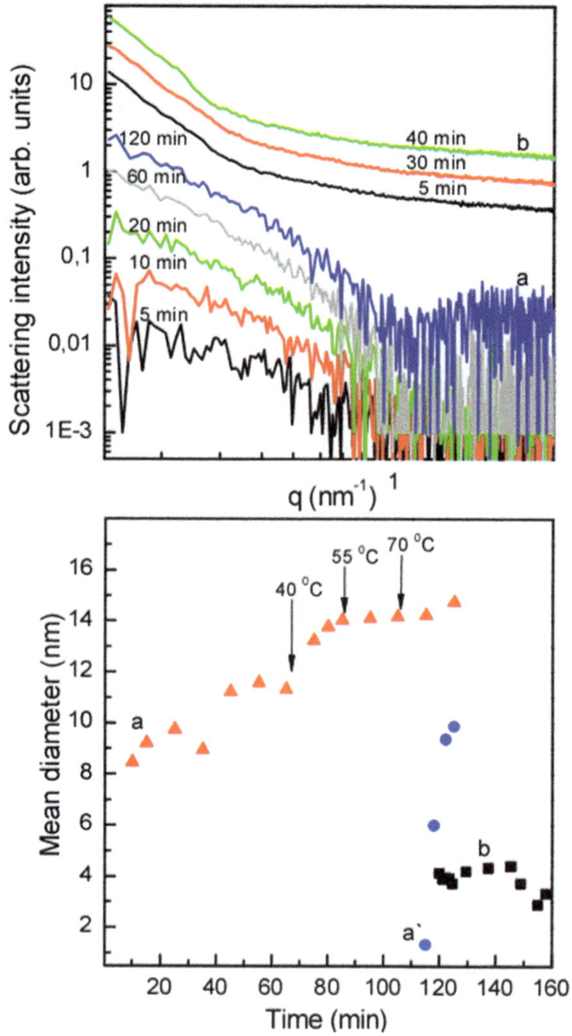

Fig. 5 SAXS curves acquired from a capillary filled with 0.3 mM $HAuCl_4$ + 1 mM Na_2S solution with programmable temperature changes (a), and that from a capillary containing the solution reacted for 2 hours under room temperature +0.05 M NaCl (b) along with time dependence of the mean diameters of the species.

analysis is not straightforward since there is a considerable contribution from bulk water and pH changes so we avoid presenting quantitative data here. It was previously established using *in situ* and *ex situ* X-ray absorption spectroscopy, XPS, and TEM[2,3] that the supersaturated solutions and precipitated material composed of Au(I) coordinated on average to less than two S atoms. In all cases, the precipitates characterized here with TEM and XPS were found to be almost identical to poorly crystalline Au(I) sulfide easily decomposing to Au clusters as described in ref. 2 and 3.

The initial "liquid clusters", several nm in size, have a low negative electrical charge and easily amalgamate into larger species up to 20–50 nm in diameter, which exhibit zeta-potentials around −40 mV impeding their further coalescence. The amount of Au–S substance in the interior is thought to be insufficient for the formation of large enough, stable "critical" nuclei requiring "liquid droplets", which are dense both in terms of the absence of internal boundaries and the enhanced concentration and chemical state of the solute. Although the 20–50 nm liquid droplets, or globules, associate into submicrometer aggregates, those are probably loose and have inner interfaces preventing mass transfer and then nucleation. The fusion of liquid intermediates and following nucleation, growth and coagulation of solid particles is expected to be faster in solutions with high background electrolyte concentrations due to weaker electrostatic repulsion.

Indeed, the addition of a strong electrolyte triggers the formation and precipitation of gold sulfide. Furthermore, a change of osmotic pressure as inert electrolyte is added can play a role of a "water pump" which renders the substance inside the liquid droplets denser. During the growth of the liquid intermediates evidenced with DLS, slow changes of the amount and the chemical state of the pre-nucleation substance were observed from the UV-vis spectra. At a certain point, both the size and optical absorption changes notably accelerate, suggesting transformation of the pre-nucleation matter. Simultaneously, SAXS fitting shows the emergence of 1.5–8 nm nanoparticles composed, most likely, of gold sulfide, followed by the coagulation of black precipitate. It is important that these phenomena proceed gradually but not instantaneously; this may be explained by a progressive transformation of the loose aggregates into the dense droplets and subsequent nucleation in the interior.

This mechanism is illustrated in Scheme 1 showing the evolution of the fluid pre-nucleation entities as in ref. 3 with amendment of the nucleation stage, which is believed to proceed within the large dense droplets, whereas smaller (or loose) fluid species still remain in the solution.

The UV-vis absorption intensities are lower and E_g magnitudes derived from the optical spectra are larger at enhanced temperatures than those at room temperature (Fig. 3), suggesting a slower formation of the pre-nucleation substance and "liquid clusters". At the same time, the experimental results imply that the lifetime of large fluid aggregates is shorter. We hypothesize that the emergence of small liquid clusters is hindered, while their fusion into large dense droplets, and then the nucleation stage, is facilitated, probably, because of more diffuse boundaries between the bulk solution and the dense liquid species.

The structures and mechanisms found for the Au–S media may be applicable for other reactions in which nucleation or(and) chemical reaction stages are effectively arrested. In particular, similar results suggesting the existence of long-

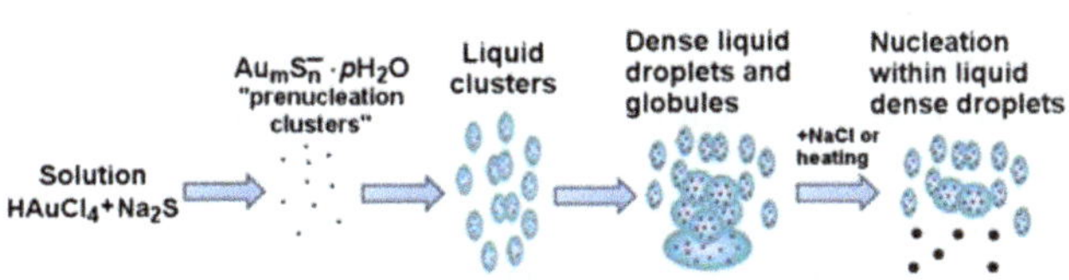

Scheme 1 The scheme exhibits the different stages in solution preceding bulk black residue deposition during the reaction studied.

lived quasi-liquid intermediates, though with some kinetic differences, were obtained in case of the reaction between H_2PtCl_6 and sodium sulfide; these data will be reported in detail elsewhere.

Conclusions

In situ time-resolved UV-vis spectroscopy, DLS, zeta-potential measurements, and synchrotron based SAXS were used in order to study pre-nucleation long-lived fluid species in the reaction of tetrachloroauric acid and sodium sulfide with the initial molar ratio of 1 : 3 at two temperatures. The initial "liquid clusters" were found to have a small negative charge and coalesce into larger species exhibiting negative zeta-potential of about 30–40 mV and associating to submicrometer aggregates. The addition of 0.05 M NaCl sharply accelerates the aggregation and nucleation. The growth of the dense liquid species proceeds slowly at elevated temperatures, although in general the formation of solid gold sulfide is faster, probably, due to easier fusion of the species into large droplets and fast nucleation.

Acknowledgements

This study was supported by the Grant MK-6489.2014.5 for young candidates of science from President of RF. We gratefully thank Prof. Heinz Amenitsch and the staff of the Austrian SAXS beamline at the ELETTRA for their precious assistance.

Notes and references

1 Y. Mikhlin, M. Likhatski, A. Karacharov, V. Zaikovski and A. Krylov, *Phys. Chem. Chem. Phys.*, 2009, **11**, 5445–5454.
2 Y. Mikhlin, M. Likhatski, Y. Tomashevich, A. Romanchenko, S. Erenburg and S. Trubina, *J. Electron Spectrosc. Relat. Phenom.*, 2010, **177**, 24–29.
3 Y. Mikhlin, A. Karacharov, M. Likhatski, T. Podlipskaya and I. Zizak, *Phys. Chem. Chem. Phys.*, 2014, **16**, 4538–4543.
4 R. D. Averitt, D. Sarkar and N. J. Halas, *Phys. Rev. Lett.*, 1997, **78**, 4217–4220.
5 A. M. Schwartzberg, C. D. Grant, T. van Buuren and J. Z. Zhang, *J. Phys. Chem. C*, 2007, **111**, 8892–8901.
6 J. Majimel, D. Bacinello, E. Durand, F. Vallée and M. Tréguer-Delapierre, *Langmuir*, 2008, **24**, 4289–4294.
7 A. M. Gobin, E. M. Watkins, E. Quevedo, V. L. Colvin and J. L. West, *Small*, 2010, **6**, 745–752.
8 D. Gebauer and H. Cölfen, *Nano Today*, 2011, **6**, 564–584.
9 P. R. ten Wolde and D. Frenkel, *Science*, 1997, **277**, 1975–1978.
10 A. Lomakin, N. Asherie and G. B. Benedek, *Proc. Natl. Acad. Sci. U. S. A.*, 2003, **100**, 10254–10257.
11 D. Kashchiev, P. G. Vekilov and A. B. Kolomeisky, *J. Chem. Phys.*, 2005, **122**, 244706.
12 P. G. Vekilov, *Nanoscale*, 2010, **2**, 2346–2357.
13 Y. Mikhlin, A. Karacharov, M. Likhatski, T. Podlipskaya, Y. Zubavichus, A. Veligzhanin and V. Zaikovski, *J. Colloid Interface Sci.*, 2011, **362**, 330–336.
14 M. Doyen, K. Bartik and G. Bruylants, *J. Colloid Interface Sci.*, 2013, **399**, 1.

15 S. Bernstorff, H. Amenitsch and P. Laggner, *J. Synchrotron Radiat.*, 1998, **5**, 1215–1221.
16 D. Pontoni, S. Finet, T. Narayanan and A. R. Rennie, *J. Chem. Phys.*, 2003, **119**, 6157–6165.
17 J. S. Pedersen, *Adv. Colloid Interface Sci.*, 1997, **70**, 171–210.

Faraday Discussions

DISCUSSIONS

Time and space resolved methods: general discussion

Wenhao Sun, Samuel Booth, Allan Myerson, Colan Hughes, Haihua Pan, Gerard Coquerel, Clement Brandel, Hugo Meekes, Marco Mazzotti, Laszlo Fabian, Simon Black, Peter Vekilov, Kevin Back, Dimitrios Toroz, Jessica Lovelock, Jan Sefcik, Ake Rasmuson, Eric Breynaert, Richard Sear, Robert Hammond, Martin Ward, Terence Threlfall, Jim De Yoreo, Roger Davey, Radoljub Ristic, Ken Lewtas, Kevin Roberts, Alan Hare, Martí Gich, Helmut Cölfen, Maxim Likhatskiy, Joop ter Horst and Nico Sommerdijk

DOI: 10.1039/c5fd90037a

Jim De Yoreo presented some slides on *in situ* AFM, TEM, dynamic force spectroscopy (DFS) and optical spectroscopy investigations of nucleation in the calcium carbonate system: The free energy barrier to homogeneous nucleation of calcite calculated within the framework of classical nucleation theory (CNT) is prohibitive, even at concentrations exceeding the solubility limits of the amorphous phases. Consistent with this analysis, during nucleation in pure solutions, in our *in situ* TEM experiments we observed direct formation of all phases, including amorphous calcium carbonate (ACC), as well as the three predominant crystalline phases: calcite, vaterite, and aragonite, even under conditions in which ACC readily forms. In addition to direct formation pathways, we observed indirect pathways in which ACC transforms to aragonite and vaterite through nucleation within or on the precursors, rather than *via* dissolution and reprecipitation. We also observed aragonate transformation to calcite, but never recorded an instance in which ACC transforms into calcite, except *via* dissolution–reprecipitation reactions.

In contrast to the situation in bulk solutions, *in situ* observations of the heterogeneous nucleation of calcium carbonate on self-assembled monolayers (SAMs) *via* either optical or AFM methods revealed the direct formation of calcite with no evidence for other phases. Moreover, analysis of nucleation rates *vs.* supersaturation gave the dependence expected classically and showed that control over nucleation arose from a moderate decrease the interfacial energy. Furthermore, the differences in nucleation kinetics on odd and even chain-length carboxyl SAMs were attributable to relative differences in these energies. In addition, DFS measurements demonstrated that the underlying source of this reduced interfacial free energy is the magnitude of the matrix-crystal binding free energy, as would be expected from a purely classical analysis.

Introduction of the charged polymer polystyrene sulfonate (PSS), which acts as a surrogate for highly polysaccharides and acidic proteins, led to formation of 10–20 nm diameter Ca-PSS globules through counter ion condensation. Titration calorimetry showed that more than 50% of the Ca ions are sequestered into these globules. Under conditions for which vaterite appeared randomly throughout the cell in the absence of PSS, the first mineral to appear in its presence was ACC, which formed only at the location of these globules. Analysis of growth rates showed that the supersaturation relative to ACC at the time of nucleation is about twice as high as that relative to vaterite when it nucleates without PSS. Hence, unlike the SAMs which controlled nucleation by promoting nucleation of calcite through the reduction of interfacial free energy, PSS controlled nucleation through ion binding and stabilization of ACC, but was still an inhibitor of nucleation.

Peter Vekilov opened a general discussion of the paper by Joop ter Horst: In the expression that uses expectancy time of detection of the first crystal to evaluate the nucleation rate, do you account for the time that the crystal takes to grow to detectable dimensions; is this time determined by the rate of crystal growth and the resolution of the chosen detection technique, typically of order one micrometer? In Kashchiev's book,[1] this time is referred to as growth time, t_G, and we have found that it may be significant, at least in the nucleation of sickle cell hemoglobin polymers.[2,3]

1. D. Kashchiev, *Nucleation. Basic theory with applications*, Butterworth, Heinemann, Oxford, 2000.
2. O. Galkin, R. L. Nagel and P. G. Vekilov, *J. Mol. Biol.*, 2007, **365**, 425–439.
3. O. Galkin and P. G. Vekilov, *J. Mol. Biol.*, 2004, **336**, 43–59.

Joop ter Horst answered: The induction time measurements in the paper are done in non-stagnant stirred 1 ml solutions. What is detected is a suspension. What we think happens, and we found compelling evidence for this,[1,2] is that a single parent crystal nucleates which continues to be suspended in the solution due to the stirring taking place. This single nucleus grows out to become a large crystal after which it undergoes secondary nucleation and a lot of fragments are created that grow out to create the suspension which is detected. Due to this single nucleus mechanism (see also Fig. 3 in our paper) there is a delay in detection: we detect the suspension and not the nucleus formation itself. We assume that this delay (we call it the growth time) is the same for all experiments under the same conditions.

1. S. S. Kadam, S. A. Kulkarni, R. Coloma Ribera, A. I. Stankiewicz, J. H. ter Horst and H. J. M. Kramer, A new view on the metastable zone width during cooling crystallization, *Chem. Eng. Sci.*, 2012, 72, 10–19.
2. S. A. Kulkarni, H. Meekes and J. H. ter Horst, Polymorphism Control through a Single Nucleation Event, *Cryst. Growth Des.*, 2014, **14**(3),1493–1499.

Ake Rasmuson commented: We are doing similar experiments using our own equipment and naked eye observation. We don't see one single crystal forming first. As soon as we can detect anything the solution it turns turbid. Our interpretation is that the growth rate is sufficiently high for the growth time to be negligible.

Alan Hare asked: Would you be able to see one crystal? How small is the smallest object you can see? Would I be able to see it?

In this context, Joop's introduction of statistics in a formal way now seems all the more important.

Ake Rasmuson replied: I think it is a fair assumption that if we cannot observe any changes in the solution for 20 min and then within a couple of seconds the entire solution becomes very turbid, the growth time to visibility is likely to be negligible for the determination of nucleation times.

The growth time evaluated by the approach of ter Horst and coworkers is based on an assumption that the first nuclei observed has nucleated at time zero and the measured time to visibility is entirely governed by growth. The growth time determined in this way sets an upper limit, but not a lower one. In my opinion the validity of this assumption needs to be further investigated. Furthermore, since the growth time tends to be short compared to the average nucleation time, and the nucleation time distribution is quite wide; this doesn't have a major influence on the evaluation.

Roger Davey said to Joop ter Horst: I want to clarify that in your equations the growth time is a parameter that can be fitted or a value assumed – correct? It is not ignored, rather it forms an integral part of the data analysis.

Joop ter Horst responded: Indeed. The fitting procedure delivers us not only the nucleation rate but also the growth time.

Peter Vekilov added: Besides the growth time discussed in my previous question to Professor Joop ter Horst, another significant contributor to the expectancy time of the first crystal is the nucleation delay time. In the classical nucleation theory, *i.e.*, for direct nucleation, this time is defined as the time required by the system to transition from one steady state, a homogeneous supersaturated solution, to another steady state, a solution in which nucleation proceeds at a steady rate. Do you account for the nucleation delay time in your model?

Joop ter Horst answered: We do not account for this time since we think this time is fast compared to the induction time as well as the time needed to cool the solution. We only use experimental data in which the supersaturation results in induction times that are more than roughly 2 min. The moment the required constant temperature is reached we thus assume that the system also reaches the stationary state.

Marco Mazzotti remarked: As shown in our own contribution, it is not possible to obtain independent information about nucleation kinetics and growth kinetics from induction time measurements only. The parameters obtained in this way result in being correlated. Can you comment on how you dealt with this problem in your work?

Joop ter Horst responded: Indeed in case of metastable zone width measurements such as measured in the paper[1] and which you use in your Faraday paper, crystal nucleation and growth kinetics are difficult to obtain independently. This is because the supersaturation changes during the experiment. The experiments in my Faraday paper are measured at constant supersaturation in small volumes resulting in large distributions of induction times. This actually means that the primary nucleation rate is small. Separate measurements indicated that indeed a single nucleus mechanism occurred,[1,2] see also Fig. 3 of our paper. A single parent crystal nucleates in the stirred solution. This single crystal grows out to become a large crystal after which it undergoes secondary nucleation and a lot of fragments are created that grow out to create the suspension which is detected. The secondary nucleation part creating the suspension seems to be fast compared to the time needed to grow the parent crystal to the size at which it undergoes secondary nucleation. This single nucleus mechanism therefore effectively splits nucleation and growth kinetics if it is assumed that the growth kinetics of the single crystal is the same for all experiments under the same conditions.

1. S. S. Kadam, S. A. Kulkarni, R. Coloma Ribera, A. I. Stankiewicz, J. H. ter Horst and H. J. M. Kramer, A new view on the metastable zone width during cooling crystallization, *Chem. Eng. Sci.*, 2012, **72**, 10–19.
2. S. A. Kulkarni, H. Meekes and J. H. ter Horst, Polymorphism Control through a Single Nucleation Event, *Cryst. Growth Des.*, 2014, **14**(3),1493–1499.

Allan Myerson remarked: There are a number of ways to perform nucleation induction time experiments. The results can vary significantly based on both experimental method and experimental technique. This include solution preparation, filtration, cleaning of containers, control of temperature, control of evaporation, head space present in a vessel, stirring *etc.* This does not mean the data is not valuable, but it must be repeatable with enough repetitions to be statistically significant and the role of the variables mentioned above have to be considered in analysis of the results.

Joop ter Horst responded: Induction time measurements and nucleation rate measurements need careful and meticulous design. Measurement procedures need to be carefully thought through and all possible sources for the mentioned variations should be eliminated as much as possible. If these can not be eliminated, the effect of the variations on the experimental results should indeed be considered.

Jim De Yoreo asked: Rather than measure inductions times, we measure nucleation rates, which do not have the uncertainty about the meaning of induction time. We can do this, because we introduce surfaces that drive nucleation to occur on them. So it is easy to count nuclei. Of course this means we get interfacial energies that include the effect of the surface. Nonetheless, the meaning of the rates is unambiguous. Must you use bulk solutions absent of interface effects to get the information you want, or could you do so by also utilizing the controls of heterogeneous nucleation?

Joop ter Horst answered: Currently in crystal primary nucleation from solutions, nuclei form on heterogeneous particles such as dust particles (heterogeneous primary nucleation). Now that we and also others have established accurate crystal nucleation rate measurement methods we will be able to investigate heterogeneous nucleation onto well-defined and functionalized heterogeneous particles and surfaces rather than on ill-defined dust particles. The addition of these heterogeneous surfaces is vital in understanding heterogeneous crystal nucleation.

Colan Hughes remarked: In Fig. 6 of your paper, you show induction time data for solutions which had been filtered before crystallization. This data set stands out as having the worst fit to the model you propose, with a much more linear dependence of $P(t)$ on t. Is this poor fit a consequence of the filtering removing heterogeneous particles and resulting in a situation where this model cannot be used? Alternately, has this particular fit simply failed to find the correct value of t_g?

Joop ter Horst replied: The fit is indeed less optimal for the filtered solutions but in our eyes still acceptable. Why the fit is of a lower quality we don't know. The filtering with the relatively large pore size would remove only part of the heterogeneous particles. After filtration therefore the solution still would contain some heterogeneous particles and we think heterogeneous nucleation continues to occur but now possibly onto less effective and/or a lower concentration of heterogeneous particles.

The model is applicable in stirred solutions in which the nucleation rate is low (roughly 1 per experimental volume per experimental time). We thus think that the same model can be applied.

Wenhao Sun said: I wanted to return to what Professor Vekilov mentioned earlier. The solution proceeds between two steady-states – the metastable supersaturated solution and the period of steady-state nucleation. The difference between these times is defined as the induction time. I was wondering if there was any equation or formulation within classical nucleation theory that relates the induction time to the classical nucleation barrier?

Joop ter Horst replied: In my paper and many other papers the induction time is defined as the time progress between the instance of constant supersaturation creation and the instance of detection of crystals. This definition follows that in the seminal book on Nucleation of Dimo Kashchiev.[1] In that book the time between the creation of a supersaturated solution and a stationary state is referred to as the time lag.

In our paper we use the induction time distributions at different supersaturations to determine nucleation rates. These nucleation rates were analyzed using Classical Nucleation Theory to come to values for the kinetic pre-exponential factor and the exponential factor (*e.g.*, Fig. 9 in the paper). This last factor is related to the nucleation work, the energy barrier for nucleation.

In literature other relations are proposed that link induction times to factors in the Classical Nucleation Theory.

1. D. Kashchiev, *Nucleation. Basic theory with applications*, Butterworth, Heinemann, Oxford, 2000.

Peter Vekilov commented: In the classical nucleation theory, *i.e.*, for direct nucleation, the delay time for nucleation is defined as the time required by the system to transition from one steady state, a homogeneous supersaturated solution, to another steady state, a solution in which nucleation proceeds at a steady rate. For direct nucleation, this time was first defined by Zel'dovich as an extension of his theory of steady-state classical nucleation.[1] Detailed analysis of this delay time by Kashchiev[2] suggest that it is inversely proportional to the rate of attachment of molecules to the nucleus, *i.e.*, roughly, to the crystal growth rate. This makes it a weak function of the supersaturation. In the case of two-step nucleation, the nucleation delay time depends on the total volume occupied the precursor clusters.[3] Since this volume is a function of the solution concentration and the supersaturation, the nucleation delay time is a sensitive indicator of the mechanism chosen by the system.[4–6]

1. Y. B. Zel'dovich, *Acta Physicochimica URSS*, 1943, **18**, 1–22.
2. D. Kashchiev, *Nucleation. Basic theory with applications*, Butterworth, Heinemann, Oxford, 2000.
3. D. Kashchiev, P. G. Vekilov and A. B. Kolomeisky, *J. Chem. Phys.*, 2005, **122**, 244706.
4. O. Galkin, R. L. Nagel and P. G. Vekilov, *J. Mol. Biol.*, 2007, **365**, 425–439.
5. O. Galkin, W. Pan, L. Filobelo, R. E. Hirsch, R. L. Nagel and P. G. Vekilov, *Biophys. J.*, 2007, **93**, 902–913.
6. O. Galkin and P. G. Vekilov, *J. Mol. Biol.*, 2004, **336**, 43–59.

Wenhao Sun responded: Professor Vekilov states that the induction time typically depends on the width of the nucleation barrier, rather than the height of the nucleation barrier. This is described by Zel'dovich in the Zel'dovich Factor.

I was surprised that this was the case, because I would expect induction times to be much shorter for heterogeneous nucleation than for homogeneous nucleation, suggesting that the nucleation barrier height, rather than the width, plays the most important role in determining the induction time. Would you elaborate?

Furthermore, the Zel'dovich factor is a pre-exponential term, and is routinely ignored, since order-of-magnitude calculations for the Zel'dovich factor ranges from 0.1 to 1, whereas the exponential term can vary by over 30 orders of magnitude, and so wouldn't it suggest that the barrier height plays a more influential role in induction time as well?

Peter Vekilov responded: The induction time is a poorly defined concept. If we assume that this is the expectancy time for the detection of the first crystal, measured from the moment that the solution became supersaturated, it will be the sum of three characteristic times

$$\Delta t_{\text{expect}} = \tau + \Delta t_{\text{nucl}} + \Delta t_{\text{growth}}.$$

In this expression, τ is the nucleation delay time, defined by Zel'dovich as the time for transition from a steady supersaturated homogeneous solution to another steady state, at which crystal nucleation proceeds at a steady rate.[1–3] The delay time $\tau = 2(3\pi Z^2 f^*)^{-1}$, where Z is the Zel'dovich factor and f^* is the frequency of attachment of molecule to the nucleus. In turn $Z = i^{*-1}(\Delta G^*/3\pi k_B T)^{1/2}$, where i^* and ΔG^* are, respectively, the number of molecules in the nucleus and the nucleation barrier, related as $\Delta G^* = 0.5 i^* \Delta\mu$; the ratio $\Delta G^*/i^{*2}$ is proportional to the width of the nucleation barrier.[1,4]

The second member of the sum, $\Delta t_{nucl} = (VJ)^{-1}$, where V is the system volume and J is the steady state nucleation rate, accounts for the expectancy time for the nucleation of the first crystal after steady state has been reached. The third additive Δt_{growth} is the time needed for the first crystal to grow to detectable dimensions; it depends on the resolution limit of the technique of observation and on the rate of growth of the crystals.

There are numerous examples in which τ is the longest of the three characteristic times. Investigating the nucleation of sickle cell hemoglobin polymers[4,5] we found that τ was in the range 10–20 s, while Δt_{nucl} and Δt_{growth} were both shorter than 1 s. Thus, assuming that $\Delta t_{expect} \approx \Delta t_{nucl}$ may be a serious misconception. This may be even further aggravated if one uses Δt_{expect}, misinterpreted as Δt_{nucl}, to determine the nucleation rate J and draw conclusions about the nucleation mechanism.

As a corollary of these considerations, since for nucleation on heterogeneous centers both i^* and ΔG^* are reduced, the Zel'dovich factor and the delay time τ are not significantly affected. Heterogeneous nucleation leads to significantly faster rates and, consequently, to shorter Δt_{nucl}. However, if τ is the major component of the expectancy time, then this time will not be affected by the presence of heterogeneous nucleation centers.

1. Y. B. Zel'dovich, *Acta Physicochimica URSS*, 1943, **18**, 1–22.
2. D. Kashchiev, *Surf. Sci*, 1969, **14**, 209–220.
3. D. Kashchiev, *Nucleation. Basic theory with applications*, Butterworth, Heinemann, Oxford, 2000.
4. B. Mutaftschiev, in *Handbook of crystal growth*, ed. D. T. J. Hurle, Elsevier, Amsterdam, 1993, vol. 1, pp. 189–247.
5. O. Galkin, R. L. Nagel and P. G. Vekilov, *J. Mol. Biol.*, 2007, **365**, 425–439.
6. O. Galkin and P. G. Vekilov, *J. Mol. Biol.*, 2004, **336**, 43–59.

Kevin Roberts commented: This discussion merits some reflection as to how active pharmaceutical ingredients (APIs) are manufactured through crystallisation processes on the industrial scale. Typically, a limited knowledge of the degree of supersaturation used for crystallisation together with a lack of its precise control can conspire to deliver product crystals characterised by a wide size distribution. In order to provide an effective formulation of the API these crystals have to be milled to provide an optimal particle size. Such intense mechanical action can result in damage to the crystal structure of the crystals and can induce the formation of lattice defects and/or lead to the formation of metastable or amorphous forms all of which would impact upon the quality and variability of drug products, hence making them far less effective.

An important future development for the industry would see the development of new processes capable of producing a sharp crystal size distribution of uniform product manufactured to an exact size. Bearing in mind that the crystal dissolution rate is driven largely through considerations of surface area, this production of precision drug particles would allow the therapeutic profile to be much narrower, enabling lower dosages without compromising efficacy.

Routes to enable the reproducible crystallisation of precision API crystals can thus represent a significant future challenge for the healthcare sector. Within this context the use of surface templates perhaps provide one potential processing option and recent work by my group[1,2] on the API entacapone have demonstrated that almost mono-sized crystal size distributions can be crystallised by this approach. This kind of approach may offer promise for the future in the above respect.

1. A. Kwokal, T. H. Nguyen and K. J. Roberts, Polymorph-directing seeding of Entacapone crystallisation in aqueous/acetone solution using a self-assembled molecular layer on Au(100), *Crys. Growth Des.*, 2009, **9**, 4324–4334.
2. A. Kwokal, D. Čavužić and K. J. Roberts, Surface adsorbed templates for directing the crystal growth of Entacapone as monitored by ATR-UVvis spectroscopy and FBRM, *Crys. Growth Des.*, 2013, **13**, 5324–5334.

Kevin Back said: Following Prof. Kevin Roberts' comment on the importance of nucleation in the manufacture of active pharmaceutical ingredients, the pharmaceutical industry are also interested in biorelevant nucleation. For example, a basic drug may dissolve in the acid environment of the stomach, but as it moves into the small intestine it encounters a pH closer to neutral, and could become supersaturated. If the drug nucleates and precipitates, this can impact on absorption of the drug and thus its bioavailability. Measuring nucleation rates in biorelevant media can give us parameters which we can use in models to predict precipitation behaviour *in vivo*, and thus aid in the computational design of our drug products.

Roger Davey commented: Induction times depend on all sorts of things but we should not use this as a reason for doing nothing. For example if we standardised our procedures (using *e.g.* Crystal16) then we could compare our results.

Allan Myerson responded: Induction time measurements can be very valuable but the data analysis and the use of the data must be carefully done with an understanding of the limitations of whatever methods and conditions were employed. In addition, statistically valid numbers of experiments must be conducted.

Richard Sear returned to the paper by Joop ter Horst: In, for example Table 1, you quote rates per unit volume. Can I ask if you have directly checked that the rates you observe are per unit volume, *e.g.*, have you doubled the volume of the droplets you work with and observed a doubling of the rate?

Joop ter Horst replied: We have investigated the effect of volume in an earlier paper.[1] We see that at larger volumes the variations in the induction time decrease. This is expected since the probability that a nucleus forms increases in larger volumes.

1. S. S. Kadam, S. A. Kulkarni, R. Coloma Ribera, A. I. Stankiewicz, J. H. ter Horst and H. J. M. Kramer, A new view on the metastable zone width during cooling crystallization, *Chem. Eng. Sci.*, 2012, **72**, 10–19.

Radoljub Ristic commented: In your paper the observables are microcrystals which can be correlated to two different types of nucleation; primary and secondary nucleation. The mechanisms of the two are different so the assigned probabilities should be different, too. For the former one you define experimentally determined probability distribution $P(t)= (M^+(t))/M$ and the cumulative probability distribution function $P(t)=1 - \exp(-JV(t - t_g))$ while for the latter one, the counterparts are not provided. Your ultimate goal is primary nucleation rate $J = zf^*C^* = zf^*C_0 \exp(-\Delta G^*/kT) = AS \exp(-B/\ln^2 S)$[1] and from it the determination of A and B by using the plot $\ln(J/S) = f(\ln^{-2}S)$ (Fig. 9). This equation is derived under the assumption that J is stationary and the solution is in a stagnant state. However, your solution is not stagnant, it had been mixed continuously and intensively (700 rpm). In other words, your system had been supplied continuously with a certain amount of external energy. It implies that this dissipative energy should have been estimated and included in the exponential of the equation originally derived for the primary crystal nucleation rate in a stagnant solution. A negligence of this energy contribution may lead to a less adequate definition of the relevant probabilities. Apart from that, the origin and the number of primary nucleated crystals remain unknown. This is probably due to a limited detecting system field of view, which needs attention, that is, more comprehensive information on the experimental set-up is needed as well as the accompanying difficulties in data collection.

The accuracy of the measurement of induction time is determined by the power of detecting system to register as earlier as possible a new phase and its origin. Bearing in mind this, I have difficulties to see any physical reason for waiting for secondary nucleation to occur from already present and naked eye visible microcrystals induced by primary nucleation. This strategy leads to a new set of events almost uncorrelated to those in primary nucleation so I am not convinced how induction times measured in the secondary nucleation mode helps an improvement of the primary nucleation rate, J. Regardless of these remarks, secondary nucleation itself is of huge interest and significance to both academic and industrial communities, and as such deserves a particular attention.

Unfortunately, the mechanism of secondary nucleation is still poorly understood and there is no quantitative description of this complexity. The thing that we know reliably at present is that this process is enhanced in a stirred solution.

There are several factors that may contribute to secondary nucleation: crystal–agitator contact, attrition process with a turbulent solution-induced mechanism, growing crystals containing dislocations, defects or inclusions are prone to secondary nucleation through the development of internal stresses which lead to crack formation and the subsequent production of breakage fragments, autocatalytic role of nucleation at the facets of parent crystals *etc.* However, we are still unable to establish a hierarchy of individual contributors and to control any of them in the course of a crystallisation process. The above set of events is entirely different compared to those for primary nucleation. In my opinion, the fundamental problem is how to define probabilities related to the above phenomena. The only thing that I am certain about, at this stage, is that any attempt to define these kinds of probabilities, without much deeper insight into the puzzle, would be highly subjective, sometimes completely arbitrary, and most often meaningless.

To summarise, I am still uncomfortable with my perception of a link between the induction times measured in the secondary nucleation mode and the experimental and cumulative probabilities defined for primary nucleation of an unstirred solution, in order to determine A and B. Could you comment?

1. D. Kashchiev, *Nucleation. Basic theory with applications*, Butterworth, Heinemann, Oxford, 2000.

Joop ter Horst responded: From your comment I deduced a number of questions I will try to answer:

Firstly: Is secondary nucleation influencing the measurement of the primary nucleation rate?

We think the nucleation rate measurements follow the mechanism we proposed in the paper and elsewhere. To deduce true nucleation rates from the measurements the single nucleus mechanism has to occur. This mechanism (explained in section 3.3 of the paper) proceeds as follows: 1. A single nucleus is formed in the stirred supersaturated solution; 2. This single nucleus grows out to a macroscopically large single crystal; 3. This single crystal, when large enough, is the source of secondary nuclei that grow further to create the suspension. If the suspension is dense enough the detection technique will be able to detect crystals.

We therefore make use of secondary nucleation to detect the single nucleus after it is grown out to a large size. Thus, the growth and secondary nucleation process delay the detection of the single nucleus. We assume that this delay in each experiment under the same conditions is equal. So, any variation in induction times would be caused by the primary nucleation rate. Primary nucleation on the one hand and crystal growth and secondary nucleation on the other hand are thus effectively uncoupled. This comes back in the equation describing the induction time probability (equation 5 in the paper) which contains the primary nucleation rate and the delay, which we defined as growth time t_g. This growth time is thus influenced by both growth rate and secondary nucleation rate. We see for instance that the fitted growth time increases with decreasing supersaturation. We have indications that if the stirrer speed is increased (which would decrease the size at which the parent crystal starts to cause secondary nucleation and possibly increase secondary nucleation rate) results in a decrease of the growth time while the nucleation rate remains roughly unaffected (unpublished work).

Secondly: What is the effect of the external energy source due to the stirring of the solution?

We did not account for any influence of fluid dynamics in the Classical Nucleation Theory.

Thirdly: Why wait for secondary nucleation to occur if the primary crystal(s) can be seen?

We did the induction time experiments on a 1 ml scale using transmission of light through the sample to detect whether crystals are present or not. We only detect crystals once the suspension is sufficiently dense. We could not visualize the parent crystals in these experiments.

We did visualize the parent crystal in 3 ml experiments. In these experiments we only were able to visualize a small part of the solution. In addition, we only could take snapshots every second. There was therefore only a small chance that the single parent crystal was captured in a picture. In our eyes it is not possible to have an accurate method for detection of this single parent crystal in a stirred solution.

Jan Sefcik commented: The effect of shear on primary nucleation can be quantitatively studied using well-controlled shear environment (such as a Couette cell) where the probability distribution of induction times is analysed using an approach proposed by ter Horst.[1] We investigated isothermal nucleation in aqueous glycine solutions[2] and within a short time of the first crystals being observed there was a very rapid increase in number of crystals present which can be attributed to secondary nucleation. Under such conditions it is possible to accurately estimate induction times of secondary nucleation, and the probability distribution of these is then interpreted using the ter Horst model, where the time of appearance of the first crystal through primary nucleation is followed by growth time t_g to the size required to induce secondary nucleation. The probability distribution was fitted to estimate the primary nucleation rate J and the growth time t_g, where both of these parameters were found to be dependent on the shear rate applied.[2]

1. S. Jiang and J. H. ter Horst, *Cryst Growth Des.*, 2011, **11**, 256–261.
2. C. Forsyth, P. A. Mulheran, C. Forsyth, M. D. Haw, I. S. Burns and J. Sefcik, *Cryst Growth Des.*, 2015, **15**, 94–102.

Radoljub Ristic answered: I am inclined to believe that my comments provide, even at this level, quite simple and sufficient physical evidence that making use of the formula for primary nucleation rate, J, derived from CNT under well-defined assumptions including the stagnancy of solution, is not applicable to the description/characterisation of the kinetic and thermodynamic parameters, A and B, respectively, for primary nucleation processes in stirred solutions. It is quite apparent that an externally disturbed crystallisation system by a mechanical energy of solution mixing is just one from a myriad of dynamic complexities. Understanding of respective and puzzling mechanisms of the creation of either primary or secondary nucleation under the above conditions still remains open to question.

Peter Vekilov turned the discussion to the presentation by Jim De Yoreo: In your presentation, you indicated that the supersaturations during your experiments is so high that they do not allow discrimination between direct and two-step nucleation. As a result, the action of both mechanism is observed. Would you speculate on which mechanism would dominate if the same experiments are performed at lower supersaturations?

Jim De Yoreo answered: What we can say is that, as we go to lower supersaturation, but still high enough to see anything happen on the timescale of our experiments, we see a tendency towards less formation of highly unstable phases and more direct formation of calcite. However, I have to emphasize again that when we speak of two-step nucleation, what we mean is formation of a macroscopic phase that then transforms to a more stable phase through a second nucleation event on or in the precursor particle. In the cases where the first phase to form is amorphous, we can not say if it is a liquid or solid. More importantly, even in the cases where we say we see one-step nucleation, our resolution is not sufficient to say whether the initial nucleus (less than ~50 nm) is crystalline, amorphous, or liquid.

Allan Myerson commented: Calcite is an important industrial material used in the paper industry and large amounts are made every day. The industry does a very good job in controlling particle size and shape but every so often they will end up with aragonite without really understanding why. Those who work in the field might use this experience to try to understand the difference in the nucleation barriers for each form and the role of impurities.

Helmut Cölfen replied: We agree that the nucleation barriers play a decisive role as well as the impurities. Looking into details of the $CaCO_3$ nucleation process would require consideration of prenucleation clusters as well[1] and of course the role of impurities is manyfold[2] since these are additives and can switch polymorphs.[3]

1. D. Gebauer, A. Volkel and H. Cölfen, *Science*, 2008, **322**, 1819–1822.
2. D. Gebauer, H. Cölfen, A. Verch and M. Antonietti, *Adv. Mater.*, 2009, **21**, 435–439.
3. D. Gebauer, A. Verch, H. G. Borner and H. Cölfen, *Cryst. Growth Des.*, 2009, **9**, 2398–2403.

Ken Lewtas communicated: Professor Myerson's comment on the industrial production of well-controlled calcium carbonate particles relates to large scale processes. For example, such particles (actually colloids) are added to almost every lubricating oil to neutralise acids formed during combustion. They are synthesised by the *in situ* reaction of carbon dioxide with lime mediated by particular surfactants which keep the particle (colloid) sizes down bellow 10 nm with a very narrow size distribution.[1] Measurement of the different crystal polymorphs at these sizes is very interesting especially as amorphous regions also seem to be present. As Alan says, this is still a subject for further studies.

1. P. J. Dowding, McBain Medal Lecture: http://www.soci.org/~/media/Files/Conference%20Downloads/2009/McBain%20Dec%2009/Dowding_McBain.ashx

Wenhao Sun asked: You stated in your presentation that if you look at the classical nucleation barrier for calcite, the required supersaturations are much too high to be feasible for direct precipitation of the final phase. My question is, since we know that calcium carbonate proceeds through many transient polymorphs, and it does so through a dissolution–reprecipitation route, could the local supersaturation near dissolving transition polymorphs result in the local ion activity being much higher than the solution average? This could provide the local supersaturations to induce formation of the final calcite phase. (This topic is also discussed in my recent work.[1])

1. W. Sun, *et al.*, Nucleation of Metastable Aragonite $CaCO_3$ in Seawater, *Proc. Nat. Acad. Sci*, 2015, **112**, 3199–3204.

Jim De Yoreo replied: What I meant by my statement is that the supersaturation required to achieve barriers to calcite nucleation that, according to classical nucleation theory (CNT), are of sufficiently small magnitude to induce such events on a reasonable timescale are also large enough to make all other calcium carbonate phases more stable than the solution. Consequently, all phases can form and we should expect that to be the case. Indeed, our liquid phase TEM experiments demonstrate this is the case. The does not imply the less stable phases are more likely to form that calcite. Whether or not that is the case depends on many more factors than the solubilities of the various phases.

In the cases where one or more of these less stable phases forms before calcite, the system is still supersaturated with respect to calcite, so it will eventually form somewhere in the solution independent of any of the other precursor particles, but perhaps not on a timescale one cares to endure. So how could the presence of these other phases promote calcite nucleation? Your suggestion is that they create a locally high supersaturation. I don't see how that is possible. The solution state will be defined by the equilibrium solubility of the precursor phases and no higher. Thus the supersaturation with respect to calcite will be lower than the original value prior to nucleation of the less stable phases. However, perhaps the interfacial energy for calcite nucleation on the less stable phases is lower than it is for nucleation in the bulk solution. Then they will provide a location of lower barrier and make the nucleation of calcite on their surfaces a faster process than nucleation of calcite out in the solution away from the precursors. If the reduction in interfacial energy is great enough, the rate of nucleation on their surfaces could even be larger than the rate of nucleation in the original highly-supersaturated solution.

Eric Breynaert opened the discussion of the paper by Nico Sommerdijk: We all invest lots of efforts in attempts to explain unexpected observations in nucleation from a purely physical point of view. Chemistry however also plays a very important role in many systems, especially in systems that are difficult to control. The transformation of ferrihydrite to magnetite is a good example of such a 'difficult' system as this system undergoes major chemical changes before or while magnetite is nucleating. In your manuscript the transformation of ferrihydrite to magnetite occurs with the intermediate appearance of green rust. Magnetite is one of the typical products observed upon partial oxidation of green rust. This in itself demonstrates the chemistry of the system is evolving, implying that it is essential to demonstrate that the chemical conditions remained constant during the 'multi-step nucleation process'. If the chemical conditions were still evolving, the multi step nucleation could simply be a series of transformations from one phase to another as result of the changing chemistry. This also relates to experimental aspects of redox measurements in such systems. How does one accurately measure the *in situ* redox conditions, sometimes also involving surfaces, which most often cannot be picked up by, for example, redox electrodes. Development of suitable methods for the determination of redox states and conditions has for example been a topic of the EU funded Recosy project[1] with partners from multiple member states that ran from 2008 to 2012.

1. Recosy project: http://www.recosy.eu/

Nico Sommerdijk responded: Dr Breynaert makes an important point. Indeed, the transformation of ferrihydrite into magnetite[1,2,3,4] which in some occasions was shown to occur *via* one or more intermediates[1,4] must involve one or more chemical transformations. This means that we must certainly take into account the kinetics of the chemical reaction(s) to describe the formation of the subsequent phases. The changing chemistry in the system goes hand in hand with changes in the pH as is demonstrated in the present paper (DOI: 10.1039/C4FD00227J) and will also affect the redox and chemical potentials. Although this complicates the analysis of the crystallization process, we still must consider the formation of every new phase – whether through precipitation from solution or through a solid state transformation – a nucleation step where both chemical kinetics and nucleation kinetics play a role. In fact the changing chemistry can be a handle for understanding a multistep crystallization pathway as was recently demonstrated for the formation of apatite from a calcium phosphate solution.[4]

1. J. Baumgartner, A. Dey, P. H. H. Bomans, C. Le Coadou, P. Fratzl, N. A. J. M. Sommerdijk and D. Faivre, *Nature Mater.*, 2013, **12**, 310–314.
2. J. J. M. Lenders, C. L. Altan, P. H. H. Bomans, A. Arakaki, S. Bucak, G. de With and N. A. J. M. Sommerdijk, *Cryst. Growth Des.*, 2014, **14**, 5561–5568.
3. J. J. M. Lenders, H. R. Zope, A. Yamagishi, P. H. H. Bomans, A. Arakaki, A. Kros, G. de With and N. A. J. M. Sommerdijk, *Adv. Func. Mater.*, 2014, **25**, 711–719.
4. W. J. E. M. Habraken, J. Tao, L. J. Brylka, H. Friedrich, L. Bertinetti, A. S. Schenk, A. Verch, V. Dmitrovic, P. H. H. Bomans, P. M. Frederik, J. Laven, P. van der Schoot, B. Aichmayer, G. de With, J. J. De Yoreo and N. A. J. M. Sommerdijk, *Nat. Commun.* 2013, **4**, 1507.

Eric Breynaert asked: What is the role of green rust in this system? One option to investigate this could be to start from sulphate based green rust GR2(SO_4^{2-}),[1] which can be obtained pure and much more easily than the GR1(Cl^-) structure with chloride.[2]

1. L. Simon, *et al.*, *Solid State Sci.*, 2003, **5**, 327–334.
2. J.-M. R. Génin and C. Ruby, *Solid State Sci.*, 2004, **6**, 705–718.

Nico Sommerdijk answered: Our hypothesis is that in the sequence described the Fe(II) rich green rust acts as a supplier of the Fe(II) required for the transformation of the primary FeH–Fe(II) particles into the more Fe(II)-rich secondary particles (Dey *et al.*, DOI: 10.1039/C4FD00227J). Indeed, synthesizing magnetite by starting from a pure sulfate-based green rust[1] would be an excellent opportunity to investigate the details of the process.

1. C. Ruby, A. Géhin, M. Abdelmoula, J.-M. R. Génin and J.-P. Jolivet, *Solid State Sci.*, 2003, **5**, 1055–1062.

Gerard Coquerel said: You have shown two pathways to produce the nanomagnetite. Are you sure you really have stoichiometric magnetite for both? Did you explore further the chemistry of the magnetite produced at a so low temperature?

Nico Sommerdijk answered: Dr Coquerel makes an excellent point here. We are aware that electron diffraction cannot distinguish between structurally very similar iron oxides that only vary in the Fe(II)/Fe(III) ratio. We therefore routinely confirm the formation of magnetite – and *e.g.* the absence of maghemite – using raman spectroscopy.[1,2,3] In this particular case these measurements were not performed and we therefore should consider the option that non-stochiometric magnetite was formed. Having said that, the saturation magnetization values of the magnetite we produce by our room temperature synthesis methods[1,2,3] are in good agreement with the literature values for magnetite. Nevertheless, we agree that the extension of the study with Mössbauer spectroscopy would be an important addition to characterize the final product.

1. J. J. M. Lenders, C. L. Altan, P. H. H. Bomans, A. Arakaki, S. Bucak, G. de With and N. A. J. M. Sommerdijk, *Cryst. Growth Des.*, 2014, **14**, 5561–5568.

2. J. J. M. Lenders, H. R. Zope, A. Yamagishi, P. H. H. Bomans, A. Arakaki, A. Kros, G. de With and N. A. J. M. Sommerdijk, *Adv. Func, Mater.*, 2014, **25**, 711–719.
3. C. L. Altan, J. J. M. Lenders, P. H. H. Bomans, G. de With, H. Friedrich, S. Bucak and N. A. J. M. Sommerdijk, *Chem. - Eur. J.*, 2015, **21**, 6150–6156.

Gerard Coquerel asked: It is well known that magnetite can bear Fe^{2+} and an excess of Fe^{3+}. There is even a metastable form of Fe_2O_3 with a spinelle structure. Did you spot different products from different routes?

Nico Sommerdijk replied: As I mentioned in my reply to your previous question, at this point – apart from diffraction data and Raman spectroscopy that are in agreement with magnetite formation – we have no further information on the exact stochiometry of the materials formed, but see no significant differences in the saturation magnetization of the products from different synthesis routes.

Gerard Coquerel opened a general discussion of the paper by Martí Gich: In the situation described in your paper, you don't have two open "hands" in the end of your process; it is as if the left hand is open and the right hand is closed (*i.e.* a fist). The morphologies of the particles are different and need to be justified by something chiral in your process.

Martí Gich answered: Exactly. The crystallization process of a given film results in the presence of one and the same type of crystal habit, which we have reasons to believe, corresponds to crystals of a definite chirality. However, I cannot think of any chiral element in the process. As is mentioned in the paper, we have evidence that a mechanism of competitive growth analogous to that described by Garcia-Ruiz[1] is at play. It is likely that the nucleation occurring at the interface between the Si substrate and the amorphous SiO_2 deposited on top of it involves the two quartz enantiomorphs. However, it is possible that the growth of one of those takes over the other due to an initial imbalance of the two populations: when adjacent crystals meet the subsequent growth is more conflicting if those are of different morphologies (that is, h1 and h2) compared to the case when the boundaries are with crystals of the same morphologies (h1 and h1 or h2 and h2). Still, returning to your question, I am aware that one can hardly justify this initial imbalance without a chiral bias, but I have no idea what this could be.

1. A. Rodriguez-Navarro, J. M. Garcia-Ruiz, *et al.*, *Eur. J. Miner.*, 2000, **12**, 609.

Hugo Meekes asked: You need dissolution and growth, repeatedly, to get pure enantiomers. You claim a melt at 900 °C in your process. This could be a necessary condition for the growth and melting/dissolution cycles.

Martí Gich responded: Indeed, the other possible mechanism that we have put forward to explain the habit (and presumably enantiomorphic) selection is a process analogous to Viedma ripening in which the enantiomorphic amplification requires repeated dissolution and recrystallization cycles. I am aware that the literature on this topic reports time scales for the process on the order of days but, importantly, all these experiments were performed at temperatures well below those at which our melt seems to form. Thus, it cannot be excluded that our system undergoes a much faster kinetics that allows the process to take place in a significantly shorter time window (on the order of minutes).

Hugo Meekes asked: If you have a melt at this temperature you would need repetitive cycles normally (as in the Viedma type of ripening paper you mention, but more particularly the temperature cycling approach,[1] but you keep things at one temperature. Therefore, this would be a very new type of ripening.

1. K. Suwannasang, A. E. Flood, C. Rougeot and G. Coquerel, *Cryst. Growth Des.*, 2013, **13**, 3498.

Martí Gich responded: Well, in fact the crystallization treatment consists in heating at 3 °C min^{-1} from room temperature to 1000 °C where an isotherm of 5 h is performed followed by a cooling at 3°C min^{-1}. We know that the melting occurs during the heating stage and this ripening would occur upon heating. To understand this mechanism it might be relevant that crystallization of the amorphous SiO_2 involves the expulsion of Sr^{2+} into the melt. This melt enriched in Sr^{2+} might dissolve the initially formed quartz.

Hugo Meekes asked: I would expect patches of left and right-handed enantiomers in your competitive growth proposal as the relevant mechanism. Do you see such patches?

Martí Gich answered: This is a good point. We have not seen patches but since we have studied the films by Atomic Force Microscopy, the areas that can be scanned are limited to squares of several tens of microns. Even though we analyzed several zones of the same film, we are trying to develop an alternative technique using a confocal microscope that will allow us to analyze larger film areas and directly measure the chiral rotation. I hope with this it will be possible to assess the existence of patches, which so far I would not exclude.

Wenhao Sun asked: It doesn't seem to me, by identification, that the exposed surfaces on the two enantiomers appear to be related by the chiral rotation along the [001] direction. If you look at the surfaces of the two enantiomers do you see similar surface structures on the exposed planes?

Martí Gich responded: I suppose that you mean that the two habits are not the mirror image of one another. We still don't have direct evidence that crystal habit h1 corresponds to one enantiomorph and h2 corresponds to the second enantiomorph but we are trying to set up techniques to obtain information on the chirality of the surfaces, one using a confocal microscope I already mentioned and another based on polarized Raman spectroscopy.

Terence Threlfall said: Are the two pictures of different morphologies or of crystals of different structures? Although two enantiomers are supposed to be equivalent in everything but their handedness there exist a handful of cases for which one enantiomer exists as a different polymorph to the other.

Martí Gich answered: We think that the two morphologies could be indicative of two different enantiomorphs but this has still not been confirmed. What we do know by X-ray diffraction studies is that the two crystal morphologies correspond to the same SiO_2 polymorph: quartz.

Jim De Yoreo remarked: Years ago, we looked at the effect of chiral impurities on the growth morphology of achircal crystals. We found that chiral molecules will bind differently to atomic steps that are crystallographically identical, but are related through a mirror symmetry and the result was that the crystal shape then became chiral. Could you be seeing the same effect in reverse in your system; that is, the achiral impurities bind preferentially to left- or right-handed crystal structures?

Martí Gich replied: The effect that you propose might explain the enatiomorphic selection. The impurity would hinder the growth of the enatiomorph to which it selectively binds; however this would not explain the selection of the second enantiomoph which is also observed.

Clement Brandel commented: From the comparison of the two supposed enantiomorph crystal morphologies, I am suspicious that these two types of crystals are not enantiomorphs (some dihedral angles are very different) but two different polymorphic forms of the same handedness. A rapid procedure to rule out this suspicion would be to have the two types of morphology analyzed by X-ray powder diffraction to confirm that both varieties are present in the same powder diffractogram, as two enantiomorphous crystals would be.

Martí Gich replied: It is possible that these two types of crystals are not enantiomorphs but I can exclude that these are two different polymorphic forms, because the films have been thoroughly analyzed by X-ray diffraction and in both cases we obtain the same type of patterns.

Robert Hammond remarked: In estimating the comparative surface-energy of the plausible crystallographic-surfaces for the two space groups *P*3121 and *P*3221 to assign the crystal habit it seems that the surfaces are assumed to be unperturbed terminations based on the bulk lattice structure. It is not unusual to observe reconstruction at such crystal surfaces to reduce surface-energy, so have you considered whether surface reconstruction could be important here?

Martí Gich answered: We did not consider it but is indeed relevant. Thank you for the comment.

Laszlo Fabian asked: How much does what you see depend on surface coverage? If you just have a few, separate small crystallites on the surface, would you see the same as with full coverage?

Martí Gich replied: You are rising a very interesting point. We have not tried to make this study but I think it would be worth doing it.

Simon Black commented: The images show very consistent sizes of the quartz crystals. How was this achieved?

Martí Gich answered: This has probably to do with the use of surfactants in the synthesis of the silica gel that is subsequently crystallized into quartz. However we do not control the size of these crystals, we can only control the porosity of the films that can be made macroporous or mesoporous.[1]

1. A. Carretero-Genevrier *et al.*, *Science*, 2013, **340**, 827.

Jessica Lovelock opened the discussion of the paper by Maxim Likhatskiy: Your model shows distinct prenucleation clusters consisting of Au(I) coordinated to sulfur atoms. What is the primary driving force behind this aggregation, and what role does the sulfur play in this process?

Maxim Likhatskiy replied: The origin of attractive forces between liquid intermediates is connected with an action of short-range van der Waals forces. A repulsive force is based on electrostatic interactions between negatively charged species (as revealed by zeta-potential measurements), with the negative charge carried by di- polysulfide ions, and S-O species, which are absorbed on the interface.

Jessica Lovelock remarked: I understand that the presence of sulfur in these early pre-nucleation clusters allows for the final particle size to be controlled (*via* mediating the addition of NaCl). My question is regarding the base case, however, when NaCl is not added – how does the nucleation process proceed normally without this external driver, and what (if anything) natively fills the role of the NaCl in expediting aggregation?

Maxim Likhatskiy replied: The reason for the aggregation is the relatively small zeta-potential of the liquid species.

Jim De Yoreo said: I can think of two reasons for getting particles in a liquid state within a solution. The first is liquid–liquid separation. The other is that the particles are small enough that their melting point below the temperature of the experiments. Are you dropping below the melting point by having particles of a sufficiently small size, perhaps with the melting point lowered even further through the addition of sulfur?

Maxim Likhatskiy responded: We deal with the liquid species the sizes of which are in range of 100–200 nm. Hence, the lowering the melting point of this species as compared with the bulk gold sulfide cannot play an important role.

Jim De Yoreo asked: Do you think there is a spinodal (a liquid phase-line) in the system?

Maxim Likhatskiy answered: Potentially, two reasons can underlay the formation of liquid intermediary species. These are (i) inhomogeneity of the solution taking place during the mixing the reactant solutions, and (ii) spinodal separation in the system. Until now we have not strong evidence to support one of these, but, in our opinion, the second one is more realistic.

Samuel Booth asked: There is an organic phase reduction of Au(III) similar to yours whereby the use of thiol to reduce $AuCl_4^-$ can form a Au(I) thiol intermediate similar to your AuS pre-nucleation clusters. The subsequent reduction to Au(0) in the organic solvent system results in the formation of a gold metal core, surrounded by a monolayer of thiol which acts as a protecting ligand. Do you think you are seeing a similar process whereby your final clusters will be made up of gold surrounded by sulphur or do you think that the sulphur will be distributed throughout your particles forming a bulk Au_2S structure?

Maxim Likhatskiy replied: In order to shed light on the problem of Au speciation within liquid species we have taken, treated and analyzed the Au $L_{2,3}$-edge X-ray absorbance spectra.[1] There one can see EXAFS and XANES spectra which clearly show that Au(I)–S bonding takes place, with strongly disordered Au_2S forming after drying in air. In general, there are, at least, two sorts of sulfur in the system. The first one is localized within the clusters $Au_mS_n \cdot {}_xH_2O$, with the second one absorbing onto the interface between the bulk of solution and the liquid domains. Possibly, the surface species of sulfur can be presented by di- and

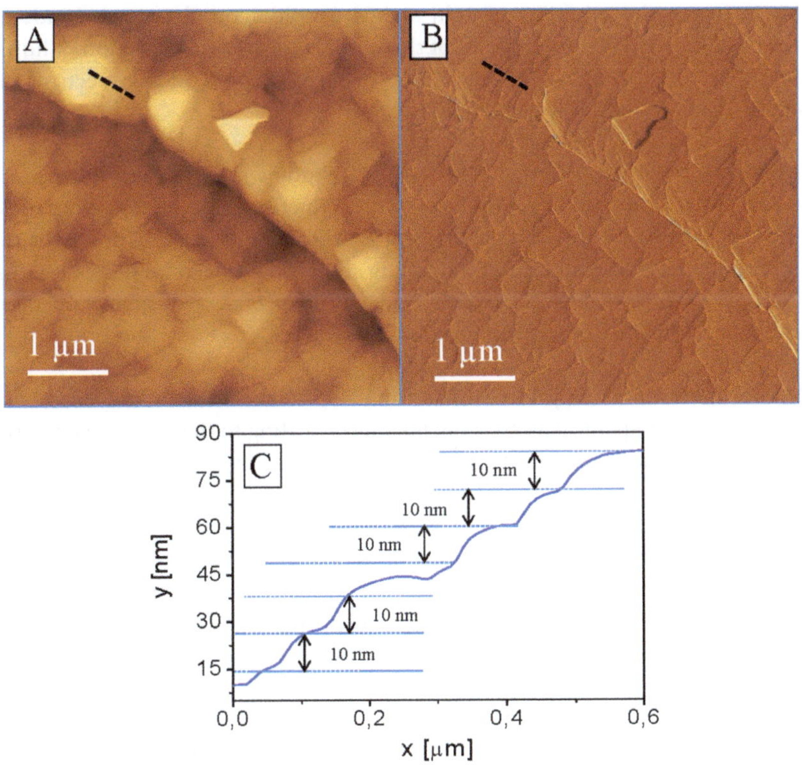

Fig. 1 Topographical (a) and phase (b) AFM images of a polycrystalline quartz film. Panel (c) shows the topographic profile along the dotted line indicated in panels (a) and (b).

polysulfide ions along with S-O species (as suggested by TEY XANES and XPS spectra), with them bearing a negative charge which stabilizes the liquid intermediates against coalescence.

1. Yu. Mikhlin, M. Likhatski, Ye. Tomashevich, A. Romanchenko, S. Erenburg and S. Trubina, *J. Electron Spectrosc.*, 2010, **177**, 24–29.

Eric Breynaert asked: What is the role of the changing redox conditions and changing speciation of sulphur and polysulphide oligomers in this system? As these experiments were not performed in anoxic conditions, the speciation of sulphur and the redox conditions can be expected to continuously evolve as function of time. Were the redox conditions or the sulphur speciation monitored during the evolution of the system? The research of the Shenberger and Barnes[1] at related conditions, although at a higher temperature, provides some insight in the complicated chemistry of the gold sulfur system, without even considering the often catalytic influence of sulphur electron donors (*e.g.* SO_3^{2-}, $S_2O_3^{2-}$, ...) on the interconversion of polythionates. Can we really draw conclusions on nucleation mechanisms in a system where one hardly controls the chemistry?

1. D. M. Shenberger and H. L. Barnes, Solubility of gold in aqueous sulfide solutions from 150 to 350 °C, *Geochim. et Cosmochim. Acta*, 1989, **53**(2), 269–278.

Maxim Likhatskiy replied: In the process of liquid intermediates formation a plethora of sulfide ions is added. All the synthesis were conducted in acid media (pH < 2), where the solubility of gold as hydrosulfide complexes was hindered thermodynamically. The main point of this work is to show an evolution of liquid species with time giving rise to nucleation disregarding chemical aspects of the problem which are actually very complicated.

Dimitrios Toroz asked: Previous theoretical studies have been performed on gold aurate ions with amino acids showing the reduction of Au(III) ions from these amino acids showing the charge distribution between the gold ions with the Au(III) complexes . Have molecular modelling studies have been carried out in this paper and to what extent have such changes been observed, giving us more fundamental information on the reaction pathway?

Maxim Likhatskiy responded: Unfortunately, no theoretical study has been conducted in the paper but all the conclusions regarding to the mechanism of the process (and its chemical aspects) have been supported by a series of microscopic and spectroscopic data.

Hugo Meekes returned to discussion of the paper by Martí Gich: It would be interesting to go to organic synthesis chemists who are increasingly interested in chiral surfaces for catalysis. You seem to produce only chiral surfaces. It would be worth speaking to these groups to determine if you really have purified enantiomers.

Martí Gich replied: Yes definitely, I would very much appreciate if you could put me in contact with those people.

Haihua Pan commented: Considering that the facets of crystallites are well defined, we can take each crystallite as single crystal as indicated by this work. But we did not see hillock or steps on the crystal surface of quartz in you AFM images, which should be expected. Could you comment on this?

Martí Gich responded: The vertical resolution depends on the surface roughness. An atomically flat substrate would allow vertical resolutions of about 0.1 nm to be achieved, enabling the observation of steps or hillocks. However, in our case the superposition of different crystallites on top of one another results in steps of about 10 nm in lateral displacements of about 100 nm (see Fig. 1) and in these conditions the vertical resolution is not enough to observe steps and hillocks.

Martin Ward communicated: On the problem of identifying which enantiomer of quartz is being formed during your experiments, I wonder if you have considered the use of circular dichroism spectroscopy? This is a technique that has been used to investigate secondary protein structure[1] as well as identifying the chirality present in the foliage of plants.[2] I bring your attention to references 2 and 3, as the method of 'diffuse reflectance circular dichroism' is one that could perhaps be applied to analysis of your quartz films.

1. P. Manavalan and W. C. Johnson, *Nature*, 1983, **305**, 831.
2. R. D. Wolstencroft, *et al.*, *Proc. Int. Astron. Union*, 2004, **213**, 149.
3. N. Asano, *et al.*, *Chem. Comm.*, 2009, 899.

Martí Gich communicated in reply: We are also trying to use linear and circular dichroism but in contrast to those references we are not doing spectroscopy but working at a fixed wavelength. However we are using a confocal microscope that will allow us to have a quite low spatial resolution (500 nm diameter spot).

Faraday Discussions

PAPER

Self-association of organic solutes in solution: a NEXAFS study of aqueous imidazole†

M. J. Thomason,[a] C. R. Seabourne,[b] B. M. Sattelle,[a] G. A. Hembury,[a] J. S. Stevens,[a] A. J. Scott,[b] E. F. Aziz[cd] and S. L. M. Schroeder*[ab]

Received 12th January 2015, Accepted 28th January 2015
DOI: 10.1039/c5fd00005j

N K-edge near-edge X-ray absorption fine-structure (NEXAFS) spectra of imidazole in concentrated aqueous solutions have been acquired. The NEXAFS spectra of the solution species differ significantly from those of imidazole monomers in the gas phase and in the solid state of imidazole, demonstrating the strong sensitivity of NEXAFS to the local chemical and structural environment. In a concentration range from 0.5 to 8.2 mol L^{-1} the NEXAFS spectrum of aqueous imidazole does not change strongly, confirming previous suggestions that imidazole self-associates are already present at concentrations more dilute than the range investigated here. We show that various types of electronic structure calculations (Gaussian, StoBe, CASTEP) provide a consistent and complete interpretation of all features in the gas phase and solid state spectra based on ground state electronic structure. This suggests that such computational modelling of experimental NEXAFS will permit an incisive analysis of the molecular interactions of organic solutes in solutions. It is confirmed that microhydrated clusters with a single imidazole molecule are poor models of imidazole in aqueous solution. Our analysis indicates that models including both a hydrogen-bonded network of hydrate molecules, and imidazole–imidazole interactions, are necessary to explain the electronic structure evident in the NEXAFS spectra.

Introduction

Volmer's concept of the nucleation stage of crystallization[1] resulted in the development of what is now known as classical nucleation theory (CNT). CNT

[a]The University of Manchester, School of Chemistry, School of Chemical Engineering and Analytical Science, Manchester, M13 9PL, UK. E-mail: slm.schroeder@outlook.com
[b]University of Leeds, School of Chemical and Process Engineering, Leeds, LS2 9JT, UK
[c]Joint Ultrafast Dynamics Lab in Solutions and at Interfaces (JULiq), Helmholtz-Zentrum Berlin für Materialien und Energie, Albert Einstein Strasse 15, 12489 Berlin, Germany
[d]Freie Universität Berlin, Arnimallee 14, 14195 Berlin, Germany

† Electronic supplementary information (ESI) available: Microhydrated cluster models – Fig. S1–S9. See DOI: 10.1039/c5fd00005j

assumes that the nucleation process occurs in two distinct steps. First, molecules in a supersaturated solution aggregate into nuclei, thereby developing an interface with the surrounding solution. The stability of these nuclei is size-dependent, reflecting the free energy balance between an interfacial tension penalty and cohesive energy stabilisation. Second, once the nuclei have grown beyond a critical size, above which the cohesive term outweighs the interfacial destabilisation, the total free energy decreases continuously as a function of size and crystal growth becomes the favourable process.[2,3] Whilst a proven and useful concept for describing and predicting nucleation and crystal growth phenomena, often quantitatively, CNT remains a description of nucleation that does not explicitly consider intermolecular interactions or the precise structural nature of the pre-crystalline state; this limits its predictive power for many systems.[4,5] Researchers are therefore using a variety of experimental techniques to obtain molecular level information about the solute species in solution, how they associate and assemble during nucleation, and whether the pre-crystalline structures relate to those in the crystalline products.[4,6–10] Examples of techniques commonly employed include nuclear magnetic resonance (NMR),[11–13] small- and wide-angle X-ray scattering (SAXS/WAXS),[14] optical microscopy,[15] and vibrational spectroscopies (infrared and Raman),[16] as well as grazing incidence X-ray diffraction (GIXD) of interfacial species.[17]

In this paper we introduce another technique for probing the molecular properties of organic solute species in concentrated solutions; namely, near-edge X-ray absorption fine-structure (NEXAFS) spectroscopy,[18,19] which is also often referred to as X-ray absorption near-edge structure (XANES). Our application of NEXAFS to concentrated solutions builds on the recent realisation that chemical shifts in atomic core level binding energies, which can be measured by X-ray photoelectron spectroscopy (XPS),[20–22] provide incisive information about the influence of hydrogen bonding and proton transfer on the structure of the organic solid state.[23–28] High resolution X-ray spectroscopies that probe core levels are generally sensitive to the local electronic structure around, and bonding by, the element that is excited in the atomic core. The X-ray absorption process underlying NEXAFS is the excitation of atomic core electrons (for light elements such as C, N and O the 1s electrons) to unoccupied valence orbitals; this can be readily interpreted by the use of molecular orbital calculations. Due to its sensitivity to unoccupied valence orbitals, NEXAFS is chemically and structurally more incisive than core level binding energy measurements by XPS, and we have recently demonstrated the level of detail that can be obtained by combining XPS, NEXAFS and density functional theory, to examine local bonding both in the organic solid state[29] and by solute species in solutions.[30] Here we apply the same conceptual framework and extend it to the analysis of local bonding by imidazole in aqueous solution.

1*H*-Imidazole (for the remainder of this paper referred to as 'imidazole') crystallises in a monoclinic crystal structure with the space group $P2_1/c$.[31–33] There are four molecules in the unit cell, and the main structural features are infinite chains of hydrogen-bonded imidazole molecules along the *c*-axis. The IUPAC numbering of the ring atoms in imidazole is indicated in Fig. 1. There is considerable interest in a deep understanding of the local interactions of imidazole in aqueous media, due to its biological importance as the side chain of the amino acid histidine. Experimental evidence for the self-association of imidazole

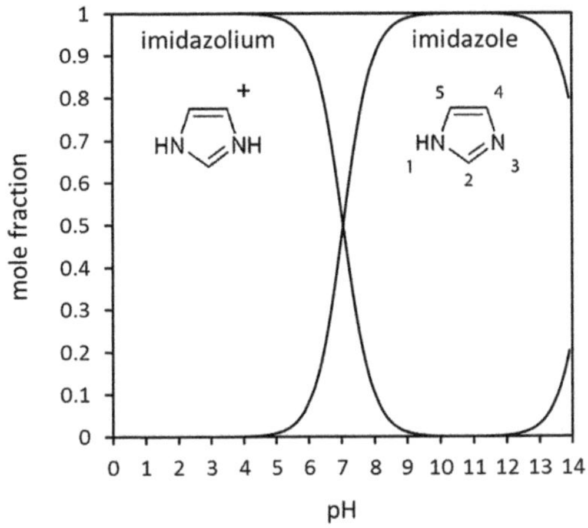

Fig. 1 Imidazole speciation in aqueous solution as a function of pH, calculated from pK_a values. Also indicated in the figure is the IUPAC numbering of the ring atoms.

in aqueous solutions at concentrations above 10^{-4} mol L^{-1} has been available for more than 70 years.[34] π–π interactions between imidazole molecules have been invoked to explain self-association through the formation of molecule stacks.[35,36] Such stack models contrast with the dominant motif of chain-formation through hydrogen bonding in solid imidazole.[32,33] Recent molecular dynamics simulations of high concentrations (>0.5 mol L^{-1}) of imidazole in aqueous solutions indicate that there is in fact a concentration-dependent balance, between π–π-assisted hydrogen-bonded stack structures and chains of hydrogen-bonded imidazole molecules[37] that are similar to those found in liquid[38] and solid[32,33] imidazole.

As a first step to experimentally probing these properties, we contrast here the N 1s NEXAFS spectra of (i) monomeric gas phase imidazole and (ii) crystalline imidazole with (iii) imidazole as a solute in concentrated aqueous solution. This approach provides an opportunity to begin building up a systematic picture of the electronic structure variations in the imidazole molecule, that arise from the three different environments. The aim of the investigations here is to establish whether available methods for electronic structure calculations correctly reproduce the experimental spectra, and whether the computational analysis of experimental core level spectra can realistically provide a quantitative insight into complex solution systems. For example, an important question arising in this context is to what extent the electronic transitions in the experimental excitation spectra are dominated by ground state ('initial state') properties of the imidazole molecules, and whether excited state properties ('final state effects') influence the spectra significantly.

Materials and methods

Imidazole was obtained from Sigma Aldrich as ACS grade ≥99% (titration) with total impurities of ≤0.2% water. The solubility of imidazole in water is very high, with saturation around 11 mol L^{-1} at 298.15 K.[39] For this study, aqueous solutions with concentrations from 0.5 to 8.2 mol L^{-1} were prepared using laboratory grade deionised water. The pH of the solutions was monitored using a micro-tipped pH electrode. Before any measurements were taken, the pH electrode was calibrated using three various buffered solutions (3, 6 and 9). As expected for a weak base, it was found that all imidazole solutions were basic, with a pH around 10.5, indicating that the solutions contained >99% neutral imidazole species. The expected speciation of imidazole as a function of pH can be calculated from the known pK_a values (7.05, 14.52) and is displayed in Fig. 1. We note that the measured pH agreed well with the expected pH (as calculated from the pK_a value) at concentrations up to approximately 1 mol L^{-1}. The experimental pH values were consistently somewhat lower (ΔpH $\sim$ 0.1) than expected at higher concentrations. This may be an indication of an increasing degree of self-association by imidazole–imidazole hydrogen bonding, which would be expected to reduce the OH^- anion concentration. However, the effect is small, and the predicted pH is close to the estimated error of the pH electrode. Furthermore, it is not clear whether the response of the pH electrode may be affected by high concentrations of hydrogen-bonding solutes.

NEXAFS measurements were undertaken at beamline U41-PGM of the BESSY-II synchrotron radiation storage ring of the Helmholtz Zentrum Berlin. The LIQUIDROM experimental chamber has been specifically designed for analysis of solution systems under vacuum or an inert helium atmosphere. Light from the storage ring passes through a differential stage and enters into the chamber through a thin Si_3N_4 membrane. To minimise radiation damage of the solution, a flow cell[40] with a peristaltic pump was used, in which the X-rays impinge onto the solution through a Si_3N_4 membrane. The N K-edge spectra were recorded *via* fluorescence/luminescence yield detection with a GaAsP 5 $\times$ 5 mm^2 photodiode.

The flow cell setup allows analysis of a series of solutions without opening the analysis chamber. During sample changes, the system was purged for 30 min with deionised water. A control measurement of the spectrum was then performed to ensure that the cell was clean. During experiments the flow rate was typically set to 40 mL min^{-1}. The flowing system was allowed to equilibrate for at least 20 min before three NEXAFS spectra, each taking about 20 min to acquire, were measured.

StoBe-DeMon ('StoBe') calculations in this paper were carried out as previously described.[40,41] The software uses a hybrid density functional theory (DFT)[42–44] with a double zeta, local spin density exchange and correlation functional by Vosko and Wilk,[45] and has been described in detail in several papers.[42,43,46–48] For the calculation of the solid state N K-edge absorption spectrum, a 6-molecule cluster cut-out from the crystal structure of imidazole was used.

Gaussian03 (ref. 49) was used to carry out single point energy calculations, minimising the total energy of the gas phase molecule by geometry optimisation. The output provided molecular orbitals, their orbital energies, and populations. For the monomer structure, a geometry-optimised imidazole monomer from Gaussian03 (ref. 49) was used, which was obtained using the B3LYP/6-31G* basis set. A geometry-optimised stack structure, of three imidazole molecules coordinated with three water molecules, was generated using Gaussian03 at the BHANDH/6-311G** level of theory.

Density functional theory (DFT) was used to predict core-level spectroscopy results. Initially, the CASTEP code was used.[50] CASTEP is a member of the linear-augmented plane-wave (LAPW) class of DFT codes. It is a pseudopotential code, whereby approximations are used to model the atomic-like core states, with the code only actively calculating valence states. Developments to the code have allowed core-level spectroscopy calculations to be carried out, as detailed in the literature.[51–53] The generalized gradient approximation (GGA)[54,55] was used throughout the calculations, applying a methodology previously developed in the context of analysing Al K-edges.[56]

The first parameter to converge was the basis set size (as defined by the kinetic energy cut-off). This was changed in intervals of 100 eV, with convergence being determined against the predicted core-level spectroscopy result (for the 'pyridine' N3 nitrogen atom), initially in the ground electronic state. Upon varying the parameter from value 'A' to value 'B', for each point on the energy axis (from −5.0 eV to 50 eV in steps of 0.05 eV) the modulus of the difference in intensity was found (with a nominal Gaussian broadening applied to the predicted results). The average percentage change relative to result A was then found across all the energy axis points. When this was less than 15% (meaning that no meaningful change could be observed by eye upon the application of a physically realistic broadening scheme) the results were considered to be converged. A similar process of convergence was carried out for the other key DFT code parameter – the quality of sampling in reciprocal space – as defined by the Monkhorst–Pack *k*-point grid.[55] Ultimately the number of *k*-points in each reciprocal space dimension was doubled until convergence was achieved. This led to parameters as detailed in Table 1.

For the CASTEP predicted spectra, supercells were constructed, and 0.5 electron core-holes used at the two nitrogen positions. It was determined that a $2 \times 2 \times 1$ supercell was sufficiently large to avoid artificial core-hole interactions (*i.e.* a minimum core-hole separation distance of 9.779 Å).

To complement the CASTEP results, the WIEN2K DFT code[57,58] was used. WIEN2K, like CASTEP is a member of the LAPW family of codes. WIEN2K works by splitting the space in the theoretical cell into 'muffin-tin' regions centred on atoms, and interstitial space, with different exact modelling methodologies being used in each of those regions. WIEN2K is an all-electron code, and therefore it is possible to predict the onset energy difference (ΔIPs) between the N1 and N3 nitrogen atoms in imidazole. In this instance, convergence was based on the determined energy onset difference between the N1 and N3 positions. The key DFT code parameters (basis set size, defined in WIEN2K by the RK_{MAX} value and the density of *k*-points in reciprocal space) were converged such that the energy difference between the two nitrogen positions was accurate to the nearest 0.1 eV.

Table 1 Details of CASTEP convergence for imidazole

Cell details[33]	Kinetic energy cut-off/eV	Minimum *k*-point separation in reciprocal space
$a = 7.732$ Å $b = 5.458$ Å $c = 9.779$ Å $\alpha = \gamma = 90°$ $\beta = 117.26°$	725	0.073 Å^{-1}

Results and discussion

Monomeric gas phase imidazole

Before embarking on the presentation and discussion of crystalline and aqueous imidazole, it is instructive to establish the origin of the features in the N K-edge core level excitation spectra of an isolated imidazole molecule. The energies of the atomic 1s core levels, the occupied molecular orbitals, and the three lowest unoccupied ('virtual') molecular orbitals obtained by a Gaussian03 ground state calculation for the gas phase monomer are all summarised in Table 2. From these data we can construct a schematic molecular orbital diagram summarising the atomic and molecular orbital energies relevant for the N K-edge NEXAFS spectrum (Fig. 2). It can be seen that the N 1s core level binding energy difference between N1 and N3 in the ground state of imidazole is 2.3 eV. Due to the aromatic nature of imidazole, the excitation of N 1s electrons to unoccupied π^* states is expected to be the dominant absorption feature in the N K-edge NEXAFS.[18] As can be seen in Table 2 and Fig. 2, there are two low-lying unoccupied π^* states available. First, transitions into the lowest unoccupied molecular orbital (LUMO) occur, through the transitions labelled N1 1s $\rightarrow$ 1π^* and N3 1s $\rightarrow$ 1π^*, and possibly an additional absorption feature due to a transition of a N 1s electrons to the 2π^* virtual orbital. In Fig. 3 the 1π^* LUMO (#19 in Table 2), is visualised together with the Gaussian-derived second- (1σ^*, #20) and third-lowest (2π^*, #21) unoccupied MOs. It can be seen that the 2π^* orbital has no density of states (DOS) at the N1 atom, so a significant N1 1s $\rightarrow$ 2π^* transition can be excluded. The 2π^* orbital has local DOS at the N3 centre, so a N3 1s $\rightarrow$ 2π^* transition may arise in

Table 2 Ground state molecular orbitals and their energies for an isolated imidazole molecule, as calculated by Gaussian03

Orbital #	Assignment	Energy/eV	
1	N1 1s	−391.602	Atomic core orbitals
2	N3 1s	−389.337	
3	C2 1s	−278.531	
4	C4 1s	−277.833	
5	C5 1s	−277.349	
6	σ	−27.449	Occupied molecular orbitals
7	σ	−23.540	
8	σ	−20.117	
9	σ	−16.638	
10	σ	−15.952	
11	σ	−15.265	
12	σ	−11.869	
13	σ	−11.761	
14	π	−11.549	
15	σ	−11.059	
16	π	−7.529	
17	σ	−7.101	
18	π(HOMO)	−6.129	
19	1π^*(LUMO)	0.921	Virtual (unoccupied) molecular orbitals
20	1σ^*	1.998	
21	2π^*	2.065	

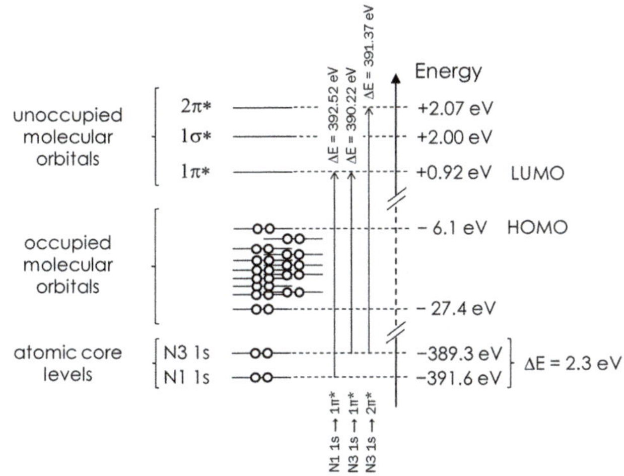

Fig. 2 Schematic representation of the Gaussian03-derived energies of atomic core levels, as well as of the occupied and unoccupied molecular orbitals in a monomeric isolated (gas phase) imidazole molecule. The N 1s core level binding energy difference and predicted energies associated with the 1s → π* transitions are also indicated. Note that the energetic positions of the occupied MOs are a schematic illustration and not an accurate reflection of the actual energies in Table 2.

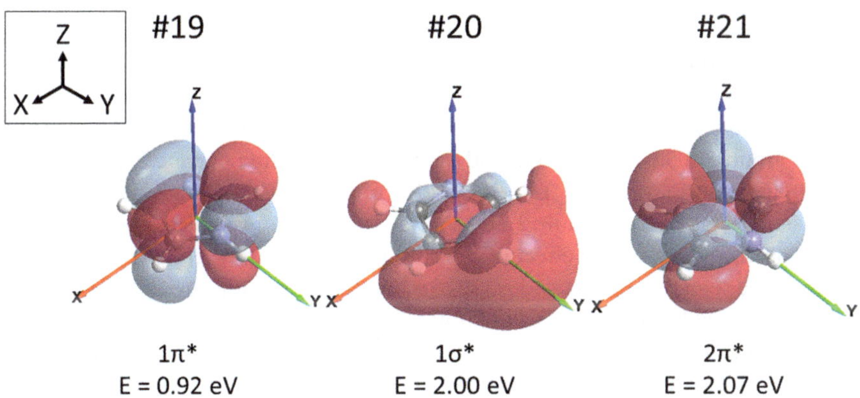

Fig. 3 The lowest unoccupied molecular orbital (LUMO) #19 (left), the second-lowest unoccupied molecular orbital #20 (middle), and the third-lowest unoccupied molecular orbital #21 (right) of imidazole. The molecule is in the *XY*-plane with the N–H bond axis aligned in the direction of the *Y* vector. Orbital energies are noted below each orbital.

the experimental spectrum. This transition is therefore also included in the scheme in Fig. 2. Electronic transitions from the N 1s core levels to the 1σ* orbital (#20) should be weak and are therefore unlikely to be evident in the experimental spectrum.

Having established the expected transitions in the N K-edge NEXAFS of an isolated imidazole molecule, we can now turn to the experimental N K-edge gas phase spectrum (Fig. 4) as previously determined by Apen *et al.*[59] using inner-shell

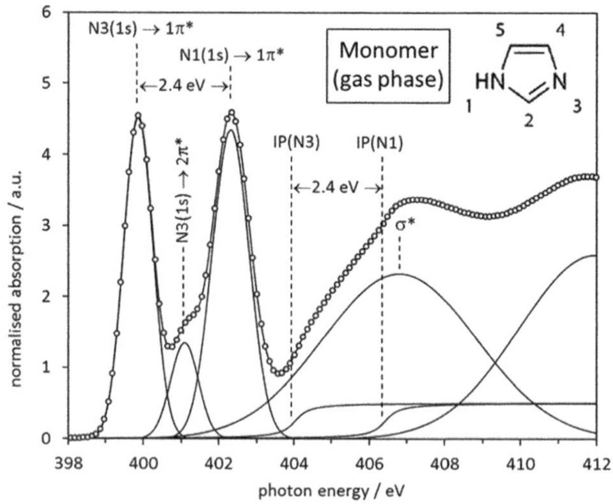

Fig. 4 Experimental ISEELS N K-edge spectrum of imidazole monomer (gas phase) species with all spectral features fitted by Gaussian and arctan functions. The origin of each feature is indicated by the annotations in the figure.

electron energy loss spectroscopy (ISEELS). All features in the spectrum have been modelled through a least-squares curve fitting procedure, using Gaussian functions for transitions to bound states and arctan edge step functions for the ionisation potentials (IPs), *i.e.*, the energies required for removal of the N 1s core electrons from the atom by photoemission into a continuum state. Beyond the IPs there are only multiple scattering resonances, such as the broad σ* shape resonance around 406.6 eV. The centroid energies of the fitted curve components in Fig. 4 are summarised in Table 3 and compared with values predicted by Gaussian03 (taken from Table 2 and Fig. 2) and a StoBe calculation of the N K-edge spectrum, taking the effect of the core hole state into account.

It can be seen that the N1 and N3 IPs for the gas phase molecule agree well with those predicted by StoBe. The Gaussian03-derived values are approximately 15 eV lower, as a result of neglecting the core hole state in the Gaussian03 ground state calculation. However, for both ground state and excited state calculations the energy difference (ΔIP) between the two IPs is in almost quantitative agreement with the experimental value (2.4 eV), with a value of 2.3 eV for Gaussian03 and 2.2 eV for StoBe.

Also clearly visible are the two distinct near-edge peaks that stem from the predicted N 1s → 1π* transitions at the N1 and N3 centres, which are almost quantitatively reproduced by the StoBe calculation. The Gaussian03 ground state calculation predicts values that are approximately 10 eV lower, due to neglecting the core hole effect. It can also be seen that there is a weak peak at 401.3 eV in the experimental spectrum arising from the predicted N3 1s → 2π* transition, which is also well reproduced by the StoBe calculation. There is excellent agreement between the IP difference and the energy difference between the 1s → 1π* transitions at N1 and N3, for experimental as well as both computational values; *i.e.*, with or without taking the core hole into account. This strongly indicates that

Table 3 Experimental ionisation potentials (IPs) and N 1s → π* transition energies for the N1 and N3 centres in a gas phase imidazole monomer, compared to predicted values from Gaussian03 and StoBe calculations

	Experim./eV	Calc. (StoBe)/eV	Calc. (Gaussian)/eV
IP(N1)	406.2	406.6	391.6
IP(N3)	403.9	404.4	389.3
ΔIP	2.4	2.2	2.3
E(N1 1s → 1π*)	402.3	402.3	392.5
E(N3 1s → 1π*)	399.9	400.1	390.2
ΔE(1π*)	2.4	2.2	2.3
E(N3 1s → 2π*)	401.3	401.2	391.4
E(2π*) − E(1π*)	1.4	1.1	1.2

there is no differential core hole relaxation effect in the isolated molecule, which had been previously invoked to explain an apparent discrepancy between experimentally determined IPs and the 1s → π* shifts.[59]

Overall, we can conclude from the analysis that the relative energetic positions of all features observed in the N K-edge spectrum are almost entirely determined by the ground state electronic properties of the molecule. The deviations between predicted and experimentally observed energetic shifts are at maximum on the order of 0.2 eV, even for the virtual 2π* state.

Finally, good agreement between experiment and theory was also achieved by the CASTEP analysis of the imidazole electronic structure. It was shown that a 15 Å theoretical 'box' was sufficiently large to simulate the molecule in the gas phase (as determined by comparing the predicted result for the N3 atom to that for smaller boxes). For the two nitrogen positions, predicted spectra were calculated, using a 0.5 electron core-hole, with spin allowed. Before comparing results with experiment,[59] the results had a lifetime-broadening scheme individually applied, before a rigid shift of +2.4 eV was applied to the N1 result and averaging of the spectra was performed.[60] The results for the two positions, normalised to the Fermi level, are shown in Fig. 5, along with comparisons to experiment. It can be

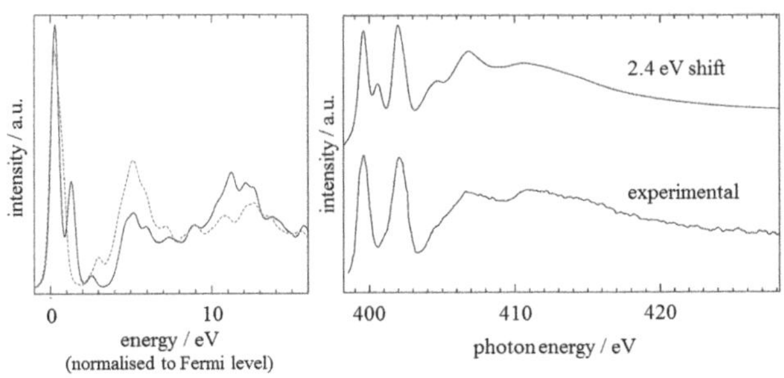

Fig. 5 Left: CASTEP monomer predictions for the N3 (black line) and N1 (blue dashed line) nitrogens. A 0.2 eV Gaussian broadening was applied and the spectra were aligned to the Fermi level. Right: Comparison of predicted N K edge with experimental data, with the spectra of the two N atoms shifted relatively by the experimental value of 2.4 eV. An energy-dependent lifetime broadening, as described in the text, was applied and the energy scale of the spectrum adjusted to match with the experimental data.

seen that the calculated spectrum based on the CASTEP analysis is in excellent agreement with the experimental spectrum.[59] The secondary peak visible at about 401 eV in the calculated spectrum is due to an overestimated transition to the 2π* MO.

Imidazole crystal

The same curve fitting procedure as for the gas phase monomer was applied to the spectrum of crystalline imidazole, which was previously determined by Apen *et al.* *via* electron-yield detection.[59] The most noticeable difference between the experimental spectra of gas phase and crystalline imidazole is a strong reduction of the energy split between the IPs and the 1s → 1π* transitions of the N1 and the N3 moieties, from 2.4 eV in the gas phase monomers (Fig. 4, Table 3) to 1.5 eV in the spectrum of the solid (Fig. 6). This difference arises from intermolecular N–H···N hydrogen bonding in the chains of imidazole molecules in the crystal structure. This causes intermolecular redistribution of electron density from N3 to N1 sites through weakening of the N1–H bond, and a partial levelling of the electronic structure difference around the two N centres in each molecule.

The spectrum predicted by StoBe for the six-molecule cluster from the imidazole crystal structure is shown in the middle of Fig. 7, displayed above the experimental spectrum. The calculated spectrum of the central imidazole molecule has an N3 1s → 1π* transition at an energy of 400.3 eV, and a small shoulder due to the 1s → 2π* transition at 401.2 eV. The N1 (NH) 1s → 1π* transition is evident as a single peak at 401.6 eV, with no noticeable shoulder contributions.

The parameters describing the features in the experimental and the calculated spectra are summarised in Table 4. The overall agreement between calculated and experimental data is good. The reduced energy split $\Delta E(1\pi^*)$ of 1.5 eV between the two 1s → 1π* transitions in the experimental data is reproduced almost

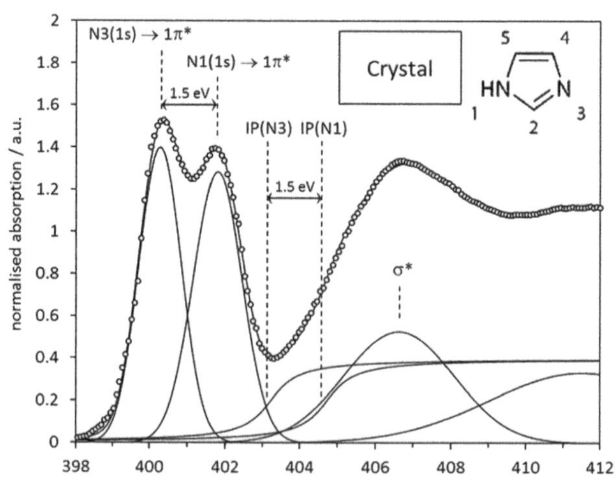

Fig. 6 Experimental electron-yield N K-edge spectrum of crystalline imidazole, with all spectral features fitted by Gaussian and arctan step functions. The origin of each feature is indicated by the annotations in the figure.

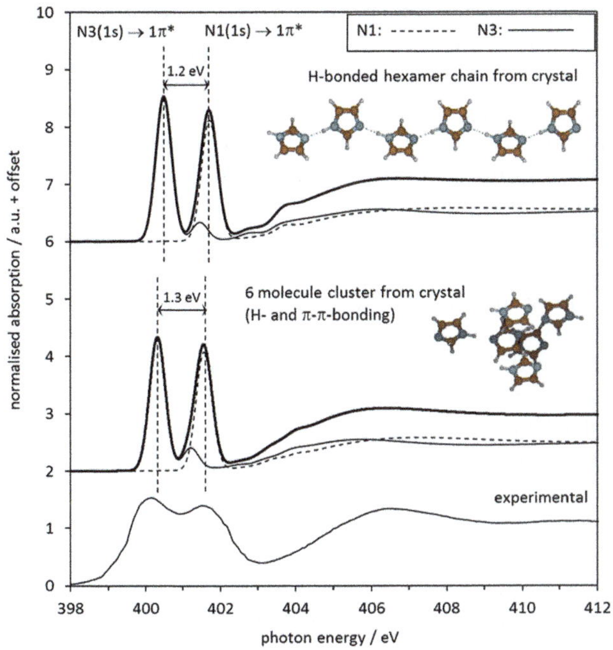

Fig. 7 Top: Middle: Calculated N K-edge NEXAFS spectrum of solid imidazole, as predicted by the StoBe code for the central molecule in a cluster of 6 molecules cut out from the crystal structure of imidazole. For both calculated spectra, the two individual absorption spectra of the two nitrogen moieties N1 and N3 in the central N atoms of the displayed molecular structures are shown alongside the combined spectrum. Bottom: Experimental observed electron-yield N K-edge spectrum from Fig. 6.

quantitatively by the StoBe cluster calculation. Comparing the StoBe results to those obtained for the gas phase monomer (Table 2) reveals that most of the reduction in the energy split between the gas and solid state is due to a decrease of 1.1 eV in the IP of N1 (its N 1s core level binding energy), from 406.6 eV to 405.5 eV. This indicates that the formation of intermolecular hydrogen bonds (N1–H··· N3) increases electron density at the N1 centre considerably, while the IP of the N3 centre is much less affected. This reflects the fact that the donation of electron density at N3 stems from the aromatic π system of the molecule, resulting in a

Table 4 Experimental IPs and 1s → 1π* transition energies for crystalline imidazole compared to the values calculated with a 6-molecule cluster calculation

	Experimental/eV	Calculated (StoBe)/eV
IP(N1)	404.7	405.5
IP(N3)	403.2	404.3
ΔIP	1.5	1.2
E(N1 1s → 1π*)	401.8	401.6
E(N3 1s → 1π*)	400.3	400.3
ΔE(1π*)	1.5	1.3

delocalisation of the electron density loss across the whole ring. In contrast, the electron density gain at the N1 centre takes place through the local N–H bond, which does not permit delocalisation of charge into the aromatic system.

Besides hydrogen bonding, additional interactions are expected in the solid state structure of imidazole. For example, π–π interactions between the sheets of hydrogen-bonded imidazole chains are expected. To examine the influence of such interactions on the spectra, we examined a chain (hexamer) of hydrogen-bonded imidazole molecules, retaining its geometry in the crystal structure. The absorption spectrum of the N1 and N3 moieties calculated by StoBe (Fig. 7, top spectrum) is almost identical to the spectrum calculated for the 3-dimensional arrangement of six imidazole molecules taken as a model of the crystal structure. This indicates that the impact of N1–H···N3 hydrogen bonding dominates the electronic structure of imidazole in the solid, with additional interactions leading only to secondary changes in the electronic structure.

It is interesting to note that recent core level binding energy measurements found an N 1s core level shift of 1.6 eV for the two nitrogen ring moieties in the neutral imidazole side chain of solid histidine.[61] This shift is comparable to the IP difference observed in the experimental and calculated NEXAFS of solid imidazole, although the hydrogen bonding experienced by the imidazole system is fundamentally different. In the crystal structure of histidine both N centres of the imidazole ring take part in hydrogen bonding, with the N1 (NH) centre donating its hydrogen to a more electronegative carboxylate acceptor (O), while the N3 (C═N) centre is accepting from the protonated primary amine group of the zwitterion.[62]

Fig. 8 shows the CASTEP-predicted absorption spectra for the N3 and N1 centres in the crystal structure. Compared to the gas phase monomer (Fig. 5) we observe a relative similarity in the predicted absorption spectra, as a consequence of the hydrogen bonding in the crystal structure levelling the differences between the nitrogen environments.

To model the onset energy difference between the two nitrogen positions, the 1.3 eV energy difference derived by StoBe was utilised, and the individual N1 and

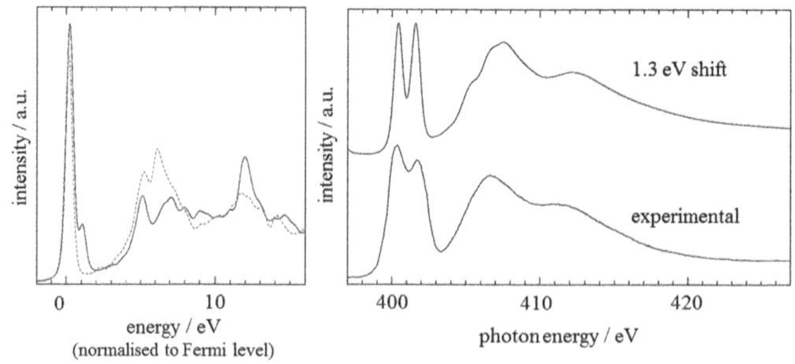

Fig. 8 Left: CASTEP results for N3 (black solid line) and N1 (blue dashed line) in the imidazole crystal. A Gaussian broadening of 0.2 eV has been applied and the calculated spectra are aligned to the Fermi level. Right: Comparison of experimental and predicted spectra, with an energetic shift of 1.3 eV between the two N 1s core level binding energies applied. The calculated spectrum includes an energy-dependent lifetime broadening, as described in the main text; the energy scale of the spectrum was adjusted to match with the experimental data.

N3 spectra were lifetime broadened as described above. As can be seen, this leads to excellent agreement with the experimental spectrum.[59]

Aqueous imidazole solutions

Shown in Fig. 9 are the obtained aqueous imidazole N K-edge NEXAFS spectra covering a concentration range from 0.50 mol L^{-1} ($\sim$100H_2O molecules per imidazole molecule) to the saturation concentration of 8.20 M ($\sim$7H_2O molecules per imidazole molecule). It can be seen that variations between the spectra as a function of concentration are minor, indicating that the average local coordination of imidazole molecules in this concentration range does not vary significantly. This observation suggests two possible scenarios. Either self-association involves only secondary interactions between individually hydrated imidazole molecules, which do not manifest themselves strongly in the N K-edge spectra; or self-association of imidazole is already dominant at the lower end of the concentration range investigated, so that any imidazole molecules added to the solution increase the volume fraction of such assemblies. Of course, the two scenarios are not mutually exclusive and may occur simultaneously.

Previous photoelectron spectroscopy investigations[41,63,64] of aqueous imidazole solutions focused on solvent interactions, *i.e.*, omitting imidazole–imidazole interactions and other self-association effects from the analysis. Insight into the electronic structure of imidazole molecules in aqueous solution as a function of pH was gained, and the sensitivity of C 1s and N 1s core level spectroscopy to pH-induced protonation of imidazole was clearly demonstrated by detection of electronically equivalent N atoms in the imidazolium cation under acidic conditions.[41,63] An attempt was made to explain the electronic structure of neutral imidazole species in solution, through calculations of the electronic properties of microhydrated gas-phase clusters comprising a single imidazole molecule and up to five water molecules.[64] This provided some insight into the effect of the nearest-neighbour water coordination shells on the imidazole molecules. The study

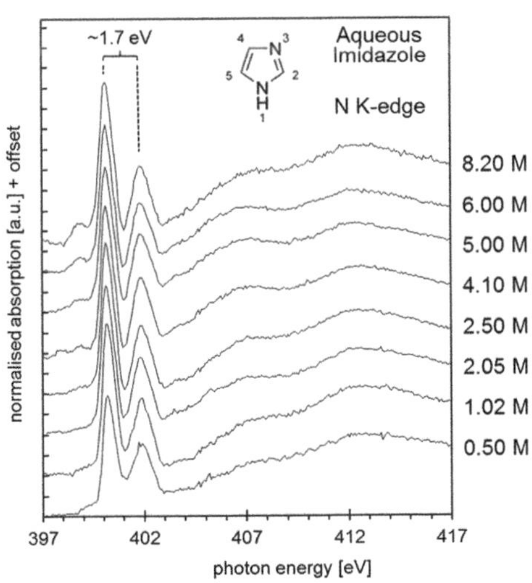

Fig. 9 N K-edge near edge X-ray absorption fine structure (NEXAFS) spectra of aqueous imidazole solutions as a function of concentration, from 0.5 mol L^{-1} to 8.2 mol L^{-1}.

concluded that longer-range effects in the solvent probably need to be taken into account to obtain a good agreement between theory and experiment.

As already observed for the gas phase monomer and the crystalline state, the solution NEXAFS spectra are dominated by the 1s → 1π* transitions of the N1 and N3 moieties in the imidazole ring, which now appear at photon energies of approximately 400.2 eV and 401.9 eV (Fig. 10). The photon energy difference of 1.7 eV is lower than the value observed for the gas phase monomer (2.4 eV) and slightly higher than in crystalline imidazole (1.5 eV). That the solution value is closer to that of the hydrogen-bonded solid state structure suggests that significant hydrogen bonding with surrounding water or imidazole molecules takes place. Since the extent of proton transfer is insignificant in the pH range of the solutions (pH ~ 10.5) it is reasonable to suggest that the observed energy difference of 1.7 eV is due to imidazole–water and/or imidazole–imidazole interactions.

The observed 1.7 eV difference between the 1s → 1π* transitions at the N1 and N3 centres matches the previously reported difference between the N 1s photoemission peaks of aqueous imidazole solutions.[41] Since the N 1s photoemission peak shifts are identical to the IP difference between the two N moieties (Fig. 10) we can conclude that the chemical shift between the two N 1s → 1π* transitions is primarily determined by the IP difference, just as observed in the above analysis of the imidazole monomer and crystal spectra, as well as in a similar analysis of the solid state of *p*-aminobenzoic acid.[29] The relative energetic positions of the transitions appear to be determined by ground state core level binding energy differences, and additional differences due to final state effects are negligible.

The effect of imidazole–water interactions was explored more systematically by calculating the N K-edge NEXAFS of the microhydrated cluster structures used previously to interpret the photoelectron spectra of imidazole in aqueous

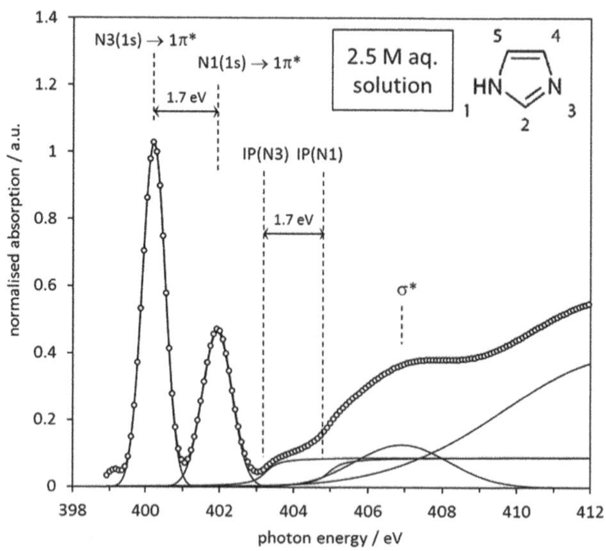

Fig. 10 Experimental N K-edge spectrum of 2.5 mol L^{-1} aqueous imidazole solution, with all spectral features fitted by Gaussian and arctan step functions. The origin of each feature is indicated by the annotations in the figure.

Table 5 Calculated N 1s core level binding energies and energy difference between the N1 ('NH') and N3 ('N=') 1s → 1π* NEXAFS bands for the imidazole N atoms in the gas phase monomer (#0, see Table 3) and the geometry-optimised gas phase imidazole–water clusters of Jagoda-Cwiklik *et al.*[64] The structures of the imidazole–water clusters are visualised in Fig. S1–S9 of the ESI

Cluster[a]		IP/eV			
		N1	N3	ΔIP/eV	ΔE(1s → 1π*)/eV
#0	imi·0H_2O	406.6	404.4	2.2	2.2
#1	imi·1H_2O	406.9	404.7	2.2	2.1
#2	imi·2H_2O	406.7	404.5	2.2	2.0
#3a	imi·3H_2O	406.5	404.7	1.8	1.8
#3b		406.0	404.3	1.7	1.6
#4a	imi·4H_2O	406.1	404.6	1.5	1.5
#4b		406.1	404.6	1.5	1.5
#4c		406.0	404.4	1.6	1.5
#4d		406.2	404.8	1.4	1.4
#5	imi·5H_2O	406.0	404.6	1.4	1.3

[a] Clusters are visualised in the ESI.

solution.[64] The structures of these clusters are displayed in the ESI (Figs. S1–S9†). Starting with a hydrogen-bonded HOH···N3 water–imidazole dimer #1, it can be seen that the effect of the hydrogen bond on the calculated NEXAFS spectrum is negligible (Table 5): the N1/N3 1s → 1π* peak energy difference is essentially the same as in the imidazole monomer (Table 3). The core level binding energies of the N3 (N=) and N1 (NH) centres remain very similar to the monomer binding energies, with only a slight shift of 0.3 eV to 406.9 eV for N1 and 404.7 eV for N3. This reflects the donation of electron density from the aromatic system to the water molecule.

The addition of another water molecule leads to cluster #2, which exhibits some subtle differences to the monomer. Although the spectral features are similar, there is a reduction of the N1/N3 1s → 1π* peak energy difference by 0.1 eV compared to the monomer. Interestingly, it appears that the N1 contribution has shifted by 0.1 eV with the addition of the extra bound water in the vicinity of the N3 centre.

Adding a third water molecule to the cluster led to equilibrium structures #3a and #3b.[64] These clusters exhibit the first significant changes compared to the monomer. The calculated N1/N3 1s → 1π* peak energy differences are reduced significantly relative to the monomer value of 2.1 eV, to 1.8 eV and 1.6 eV, respectively. These values, along with the corresponding N 1s core level binding energy differences of 1.8 eV and 1.7 eV, are actually in good agreement with the NEXAFS data, and with the N 1s binding energies previously reported.[41] However, this seemingly excellent agreement should not be overinterpreted, for the following reasons. First, the electronic structure of these clusters does not satisfactorily model the vertical ionisation potentials.[64] Second, it seems unlikely that imidazole in aqueous solution is solvated by only three water molecules. Third, we observed that adding more water molecules to the microhydrated clusters continues the trend of reducing the IP difference between the N1 and the N3 centres. As can be seen in Table 5, adding additional water molecules to yield hydrate shells with 4 (clusters #4a–#4d) and 5 (cluster #5) water molecules has the effect of further reducing the N1/N3 1s → 1π* peak energy difference, with the 5 water hydration shell producing a peak split of 1.3 eV and an IP difference of 1.4 eV. These are both significantly below the experimentally observed value of 1.7 eV. In fact, the system approaches the values calculated for the extensively hydrogen-bonded crystal, suggesting that hydrogen bonding of monomeric imidazole to water molecules alone does not correctly represent the structure of the system. Inclusion of larger hydration shells leads to a stronger than experimentally observed effect on the levelling of the N 1s core level energy difference between the two nitrogen centres. It may be speculated that one contribution to this result may be the asymmetric and incomplete coordination of the imidazole molecules in models invoking so few water molecules. In real solutions, the electron density variations induced by hydrate coordination from one side of the molecule are counterbalanced by coordination from the other, leading perhaps to an overall weaker net effect on the electronic structure of the central imidazole molecule. It remains to be examined whether this and other possible effects (such as longer range polarisation of the hydrogen bonded water network) in more extended hydration clusters may weaken hydrogen bonding to the imidazole molecule.

There is, of course, strong previous evidence that self-association of imidazole takes place in the range of aqueous concentrations investigated here. It is likely that either secondary interactions between hydrated clusters take place, or that even direct imidazole–imidazole interactions exist in solution, perhaps in hydrogen-bonded chain-and-stack structures such as those recently predicted by molecular dynamics simulations.[37] Indeed, a previous X-ray scattering analysis of aqueous imidazole solutions suggested that self-association involves the formation of molecule stacks held together by hydrogen-bonding through water molecules.[35] To examine whether such a model would lead to a more correct reproduction of the NEXAFS data, we set up a more complex geometry-optimised structure model in Gaussian03. A geometry-optimised stack structure of three imidazole molecules was generated at the BHANDH/6-311G** level of theory, involving three water molecules coordinating the central imidazole molecule. The resulting structure and its predicted absorption spectrum are shown in Fig. 11. It can be seen that the geometry-optimised cluster exhibits hydrogen bonding interactions for one of the water molecules linking two of the imidazole species, just as in the model previously generated by X-ray scattering.[35] The energy difference between the N 1s → 1π* resonances in the spectra is 1.5 eV, significantly more in agreement with experiment than the microsolvated cluster models involving only one imidazole molecule.

This result suggests that self-association involves complex synergistic hydrogen bonding interactions mediated by water molecules alongside imidazole–imidazole interactions. This model is compatible with the idea of π–π-assisted hydrogen-bonded stack structures, in competition with chains of hydrogen-bonded imidazole molecules[37] that are similar to those found in liquid[38] and solid[32,33] imidazole. We should mention that the calculated absorption spectra of the central imidazole in these structures are quite sensitive to conformational detail in the evaluated cluster, resulting in varying degrees of agreement with experimental data. Clearly, more complex structure models, including more water and imidazole molecules, need to be more systematically

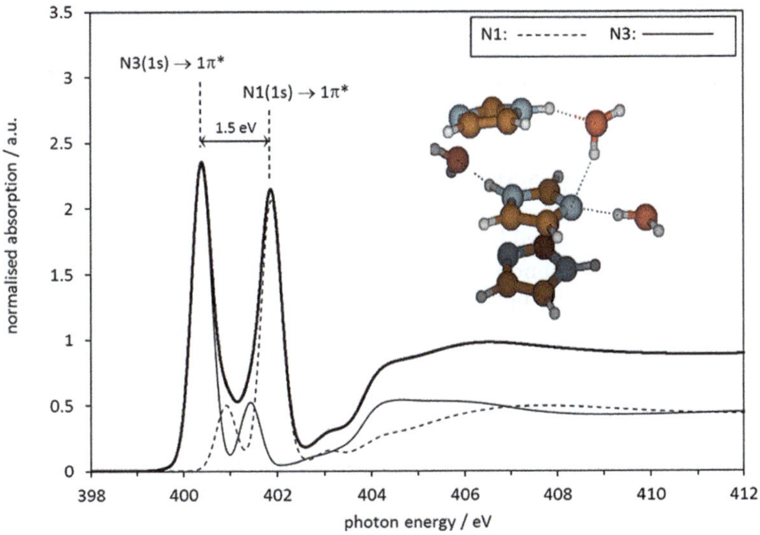

Fig. 11 Calculated N K-edge absorption spectrum for the central molecule in a micro-hydrated Gaussian 03 optimised stack of 3 imidazole molecules.

evaluated before firm conclusions about the limitations of the micro-cluster system modelling approach can be drawn. Moreover, combination with other techniques sensitive to local structure in solution (X-ray and neutron scattering, NMR, vibrational spectroscopies, also C K-edge NEXAFS) should be explored to generate a more complete picture. However, the insight already obtained by the modelling of the NEXAFS N K-edge data, and the observed sensitivity of NEXAFS to local structure, provide some confidence that the technique could start playing a role as a tool for validating computationally derived structural predictions. In the context of nucleation studies, a particularly valuable objective for further development will be the setting up of an experimental infrastructure that permits studies of supersaturated solutions.

Conclusions

We have carried out a feasibility study examining the possibility of obtaining N K-edge near-edge X-ray absorption fine-structure (NEXAFS) spectra of a solute at high concentrations in aqueous solution. NEXAFS has been shown to be able to distinguish between different chemical environments of imidazole, including gas phase, solid state and the solution state. We have shown that the analysis of known structures (monomer, crystal) provides valuable insight for unravelling the more complex origins of electronic structure variations in the solution system. We have also shown that different types of electronic structure calculations (Gaussian, StoBe, CASTEP) can be used to provide a consistent interpretation of NEXAFS data, which, especially in combination with information from other experimental techniques, should enable more incisive analytical approaches to determining the structure and properties of organic solutes in solution. The results obtained so far have confirmed that microhydrated clusters with a single imidazole molecule are poor models for aqueous imidazole, and that more complex models, which include hydrogen-bonded hydrate molecules as well as imidazole–imidazole interactions lead to better agreement with experimental data.

Acknowledgements

This research was supported by EPSRC grants GR/S85801/01, EP/D023807/1, GR/R59069/01, EP/I014446/1 and EP/I013563/1. MJT gratefully acknowledges support by EPSRC through a DTA PhD studentship. We thank the Helmholtz Zentrum Berlin (HZB) for a BESSY-II beamtime award (2007_61140). This work made use of the facilities of N8 HPC, provided and funded by the N8 consortium and EPSRC (Grant no. EP/K000225/1). The Centre is coordinated by the Universities of Leeds and Manchester. The authors would also like to acknowledge the usage of the ARC1 computing facility at the University of Leeds, and the usage of Hartree Centre resources for this work. The STFC Hartree Centre is a research collaboration funded by the UK's investment in e-Infrastructure. CRS would like to acknowledge the support of the High Performance Computing (HPC) team at the University of Leeds, for their design of the advanced research computing 1 (ARC1) facility, and assistance in code compilation and optimization. We thank Dr Luanga N. Nchari and Dr Angela M. Beesley for assistance during the NEXAFS measurements. We thank Professor Pavel Jungwirth for sharing the micro-hydrated cluster model structures.

Notes and references

1 M. Volmer, in *Kinetik der Phasenbildung*, Steinkopff, Leipzig, Germany, 1939.
2 P. G. Debenedetti, *Nature*, 2006, **441**, 168–169.
3 A. C. Pan, T. J. Rappl, D. Chandler and N. P. Balsara, *J. Phys. Chem. B*, 2006, **110**, 3692–3696.
4 R. J. Davey, S. L. M. Schroeder and J. H. ter Horst, *Angew. Chem., Int. Ed.*, 2013, **52**, 2166–2179.
5 D. W. Oxtoby, *Acc. Chem. Res.*, 1998, **31**, 91–97.
6 D. Kashchiev and G. M. van Rosmalen, *Cryst. Res. Technol.*, 2003, **38**, 555–574.
7 D. Kashchiev, in *Nucleation: Basic Theory with Applications*, Butterworth-Heinemann, Oxford, 2000.
8 P. G. Vekilov, *Cryst. Growth Des.*, 2010, **10**, 5007–5019.
9 D. Gebauer and H. Cölfen, *Nano Today*, 2011, **6**, 564–584.
10 J. Bernstein, in *Polymorphism in Molecular Crystals*, Clarendon, Oxford, 2002.
11 A. Spitaleri, C. A. Hunter, J. F. McCabe, M. J. Packer and S. L. Cockroft, *CrystEngComm*, 2004, **6**, 489–493.
12 C. A. Hunter, C. M. R. Low, M. J. Packer, S. E. Spey, J. G. Vinter, M. O. Vysotsky and C. Zonta, *Angew. Chem., Int. Ed.*, 2001, **40**, 2678–2682.
13 J. Bernstein, R. J. Davey and J. O. Henck, *Angew. Chem., Int. Ed.*, 1999, **38**, 3441–3461.
14 H. G. Alison, R. J. Davey, J. Garside, M. J. Quayle, G. J. T. Tiddy, D. T. Clarke and G. R. Jones, *Phys. Chem. Chem. Phys.*, 2003, **5**, 4998–5000.
15 P. E. Bonnett, K. J. Carpenter, S. Dawson and R. J. Davey, *J. Chem. Soc., Chem. Commun.*, 2003, 698–699.

16 R. J. Davey, G. Dent, R. K. Mughal and S. Parveen, *Cryst. Growth Des.*, 2006, **6**, 1788–1796.
17 I. Weissbuch, M. Lahav and L. Leiserowitz, *Cryst. Growth Des.*, 2003, **3**, 125–150.
18 J. Stöhr, in *NEXAFS Spectroscopy (Springer Series in Surface Sciences 25)*, Springer, Berlin, Heidelberg, 1992.
19 G. Hähner, *Chem. Soc. Rev.*, 2006, **35**, 1244–1255.
20 *Practical Surface Analysis, Auger and X-ray Photoelectron Spectroscopy*, ed. D. Briggs and M. P. Seah, John Wiley & Sons, Chichester, New York, Brisbane, Toronto, Singapore, 2 edn, 1990, vol. 1.
21 J. W. Niemantsverdriet, in *Spectroscopy in Catalysis*, VCH, Weinheim, 1993.
22 S. Hüfner, in *Photoelectron Spectroscopy. Principles and Applications*, Springer Verlag, Berlin, Heidelberg, New York, 2 edn, 1996.
23 J. S. Stevens, S. J. Byard and S. L. M. Schroeder, *Cryst. Growth Des.*, 2010, **10**, 1435–1442.
24 J. S. Stevens, S. J. Byard and S. L. M. Schroeder, *J. Pharm. Sci.*, 2010, **99**, 4453–4457.
25 J. S. Stevens, S. J. Byard, C. A. Muryn and S. L. M. Schroeder, *J. Phys. Chem. B*, 2010, **114**, 13961–13969.
26 J. S. Stevens, S. J. Byard, C. C. Seaton, G. Sadiq, R. J. Davey and S. L. M. Schroeder, *Angew. Chem., Int. Ed.*, 2011, **50**, 9916–9918.
27 J. S. Stevens, S. J. Byard, E. Zlotnikov and S. L. M. Schroeder, *J. Pharm. Sci.*, 2011, **100**, 942–948.
28 J. S. Stevens, S. J. Byard, C. C. Seaton, G. Sadiq, R. J. Davey and S. L. M. Schroeder, *Phys. Chem. Chem. Phys.*, 2014, **16**, 1150–1160.
29 J. S. Stevens, C. R. Seabourne, Ch. Jaye, D. A. Fischer, A. J. Scott and S. L. M. Schroeder, *J. Phys. Chem. B*, 2014, **118**, 12121–12129.
30 J. S. Stevens, A. Gainar, E. Suljoti, J. Xiao, R. Golnak, E. F. Aziz and S. L. M. Schroeder, *Chem.–Eur. J.*, 2015, **21**, DOI: 10.1002/chem.201405635.
31 G. Will, *Nature*, 1963, **198**, 575.
32 S. Martinez, *Acta Crystallogr.*, 1966, **20**, 783.
33 B. M. Craven, R. K. Mcmullan, J. D. Bell and H. C. Freeman, *Acta Crystallogr., Sect. B: Struct. Sci.*, 1977, **33**, 2585–2589.
34 W. Hückel, J. Datow and E. Simmersbach, *Z. Phys. Chem., Abt. A*, 1940, **186**, 129–179.
35 L. Gontrani, R. Caminiti, L. Bencivenni and C. Sadun, *Chem. Phys. Lett.*, 1999, **301**, 131–137.
36 F. Peral and E. Gallego, *J. Mol. Struct.*, 1997, **415**, 187–196.
37 S. Y. Liem, M. S. Shaik and P. L. A. Popelier, *J. Phys. Chem. B*, 2011, **115**, 11389–11398.
38 M. S. Shaik, S. Y. Liem, Y. N. Yuan and P. L. A. Popelier, *Phys. Chem. Chem. Phys.*, 2010, **12**, 15040–15055.
39 U. Domanska, M. K. Kozlowska and M. Rogalski, *J. Chem. Eng. Data*, 2002, **47**, 456–466.
40 E. F. Aziz, PhD Dissertation, Freie Universität Berlin, 2007.
41 D. Nolting, N. Ottosson, M. Faubel, I. V. Hertel and B. Winter, *J. Am. Chem. Soc.*, 2008, **130**, 8150–8151.
42 L. Triguero and L. G. M. Pettersson, *Surf. Sci.*, 1998, **398**, 70–83.

43 L. Triguero, L. G. M. Pettersson, B. F. Minaev and H. Agren, *J. Chem. Phys.*, 1998, **108**, 1193–1205.
44 K. Herrmann, L. G. M. Pettersson, M. E. Casida, C. Daul, A. Goursot, A. Koester, E. Proynov, A. St-Amant and D. R. Salahub, *StoBe-deMon version 2.2*, Berlin, Stockholm, 2006.
45 S. H. Vosko and L. Wilk, *Phys. Rev. B*, 1980, **22**, 3812–3815.
46 C. Kolczewski, R. Püttner, O. Plashkevych, H. Agren, V. Staemmler, M. Martins, G. Snell, A. S. Schlachter, M. Sant'Anna, G. Kaindl and L. G. M. Pettersson, *J. Chem. Phys.*, 2001, **115**, 6426–6437.
47 L. Triguero, L. G. M. Pettersson and H. Agren, *J. Phys. Chem. A*, 1998, **102**, 10599–10607.
48 L. Triguero, L. G. M. Pettersson and H. Agren, *Phys. Rev. B*, 1998, **58**, 8097–8110.
49 M. J. Frisch, G. W. Trucks, H. B. Schlegel, G. E. Scuseria, M. A. Robb, J. R. Cheeseman, J. A. Montgomery Jr, T. Vreven, K. N. Kudin, J. C. Burant, J. M. Millam, S. S. Iyengar, J. Tomasi, V. Barone, B. Mennucci, M. Cossi, G. Scalmani, N. Rega, G. A. Petersson, H. Nakatsuji, M. Hada, M. Ehara, K. Toyota, R. Fukuda, J. Hasegawa, M. Ishida, T. Nakajima, Y. Honda, O. Kitao, H. Nakai, M. Klene, X. Li, J. E. Knox, H. P. Hratchian, J. B. Cross, V. Bakken, C. Adamo, J. Jaramillo, R. Gomperts, R. E. Stratmann, O. Yazyev, A. J. Austin, R. Cammi, C. Pomelli, J. W. Ochterski, P. Y. Ayala, K. Morokuma, G. A. Voth, P. Salvador, J. J. Dannenberg, V. G. Zakrzewski, S. Dapprich, A. D. Daniels, M. C. Strain, O. Farkas, D. K. Malick, A. D. Rabuck, K. Raghavachari, J. B. Foresman, J. V. Ortiz, Q. Cui, A. G. Baboul, S. Clifford, J. Cioslowski, B. B. Stefanov, G. Liu, A. Liashenko, P. Piskorz, I. Komaromi, R. L. Martin, D. J. Fox, T. Keith, M. A. Al-Laham, C. Y. Peng, A. Nanayakkara, M. Challacombe, P. M. W. Gill, B. Johnson, W. Chen, M. W. Wong, C. Gonzalez and J. A. Pople, *Gaussian 03, Revision C.02*, Wallingford, CT, Gaussian, Inc., 2004.
50 S. J. Clark, M. D. Segall, C. J. Pickard, P. J. Hasnip, M. J. Probert, K. Refson and M. C. Payne, *J. Crystallogr.*, 2005, **220**, 567–570.
51 S. P. Gao, C. J. Pickard, A. Perlov and V. Milman, *J. Phys.: Condens. Matter*, 2009, **21**, 104203.
52 S. P. Gao, C. J. Pickard, M. C. Payne, J. Zhu and J. Yuan, *Phys. Rev. B*, 2008, **77**, 115122.
53 C. J. Pickard, PhD thesis, University of Cambridge, 2007.
54 J. D. Pack and H. J. Monkhorst, *Phys. Rev. B*, 1977, **16**, 1748–1749.
55 H. J. Monkhorst and J. D. Pack, *Phys. Rev. B*, 1976, **13**, 5188–5192.
56 C. R. Seabourne, A. J. Scott, R. Brydson and R. J. Nicholls, *Ultramicroscopy*, 2009, **109**, 1374–1388.
57 P. Blaha, K. Schwarz, P. Sorantin and S. B. Trickey, *Comput. Phys. Commun.*, 1990, **59**, 399–415.
58 K. Schwarz, P. Blaha and S. B. Trickey, *Mol. Phys.*, 2010, **108**, 3147–3166.
59 E. Apen, A. P. Hitchcock and J. L. Gland, *J. Phys. Chem.*, 1993, **97**, 6859–6866.
60 P. Moreau, F. Boucher, G. Goglio, D. Foy, V. Mauchamp and G. Ouvrard, *Phys. Rev. B*, 2006, **73**, 195111.
61 J. S. Stevens, A. C. de Luca, M. Pelendritis, G. Terenghi, S. Downes and S. L. M. Schroeder, *Surf. Interface Anal.*, 2013, **45**, 1238–1246.

62 P. Edington and M. M. Harding, *Acta Crystallogr., Sect. B: Struct. Sci.*, 1974, **30**, 204–206.

63 B. Jagoda-Cwiklik, P. Slavícek, D. Nolting, B. Winter and P. Jungwirth, *J. Phys. Chem. B*, 2008, **112**, 7355–7358.

64 B. Jagoda-Cwiklik, P. Slavícek, L. Cwiklik, D. Nolting, B. Winter and P. Jungwirth, *J. Phys. Chem. A*, 2008, **112**, 3499–3505.

Faraday Discussions

PAPER

Urea homogeneous nucleation mechanism is solvent dependent

Matteo Salvalaglio,*[abc] Marco Mazzotti[a] and Michele Parrinello[bc]

Received 25th November 2014, Accepted 9th January 2015
DOI: 10.1039/c4fd00235k

The composition of the mother phase plays a primary role in crystallization processes, affecting both crystal nucleation and growth. In this work, the influence of solvents on urea nucleation has been investigated by means of enhanced sampling molecular dynamics simulations. We find that, depending on the solvent, the nucleation process can either follow a single-step or a two-step mechanism. While in methanol and ethanol a single-step nucleation process is favored, in acetonitrile a two-step process emerges as the most likely nucleation pathway. We also find that solvents have a minor impact on polymorphic transitions in the early stages of urea nucleation. The impact of finite size effects on the free energy surfaces is systematically considered and discussed in relation to the simulation setup.

1 Introduction

Fine chemicals and active pharmaceutical ingredients are typically produced in crystalline form. In such cases it is of great importance to control both polymorphism and particles' morphology. Both these characteristics, in fact, contribute crucially to the definition of key properties such as dissolution kinetics[1] and bioavailability.[2] Crystallization of organic molecules is typically carried out in solution. Addressing the role of solvents and predicting their impact on the process is therefore crucial for the control and design of crystallization. Solvents, in fact, can influence both the crystal morphology and the bulk crystal structure.[3,4] Crystallization proceeds in two subsequent stages: nucleation and growth. Nucleation is the process of formation of the first stable embryonic structure of a crystal, that eventually grows to become a macroscopic particle.[5] Understanding the effect of the mother phase composition on crystal growth might allow the steering of the crystallization process towards a predetermined

[a]*ETH Zurich, Institute of Process Engineering, Soneggstrasse 3, CH-8092 Zurich, Switzerland. E-mail: salvalaglio@ipe.mavt.ethz.ch*
[b]*Facoltà di Informatica, Istituto di Scienze Computazionali, Università della Svizzera Italiana, CH-6900 Lugano, Switzerland*
[c]*Department of Chemistry and Applied Biosciences, ETH Zurich, CH-8092 Zurich, Switzerland*

outcome. As such, a large body of experimental and theoretical work deals with the impact of solvents and additives on crystal growth.[6–8]

In this work we focus on nucleation, with the aim of uncovering the effect of solvents on the nucleation mechanism by means of molecular simulations. Studying nucleation from solution represents a challenge for both experiments and simulations.[9,10] On the one hand, nucleation is initiated at the molecular scale, thus making its direct experimental investigation extremely challenging. On the other hand, nucleation is a paradigmatic example of a rare event, characterized by timescales that often more than not exceed those accessible with molecular simulations. Therefore, while molecular modelling techniques display the clear advantage of providing an atomistic resolution of nucleation, timescale limitations still need to be properly addressed.[11] To this aim, a variety of enhanced sampling methods based on molecular dynamics have been proposed.[12–21] Taking advantage of these tools, much work has been devoted to the investigation of homogeneous nucleation in simple model systems, like Lennard-Jones particles or hard spheres.[22–24] Recently, however, attention has shifted to more complex and more practically relevant systems, such as organic molecules in solution.[11,25–34] In this work we use Well Tempered (WT) metadynamics, a state-of-the-art enhanced sampling method that allows the investigation of nucleation events in explicit solution.[12,35–38]

In particular, we focus our study on the effect of organic solvents, such as acetonitrile (ACN), methanol (MeOH) and ethanol (EtOH), on the nucleation of urea crystals. Urea, despite being a simple organic molecule, exhibits a rich crystallization behaviour that has been extensively investigated using both experimental and computational studies.[8,29,33,39–44] It has emerged that different crystal faces grow following different mechanisms,[8,41] and that solvents and additives may substantially affect urea crystal morphologies, that can range from needle-like particles to compact tetrahedra.[40,43,44] A two-step nucleation mechanism has been identified as the most probable pathway for homogeneous nucleation of urea from aqueous solution.[33] Moreover, when analysing urea nucleation both from the melt[29] and from aqueous solution,[33] it has been found that two polymorphs compete in the early stages of nucleation: form I, the experimentally known urea crystal structure, and form II, a metastable polymorph. In this paper we investigate the effect of solvents on the nucleation mechanism of urea, focusing our attention on both the competition between polymorphs and on the dominant nucleation mechanisms. Similarly to the nucleation of liquid droplets from a vapour,[45–47] the simulation of crystal nucleation from solution is heavily affected by finite-size effects.[33,48] In order to deal with these effects, we shall apply the strategy discussed in detail in ref. 33.

2 Theory and methods

2.1 Reversible work of nucleation in a finite-sized system

Molecular dynamics (MD) simulations of crystallization processes in solution are conventionally performed in the isothermal–isobaric ensemble, in which temperature, pressure and number of atoms are kept constant. When nucleation takes place in a simulation box, the chemical potential of the mother solution inherently couples with the size of the nucleus, as described in depth in ref. 33. This effect is reflected in the free energy profile associated with the formation of a finite sized crystal nucleus of size N_c. In ref. 33, the reversible work for the formation of a crystalline nucleus of size N_c in a confined system is thus derived as:

$$\Delta G_{l\to c} = -N_c k_B T \ln\left(\frac{\gamma x}{\gamma^* x^*}\right) + \sigma' N_c^{2/3} + N_{tot} k_B T \ln\left(\frac{\gamma x}{\gamma_0 x_0}\right) + N_s k_B T \ln\left(\frac{\gamma_s(1-x)}{\gamma_{s,0}(1-x_0)}\right) \quad (1)$$

where: N_c is the number of molecules belonging to the crystal phase, k_B is the Boltzmann constant, T is the temperature, γ_i are the activity coefficients, x is the actual molar fraction of the solute, x^* is the molar fraction of the solute at equilibrium with the solid, x_0 is its molar fraction in absence of any crystal-like nuclei, N_s is the number of solvent molecules, N_{tot} is the total number of solute molecules in the system, and σ' is the effective surface energy that takes into account both the surface tension and the shape of the nucleus.

For macroscopic systems, where $x \simeq x_0 = \text{const}$, this expression reduces to the free energy profile derived within the context of Classical Nucleation Theory (CNT) for a system at fixed composition:

$$\Delta G_{CNT}(N_c) = -N_c k_B T \ln\left(\frac{\gamma_0 x_0}{\gamma^* x^*}\right) + \sigma' N_c^{2/3} \quad (2)$$

The free energy profiles reported in eqn (1) and (2), allow the determination of the thermodynamic stability of crystal nuclei as a function of their size N_c in a finite-sized or an infinite system, respectively.[5]

The most relevant consequence of the confinement effect is that the stability of the solution does not depend only on supersaturation $S = \gamma_0 x_0/\gamma^* x^*$, but also on the system size.[33,48] Three regimes (labelled A, B, and C) can be identified as a function of the simulation box volume and of the initial composition of the liquid phase for $S > 1$ (see Fig. 1):

A: the solution is supersaturated; however, depletion effects render the formation of a crystal nucleus unfavourable. In this case, the free energy profile described by eqn (1) does not display stationary points, a critical nucleus thus cannot be identified and any state comprising a crystal-like particle is unstable with respect to dissolution.

B: the absolute minimum is still the homogeneous solution; however, the ΔG curve is not monotonous, exhibiting both a local maximum and a minimum. The latter is a state in which a crystallite is in metastable equilibrium with the solution. The maximum instead identifies the size of the critical nucleus that leads to the formation of the metastable state.

C: here the curve has a shape similar to that in case B, but the local minimum corresponding to the small crystallite in equilibrium with the solution becomes the stable state, towards which the homogeneous solution can evolve. In this case as well, the critical nucleus size is defined by the location of the maximum in the free energy profile.

Examples of typical free energy profiles in the three cases, together with the domains in the space of the initial composition x_0 and the system volume, are

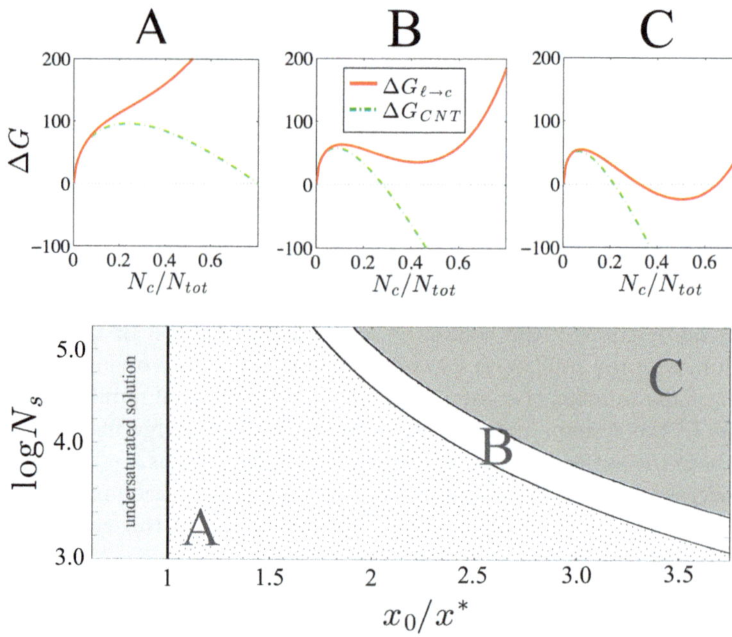

Fig. 1 Upper row: Representation of characteristic free energy profiles calculated in regimes A, B, and C. Lower row: existence domains for regimes A, B, and C as a function of supersaturation x_0/x^* and the number of solvent molecules N_s. Once x_0 is fixed the number of solvent molecules N_s determines the system volume, as $V \propto (N_{tot} + N_s) = (1 - x_0)^{-1}N_s$.

reported in Fig. 1. A detailed description and derivation of the nucleation free energy profile in a confined system is reported in ref. 33.

From the free energy profiles defined in eqn (2) and (1) we introduce the correction term:

$$\begin{aligned}\Delta G_{\text{corr}}(N_c) &= \Delta G_{\text{CNT}} - \Delta G_{1\to c}\\ &= -(N_{\text{tot}} - N_c)k_BT\ln\left(\frac{\gamma x}{\gamma_0 x_0}\right) - N_s k_B T\ln\left(\frac{\gamma_s(1-x)}{\gamma_{s,0}(1-x_0)}\right)\end{aligned} \quad (3)$$

This expresses the difference in free energy between a finite size system and an infinitely large one at constant supersaturation.[33,46] In order to extract from finite sized simulations information in the limit of a macroscopic system, we shall thus weight each configuration considering the correction term ΔG_{corr}.

The previous discussion has important consequences for simulations. In fact, for computationally accessible system sizes we typically fall in case A, where neither stable nor metastable states other than the solution are possible. The states in which the crystal has nucleated are to be found in the tail of the probability distribution, and practically cannot be observed with unbiased simulations. Thus, if we want to explore these states we need enhanced sampling. In this work we use WT metadynamics, an enhanced sampling method based on the deposition of a history-dependent bias potential, which enhances the fluctuations in a coarsened phase space defined by collective variables (CVs). In the following paragraphs, details of the CVs definition as well as simulation details are provided. The interested reader can refer to a recent review[49] for a detailed description of the WT metadynamics method details, and to ref. 29 and 38 for examples of WT metadynamics simulations applied to nucleation processes.

2.2 Collective variables

A key ingredient of metadynamics is the definition of collective variables (CVs), the coordinates of the coarsened space in which the metadynamics bias potential is adaptively constructed.[49] Collective variables should be able to capture the slow degrees of freedom associated to the rare transformation that is investigated. In this work, WT metadynamics simulations were carried out using a set of collective variables inspired by the study of urea growth from solution,[8,44] as described in detail in a recent work on urea crystal nucleation from the melt.[29] The CVs were implemented in PLUMED-1.3.[50] The CVs are formulated with the aim of accounting for both local density and the mutual orientation of urea molecules. The general expression for the CVs, discussed in detail in ref. 29, follows:

$$S = \frac{1}{N}\sum_{i=1}^{N} \Gamma_i \tag{4}$$

where N is the total number of solute molecules in the system and Γ_i is an estimate of the degree of order for the single solute molecule, defined with respect to a specific orientation axis of the urea molecule. In analogy to ref. 29 and 33, we use two CVs: $S_1 = \frac{1}{N_{\text{tot}}}\sum_{i=1}^{N}\Gamma_{CO}(i)$ and $S_2 = \frac{1}{N_{\text{tot}}}\sum_{i=1}^{N}\Gamma_{NN}(i)$, where $\Gamma_{CO}(i)$ and $\Gamma_{NN}(i)$ account for the degrees of local order with respect to the CO and NN axes within the urea molecule. $\Gamma_{CO}(i)$ and $\Gamma_{NN}(i)$ are defined as:

$$\Gamma_{CO}(i) = \frac{\rho_i}{n_i}\sum_{j=1}^{N} f_{ij}\left(e^{-\frac{(\vartheta_{ij}-\theta_1)^2}{2\delta^2}} + e^{-\frac{(\vartheta_{ij}-\theta_2)^2}{2\delta^2}}\right) \tag{5}$$

$$\Gamma_{NN}(i) = \frac{\rho_i}{n_i}\sum_{j=1}^{N} f_{ij}\left(e^{-\frac{(\vartheta_{ij}-\theta_1)^2}{2\delta^2}} + e^{-\frac{(\vartheta_{ij}-\theta_2)^2}{2\delta^2}}\right) \tag{6}$$

where:

- n_i is the coordination number of molecule i.
- ρ_i is a switching function defined in the space of the coordination number, that decreases from one to zero when the number of neighbours of the ith urea molecule is less than a threshold value. As in ref. 29, a threshold of four neighbours within a cut-off distance of 0.6 nm has been used.
- ϑ_{ij} is the angle between the CO or the NN axis of the i and j urea molecules.
- f_{ij} is a function weighing the contribution of the relative orientation of neighbouring urea molecules with their Cartesian distance.
- θ_1, θ_2 represent characteristic orientations for the CO and NN vectors.[29] In eqn (5) $\theta_1 = 0°$, $\theta_2 = 180°$, while in eqn (6) $\theta_1 = 90°$, $\theta_2 = 270°$.
- δ is the standard deviation of the Gaussian functions centered in θ_1 and θ_2. In both eqn (5) and (6) $\delta = 27.5°$.

A detailed description of the terms ρ_i and f_{ij} can be found in ref. 29.

2.3 Analysis tools

The addition of a bias potential with WT metadynamics[12] alters the statistical weight with which configurations are sampled. In order to restore canonical sampling, a reweighing procedure needs to be put in place.[36,37] This allows the analysis of the nucleation mechanisms, by plotting the free energy surface (FES) relative to variables different from S_1 and S_2. In particular we found the introduction of the variables (n, n_o) and (N_I, N_{II}) illuminating.

The first pair considers the number of molecules in the largest connected cluster. The difference between them is that to compute n, we link molecules on the basis of their distance and on the density of their local environment, irrespective of their relative orientation, whereas in n_o the relative orientation is also taken into account. Thus n_o represents the size of the largest cluster that is crystal-like, and $n > n_o$. In order to calculate n, we cluster molecules that have at least four neighbours within a distance of 0.6 nm. The connectivity criterion for n_o, on the other hand, has the additional condition that the order parameter Γ_{CO} is greater than 0.5.

The variables N_I and N_{II} give the number of molecules in the largest crystalline cluster, with the structures of form I and form II, respectively. To discriminate between form I and form II molecules, both molecular order parameters Γ_{CO} and Γ_{NN} were considered. Molecules were assigned to a form I configuration if $\Gamma_{CO} > 0.5$ and $\Gamma_{NN} > 0.3$, and to a form II configuration if $\Gamma_{CO} > 0.5$ and $\Gamma_{NN} \leq 0.3$. The choice of these threshold values is based on the relative position of basins identifying form I and form II configurations, in the FES obtained for urea nucleation from its melt in Fig. 3 of ref. 29.

Having defined a connectivity criterion, we identify the clusters using a *depth-first* algorithm, similar to the approach used in ref. 51.

2.4 Simulation details

Forcefield. In order to meet the requirement of an extensive sampling, we have chosen to simulate our systems with a classical forcefield. Similarly to what has been done in previous works,[8,29,33,44] both urea and solvent molecules have been explicitly represented using the Generalized Amber Force Field.[52–54]

Table 1 WT metadynamics simulations of urea nucleation in MeOH, EtOH, and ACN. Simulation conditions and parameters obtained from the fitting of the WT metadynamics FES

	MeOH	EtOH	ACN
T [K]	300	300	300
x_0	0.0978	0.1348	0.1245
N_s	2767	1926	2110
Time [ns]	1×10^3	1×10^3	1×10^3
x^*	0.039	0.062	0.0019
σ' [kJ mol^{-1}]	58.15	29.05	75.05
N^* at $x = x_0$	4624	991	108

Simulations setup. Three simulations were carried out in order, to study urea nucleation explicitly in ACN, MeOH, and EtOH. All the systems have been prepared by distributing 300 urea molecules in an appropriate number of solvent molecules (see Table 1). Urea molecules were dispersed in solvent using the *genbox* utility of the gromacs simulation package, and three dimensional periodic boundary conditions were applied. Production runs were performed at 300 K and 1 bar. The metadynamics production runs were carried out for approximately one microsecond each. Each system was firstly minimized with the conjugate gradient algorithm, with a tolerance on the maximum force of 200 kJ mol^{-1} nm^{-1}. A 20 ns unbiased NPT equilibration of the systems was carried out prior to performing the metadynamics simulations. All the simulations were carried out in the isothermal–isobaric ensemble at 1 bar, using the Bussi–Donadio–Parrinello stochastic thermostat[55] and the isotropic Parrinello–Rahman barostat[56] as implemented in GROMACS 4.5.4.[57] The particle-mesh Ewald approach was used to calculate long range electrostatic interactions, with a cutoff of 1 nm.[58] The LINCS algorithm was applied at each step to preserve the bond lengths, with 1×10^{-5} tolerance and 4 iterations.[59] The time step used in the production run was 2×10^{-3} ps. In our simulations the bias deposition stride was set to be 1 ps; the height of the Gaussian bias rate was set to $\approx 2k_BT$. The width of the Gaussian bias rate was set to 2×10^{-2}. In all simulations the WT algorithm was applied with a bias factor of 200.[12,29]

3 Results

3.1 Reversible work of crystal nucleation

The expression of the free energy of nucleation in a closed, confined system formulated in eqn (1) has been derived under the hypotheses that: (i) only one nucleus is formed, (ii) surface tension can also be defined for small crystal-like clusters, and (iii) the nucleus possesses a well-defined crystal structure. As discussed at length in ref. 33, these hypotheses are considered only to derive an analytical expression for the reversible work of formation of a nucleus, and do not reflect the complexity of the system evolution at the molecular scale.

In this first paragraph we report the free energy profiles calculated from WT metadynamics for the nucleation of a single cluster of urea molecules in a crystal-like configuration. We compare these free energy profiles with those derived from the analytical expression reported in eqn (1). In Table 1, the fitted values of σ' and x^* are reported.

It can be seen in Fig. 2 that the finite size expression of eqn (2) fits the data for all three solvents rather well. Together with our previous results,[33] this provides further validation of this equation. In the right column of Fig. 2, we report the solvent-specific location in the parameter space of domains A, B, and C described in the previous section. Their relative positions and extensions are clearly solvent-dependent, reflecting the differences in x^* and σ'.

The fitted values of x^* display a solubility for urea in ACN at least 20 times lower than in alcohols, in agreement with the experimental observations (Table 2). The relative solubility of urea in MeOH and EtOH appears instead to rank in the opposite order, with respect to the experimental values. This discrepancy reflects the typical energy accuracy of the molecular modelling approach used in our simulations.

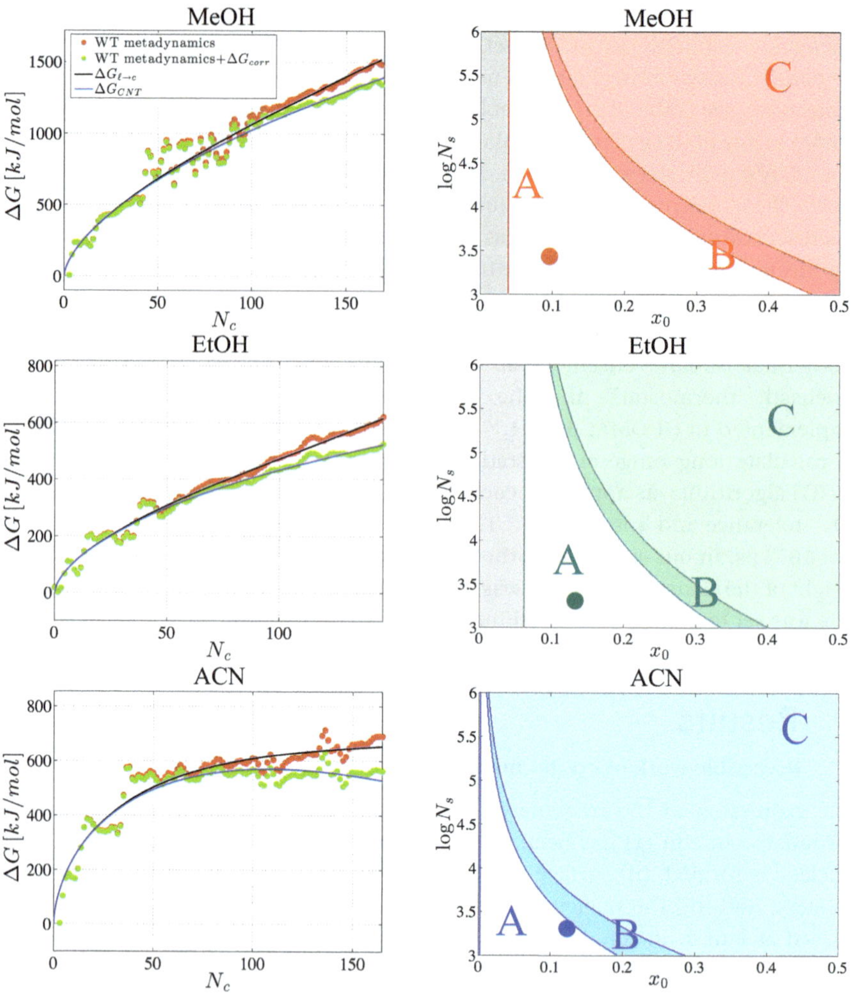

Fig. 2 Left column: Comparison between the analytical expression of the free energy of nucleation of a crystal-like cluster calculated with parameters reported in Table 1, and the free energy profile computed from WT metadynamics simulations for the nucleation of a single urea cluster in solution. The analytical expression consistently describes the free energy profile obtained from simulations, and the correction term reported in eqn (3) allows the free energy profile to be obtained in the limit of a macroscopic system at constant composition $x = x_0$. Right column: Regimes A, B, and C in the domain of x_0 and N_s. The point corresponding to the simulation conditions is highlighted with a circle. Systems prepared with similar initial conditions in the x_0, N_s space may be affected by drastically different finite size effects, depending on the choice of solvent.

Consider, for instance, the thermodynamic driving force to crystallization, $\Delta\mu_{\rm ref} = k_{\rm B}T\ln(x_{\rm ref}/x^*)$, calculated with respect to a common reference state at composition $x_{\rm ref}$ (Table 2). It can be noted that the discrepancy between the calculated and experimental $\Delta\mu$ values, as well as the difference in $\Delta\mu$ between MeOH and EtOH, is of the order of $k_{\rm B}T$, the typical level of accuracy expected from classical forcefields. The difference in $\Delta\mu_{\rm ref}$ between alcohols and ACN is instead

Table 2 Solubility values obtained from simulations compared with their experimental counterparts. To estimate the error in the $\Delta\mu$,a common reference state at $x_{ref} = 0.3$ has been chosen. The value of x_{ref} has been such that the reference state is supersaturated with respect to all solvents, *i.e.* $x_{ref} > \max. x^*$. The particular choice of the reference state does not influence the absolute error, reported in the last column

	x^* calc.	x^* exp.	$\Delta\mu_{ref}$ calc. (kJ mol^{-1})	$\Delta\mu_{ref}$ exp. (kJ mol^{-1})	$\left\lvert\frac{\Delta\mu_{ref}^{exp} - \Delta\mu_{ref}^{calc}}{k_BT}\right\rvert$
MeOH	0.039	0.107[a]	−5.13	−2.57	1.02
EtOH	0.062	0.045 (ref. 60)	−3.94	−4.76	0.33
ACN	0.0019	0.0035[a]	−12.65	−11.13	0.61

[a] Data obtained from the Open Notebook Science Solubility Challenge as of November 25, 2014.[61]

much larger than k_BT, leading to significant differences in solubility. It is interesting to note that the highest discrepancy with respect to experimental data occurs in the case of the most polar solvent, pointing towards the lack of polarizability as a possible source of error.

Inserting these values in eqn (2) allows the reconstruction of the theoretical free energy profile for an infinite system at constant composition x_0 (see Table 1). Under these conditions, we find critical nuclei sizes (N^*) that far exceed the size of the simulation box for MeOH and EtOH (see Table 1), thus implying that in the simulations of these two cases we are observing only the first steps of the nucleation process. For ACN, on the other hand, the critical nucleus for an infinite system at $x = x_0$ is estimated to be around 108 molecules. The differences in nucleus size estimated for ACN with respect to MeOH and EtOH can be rationalised by analysing the dependence of N^* on concentration, as predicted by CNT. Under the hypothesis of a spherical nucleus, the critical size can be written as[5] $N^* = 32\pi v_0^2\sigma^3/\Delta\mu^3$, where v_0 is the molecular volume, σ is the surface tension, and $\Delta\mu$ is the difference in chemical potential between the solution and the crystal. This means that the critical nucleus size for crystals of the same substance scales as $N^* \propto (\sigma/\Delta\mu)$.[3] The term at the denominator can be straightforwardly expressed as: $\Delta\mu = k_BT\ln(x_0/x^*)$, assuming unitary activity coefficients. The estimate of surface tension σ, however, requires the formulation of a simplified model. Typically, theoretical expressions adopt the general functional form:[62–64] $\sigma \propto \ln(C_s/C_l)$, where C_s is the solute concentration in the solid and C_l is its concentration in solution. As C_s is constant by definition and C_l is very similar in the three simulations, σ is within the same order of magnitude for all three solvents. In agreement with the simulation results, the dependence of N^* on the solvent results is dominated by $\Delta\mu$, and therefore by the supersaturation ratio x_0/x^*.

3.2 The nucleation mechanism is solvent-dependent

Single step *vs.* two step nucleation. As mentioned in the previous paragraph, explicit solvent molecular simulations allow mechanistic information on the nucleation process to be obtained, helping to decrypt its inherent complexity. In this paragraph we analyse the WT metadynamics simulations, in order to estimate whether urea homogeneous nucleation is a single-step or a multiple step process in each of the solvents investigated. To this aim, we proceed as in ref. 33 for the nucleation of urea in aqueous solution, *i.e.* projecting the free energy surface obtained from metadynamics simulations in the space defined by variables n and n_o (see the Methods section). By definition, $n \geq n_o$, and thus the free energy surface assumes a triangular shape, with finite values only above the $n = n_o$ line. In order to account for finite size effects, the free energy obtained from WT metadynamics has been corrected with the term reported in eqn (3), calculated solely as a function of n_o. In other words, this means that we are considering only the crystal phase as a separate phase that depletes the mother solution. Thus molecules belonging to locally dense, disordered clusters are still considered as belonging to the liquid phase. This assumption is equivalent to considering disordered clusters as the result of local solute density fluctuations, rather than a separate metastable phase. In the (n, n_o) space a perfectly homogeneous solution is at the origin, while in the upper right corner states characterised by a large ordered clusters can be identified, and in the upper left corner large disordered clusters are represented. In this space, a typical single-step nucleation mechanism corresponds to an ideal path along the diagonal, where every constituent in the forming nucleus is crystal-like. In Fig. 3, the free energy surfaces $F(n_o, n)$ computed from WT metadynamics simulations of urea nucleation in methanol, ethanol, and acetonitrile are reported. For each of these surfaces, the most probable pathway connecting the homogeneous solution and the largest crystal-like cluster sampled during the simulation is reported. The pathway is constructed as the locus of the minima in $F(n)|_{n_o}$ the free energy profile in n at fixed n_o and plotted as a function of n_o. Each of the points belonging to such pathway represents the most probable state with respect to n, at a fixed n_o.

From a comparison of the most probable pathways in the (n_o, n) space in the case of MeOH and ACN it can be noted that in the former the evolution towards larger nucleus sizes proceeds following the diagonal, while in the latter the most probable pathway proceeds at first across states characterised by a low degree of order, and then evolves towards more ordered states. The mechanism of crystal nucleation in MeOH can thus be described as a single-step process, during which the crystal phase emerges from solution without intermediates: in this case, in fact, small crystal-like clusters are self-assembled from solution. It is important to note that in this case the formation of disordered clusters is still possible, however the nucleation of crystal-like phase within disordered clusters does not appear to be the most favourable pathway towards the formation of a cluster with a crystal-like structure. On the contrary, in acetonitrile the formation of an ordered phase within disordered clusters represents the most favourable pathway. In analogy with the mechanism of nucleation observed in water,[33] the most probable nucleation mechanism involves two steps: the formation of locally dense disordered urea clusters, and the subsequent nucleation of a crystal phase within the dense clusters. In Fig. 3, these two alternative mechanisms are illustrated with configurations extracted from the simulations in methanol and acetonitrile. It is interesting to note that the FES obtained in ethanol exhibits a competition between the two mechanisms; while the single-step process still seems to dominate, the most favourable pathway coincides less precisely with the diagonal. Moreover, a secondary transition pathway associated to the formation of crystalline domains within disordered clusters is also present.

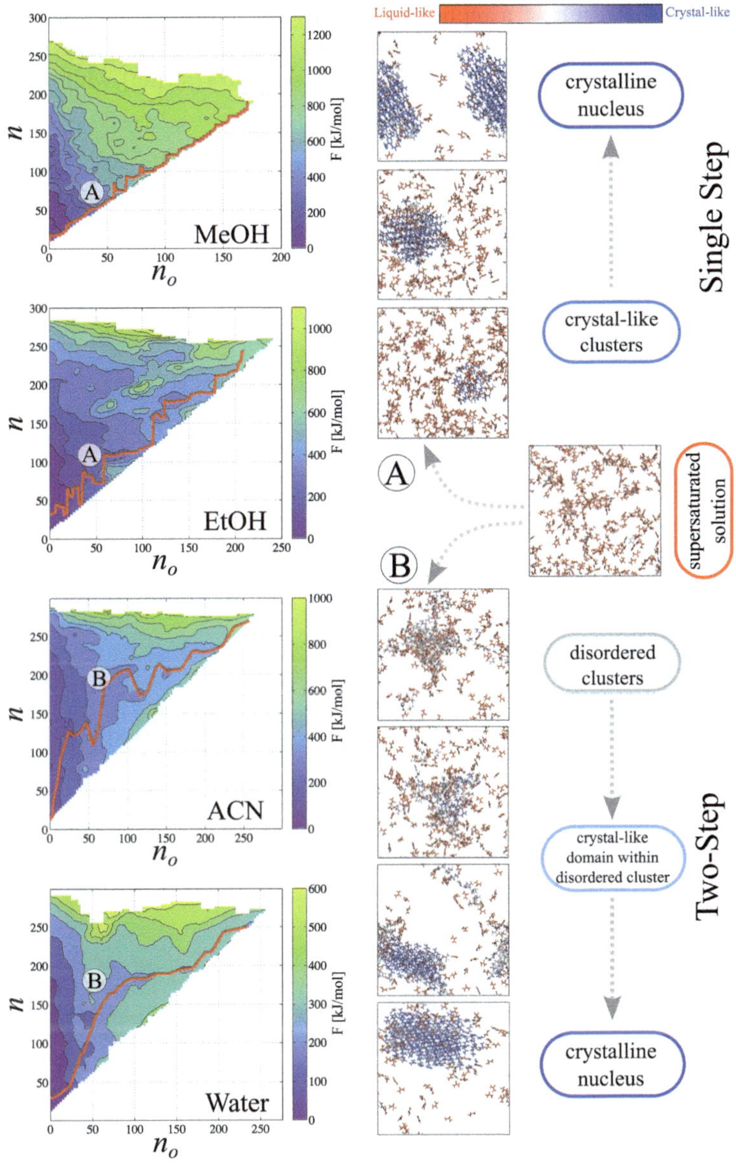

Fig. 3 $F(n_0, n)$ computed from the nucleation simulations in methanol, ethanol, acetonitrile and water (reported from ref. 33). It can be observed that the most favourable process follows a single-step mechanism in the case of methanol and ethanol (A), and a two step process in the case of acetonitrile and water (B). Differences in the two mechanisms are shown on the right, with configurations extracted from the trajectories obtained in methanol and acetonitrile, where urea molecules are coloured according to their respective degree of crystallinity expressed through their Γ_{CO} value, ranging from red for liquid-like molecules, to blue for crystal-like molecules. In the snapshots, periodic boundary conditions apply.

Crystal structure of the nuclei. In our analysis of urea nucleation, both from aqueous solution and from the melt, we have identified two polymorphic structures appearing in the early stages of the nucleation process.[29,33] The structures observed are the result of a replication in the three-dimensional space of the intermolecular interaction motifs displayed by *head-to-tail* and *cyclic* dimers.[65] The former gives rise to the experimental crystal structure (form I), while the latter

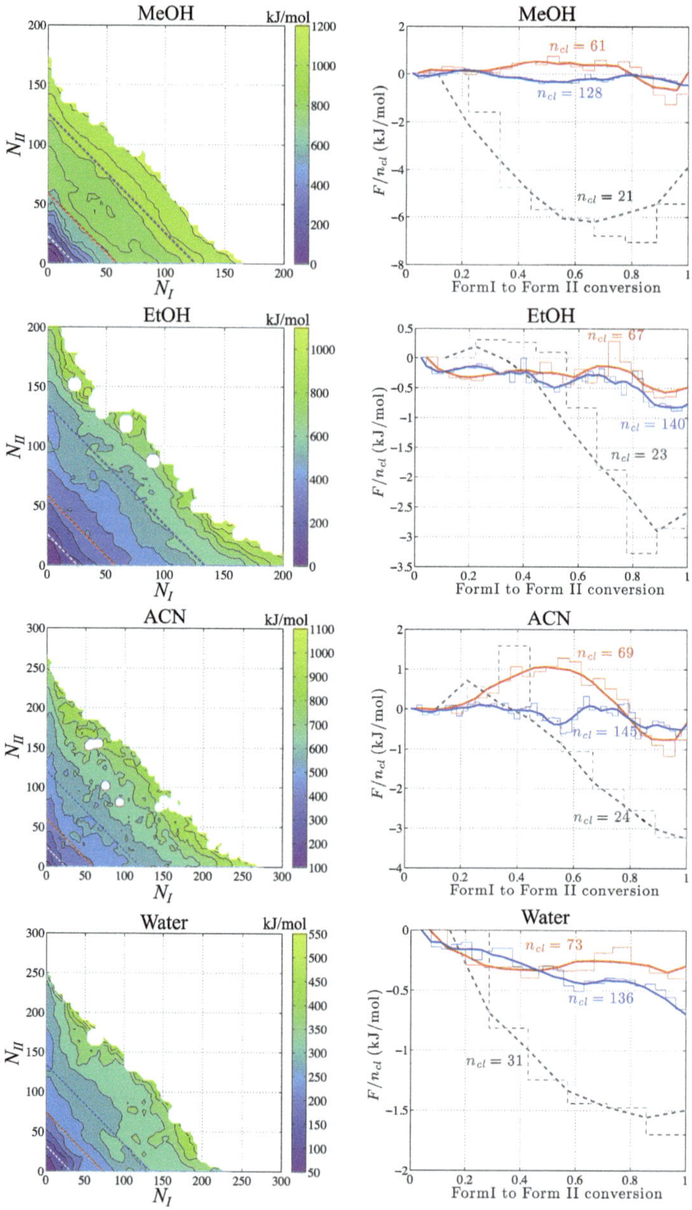

Fig. 4 Left column: Contour plots of the free energy surfaces obtained for urea nucleation in methanol, ethanol, acetonitrile, and water (reported from ref. 33) in the space defined by N_I and N_{II}, the number of molecules in the largest form I or form II cluster (see the Methods section for further details). Right column: Free energy profiles associated with the conversion of a crystal-like cluster from form I to form II, for three increasing cluster sizes. In each curve, the number of constituents n_{cl} is constant. Curves are extracted as cuts of the FES, reported on the right and normalized by the number of constituents.

produces a distorted structure (form II), in which the relative orientation between carbonyl bonds of neighbouring molecules is preserved, while the relative orientation of the axis joining the two nitrogen atoms is changed.[29,33] In the previous paragraph we focused our analysis on the nucleation pathway in relation to the development of crystalline clusters. In doing so, we considered nuclei belonging to both polymorphs as crystalline structures contributing to the value of n_{o}.

Now we go a step further and analyse the relative stability of the two polymorphs, as a function of the cluster size. As shown in our previous work on urea nucleation from the melt,[29] in the limit of an infinite crystal form I is the most stable polymorph in a temperature range extending from 300 K to the melting temperature. This means that the polymorphic transformation between form I and form II can be considered monotropic. However, for small finite-sized crystal nuclei, surface effects play an important role, and the relative stability between form I and form II clusters becomes a size-dependent property.

In the left column of Fig. 4, the FES in the (N_{I}, N_{II}) space is reported for urea nucleation in methanol, ethanol, acetonitrile, and water.[33] In the right column of Fig. 4, the free energy profiles associated to the conversion of a cluster of constant size from form I to form II are displayed. It can be observed that the nucleation of small form II clusters appears to be favoured with respect to form I clusters. The difference in the relative stability of the two forms decreases with an increase in the cluster size. This indicates that for larger clusters the surface energy contributions become gradually less important, and the relative stability of the two forms tends to the limit of an infinite crystal. This effect of the size-dependent stability of different polymorphs has also been recently hypothesised for the nucleation of NaCl in water solution,[30] and is compatible with the Ostwald empirical rule of stages[66] whereby the least stable polymorph is the first appearing during the nucleation process.

Let us now analyse the features of the free energy profile associated with the conversion between form I and form II clusters. In Fig. 4 it can be seen that, for large cluster sizes, the transition between form I and form II becomes a barrierless process. We interpret this feature as a consequence of the mechanism of

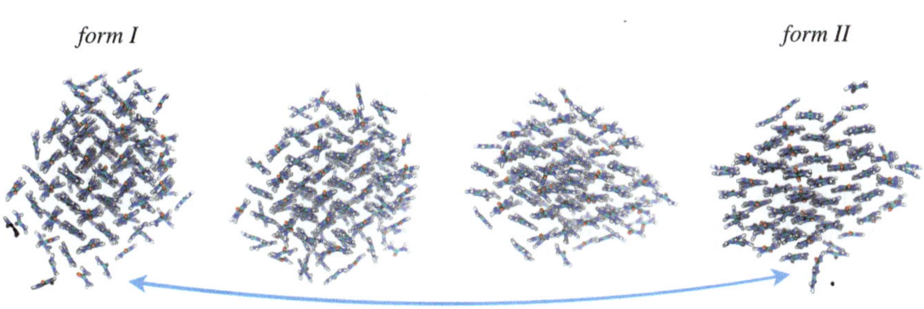

Fig. 5 Form I to form II polymorphic transitions observed in clusters obtained from simulations of urea nucleation in aqueous solution.[33] Intermediate structures visited during the polymorphic transition are shown in order, to highlight that the lattice distortion does not involve conformational changes of urea molecules, but simply a reorientation of urea chains within the crystal lattice. This transition mechanism is observed for all solvents investigated.

conversion between the two forms. Such a transition in fact involves a distortion of the crystal lattice, taking place without relevant conformational rearrangements of the solute molecules or a dissolution step, as shown in Fig. 5. The form I to form II transition is thus not a solvent-mediated process, and can take place due to the characteristics of the urea crystal bulk.[29,33] An analogous polymorph interconversion mechanism has been observed by Peters[26] in the case of charged colloids crystals. Also in this case the nature of the colloidal particles is such that conformational changes are not required and the polymorph transformation proceeds *via* a seemingly barrier-less distortion of the lattice, without needing a dissolution step.

Consistently with such an interpretation, the barrier-less polymorphic transformation of small urea clusters is observed consistently across all solvents considered in this study, thus confirming that the interconversion of crystal-like urea clusters from form II to form I is dominated by the bulk properties of the crystal.

4 Conclusions

In this work, taking advantage of the sampling of nucleation events made possible by WT metadynamics, we have investigated urea nucleation from methanol, ethanol and acetonitrile solutions. The main finding of our analysis concerns the effect of solvent on the nucleation mechanism. Whilst in acetonitrile a two-step nucleation process emerges as the most likely mechanism, in methanol and ethanol a classical, single-step process dominates. In acetonitrile the crystal phase is nucleated within amorphous urea clusters, whereas in methanol small crystal-like clusters are formed directly from solution.

On the contrary, the competition between urea polymorphs is confirmed to be present with similar features in all three solvents. In all three cases, metastable form II clusters are favored over form I clusters, for cluster sizes reaching up to approximately 50 molecules. For larger cluster sizes, the free energy difference between the two forms tends to progressively level off, with the transition between the two crystal forms becoming essentially barrier-less. These findings indicate that, unlike the nucleation mechanism which is heavily affected by the solvent, the polymorph interconversion is dominated by the bulk structure of the urea polymorphs.

To complete our analysis, we show that finite size effects associated with solution depletion are not solely a function of the initial composition (x_0) and system size ($\propto N_S$) but also crucially depend on solubility and surface tension and thus on solvent. These effects lead to significant differences in the shape of the free energy profiles calculated for systems prepared in similar x_0 and N_s conditions, but characterized by different solvents.

Acknowledgements

The authors acknowledge the computational resources provided by the Swiss Center for Scientific Computing (CSCS) and the Brutus Cluster at ETH Zurich. M.P. acknowledges the European Union grant ERC-2009-AdG-247075 and the NCCR MARVEL project for funding.

References

1 R. C. Snyder, S. Veesler and M. F. Doherty, *Cryst. Growth Des.*, 2008, **8**, 1100–1101.

2 N. Variankaval, A. S. Cote and M. F. Doherty, *AIChE J.*, 2008, **54**, 1682–1688.

3 I. Weissbuch, L. Addadi, M. Lahav and L. Leiserowitz, *Science*, 1991, **253**, 637–645.

4 I. Weissbuch, R. Popovitz-Biro, M. Lahav and L. Leiserowitz, *Acta Crystallogr., Sect. B: Struct. Sci.*, 1995, **51**, 115–148.

5 D. Kashchiev, *Nucleation, Basic Theory with Applications*, Elsevier, 2000.

6 J. P. Sizemore and M. F. Doherty, *Cryst. Growth Des.*, 2009, **9**, 2637–2645.

7 T. Vetter, M. Mazzotti and J. Brozio, *Cryst. Growth Des.*, 2011, **11**, 3813–3821.

8 M. Salvalaglio, T. Vetter, F. Giberti, M. Mazzotti and M. Parrinello, *J. Am. Chem. Soc.*, 2012, **134**, 17221–17233.

9 R. J. Davey, S. L. M. Schroeder and J. H. TerHorst, *Angew. Chem., Int. Ed.*, 2013, **52**, 2166–2179.

10 A. S. Myerson and B. L. Trout, *Science*, 2013, **341**, 855–856.

11 J. Anwar and D. Zahn, *Angew. Chem., Int. Ed.*, 2011, **50**, 1996–2013.

12 A. Barducci, G. Bussi and M. Parrinello, *Phys. Rev. Lett.*, 2008, **100**, 020603.

13 A. Laio and M. Parrinello, *Proc. Natl. Acad. Sci. U. S. A.*, 2002, **99**, 12562–12566.

14 P. G. Bolhuis, C. Dellago and D. Chandler, *Proc. Nat. Acad. Sci. U. S. A.*, 2000, **97**, 5877–5882.

15 A. F. Voter, *J. Chem. Phys.*, 1997, **106**, 4665–4677.

16 H. Grubmüller, *Phys. Rev. E: Stat. Phys., Plasmas, Fluids, Relat. Interdiscip. Top.*, 1995, **52**, 2893–2906.

17 T. Huber, A. E. Torda and W. F. van Gunsteren, *J. Comput.-Aided Mol. Des.*, 1994, **8**, 695–708.

18 R. J. Allen, C. Valeriani and P. R. ten Wolde, *J. Phys.: Condens. Matter*, 2009, **21**, 463102.

19 C. Snow, H. Nguyen, V. Pande and M. Gruebele, *Nature*, 2002, **420**, 102–106.

20 D. Hamelberg, C. A. F. de Oliveira and J. A. McCammon, *J. Chem. Phys.*, 2007, **127**, 155102.

21 A. Kawska, J. Brickmann, R. Kniep, O. Hochrein and D. Zahn, *J. Chem. Phys.*, 2006, **124**, 024513.

22 P. R. ten Wolde and D. Frenkel, *Science*, 1997, **277**, 1975–1978.

23 S. Auer and D. Frenkel, *Nature*, 2001, **409**, 1020–1023.

24 F. Trudu, D. Donadio and M. Parrinello, *Phys. Rev. Lett.*, 2006, **97**, 105701.

25 E. E. Santiso and B. L. Trout, *J. Chem. Phys.*, 2011, **134**, 064109.

26 B. Peters, *J. Chem. Phys.*, 2009, **131**, 244103.

27 G. A. Tribello, F. Bruneval, C. C. Liew and M. Parrinello, *J. Phys. Chem. B*, 2009, **113**, 11680–11687.

28 D. Zahn, *Phys. Rev. Lett.*, 2004, **92**, 040801.

29 F. Giberti, M. Salvalaglio, M. Mazzotti and M. Parrinello, *Chem. Eng. Sci.*, 2015, **121**, 51–59.

30 F. Giberti, G. Tribello and M. Parrinello, *J. Chem. Theory Comput.*, 2013, **9**, 2526–2530.

31 B. Knott, V. Molinero, M. Doherty and B. Peters, *J. Am. Chem. Soc.*, 2012, **134**, 19544–19547.

32 M. Salvalaglio, F. Giberti and M. Parrinello, *Acta Crystallogr., Sect. C: Cryst. Struct. Commun.*, 2014, **70**, 132–136.
33 M. Salvalaglio, C. Perego, F. Giberti, M. Mazzotti and M. Parrinello, *Proc. Natl. Acad. Sci. U. S. A.*, 2015, **112**, E6–E14.
34 P. Ectors, P. Duchstein and D. Zahn, *Cryst. Growth Des.*, 2014, **14**, 2972–2976.
35 J. F. Dama, M. Parrinello and G. A. Voth, *Phys. Rev. Lett.*, 2014, **112**, 240602.
36 M. Bonomi, A. Barducci and M. Parrinello, *J. Comput. Chem.*, 2009, **30**, 1615.
37 P. Tiwary and M. Parrinello, *J. Phys. Chem. B*, 2015, **119**, 736–742.
38 F. Giberti, M. Salvalaglio and M. Parrinello, *IUCrJ*, 2015, **2**, 256–266.
39 R. Docherty, K. Roberts, V. Saunders, S. Black and R. Davey, *Faraday Discuss.*, 1993, **95**, 11–25.
40 V. Bisker-Leib and M. Doherty, *Cryst. Growth Des.*, 2001, **1**, 455–461.
41 S. Piana, M. Reyhani and J. D. Gale, *Nature*, 2005, **438**, 70–73.
42 S. Piana and J. D. Gale, *J. Am. Chem. Soc.*, 2005, **127**, 1975–1982.
43 X. Y. Liu, E. S. Boek, W. J. Briels and P. Bennema, *Nature*, 1995, **374**, 342–345.
44 M. Salvalaglio, T. Vetter, M. Mazzotti and M. Parrinello, *Angew. Chem., Int. Ed.*, 2013, **52**, 13369–13372.
45 D. Reguera, R. K. Bowles, Y. Djikaev and H. Reiss, *J. Chem. Phys.*, 2003, **118**, 340–353.
46 J. Wedekind, D. Reguera and R. Strey, *J. Chem. Phys.*, 2006, **125**, 214505.
47 J. W. W. P. Schmelzer and A. S. Abyzov, *J. Non-Cryst. Solids*, 2014, **384**, 2.
48 R. Grossier and S. Veesler, *Cryst. Growth Des.*, 2009, **9**, 1917–1922.
49 A. Barducci, M. Bonomi and M. Parrinello, *Wiley Interdiscip. Rev.: Comput. Mol. Sci.*, 2011, **1**, 826–843.
50 M. Bonomi, D. Branduardi, G. Bussi, C. Camilloni, D. Provasi, P. Raiteri, D. Donadio, F. Marinelli, F. Pietrucci, R. A. Broglia and M. Parrinello, *Comput. Phys. Commun.*, 2009, **180**, 1961–1972.
51 G. C. Sosso, G. Miceli, S. Caravati, F. Giberti, J. Behler and M. Bernasconi, *J. Phys. Chem. Lett.*, 2013, **4**, 4241–4246.
52 W. Cornell, P. Cieplak, C. Bayly, I. Gould, K. Merz, D. Ferguson, D. Spellmeyer, T. Fox, J. Caldwell and P. Kollman, *J. Am. Chem. Soc.*, 1995, **117**, 5179–5197.
53 J. Wang, R. Wolf, J. Caldwell, P. Kollman and D. Case, *J. Comput. Chem.*, 2004, **25**, 1157–1174.
54 D. Van der Spoel, P. J. van Maaren and C. Caleman, *Bioinformatics*, 2012, **28**, 752–753.
55 G. Bussi, D. Donadio and M. Parrinello, *J. Chem. Phys.*, 2007, **126**, 014101.
56 M. Parrinello and A. Rahman, *J. Appl. Phys.*, 1981, **52**, 7182–7190.
57 B. Hess, C. Kutzner and E. Lindahl, *J. Chem. Theory Comput.*, 2008, **4**, 435–447.
58 T. Darden, D. York and L. Pedersen, *J. Chem. Phys.*, 1993, **98**, 1008910092.
59 B. Hess, H. Bekker, H. J. C. Berendsen and J. G. E. M. Fraaije, *J. Comput. Chem.*, 1997, **18**, 14631472.
60 F.-M. Lee and L. E. Lahti, *J. Chem. Eng. Data*, 1972, **17**, 304–306.
61 J.-C. Bradley, C. Neylon, R. Guha, A. Williams, B. Hooker, A. S. I. D. Lang, B. Freisen, T. Bohinski, D. Bulger, M. Federici, J. Hale, J. Mancinelli, K. B. Mirza, M. J. Moritz, D. Rein, C. Tchakounte and H. T. Truong, *Open notebook science challenge: solubilities of organic compounds in organic solvents*, ONS Books, 2010
62 A. E. Nielsen and O. Söhnel, *J. Cryst. Growth*, 1971, **11**, 233–242.
63 A. Mersmann, *J. Cryst. Growth*, 1990, **102**, 841–847.

64 J. Christoffersen, E. Rostrup and M. R. Christoffersen, *J. Cryst. Growth*, 1991, **113**, 599–605.
65 G. ÖzpInar, W. Peukert and T. Clark, *J. Mol. Model.*, 2010, **16**, 1427–1440.
66 R. A. Van Santen, *J. Phys. Chem.*, 1984, **88**, 5768–5769.

Faraday Discussions

PAPER

Influence of solvent on crystal nucleation of risperidone†

Donal Mealey,[a] Jacek Zeglinski,*[a] Dikshitkumar Khamar[a] and Åke C. Rasmuson*[ab]

Received 17th November 2014, Accepted 3rd February 2015
DOI: 10.1039/c4fd00223g

Over 2100 induction time experiments were carried out for the medium-sized, antipsychotic drug molecule, risperidone in seven different organic solvents. To reach the same induction time the required driving force increases in the order: cumene, toluene, acetone, ethyl acetate, methanol, propanol, and butanol, which reasonably well correlates to the interfacial energies as determined within classical nucleation theory. FTIR spectroscopy has been used to investigate any shifts in the spectra and to estimate the interaction of solute and solvent at the corresponding site. The solution condition has also been investigated by Density Functional Theory (DFT) calculations over (1 : 1) solvent–solute binding interactions at 8 different sites on the risperidone molecule. The DFT computational results agree with the spectroscopic data suggesting that these methods do capture the binding strength of solvent molecules to the risperidone molecule. The difficulty of nucleation correlates reasonably to the DFT computations and the spectroscopic measurements. The results of the different measurements suggest that the stronger the solvent binds to the risperidone molecule in solution, the slower the nucleation becomes.

1. Introduction

Crystal nucleation from solution is a key step in crystallization processes and solution crystallization is widely used in the pharmaceutical, food, chemical and agrochemical industries. Crystal nucleation is influenced by the solvent and within the Classical Nucleation Theory (CNT) framework this is captured in the interfacial energy[1,2] and in the pre-exponential factor.[3] The influence on the kinetics of nucleation is assumed to be partly related to an influence on the type and the size of molecular assemblies (clusters) in the solution. Solution structuring of solute molecules has been investigated by various spectroscopic

[a]*Materials and Surface Science Institute, Chemical and Environmental Science Department, University of Limerick, Limerick, Ireland. E-mail: ake.rasmuson@ul.ie; jacek.zeglinski@ul.ie; Tel: +353 61 23 46 17; +353 61 23 41 57*

[b]*Department of Chemical Engineering and Technology, KTH Royal Institute of Technology, Stockholm, Sweden*

† Electronic supplementary information (ESI) available. See DOI: 10.1039/c4fd00223g

methods (IR,[4] Raman,[5] NMR,[6] UV-visible[7]) to establish a link with the crystal structure of the nucleating solid phase. In some cases such a link could be established and in others cases not.[8,9] In addition to the type of aggregates[6,10] in the solutions, the type of solvent–solute interaction can also dictate the morphology[11] as well as the kinetics of nucleation.[12] Lohani *et al.*[13] showed the type of solvent–solute interaction affecting the diffusion coefficient, resulting in the preferential crystallization of one form over another. The hydrogen bonding capabilities of a solvent and the nature of interaction prevailing between solute and solvent resulting in the nucleation of a particular form has also been a subject of a number of studies.[5,14]

Application of the CNT framework on experimentally measured nucleation rate data helps to determine the nucleation work and the interfacial energy in different solvents. The influence of solvent on the ease or difficulty of nucleation can be used to probe the nucleation process on the molecular level. Our previous work[12] on a small molecule, salicylic acid (138.12 g mol^{-1}) reveals the role of the solvent in the nucleation of this compound. By employing solution calorimetry, vibrational spectroscopy and quantum-chemical calculations over one-to-one solvent–solute molecule interactions as well as over the complete first solvation shell, it was consistently observed that the stronger the solvent binds to salicylic acid, the greater is the difficulty of nucleation. On a similar note, the effect of solvent was also recently highlighted by Sullivan *et al.*[15] in which the authors suggest the process of desolvation affects the kinetics of nucleation and thus governs the nucleation process. Zhao *et al.*[16] studied the effect of solvent on the crystallization of nesquehonite ($MgCO_3 \cdot 2H_2O$). They reported faster nucleation kinetics with the addition of dimethyl formamide (DMF) to the water which they attributed to the kinetic factors accelerating the cluster aggregation in a non-classical mechanism.

In spite of the critical importance of the choice of solvent in solution crystallization, there is only a limited amount of nucleation rate data available for medium- to large-sized organic molecules in a range of organic solvents. Much of the previously published work is carried out either in one solvent only or for small-sized molecules only. The size and complexity of the solute molecule can certainly add complications to an already complex nucleation process. In this work, we use a medium sized (410.49 g mol^{-1}), fairly complex and conformationally flexible pharmaceutical molecule (API), risperidone, and use the influence of the solvent as a probe to investigate the mechanisms of nucleation. The results are analysed directly without the use of the CNT framework as well as within that framework. The strength of solvent–solute interaction is quantified using DFT calculations and is examined by solution spectroscopy for particular

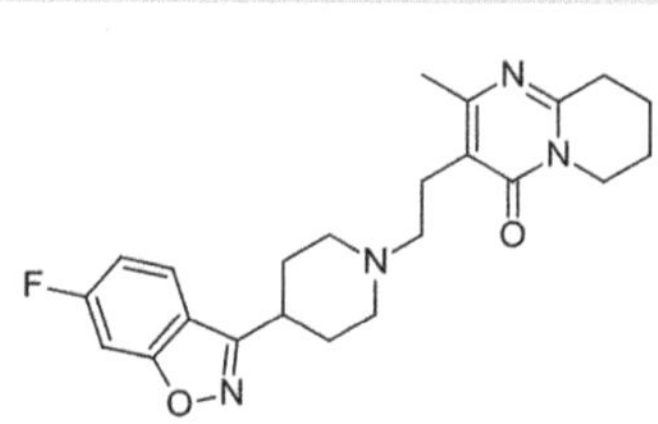

Fig. 1 Molecular structure of risperidone.

interaction sites. The molecular structure of risperidone is shown in Fig. 1. There are two reported polymorphs of risperidone and both structures, form I (WASTEP) and form II (WASTEP01) are reported in the Cambridge Structure Database.

2. Materials and methods

2.1. Materials

Risperidone (>99.95%), form I, was kindly provided by Janssen Pharmaceuticals, Ireland, and used without further purification. All solvents were purchased from VWR International; methanol (99.8% purity by volume), 1-propanol (99.8%), 1-butanol (99.8%), ethyl acetate (99.8%), acetone (99.8%), toluene (99.8%), and cumene (99.8%).

2.2. Induction time measurement

The induction time of nucleation of risperidone was measured in seven solvent systems with a saturation temperature of 50 °C. Over 2160 nucleation events were observed at 27 different conditions.

Stock solutions were prepared in 500 ml sealed glass bottles by weighing appropriate amounts of risperidone and solvent according to previously published solubility data.[9] An equilibration period of 12 h at 55 °C was allowed, with agitation provided with a PTFE (polytetrafluoroethylene) coated magnetic stirrer (38 × 10 mm, 3 × 11 mm pivot ring) at 500 rpm. Solutions were dispensed *via* preheated syringes and 0.2 μm solvent compatible filters (PTFE or Nylon), into twenty 20 ml glass vials (70 × 20 mm, VWR International). A magnetic stirring bar (13 × 3 mm) was added to each vial prior to sealing with a plastic screw cap with a PTFE seal. Solutions were then subjected to a second equilibration period at 55 °C overnight before the crystallization experiments were performed.

An experimental apparatus identical to that described by Mealey *et al.*[17] was employed to simultaneously monitor 20 individual nucleation experiments. Solutions were transferred from bath A at 55 °C to bath B held at a temperature where the solutions were supersaturated (T_{supersat}). Agitation was provided at 400 rpm *via* a 60 point submersible magnetic stirring plate (2Mag). Immediately after the transfer, the time to nucleate (temperature equilibration time + isothermal nucleation time) was measured with a high definition video camera observing the transition from a clear to cloudy solution. The approach to temperature equilibrium was measured separately with a control tube of pure solvent and an *in situ* calibrated temperature probe (Dostmann P600). Once all tubes had nucleated they were transferred back to water bath A where complete dissolution occurred and equilibration was allowed overnight. The cycle was repeated four times until approximately 80 nucleation events were recorded for a given driving force and solvent.

To avoid the possible effects of degradation of the risperidone molecule each batch of 20 tubes were subjected to just four temperature cycles. It was observed that upon extended time at elevated temperature an obvious colour change would occur in the solutions. A new batch of tubes was prepared for each separate driving force. Along with monitoring solutions visually, after four cycles solutions were spotted onto silica gel thin layer chromatography (TLC) plates using methanol as a mobile phase and viewed under UV light to ensure there were no detectable impurities.

Solid samples were isolated by filtration, moments after nucleation for both an early nucleation and a late nucleation event, at both high and low driving forces. The polymorphic form was determined by reflectance powder X-ray diffraction (PXRD, Philips PANalytical X'Pert MP PRO) in the range 5–35° (2θ) with a wavelength of 1.54 Å (40 kV, 35 mA). In all cases the solid phase analysed was pure form I. The theoretical PXRD patterns generated from the crystallographic information files for both form I and form II along with the samples taken from risperidone nucleation in toluene are shown in the ESI.†

2.3. Computational methods

Density Functional Theory (DFT) calculations have been applied to investigate (1 : 1) solvent–solute pair interactions. For this purpose, a risperidone molecule has been extracted from the crystal lattice of its form I (WASTEP) and optimized in isolation at the B97-D3/6-31G(d,p) level. The molecular geometry changed only slightly upon gas-phase optimization, preserving the original crystal-like conformation (*cf.* Fig. S4, ESI†). The seven solvents used in our experimental work can be divided into three groups: polar protic (alcohols), polar aprotic (acetone and ethyl acetate), and nonpolar (toluene and cumene). Three solvents (one from each group), representing different polarity and H-bond capability, were chosen for the modelling part, *i.e.* 1-propanol, acetone, and toluene. These solvents posses different bonding mechanisms. 1-Propanol is both a H-bond donor and acceptor, acetone is only capable of accepting H-bonds, while toluene is not capable of making strong H-bonds, although it can interact through π-electrons located over its benzene ring. The (1 : 1) solvent–solute interactions are probed at eight sites of the risperidone molecule chosen with aid of the electrostatic potential map (EPM) as shown in Fig. 2. The EPM indicates higher electron deficiency (dark blue colour) at the hydrogens labelled H_1, H_2, and H_3, while the electron rich regions, having H-bond accepting capabilities, are highlighted in red/yellow at the heteroatoms O_1, O_2, and N_1–N_3 (Fig. S2, ESI,† shows more projections of the EPM). Based on the electron distribution, one can expect that in the crystal, the higher electron deficiency at the ring 1 hydrogens will be complemented by neighbouring electron rich polar atoms resulting in relatively strong intermolecular bonding. This can indeed be observed in the risperidone crystal structure (*cf.* Fig. 2b) and thus validates the electrostatic potential map presented.

The considered eight sites represent all the important binding features of the risperidone molecule such as interaction at (i) polar sites containing O and N heteroatoms (H-bond acceptors; sites 3, 5, and 7), (ii) non-polar C–H-terminated sites 1 and 8, and (iii) the connected rings 1–2 (site 4) and 4–5 (site 6). In addition, for all the seven solvents we have modelled the (1 : 1) risperidone–solvent associates at site 5; this site combines two electron-rich atoms, *i.e.* carbonyl oxygen and tertiary amine nitrogen, and thus it is expected to yield strong interactions. Another reason for choosing site 5 was the ability of comparing the calculated solvent–solute binding energies of the risperidone carbonyl group across all the seven solvents with the respective experimentally determined C=O shifts in the FTIR spectrum.

The equilibrium geometries of the (1 : 1) dimers are calculated with a B97-D3 Grimme's functional,[18] and a Gaussian-type 6-31G(d,p) basis set.[19] Binding energies in the dimers are calculated for the lowest energy configurations, after

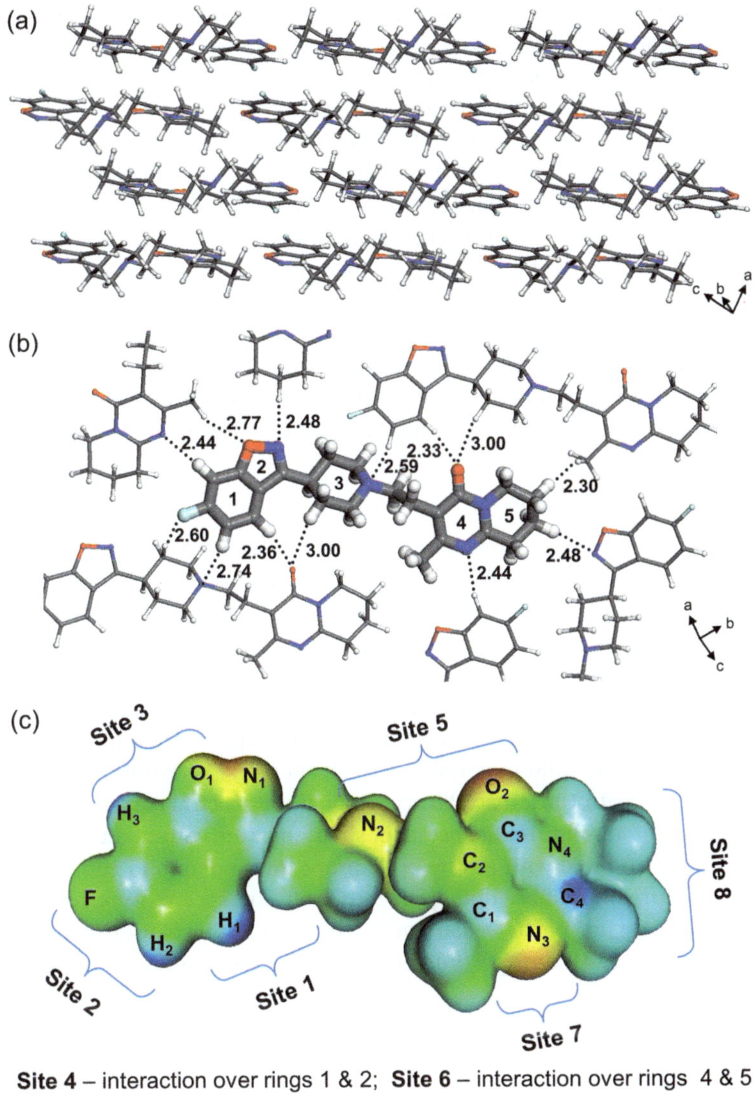

Fig. 2 Molecular arrangement in the risperidone crystal (a); chief interactions and intermolecular distances (in Angstroms) between molecules neighbouring in a monolayer (b); electrostatic potential isosurface and definition of interaction sites in risperidone molecule for probing solvent–solute bonding (c) (blue – positive, red – negative, green – neutral potential).

probing a set of sensible molecular orientations, using a double hybrid B2PLYP-D3 functional,[20] which combines exact Hartree–Fock exchange with an MP2-like correlation and long-range dispersion corrections; here we use a basis set of quadruple-ζ valence quality (def2-QZVP).[21] The binding energy is calculated as follows:

$$\Delta E_{\mathrm{bind}} = E_{\mathrm{A-B}} - (E_{\mathrm{A}} + E_{\mathrm{B}}) \qquad (1)$$

where E_{A-B} is the energy of a dimer and E_A and E_B are the energies of isolated monomers A and B in fully relaxed gas phase geometries.

Conformational analysis of the risperidone molecule has been carried out at the DFT B97-D3/6-31G(d,p)//B2PLYP-D3/def2-QZVP level (gas-phase, 0 K), with energy barriers to rotation calculated for four rotational centres (single bonds) of the risperidone molecule (Fig. S5, ESI†). We restricted our calculations to electronic energies only (corrections for zero point energy and other vibrational contributions are not included). A relaxed potential energy surface (PES) scan over 360°, with the step interval of 5°, has been performed for the respective dihedral angles. Four new low energy conformers were identified upon PES scanning and their stabilities have been compared relative to the relaxed crystal-like conformation.

Calculations were performed using the GAUSSIAN 09 package.[22] Electrostatic potential maps were generated with the MOLEKEL 5.4 software.[23]

2.4. Spectroscopic methods

Solution spectroscopy was carried out for risperidone in all the seven solvents that were used for nucleation work. Raman spectra were collected using Kaiser Raman Rxn2 analyser with an Invictus 785 nm excitation laser. For liquid samples, a short focus immersion probe of 1/4 inch diameter was used and the solid samples were analysed using a non-contact optic probe. Each spectrum was collected for a minimum of 15 s exposure time and five accumulations in the spectral region of 3400–200 cm^{-1} using Mettler Toledo iC Raman software version 4.1.

IR spectrum of the solid material was collected using Perkin Elmer spectrum 100 spectrometer with Universal ATR accessory in the spectral region of 4000–650 cm^{-1} with 4 cm^{-1} resolution. IR spectra of solutions were collected using Mettler Toledo ReactIR 10 fitted with a fibre optic probe and a diamond composite tip. Due to the nature of the fibre optic probe, the solution spectra were collected in the fingerprint region of 2000–650 cm^{-1} with 4 cm^{-1} resolution using iC IR software version 4.3. All the spectral data were collected at ambient temperature (20–22 °C).

3. Results

3.1. Rate of nucleation in different solvents

Induction time distributions from toluene (a) and ethyl acetate (b) are shown in Fig. 3.

The use of various distribution functions to examine induction time data has previously been examined.[17,24] Here we have used the lognormal cumulative distribution function (LCDF) and the exponential-based function of Jiang and ter Horst.[24] Both distributions provide reasonably similar fits to the data, and results are collected in Table 1. The calculated nucleation rates using both methods display similar trends with driving force. Fitting the LCDF produced coefficients of determination greater than 0.97 in all instances while the fits of the exponential function were similar with one exception resulting in a coefficient of determination of 0.94.

The fitting to the LCDF includes no assumption of a particular mechanism of nucleation. By examining the relation of the induction time distributions to the

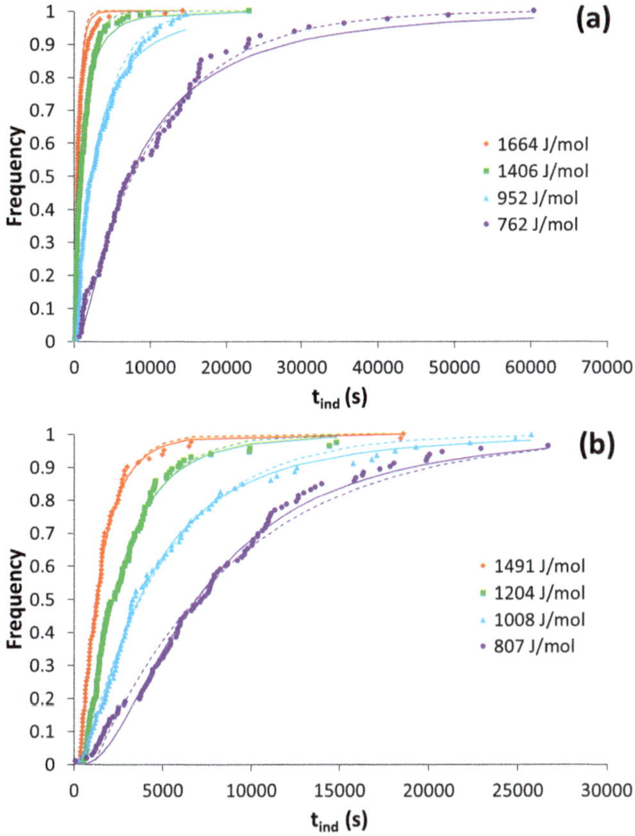

Fig. 3 Induction time distributions from toluene (a) and ethyl acetate (b) at various driving forces. Also shown is the fit of the LCDF (solid lines) and the exponential distribution function[24] (dashed lines).

driving force in Fig. 4 across the various solvents, it is obvious that nucleation is faster in non-polar (toluene and cumene) and polar aprotic (acetone and ethyl acetate) solvents than in the alcohols (1-propanol, methanol, and butanol). To reach equal median induction time, in the range where data is available for all solvents, the driving force required increases in the order: cumene, toluene, acetone, ethyl acetate, methanol, propanol and butanol (Table 2).

Based on the geometric mean induction times reported in Table 1, Fig. 5 presents the standard plot for evaluation of the induction time data within the classical nucleation theory. By relating the induction time (t_{ind}) to the nucleation rate (J) it is possible to estimate the pre-exponential factor (A) and the interfacial energy (φ) using eqn (2) and (3):

$$\frac{1}{t_{\text{ind}}V} = J = A\exp\left(\frac{-\Delta G_{\text{crit}}}{RT}\right) \tag{2}$$

$$\Delta G_{\text{crit}} = \frac{16\pi N_{\text{A}}\varphi^3 \upsilon^2}{3(RT\ln S)^2} \tag{3}$$

Table 1 Parameters describing induction time distributions of risperidone in various solvents from fitting the lognormal distribution function (geometric mean (η^*), geometric standard deviation (σ^*), calculated nucleation rate (J), and size of the critical nucleus (N^*), or exponential based function[24] (nucleation rate (J^a))

Solvent	$RT \ln S$ [J mol^{-1}]	η^* (s) $\times/\sigma^*$	J (m^{-3} s^{-1})	N^*	J^a (m^{-3} s^{-1})
Cumene	757	4682 ×/3.75	10.68	24	6.67
	940	1343 ×/1.93	37.25	13	43.48
	1120	732 ×/1.83	68.29	7	87.22
Toluene	762	7397 ×/2.84	6.76	23	4.78
	952	2286 ×/3.24	21.87	12	14.28
	1406	886 ×/3.15	56.44	4	38.00
	1664	489 ×/2.19	102.28	2	105.12
Acetone	526	18374 ×/3.22	2.72	78	1.99
	858	5528 ×/2.36	9.05	18	7.43
	1074	2472 ×/2.02	20.23	9	18.25
	1287	1036 ×/2.17	48.28	5	46.65
Ethyl acetate	807	6914 ×/2.21	7.23	15	6.05
	1008	3666 ×/2.55	13.64	8	10.78
	1204	2247 ×/2.20	22.25	5	21.16
	1491	1265 ×/2.17	39.51	2	37.46
Methanol	1338	3166 ×/2.07	15.79	9	14.66
	1553	2289 ×/2.68	21.85	6	16.42
	1897	1514 ×/3.45	33.02	3	20.20
	2224	673 ×/2.41	74.31	2	64.33
1-Propanol	1072	16549 ×/3.52	3.02	19	1.81
	1632	3216 ×/3.43	15.55	5	9.54
	2013	1172 ×/2.76	42.65	3	31.99
	2378	654 ×/2.39	76.48	2	68.52
Butanol	1377	4362 ×/2.86	11.46	6	8.41
	1654	2963 ×/2.51	16.88	4	13.84
	2025	2262 ×/2.52	22.10	2	18.97
	2370	1239 ×/2.01	40.37	1	42.53

where N_A is the Avogadro number, R is the gas constant, T is temperature, ΔG_{crit} is the nucleation work assuming a spherical nucleus, V is the volume of the solution, υ is the molar volume of risperidone in the cluster, and $RT \ln S$ is the supersaturation driving force where S is the supersaturation ratio in terms of the mole fraction. The solid liquid interfacial energies and the preexponential factors are reported in Table 3.

No direct relation can be found between the interfacial energy in the different solvents and the solid–liquid solubility, regardless of units of the latter which agrees with our previous findings.[3,12,25] There is neither a clear correlation with the Hansen solubility parameter nor with the viscosity of the solvent.

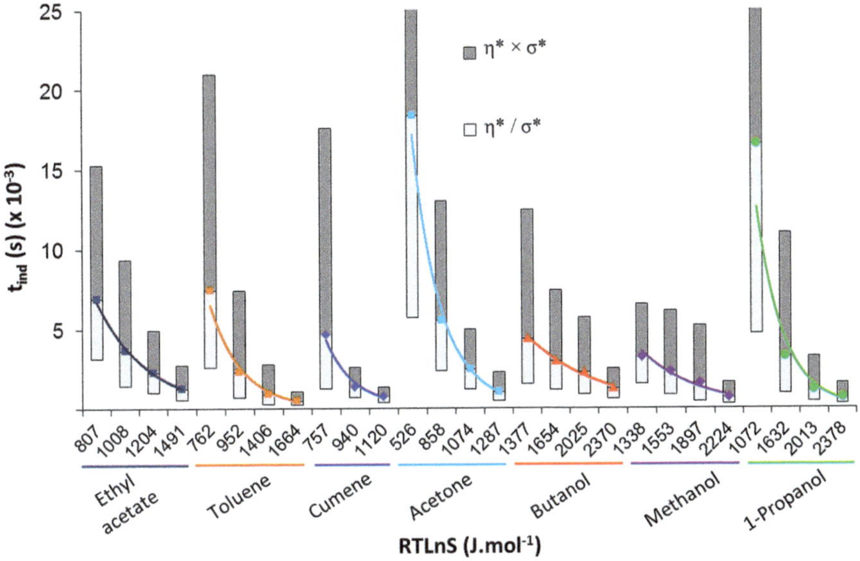

Fig. 4 Induction time distributions. Data points represent the geometric mean (η^*) of the LCDF while the shaded regions represent one geometric standard deviation (σ^*) from this mean.

3.2. DFT calculations

By running the potential energy surface scan over the dihedral angles of the four single bonds that serve as rotational centres, we have found that the relevant energy barriers to interconversion between the different conformational isomers range from 10 kJ mol^{-1} to 26 kJ mol^{-1} (ESI†). There is one new isomer identified for each rotational centre being structurally different from those constituting known conformations in the crystal structure of risperidone form I (and form II). Three of the new conformers are linear and the fourth is U-shaped. Two of the linear conformers are slightly more stable (lower in energy by 2–3 kJ mol^{-1}), while the U-shaped molecule is higher in energy by 6.3 kJ mol^{-1} as compared to the

Table 2 A quantified measure of the ease of nucleation of risperidone in 7 different solvents

Solvent	Driving force (RT ln S, J mol^{-1}) required to reach median induction time 2000 s	3000 s	Driving force order
Cumene	854	784	1
Toluene	1015	841	2
Acetone	1055	954	3
Ethyl acetate	1288	1112	4
Methanol	1639	1370	5
Propanol	1829	1657	6
Butanol	1976	1671	7

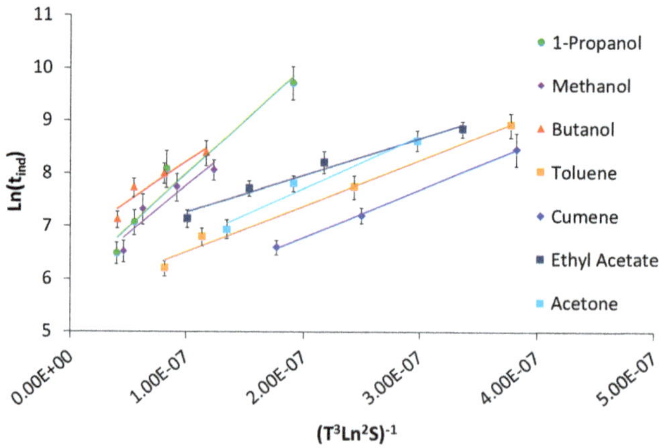

Fig. 5 Log geometric mean induction times plotted against driving force to determine the interfacial energy and preexponential factor within classical nucleation theory.

crystal-like starting conformer; this suggests the least favourable geometry. It can be assumed that only kinetically stable conformers, *i.e.* those separated by higher energy barriers, could significantly affect the nucleation process. The relatively low rotational barriers along with small energy differences of the three gas-phase linear conformations *vs.* the crystal structure conformations indicate absence of one kinetically stable conformation that could be unambiguously chosen as a model conformation to study interactions of risperidone with solvent molecules. For this reason, the linear conformation appearing in the crystal structure of risperidone is used consistently in our modelling work. The calculated low energy barriers to rotation indicate high conformational flexibility of the risperidone molecule and suggest a variety of different conformations are present in solution.

The relatively large and complex risperidone molecule can possess a variety of binding configurations with solvent molecules. Here we consider eight binding sites, where the interactions at the risperidone molecule are probed with three

Table 3 Experimental solid–liquid interfacial energy and preexponential factor of risperidone nucleation in the different solvents. Also given is the solubility as mole fraction (MF) and mol l^{-1} at 25 °C, as well as the Hansen solubility parameter χ_{12} and the solvent viscosity

Solvent	Driving force order from Table 2	Interfacial energy (mJ m^{-2})	Pre-exponential factor, A (m^{-3} s^{-1})	Solubility at 25 °C (MF × 10^3)	Solubility at 25 °C (mol l^{-1})	χ_{12}	Solvent viscosity (mPa S)
Cumene	1	1.72	348	2.60	18.68	2.05	0.78
Toluene	2	1.70	181	6.67	63.17	1.43	0.59
Acetone	3	1.77	161	1.75	23.79	1.18	0.31
Ethyl acetate	4	1.58	71	2.16	22.01	1.32	0.426
Methanol	5	2.18	134	3.96	98.12	1.44	0.59
1-Propanol	6	2.25	129	3.05	40.86	1.35	1.94
Butanol	7	2.04	61	3.47	38.01	1.46	2.54

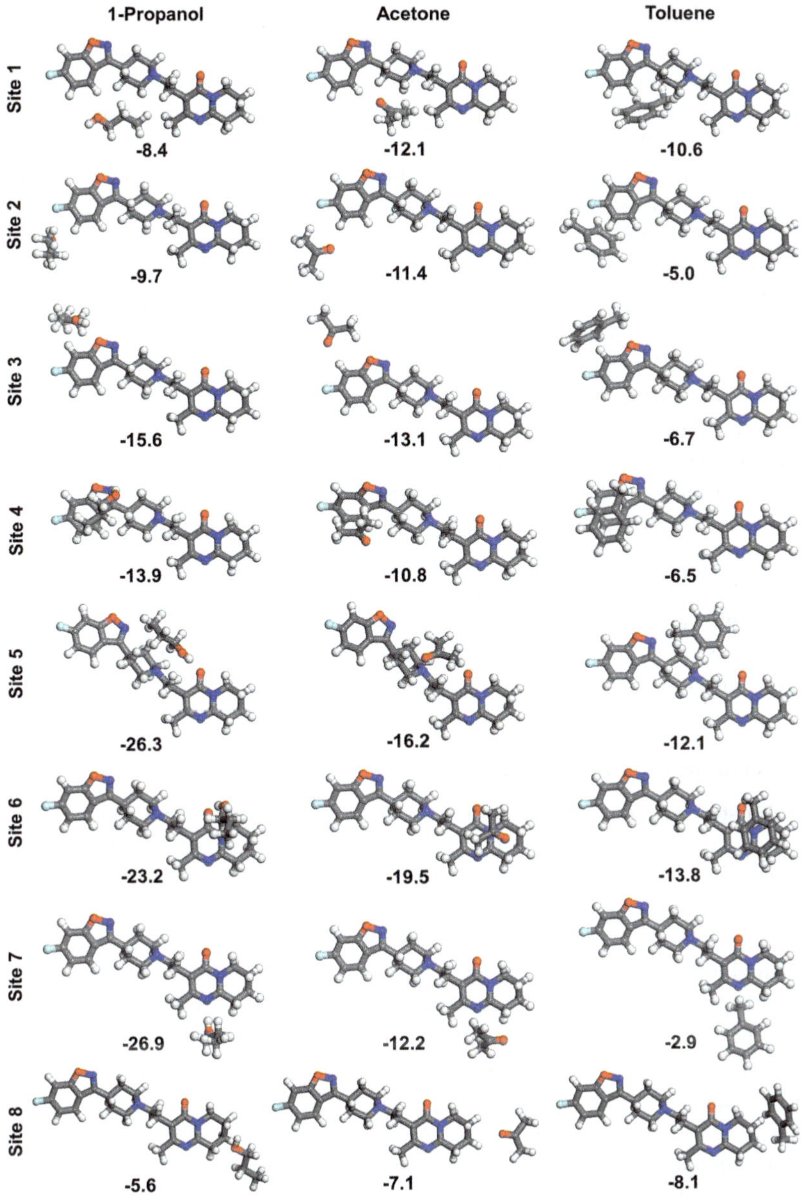

Fig. 6 Optimized geometry of risperidone–1-propanol, risperidone–acetone, and risperidone–toluene (1 : 1) molecular associates (B97-D3/6-31G(d,p) level). Binding energy in kJ mol^{-1}, calculated at B2PLYP-D3/def2-QZVP level. Hydrogen – white, carbon – grey, oxygen – red, nitrogen – blue, fluorine – pale blue.

representative solvents, *i.e.* 1-propanol, acetone and toluene. The DFT binding energy is calculated for each of the 24 solvent–solute (1 : 1) heterodimers (Fig. 6). One solvent molecule can interact with a number of atoms of the risperidone molecule. In case of site 1 and site 8, only C–H hydrogens are involved in bonding. The other sites feature a combined interaction of hydrogens and heteroatoms.

Our calculations indicate that 1-propanol does not establish strong bonding at the nonpolar sites 1 and 8 of risperidone, being up to 30% weaker than the respective bonding to acetone and toluene. However, the propanol molecule binds stronger at sites 3–7, where it forms H-bonds with polar heteroatoms of the solute molecule. The strongest binding of -26.9 kJ mol^{-1} and -26.3 kJ mol^{-1} is observed for site 7 and 5, respectively. The interaction at site 2, which combines binding to hydrogen and fluorine ring atoms, is relatively weak and decreases in order: acetone > 1-propanol > toluene. Acetone and toluene molecules yield the strongest binding of -19.5 kJ mol^{-1} and -13.8 kJ mol^{-1}, respectively, when placed over rings 4–5 (site 6). The solvent–risperidone (1 : 1) binding energies averaged over the eight sites form an order: 1-propanol (-16.2 kJ mol^{-1}) > acetone (-12.8 kJ mol^{-1}) > toluene (-8.2 kJ mol^{-1}).

When comparing all the eight interaction sites it appears that site 5 is amongst the strongest interacting sites. This site provides a set of polar groups, including carbonyl oxygen and tertiary amine nitrogen, which result in strong binding with a solvent molecule. In order to compare strength of the risperidone–solvent interaction for all the seven solvents, the site 5 was chosen, and the respective binding energies are shown in Fig. 7.

The binding energies calculated at site 5 form three distinct groups of comparable values: (i) alcohols, yielding very similar and the strongest binding interactions, (ii) acetone and ethyl acetate, with moderate binding, and (iii) cumene and toluene, being relatively weakly bound to the risperidone molecule.

3.3. Solution spectroscopy

Solid state spectra. IR and Raman spectra of risperidone form I are shown in Fig. 8. In the fingerprint region carbonyl stretching appears as a strong peak at 1643 cm^{-1} in the IR spectra, whereas in the Raman spectra it is only of moderate intensity. In the crystal structure of form I the carbonyl is attached to three different C–H sites through weaker interaction and the interaction at the carbonyl site roughly corresponds to the interaction at site 5 of the risperidone molecule (Fig. 2). Apart from the dominant carbonyl peak, the fingerprint region shows a strong peak at 1534 cm^{-1} corresponding to mainly $\nu(C_4{=}N_3,\ C_1{=}C_2)$[26] in the pyrimidine ring (ring 4) and also some contribution from $\nu(C_1{-}N_3)$ of the same ring (*cf.* Fig. 2). Functional groups responsible for this vibration frequency correspond to site 7 as shown in Fig. 2.

Solution spectra. For solution spectroscopy of this compound, IR spectroscopy was found to be more suitable than Raman spectroscopy due to the low solubility in the different solvents leading to a weak solute signal in the Raman spectra. The biggest difference in the solution spectra was observed for the carbonyl peak. In case of alcoholic solutions this peak is marginally shifted to 1648 cm^{-1} followed by higher shifts in acetone and further shifts in ethyl acetate, toluene and cumene (Fig. 9). The shift in the carbonyl frequency follows the polarity order[27] of the solvents: alcohols > acetone > ethyl acetate > methyl benzenes (toluene and cumene). The carbonyl stretching frequency, $\nu(C{=}O)$ is sensitive to hydrogen bonding and other solvent–solute interactions due to the effect of interacting species on the double bond characteristics of the carbonyl bond. The $\nu(C{=}O)$, has often been used as a probe to study the solvent–solute interaction[12,28,29] and generally the stronger the solvent interacts, the lower the frequency at which it

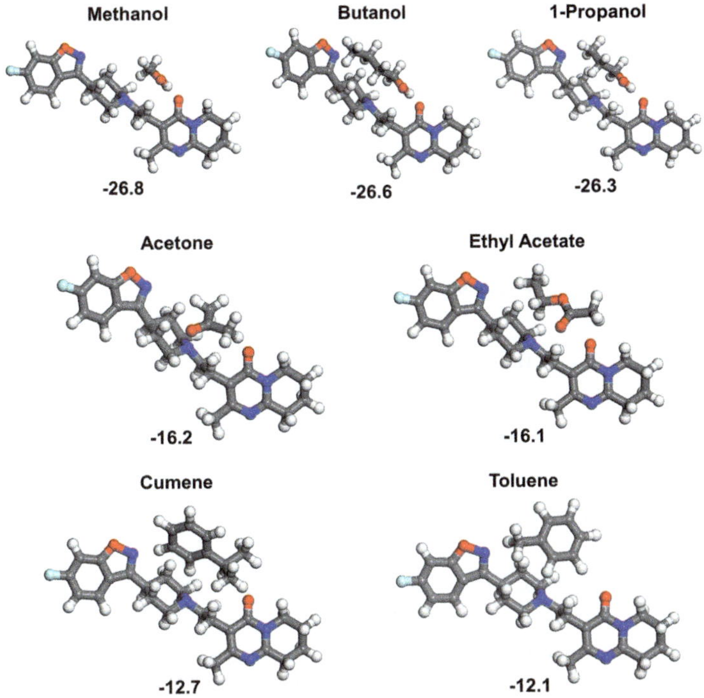

Fig. 7 Risperidone–solvent (1 : 1) associates at site 5, with all the seven solvents used in the experiment (B97-D3/6-31G(d,p) level). Binding energy in kJ mol^{-1}, calculated at B2PLYP-D3/def2-QZVP level.

absorbs. The shifts of the carbonyl frequency suggest that at site 5 of the risperidone molecule (Fig. 2) the alcohols are strongly bound followed by acetone and ethyl acetate, and the methyl benzenes (toluene and cumene) are relatively weakly interacting at this site.

Similar to the carbonyl peak, shifts were also observed for the peak (1534 cm^{-1}) corresponding to site 7. This peak at 1534 cm^{-1} (in solid material) moves to somewhat higher frequency in alcoholic solutions (1536 cm^{-1}) followed by slightly higher frequency in acetone solution (1538 cm^{-1}) and further higher frequency (1540 cm^{-1}) in toluene, cumene and ethyl acetate solutions. The

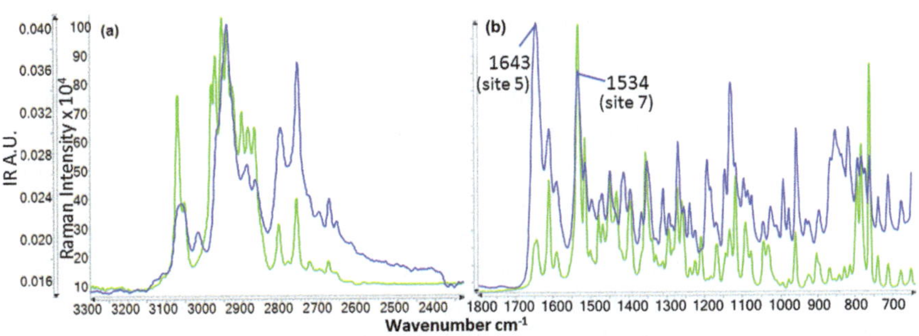

Fig. 8 IR (blue) and Raman (green) spectrum of risperidone Form I (solid).

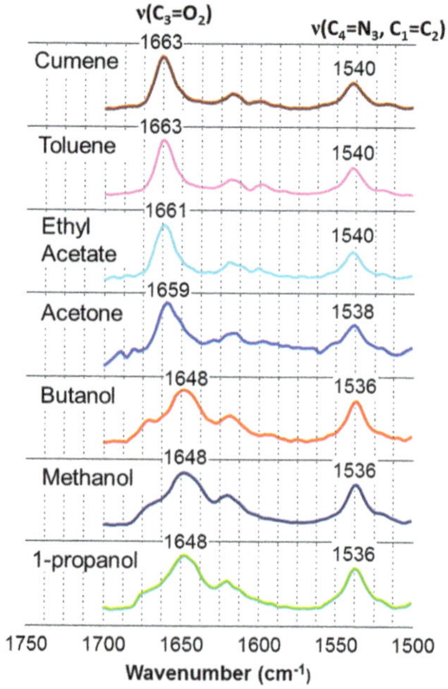

Fig. 9 IR spectra of risperidone showing the peak shift observed in different solvents. The solvent spectrum is subtracted from the solution spectra.

relative shift difference was smaller than that observed for the carbonyl peak which is expected considering a higher force constant for carbonyl stretching frequency over the frequency responsible for the peak at 1534 cm^{-1} in IR. The concentration effect on the spectra was also analysed for toluene, 1-propanol, methanol and acetone solutions: there was no clear difference in the spectra due to the different concentrations. Details of concentration and the wavenumber accuracy are shown in the ESI† file.

4. Discussion

The induction time experiments reveal that in order to reach the same induction time the required driving force increases in the order: cumene < toluene < acetone < ethyl acetate < methanol < propanol < butanol. We will refer to this order as increasing difficulty of nucleation, and point out that this observation is directly extracted from the experimental work without assumption of a particular theory. This work further shows that if the data are processed within the framework of the classical nucleation theory the obtained interfacial energies increase in a similar but not identical order. Overall, the difference in interfacial energy of risperidone depending on the solvent is rather limited, spanning from 1.58 to 2.25 mJ m^{-2} for the seven solvents investigated. In the alcohols the value is clearly higher than in the other four solvents, but within these two groups of solvents the differences are relatively small. With respect to the experimental uncertainties we do not see a big discrepancy between the driving force order and the interfacial energy order even though we do want to point out the difference with respect to ethyl acetate in particular. Furthermore, since the nucleation rate in the classical nucleation

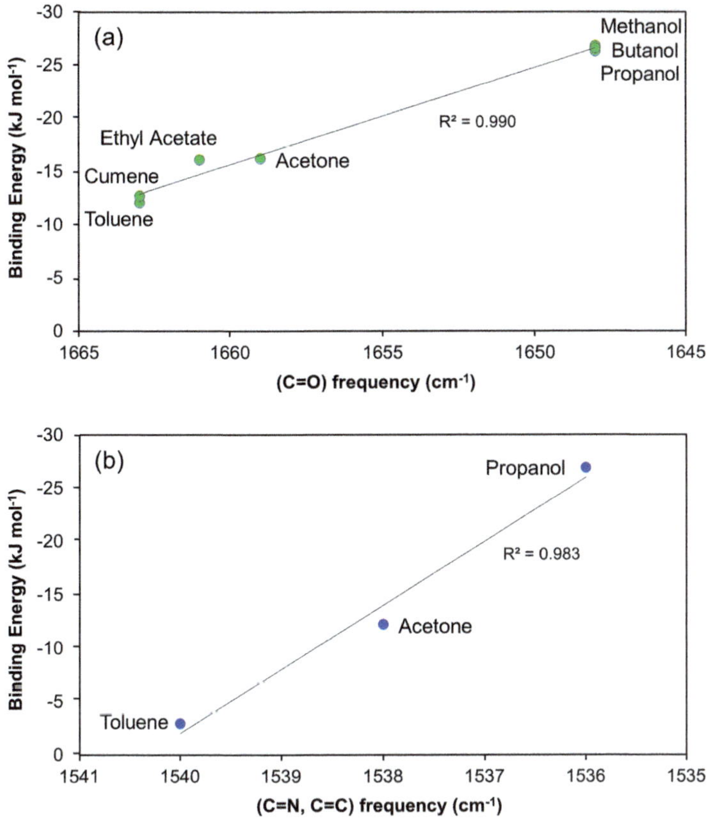

Fig. 10 Cross-validation of solvent–risperidone interactions: DFT binding energy *vs.* FTIR vibrational frequency at (a) carbonyl ν(C=O) (site 5) for all the 7 solvents and (b) ν(C=N + C=C) in ring 4 (site 7) for selected 3 solvents.

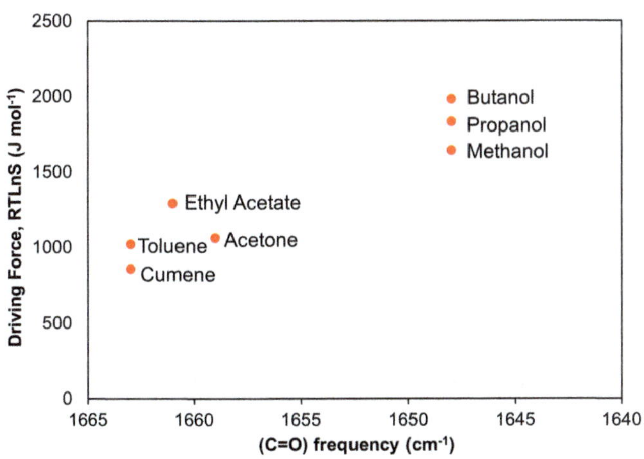

Fig. 11 Relationship between driving force required to reach a median induction time of 2000 s and the FTIR vibrational frequency of the risperidone carbonyl group, site 5.

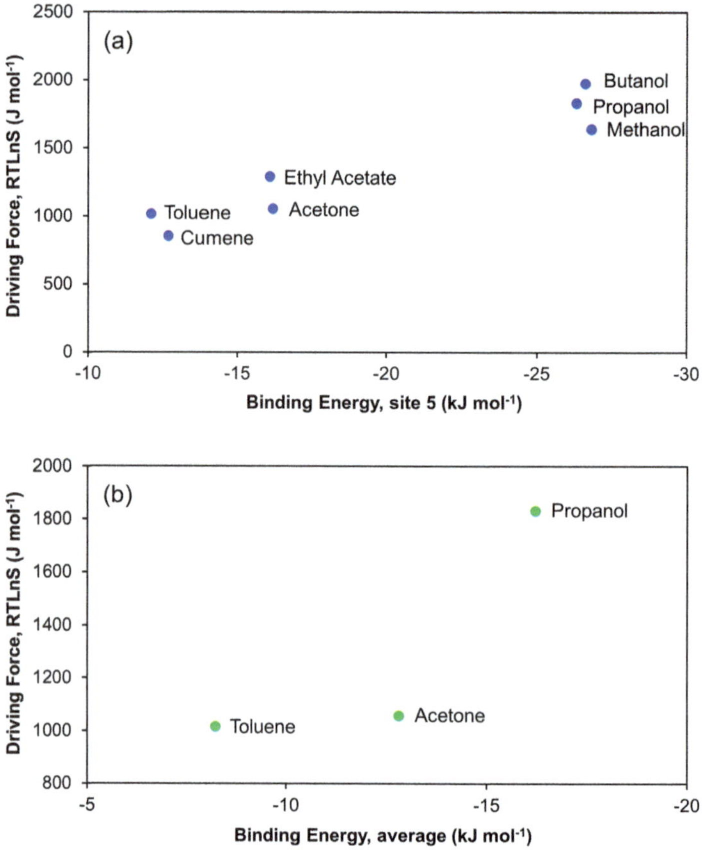

Fig. 12 Relationship between driving force at median induction time of 2000 s and DFT binding energy: (a) for seven solvents at site 5, and (b) averaged over eight interaction sites considering three representative solvents.

theory depends on the preexponential factor as well as the nucleation work, the order of nucleation according to the driving force scale doesn't necessarily have to correlate with the interfacial energy order.

Risperidone is a relatively large molecule (410.49 g mol^{-1}) that possesses various functional groups of different polarity, serving as sites for intermolecular interactions. Because of this DFT simulation of the complete first solvation shell[10] becomes too demanding and we settled for simulating a number of one-to-one molecule solvent–solute pairs. In Fig. 10 the spectroscopic results are compared with the DFT-computed solvent–risperidone binding energies at the carbonyl group (Fig. 10a) site 5 and the pyrimidine ring (Fig. 10b) site 7. Obviously, there is a good correlation in the data from the two methods: with increasing binding energy the carbonyl peak and the pyrimidine peak both shift towards lower frequencies. The results of both methods suggest that the binding to the non-polar solvents cumene and toluene, is quite weak, is somewhat stronger to the aprotic solvents ethyl acetate and acetone, and clearly stronger than the alcohols. Since the carbonyl site shows a relatively larger shift in solution spectra, as compared to the (C=N, C=C) vibrations of ring 4, the carbonyl group appears a relatively more sensitive probe for investigating the solvent–solute interactions in solution.

In Fig. 11 we examine the correlation between the driving force required to reach a median induction time of 2000 s and the FTIR carbonyl peak frequency. The overall trend is that the required driving force increases with decreasing peak frequency suggesting that the crystal nucleation of risperidone becomes gradually more difficult the stronger the solvent binds to the carbonyl site. The correlation rather clearly distinguishes between the non-polar solvents, the aprotic solvents and especially the alcohols. It must be noted though that the correlation in Fig. 11 shows strength of solvent–solute interaction only at one site. However a similar correlation could be observed for the shift in the pyrimidine ring vibrations affecting site 7 as well. Unfortunately, in the solution IR spectroscopy only certain sites of interactions can be probed, and the resolution is somewhat limited for solvents bearing similar structural features such as the group of alcohols and the pair of toluene and cumene. This points out certain limitations of FTIR spectroscopy in such work.

Fig. 12a shows, as expected, that the nucleation driving force correlates nicely with the DFT computed (1 : 1) binding energy to the carbonyl group at site 5. At the sites where the DFT binding energies are the strongest, *i.e.* sites 5, 6, and 7, we also observe a correlation between binding energy and the ease of nucleation for the three considered solvents. It is noteworthy that, for five sites out of the eight considered, the binding energies follow the order of difficulty of

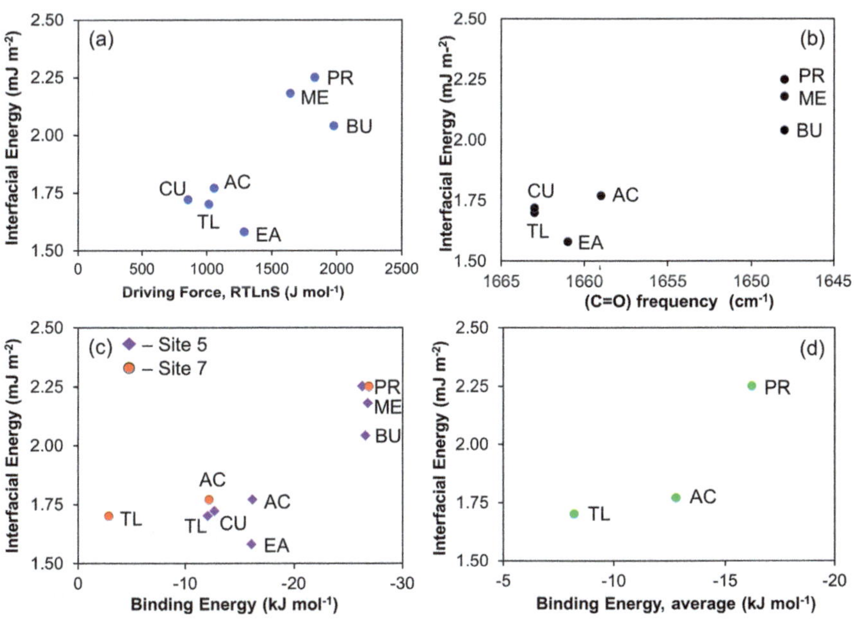

Fig. 13 Relationship between interfacial energy and (a) driving force at median induction time of 2000 s, (b) FTIR vibrational frequency of risperidone carbonyl group, (c) DFT binding energy of solvent–solute at site 5 (purple diamonds) and site 7 (red dots), and (d) DFT binding energy of solvent–solute as averaged over eight interaction sites. Cumene (CU), toluene (TL), acetone (AC), ethyl acetate (EA), methanol (ME), propanol (PR), and butanol (BU).

nucleation. The three sites where there is no such a correlation are those where the solvent–solute interactions are relatively weak, as they feature either relatively non-polar C–H hydrogens (site 1 and 8) or a combined interaction of hydrogen and fluorine (site 2). For the three solvents where we have DFT-computed binding energies to all 8 sites, we have calculated a mean interaction energy, and the correlation to the nucleation driving force is shown in Fig. 12b. Based on these three solvents only the correlation is not overly impressive, since the nucleation driving force required in acetone for 2000 s induction time is almost the same as for toluene. It is noteworthy though that at 3000 s induction time (Table 2), there is a somewhat more clear difference in driving force between toluene and acetone.

Since often the nucleation propensity is characterised by the interfacial energy only, for completeness we also show the corresponding correlations in Fig. 13. Fig. 13a shows the correlation between the interfacial energy and the driving force required to induce the same induction time. Overall, the difference between the alcohols and the other solvents is distinguished but the more detailed order differs, and in particular a discrepancy is observed in case of ethyl acetate, where the lowest interfacial energy of 1.58 mJ m^{-2} is calculated, despite the driving force being somewhat in the mid-range (*cf.* Fig. 4 and Table 2). The correlations of interfacial energy with carbonyl frequency (Fig. 13b), the DFT calculated binding energies at sites 5 and 7 (Fig. 13c), and the average binding energy (Fig. 13d), are similar to that shown in Fig. 13a. Overall, we observe that the strength of solvent–solute interaction affects the nucleation process, such that the stronger the solvent binds to the solute, the greater the difficulty of nucleation becomes. This very much follows the observations made for salicylic acid,[13] further supporting the hypothesis that desolvation of the solvent molecules during the process of formation of crystal nuclei can be a key step in explaining the influence of the solvent on the nucleation process.

5. Conclusions

The nucleation of risperidone depends on the solvent. To reach the same induction time the required driving force increases in the order: cumene, toluene, acetone, ethyl acetate, methanol, propanol, and butanol. This order overall corresponds to the order of increasing interfacial energy as determined within the classical nucleation theory, but there are deviations in the detailed order. The frequency of the FTIR spectroscopy peaks corresponding to the carbonyl and pyrimidine group vibrations decreases in the polarity order of the three solvent groups. DFT (1 : 1) solvent–solute binding energies depend significantly on the binding site for all three solvents investigated. For the carbonyl site the binding energy increases essentially in the same order as the order of increasing driving force. A good correlation between spectroscopic data and DFT simulations provides validation for the relevance of these techniques in characterising solvent–solute interactions. The results of this work support the hypothesis that the influence of the solvent on the crystal nucleation is related to the desolvation of the risperidone molecules. The stronger the solvent binds to the risperidone molecule in solution, the more difficult the desolvation and the slower the nucleation becomes.

Acknowledgements

The support of the Science Foundation Ireland (10/IN.1/B3038) is gratefully acknowledged. J. Zeglinski acknowledges the Science Foundation Ireland (SFI) and Higher Education Authority funded Irish Centre for High End Computing (ICHEC) for access to computational facilities, and Å. C. Rasmuson the support of the Swedish Research Council (621-2010-5391).

References

1 R. A. Granberg, C. Ducreux, S. Gracin and Å. C. Rasmuson, *Chem. Eng. Sci.*, 2001, **56**, 2305–2313.
2 S. Teychene and B. Biscans, *Cryst. Growth Des.*, 2008, **8**, 1133–1139.
3 M. Svärd and Å. C. Rasmuson, *Cryst. Growth Des.*, 2013, **13**, 1140–1152.
4 R. J. Davey, G. Dent, R. K. Mughal and S. Parveen, *Cryst. Growth Des.*, 2006, **6**, 1788–1796.
5 S. A. Kulkarni, E. S. McGarrity, H. Meekes and J. H. ter Horst, *Chem. Commun.*, 2012, **48**, 4983.
6 C. A. Hunter, J. F. McCabe and A. Spitaleri, *CrystEngComm*, 2012, **14**, 7115.
7 R. J. Davey, N. Blagden, S. Righini, H. Alison, M. J. Quayle and S. Fuller, *Cryst. Growth Des.*, 2001, **1**, 59–65.
8 C. S. Towler and L. S. Taylor, *Cryst. Growth Des.*, 2007, **7**, 633–638.
9 R. J. Davey, S. L. Schroeder and J. H. ter Horst, *Angew. Chem., Int. Ed. Engl.*, 2013, **52**, 2166–2179.
10 A. Mattei, X. Mei, A. Miller and T. Li, *Cryst. Growth Des.*, 2013, **13**, 3303–3307.
11 M. Lahav and L. Leiserowitz, *Cryst. Growth Des.*, 2006, **6**, 619–624.
12 D. Khamar, J. Zeglinski, D. Mealey and Å. C. Rasmuson, *J. Am. Chem. Soc.*, 2014, **136**, 11664–11673.
13 S. Lohani, I. V. Nesmelova, R. Suryanarayanan and D. J. W. Grant, *Cryst. Growth Des.*, 2011, **11**, 2368–2378.
14 W. Du, Q. Yin, J. Gong, Y. Bao, X. Zhang, X. Sun, S. Ding, C. Xie, M. Zhang and H. Hao, *Cryst. Growth Des.*, 2014, **14**, 4519–4525.
15 R. A. Sullivan, R. J. Davey, G. Sadiq, G. Dent, K. R. Back, J. H. ter Horst, D. Toroz and R. B. Hammond, *Cryst. Growth Des.*, 2014, **14**, 2689–2696.
16 L. Zhao, C. Zhu, J. Ji, J. Chen and H. H. Teng, *Geochim. Cosmochim. Acta*, 2013, **106**, 192–202.
17 D. Mealey, D. M. Croker and Å. C. Rasmuson, *CrystEngComm*, 2015, DOI: 10.1039/C4CE01428F.
18 S. Grimme, S. Ehrlich and L. Goerigk, *J. Comput. Chem.*, 2011, **32**, 1456–1465.
19 V. A. Rassolov, M. A. Ratner, J. A. Pople, P. C. Redfern and L. A. Curtiss, *J. Comput. Chem.*, 2001, **22**, 976–984.
20 L. Goerigk and S. Grimme, *J. Chem. Theory Comput.*, 2011, **7**, 291–309.
21 F. Weigend and R. Ahlrichs, *Phys. Chem. Chem. Phys.*, 2005, **7**, 3297–3305.
22 M. J. Frisch, G. W. Trucks, H. B. Schlegel, G. E. Scuseria, M. A. Robb, J. R. Cheeseman, *et al.*, *Gaussian 09, Revision D.01*, Gaussian Inc., Wallingford, CT, 2013, 24.
23 S. Portmann and H. P. Luthi, *Chimia*, 2000, **54**, 766–770.
24 S. Jiang and J. H. ter Horst, *Cryst. Growth Des.*, 2011, **11**, 256–261.
25 H. Y. Yang and Å. C. Rasmuson, *Cryst. Growth Des.*, 2013, **13**, 4226–4238.

26 A. Alparone, *Spectrochim. Acta, Part A*, 2011, **81**, 631–639.
27 C. Reichardt and T. Welton, *Solvents and Solvent Effect in Organic Chemistry*, Wiley-VCH Verlag GmbH & Co. KGaA, 2011, pp. 425–490.
28 R. A. Nyquist, in *Interpreting Infrared, Raman, and Nuclear Magnetic Resonance Spectra*, ed. R. A. Nyquist, Academic Press, San Diego, 2001, pp. 165–203.
29 S. I. Nam, E. S. Min, Y. M. Jung and M. S. Lee, *Bull. Korean Chem. Soc.*, 2001, **22**, 989–993.

Faraday Discussions

PAPER

A design equation for low dosage additives that accelerate nucleation

Geoffrey G. Poon,[a] Stefan Seritan[a] and Baron Peters*[ab]

Received 18th November 2014, Accepted 7th January 2015
DOI: 10.1039/c4fd00226a

Additives are used to control nucleation in many natural and industrial environments. However, the mechanisms by which additives inhibit or accelerate solute precipitate nucleation are not well understood. We propose an equation that predicts changes in nucleation barriers based on the adsorption properties and concentrations of trace additives. The equation shows that nucleant efficacy depends on the product of an adsorption equilibrium constant and the reduction in interfacial tension. Moreover, the two factors that determine the potency of additives are related to each other, suggesting that assays of just one property might facilitate additive design. We test the design equation for a Potts lattice gas model with surfactant-like additives in addition to solutes and solvents.

1 Introduction

Metastable phases of matter include liquids that can be supercooled without freezing, vapors that can be supercooled without condensing, and solids that can exist as multiple polymorphs. Some metastable phases can survive for a long time, because the more stable phase cannot form without first surmounting an activation barrier for nucleation. Nucleation is a stochastic process that forms the first growing embryo of the more stable phase. Nucleation and growth from single-component systems have been extensively studied, but there has been relatively little theoretical and computational work on solute precipitate nucleation. Even fewer studies have focused on nucleation in the presence of additives.

Additives are important in nucleation and growth in many natural and industrial processes, including ice formation,[1–5] pharmaceutical crystallization,[6–8] biomineralization,[9–16] and material synthesis.[17–20] For example, pharmaceutical companies use nucleants, growth promoters, and inhibitors to drive the selective crystallization of the desired polymorph.[21–24] Oil and gas companies invest millions of dollars in inhibitors to prevent methane hydrates from clogging pipelines.[25–27] Others have investigated additives to accelerate gas hydrate formation, for gas storage at moderate temperatures and pressures.[28–31] In

[a]Department of Chemical Engineering, University of California, Santa Barbara, California 93106, USA. E-mail: baronp@engineering.ucsb.edu
[b]Department of Chemistry, University of California, Santa Barbara, California 93106, USA

biology, salts, metabolytes, and proteins prevent the formation of urinary stones[32,33] and ice in Antarctic fish.[34–36] Therefore, there is substantial scientific and industrial interest in designing additives for controlling nucleation and growth. Additives are currently discovered by trial-and-error experiments rather than focused searches guided by physical models.

Classical nucleation theory (CNT) predicts that nucleation rates are primarily dependent on two competing factors: (1) the bulk chemical potential difference between the stable and metastable phases, and (2) the interfacial free energy between the two phases.[37,38] The first factor is the nucleation driving force and is often written in terms of supersaturation (*i.e.* $\Delta\mu = k_BT \ln S$). The second factor, interfacial free energy, is directly responsible for the nucleation barrier. Although the nucleation barrier can be adjusted by modulating supersaturation, we are primarily interested in additives that selectively bind to the nucleus and lower the interfacial free energy. Surface-active additives can potentially have a strong effect on nucleation at trace concentrations.

Most experimental and theoretical research on trace additives focuses on crystal growth, not nucleation. For example, Weissbuch and coworkers have experimentally studied "tailor-made" stereospecific promoters and inhibitors that drive the selective crystallization of polymorphs.[39,40] Ward and coworkers studied molecular inhibitors that bind stereospecifically to growing crystal faces and suppress growth.[33,41] Storr *et al.*[26] and Anderson *et al.*[25] used molecular simulations to compare the binding energies of hydrate growth inhibitors. All of these works investigate the effect of additives on large crystal surfaces, instead of the microscopic clusters present during the nucleation stage.

A few recent theoretical studies investigated the effect of additives on nucleation, but these did not consider the interfacial and solubility contributions to the free energy barrier separately. Anwar *et al.* used short unbiased molecular simulations of a ternary Leonard Jones (LJ) system (*i.e.* solute–solvent–additive) to compare nucleation rates for different additive interaction and size parameters.[42] Duff *et al.* developed an alchemical transformation approach to compare the effects of NaCl on the interfacial free energy of nuclei for two polymorphs of glycine.[43] However, Duff *et al.* did not compute the effect of salt on the solubility of glycine in solution, which is a separate and important effect of adding salt.

This paper proposes a theoretically-motivated additive design equation that applies to low dosage nucleation promoters that strongly interact with the solution–precipitate interface. The equation is motivated by (1) the barrier's strong dependence on interfacial tension, (2) the drop in interfacial tension with increasing additive interfacial coverage, and (3) the increase in coverage with increasing additive concentration. It predicts a drop in barrier height proportional to the trace additive concentration. The proportionality constant is the product of two measurable quantities: an equilibrium constant for adsorption and the reduction in interfacial tension per unit coverage. These two quantities are related to each other, so a method to quickly determine either property for a set of trial additives could help design potent nucleation promoters.

2 Theory

CNT predicts that the reversible work to grow a nucleus from $n = 1$ building units to a size n is:[37,38,44–47]

$$F(n) = -n\Delta\mu + \gamma a\phi n^{2/3} \tag{1}$$

where the free energy $F(n)$ depends on the chemical potential difference between the metastable and stable phases ($\Delta\mu$), the interfacial tension (γ), the area per adsorption site (a), and a shape factor (ϕ) such that $N_{\text{site}} = \phi n^{2/3}$ is the number of adsorption sites on the interface. The qualitative dependence of free energy on cluster size in eqn (1) is supported by many previous simulations.[48–57]

According to eqn (1), the free energy barrier scales with the cube of the interfacial tension:

$$F^{\ddagger} = \frac{4}{27}\left[\frac{(\gamma a\phi)^3}{\Delta\mu^2}\right] \tag{2}$$

where $F^{\ddagger}$ is the work required to create a critical nucleus of size $n^{\ddagger} = (2\gamma a\phi/3\Delta\mu)^3$ at which $\partial F/\partial n = 0$. While the free energy barrier is strongly dependent on the driving force, large loadings of additives would generally be required to alter the driving force. By contrast, additives that selectively interact with the interface can lower the interfacial tension, and therefore promote nucleation at low concentrations. CNT predicts that the change in the free energy barrier due to a change in the interfacial tension when the driving force is held constant is:

$$\frac{F^{\ddagger} - F_0^{\ddagger}}{F_0^{\ddagger}} = \left(\frac{\gamma}{\gamma_0}\right)^3 - 1 \tag{3}$$

where the subscript 0 indicates the reference value with no additives present.

In the simplest case, we suppose that additive adsorption follows the Langmuir isotherm:

$$\theta = \frac{Kx}{1 + Kx} \tag{4}$$

where θ is the fraction of adsorption sites occupied by additives, K is the Langmuir constant, and x is the additive concentration.[58,59] Langmuir type adsorption assumes localized adsorption onto uniform surface sites and no interaction between adsorbed molecules, which are both reasonable assumptions for dilute adsorbates. In the dilute limit (*i.e.* $Kx \ll 1$), eqn (4) is simplified to a relationship resembling Henry's law:[58,59]

$$\theta \approx Kx \tag{5}$$

where K is now just the equilibrium constant of adsorption.

The change in interfacial energy upon addition of additives, $\gamma - \gamma_0$, can be related to additive coverage on the interface θ, as depicted in the thermodynamic cycle of Fig. 1:

$$[\gamma(\theta) - \gamma_0]a = \theta\Delta F^{\circ}_{\text{ads}} \tag{6}$$

where $\Delta F^{\circ}_{\text{ads}}$ is the average interfacial free energy change for adsorption at the reference state and θ is the fraction of adsorption sites occupied by an additive. Since $\Delta F^{\circ}_{\text{ads}} = -k_BT\ln K$,[58] eqn (6) can be simplified to:

$$\gamma(\theta) - \gamma_0 = -\frac{K\ln K}{\beta a}x\ \ldots \tag{7}$$

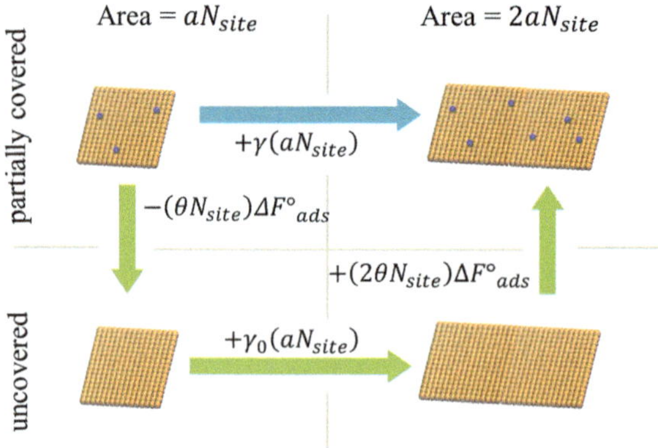

Fig. 1 The reversible work to double the area of a partially covered surface (blue path) is equivalent to the work to (1) remove the additives from the surface, (2) double the surface area, and (3) adsorb the same fraction of additives on the surface (green path).

which, in agreement with the Gibbs adsorption isotherm,[58,59] shows that interfacial tension decreases if additives favorably adsorb to the interface. Eqn (7) is also consistent with the linear reduction in surface tension for small concentrations of surfactants.[60]

Note, however, that eqn (6) and (7) are only valid in the dilute limit when the adsorbed additives do not interact. The model must break down for sufficiently large θ, else a negative γ would result. Additionally, eqn (7) ignores the effects of additives that can dissolve into the solid precipitate, and additives that preferentially bind at surface defects like kinks, steps, and pits.

For our idealized model in which dilute additives sparingly cover smooth surfaces of the nuclei, eqn (3) and (7) can be combined to obtain:

$$\frac{F^{\ddagger}(x) - F_0^{\ddagger}}{F_0^{\ddagger}} \approx -3\frac{K \ln K}{\beta a \gamma_0}x \tag{8}$$

where $\beta = 1/k_BT$. Eqn (8) predicts relative changes in free energy barriers with additive concentration. Instead of the equilibrium adsorption constant K, the change in free energy barrier can equally be written entirely in terms of $\gamma - \gamma_0$. Both forms are potentially useful for screening additives, *e.g.* by computing the adsorption equilibria K, or by measuring changes in the contact angles $\gamma - \gamma_0$, for trial additives.

Note that nucleation rates depend on free energy barriers as $J = A\exp[-\beta\Delta F^{\ddagger}]$. Assuming that the prefactor is only weakly dependent on additive concentration, eqn (8) further suggests that:

$$\ln[J(x)/J_0)] \approx \frac{3F_0^{\ddagger}}{a\gamma_0}(K \ln K)x \tag{9}$$

where $J(x)$ and J_0 are nucleation rates, with and without additives respectively. Using eqn (2), the quantity $3F_0^{\ddagger}/a\gamma_0$ can be written as $\phi(2\gamma_0 a\phi/3\Delta\mu)^2$. Thus, assuming spherical nuclei and given values for $\Delta\mu$ and K, eqn (9) offers an intriguing new method for extracting the bare interfacial tension (γ_0) from the slope of $\ln[J/J_0]$, as a function of additive concentration (x). Eqn (9) also conveniently separates the effect of non-additive properties ($3F_0^{\ddagger}/a\gamma_0$) from additive-specific properties ($K\ln K$), and suggests that the efficacy of nucleation promoters can be compared using ratios of $K\ln K$.

3 Simulation methods

3.1 Ternary Potts lattice gas model

Lattice models enable the investigation of metastable solutions independent of system-specific chemical details.[61,62] They are frequently used to test new rare-event algorithms and to provide general insight into nucleation, but are primarily used on single component systems. A number of studies have combined lattice gas simulations with theoretical analyses based on CNT.[62–65] For example, Pan and Chandler studied the transition state ensemble of the critical nuclei of a lattice gas model, and confirmed that cluster size is a reasonable reaction coordinate for nucleation.[49] Sear used lattice gas models to investigate heterogeneous nucleation in pores of different shapes and sizes.[66–68]

The Potts model is a generalization of the lattice gas model with each Potts orientation representing a particular phase or component,[69,70] and has been used to study structural transitions in solids.[71,72] Peters and coworkers developed a binary Potts lattice gas (PLG) model, with a lattice gas-like degree of freedom to distinguish between lattice sites occupied by solutes or solvents, and a Potts-like degree of freedom to represent the orientation of molecules or lattice vectors at each lattice site.[56,73] The PLG model is a minimal nearest-neighbor model for orientation-specific interactions between solute–solute, solute–solvent, and solvent–solvent pairs. The interaction parameters can be tuned to obtain phase diagrams that resemble those of real binary mixtures. The PLG and closely related models have now been used in studies of phase equilibria, self-assembly, and several nucleation processes.[56,73–78]

The model in this work starts with the PLG and incorporates additives as a third component. Each lattice site is therefore occupied by either a solute, solvent, or additive (labeled species $k = 1$, 2, and 3 respectively), having one of $Q = 24$ possible orientations. For a cubic lattice, $Q = 24$ represents the most asymmetric case,[79] but smaller values of Q can be used to represent more symmetric molecules.[57,80] The PLG Hamiltonian takes solute–solvent interactions as the zero of energy and sums over all nearest neighbor solute–solute, solvent–solvent, and additive–solute interactions. Using the notation $\langle i,j\rangle$ to denote nearest neighbor pairs, the Hamiltonian is:

$$\beta H = \beta H_0 + \beta\Delta H \tag{10}$$

where $\beta\Delta H$ and βH_0 are the additive and additive-free Hamiltonians, respectively. The solute–solute and solvent–solvent interactions are those of the standard PLG model:

$$\beta H_0 = -\sum_{k=1}^{2}\sum_{\langle i,j\rangle}\delta_{m(i),k}\delta_{m(j),k}\left[\beta G_k + \beta A_k\delta_{s(i),s(j)}\right] \tag{11}$$

where $m(i)$ represents the species at lattice site i, $s(i)$ represents the local orientation of the molecule at lattice site i, and βG_k and βA_k are the species-specific and orientation-specific interactions, respectively. βG_1 and βG_2 stabilize solute–solute and solvent–solvent pairs, respectively, and primarily control the solubility of solute in the solvent-rich phase and solvent in the solute-rich phase. Like neighbors with matching orientations are further stabilized by βA_1 and βA_2. For this study, we use the following PLG Hamiltonian parameters: $\beta G_1 = \beta G_2 = \beta A_1 = 1.25$ and $\beta A_2 = 0$. This choice of PLG parameters results in a weakly soluble solute and a relatively pure solute precipitate, identical to the binary PLG model shown in Fig. 1a of Duff and Peters at $k_BT = 0.8$.[56,79]

The additives mimic amphiphilic molecules through interactions that favor adsorption at the solute precipitate–solution interface. Additive–solvent and additive–additive interactions are equivalent to the zero energy solute–solvent interactions, making the additives sparingly soluble in solution and preventing them from aggregating. Additives interact with solutes only along the direction of their orientation:

$$\beta \Delta H = -\sum_{\langle i,j \rangle} \delta_{m(i),\,3}\delta_{m(j),\,1}\delta_{g(i),\,n(j,\,i)}\beta A_3 \tag{12}$$

where $n(j, i)$ is the neighbor index of lattice site j with respect to lattice site i, which varies from 1 to 6 for a cubic lattice, and $g(i)$ indexes the neighbor to which the orientation of the additive at lattice site i points. For a cubic lattice, $g(i) = \text{mod}[Q, 6] + 1$ (*i.e.* 1 plus the remainder of $Q/6$) also varies from 1 to 6. Therefore, only additives whose orientation points toward a solute experience a favorable interaction (see Fig. 2). This directional additive–solute interaction mimics the attraction between the head/tail group of surfactants and the interface; it prevents the nucleating phase from having many additive inclusions, even though their interactions are strong. For this study, we used five different values of βA_3 to investigate the effect of additive binding strength: $\beta A_3 = 2.5$, 2.875, 3.125, 3.375, and 3.75.

3.2 Simulation details

In a small closed simulation with fixed composition, a solute nucleus depletes the solute from its surroundings and consumes the driving force for its own growth.[47,81–84] Because of these finite size effects, solute precipitate nucleation is best simulated in grand or semigrand (open) ensembles.[47,56,73,85] We use semigrand canonical Monte Carlo (SGMC) (or NT$\{\mu_i - \mu_r\}$ ensemble) simulations[86] to maintain differences in the chemical potentials of each species and that of the solvent reference species. Our SGMC moves include (1) local and non-local swaps, (2) orientation flips, and (3) semigrand identity changes. The first two move types are accepted or rejected using the standard canonical Metropolis criterion. The

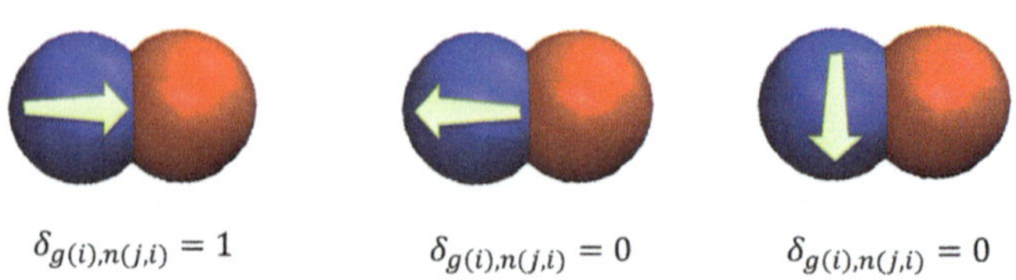

Fig. 2 The additive (blue) only interacts with a neighboring solute if its orientation points toward a solute (red), *i.e.* if $g(i) = n(j, i)$.

acceptance probability for identity changes from species i to j involves an additional fugacity ratio prefactor.[86]

Free energy landscapes along the nucleus size coordinate n are computed as:

$$\beta F(n) = -\ln \frac{\langle N(n)\rangle}{\langle N(1)\rangle} \tag{13}$$

where $\langle N(n)\rangle$ is the average number of nuclei composed of n solute monomers in the simulation box.[47,51,52] The size n of each nucleus is computed by counting contiguously neighboring solutes. However, it is computationally impractical to calculate $F(n)$ using unbiased Boltzman sampling for high barrier processes, due to the exponentially vanishing probability of visiting higher free energy states. We overcome this issue with umbrella sampling, which imposes a set of artificial biasing potentials to improve the sampling of less probable configurations.[87–89] We use windows with widths of three units along the nucleus size n-coordinate with hard walls at the window edges. Each umbrella window is sampled for at least 500 000 SGMC sweeps to guarantee sufficient sampling.

3.2.1 Compositions of equal driving force. The nucleation free energy barrier depends on both the driving force (or supersaturation) and the interfacial free energy. To isolate the effects of the additive on the interfacial free energy, we prepared compositions of equal driving force (Fig. 3). The driving force is measured by simulating slab growth in a long elongated cell (in this case a 10 × 10 × 80 lattice), which maximizes the volume-to-surface ratio for a two-phase system. The change in free energy for layer-by-layer slab growth is $F_{\text{slab}}(n) - F_{\text{slab}}(n_0) = -(n - n_0)\Delta\mu$. The driving force is therefore:[90]

$$\Delta\mu = -\frac{\partial F_{\text{slab}}}{\partial n} \tag{14}$$

The set of compositions of constant driving force can be calculated by the following iterative process:

1. Simulate slab growth using SGMC and calculate the dimensionless driving force with no additive ($\beta\Delta\mu_0$), using eqn (14).

2. Adjust the additive–solvent (f_3/f_2) and solute–solvent (f_1/f_2) fugacity ratio, and repeat step 1 until the driving force is approximately $\beta\Delta\mu_0$.

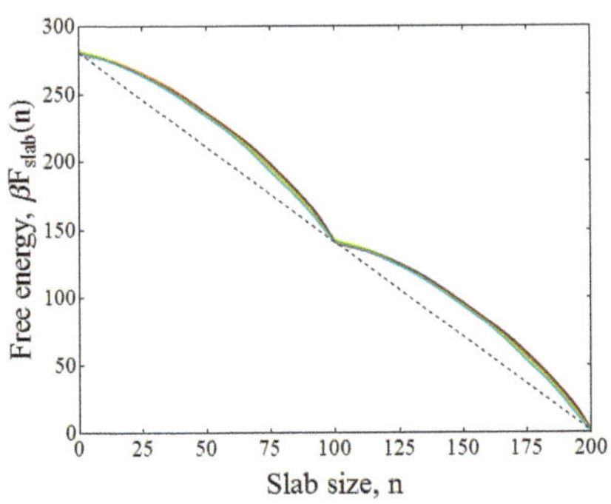

Fig. 3 The free energy landscape of solute slab growth at different fugacity ratios with the same driving force for $\beta A_3 = 3.75$. The driving force $\Delta\mu$ is estimated by the common tangent through troughs (dashed line).

3. Repeat step 2 for each solute–solvent–additive composition.

Increasing f_1/f_2 will increase the solute concentration and therefore the driving force. Increasing f_3/f_2 will increase the additive concentration and therefore increase coverage, lower interfacial tension, and lower barrier height. The four sets of fugacity ratios we use that have the same driving force are: $f_1/f_2 = 2.25$ and $f_3/f_2 = 0$, $f_1/f_2 = 2.25$ and $f_3/f_2 = 0.01$, $f_1/f_2 = 2.258$ and $f_3/f_2 = 0.03$, and $f_1/f_2 = 2.26$ and $f_3/f_2 = 0.05$. These fugacity ratios are used for each metastable state in all other simulations. The composition of each metastable state is determined using unbiased 500 000 sweep-long SGMC simulations on a 64 × 64 × 64 periodic lattice.

3.2.2 Free energy barriers and interfacial tension. We use umbrella sampling to compute the work required to grow a nucleus from size 1 to n on a 32 × 32 × 32 periodic lattice.[87–89,91] Since the probability of observing more than one cluster of size $n > 4$ in our simulation box is negligible, $N(n)$ for values of $n > 4$ is approximated by the probability that the largest cluster in the system is of size n.[51,52] CNT fits of the free energy landscape with eqn (1) are used to estimate the nucleation barriers and interfacial tension.

3.2.3 Additive adsorption. SGMC simulations are used to estimate the adsorption constant. Simulations are done on a 64 × 64 × 64 lattice with a 64 × 64 solute sheet. The sheet is not allowed to grow, dissolve, or change any of its Potts degrees of freedom. Coverage is estimated by averaging the fraction of lattice sites neighboring the solute sheet that is occupied by an additive,[92] from at least 500 000 SGMC sweeps.

4 Results and discussion

Fig. 4a shows that the average additive coverage θ for a range of the average additive mole fraction x agrees with the predicted Henry's law behavior at low concentrations. We obtain the adsorption constant K by fitting additive coverage θ to eqn (5). Fig. 5 shows the free energy $F(n)$ for the four compositions of equal driving force for $\beta A_3 = 3.75$, which is qualitatively similar for other values of βA_3. Interfacial tension $\beta a\gamma\phi$ is obtained by fitting $F(n)$ to eqn (1) for each additive

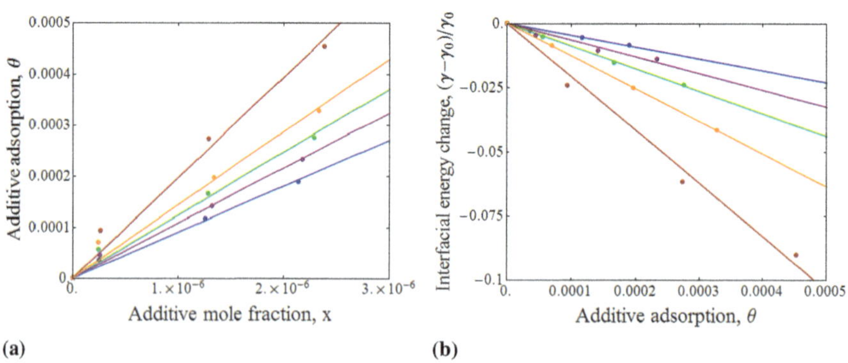

Fig. 4 Proportional relationships between (a) additive coverage θ and additive concentration x and (b) change in interfacial free energy $(\gamma - \gamma_0)/\gamma_0$ and θ. Lines are linear fits constrained to pass through the origin for additive–solute interactions $\beta A_3 = 2.5$ (blue), 2.875 (purple), 3.125 (green), 3.375 (orange), and 3.75 (red).

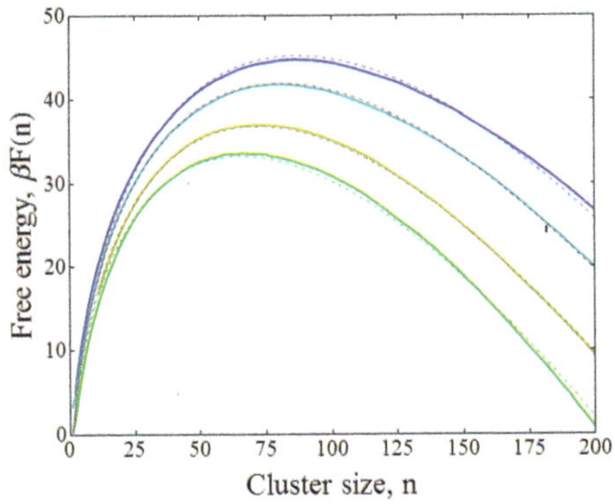

Fig. 5 The reversible work $\beta F(n)$ to create a nucleus of size n at different additive concentrations for $\beta A_3 = 3.75$. The dotted lines are CNT fits with the constraint that the driving force at each additive concentration is the same.

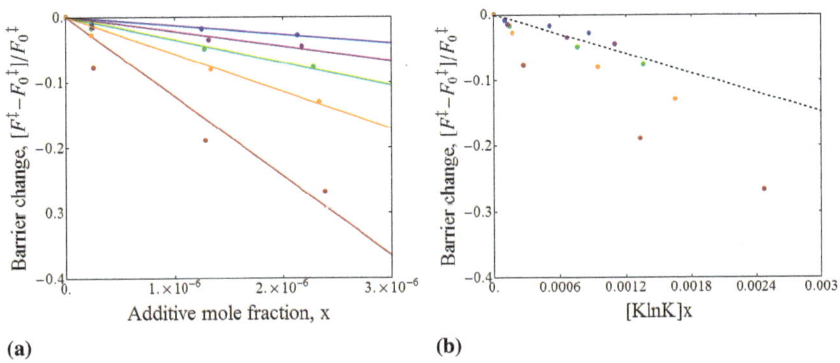

Fig. 6 While (a) changes in free energy barriers are approximately proportional to additive concentration, (b) only additives with small values of βA_3 are consistent with eqn (8). The lines are linear fits that pass through the origin for additive–solute interactions $\beta A_3 = 2.5$ (blue), 2.875 (purple), 3.125 (green), 3.375 (orange), and 3.75 (red). The dashed line is for the combined $\beta A_{13} = 2.5$, 2.875, and 3.125 data.

concentration, with the constraint that the driving force at each additive concentration is the same. The proportional relationships in Fig. 4a and b are consistent with our theoretical prediction that increased adsorption will lower interfacial tension.

For the most weakly binding additives, Fig. 6b validates the approximately proportional relationship in eqn (8) between the free energy barrier and $(K \ln K)x$. The discrepancy for additives with large values of K may be due to a breakdown in the infinitely dilute additive assumption, due to increased interaction with solute in solution. Even at large values of K, Fig. 6b suggests that ratios of $K \ln K$ can still be used to predict nucleant efficacy, albeit qualitatively.

5 Concluding remarks

We developed an equation starting from classical nucleation theory to predict the effects of additive binding strength and additive concentration on nucleation barriers. The rate is modulated by a product of the interfacial free energy reduction per unit coverage and the equilibrium constant for additive binding (K). These two quantities are closely related to each other, so it may be possible to screen for potent additives by comparing ratios of $K \ln K$ for trial additives. Promising nucleants can also be screened using contact angle measurements with and without trial additives. Trace additives with higher values of $K \ln K$ are more effective at reducing surface tension, lowering the free energy barrier, and accelerating nucleation. Simulations on a ternary Potts lattice gas model confirm that the potency of a trace additive for barrier reduction increases with $K \ln K$. Further investigations are needed to test the validity of these results for molecular additives, solutes, and solvents.

Acknowledgements

G.G.P acknowledges the National Science Foundation (NSF) for support through the Graduate Research Fellowship under DGE 1144085 and through CBE-1125235. S.S. acknowledges the Summer Undergraduate Research Fellowship (SURF) at the University of California, Santa Barbara for support. B.P. acknowledges the NSF Career Award 955502 for support. We thank Valeria Molinero and Sean Cray for many helpful discussions.

References

1 A. Hudait and V. Molinero, *J. Am. Chem. Soc.*, 2014, **136**, 8081–8093.
2 G. Bullock and V. Molinero, *Faraday Discuss.*, 2013, **167**, 371–388.
3 C. Budke, A. Dreyer, J. Jaeger, K. Gimpel, T. Berkemeier, A. S. Bonin, L. Nagel, C. Plattner, A. L. DeVries, N. Sewald and T. Koop, *Cryst. Growth Des.*, 2014, **14**, 4285–4294.
4 P. A. Alpert, J. Y. Aller and D. A. Knopf, *Atmos. Chem. Phys.*, 2011, **11**, 5539–5555.
5 T. Koop and B. Zobrist, *Phys. Chem. Chem. Phys.*, 2009, **11**, 10839–10850.
6 R. J. Davey, N. Blagden, G. D. Potts and R. Docherty, *J. Am. Chem. Soc.*, 1997, **119**, 1767–1772.
7 N. Blagden and R. J. Davey, *Cryst. Growth Des.*, 2003, **3**, 873–885.
8 A. Llinas and J. M. Goodman, *Drug Discovery Today*, 2008, **13**, 198–210.
9 V. Pipich, M. Balz, S. E. Wolf, W. Tremel and D. Schwahn, *J. Am. Chem. Soc.*, 2008, **130**, 6879–6892.
10 Y. Levi, S. Albeck, A. Brack, S. Weiner and L. Addadi, *Chem.–Eur. J.*, 1998, **4**, 389–396.
11 M. F. Butler, N. Glaser, A. C. Weaver, M. Kirkland and M. Heppenstall-Butler, *Cryst. Growth Des.*, 2006, **6**, 781–794.
12 F. C. Meldrum and H. Coelfen, *Chem. Rev.*, 2008, **108**, 4332–4432.
13 K. Naka and Y. Chujo, *Chem. Mater.*, 2001, **13**, 3245–3259.
14 S. Elhadj, J. J. De Yoreo, J. R. Hoyer and P. M. Dove, *Proc. Natl. Acad. Sci. U. S. A.*, 2006, **103**, 19237–19242.
15 D. Gebauer, H. Coelfen, A. Verch and M. Antonietti, *Adv. Mater.*, 2009, **21**, 435–439.
16 D. C. Popescu, M. M. J. Smulders, B. P. Pichon, N. Chebotareva, S.-Y. Kwak, O. L. J. van Asselen, R. P. Sijbesma, E. DiMasi and N. A. J. M. Sommerdijk, *J. Am. Chem. Soc.*, 2007, **129**, 14058–14067.

17 C. Brinker and G. Scherer, *Sol–gel Science: The Physics and Chemistry of Sol–gel Processing*, Academic Press, 1990.
18 R. F. Lobo, S. I. Zones and M. E. Davis, *J. Inclusion Phenom. Mol. Recognit. Chem.*, 1995, **21**, 47–78.
19 Q. S. Huo, D. I. Margolese and G. D. Stucky, *Chem. Mater.*, 1996, **8**, 1147–1160.
20 Y. Kubota, M. M. Helmkamp, S. I. Zones and M. E. Davis, *Microporous Mater.*, 1996, **6**, 213–229.
21 O. Galkin and P. G. Vekilov, *Proc. Natl. Acad. Sci. U. S. A.*, 2000, **97**, 6277–6281.
22 P. G. Vekilov, *Cryst. Growth Des.*, 2010, **10**, 5007–5019.
23 V. Y. Torbeev, E. Shavit, I. Weissbuch, L. Leiserowitz and M. Lahav, *Cryst. Growth Des.*, 2005, **5**, 2190–2196.
24 R. P. Sear, *J. Chem. Phys.*, 2001, **115**, 575–579.
25 B. J. Anderson, J. W. Tester, G. P. Borghi and B. L. Trout, *J. Am. Chem. Soc.*, 2005, **127**, 17852–17862.
26 M. T. Storr, P. C. Taylor, J. P. Monfort and P. M. Rodger, *J. Am. Chem. Soc.*, 2004, **126**, 1569–1576.
27 A. K. Sum, C. A. Koh and E. D. Sloan, *Ind. Eng. Chem. Res.*, 2009, **48**, 7457–7465.
28 J. S. Zhang, S. Lee and J. W. Lee, *Ind. Eng. Chem. Res.*, 2007, **46**, 6353–6359.
29 J. S. Zhang, C. Lo, P. Somasundaran, S. Lu, A. Couzis and J. W. Lee, *J. Phys. Chem. C*, 2008, **112**, 12381–12385.
30 A. Fazlali, S.-A. Kazemi, M. Keshavarz-Moraveji and A. H. Mohammadi, *Energy Technol.*, 2013, **1**, 471–477.
31 S.-Q. Hao, S. Kim, Y. Qin and X.-H. Fu, *Environ. Chem. Lett.*, 2014, **12**, 341–346.
32 H. X. Yuan and J. M. Ouyang, *Prog. Chem.*, 2006, **18**, 573–578.
33 J. D. Rimer, Z. An, Z. Zhu, M. H. Lee, D. S. Goldfarb, J. A. Wesson and M. D. Ward, *Science*, 2010, **330**, 337–341.
34 A. L. Devries, *Science*, 1971, **172**, 1152–1155.
35 J. G. Duman, *Annu. Rev. Physiol.*, 2001, **63**, 327–357.
36 F. Sicheri and D. S. C. Yang, *Nature*, 1995, **375**, 427–431.
37 D. Kashchiev and G. M. van Rosmalen, *Cryst. Res. Technol.*, 2003, **38**, 555–574.
38 D. Kashchiev, *Nucleation: Basic Theory With Applications*, Butterworth-Heinemann, Oxford, 2000.
39 I. Weissbuch, M. Lahav and L. Leiserowitz, *Cryst. Growth Des.*, 2003, **3**, 125–150.
40 I. Weissbuch, V. Y. Torbeev, L. Leiserowitz and M. Lahav, *Angew. Chem., Int. Ed.*, 2005, **44**, 3226–3229.
41 L. N. Poloni and M. D. Ward, *Chem. Mater.*, 2014, **26**, 477–495.
42 J. Anwar, P. K. Boateng, R. Tamaki and S. Odedra, *Angew. Chem., Int. Ed.*, 2009, **48**, 1596–1600.
43 N. Duff, Y. R. Dahal, J. D. Schmit and B. Peters, *J. Chem. Phys.*, 2014, **140**, 014501.
44 R. Becker and W. Doring, *Ann. Phys.*, 1935, **24**, 719–752.
45 M. Volmer and A. Weber, *Z. Phys. Chem.*, 1926, **119**, 277–301.
46 J. Frenkel, *Kinetic Theory of Liquids*, Dover, 1955.
47 V. Agarwal and B. Peters, *Solute Precipitate Nucleation: A Review of Theory and Simulation Advances*, John Wiley & Sons, New Jersey, USA, 2014, vol. 155.
48 S. Auer and D. Frenkel, *Nature*, 2001, **409**, 1020–1023.
49 A. C. Pan and D. Chandler, *J. Phys. Chem. B*, 2004, **108**, 19681–19686.
50 S. Punnathanam and P. A. Monson, *J. Chem. Phys.*, 2006, **125**, 024508.

51 L. Maibaum, *Phys. Rev. Lett.*, 2008, **101**, 019601.
52 L. Maibaum, *Phys. Rev. Lett.*, 2008, **101**, 256102.
53 S. L. Meadley and F. A. Escobedo, *J. Chem. Phys.*, 2012, **137**, 074109.
54 I. Saika-Voivod, P. H. Poole and R. K. Bowles, *J. Chem. Phys.*, 2006, **124**, 224709.
55 E. Sanz, C. Vega, J. R. Espinosa, R. Caballero-Bernal, J. L. F. Abascal and C. Valeriani, *J. Am. Chem. Soc.*, 2013, **135**, 15008–15017.
56 N. Duff and B. Peters, *J. Chem. Phys.*, 2009, **131**, 184101.
57 B. C. Knott, N. Duff, M. F. Doherty and B. Peters, *J. Chem. Phys.*, 2009, **131**, 224112.
58 J. Kipling, *Adsorption from solutions of non-electrolytes*, Academic Press, 1965.
59 D. Ruthven, *Principles of Adsorption and Adsorption Processes*, Wiley, 1984.
60 S. Safran, *Statistical Thermodynamics of Surfaces, Interfaces, and Membranes*, Addison-Wesley Pub, 1994.
61 K. Binder and H. Muller-Krumbhaar, *Phys. Rev. B: Solid State*, 1974, **9**, 2328–2353.
62 D. Stauffer, A. Coniglio and D. W. Heermann, *Phys. Rev. Lett.*, 1982, **49**, 1299–1302.
63 J. Schmelzer and D. P. Landau, *Int. J. Mod. Phys. C*, 2001, **12**, 345–359.
64 V. A. Shneidman, K. A. Jackson and K. M. Beatty, *J. Chem. Phys.*, 1999, **111**, 6932–6941.
65 F. Schmitz, P. Virnau and K. Binder, *Phys. Rev. E: Stat., Nonlinear, Soft Matter Phys.*, 2013, **87**, 053302.
66 R. P. Sear, *J. Phys. Chem. B*, 2006, **110**, 4985–4989.
67 R. P. Sear, *J. Phys.: Condens. Matter*, 2012, **24**, 052205.
68 A. J. Page and R. P. Sear, *Phys. Rev. Lett.*, 2006, **97**, 065701.
69 F. Y. Wu, *J. Phys. A: Math. Gen.*, 1981, **14**, L39–L44.
70 F. Y. Wu, *Rev. Mod. Phys.*, 1982, **54**, 235–268.
71 R. Devanathan, L. Van Brutzel, A. Chartier, C. Gueneau, A. E. Mattsson, V. Tikare, T. Bartel, T. Besmann, M. Stan and P. Van Uffelen, *Energy Environ. Sci.*, 2010, **3**, 1406–1426.
72 V. Tikare, E. A. Holm, D. Fan and L. Q. Chen, *Acta Mater.*, 1998, **47**, 363–371.
73 V. Agarwal and B. Peters, *J. Chem. Phys.*, 2014, **140**, 084111.
74 D. Bellucci, V. Cannillo and A. Sola, *Ceram. Int.*, 2010, **36**, 1983–1988.
75 L. O. Hedges and S. Whitelam, *Soft Matter*, 2013, **9**, 9763–9766.
76 J. Grant, R. L. Jack and S. Whitelam, *J. Chem. Phys.*, 2011, **135**, 214505.
77 W. M. Jacobs, D. W. Oxtoby and D. Frenkel, *J. Chem. Phys.*, 2014, **140**, 204109.
78 T. K. Haxton and S. Whitelam, *Soft Matter*, 2013, **9**, 6851–6861.
79 N. Duff and B. Peters, *J. Chem. Phys.*, 2010, **132**, 129901.
80 B. C. Knott, M. F. Doherty and B. Peters, *J. Chem. Phys.*, 2011, **134**, 154501.
81 D. Reguera, R. K. Bowles, Y. Djikaev and H. Reiss, *J. Chem. Phys.*, 2003, **118**, 340–353.
82 R. Grossier and S. Veesler, *Cryst. Growth Des.*, 2009, **9**, 1917–1922.
83 J. Wedekind, D. Reguera and R. Strey, *J. Chem. Phys.*, 2006, **125**, 214505.
84 A. S. Abyzov and J. W. P. Schmelzer, *J. Non-Cryst. Solids*, 2014, **384**, 8–14.
85 R. Ni, F. Smallenburg, L. Filion and M. Dijkstra, *Mol. Phys.*, 2011, **109**, 1213–1227.
86 D. A. Kofke and E. D. Glandt, *Mol. Phys.*, 1988, **64**, 1105–1131.
87 G. M. Torrie and J. P. Valleau, *J. Comput. Phys.*, 1977, **23**, 187–199.

88 S. Kumar, D. Bouzida, R. H. Swendsen, P. A. Kollman and J. M. Rosenberg, *J. Comput. Chem.*, 1992, **13**, 1011–1021.
89 S. Auer and D. Frenkel, *Annu. Rev. Phys. Chem.*, 2004, **55**, 333–361.
90 M. Salvalaglio, T. Vetter, M. Mazzotti and M. Parrinello, *Angew. Chem., Int. Ed.*, 2013, **52**, 13369–13372.
91 B. Roux, *Comput. Phys. Commun.*, 1995, **91**, 275–282.
92 P. Rinaldi, F. Bulnes, A. J. Ramirez-Pastor and G. Zgrablich, *Surf. Sci.*, 2008, **602**, 1783–1794.

Faraday Discussions

PAPER

The many roles of mellitic acid during barium sulfate crystallization†

Matthew P. Boon,[a] Sandra Freeman,[a] Mark I. Ogden,[a] Allan Oliveira,[a] William R. Richmond,[a] Brian W. Skelton[b] and Franca Jones*[a]

Received 7th November 2014, Accepted 25th November 2014
DOI: 10.1039/c4fd00211c

The various roles of mellitic acid during barium sulfate crystallization from nucleation to mesocrystal formation are explored and elucidated.

Introduction

Classical crystallization has provided a basis for understanding the process of how random hydrated ions in solution come together and form ordered crystalline solids. Knowing crystallization theory can, in principle, help us understand how the minerals form on our earth, how nature forms various biominerals in living organisms and how we may make particles to suit our technological needs. Unfortunately, classical crystallization theory cannot explain many crystallization phenomena.[1,2] In biomineralization, amorphous precipitates are often formed, rather than crystalline nuclei.[3,4] Quite recently, the stabilisation of nanoparticles that aggregate and then form mesocrystals has shown another non-classical crystallization mechanism.[5] In mesocrystal formation growth beyond the nanoscale does not occur; rather the micron-sized particles are formed by the self-assembly of the stabilised nanoparticles into larger structures.[6,7]

There has been much work conducted in the area of mesocrystals and their formation in the presence of polymeric species.[8–11] We have, in the past, shown that small molecules can stabilise barium sulfate and that this is most likely due to self-assembly of the organic molecule[12–14] resulting in the steric stabilisation required. Formation of meso-crystals is also possible by charge stabilisation of nanoparticles[15,16] rather than steric stabilisation since the general mechanism

[a]*Department of Chemistry, Curtin University, Bentley, WA 6102, Australia. E-mail: F.Jones@curtin.edu.au*
[b]*Centre for Microscopy, Characterisation and Analysis, The University of Western Australia, Crawley, Western Australia 6009, Australia*

† Electronic supplementary information (ESI) available: Complexation as measured by conductivity, barium mellitate crystal structure images, surface roughness shown by line section of height image in AFM, turbidity measured by absorbance at different mellitic acid concentrations, DLS counts measured from timed crystallization experiments and aggregation of particles at high mellitic acid concentration. CCDC 1032712. For ESI and crystallographic data in CIF or other electronic format see DOI: 10.1039/c4fd00211c

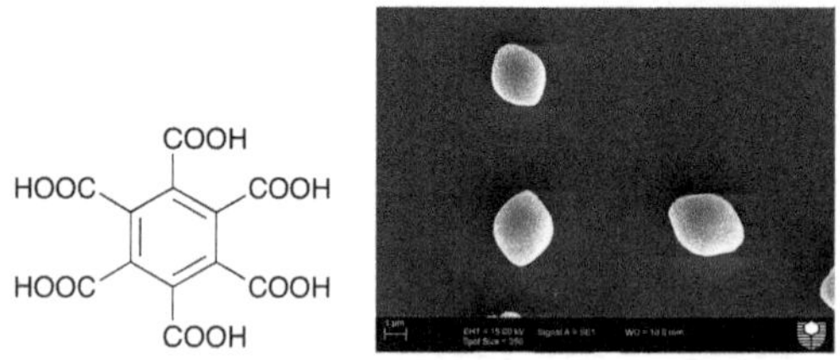

Fig. 1 Schematic of mellitic acid (left) and SEM image of baryte solids formed when 0.03 mM mellitic acid is present.

relies on any form of stabilization of the initial nanoparticle before formation of the mesocrystal and/or fusion to an iso-oriented crystal.

In this work, barium sulfate is used as our crystallizing solid phase. Baryte (barium sulfate solid) has several properties which make it useful as a model system; the first is that the ionic species are known and do not alter speciation significantly even with moderate pH changes, the other is that only one polymorph of barium sulfate is formed (baryte) even when conditions are significantly altered.

Previous work has shown mellitic acid to be an inhibitor of barium sulfate crystallization, which led to the formation of mesocrystals (see Fig. 1) of barium sulfate.[17] Thus, an important aspect of this study was to determine how mellitic acid is involved in the formation of the meso-crystal. Additionally, infrared studies on this system showed that mellitic acid was intimately associated with the initial nucleation of barium sulfate particles from solution.[18] Thus, mellitic acid was also influencing the nucleation behaviour of barium sulfate. For this reason, the system was investigated to determine what impact the mellitic acid had on all aspects of crystallization and particle formation and to rationalise those impacts based on chemistry and complexation behaviour. The results emphasise that impurities have far reaching impacts that are not always fixed to one particular stage of crystal formation.

Experimental

Materials

The materials used in this study were AR grade, from Ajax Chemicals, Sigma Aldrich or BDH and were dissolved to the required concentrations using filtered (0.2 μm), ultrapure (>18 MΩ cm resistivity) water.

Methods

Morphology and electron microscopy. Batch crystallization was undertaken to observe the impact on morphology; the method used was identical to that previously reported in ref. 17. Samples from timed experiments were used for transmission electron microscopy investigations as reported in ref. 17 and 18. Conditions are detailed in the ESI.†

Turbidity. The effect of mellitic acid on the nucleation behaviour of barium sulfate during crystallization was determined by its impact on turbidity. The induction time (time taken for the turbidity to rise above zero) was determined at different mellitic acid concentrations. Induction time changes reflect nucleation rate changes, which in turn can give information about the surface free energy of the nuclei (in classical nucleating systems). The turbidity was measured using a UV-vis instrument (GBC, UV-vis 916 instrument) operating at 900 nm to limit absorption of the light by the solution and dissolved species. The crystallization reaction, equilibrated to temperature, involved 200 mL ultrapure water, 0.5 mL barium chloride solution (0.1 M) and 0.5 mL sodium sulfate solution (0.1 M) stirred at 150 rpm with a four-pronged glass impellor. If mellitic acid solution (1 g L^{-1} stock solution) was added, the amount of ultrapure water was adjusted to maintain a constant final volume. This solution was pumped through a flow-through quartz cuvette and the absorbance reading through the cell constantly monitored. The flow rate through the cell was 67 mL min^{-1} and this was achieved using a Masterflex® peristaltic pump and Tygon® tubing.

AFM. The AFM procedure is the same as that described in ref. 19. A mineralogical sample of barite was freshly cleaved prior to each experiment and the supersaturation index was SI = 25. The supersaturation calculated in this manuscript uses the PHREEQC program[20] to calculate the ion activity product (IAP), and then this is used to determine the supersaturation index such that:

$$\mathrm{SI} = (\mathrm{IAP}/K_{\mathrm{sp}}) \tag{1}$$

A syringe pump (driven at 0.2 mL min^{-1}) was used to deliver the solutions to a fluid cell while the AFM software collected images continuously, thereby monitoring crystallization growth *in situ*. Growth under conditions *sans* mellitic acid were first obtained so that all subsequent data could be normalised.

Confocal Raman. Confocal Raman spectra and maps were obtained on an Alpha 300SAR confocal Raman microscope (WITech, Germany) equipped with a 100× objective (numerical aperture 0.9). Spectra were obtained with a frequency doubled NdYAG laser having a final excitation wavelength of 532 nm and using an integration time of 50 ms.

Zeta potential and size. The zeta potential and particle counts (by dynamic light scattering, DLS) were measured using a Malvern NanoZeta ZS in a disposable zeta cell. Particles were measured in the solution in which they were formed. This consisted of preparing 20 mL solutions containing 50 μL of the barium chloride (0.1 M) with the required mellitic acid (with adjusted water volume if necessary to maintain 20 mL total volume). Sodium sulfate (50 μL, 0.1 M) solution was added to commence crystallization and the vial was then left for three days. After this time, the vial was sonicated gently in a bath sonicator to disperse the particles and an aliquot taken for measurement. For the timed experiments the reactor and conditions used in the turbidity experiments were replicated. At pre-determined intervals an aliquot was taken and the zeta potential or size measured. Although size measurements were taken during the timed experiments the data obtained was used to determine the particle counts.

Degree of complexation between mellitic acid and barium ions. The extent to which mellitic acid complexed barium ions in solution at the same pH as in the conductivity, morphology, AFM and zeta potential experiments was determined in two ways. Firstly, a Ba ion selective electrode (ISE) was used to see whether the addition of mellitic acid to a known barium ion concentration resulted in a lowering of the barium ion activity. Secondly, the barium chloride conductivity was compared to the barium ion conductivity with mellitic acid to see whether there was any change in the mobility of the solution measured.

Crystallography. The solid formed by complexation of barium ions with mellitate ions was obtained by adjusting the mellitic acid pH to >9 using sodium hydroxide. Barium chloride (0.1 M) solution was subsequently added at 3 mol equivalents. Initially, a white cloudy precipitate forms (presumably barium hydroxide) which overnight transforms to clear long needles appropriate for single crystal X-ray structure determination.

Crystallographic data for the structure were collected at 150(2) K on an Oxford Diffraction Gemini diffractometer fitted with Mo Kα radiation. Following multi-scan absorption corrections and solution by direct methods, the structure was refined against F^2 with full-matrix least-squares using the program SHELXL-97.[21] Several peaks in later maps were modelled as solvent water molecules with site occupancy factors constrained to 0.5 or 1.0 after trial refinement. Water molecule hydrogen atoms were not included in the model. All remaining hydrogen atoms were added at calculated positions and refined by use of riding models with isotropic displacement parameters based on those of the parent atoms. Anisotropic displacement parameters were employed throughout for the non-hydrogen atoms. Full details of the structure determination have been deposited with the Cambridge Crystallographic Data Centre.†

Crystal data: $C_{12}H_{28}Ba_3O_{26}$; $M = 1000.36$, monoclinic, $I2/a$; $a = 13.4109(2)$, $b = 16.2568(4)$, $c = 27.9028(6)$ Å, $V = 6008.7(2)$ Å^3, $Z = 8$; $D_c = 2.212$ g cm^{-3}, $\mu = 3.988$ mm^{-1}, crystal size = 0.30 × 0.035 × 0.03 mm^3. 59 752 reflections collected, 15 427 unique ($R_{int} = 0.0816$), max./min. transmission = 1.00/0.95. Data/restraints/parameters = 9865/0/375; GoF = 1.039. $R_1 = 0.0611$ ($I > 2\sigma(I)$), wR_2 = 0.1681 (all data). Largest diff. peak and hole = 4.206 and −1.427 e Å^{-3}. CCDC no. 1032712.

Thermogravimetry. TGA analysis was conducted on a TA Instruments SDT 2960 simultaneous DSC-TGA from ambient to 900 °C at 5 °C per minute in air at a flow rate of 40 mL min^{-1} using nitrogen gas. A known weight of sample (15–50 mg) was heated in a platinum pan for each measurement. The temperature of the instrument was calibrated against the melting points of indium, zinc, tin, silver and gold. The balance was calibrated over the temperature range with standard alumina weights as provided by the vendor.

Results

Complex formation and analysis by single crystal XRD

The ability of organic molecules to complex ions in solution is important to the process of crystallization. This is because complexation will lead to changes in the activity of ions and so must be taken into consideration when determining supersaturation. Additionally, by understanding the solution interactions of the ion with the organic molecule it is hoped to better understand the surface interactions.

It was found that mellitic acid complexed insignificant amounts of barium at pH 6 as determined by either the use of the barium ISE or conductivity (conductivity data shown in ESI, S1†) up to concentrations of 10 ppm or 0.03 mM mellitic acid.

Fig. 2 SEM images of (a–c) barium mellitate particles and (d) barium sulfate particles formed after heating barium mellitate in the presence of sulfate to 60 °C (see text for further details).

Crystals of barium mellitate were obtained and single crystal XRD data taken for structure determination. Barium mellitate was found to readily crystallize from mellitic acid solutions of pH > 9 at concentrations as low as 0.03 mM. When sulfate solution was added, no barium sulfate precipitated despite the supersaturation for barium sulfate being ~16 even when complete complexation to mellitic acid was assumed. The barium mellitate particles were long and needle-like as can be seen in Fig. 2a and c. The ends did not appear to have sharp well defined faces (Fig. 2b) and had glassy type breakages. When the slurry containing barium mellitate particles and sulfate in solution was heated to 60 °C overnight, barium sulfate particles were seen to precipitate (rhombic particle in Fig. 2d).

The barium mellitate crystals were of sufficient quality for a single crystal X-ray structure determination, the results of which were consistent with a formulation of $[Ba_3(mellitate)(H_2O)_{10}]\cdot 4H_2O$. There are three unique barium atoms in the structure, all of which are nine-coordinate with the coordination sphere consisting of carboxylate O atoms and water molecules, three H_2O molecules around Ba1 and Ba3 and four around Ba2. Ba1 and Ba3 form two seven membered chelate rings, bridging adjacent carboxylate groups of the same ligand, as well as a four membered chelate ring to both O atoms of one carboxylate group. Ba2 is similar, except that only one seven-membered chelate ring is found, the other being replaced by a unidentate mellitate ligand and the additional water O atom (Fig. S2†). As a result, the mellitate anion is bridged across all carboxylate groups aside from group 1 to 6, with groups 1, 3 and 5 also bound as O,O′ bidentate (Fig. S3†).

The structure consists of a 3-D polymer with large channels running through the lattice in the *a* direction (Fig. 3, occupying water molecules not shown). The Ph ring lies almost in the *bc* plane, the dihedral angle being 8.3 degrees. The Ba atoms bridge between ligands created by the crystallographic *a* glide forming stacks along the *a*-axis, (centroid–centroid = 6.7 Å) (this is seen in the cell plot down *a*, ESI S4†). There are also bonds from the Ba atoms to the O atoms of other

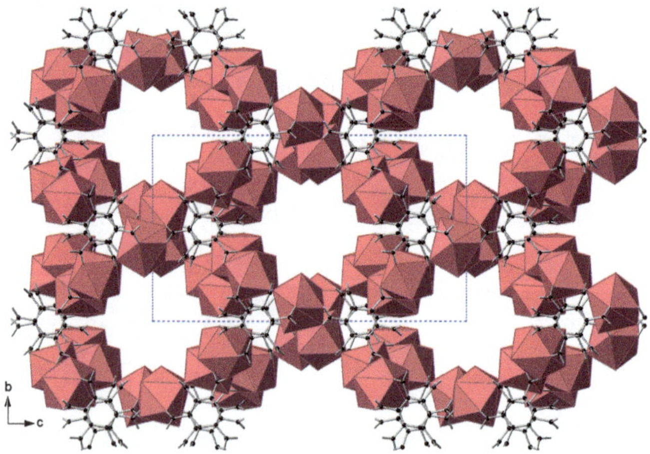

Fig. 3 Crystal packing in the $[Ba_3(mellitate)(H_2O)_{10}]\cdot 4H_2O$ crystal structure. BaO_9 centres are represented as polyhedral, and water molecules occupying the channels have been omitted for clarity.

ligands generated by the crystallographic 2_1 screw along the b axis and by the inversion centres, which create the 3D coordination polymer.

TGA

The results from the thermogravimetric analysis of the solids are shown in Fig. 4. Clearly, little mass is lost from pure barium sulfate at temperatures ≤800 °C. The barium mellitate complex loses water up until 100–150 °C and decomposes above 650 °C. The water loss from barium mellitate is calculated to be ~25.8 wt%. This is only marginally higher than 25.2 wt% expected from the single crystal data and

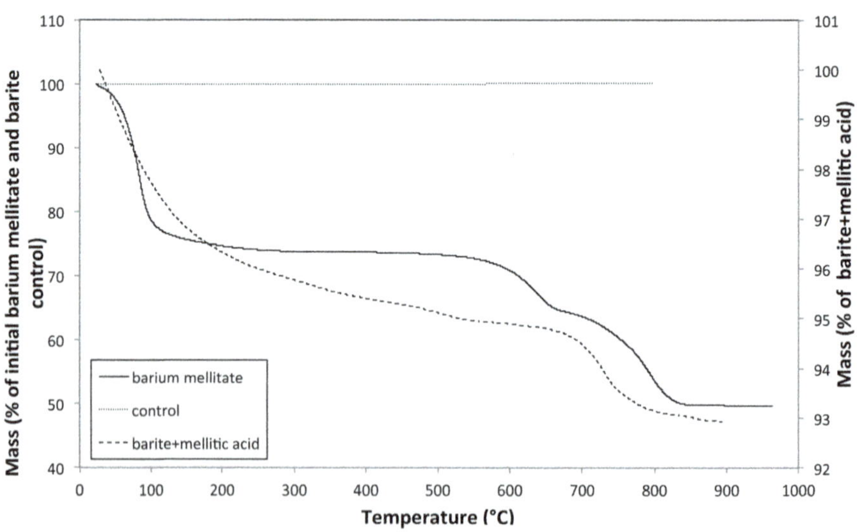

Fig. 4 TGA curves for pure barium sulfate, barium mellitate and barium sulfate formed in the presence of mellitic acid.

may suggest some residual surface adsorbed water. The loss of water also suggests that the water found in the barium mellitate structure is weakly bound, being driven off at temperatures less than 200 °C.

The barium sulfate particles formed in the presence of mellitic acid at 0.03 mM do not show a plateau in the mass loss above 100 °C. Given that pure barium sulfate does not have a mass loss in this region and that water is being lost for barium mellitate, this would suggest a range of water environments (of differing energy) in this system. In fact, one would additionally conclude that, generally, more energy is required to drive out this water from the barium sulfate structure when formed in the presence of the mellitic acid. It is found that approximately 7.1 wt% of mass is lost for the baryte particles formed in the presence of mellitic acid. If we assume, just as found in the crystal structure of the complex, that for every molecule of mellitic acid that is present, 14 molecules of water are also present this would equate to 0.03 moles mellitate per mole $BaSO_4$. This is despite the fact that the solution molar ratio of mellitic acid to barium ions was roughly 1.2 : 1. Thus, only a small proportion of the mellitic acid adsorbs or incorporates into the solid structure. In order to determine whether the mellitic acid is mainly adsorbed on the surface or is also incorporated into the particles, the maximum amount of mellitic acid that could be surface adsorbed was calculated (by assuming the mellitate is not lying flat). For a 50 nm particle, adsorption on the surface would account for only half the mellitate found in the sample. Thus, it appears that the mellitate ion can both adsorb on the surface of the particles and/or incorporate into the barium sulfate solids.

Zeta potential

The zeta potential results for the barium sulfate particles formed in the absence of mellitic acid show a surface potential of less than −5 mV, confirming that these

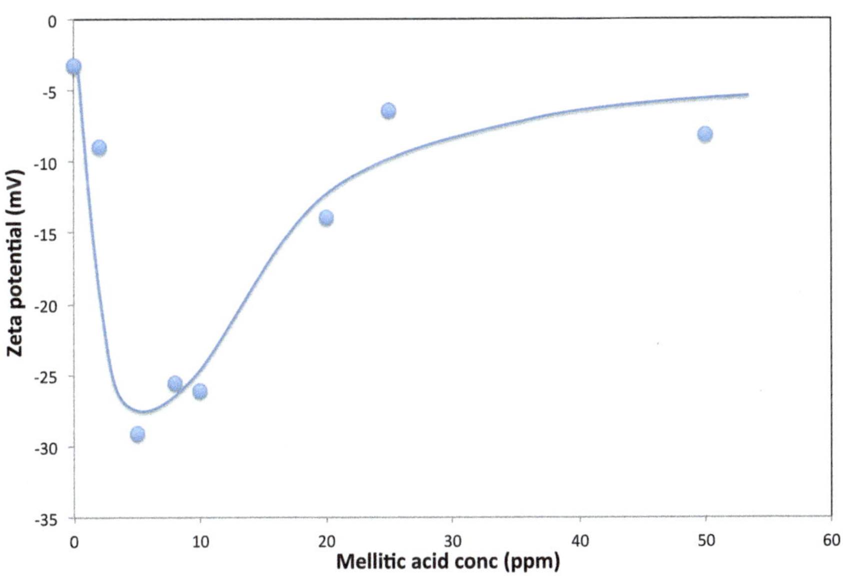

Fig. 5 Zeta potential measured on barium sulphate particles after 3 days in the solution in which they were formed (line drawn to aid reader only).

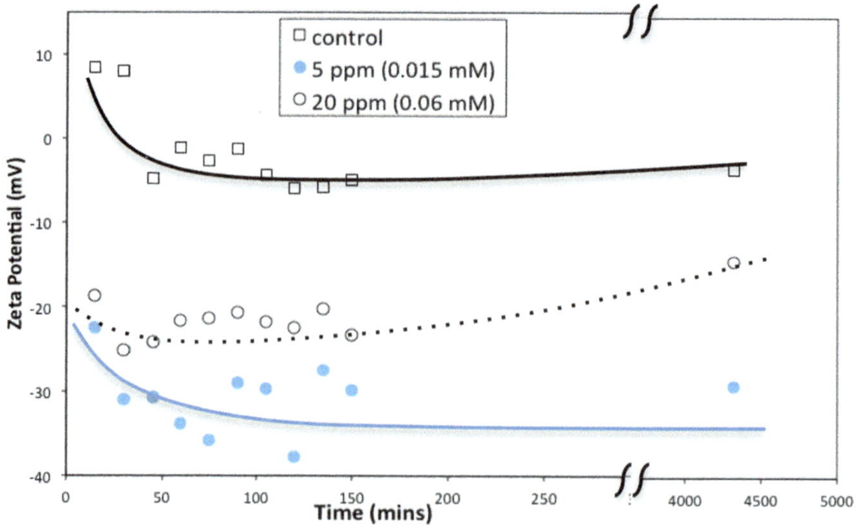

Fig. 6 Zeta potential measured on barium sulphate particles *versus* time in the solution in which they were formed for 0, 5 and 20 ppm mellitic acid present (lines drawn to aid reader only).

particles essentially carry little to no surface charge (Fig. 5). Fig. 5 shows the zeta potential measured on barium sulfate particles after three days suspended in the solution in which they were prepared. From this figure it is observed that the zeta potential reaches a minimum value of ~-30 mV before increasing dramatically to ~-10 mV at higher mellitic acid concentrations. Thus, at the minimum zeta potential the barium sulfate particles are on that borderline region between stability and coagulation (it is a common rule of thumb that slurries with surface charges larger, in absolute terms, than ±30 mV are stable). Thus, between 5–10 ppm mellitic acid, the nanoparticles could be charge-stabilised. This raises an interesting point. If the particles are within the surface charge range where coagulation is expected, why is it that mesocrystals are still able to form? What is the nanoparticle stabilisation mechanism that allows them to self-assemble outside of this concentration range?

Perhaps, the answer lies with how the zeta potential evolves over time. At 5 ppm (0.015 mM) mellitic acid, it is clearly seen that the zeta potential of baryte particles decreases quickly over the first 30–60 minutes and then plateaus on longer time scales (Fig. 6). The behaviour is similar for the control even if the magnitudes are different. However, when 20 ppm (0.06 mM) mellitic acid was present, initially a slightly lower minimum was found (~-25 mV) but this increased over time to ~-14 mV. Thus, at higher mellitic acid concentrations the initial surface charge on the barium sulfate particles is only slightly less negative while the steady state value after three days can be significantly less negative.

This may point to the mesocrystal formation occurring soon after formation of the nanoparticles when the zeta potential is at its most negative.

AFM

AFM of the growing barium sulfate (001) face in the presence of mellitic acid showed an interesting phenomenon whereby the surface became significantly

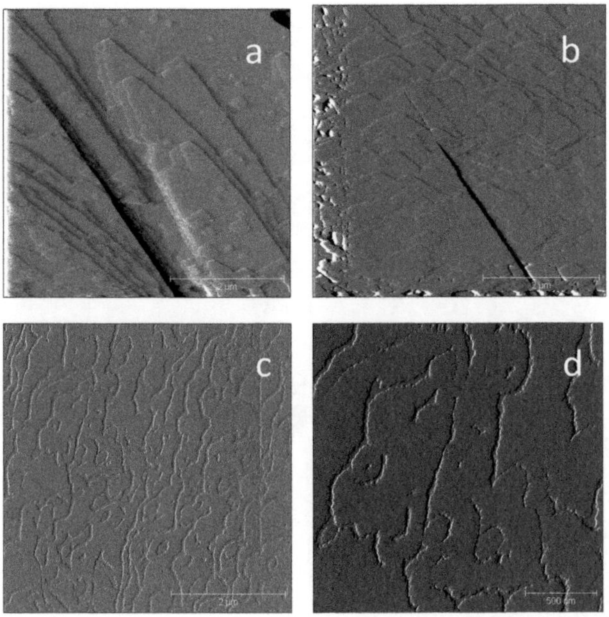

Fig. 7 AFM deflection images of the barium sulfate (001) face growing with (a) no mellitic acid present, (b) 0.1 ppm (c) 0.2 ppm and (d) 0.2 ppm mellitic acid present at a higher magnification. Scale bars are 2 μm except for (d) where it is 500 nm.

roughened compared to the control (Fig. 7a, e and d and the line profile shown in the ESI, S4†). This made measurement of growth rates and nucleation rates for the 2D islands almost impossible.

In order to determine the distribution of mellitic acid on the surface of the growing particle, confocal Raman mapping was conducted. The amount of mellitic acid on the surface was not sufficient to observe its Raman spectrum over and above that of the barium sulfate Raman spectrum (Fig. 8d), however, a broad fluorescence band was observed that was not seen in the control. Mapping of the intensity of the fluorescence proved more sensitive than the Raman spectra and is shown in Fig. 8c. The fluorescence was observed on the entire surface but some areas were more fluorescent than others as seen in Fig. 8c. Assuming that the intensity of the fluorescence is related to the quantity of adsorbed mellitic acid, it is clear that the mellitic acid is non-uniformly distributed on the surface of the barium sulfate. This distribution of the mellitic acid is broadly consistent with the observations from AFM.

Discussion

The mellitic acid is known to be an integral part of the initially forming nuclei of baryte particles as determined from infrared data previously obtained.[18] Moreover, these particles are initially amorphous and transform over time to crystalline solids[18] confirming a non-classical pathway for formation of the nuclei. Standard turbidimetric data suggested that the nucleation rate was significantly reduced in the presence of mellitic acid (see ESI, S5†), however, more sensitive

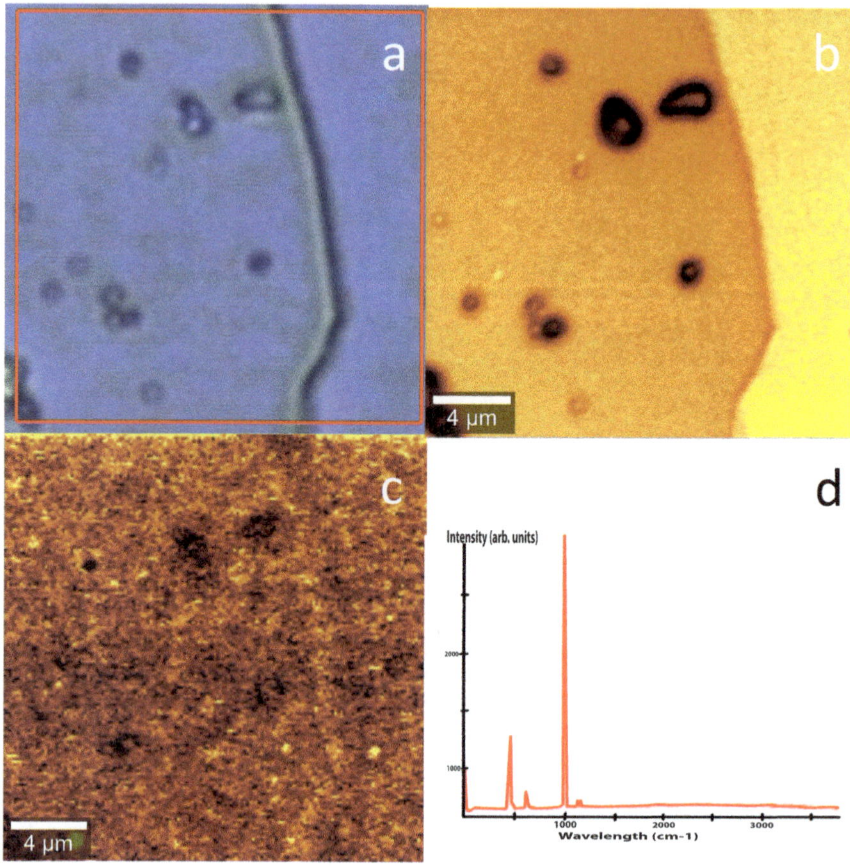

Fig. 8 Raman confocal mapping experiment. (a) Optical image of area mapped outlined by red square, (b) map associated with the 990–1000 cm^{-1} sulfate peak, (c) map associated with the fluorescence intensity and (d) typical Raman spectrum observed for the area.

DLS data did not show decreased particle counts in the presence of mellitic acid nor was the onset of the high particle counts delayed when mellitic acid was present (see ESI, S6†). This can be rationalised by the fact that the standard turbidimetric method fails to detect particles less than 400 nm,[22] thus 3D nucleation may not be inhibited at all. In addition, the roughness observed in the AFM images when mellitic acid is present suggests that 2D nucleation is promoted. Thus, while complexation of barium with mellitate is not expected to be significant, the presence of mellitic acid appears to promote the 2D nucleation rate of barium sulfate.

From the complex formed between barium and mellitate ions it is clear that water is a significant part of the structure and that significant amounts of water are also found in the baryte solids formed. Thus, the formation of amorphous barium sulfate nuclei can be reconciled with the substantial amount of water present meaning the crystallization process would involve removal of this water. This would also correlate with the internal porosity and ageing effects seen for the

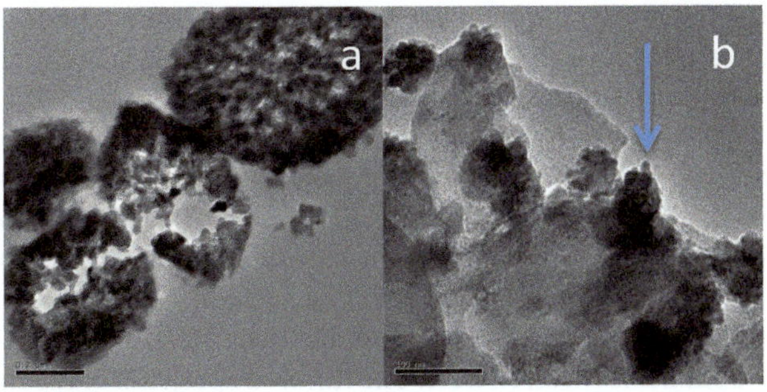

Fig. 9 TEM images of barium sulfate formed in the presence of mellitic acid after (a) 24 h and (b) 5 hours (blue arrow highlights one of many smaller particles, which appear to make up the 50 nm particles).

particles (where some fusion of the mesocrystal form is observed, see Fig. 9a) as well as the TGA analysis. From the infrared data we have seen that the organic molecule de-protonates on adsorption[18] (mellitic acid FTIR peaks are consistent with the fully de-protonated mellitate ion, L^{6-}). This behaviour (de-protonation on adsorption) has been previously observed for other carboxylate systems.[23] In addition, the promotion of 2D nucleation by mellitic acid may be due to the association of the barium ions to the mellitate ions after adsorption, especially since dehydration of the barium ions is known to be a rate-limiting step for baryte crystallization.[24,25] It is also seen that while the 'particles' of baryte are ~50 nm in diameter they appear to be made of smaller particles still, of ~10 nm in diameter (see Fig. 9b).

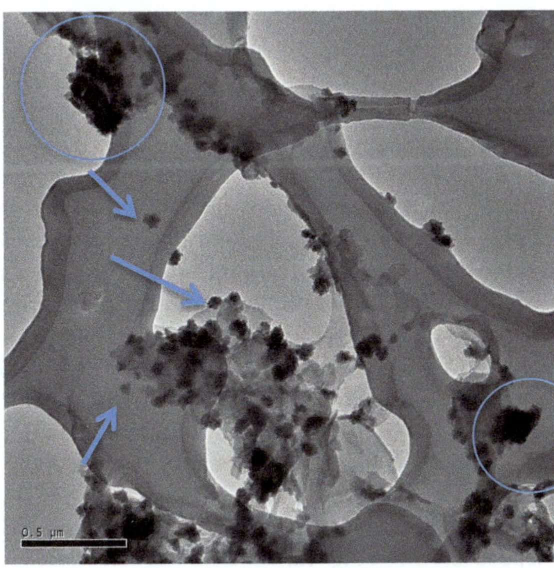

Fig. 10 TEM image of solids obtained after 5 hours during crystallization of barium sulfate in the presence of 0.03 mM mellitic acid. Blue arrows highlight ~50 nm nanoparticles, circles highlight self-assembled aggregates.

Fig. 10 shows that the nanoparticles do not change size or shape appreciably over time, but rather simply aggregate, suggesting that this is not a dissolution–recrystallization process but a solid-state transformation.

After nucleation, the ~50 nm particles self-assemble into almost rhombohedral structures (the evolution of the shape is discussed later). This is at a time when the zeta potential is found to be at its most negative but is mellitic acid concentration dependant. The fact that the surface is negative implies an excess of sulfate or an excess of mellitate ions on the surface. As mellitic acid concentration alters the zeta potential it is reasonable to suggest that the mellitic acid rather than the sulfate is involved here. The infrared spectrum of the mellitic acid during barium sulfate crystallization[18] is consistent with the L^{6-} species[26] meaning that all the carboxylate groups are de-protonated. Thus adsorption of mellitate ions increases the zeta potential. Clearly, these adsorbed ions are not fully screened by barium ions or the surface charge would be more positive. As the concentration of mellitic acid is increased, further adsorption of mellitate would be expected, however further increases in mellitic acid show a decrease in zeta potential. This can only be reconciled if the mode in which mellitate adsorbs changes, either by attracting more barium ions to it (screening the charge) or by changing the physical form in which it adsorbs that results in more carboxylates being incorporated into the growing surface.

At first it may appear surprising that the mellitic acid adsorbs as the completely deprotonated ion. This can be understood in terms of the speciation of the mellitic acid, which at pH 6 while not completely de-protonated has up to 26% as the L^{6-} and 62% L^{5-} ligand.[26] Thus, adsorption could alter the solution equilibrium such that more and more L^{6-} is formed as it is removed by the adsorption process until adsorption is no longer favoured. The presence of even small amounts of L^{6-} ligand may also explain the 2D nucleation promotion. If complexation with barium ions is preferred with the fully de-protonated ligand, the presence of this small amount of L^{6-} may promote the de-watering of the barium ion even if it does not significantly impact on the activity of the barium ions in solution.

What is the mesocrystal shape? From the SEM images, the morphology of the particles formed changes only slightly with the concentration as shown in Fig. 11 when mellitic acid is present at 0.075 mM. As concentration increases, the particles become more elongated and more inter-grown solids are seen. In addition, a second population of very small almost rhombic particles begin to appear. The shape of the mesocrystals now becomes clear; although very rounded, the self-assembly process is to form essentially rhombic particles (compare these small mesocrystals to the rhombic particle in Fig. 2d). This shape is, in essence, a flattened version of the thermodynamic shape.[27,28] This interpretation is supported by measuring the angles of the less elongated particles which show an obtuse angle of ~105° compared to the theoretical obtuse angle for the rhombohedron of 101.7°. The elongated particles appear to be developing (010) faces (the flattened ends) and the obtuse angle here is measured to be ~130°, which may imply the (210) faces are becoming less well defined over time. A schematic of the mesocrystal shape is shown in Fig. 12.

Finally, even after mesocrystal formation, the aggregation state also changes at high mellitic acid concentration due to the low zeta potential on the surface of the particles (as can be seen in the ESI, S7†).

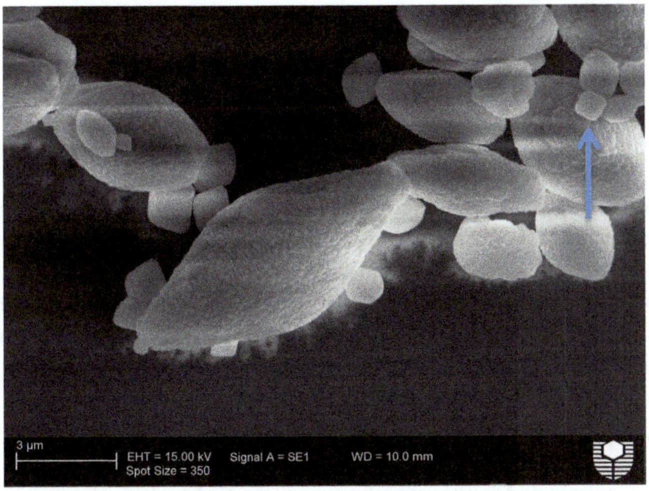

Fig. 11 SEM image of barium sulphate mesoparticles formed in the presence of 0.075 mM mellitic acid. Blue arrow highlights a rhombic aggregate.

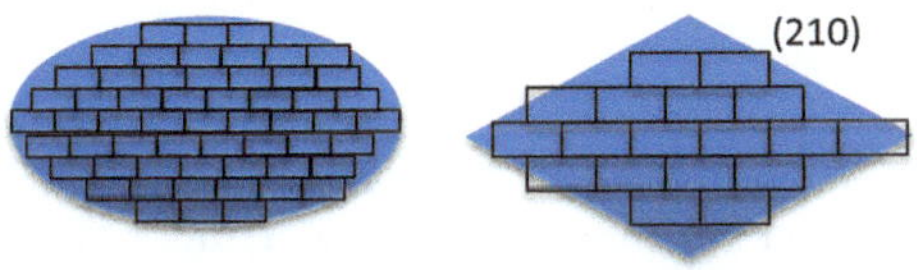

Fig. 12 Schematic of mesocrystal growth showing the 210 diagonal faces in the rhombic particle when viewed from the top.

Conclusions

Mellitic acid can be seen to be an intimate partner during the crystallization of the baryte solids. It appears that it does not inhibit 3D nucleation despite the turbidity results, leading to the formation of small 50 nm particles, which are themselves aggregates of even smaller particles. At this stage the mellitic acid associated with the solids is the fully de-protonated ligand and results in a minimum surface charge. The presence of mellitic acid also limits the growth of the particles. It can be seen in TEM images that once the particles have approached the 50 nm size, no further growth occurs and aggregation leads to larger assemblies. These assemblies appear to orient themselves into the thermodynamically stable morphology albeit with a short *c* axis. At high mellitic acid concentration, these assemblies also coagulate due to the low surface charge.

While turbidimetric studies showed a long induction time even with low mellitic acid concentration, implying 3D nucleation inhibition, this may not be the case as the same trend is not observed in DLS measurements. AFM images show many small particles deposited on the surface, too numerous to count, showing 2D nucleation promotion. The TEM images also support the formation of small nanometric particles of baryte. The promotion of 2D nucleation is most probably due to association of the barium ions with the mellitate ion (which exists in solution at low concentration) resulting in at least partial de-watering of the barium ion.

Of particular interest is that the surface charge measured for the nanoparticles, while repulsive, is not large, thus surface charge may direct mesocrystal formation but one would assume only for those mellitic acid concentrations between 5–10 ppm. Above or below these concentrations random coagulation would be expected. It is possible that for short times the surface charge on the 50 nm particles may be sufficient to induce mesocrystal formation, however, the observation that the mesocrystal morphology is very similar to the thermodynamic morphology may suggest an oriented attachment mechanism unrelated to the surface charge.

Acknowledgements

The authors acknowledge the use of equipment and technical assistance of the Curtin University Electron Microscope Facility, which has been partially funded by the University, State and Commonwealth Governments, Peter Chapman for providing the thermogravimetric data and Tom Becker for use of, and technical assistance with, the AFM facility at Curtin University.

References

1 D. W. Oxtoby, *J. Phys.: Condens. Matter*, 1992, **4**, 7627–7650.
2 D. Erdimemir, A. Y. Lee and A. S. Myerson, *Acc. Chem. Res.*, 2009, **42**, 621–629.
3 L. Addadi, S. Raz and S. Weiner, *Adv. Mater.*, 2003, **15**, 959–970.
4 M. J. Olszta, D. J. Odom, E. P. Douglas and L. B. Gower, *Connect. Tissue Res.*, 2003, **44**, 32–334.
5 M. Niederberger and H. Cölfen, *Phys. Chem. Chem. Phys.*, 2006, **8**, 3271–3287.
6 T. Wang, A. Reinecke and H. Cölfen, *Langmuir*, 2006, **22**, 8986–8994.
7 T. Wang and H. Cölfen, *Langmuir*, 2006, **22**, 8975–8985.
8 S. H. Yu, M. Antonietti, H. Cölfen and J. Hartmann, *Nano Lett.*, 2003, **3**, 379–382.
9 L. Qi, H. Cölfen, M. Antonietti, M. Li, J. D. Hopwood, A. J. Ashley and S. Mann, *Chem. - Eur. J.*, 2001, **7**, 3526–3531.
10 L. Qi, H. Cölfen and M. Antonietti, *Chem. Mater.*, 2000, **12**, 2392–2403.
11 L. Qi, H. Cölfen and M. Antonietti, *Angew. Chem., Int. Ed.*, 2000, **39**, 604–607.
12 A. Baynton, T. Radomirovic, M. I. Ogden, C. L. Raston, W. R. Richmond and F. Jones, *CrystEngComm*, 2011, **13**, 109–112.
13 A. Baynton, M. I. Ogden, C. L. Raston and F. Jones, *CrystEngComm*, 2012, **14**, 1057–1062.
14 M. I. Ogden, C. L. Raston, T. Radomirovic and F. Jones, *Cryst. Growth Des.*, 2014, **14**, 1419–1429.
15 M. Li, H. Schnablegger and S. Mann, *Nature*, 1999, **402**, 393–395.
16 A. Ahniyaz, Y. Sakamoto and L. Bergstrom, *Proc. Natl. Acad. Sci. U. S. A.*, 2007, **104**, 17570–17574.
17 S. R. Freeman, F. Jones, M. I. Ogden, A. Oliviera and W. R. Richmond, *Cryst. Growth Des.*, 2006, **6**, 2579–2587.
18 F. Jones, *CrystEngComm*, 2012, **14**, 8374–8381.

19 A. Baynton, T. Becker, B. D. Chandler, F. Jones, M. I. Ogden, T. Radomirovic and G. K. H. Shimizu, *CrystEngComm*, 2011, **13**, 1090–1095.
20 D. L. Parkhurst and C. A. Appelo, *Users guide to PHREEQC (version 2) – a computer program for speciation, batch reaction, one dimensional transport, and inverse geochemical calculations*, U.S Geological Survey Water-Resources Investigation Report, 1999, pp. 99–4259.
21 G. M. Sheldrick, *Acta Crystallogr., Sect. A: Found. Crystallogr.*, 2008, **64**, 112–122.
22 T. Radomirovic, P. Smith and F. Jones, *Int. J. Miner. Process.*, 2013, **118**, 59–64.
23 F. Jones, P. Jones, M. I. Ogden, W. R. Richmond, A. L. Rohl and M. Saunders, *J. Colloid Interface Sci.*, 2007, **316**, 553–561.
24 S. Piana, F. Jones and J. D. Gale, *CrystEngComm*, 2007, **9**, 1187–1191.
25 S. Piana, F. Jones and J. D. Gale, *J. Am. Chem. Soc.*, 2006, **128**, 13568–13674.
26 B. B. Johnson, S. Sjöberg and P. Persson, *Langmuir*, 2004, **20**, 823–828.
27 F. Jones and A. L. Rohl, *Mol. Simul.*, 2005, **31**, 393–398.
28 F. Jones, W. R. Richmond and A. L. Rohl, *J. Phys. Chem. B*, 2006, **110**, 7414–7424.

Faraday Discussions

PAPER

Modelling the stochastic behaviour of primary nucleation

Giovanni Maria Maggioni and Marco Mazzotti*

Received 9th December 2014, Accepted 23rd December 2014
DOI: 10.1039/c4fd00255e

We study the stochastic nature of primary nucleation and how it manifests itself in a crystallisation process at different scales and under different operating conditions. Such characteristics of nucleation are evident in many experiments where detection times of crystals are not identical, despite identical experimental conditions, but instead are distributed around an average value. While abundant experimental evidence has been reported in the literature, a clear theoretical understanding and an appropriate modelling of this feature is still missing. In this contribution, we present two models describing a batch cooling crystallisation, where the interplay between stochastic nucleation and deterministic crystal growth is described differently in each. The nucleation and growth rates of the two models are estimated by a comprehensive set of measurements of paracetamol crystallisation from aqueous solution in a 1 mL vessel [Kadam *et al.*, *Chemical Engineering Science*, 2012, **72**, 10–19]. Both models are applied to the cooling crystallisation process above under different operating conditions, *i.e.* different volumes, initial concentrations, cooling rates. The advantages and disadvantages of the two approaches are illustrated and discussed, with particular reference to their use across scales of nucleation rate measured in very small crystallisers.

1 Introduction

The importance of being able to describe and predict the course of primary nucleation in crystallisation processes cannot be overestimated. This is however challenging, both because monitoring and characterizing nucleation events experimentally is always difficult, and because scaling up information and models obtained in the lab to the scales of industrial processes is often not accurate enough.

Critical nuclei and sub-critical clusters are in fact nano-scale objects, which can not normally be detected when formed upon nucleation, but only after growing to a size large enough to occupy a significant volume fraction of the suspension. Hence, the detection (or induction) time, t_d, follows the nucleation time, t_n, in a time interval that depends not only on the nucleation rate but also on

ETH Zurich, Institute of Process Engineering, Sonneggstrasse 3, Zurich, Switzerland. E-mail: marco.mazzotti@ipe.mavt.ethz.ch; Tel: +41 44 632 24 56. Fax: +41 44 632 11 41

the growth rate, *i.e.* the speed at which newly formed nuclei grow while consuming the solute in the mother liquor.

Moreover, nucleation is on the one hand affected by mixing, heat transfer limitations and materials, whose characteristics vary when scaling up, and on the other hand exhibits poor reproducibility, in ways that are difficult to rationalise and control.

There are two interpretations of such poor reproducibility. When viewing nucleation and growth as deterministic phenomena, it is attributed to a high sensitivity to experimental conditions, which are difficult to reproduce exactly in repeated experiments. Such difficulty is well known to experimentalists, and is indeed an issue to be taken into serious consideration.

Otherwise, variability in the outcome of nucleation experiments is attributed to an intrinsic stochastic behaviour in the formation of critical nuclei, of which there is abundant experimental evidence (see the references reported in Table 1). This manifests itself in the fact that nucleation times (or induction times) are distributed statistically when many identical experiments are repeated, something that is typically only possible using vessels of very small size, namely in the nanolitre to millilitre range.

Table 1 Overview of relevant literature on stochastic nucleation from solution, and the key features of each study. The category "Stochastic process" applies to the cases where stochastic nucleation is observed but no further specification about its nature is provided

Literature Overview				
Reference	Technique	Volume	Detection	Interpretation
1 and 2	Microfluidic chip	2–250 nL	Optical microscopy	Poisson process
3		0.1–1000 μL		
4 and 5	Microfluidic chip	2–150 nL	Light microscopy	Poisson process
6	Microfluidic chip	5 μL	Optical microscopy	Poisson process
7				
8	LQELT	10–20 nL	Light-polarisation	Stochastic process
9			Voltage Measurements	Poisson process
10	Crystallisation platform	1.5 mL	Transmittivity	Poisson process
11	Various reactors	10–1000 mL		Poisson process
12	Crystallisation platform	10–1000 mL		Poisson process
13 and 14	Crystallisation platform	6 mL	Optical microscopy	Stochastic process
15	Various reactors	15–150–500 mL	Direct observation	Stochastic process
16	Crystallisation platform	5 mL		Poisson process
17		15 mL		Stochastic process
18		30 mL		Poisson process
19	Crystallisation platform	1–2 mL	Transmittivity	Poisson process
20	Various reactors	1.8–250 –1000 mL	Turbidity	Stochastic process

The conventional interpretation is that the formation of a critical nucleus is a rare event, which can be classified as a Poisson process. As a consequence, the number of nuclei obeying a Poisson distribution, and the nucleation times are distributed as follows:

$$P(t) = 1 - \exp\left(-\int_0^t k(t')\mathrm{d}t'\right) \tag{1}$$

where t is time; $P(t)$ is the probability that nucleation (the formation of the first stable nucleus) occurs between $t = 0$ and t; and $k(t)$ is the rate constant of the Poisson process, which is called homogeneous if k is time independent and inhomogeneous otherwise. The rate constant k multiplied by a generic time interval Δt gives the probability that an event, *i.e.* the formation of a new nucleus, occurs in that time interval. Note that in the previous equation as well as in the following ones, we assume that the system is saturated or supersaturated from $t = 0$ onwards.

In the present context k equals the product of the volume of the solution, V, where nucleation can occur, and of the rate of nucleation, J, given in number of particles per unit time and unit volume, as derived within the classical nucleation theory (CNT). A homogeneous Poisson process corresponds to nucleation of a solution at a constant volume V and constant nucleation rate, J, *i.e.* constant supersaturation and constant temperature. An inhomogeneous Poisson process corresponds to nucleation triggered by cooling of the solution or solvent evaporation, *i.e.* under conditions where supersaturation or temperature or volume, or combinations thereof, change over time. The average nucleation time, $t_\mathrm{n}^\mathrm{avg}$, and the standard deviation of its distribution, s_n, are calculated through the moments of the derivative $\mathrm{d}P/\mathrm{d}t$ of the cumulative distribution (1) as:

$$t_\mathrm{n}^\mathrm{avg} = \int_0^\infty tk(t)\exp\left(-\int_0^t k(t')\mathrm{d}t'\right)\mathrm{d}t \tag{2}$$

$$s_\mathrm{n} = \left(\int_0^\infty \left(t - t_\mathrm{n}^\mathrm{avg}\right)^2 k(t)\exp\left(-\int_0^t k(t')\mathrm{d}t'\right)\mathrm{d}t\right)^{1/2} \tag{3}$$

Note that in the case of a homogeneous Poisson process, $t_\mathrm{n}^\mathrm{avg} = s_\mathrm{n} = 1/k = 1/(JV)$. This shows that not only the average nucleation time, but also the standard deviation decrease with volume – hence the expected variability in the measurements of nucleation times.

From this interpretation a seemingly reassuring picture seems to emerge: the stochastic nature of nucleation fades away as V increases, and the variability in measured nucleation times vanishes accordingly. It is also worth noting that in the frame of a deterministic description of nucleation, where JV gives the deterministic rate of formation of new crystals per unit time in the volume V, the nucleation time t_n can be predicted as the time when the first nucleus is deterministically formed. This is the time at which the following integral, which gives the number of crystals that have formed, equals 1:

$$\int_0^{t_\mathrm{n}} J(t)V(t)\mathrm{d}t = 1 \tag{4}$$

At constant nucleation rate and volume, $t_n = 1/(JV)$, *i.e.* the nucleation time in the deterministic interpretation of nucleation equals the average nucleation time of eqn (1) in the stochastic description (at constant supersaturation and volume). The crucial difference is, of course, that in the former interpretation all nucleation experiments are perfectly reproducible at any scale, whereas in the latter description the nucleation times and hence experimental outcomes are distributed, their degree of variability decreasing monotonically with increasing system volume.

It is worth noting that a stochastic behaviour in smaller volumes and a deterministic behaviour in larger volumes are consistent with the mononuclear mechanism and the polynuclear mechanism, respectively, used by Kashchiev, Verdoes and van Rosmalen to describe nucleation and to derive expressions for the experimental induction time in their seminal paper.[1]

Unfortunately, there are several pieces of evidence that urge us to challenge the picture of nucleation above.

First, experimental evidence indicates that nucleation is also poorly reproducible in large vessels, *i.e.* where the standard deviation of nucleation times should be so small as to be effectively negligible.[2,3]

Secondly, the description of nucleation as a Poisson process is correct only if the evolution of the system at any time does not depend on previous stochastic events. This is true only for the formation of the first nucleus, but not for the next ones, particularly when the further growth of nuclei that were previously formed contributes to the depletion of supersaturation.

Finally, the nucleation time is generally not observable. What one can measure is the detection time t_d, which, although differently defined depending on the system and the measurement apparatus, is typically the time when a certain fraction of the solution volume is occupied by the newly formed crystals. In this case, one needs information about the crystal growth rate in order to quantitatively interpret the outcome of the experiments. Alternatively, and typically in microfluidics, one uses optical microscopy to look at many micro-crystallisers and to measure in how many of them was at least one particle formed. Both types of information can be used to build statistics, but they present different challenges when one wants to interpret the data quantitatively using a model.

Though in many ways successful from an experimental or a theoretical perspective, the papers listed in Table 1 have not yet answered a question of key importance. Can the nucleation kinetics estimated in small volumes be used to describe nucleation when scaling up the process? Although we are fully aware that when scaling up, many features of the crystalliser and the process can lead to deviations and differences, we do believe that there might be a further explanation to justify the difficulties experienced and reported in the literature, *e.g.* in a recent study carried out by a group at the Technical University of Delft (TU Delft). We believe that there is a need to find ways to account for the stochastic behaviour observed at smaller scale in the description of nucleation at larger scale, and this is indeed the objective of this paper. First, we will summarise the results reported earlier, that will be used as backbone and benchmark for our analysis. Then, we will use two different models to describe those data. The first model, though similar to that proposed by the TU Delft group, seems to us to be physically more consistent; the second one, to the best of our knowledge, is novel and attempts to also incorporate stochasticity in the description of nucleation processes where a large number of crystals are formed, in an arbitrary large volume and non-constant temperature. Finally, we will discuss our findings and draw conclusions.

2 Background

2.1 Literature review

The relevant literature for this work, *i.e.* concerning stochastic effects in nucleation, is summarised in Table 1, where for each paper the system volume used, the detection technique applied and the interpretation of the experiments provided are reported.

Detection techniques can be divided into two main categories. The first relies on direct observation of crystals *via* optical microscopy, where the detection of crystals depends on the resolution of the optical instrument.[4–14] The second category encompasses the experiments where a macroscopic system property (such as transmittivity or turbidity), which is related to the volume fraction of the crystals, is measured.[2,15–24] Crystals are detected, and hence the detection time is measured, when such volume fraction exceeds a specific threshold value. This latter technique suffers from two weaknesses: first, the quantitative relationship between the solid volume fraction and, for instance, turbidity is difficult to calibrate; secondly, the threshold value is extremely difficult to estimate and might well be dependent on a series of factors, *e.g.* instrument sensitivity, system intrinsic properties, fluid-dynamic effects *etc.* Observation using optical microscopy is typically utilized in very small volumes, whereas macroscopic measurements are used in larger volumes.

The time associated with the observation of crystals in the system, obtained either by direct observation or by indirect measurements, can be used to quantify nucleation kinetics. To this purpose, a physical and mathematical model of the entire crystallisation process must be provided. Such a model, on the one hand, needs to account for the stochastic nature of nucleation itself, *e.g.* assigning statistics to describe the distribution of experimental data; and, on the other hand, needs to account for the combined action of nuclei formation, which is stochastic, and growth until detection, which is deterministic.

Assumptions concerning the growth process are hence necessary to interpret the experimental data. In microscopic systems, it is often possible to assume that growth is much faster than nucleation, so that the time between particle formation and detection is negligible, *i.e.* $t_n \approx t_d$.[9–14] Alternatively, also in microscopic volumes, one can decouple nucleation and growth and analyse them independently, for example by applying the so-called double-pulse technique.[4–8]

An example of a microfluidic study where an optical technique has been used is the work of the group at the University of Illinois at Urbana-Champaign that performed isothermal experiments on the crystallisation of paracetamol and lysozyme from evaporating aqueous micro-droplets[11]. Assuming that growth is infinitely fast, and hence that the statistics of eqn (1) can be directly applied to detection times, they could estimate the parameters of the following classical nucleation rate expression:

$$J(S) = q_1 S c^* \exp\left(-\frac{q_2}{\ln^2 S}\right) \tag{5}$$

where c^* is the solubility at the experimental temperature, $S = c/c^*$ is the supersaturation, and q_1 and q_2 are constant parameters. They showed that the estimated nucleation rate exhibited good predictive capabilities when changing the operating conditions (but not the droplets' volume).

2.2 TU Delft study

The study carried out by the group at the TU Delft will be used in the following as a benchmark.[2] They carried out a large number of experiments on the cooling crystallisation of paracetamol from aqueous solutions in vessels of different volumes. In particular, in a 1 mL crystalliser they explored four different initial paracetamol concentrations, always starting the linear cooling process from the corresponding saturation temperature. Then, they carried out systematic experiments in larger crystallisers, namely of volumes of 500, 700, 900 and 1000 mL. In each experiment the detection time was measured, and cumulative distribution curves of detection times were obtained at each initial concentration and for each crystalliser volume. The two sets of data at the lowest and highest concentration explored, namely 15 g L^{-1} and 47 g L^{-1} (the others being 22 g L^{-1} and 32 g L^{-1}) are reported in Fig. 1, where the cumulative probability is plotted as a function of the temperature change between time zero and the detection time, ΔT. It can be observed that the qualitative behaviour of the cumulative probability curves is in accordance with eqn (2) and (3), *i.e.* the average value of ΔT and hence the detection time, and the standard deviation decrease with increasing volume.

The data at 1 mL were then used to estimate the parameters of the model that was finally applied to predict the data measured in larger volumes. The model assumed that nucleation occurred through a mononuclear mechanism and was followed by growth of the single crystal to a threshold size r_a, where it was finally shattered by attrition, causing a burst of secondary crystals which were immediately detected. Therefore the detection time, t_d, is the sum of the nucleation time, t_n, and of the deterministic growth time, t_g. The latter is equal to the ratio between the threshold size r_a and a growth rate, G, which is assumed to be the same for all experiments at a given initial concentration. The nucleation time t_n, on the other hand, is assumed to be distributed according to the Poisson distribution of eqn (1), with $k(t) = VJ(t)$; the nucleation rate depends on time due to cooling and is given by the following expression:

$$J = a \exp\left(-\frac{b}{\ln^2 S}\right) \tag{6}$$

where a and b are in principle temperature dependent, but are assumed to be constant for each initial concentration level, although different for the different initial concentrations. The TU Delft group related the threshold size to a growth time, t_g using: $t_g = r_a/G$, with $G = k_g(S - 1)$, the crystal growth rate. After determining the threshold size through experiments, they estimated the three parameters k_g, a, and b for each of the four sets of 1 mL experiments, thus obtaining twelve parameters in total (see Tables 1 and 2 in the original paper[2]).

Although the agreement between the model results and the experimental results is quite satisfactory, there are a number of issues. Firstly, assuming that rates of nucleation and growth are temperature independent is clearly inconsistent with the experimental evidence, where metastable zone widths between

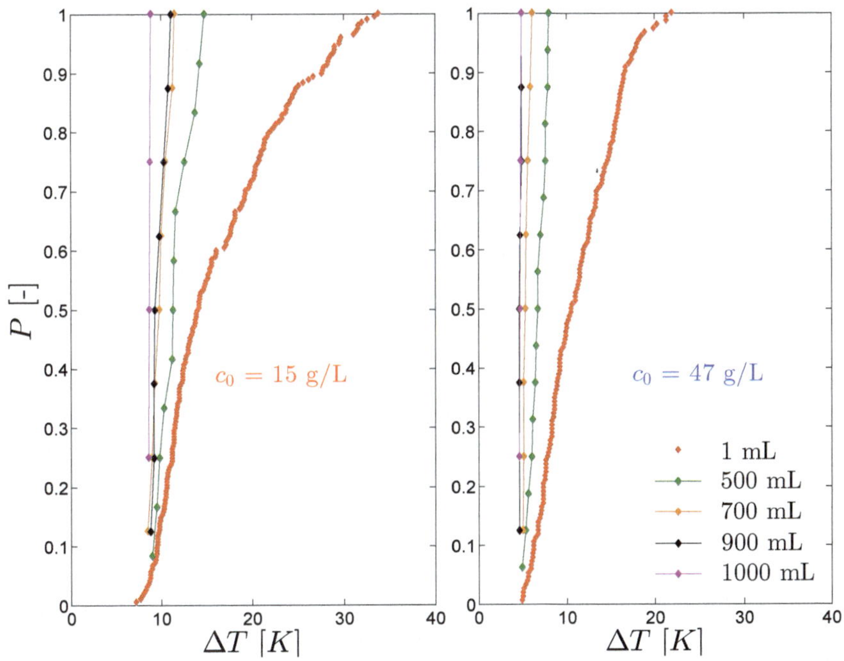

Fig. 1 The TU Delft experimental data of detection times, expressed as cumulative nucleation probability, P, *vs.* the extent of cooling, ΔT, for the cooling crystallisation of paracetamol from aqueous solution.[2] The experiments shown here represent two different initial concentrations, 15 g L^{-1} (left) and 47 g L^{-1} (right), in crystallisers of five different volumes. The lines connecting the data at larger volumes (from 500 to 1000 mL) are drawn as guides for the eye.

20 and 30 K are observed. Secondly, using different sets of parameters for different initial concentrations and temperatures allows for a rather empiric account of temperature effects, which makes it impossible to expect that the model parameters can be used in a predictive manner, *e.g.* to describe the effect of different initial concentrations or different cooling rates. Finally, there is no quantitative agreement with the measurements from the larger crystallisers ($V = 500$, 700, 900 and 1000 mL); as already noted by the authors, the Poisson model predicts a much faster transition to a deterministic behaviour with increasing volume than the one experimentally observed.

The difference between the model based on the 1 mL experiments and the measurements for larger volumes could be for several different reasons. For example, the mixing in the vessel might not be good enough to assure that the desired temperature profile is reproduced at all the tested scales. It could also be possible that hot or cold spots form locally in the larger crystallisers, thus locally altering the nucleation rate. Another explanation could be that, by changing the reactor, a different surface is provided to the system, which may change the nucleation mechanism. Moreover, the different batches might have been influenced by the presence of dust or other impurities in different amounts, hence preventing the possibility of a straightforward scaling. Finally, the detection limit determined in small volumes might not be appropriate in larger volumes. This was explicitly discussed in the frame of the theory of nucleation based on the mononuclear and polynuclear mechanisms, but was not considered in the work of the TU Delft group.

3 Modelling stochastic nucleation in crystallisation

Our goal is to analyse the different approaches to modelling crystallisation processes at different scales while accounting for stochastic nucleation. We will try and improve on the TU Delft approach, and use their experimental data as a case study and as a benchmark. More specifically, we will present and discuss two different models that differ in how nucleation is described, but that are based on common assumptions that are discussed in the following.

We describe a crystallising system of constant volume V undergoing batch cooling crystallisation, starting from an initial temperature T_0 where the solution is saturated and clear. During cooling, an arbitrary number N of particles can form. The size of the particles, having a single characteristic dimension L, is distributed according to the 1D-particle size distribution (PSD) $f(t, L)$; the quantity $f(t, L)\mathrm{d}L$ is the concentration of particles of a size between L and $L + \mathrm{d}L$; the jth-order moment of the PSD, $\phi_j(t)$, is defined as:

$$\phi_j(t) = \int_0^\infty L^j f(t, L)\mathrm{d}L \tag{7}$$

The linear cooling process is characterized by the following relationship:

$$T(t) = T_0 - \beta t \tag{8}$$

where the cooling rate β [K min^{-1}] is constant. Note that the initial solute concentration c_0 equals the solubility at the initial temperature, *i.e.* $c^*(T_0)$.

The zero-th order moment of the PSD, $\phi_0(t)$, is the number of particles per unit volume, hence:

$$N(t) = V\phi_0(t) \tag{9}$$

The third order moment of the PSD, $\phi_3(t)$ multiplied by k_v, represents the volume fraction of solids in the suspension; through the mass balance it is linked to the solute concentration $c(t)$ and to its initial concentration, c_0:

$$c_0 = c^*(T_0) = c(t) + k_v\rho_c\phi_3(t) \tag{10}$$

where ρ_c is the crystal density [kg m^{-3}], and k_v is the shape factor. In agreement with the literature,[2,3] the detection time t_d is the time when $k_v\phi_3(t)$ reaches a threshold value α, *i.e.*:

$$\frac{V_c(t_d)}{V} = k_v\phi_3(t_d) = \alpha \tag{11}$$

In the following we will always use the same value for such a threshold, namely $\alpha = 10^{-4}$.

In this work, as an exemplification but without loss of generality, we use a nucleation rate, J, and a growth rate, G, based on the classical nucleation theory and on the (size independent) birth and spread growth model, respectively.[3,25] They are expressed in such a way that their temperature dependence is explicitly accounted for:

$$J(S,T) = A_0 S \exp\left(-\frac{A_1}{T}\right)\exp\left(-\frac{B}{T^3 \ln^2 S}\right) \tag{12}$$

$$G(S,T) = K_0 \exp\left(-\frac{K_1}{T}\right)(S-1)^{2/3}(\ln S)^{1/6} \exp\left(-\frac{K_2}{T^2 \ln S}\right) \tag{13}$$

Both rates depend on three independent parameters each, namely A_0, A_1 and B for the nucleation rate, and K_0, K_1 and K_2 for the growth rate, and on the supersaturation, defined as $S = c/c^*(T)$, using an empirical temperature dependent solubility – namely that of paracetamol in water – which is given by the following polynomial expression:[26]

$$c^*(T) = \sum_{i=0}^{4} \Gamma_i (T - 273.15)^i \tag{14}$$

The parameters Γ_i have been fitted on experimental data from the literature.[2] See Table 2 for more details.

3.1 Model 1: stochastic onset of nucleation

In Model 1, we assume that the first nucleation event, *i.e.* the formation of the first nucleus, occurs as a Poisson process. That is, it follows the statistics of eqn (1), and triggers the start of a deterministic process, through which new nuclei are formed while existing ones grow; the deterministic part of the process can be modelled using a deterministic population balance equation (PBE). Additionally, we assume perfect mixing, no secondary nucleation, no agglomeration and no breakage. Such a model is consistent with the mononuclear nucleation mechanism, whereby in small volumes and for slowly nucleating substances, the appearance of the first nucleus triggers a change of regime, *i.e.* a transition from a metastable state to a deterministic evolution.[1] It also bears an obvious similarity

Table 2 The coefficients and physical parameters used in the present work

Parameter	Value
β [K min^{-1}]	0.5
ρ_c [kg m^{-3}]	1260
k_v [-]	$\pi/6$
Γ_0 [g L^{-1}]	7.147
Γ_1 [g L^{-1} K^{-1}]	1.986×10^{-1}
Γ_2 [g L^{-1} K^{-2}]	5.048×10^{-3}
Γ_3 [g L^{-1} K^{-3}]	-1.273×10^{-4}
Γ_4 [g L^{-1} K^{-4}]	3.0188×10^{-6}

to the simplified model proposed by the TU Delft group which is summarised in section 2.2.[2]

The process is described as follows. The cooling crystallisation process starts at $t = 0$, with the temperature following the linear cooling profile of eqn (8) and the solute concentration constant at $c = c_0$; no crystals are present. At time t_n the first nucleus appears; since nucleation is assumed to be a non-homogeneous Poisson process, the statistics of t_n follow eqn (1), given in the form:

$$P(t_n) = 1 - \exp\left(-V\int_0^{t_n} J(t')\mathrm{d}t'\right) \tag{15}$$

with J given by eqn (12). After time t_n, the system evolves in a deterministic manner following the equations reported below, whose initial conditions are assigned at $t = t_n$. Note that in term of PSD, and its moments, at t_n only one crystal of negligible size is present.

To describe the evolution of the crystals from t_n on, we use a 1D-PBE or, alternatively but equivalently, the associated *Equations of Moments* (MoM equations).[27,28] The standard 1D-PBE for a batch crystalliser where growth is size independent is:

$$\frac{\partial f}{\partial t} + G\frac{\partial f}{\partial L} = 0 \tag{16}$$

with G calculated through eqn (13), and initial and boundary conditions given by:

$$\begin{aligned} f(t = t_n, L) &= \frac{\delta(t - t_n, L)}{V} \\ f(t, L = 0) &= \frac{J}{G} \end{aligned} \tag{17}$$

The corresponding MoM equations are:

$$\frac{\mathrm{d}\phi_j}{\mathrm{d}t} = J\delta_{j0} + jG\phi_{j-1} \quad j = 0,\ 1,\ 2,\ \ldots \tag{18}$$

where δ_{j0} is the Kronecker symbol and $\phi_{-1}(t) = 0$; the initial conditions are:

$$\begin{aligned} \phi_0(t_n) &= \frac{1}{V} \\ \phi_j(t_n) &= 0 \quad j > 0 \end{aligned} \tag{19}$$

Since here we are interested only in computing the values of t_d according to eqn (11) and in some average properties, we solve the MoM eqn (18) for $j = 0$, 1, 2, 3.

It is rather clear that this model, which might be realistic in a very small volume of solution, loses credibility when the suspension volume increases. It is in fact counter-intuitive that the formation of a critical nucleus in a specific position in the solution volume immediately impacts the course of nucleation in the whole volume and renders the formation of further nuclei fully deterministic. Therefore, the sharp separation between a stochastic nucleation mechanism for the first crystal and a deterministic growth mechanism (as assumed here and in the TU Delft model) seems to be unphysical; this observation motivates us to pursue an alternative description where stochasticity and determinism coexist, at least during the early stages of the crystallisation process.

3.2 Model 2: stochastic primary nucleation

In the case of Model 2, we assume that all nuclei, not only the first one as in Model 1, are formed stochastically. After formation, each nucleus grows independently and deterministically, thus giving its deterministic contribution to the depletion of the supersaturation in solution. In such a framework, the time when older nuclei were stochastically born influences the conditions at which future nuclei are stochastically born.

Such effect of the past on the future leads to a new type of stochastic process, namely the so called *self-exciting point process*,[29] which can be viewed as a generalization of a non-homogeneous Poisson process (see eqn (1)) where the function $k(t)$ is stochastic, as it depends on a sequence of past stochastic events. In the case of self-exciting point processes in fact, the distribution of the time intervals between two subsequent nucleation events is the same as in a Poisson process, but the relevant value of $k(t)$ is not only time-dependent but also stochastic. In other words $k(t)$ depends on more than time t only. Therefore the self-exciting point process can be understood as a step-by-step non-homogeneous Poisson process. Therefore the probability, P_1, that at least one new nucleus forms in the time interval $\Delta t = t_i - t_{i-1}$ is the complement to the probability that in the same time interval no new nuclei form, P_0; one can thus write:

$$P_1(t_{i-1};t_i) = 1 - P_0(t_{i-1};t_i) = 1 - \exp\left(-V\int_{t_{i-1}}^{t_i} J(t')\mathrm{d}t'\right) \tag{20}$$

where the time dependence of the nucleation rate, J, is affected by the formation of nuclei in the previous time interval from 0 to $t = t_i$.

Each new crystal, which is born at $t = t_{n,k}$ and is labelled k ($\forall k = 1, 2, \ldots$), grows independently according to the following deterministic equation:

$$\begin{aligned} L_k &= 0 \quad t \le t_{n,k} \\ \frac{\mathrm{d}L_k}{\mathrm{d}t} &= G(S,T) \quad t > t_{n,k} \end{aligned} \tag{21}$$

The population of crystals is in this case discrete and the crystal volume fraction, which is proportional to its third moment, is given by:

$$\frac{V_c(t)}{V} = k_v\phi_3(t) = \frac{k_v\sum_{j=1}^{N(t)} L_j^3(t)}{V} \tag{22}$$

with $N(t)$ the number of crystals at time t. With this definition, the mass balance of eqn (10), and the detection condition, eqn (11), can be computed. Note that $N(t)$ is a monotonically increasing staircase function, with $N(0) = 0$ and all step increments equal to 1.

Modelling primary nucleation as a self-exciting point process has two important consequences. First, from a physical point of view, the self-exciting point process of Model 2 applied to the formation of all crystals in the system should exhibit a higher degree of variability than the Poisson process of Model 1, which applies only to the first nucleus. This is a straightforward consequence of the coexistence of stochastic and deterministic mechanisms during the formation and growth of all the particles in suspension. It should nevertheless be noticed that, in line with experimental evidence, the overall variability still reduces with increasing volume, since the probability of particle formation is governed by eqn (20), where the volume appears in the exponent and with a negative sign.

The second consequence is of a numerical nature: the self-exciting processes can be simulated exactly when their intensity rate $k(t)$ is known, as in our case, by applying an inversion algorithm. Starting from the initial condition, $N_0 = 0$, we generate a uniformly distributed random number and find the time corresponding to it drawn from the exponential distribution given in (20). That is, we compute the time interval Δt that elapses before the first nucleation event occurs ($\Delta t = t_{n,1} - 0$). Afterwards we update the system, which now counts one particle, and compute the new time interval for the nucleation of the second particle, $\Delta t = t_{n,2} - t_{n,1}$; this time we also solve eqn (21) and (22) at each time step, while using eqn (20). The procedure is repeated until the detection condition is met. This simulation technique is fundamentally different from the Monte Carlo algorithm suggested by Biscans *et al.*,[10] based on a rejection method. In a Monte Carlo algorithm the integration step, Δt, determines the number of times the rejection process is called. Too fine or too coarse choices can over- or under-estimate the number of nuclei that form. A large number of simulations is needed simply to validate the correct time step, seriously limiting the applicability of this specific method for regressing nucleation parameters.

4 Results

In this section the results obtained with the two models described above are reported. First, Model 1 is used to regress the experimental data from TU Delft; the quality of the fitting, its predictive capabilities and the correlation among the parameters of the nucleation and growth rate expressions are discussed. Then the obtained parameters are used in Model 2; similarities and differences between the two models are analysed, and a simulation illustrating the effect of changing the cooling rate is reported and discussed. Since the focus is on stochastic nucleation and scaling from small to large volumes, all results will be presented so that the reader is informed not only of the average properties of the crystallisation process but also their statistic distribution.

4.1 Model 1: estimation of parameters

Model 1 has been applied to estimate the six parameters in the nucleation and growth rate expressions of eqn (12) and (13) by fitting the experimental detection times measured by the TU Delft group in the 1 mL crystalliser at two different initial concentrations – namely, 15 and 47 g L^{-1} – with the calculated ones. The values of the six estimated parameters are reported as *Reference kinetics* in the second column of Table 3. Fig. 2 shows the experimental cumulative distributions of detection times and the corresponding calculated distributions, as well as the calculated distributions of nucleation times.

A few remarks are worth making. The model predicts a detection time, which is somewhere from 0 s up to about 400 s larger than the nucleation time. This

Table 3 Kinetic parameters obtained from fitting the 1 mL crystalliser data of TU Delft[2]

Parameter	Reference kinetics	Faster growth	Slower growth
A_0 [m^{-3} s^{-1}]	7.55×10^{3}	8.410×10^3	5.5×10^3
A_1 [K]	660	570	630
B [K^3]	7.35×10^5	2.15×10^6	1.3×10^5
K_0 [m s^{-1}]	1.67×10^{-4}	1.67×10^{-3}	8.35×10^{-5}
K_1 [K]	1.0×10^3	1.0×10^3	1.0×10^3
K_2 [K^2]	3.22×10^3	3.22×10^3	3.22×10^3

indicates the importance of growth in these crystallisation experiments on paracetamol from aqueous solution. It is also worth noting that due to cooling, the later the nucleation, the lower the temperature and the higher the supersaturation at which it occurs; higher supersaturation (that speeds growth up) prevails on lower temperatures (that slows it down), thus leading to a decrease in growth times for larger values of t_n (from about 400 s, when nucleation is the fastest, to essentially 0 s, *i.e.* instantaneous growth to the detection size, when $t_n \approx 4000$ s).

The agreement between the measured and simulated detection times is rather good, not only at the initial concentrations of 15 g L^{-1} and 47 g L^{-1}, but also at 22 g L^{-1} and at 32 g L^{-1} (data not shown here). This is an important result, because such a good fitting has been obtained with temperature dependent rate models for nucleation and growth, where only six parameters have been

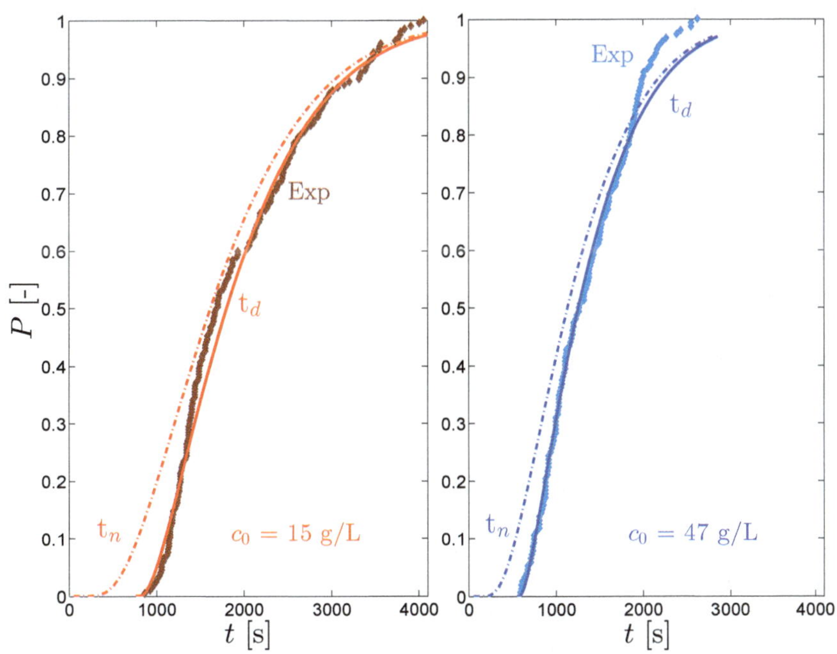

Fig. 2 Results of Model 1 and experimental data of TU Delft[2] in a 1 mL vessel. Symbols: experimental data; solid lines: detection time probability curves; dashed lines: nucleation time probability curves (given by eqn (1)).

estimated. The resulting model can reliably be used to predict crystallisation at different initial concentrations and for different cooling rates, something not possible with the TU Delft model and their twelve estimated parameters.

When looking in more detail, at $c_0 = 47$ g L^{-1} a deviation between the experiments and simulations can be observed at values of the detection time beyond about 1800 s. In fact, the experimental data for $t_d > 1800$ s seem to follow a different trend with respect to those for $t_d < 1800$ s. Considering that supersaturation increases upon cooling, a change of nucleation mechanism might explain the deviation, particularly considering that the supersaturation is as high as 1.8 when nucleation leading to $t_d = 1800$ s occurs. However, more experimental data would be needed to demonstrate this property thoroughly.

4.2 Model 1: effect of volume

Fig. 3 illustrates results for larger crystalliser volumes, namely 500, 700, 900, and 1000 mL; each diagram refers to a specific volume as indicated, and shows the experimental cumulative distributions of detection times (for the two initial concentrations, namely 15 g L^{-1} and 47 g L^{-1}, associated to larger and smaller detection times, respectively), and the corresponding distributions of both detection times and nucleation times calculated using Model 1 with the *Reference kinetics* parameters.

It is quite clear that in general the model is not able to describe the experiments, as happens in the case of the model developed by the TU Delft group.[2] The model is unable to consistently predict the average detection times, and underestimates their variability; the experiments exhibit a much larger variability than predicted by the model. This behaviour might be due to a model mismatch that makes the extrapolation to larger volumes infeasible, or to the same reasons spelled out above with reference to the attempts made by the TU Delft group (see section 2.2). We also tried to specifically fit the measurements in larger volumes with a set of *ad hoc* parameters different to the *Reference kinetics* in Table 3, but without achieving satisfactory results.

4.3 Model 1: correlation among parameters

The set of parameters indicated as *Reference kinetics* in Table 3 have been obtained by minimizing the deviation between experiments and simulations using a standard optimizer. By analysing the results, one realizes that the objective function for this problem is rather flat, and the minimum rather shallow. In fact, the estimated model parameters are correlated; in other words, very similar distributions of detection times as the one plotted in Fig. 2 can be obtained by increasing the growth rate and reducing the nucleation rate, or *vice versa*.

To demonstrate this, we have selected two different sets of values for the growth rate parameters, and have estimated the nucleation rate parameters, to minimize the error between the experiments and simulations. As reported in Table 3, the growth rate constant K_0 has been set to a value ten times larger (column 3, *Faster growth*) and to a value two times smaller (column 4, *Slower growth*) than that of the *Reference kinetics*, whereas K_1 and K_2 have been kept unchanged. After fitting, the nucleation rate parameters A_0, A_1 and B attained the new values also reported in Table 3.

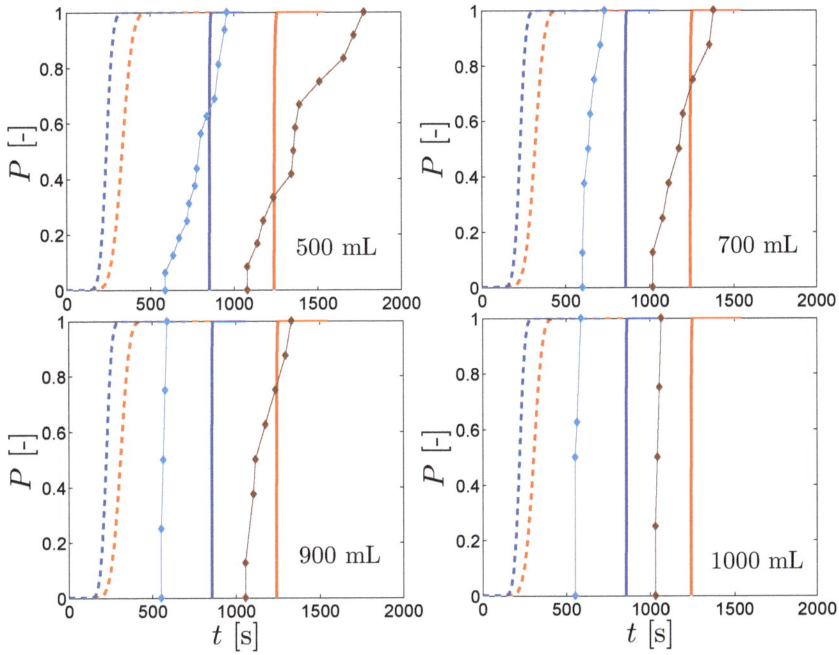

Fig. 3 Results of Model 1 and experimental data of TU Delft,[2] in 4 different volumes. Initial concentrations are 15 (red) and 47 (blue) g L^{-1}. Symbols: experimental data; solid lines: detection time probability curves; dashed lines: nucleation time probability curves (given by eqn (1)).

Fig. 4 illustrates the results, by showing the experimental detection time distributions at the two concentrations considered in the 1 mL crystalliser, together with the same distributions calculated with the three different sets of parameters; it is clear that the fitting accuracy is quite comparable. In the insets, the distributions of nucleation times for the two initial concentration levels (the larger and smaller concentrations on the l.h.s. and on the r.h.s., respectively), as calculated with the three sets of parameters are also plotted. One can readily observe that indeed, nucleation times are the longest for the set of *Faster growth* parameters, whereas the *Reference kinetics* yields intermediate values of t_n.

In conclusion, estimating nucleation and growth rate parameters just from statistics of detection times is not possible, because the two groups of parameters are highly correlated. This is a mathematical property reflecting the physical reality that both nucleation and growth contribute to the determination of detection times. To overcome this problem, either independent growth measurements or additional, different pieces of information (*e.g.* the number of particles or the average particle size at a time t) are required.

Interestingly though, the simulation results indicate that, if in 1 mL the different sets of kinetic parameters yield very similar distributions of detection times, in larger volumes the corresponding distributions are very distinct. This is illustrated in Fig. 5 for the case of $c_0 = 47$ g L^{-1}, where the cumulative distributions of nucleation and induction times obtained for the three sets of kinetic parameters are plotted and compared with the experimental distributions. It is

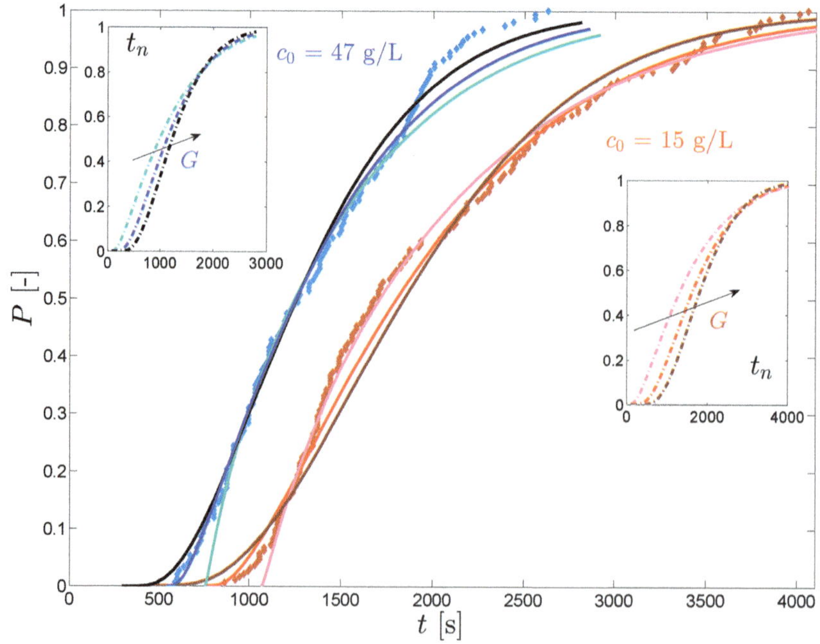

Fig. 4 Results of Model 1 with three different sets of kinetic parameters (*cfr.* Table 3) and experimental data of TU Delft[2] in 1 mL. In the main figure, symbols: experimental data; solid lines: detection time probabilities. In the figure insets, dashed lines: nucleation time probability (given by eqn (1)). For $c_0 = 15$ and 47 g L^{-1} (red and blue shades, respectively), G increases from the lightest to the darkest colour.

worth noting that by and large the *Faster growth* parameters do a better job than the two other sets. Although one might be tempted to argue that the volume effect could be used to identify the best among different sets of parameters, in reality one should be cautious and rule out first the possible causes of differences in measured detection times when scaling up the system volume, as discussed in section 2.2.

4.4 Model 2: comparison with Model 1

Let us now consider Model 2; for the sake of comparison with Model 1, we have used the *Reference kinetics* set of parameters in Table 3 rather than estimating a new set of kinetic parameters. We have carried out simulations of nucleation and growth with both models, starting from the two reference initial concentration values, for a number of system volumes between 1 mL and 1000 mL. Each simulation using Model 1 or Model 2 has been performed independently, and for each volume V at each initial concentration c_0, at least one thousand simulations have been carried out in order to build reliable statistics.

The key differences between the results obtained with the two models can be appreciated by examining Fig. 6, which plots the total number of particles present at the detection time, *i.e.* $N(t_d)$, for simulations carried out at different initial concentrations, c_0, and different system volumes, V. Let us focus on the results for $c_0 = 15$ g L^{-1} and $V = 100$ mL; similar considerations can be made for all other

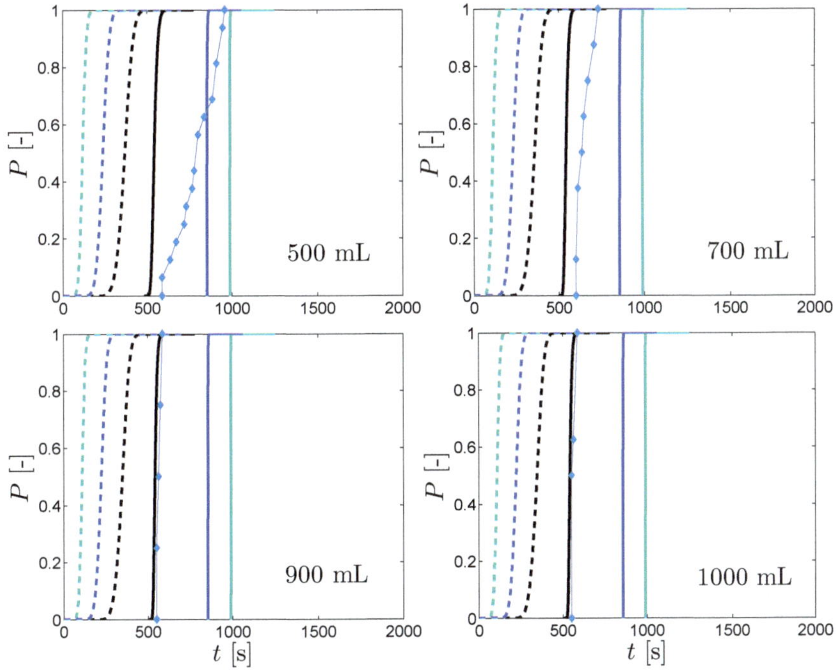

Fig. 5 Results of Model 1 with 3 different sets of kinetic parameters (*cfr.* Table 3) and experimental data, in 4 different volumes, for $c_0 = 47$ g L^{-1}. Symbols: experimental data of TU Delft;[2] solid lines: detection time probabilities (given by Model 1); dashed lines: nucleation time probability (given by eqn (1)). *G* increases in colour from light blue to black.

cases. There are two sets of points: the aligned brown (darker) symbols on the l.h.s. obtained through Model 1, and the red (lighter) symbols, forming a cloud on the r.h.s. and calculated using Model 2.

In the case of Model 1, the stochasticity of nucleation yields a range of detection times and numbers of particles as discussed above; however, for each stochastic value of the nucleation time, there is only one value of the detection time and one number of particles, because after the birth of the first nucleus the evolution of the system is deterministic.

In the case of Model 2, on the contrary, the same detection time may correspond to different numbers of particles, and *vice versa*; this is indeed a consequence of the co-existence of stochastic nucleation and deterministic growth after the formation of the first nucleus.

It is also worth observing that, as a consequence of the different structure of the two models, the average detection time and number of particles obtained with Model 2 are larger than those calculated using Model 1. On the one hand, the variability of the detection times decreases with increasing volume for both models, *i.e.* the self-exciting point process of Model 2 behaves in this respect like the inhomogeneous Poisson process of Model 1. On the other hand, the variability (or the standard deviation) of the Model 2 statistics for detection times is larger than that of Model 1.

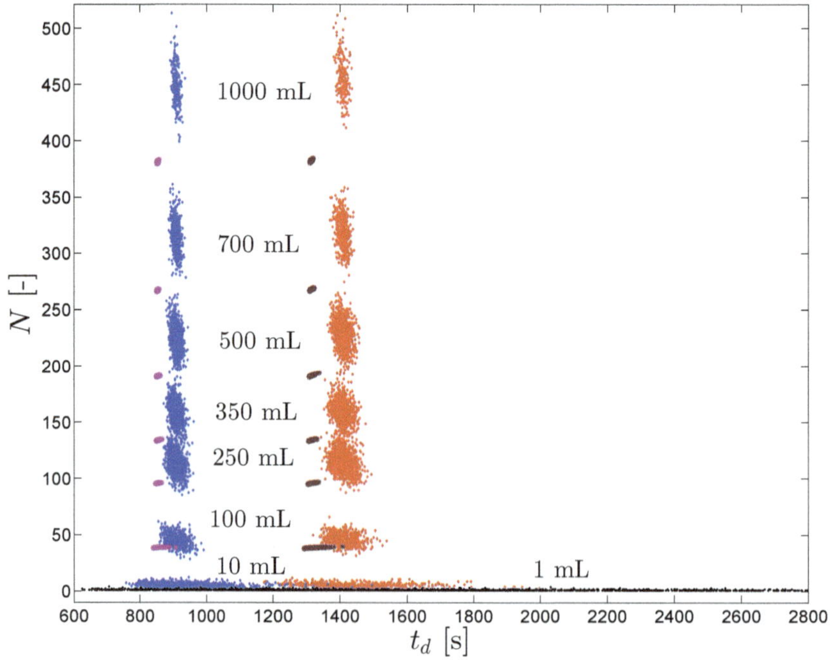

Fig. 6 Results of simulations of Model 2 given as number of crystals, N, formed at the detection time, t_d, for different volumes. Two different initial concentrations, 15 and 47 g L^{-1}, are represented in red and blue, respectively, for volumes from 10 to 1000 mL. Results of simulations in 1 mL volume, largely overlapping for both initial concentrations, are all represented in black. Results of simulations of Model 1 for 100, 250, 350, 500, 700, and 1000 mL are indicated as violet and brown points (for $c_0 = 47$ g L^{-1} and $c_0 = 15$ g L^{-1}, respectively).

Considering the statistics of the number of particles, the variability of the number of particles calculated with Model 2 increases a great deal with increasing volume. It can also be noted that for $V = 1$ mL, *i.e.* the volume on which the quantitative study of the TU Delft group is based, only a very small number of particles are formed, typically less than five, which corresponds to a very broad range of detection times (see Fig. 2 for the cumulative probability distribution of the detection times measured and calculated with Model 1 under the same conditions); from $V = 250$ mL on, the number of particles exceeds about one hundred. This result gives a quantitative confirmation and a visual illustration of how, at different system volumes, a mononuclear or a polynuclear mechanism could be the better way to describe nucleation.[1] It also shows that the mononuclear mechanism implicit in the TU Delft model, and in our Model 1, is inadequate to describe volumes above 10 mL, in this case.

The same results are illustrated in a different manner for the same two initial concentration values in Fig. 7, which shows as a function of the volume, V, and the mean detection time, t_d, together with error bars, that are calculated by adding and subtracting from the mean t_d the value of the relevant standard deviation of the t_d distribution. It is worth noting that the two models behave very similarly when $V = 1$ mL. This is expected, since at that volume only a few particles are

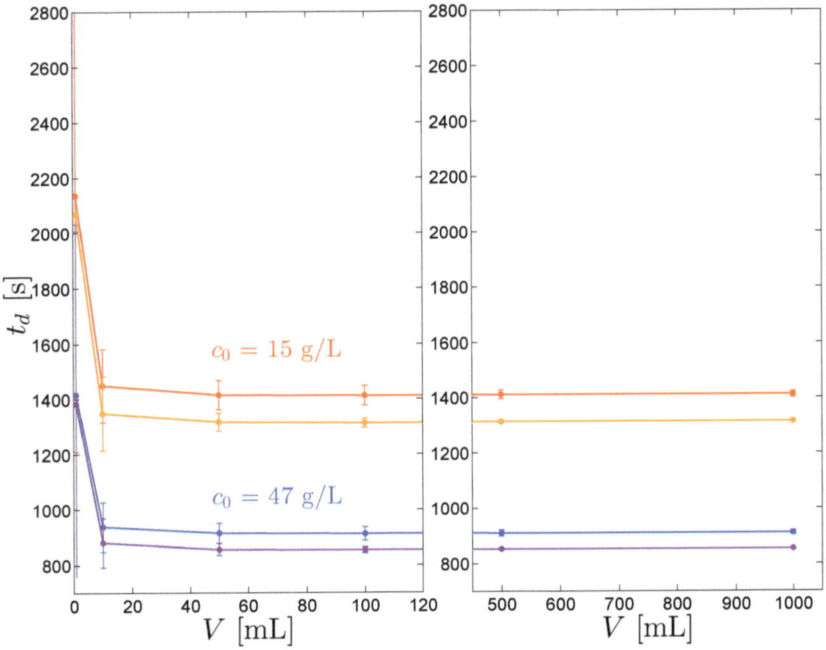

Fig. 7 Results of simulations of Model 2, $c_0 = 47$ 15 g L^{-1} (blue), and $c_0 = 15$ g L^{-1} (red), and Model 1, $c_0 = 47$ g L^{-1} (violet), and $c_0 = 15$ g L^{-1} (orange), given as average values of detection times, t_d. In the left figure, the volume ranges from 0 to 120 mL; in the right one, from 450 to 1100 mL. The error bars indicate the standard deviation of the corresponding average t_d value.

formed; since the first nucleus follows the same statistics for both models, with such a small number of particles to be formed, the differences between the results obtained with the two models can only be very small.

4.5 Model 2: sensitivity to kinetic parameters

Simulations using Model 2 have been repeated for the *Faster growth* parameters in Table 3, *i.e.* the parameters that, when using Model 1, exhibited the best accuracy in describing measurements at higher system volumes (see Fig. 5). Fig. 8 shows as a function of the volume, V, the mean detection time, t_d, together with error bars, calculated as in Fig. 7. For each initial concentration value two curves are plotted, obtained using the two different sets of kinetic parameters. Moreover, the average detection times observed in the TU Delft experiments in crystallisers with $V = 500$ mL, 700 mL, 900 mL and 1000 mL are plotted.

As already observed using Model 1, the two sets of parameters yield very similar results when $V = 1$ mL, but progressively diverge as the volume increases. At both concentrations the *Faster growth* parameters yield lower values of average detection time (about 500 s less and 300 s less at 15 g L^{-1} and at 47 g L^{-1}, respectively), and larger values of standard deviation (which is typical for slower nucleation rates, as discussed in the Introduction when summarizing the main features of a homogeneous Poisson process), than the *Reference kinetics* parameters.

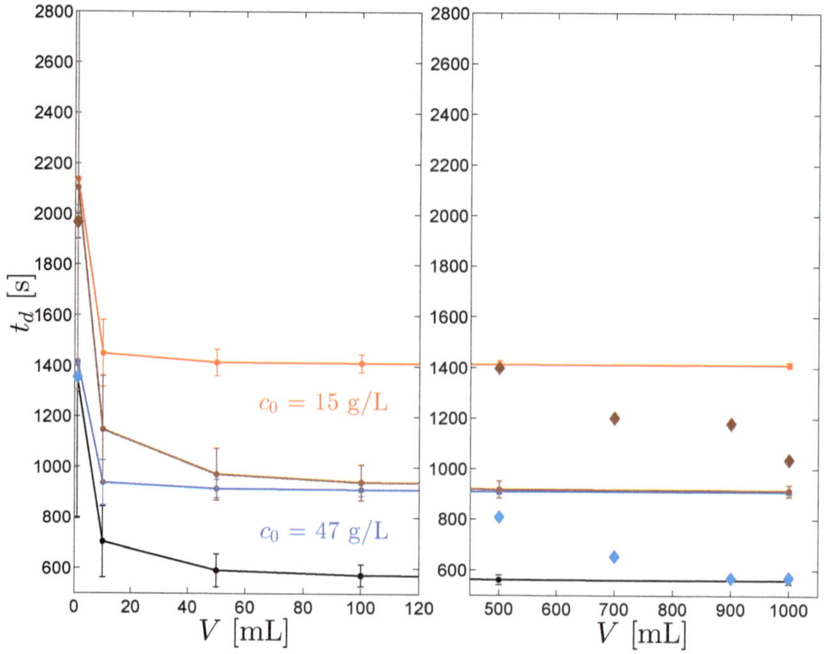

Fig. 8 Comparison of simulations of Model 2, in 1 mL crystalliser, given as average detection times, t_d. Two different sets of parameters, *Reference kinetics* (blue and red, $c_0 = 47$ g L^{-1} and $c_0 = 15$ g L^{-1}, respectively) and *Faster growth* (black and brown, $c_0 = 47$ g L^{-1} and $c_0 = 15$ g L^{-1}, respectively) have been used (see Table 3). In the left figure, the volume ranges from 0 to 120 mL; in the right one, from 450 to 1100 mL. The symbols indicate the experimental data of TU Delft[2] at the corresponding volumes. The error bars indicate the standard deviation of the corresponding average t_d value. The overlapping of the brown and blue line observed here is a coincidence of this particular system, and does not represent a specific feature of the model.

When comparing the simulation results and the experimental data in Fig. 8, one notices that the experimental data fall in the region delimited by the two curves calculated with the *Reference kinetics* and *Faster growth* parameters; the former set works better in the 500 mL crystalliser, whereas the latter works better in the 900 mL and 1000 mL crystallisers. Even though the standard deviation calculated with Model 2 is larger than that predicted by Model 1, it is still much smaller than the experimental one.

In conclusion, although Model 2 is a clear improvement with respect to Model 1 in terms of physical consistency, it is not possible to demonstrate that Model 2 is able to completely capture the effects of changing from very small, microfluidic volumes to larger ones, based only on the TU Delft experimental results from large crystallisers. It is clear that, to this aim, more experiments which follow a well designed experimental protocol will have to be performed.

4.6 Effect of cooling rate

Being intrinsically temperature dependent, both Model 1 and Model 2 can be used to study the effects of the cooling rate on the crystallisation process, the nucleation stochasticity, and the metastable zone width (MZW), *i.e.* the region of the phase diagram where the solution is supersaturated but nucleation is kinetically hindered.

We have investigated these effects using Model 2 as illustrated in Fig. 9, where the average difference between the temperatures at time zero and at the detection time, $\Delta T(t_d)$, *i.e.* the observed MZW, is plotted with its variability (error bars) *vs.* the system volume. These have been calculated using the parameters corresponding to the *Reference kinetics*, for an initial concentration of 47 g L^{-1} and three different cooling rates, namely $\beta = 0.25$ K min^{-1}, 0.5 K min^{-1} (the value used in all simulations reported so far), and 1 K min^{-1}. Therefore, the data corresponding to $\beta = 0.5$ K min^{-1} have already been reported in terms of average detection time, both in Fig. 7 and in Fig. 8. Such average detection times at large volumes are also reported in Fig. 9 beside the corresponding $\Delta T(t_d)$ *vs.* V curve.

As expected based on physical intuition, larger cooling rates lead to larger values of the MZW (and of its variability) and correspondingly smaller values of the detection time. This is due to the fact that the MZW is determined by competition between a supersaturation generating mechanism, *i.e.* cooling, and a supersaturation depleting process, namely particle formation and growth. Changing the cooling rate affects the former but not the latter, thus yielding larger values of MZW. The reduction of the detection time is a mere consequence of the

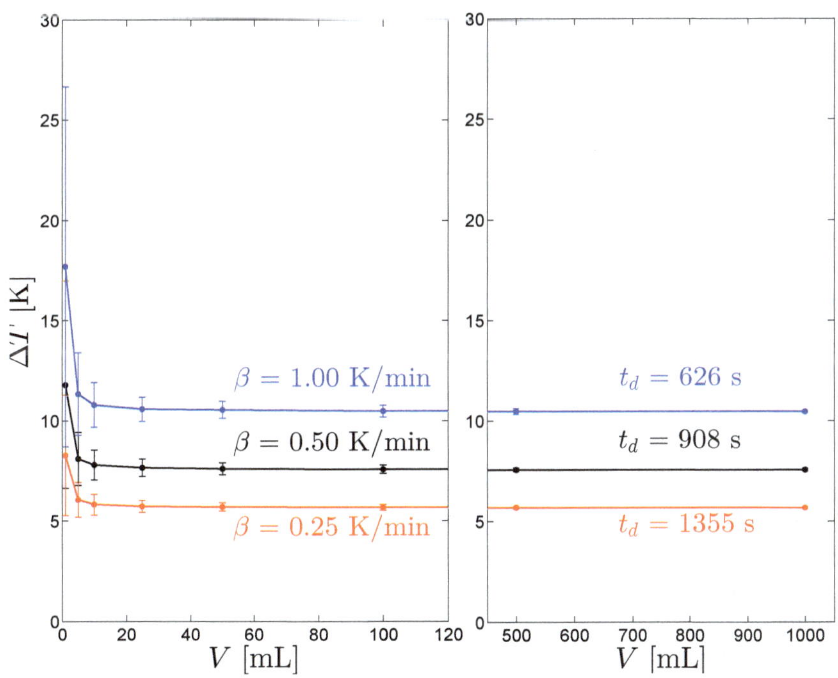

Fig. 9 Results of simulations of Model 2, at different volumes, for $c_0 = 47$ g L^{-1}, with three different cooling rates, β, reported as average metastable zone width, ΔT. In the right-hand-side box, at larger volumes, the asymptotic values of the average t_d for each cooling rate are indicated. In the left figure, the volume ranges from 0 to 120 mL; in the right one, from 450 to 1100 mL. The error bars indicate the standard deviation of the corresponding average ΔT value.

fact that the same temperature change requires less time when the cooling rate is larger.

Our time-dependent model is also able to describe the effect of the stochastic nature of nucleation on the MZW in very small volumes; namely that the MZW and its variability are larger in very small volumes than in large volumes. In volumes beyond a system-specific threshold value, the MZW reaches an asymptotic value for each cooling rate, and in principle its variability vanishes.

5 Discussion and conclusions

This contribution focuses on studying the variability of detection times observed in cooling crystallisation experiments. Such variability has been assumed to be primarily the consequence of an intrinsic stochastic nature of nuclei formation: nucleation is thus mathematically defined as a stochastic process.

Based particularly on the data of paracetamol crystallisation in water, measured by the TU Delft group[2] in small, well-mixed 1 mL reactors, we have developed two different models (section 3.1 and 3.2) combining the statistical formation of primary nuclei with their deterministic growth to fully-grown, detectable crystals. Both models not only account for dynamic profiles of temperature and concentration, but they can also be easily modified to describe systems with a non-constant volume. Consequently, they can be used directly to analyse isothermal as well as non-isothermal nucleation experiments, and crystallisation where the driving force is not due to cooling, but is due instead to, *e.g.*, solvent evaporation or anti-solvent addition.

Model 1 (section 3.1) assumes that the stochastic nature of nucleation, modelled as a Poisson process, influences only the very beginning of the crystallisation, *i.e.* the formation of the first particle. The subsequent nuclei formation, meanwhile, is perfectly deterministic and is described by a population balance equation or, equivalently, by the method of moments. The effect of nucleation stochasticity determines only the time when the system shifts to the deterministic regime. Model 2 (3.2), on the contrary, considers that the formation of all nuclei, not only the first, is stochastic, whereas their growth is deterministic.

As suggested by physical considerations and experimental evidence, the magnitude of detection time variance reduces with increasing system volume in both models. However, due to the different stochastic process used to describe nucleation, the second model always exhibits a larger detection time variability.

Model 1 has been used for regressing nucleation and growth kinetic parameters, whose correlation has also been investigated. We concluded that a unique set of parameters cannot be obtained from one series of experiments, if only the information concerning the detection times in a single volume is provided. Under these conditions, independent growth measurements are necessary to avoid parameter correlation. Nevertheless, both models show that different sets of kinetic parameters, which yield very similar detection times at the specific volume at which they have been obtained, produce clearly different results when the system volume changes. The correct kinetics could thus be discriminated by simply repeating detection time experiments at different system sizes.

In conclusion, the two models developed and studied in this contribution allow us detailed analysis of nucleation experiments at different initial conditions and with different temperature profiles. The models provide a general framework to describe the effect of volume on nucleation, while also accounting for crystal growth. Assuming scale-independent kinetics, they can thus be used to simulate the behaviour of the system at different volumes, and therefore gain a better understanding of the effects of scaling-up. Thus, the two models allow us to assess the feasibility of extrapolating kinetic information obtained in small, controlled volumes to larger values, by directly comparing experimental data and simulation results at different scales.

General mathematical descriptions are necessary to better understand the physics of nucleation, and to obtain reliable kinetic parameters for process design and control. We believe that the models presented in this work could be a valid contribution towards the development of sound and reliable protocols for designing, analysing, and interpreting nucleation experiments.

Acknowledgements

The authors would like to thank Joop H. ter Horst and Somnath S. Kadam of the TU Delft for kindly providing the raw data for the TU Delft study[2] used in this work.

References

1 D. Kashchiev, D. Verdoes and G. van Rosmalen, *J. Cryst. Growth*, 1991, **110**, 373–380.
2 S. S. S. Kadam, S. S. a. Kulkarni, R. Ribera, A. I. A. Stankiewicz, J. H. J. ter Horst, H. J. H. Kramer and R. Coloma Ribera, *Chem. Eng. Sci.*, 2012, **72**, 10–19.
3 C. Lindenberg and M. Mazzotti, *J. Cryst. Growth*, 2009, **311**, 1178–1184.
4 O. Galkin and P. G. Vekilov, *J. Phys. Chem. B*, 1999, **103**, 10965–10971.
5 P. Laval, J.-B. Salmon and M. Joanicot, *J. Cryst. Growth*, 2007, **303**, 622–628.
6 M. Ildefonso, N. Candoni and S. Veesler, *Org. Process Res. Dev.*, 2012, **16**, 556–560.
7 M. Ildefonso, N. Candoni and S. Veesler, *Cryst. Growth Des.*, 2013, **13**, 2107–2110.
8 Z. Hammadi, N. Candoni, R. Grossier, M. Ildefonso, R. Morin and S. Veesler, *C. R. Phys.*, 2013, **14**, 192–198.
9 S. Teychené and B. Biscans, *Cryst. Growth Des.*, 2011, 4810–4818.
10 S. Teychené and B. Biscans, *Chem. Eng. Sci.*, 2012, **77**, 242–248.
11 L. Goh, K. Chen, V. Bhamidi, G. He, N. C. S. Kee, P. J. A. Kenis, C. F. I. I. I. Zukoski and R. D. Braatz, *Cryst. Growth Des.*, 2010, **10**, 2515–2521.
12 K. Chen, L. Goh, G. He, P. Kenis, C. Zukoski and R. Braatz, *Chem. Eng. Sci.*, 2012, **77**, 235–241.
13 A. F. Izmailov, A. S. Myerson and S. Arnold, *J. Cryst. Growth*, 1999, **196**, 234–242.
14 D. Knezic, J. Zaccaro and A. S. Myerson, *J. Phys. Chem. B*, 2004, **108**, 10672–10677.
15 S. Jiang and J. H. ter Horst, *Cryst. Growth Des.*, 2011, **11**, 256–261.
16 D. Randall, J. Nathoo, F. Genceli-Güner, H. Kramer, G. Witkamp and a. E. Lewis, *Chem. Eng. Sci.*, 2012, **77**, 184–188.
17 S. a. Kulkarni, S. S. Kadam, H. Meekes, A. I. Stankiewicz and J. H. ter Horst, *Cryst. Growth Des.*, 2013, **13**, 2435–2440.

18 R. Sullivan, R. Davey and G. Sadiq, *Cryst. Growth Des.*, 2014, **14**, 2689–2696.
19 O. Pino-García and A. Rasmuson, *Ind. Eng. Chem. Res.*, 2003, **42**, 4899–4909.
20 O. Pino-García and A. Rasmuson, *Cryst. Growth Des.*, 2004, **4**, 1025–1037.
21 F. L. Nordström, M. Svärd and A. k. C. Rasmuson, *CrystEngComm*, 2013, **15**, 7285.
22 Y. Huaiyu, A. Rasmuson and H. Yang, *Cryst. Growth Des.*, 2013, **10**, 4226–4238.
23 M. Svärd, F. Nordström, E. Hoffmann and A. Rasmuson, *Cryst. Growth Des.*, 2013, **15**, 5020–5031.
24 M. Kuhs, J. Zeglinski and A. Rasmuson, *Cryst. Growth Des.*, 2014, **14**, 905–915.
25 D. Kashchiev, *Nucleation, Basich Theory with Applications*, Butterworth-Heinemann, 2000.
26 C. Lindenberg and M. Krattli, *Cryst. Growth Des.*, 2009, **9**, 1124–1136.
27 A. Randolph and M. Larson, *Theory of Particulate Processes*, Academic Press, London, 1971.
28 D. Ramkrishna, *Population Balances: Theory and Applications to Particulate Systems in Engineering*, 2000, p. 355.
29 D. Snyder and M. Miller, *Random Point Processes in Time and Space*, Springer-Verlag, New York, 1991, p. 481.

Faraday Discussions

DISCUSSIONS

Solvent and additive interactions as determinants in the nucleation pathway: general discussion†

Changquan Calvin Sun, Wenhao Sun, Sarah Price, Colan Hughes, Joop Ter Horst, Stéphane Veesler, Ken Lewtas, Allan Myerson, Haihua Pan, Gerard Coquerel, Joost van den Ende, Hugo Meekes, Marco Mazzotti, Ian Rosbottom, Francis Taulelle, Simon Black, Alasdair Mackenzie, Sophie Janbon, Peter Vekilov, Terence Threlfall, Thomas Turner, Kevin Back, Herma Cuppen, Dimitrios Toroz, Jan Sefcik, Jessica Lovelock, Robert Hammond, Nadine Candoni, Elena Simone, Martin Ward, Celso Aparecido Bertran, Thomas Vetter, Richard Sear, Jim De Yoreo, Roger Davey, Jamshed Anwar, Erik Santiso, David T. Wu, Kevin Roberts, Baron Peters, Sven Schroeder, Franca Jones, Ake Rasmuson, Helmut Cölfen, Jacek Zeglinski and Matteo Salvalaglio

DOI: 10.1039/c5fd90038g

Sarah Price opened a general discussion of the paper by Sven Schroeder: I have been generating the thermodynamically plausible crystal structures of organic molecules for many years, and back in 2004 we did a crystal structure prediction (CSP) study on imidazole[1] and found that it was relatively straightforward. Following your paper, we have reclassified the low energy structures according to the tilt within the hydrogen-bonded chain and the relative direction of the chains. Although the observed structure was the global minimum, two other structures with a displacement of otherwise identical layers are very close in energy. Do you think that if imidazole had crystallised in one of these alternative structures it would be distinguishable by NEXAFS? This would be a very sensitive test of whether NEXAFS combined with CSP could be used in characterising crystal structures.

1. S. L. Price, B, Patel, P. Pridhanani-Jethani and A. Torrisi, Crystal structure prediction and polymorphism - some mutual insights, *Transactions of the American Crystallography Association*, 2004, **39**, 1–12.

Sven Schroeder replied: We are working towards this aim. We have so far established that NEXAFS is extremely sensitive to variations in speciation in

† Electronic Supplementary Information available: copies of the slides presented at the meeting by Wenhao Sun on the subject of metastable aragonite $CaCO_3$ nucleation in seawater.

solution, because it probes virtual orbitals,[1,2] as well as bond lengths (through multiple scattering resonances[3]) and bond angles,[2] very incisively. With better experimental control of supersaturation conditions, we should be able to characterise the structural evolution of systems in the metastable zone, as a function of supersaturation. Microjet and droplet jet systems permitting such studies are currently under development. Combination with NMR as well as X-ray/neutron scattering studies of supersaturated solutions (pair distribution function analysis) could be extremely powerful. Even more incisive characterisation of molecular dynamics in solution should become achievable with X-ray free electron laser (XFEL) sources in the future.

1. J. S. Stevens, L. K. Newton, C. Jaye, C. A. Muryn, D. A. Fischer and S. L. M. Schroeder, Proton Transfer, Hydrogen Bonding, and Disorder: Nitrogen NEXAFS and XPS of Bipyridine-Acid Salts and Co-crystals, *Cryst. Growth Des.*, 2015, **15**, 1776–1783.
2. J. S. Stevens, C. R. Seabourne, C. Jaye, D. A. Fischer, A. J. Scott and S. L. M. Schroeder, Incisive Probing of Intermolecular Interactions in Molecular Crystals: Core Level Spectroscopy Combined with Density Functional Theory, *J. Phys. Chem. B*, 2014, **118**, 12121–12129.
3. J. S. Stevens, A. Gainar, E. Suljoti, J. Xiao, R. Golnak, E. F. Aziz and S. L. M. Schroeder, Chemical Speciation and Bond Lengths of Organic Solutes by Core Level Spectroscopy: pH- and Solvent-Influence on p-Aminobenzoic Acid,*Chem. - Eur. J.*, 2015, DOI: 10.1002/chem.201405635

Sarah Price said: My interest in nucleation is that we cannot have a genuine method of computational polymorph prediction until our understanding of nucleation has developed sufficiently to have a method of distinguishing which of the CSP generated thermodynamically plausible crystal structures can actually nucleate and grow into long-lived polymorphs. At the moment, we can just look at the difference in the packing of the low energy crystal structures and guess which motifs are sufficiently different that they might nucleate and grow, rather than transform at some stage into a slightly lower energy related structure (see Fig. 1). In the case of imidazole, this is asking whether the stacking of the hydrogen bonded ribbons and tilt angles of the molecules within the ribbons can readily change during nucleation and growth. This is assuming that it would be difficult for the direction of the hydrogen bonding chain to change once formed, though it conceivably might be easier by proton transfer than by breaking and remaking hydrogen bonds. Do you think we will be able to get an experimental handle on what sort of rearrangement processes are going on during nucleation?

Sven Schroeder answered: This would be a really interesting study. We have recently shown how sensitively NEXAFS probes the location of protons in hydrogen bonds present in the organic solid state.[1,2] As DFT electronic structure calculations reproduce the experimentally observed NEXAFS variations, it may be possible that a combination of NEXAFS, crystal structure prediction and local electronic structure calculations can determine the structure of molecular crystals.

1. J. S. Stevens, L. K. Newton, C. Jaye, C. A. Muryn, D. A. Fischer and S. L. M. Schroeder, Proton Transfer, Hydrogen Bonding, and Disorder: Nitrogen NEXAFS and XPS of Bipyridine-Acid Salts and Co-crystals, *Cryst. Growth Des.*, 2015, **15**, 1776–1783.
2. J. S. Stevens, C. R. Seabourne, C. Jaye, D. A. Fischer, A. J. Scott and S. L. M. Schroeder, Incisive Probing of Intermolecular Interactions in Molecular Crystals: Core Level Spectroscopy Combined with Density Functional Theory, *J. Phys. Chem. B*, 2014, **118**, 12121–12129.

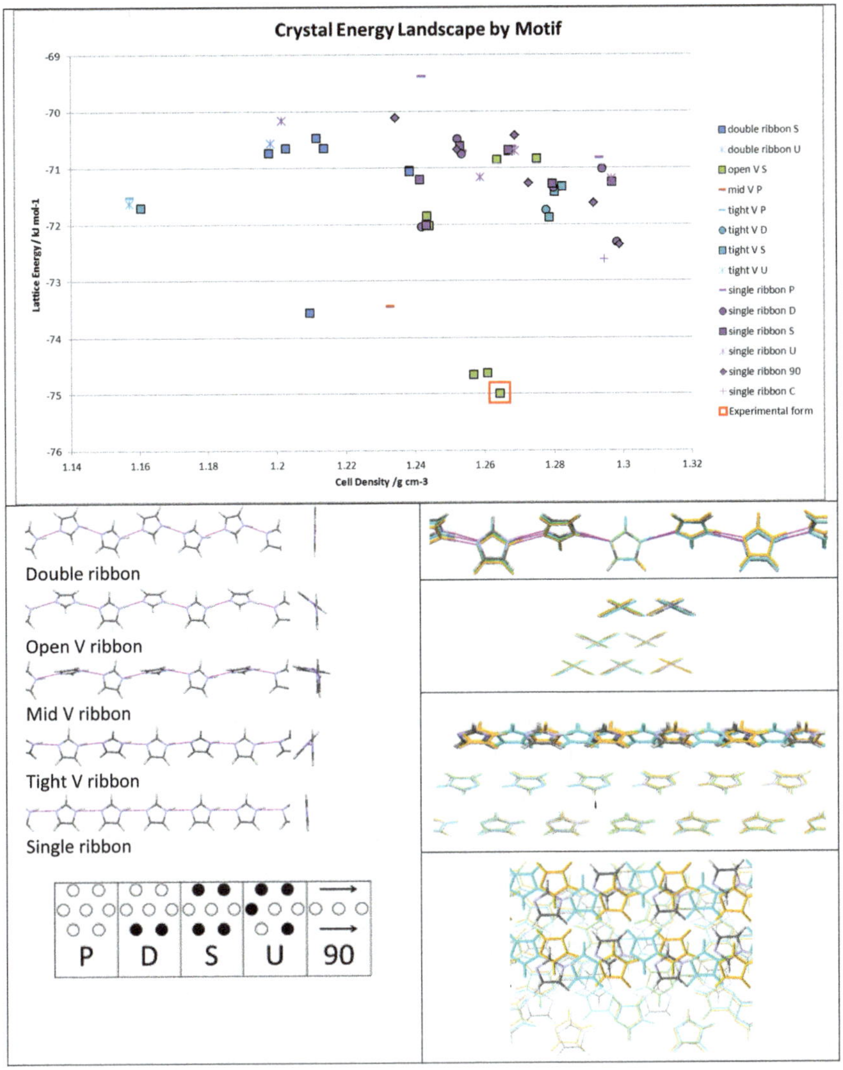

Fig. 1 The crystal energy landscape of imidazole. Each point represents a crystal structure, with the symbol depicting its ribbon structure (shown mid left) and the ribbon packing (shown bottom left, with open and filled circles denoting ribbons in opposite directions). The three lowest energy structures are overlaid on the right, with the experimental, lowest energy structure coloured by element, and the structure which are +0.34 and +0.37 kJ mol_{-1} higher in energy in cyan and orange, respectively.

Kevin Roberts commented: Crystallisation from the solution or vapour state involves the transformation of solvated or isolated solute monomers, respectively, into a packed crystal structure, where there can be confrontation between the intramolecular forces, that define the conformation of an isolated molecule, and the intermolecular forces, which define the packing of the molecules into the solid-state structure. Thus, when molecules have flexible constituent groups, efficient packing can result in the molecular conformations within the crystal structure adopting slightly "unstable" conformations in order to achieve the highest packing density and hence structural stability. Conversely, this implies that the crystallisation step can involve a conformational change as the molecule is accommodated within the crystal lattice.[1]

Mindful that the above analysis reflects the molecular scale processes taking place at the micro- and macroscale level, the question thereafter follows with respect to what happens when the crystal size approaches that typified by the nanoscales that are consistent with nucleation and the early stages of the growth process after facetting. In such circumstances the surface-to-volume ratio of such small crystals will be high, with many intermolecular synthons being unsaturated due to surface termination. Molecular and synthonic modelling of these states reveal that polymorphic stabilities can become much closer[2] or even switch over, with "metastable" forms becoming "stable" at the nanocrystal size scales and vice versa.[3] Further examination of the cluster disorder with respect to size has shown that disorder is higher in the surface regions than in the bulk[2] and that the "stable" polymorphs, at a given size, tend to have less disorder with respect to the bulk structures when compared to the "metastable" structures.[3] Such disorder has also been directly correlated to the crystallisability of a material, as revealed through careful measurement of the metastable zone width.[4]

Reflecting on the above scenario, what then are the potential implications of size-dependent conformational flexibility and by implication, polymorphic stability, in terms of our ability to reliably predict crystal structures? The latter mindful that such predictions are usually based upon periodic boundary conditions, which neglect the fact that a crystal in its early stages of development might deviate from the regular 3-D structures adopted by larger crystals.

1. K. J. Roberts, R. Docherty, P. Bennema and L. A. M. J. Jetten, The importance of considering growth-induced conformational change in predicting the morphology of benzophenone, *J. Phys. D: Appl. Phys.*, 1993, **26**(8B), B7–B21
2. R. B. Hammond, K. Pencheva and K. J. Roberts, An examination of the polymorphic stability and molecular conformational flexibility as a function of crystal size associated with the nucleation and growth of benzophenone, *Faraday Discuss.*, 2007, **136**, 87–102.
3. R. B. Hammond, K. Pencheva and K. J. Roberts, Structural variability within, and polymorphic stability of, nano-crystalline molecular clusters of L-glutamic acid and D-mannitol, modelled with respect to their size, shape and 'crystallisability', *CrystEngComm*, 2013, **14**, 1069–1082.
4. H. Jiang, An Examination of the Molecular-scale Effect of Tailor-made Additives on the Nucleation and Growth and Crystal Properties of Long-chain Alkyl Methyl-Esters, PhD thesis, University of Leeds, 2012.

Sarah Price replied: Currently, most so-called "crystal structure prediction" methods are only searching for the thermodynamically plausible, infinite, perfect crystal structures, to give the crystal energy landscape. This often contains far more structures than known polymorphs, raising questions[1] as to whether the structures could be realised, given the right experimental conditions - which may include heterogeneous nucleation by additives - or whether they are artefacts of the many approximations in evaluating the relative thermal stability of different structures. Indeed, one of the biggest difficulties in getting the relative lattice energies correct is to be able to use a computational method that reflects the balance you describe between the intramolecular energy penalty for changing conformation, and the improvement in the intermolecular energy from more effective packing.[2] So, being able to model the nucleation and growth of a crystal structure, including how the conformation changes between molecules in the solution or melt, on the surface and in the bulk crystal, would be an ideal method of predicting the crystal structure as you suggest. However, the work by Matteo Salvalaglio *et al.* (DOI: 10.1039/c4fd00235k) is leading the way in this endeavour and shows many of the challenges. Unfortunately, molecular dynamics simulations require force-fields, and in general these are not yet accurate enough to even get the right order of stability of polymorphs.[3] In addition, the simulation needs to include all factors that affect the nucleation, *i.e.* it must explicitly model any surfaces, and additives/impurities that play a role. Hence, as this Discussion meeting shows, we do not yet have a sufficiently good understanding of all the factors that influence crystallisation and polymorphic form to include these in a Molecular Dynamics simulation of nucleation and growth to predict the crystal structure. Hence, at the moment, computed crystal energy landscapes only suggest the range of conformations that can form thermodynamically competitive structures, and provide a necessary test of whether a force-field is realistic enough for simulating nucleation and growth.[4]

1. S. L. Price, Why don't we find more polymorphs? *Acta Crystallogr., Sect. B: Struct. Sci., Cryst. Eng. Mater.*, 2013, **69**, 313–328.
2. S. L. Price, Predicting crystal structures of organic compounds, *Chem. Soc. Rev.*, 2014, **43**, 2098–2111.
3. C. Bernardes and A. Joseph, Evaluation of the OPLS-AA Force Field for the Study of Structural and Energetic Aspects of Molecular Organic Crystals, *J. Phys. Chem. A*, 2015, **119**, 3023–3034.
4. O. G. Uzoh, P. T. A. Galek and S. L. Price, Analysis of the conformational profiles of fenamates shows route towards novel, higher accuracy, force-fields for pharmaceuticals, *Phys. Chem. Chem. Phys.*, 2015, **17**, 7936–7948.

Alasdair Mackenzie asked: Following on from your comment about looking at solution interaction with your calculated crystal structures, would it be possible to do the following:

You calculate many structures as being stable at 0 K, but most are higher in energy than the crystallographically determined ones. You have also stated that they fall into groups with similar structural motifs with slightly different intramolecular arrangements. If you were to add thermal energy into your higher energy structure calculations, would they rearrange into the stable ones of the similar group?

If so, could you focus on the highest energy structure in each group as the one to be stabilised by solvent interactions, which then transitions to the stable arrangement once formed?

Sarah Price answered: Yes, in general, we only need to consider the most stable of each group of structures (*i.e.* those we think would be the free energy minima if we could do a dynamic simulation at the crystallisation temperature) as potential polymorphs, and think whether the choice of solvent or other crystallisation conditions might favour the nucleation.[1] In the case of imidazole, I do think that many structures which differ only in a translation of the stacking of the hydrogen bonded chains, or with different tilt angles may well become the same structure if the thermal energy could be added in a molecular dynamics simulation. However, it seems unlikely that the direction of the hydrogen bonding chains could so readily change, and I'm unsure whether it is practically possible to change the crystallisation conditions of imidazole sufficiently to produce, for example, a polar crystal.

1. S. L. Price, Why don't we find more polymorphs? *Acta Crystallogr., Sect. B: Struct. Sci., Cryst. Eng. Mater.*, 2013, **69**, 313–328.

Simon Black returned to the discussion of the paper by Sven Schroeder: Would you expect imidazole to form hydrates, and is there any evidence that it does?

Sven Schroeder responded: I am not sure I can give an entirely conclusive answer. Both experimental NEXAFS as well as our DFT calculations indicate that imidazole-imidazole hydrogen bonding is stronger than imidazole–water hydrogen-bonding. The cohesive energy of the imidazole crystal is additionally determined by the π–π interactions between the hydrogen-bonded imidazole chains. The π–π stacking seems to destabilise the intra-chain hydrogen bonding only very slightly. These observations suggest that retaining water molecules in an ordered structure is unfavourable. With regard to evidence that hydrates are formed, we have examined aqueous imidazole solutions up to near-saturation concentration, not only by NEXAFS but also by viscosity measurements and Raman spectroscopy, and have not found any significant variations. This suggests that water is not strongly structured even when solutions contain only about 5 water molecules per dissolved imidazole molecule. So interaction with water molecules appears to be comparatively weak.

Sarah Price added: This is consistent with there being no imidazole hydrate structures in the Cambridge Structural Database, and only four structures (KAXJAW, MAFRAN, WIWHUH and XEMCIB) which contain a much larger molecule, solvated by both imidazole and water.

Terence Threlfall commented: Desiraju showed *via* analysis of the Cambridge Structural Database that hydrates were only likely to form in crystals from organic compounds where the ratio of donors to acceptors was less than 1.[1]

1. G. R. Desiraju, Hydration in organic crystals: prediction from molecular structure, *J. Chem. Soc., Chem. Commun.*, 1991, 426–428

Hugo Meekes opened a general discussion of the paper by Matteo Salvalaglio: First of all, my compliments; this is really nice work. In view of the preceeding discussion it is nice that you showed a two-step mechanism for polymorphic forms. You mentioned that you observe various metastable forms of up to 50 molecules, which transform, almost without a barrier, to a stable form. That would mean that this is a very nice way of showing the Ostwald's rule of stages. In crystallization experiments we do not often see a metastable form transforming to a stable form within the crystal. Usually the transformation proceeds through nucleation of the stable form, combined with dissolution of the metastable form. In your simulations you probably would need more computation time for the stable form to nucleate in the solution and the metastable form to dissolve. Do you think you would be able to see such a process with your technique?

Matteo Salvalaglio answered: Thank you very much for your comment. In a macroscopic system, one can imagine that large nuclei will grow at the expense of the smaller clusters through Ostwald ripening. Therefore it is reasonable to think that, when a stable form I cluster is formed through a solid-solid transition, it then grows at the expenses of smaller clusters, regardless of their internal structure. While we show that a solid–solid transition between form I and form II is possible for urea, we imagine that in macroscopic systems where a population of clusters develops, both solid–solid and solvent mediated transformations will take place.

In order to observe such processes directly in a simulation, we would need to construct model systems large enough to contain at least a stable crystal-like cluster together with a population of smaller clusters. While this is a challenge for urea, it can be done for poorly-soluble larger organic molecules. In such cases we observe an Ostwald-like ripening process taking place during simulations in the 100 ns timescale.

Hugo Meekes asked: If you look at the transition inside the cluster, can you see whether it starts on the surface of the cluster?

Matteo Salvalaglio answered: Yes, it is just a qualitative observation. However, the transition seem to be typically initiated on the surface of the clusters.

David T. Wu commented: In Figure 3, you show the nucleation simulation's progress measured with two order parameters, n and n_o, and describe progress off the diagonal as corresponding to a two-step process. However, shouldn't we define the process as being one- or two-step in terms of whether there is a well-defined intermediate, *i.e.* whether there are two free energy barriers? For instance, if there is only one barrier for a reaction coordinate that curves off the diagonal, wouldn't that be better considered a one-step process that involves both order parameters concomitantly?

Matteo Salvalaglio replied: This question highlights a very important point that I believe is relevant, not just to this paper, but probably also to a more general discussion about nomenclature in two-step nucleation mechanisms. In our case, when the most favoured pathway departs from the diagonal it means that disordered intermediates exist. Even if they are not characterised by a metastable basin in free energy, such structures possess a finite, non-negligible lifetime.

We are therefore referring to a two-step process when nucleation of ordered domains takes place within such structures, regardless of the fact that they cannot be considered a metastable phase, but rather the transient product of solute density fluctuations. Regardless of the character of the intermediates, the process that we refer to as two-step is qualitatively different from the classical description of a single-step process, whereby for each cluster size the system can be considered as already possessing its final crystal structure.

Dimitrios Toroz remarked: To concentrate on the role of the solvent in order to better understand the two-step to one-step mechanism, it would be interesting to reveal from the WT-MD trajectories fundamental information on the reaction pathway that occurs between urea and the alcohols and the acetonitrile respectively. This would provide a better understanding of the reaction mechanism (solute–solute, solute–solvent and solvent–solvent) that is taking place.

Your paper nicely presents the idea that depending on the solvent, the nucleation process can either follow a one-step or two-step mechanism. My question is whether, by tuning the concentration of the system, any difference on the step mechanism for the nucleation process can be observed or not.

Matteo Salvalaglio responded: We cannot rule out this possibility, and it could indeed be interesting to investigate this aspect by taking advantage of algorithms such as CμMD.[1].

1. C. Perego, M. Salvalaglio and M. Parrinello, Molecular dynamics simulations of solutions at constant chemical potential, *J. Chem. Phys.*, 2015, **142**, 144113.

Robert Hammond communicated: I would like to draw together a theme from the question posed by Professor Sarah Price at the beginning of this session with an aspect of the paper presented by Dr Matteo Salvalaglio.

Professor Price asked, in reference to imidazole - a question on Professor Schroeder's paper (DOI: 10.1039/c5fd00005j) - whether it is possible from experiments (for example, from solution state NEXAFS) to get a clearer idea which of the often numerous different, fully periodic, bulk crystal structures representing possible polymorphs of a material, as predicted from first principles calculations, and lying within a relatively small energy range, will be manifested in nature through the process of nucleation. In the sense that Dr Salvalaglio has suggested, through his computational experiments using molecular dynamics, that a metastable form of urea - form II - is suggested, and that the cluster structures he has predicted could be used, for example, to predict NEXAFS spectra, could he comment on whether form II, if it were to continue to grow into a bulk crystal, would represent a distinct structure in terms of the implied symmetry elements (space group) and unit cell parameters? In this context, could he speculate on the likely energy difference between the two polymorphs, if they could both be obtained as bulk structures?

Additionally, could Dr Salvalaglio comment on my observation that the cluster of form II depicted in Figure 5 of his paper appears to be faceted, whereas the cluster of form I does not?

Matteo Salvalaglio communicated in reply: The metastable intermediate that we observe in our calculations represents a stable periodic structure with a distinct unit cell. We have evaluated the energy difference between form I and form II as ~10 kJ mol^{-1} (ref. 1). Clusters depicted in Fig. 5 are just a representative snapshot of form I and form II structures, extracted from a metadynamics simulation in water. They both show facets to a similar extent. It is worth mentioning that the form I clusters expose predominantly {110} faces, in agreement with the experimentally known morphology of urea in water.[2]

1. F. Giberti, M. Salvalaglio, M. Mazzotti and M. Parrinello, Insight into the nucleation of urea crystals from the melt, *Chem. Eng. Sci.*, 2015, **121**, 51–59.
2. M. Salvalaglio, T. Vetter, F. Giberti, M. Mazzotti, and M. Parrinello, Uncovering Molecular Details of Urea Crystal Growth in the Presence of Additives, *J. Am. Chem. Soc.*, 2012, **134**, 17221–17233.

Joost van den Ende communicated: In figure 4 there are some voids in the N_I *vs.* N_{II} graphs, for example in the case of EtOH. How should these voids be interpreted?

Matteo Salvalaglio communicated in reply: These FES are the result of reweighing the free energy obtained during WT metadynamics as a function of the two coordinates N_{I} and N_{II}. Voids in those plots are simply related to a local lack of statistics.

Joop Ter Horst remarked: You stated that the solvent is responsible for whether the two-step mechanism or the conventional mechanism is observed. In ethanol and methanol, you do not see the two-step mechanism but you do in the other solvent used. You also mentioned the nucleus size from your simulations: in the other solvent this is relatively small, while in ethanol and methanol the nucleus size is an order of magnitude larger. This indicates that in ethanol and methanol the supersaturation ratio is relatively low, while it is higher in the other solvent. Therefore, could supersaturation also be responsible for whether we observe the two-step or conventional mechanism?

Matteo Salvalaglio replied: The dependence of the nucleation mechanism on the solvent is the result of the interplay between surface tension, solubility of the crystal phase, and the stability/lifetime of the amorphous intermediates. These three aspects are certainly correlated and, based on our work, we cannot rule out a dependence of the nucleation mechanism on supersaturation alone. It would indeed be interesting to isolate such supersaturation dependence. We think that the correct way of approaching this problem would be to choose a solvent and simulate nucleation while modulating the concentration of the liquid phase, as I briefly mentioned earlier in response to Dr Toroz.

It is worth highlighting that carrying out nucleation simulations in which supersaturation is precisely controlled is far from trivial, requiring computational schemes such as the one proposed in a recent work from Perego *et al.*[1]

1. C. Perego, M. Salvalaglio and M. Parrinello, Molecular dynamics simulations of solutions at constant chemical potential, *J. Chem. Phys.*, 2015, **142**, 144113.

Sophie Janbon communicated: You showed that urea homogeneous nucleation mechanism is solvent dependent. Do you think that the nucleation mechanism may also depend on supersaturation level?

Matteo Salvalaglio communicated in reply: It can be definitely possible, as discussed with Prof. Ter Horst, and we cannot rule it out on the basis of our work.

Stéphane Veesler asked: if I understand fully, you have shown or obtained two possible mechanisms, one-step or two-step, according to the nature of the solvent. What experiments can you imagine doing to test these, and what parameters would you measure?

Matteo Salvalaglio responded: In order to detect clusters in the nanometer size range, one would need an experimental technique possessing a detection limit on that scale. The technique illustrated by Cölfen in his contribution to this Discussion (DOI: 10.1039/c4f00269e) displays such a characteristic. Nevertheless, I believe that independently of the detection technique of choice, one should be able to localise in the space where the nucleation event occurs. In this regard, the microfluidic approaches developed by your group (DOI: 10.1039/c4fd00274a) may be extremely useful.

Nadine Candoni communicated: Simulation results obtained on small cluster sizes do not seem easy to observe experimentally. However, for large cluster sizes the simulations show that the transition between form I and form II becomes a barrier-less process. Has this transition been observed experimentally?

Matteo Salvalaglio communicated in reply: To the best of our knowledge, such a transition has never been observed experimentally. Nevertheless, there are other computational results[1] which indicate that that solid–solid polymorphic transitions may become barrier-less when the molecular structure of the solute has few conformational degrees of freedom, and the polymorphic transition is associated to a distortion of the crystal lattice.

1. B. Peters, Competing nucleation pathways in a mixture of oppositely charged colloids: Out-of-equilibrium nucleation revisited, *J. Chem. Phys.*, 2009, **131**, 244103.

Roger Davey opened a general discussion of the paper by Ake Rasmuson: Do I understand it correctly, that you quote supersaturations as RTlnS because your data is not measured isothermally?

Ake Rasmuson responded: We report driving forces as RTlnS mainly because it gives units of J mol^{-1}, which we feel are useful units for comparing one system to another. In many experiments, we used the same vials with their content undergoing repeated nucleations. In that case, it facilitates the experimental work if the various supersaturations are created by cooling to different temperatures, instead of using different concentrations and cooling to the same temperature. Usually, the temperature difference is quite small between our experiments at different driving forces. In addition, ideally we would like to keep concentration and temperature constant, since both appear in the pre-exponential factor. However, we can't then generate different supersaturations.

Roger Davey said: From your data you obtain values of the pre-exponential factor, *A*. Why do not use this to correlate with key parameters, so as to elucidate the rate determining step?

Ake Rasmuson replied: This is a good suggestion that we will follow. However, the pre-exponential determinations must be regarded as quite uncertain, and this will influence the reliability of the analysis.

Richard Sear asked: In, for example, Table 1, you quote rates as per unit volume. Can I ask if you have directly checked that the rates you observe are per unit volume; *e.g.*, have you doubled the volume of the droplets you work with, and observed a doubling of the rate? It is just that if nucleation is occurring on a surface, then the rate may vary not with volume, but with the area of that surface.

In addition, you vary the solvent in your interesting studies of nucleation. If nucleation is occurring on a surface, then changing the solvent could change the nucleation behaviour, not only by changing how the molecules interact with each other in the solvent, but also by changing how these molecules, and hence the nucleus, interacts with this surface. Classical nucleation theory predicts that the barrier to nucleation on a surface depends on both the nucleus/surface and the solvent/surface interfacial tensions.[1] The solvent/surface interfacial tension will of course change when you change the solvent. Do you think that nucleation is heterogeneous? If so, could changes in solvent/surface interactions be contributing to the changes in nucleation you see when you change the solvent?

1. P. G. Debenedetti, *Metastable Liquids*, Princeton University Press, Princeton, 1996.

Jacek Zeglinski responded: We have not done experiments with different volumes for this system, so the units are only to be understood as a way of normalizing the data. However, the nucleation appears to occur in the bulk of the solution according to the video recordings, and is not particularly related to the liquid surface. The main reason for this is likely to be that the vapor phase above the liquid is saturated by the solvent, since the vials are completely capped.

Marco Mazzotti commented: In interpreting your measurements you are assuming that growth is instantaneous, and that the induction time coincides with the nucleation time, so you can ignore growth and growth rate in your analysis. This is a critical assumption in general, and in particular in your case, where you are considering nucleation of the same solute from different solvents. It is reasonable to assume that growth is instantaneous in different ways in different solvents, *i.e.* that your assumption might be more or less critical in different solvents. Can you comment on this, and on how it affects the outcome of your analysis?

Jacek Zeglinski replied: In our work we assume the growth time to be negligible, compared to the time of nucleation and considering the induction time distribution. Even though there might be variations in the growth time we expect that the variation in nucleation induction time is much greater. The variation in nucleation induction time leads to uncertainty in the estimation of the mean induction time. With respect to this uncertainty, and the magnitude of the average induction time, we believe that the growth time can be neglected in this study. However, of course it depends on how the experiment develops and hence on the system. It appears to be a fair assumption that if we cannot observe any changes in the solution for *e.g.* 20 min, and then within a couple of seconds the entire solution becomes very turbid, the growth time to visibility is likely to be short and negligible for the determination of average nucleation time. The reason for fast growth is, of course, a comparatively high supersaturation.

Simon Black addressed Roger Davey and Ake Rasmuson: The barrier to nucleation is a free energy barrier.[1] When a solvent molecule detaches from a solute molecule, there is a gain in entropy, which could be estimated, for example, from the entropy of melting of the solvent. What is the resulting free energy change for solvent–solute separation?

1. S. Black, Simulating nucleation of molecular solids, *Proc. R. Soc. A*, 2007, **463**, 2799–2811.

Ake Rasmuson communicated in reply: Unfortunately, this is something we haven't yet investigated.

Roger Davey added: I am not sure what this question is really getting at. Clearly the question of the entropy change accompanying nucleation is important.

Thomas Turner asked: The molecular dynamics simulations presented in this work consider only 8 binding sites for the solvent to the risperidone molecule. Although the results presented show good agreement between binding energy at the carbonyl group and difficulty of nucleation in the various solvents, is there a computational method which considers the full energy landscape of such a large solute, with the aim to calculate free energies of solvation, for example? This could then give a more accurate representation of the solvation environment of this molecule.

Jacek Zeglinski responded: The full energy landscape, *i.e.* the full three-dimensional space of interaction of the solvent molecules surrounding a solute molecule, can be modelled. In our recent paper[1] we proposed a methodology for computing the solvation energy of salicylic acid in different solvents, using high-level DFT method and a solvation shell model. It would be also possible to estimate the solute molecule–bulk solvent interactions by employing force-field molecular dynamics. We have done this for a number of organic molecules, and will include the relevant methodology in our successive papers.

1. D. Khamar, J. Zeglinski, D. Mealey and Å. C. Rasmuson, Investigating the Role of Solvent–Solute Interaction in Crystal Nucleation of Salicylic Acid from Organic Solvents, *J. Am. Chem. Soc.*, 2014, **136**, 11664-11673

Simon Black addressed Jacek Zeglinski and Ake Rasmuson: Nucleation rates are sometimes estimated using interfacial tensions that are inferred from solubility data. Please comment on how these data challenge this practice.

Ake Rasmuson replied: The origin of this stems from the work of Nielsen on mostly inorganic compounds in aqueous solution, and a later theoretical analysis by Mersmann using regular solution theory. We have evaluated this for individual organic compounds in different solvents, but have not been able to find a general validity of this approach. No reasonable correlation was found for risperidone (as presented here), salicylic acid[1] or salicylamide.[2] Some correlation is found for m-hydroxy benzoic acid.[3] Reasonable correlation is found for butyl, propyl and ethyl paraben.[4,5]

1. D. Mealey, D. Croker and Å. C. Rasmuson, Crystal nucleation of salicylic acid in organic solvents; *CrystEngComm*, 2015, DOI: 10.1039/c4ce01428f.
2. F. Nordström, M. Svärd and Å. C. Rasmuson, Primary nucleation of salicylamide: influence of process conditions and solvent, *CrystEngComm*, 2013, **15**, 7285–7297.
3. M. Svärd, and Å. C. Rasmuson, m-Hydroxybenzoic Acid: Quantifying Thermodynamic Stability and Influence of Solvent on the Nucleation of a Polymorphic System, *Cryst. Growth Des.*, 2013, **13**, 1140–1152.
4. H. Yang, M. Svärd, J. Zeglinski and Å. C. Rasmuson, Influence of Solvent and Solid State Structure on Nucleation of Parabens, *Cryst. Growth Des.*, 2014, **14**, 3890–3902.
5. H. Yang and Å. C Rasmuson, Nucleation of butyl paraben in different solvents, *Cryst. Growth Des.*, 2013, **13**, 4226–4238.

Francis Taulelle remarked: If you were considering using techniques used in NMR, especially with peptides, and want to know if site one is connected to site two *etc.*, you can look at the residual dipolar interactions; several bidimensional spectra allow the establishment of such proximity correlations. The site-to-site proximity count allows us to count the number of links in the crystalline network formed. Additionally, with the actual residual dipolar coupling value, the presence and role of the solvent in the linkage of nodes of the crystalline network can be investigated.

Jan Sefcik returned to the discussion of the paper by Matteo Salvalaglio: I have a question about the free energy profiles shown in Figure 2 in your paper. You plot them for fully crystalline clusters going along the diagonal in Figure 3. But you also say that the preferred pathway wouldn't be along this diagonal. Why is it then relevant to plot this, and what would you get by plotting free energy profiles along other, more favourable, pathways?

In addition, you are fitting classical nucleation theory parameters for the initial part of the free energy profile curves that are accessible computationally, but you then extrapolate these 10 times up the cluster size to get the critical nucleus. This may be fine but it is a long extrapolation, so how relevant is this if it is not following the most favourable pathway?

In your Figure 2c for acetonitrile, the free energy profile seems to be flat from about 40 upwards. This is very interesting; do you have any explanation for that? You are fitting it with a maximum but it appears to actually be constant for a wide range of cluster sizes.

Matteo Salvalaglio responded: The plot in Fig. 2 is aimed at showing that the finite-size effects on the free energy profile can be understood and quantitatively evaluated using Eq. 1. Such an equation is derived from CNT and thus we compare it with free energy profiles that comply with the CNT hypotheses of having a crystal-like structure for all cluster sizes. Such a comparison is performed in Fig. 2, and the fitting of CNT parameters is therefore only relevant to display that indeed, the depletion effects can be quantitatively evaluated. Plotting the free energy profiles on more favourable pathways will show a shape different from CNT, due to the fact that the internal structure of the clusters will depend on the progress along the coordinate, but will indeed display the same dependence on the depletion effects that are quantitatively captured by fitting the profiles in Fig. 2.

In fact, it is worth highlighting that even if the fitting of the curves in Fig. 2 allows us to extract CNT-like parameters from the simulations, the correction term expressed in Eq. 3 does not require their knowledge.

The CNT profile is an analytical approximation of the free energy profile reweighed from simulations. The deviation of the latter from the expected maximum behaviour is not large enough to be considered a significant feature in this context.

Elena Simone communicated: In the modelling results for the polymorphic transformation of urea, the authors find that the transformation is not solvent mediated and that dissolution of the solid does not happen. Does this mean that the transformation would take exactly the same time in all solvents? Are the authors aware of any other compounds that show this behaviour (from literature or experimental work)?

Matteo Salvalaglio communicated in reply: This is indeed a very good point. If the transformation gradually becomes a barrier-less process, the exponential dependence of the rate from the energy barrier gradually vanishes. Therefore, the polymorphic transformation should proceed with rates that are within the same order of magnitude in all solvents. Such a transformation takes place in nuclei that contain hundreds of molecules. Although we refer to such clusters as "large", they are technically in the nanometer range, and we are not aware of experimental works displaying this behaviour. Similar computational results have been obtained by Peters[1] in his computational study of nucleation in charged colloids.

1. B. Peters, Competing nucleation pathways in a mixture of oppositely charged colloids: Out-of-equilibrium nucleation revisited, *J. Chem. Phys.*, 2009, **131**, 244103.

Joop Ter Horst remarked: What about heterogeneous nucleation? That is the kind of nucleation that is occurring in practice and that we are interested in. Can we simulate, measure or account for this?

Sven Schroeder answered: One could deliberately add seeds, *e.g.* nanoparticles, into the system and monitor their interaction with the solute molecules while carefully adjusting the supersaturation. One could do such measurements for solutes with absorption edges at higher X-ray energies (*e.g.* for transition metal nanoparticle nucleation) by localising nucleation events at a liquid–liquid interface[1]. For organic systems, one would have to construct a sample environment permitting such measurements with soft X-rays, but in principle this should be possible.

1. Y. Gründer, J. F. W. Mosselmans, S. L. M. Schroeder and R. A. W. Dryfe, In Situ Spectroelectrochemistry at Free-Standing Liquid–Liquid Interfaces: UV-vis Spectroscopy, Microfocus X-ray Absorption Spectroscopy and Fluorescence Imaging,*J. Phys. Chem. C*, 2013, DOI: 10.1021/jp312060f.

Matteo Salvalaglio added: We agree on the fact that understanding both homogeneous and heterogeneous nucleation is important. Heterogeneous nucleation can be readily simulated with molecular modelling techniques considering simplified surface models. Nevertheless, building quantitatively representative surface models still remains a challenge. We are all aware, in fact, that the nature of the particles triggering heterogeneous nucleation in real systems is typically unknown. Even when it is known, it is still far from trivial to design reliable structural models of the particle surfaces, especially when their nature is amorphous.

Herma Cuppen commented on the paper by Matteo Salvalaglio: I find this very impressive work. My question relates to your collective variables (CV). These are a sum of many coordinates; how do you uniquely propagate your CVs, and does this propagation choice affect the outcome of your simulations? I can imagine that, for the onset of crystallization, it matters whether you start with a very small nucleus of near perfect alignment or you gradually change the full cluster.

Matteo Salvalaglio answered: Thank you very much for your comment. As I mentioned earlier in my answer to Mr van den Ende, the CVs that we use have been designed in order to approximate the number of molecules in a crystal-like configuration, and are expressed as the sum of functions computed over all the atoms in the simulation box.

The potential affects the dynamics through the introduction of an artificial force, that is computed as the derivative of the bias potential with respect to the position of each atom. In this there is no ambiguity, and our biasing coordinates do not differ from conventional CVs such as coordination number, or the total number of H-bonds.

By construction, the CVs that we use as biasing coordinates (S1 and S2 in the paper) do not explicitly include any notion of clusters within the simulation box. This means that such CVs have a certain degree of degeneracy: for instance, configurations characterised by the same value of S1 and S2 may be characterised by a different number of clusters in the system. This is especially true when the clusters are small, *i.e.* in the "solution" basin. As the clusters grow in size, this degeneracy becomes less important.

In order to solve such issues we are *a-posteriori* clustering the configurations sampled with WT metadynamics, and then reweighing the free energy surface as a function of the variables explicitly referred to the largest cluster in the system (n, n_o, N_I, N_{II}).

Ken Lewtas addressed Matteo Salvalaglio and Marco Mazzotti: Your work is very interesting and thought provoking. From your paper, it is clear that the motivation behind this work is in the areas of pharmaceuticals and fine chemicals. I would like to draw your attention to another area of research which is very environmentally important, the question of the reduction of nitrogen oxides in vehicle emissions. NO_x is removed by a process called Selective Catalytic Reduction (SCR) which uses urea. Unfortunately, the crystallisation of urea often hinders the activity of this process. Perhaps your fundamental work can add to the development of a solution for this problem, rather than using a more empirical approach.

Marco Mazzotti responded: Our research program aims at understanding how macroscopic stochasticity can be characterised and how it affects processes at different scales, where nucleation occurs. We believe it is very important to analyse real systems, such as the one you have mentioned, in order to ascertain the role played by the stochastic behaviour of nucleation. As such, our models and tools would indeed be helpful to assess whether undesired precipitation in different types of devices and processes might be controlled, and how.

Matteo Salvalaglio added: Thank you for your comment and suggestion. Moving from the understanding of the fundamentals of urea homogeneous nucleation, to investigating urea nucleation in a SCR device, requires a leap in complexity that honestly appears quite intimidating. Nevertheless, computational techniques may be used in this field to enable a preliminary screening of nucleation, and investigate growth inhibitors that may be helpful to reduce the crystallisation of urea and thus improve the process efficiency.

Colan Hughes asked: You said that the form II urea clusters that you see in your simulations are too small for our solid-state NMR techniques to detect, at around 50 molecules. This size limitation is based on the correlation time (t_c) for the tumbling of the particles, which we have linked to the size using a number of assumptions. Therefore, I wondered if you could estimate the correlation time for these particles directly from your MD simulations, to assess whether or not such particles would be tumbling too rapidly, with $t_c < 10$ μs?

Matteo Salvalaglio answered: In principle, one could estimate the tumbling rate of clusters of arbitrary size directly from MD simulation. We haven't yet directly performed such a calculation; however, we can use the Stokes relation to compute the characteristic tumbling time for a spherical object in the size range of interest. For clusters with a radius in the nanometer range, such estimate suggests tumbling times of the order of the hundreds of nanoseconds.

Roger Davey commented: Could Schroeder's experimental technique for the study of solution species provide useful input to Salvalaglio's simulations?

Sven Schroeder replied: Urea is an excellent candidate for probing by soft X-ray spectroscopy because it contains several atoms (C, N, O) that can be probed by the instrument, and all three should be sensitive to the intermolecular interactions experienced by the urea molecules. It would be interesting to carry out a comparative study of urea nucleation from alcohols and from water, to provide experimental evidence for the predicted crystalline nuclei formed by one-step nucleation and the less ordered nuclei formed in the more complex nucleation process from water and acetonitrile.

Matteo Salvalaglio added: The technique described by Schroeder in his paper would indeed be a useful tool to experimentally investigate the dominant intra-molecular interactions in solution, and provide information on the relevant interaction motifs that should be observed in simulation.

On the other hand, enhanced sampling simulations also provide ensembles of molecular structures that can be analysed in order to interpret experimental spectra.

Changquan Calvin Sun addressed Matteo Salvalaglio: If you take out the solvent altogether, you are studying the nucleation of a supercooled liquid. What would be the possible outcome of this modelling? Would density fluctuation in the melt still occur? If so, would a two-step nucleation mechanism still be possible?

Matteo Salvalaglio answered: We have investigated urea nucleation from its melt as a first step to validate our computational approach.[1] Local density fluctuations certainly also occur in that case, however it is probably not proper to describe it as a two-step process. A two-step process, in fact, is characterised by a first stage in which solute molecules come together, forming disordered intermediates. Within those intermediates the nucleation of a crystal phase takes place. In the melt, this is the sole relevant step. Local density fluctuations in the melt, as well as in the disordered intermediate, may nevertheless characterise the ordering process. For further examples of simulation studies of the nucleation of organic crystals from the melt, please refer to the work of Santiso's group.[2]

1. F. Giberti, M. Salvalaglio, M. Mazzotti and M. Parrinello, Insight into the nucleation of urea crystals from the melt, *Chem. Eng. Sci.*, 2015, **121**, 51-59.
2. M. Shah, E .E. Santiso and B. L. Trout, Computer Simulations of Homogeneous Nucleation of Benzene from the Melt, *J. Phys. Chem. B*, 2011, **115**, 10400.

Erik Santiso remarked: This is really a comment on the discussion of Matteo's paper, where the issue of a two-step nucleation mechanism being possible in melt nucleation was raised. In work we have done, it is sometimes observed that the system goes through an apparent metastable state when following the minimum free energy path between the liquid and the solid (for nucleation of organic crystals). However, given all the difficulties in deconvoluting the effect of the periodic boundary conditions, and the physics of the process in such simulations, it would be far-fetched to conclude that this an indication of a two-step mechanism. Much bigger simulations would be needed to address this.

Kevin Roberts commented: With regards to the solvent effect, it's worthwhile remembering that urea is quite an unusual molecule, as it only has 8 atoms but is nonetheless a solid due to its large amount of hydrogen bonds. Like water, it is both a hydrogen bond acceptor and donor. Additionally, its van der Waal's contributions are surprisingly low. When it is solvated with ethanol and methanol, urea's crystallisation behaviour is quite different to its behaviour with other solvents such as water.

It's worth noting here that it is perhaps not just at the hydrogen bonds in the structure, but when the molecule has a low molecular weight, that there may be value in looking at the van der Waal's forces too.

Matteo Salvalaglio replied: Thank you for your comment. In order to interpret differences in the urea nucleation mechanism moving from alcohols to water solutions, we are planning to investigate the structure making/breaking character of urea in each of these solvents. As you point out, it would indeed be insightful to complement such analyses with a breakdown of the contributions to solvation due to hydrogen bonding and van der Waals's forces.

Roger Davey opened a general discussion of the paper by Ake Rasmuson: In the analysis of the induction time method, I notice that Ter Horst uses a Poisson distribution, while you fit with a log-normal. I understand that the Poisson distribution reflects a random process in which events are unrelated. What is your physical justification for using a log-normal distribution?

Ake Rasmuson responded: The justification is simply that, overall, the log-normal function gives a better fit to the data.

Ian Rosbottom addressed Ake Rasmuson and Jacek Zeglinski: It might be an idea to compare the isothermal data to polythermal data. Since you are empirically comparing the ease of crystallisation to induction times, you are obviously aware that the crystal spends a reasonable amount of time growing to the detection point, and if two of the solvents tested have significantly different solubilities, then one system may be significantly more de-supersaturated than the other, resulting in slower growth. Conversely, if you check metastable zone widths then the continuous cooling would keep the driving force up, and possibly the de-supersaturation might be less of an issue. Do you think it would be worth checking if the trends are the same between isothermal and polythermal data, with respect to the ease of crystallisation?

Ake Rasmuson answered: In the induction time work we do compare the results at equal driving forces; however, it is with the approximation that the activity coefficient ratio can be neglected. It is not the growth rate that is a limiting factor, since in induction time measurements the solution becomes immediately turbid as soon as any change can be observed in the solution. For salicylic acid we have made a detailed analysis and comparison of induction time and metastable zone width results. The work has recently been accepted for publication.[1] The comparison of metastable zone width and induction time data is governed by the rate of nucleation (induction time) in relation to the rate of supersaturation generation in the metastable zone width measurements.

1. D. Mealey, D M Croker and A Rasmuson, Crystal nucleation of salicylic acid in organic solvents; *CrystEngComm*, 2015, DOI: 10.1039/c4ce01428f

Nadine Candoni communicated: In the determination of the pre-exponential factor A, is it more appropriate to compare $\ln(A)$ values, because the kinetic pre-exponential factor is extracted from a plot of the logarithm of J as a function of the inverse of the square logarithm of supersaturation S? Moreover, $\ln(A)$ is extrapolated to infinite supersaturation, outside of the experimentally accessible range: it is a domain where the classical nucleation theory fails to describe nucleation. For instance, when$S \rightarrow \omega$ the classical nucleation theory predicts $J \rightarrow A$.

Ake Rasmuson communicated in reply: Sorry, I don't see how comparing ln$A$ instead of A would help. I fully agree that determination of A-values tends to be uncertain; however, there is a clear theory developed for physical interpretation of A-values, and I fail to see how using ln$A$ instead would be beneficial.

Dimitrios Toroz asked: The binding sites of the target molecule with some of the solvents were quite close in energy (in a range of 0.2 kJ mol^{-1} is some cases), so the question is: what is the effect of the counterpoise correction energies (although they might not be crucial for such a system)? The second query is the degree of agreement between the simulated IR spectrum from the vibrational frequencies calculations with the spectrum obtained from the experimental data.

Jacek Zeglinski replied: We did not correct our DFT binding energies for the basis set superposition error (counterpoise error). This procedure is not recommended in light of recent benchmark studies of density functional methods by Goerigk and Grimme.[1] To calculate accurate energies we used a large basis set of quadruple quality (def2QZVP) and a double-hybrid functional, B2PLYP-D3.

Our benchmark studies indicated that the B97-D functional, in combination with the 6-31G(d,p) basis set, gives vibrational spectra that match experimental spectra more closely than those obtained with the popular B3LYP functional. The degree of agreement would certainly depend on the type and complexity of a system. We found a very good agreement for some chief vibrational modes in the salicylic acid dimer.

1. L. Goerigk and S. Grimme, A thorough benchmark of density functional methods for general main group thermochemistry, kinetics, and noncovalent interactions, *Phys. Chem. Chem. Phys.*, 2011, **13**, 6670-6688.

Peter Vekilov opened a general discussion of the paper by Baron Peters: I enjoyed your paper; the premise that one can use CNT to predict the responses of the system to parameters of interest is very appealing. I have a question about the assumption of adsorption equilibrium between a critical cluster and impurities in the solution. Adsorption may be slowed down because of the slow rate of two processes: diffusion in the solution, and adsorption. How will the predictions of your model change if adsorption is not at equilibrium, but is transient?

Baron Peters replied: Thank you, Professor Vekilov, for this excellent question. Our model assumed a mean field additive coverage (from the Langmuir isotherm) at each nucleus size. In principle, both the size of the nucleus and the number of adsorbed additives are dynamically fluctuating variables. We are currently developing a generalized model of the free energy, dynamics, and steady-state fluxes on the two-dimensional landscape of nucleus size and additive coverage. This new model should be able to provide answers to the questions raised by you and others at this meeting.

Joop Ter Horst remarked: Additives can promote nucleation. You account for the additive to influence the interfacial energy, and herewith the energy barrier for crystal nucleation. However, if the additive takes part in the cluster formation process, it should be considered to be a nucleation centre or heterogeneous particle. This should be accounted for in the pre-exponential factor of the heterogeneous crystal nucleation rate equation of Classical Nucleation Theory, in which the concentration of nucleation sites comes back. The effect of an additive promoting nucleation can therefore be twofold: a lowering of the nucleation work and a change in the concentration of nucleation sites. Did you not account for this?

Baron Peters replied: The theory attempts to describe something very different from heterogeneous nucleation (HEN). In HEN, the nucleus is not spherical, there are typically three types of interfaces, and heterogeneous rates are expressed per area per time, or per site per time. The process that I am describing would still occur in a clear homogeneous solution, with compact nuclei, one type of interface, and a rate per volume per time. As you note, additives will influence the prefactor, but perhaps not in the way that you suggest.

You are thinking of each additive molecule as a nucleation site, and therefore suggesting that the "concentration of monomers" in the prefactor should become the "concentration of additives". It is an appealing picture, but recall that the free energy barrier in classical nucleation theory (CNT) is $F(n) = -k_BT \ln[N(n)/N(1)]$ where $N(n)$ and $N(1$ are concentrations of n-mers and of monomers, respectively. Upon exponentiation, the concentration of monomers in the free energy expression cancels with that in the prefactor. What is left is a Zeldovich factor and a mobility multiplied by the concentration of n-mers at equilibrium. My point is that, if you write the CNT prefactor in terms of additive concentration, then a consistent treatment would also compute the free energy as a function $F(m)$, where m is the number of adsorbed additives on a nucleus. The barrier as a function of adsorbed additives would no longer have the CNT shape. Perhaps more importantly, a theory built around $F(m)$ would, by definition, implicitly assume that nucleus size is not dynamically important.

Several questions from the audience suggest important effects from the dynamics of additive coverage. I interpret these to mean that our reaction coordinate (we used nucleus size n alone) might need extension to also include contributions from additive coverage. We are working on a generalized theory now, in which both n and m are important variables. If both variables turn out to be important, then the new theory will no longer look like CNT. It will instead look like the theory of my colleague.[1] His multidimensional framework is theoretically appealing, but less easily used for experimental interpretation than CNT. Therefore, I am hopeful that the reaction coordinate remains nucleus size alone as we have assumed.

1. J. S. Langer, *Ann. Phys. (Amsterdam, Neth.)*, 1969, **54**, 258–275.

Stéphane Veesler remarked: As I was taught by Dr Boistelle, and to the best of my knowledge, there is no additive which can accelerate nucleation whilst keeping the nucleation mechanism unchanged.

What I meant is that an additive or impurity can, for instance, change the mechanism of nucleation from homogeneous to heterogeneous, thereby acting as a catalyst.

In your model, you have said that the prefactor in the J equation is weakly dependent on additive concentration. But most of the experiment shows a drastic diminution of J with additive concentration (when these have an effect - the effect is related to absorption). If we define the prefactor as the kinetic term and the exponential term as the thermodynamic one, we can say that there is always a competition between kinetic and thermodynamic, and most of the time the kinetic won. That is to say, the decrease in surface energy is counterbalanced by the decrease by orders of magnitude of the prefactor term.

So, using your model of adsorption, could you model the effect of additives on the prefactor, instead of keeping it constant?

Baron Peters answered: Thank you, Prof. Veesler, for this question. Our theoretical expressions do ignore changes in the prefactor, but note that our simulations and mean first passage time calculations do not. With that in mind, note that the predicted relationship between additive concentration and free energy barriers is good, while the predicted relationship between rates and additive concentration is not as good. The additional source of error may indeed be effects of the additives on prefactors. As Prof. Davey illustrated in his lecture, there is a relationship between growth and nucleation prefactors. Therefore, models of growth inhibition may provide a starting point to understand the effects of additives on nucleation prefactors.

There are cases where additives accelerate nucleation without changing the mechanism from homogeneous to heterogeneous nucleation: detergents accelerate bubble nucleation from carbonated water. Salt accelerates the nucleation of gamma glycine from water because of double layer formation around the nuclei.[1] Minor Zr components of an Al-Zr-Sc alloys facilitate nucleation and concentrate at the interfaces between Sc precipitates and the Al matrix, *cf.* the work by G. Martin and coworkers.

1. Duff *et al. J. Chem. Phys.*, 2013

Richard Sear commented: I like your idea and model. I am happy that a dilute species that binds to the interface of the nucleus with a free energy of a few kT will lower the surface tension, and hence the barrier to homogeneous nucleation. What about heterogeneous nucleation? There we have three interfaces (nucleus/solution, nucleus/surface and solution/surface) and the rate is sensitive to the sizes of all three. For example, to take an extreme case, if the surface-active molecule makes the nucleus wet the surface, it should abolish the barrier to nucleation and so dramatically increase the rate. It could also potentially make the nucleus–surface interaction less favourable and then could actually decrease the rate. Can I ask if you agree with those two statements? And do you know how we could test for this effect in experiments?

Baron Peters answered: Thank you, Professor Sear, for the astute comment. As you've noted, heterogeneous nucleation presents at least three interfaces which could be affected by the additive: nucleus/solvent, nucleus/surface, and solvent/surface. Combining the considerations in our model with the Dupre–Young equation does suggest that an additive which binds strongly to the solvent/surface interface could either accelerate or decelerate heterogeneous nucleation. The effect of additives on heterogeneous nucleation might be measured by studying nucleation on atomistically smooth surfaces, like mica.

Sarah Price remarked: I would like just to check that you are modelling the effect of an additive on the surface tension of an existing nucleus. Is this very distinct from the case where the additive is in the centre of the nucleus, and is the cause of the nucleation?

Baron Peters answered: Our model only considers additives that bind to the exterior surface of a nucleus. There are simulations of the case where molecular scale templates end up in the center of the nucleus.[1] The structure directing agents in zeolite synthesis may also be an example.

1. See, for example, S. Jungblut and C. Dellago, Crystallization on prestructured seeds, *Phys. Rev. E.*, 2013, **87**, 012305.

Wenhao Sun asked: I'd first like to defend the main hypothesis in your paper from two earlier comments: 1.) Someone said that additives tend to affect the nucleation rate by affecting the nucleation prefactor. I find this unlikely, since the prefactor reflects ion diffusivity in the solution, which should not be strongly influenced by additives. The prefactor also reflects ion attachment rates, which may be influenced by additives, but only if additives strongly adsorb and "poison" the growth of the surface. This would be an inhibitory effect, rather than an accelerating effect. Any acceleration of nucleation resulting from adsorbing additives should not, therefore, reflect any influence from ion attachment rates. 2.) Someone said that additives typically serve as heterogeneous nucleation centers, rather than as surface energy modifiers. While additives certainly could act as heterogeneous nucleation centers, particularly large dust particles, a chemisorption effect onto surfaces should modify the surface energy and stabilize nascent nuclei. I also appreciate the comment by Baron Peters, who said that additives that stabilize a smaller critical nucleus radius would allow smaller fluctuations to lead to nucleation. How the additives interact with the pre-critical sized nuclei doesn't matter, as those dissolve anyway. But the ones that do make it can continue to grow, and those are the ones that we see.

Baron Peters answered: On the first remark, which was also alluded to by Joop ter Horst earlier, our assumption that prefactors are unaffected is an approximation. I say this because the attachment kinetics can be influenced by things other than diffusion, *e.g.* solvent displacement can be associated with barriers. Note that the barrier reductions in our paper follow the predicted trends better than do the relative rates $J(x)/J_0$. The deviations from the expected trend for strongly bound additives may already be a poisoning effect.

I especially appreciate your second remark. I think the skepticism is centered on a feeling that nucleus formation is too fast for additives to "find" the nucleus. To this, I would put forth two arguments. First, the factor of x in equation (9) accounts for the fraction of surface sites on the nucleus that will be found by additives. As additives become more rare, it does become possible for nuclei to form before additives can find them. But equation (9) accounts for that too. The reduction in $\ln[J(x)/J_0]$ is also proportional to the "naked" free energy barrier $F_0^{\ddagger}$. Thus, rare additives won't accelerate nucleation *if it was already fast*.

The process of nucleation in the presence of additives is dynamically complex, but if there is a large barrier then everything to its left is approximately at equilibrium. Quasi-equilibrium does break down *ca.* kT from the barrier top and beyond, but the equilibrium properties give the population at the barrier top. Remarkably, this population is only a factor of order one from the population of transition states at steady state. These quasi-equilibrium theories are amazingly powerful.

A related point which no one raised is this: The simulations include an equilibrium ensemble of all possible surface coverages. The theoretical model makes a sort of mean field approximation by using the Langmuir isotherm. That approximation is another possible reason for the discrepancies between theory and simulation at high K. Thank you for these comments, Dr Sun.

Wenhao Sun remarked: I wanted to ask next - your model reflects the effect of spectator ion additives, but could it also work for pH? There is some evidence in minerals that pH can influence polymorph selection. Should a polymorph more preferentially adsorb H^+ or OH^-, couldn't that also be a nucleation accelerant, *vs.* competing polymorphs? As a follow-up question: when charged ions are adsorbed onto a surface, this results in a charged surface, which often structurally manifests itself as an electrochemical double layer. How would an electrochemical double layer change the thermodynamics of the surface energy? How could we quantify this?

Baron Peters answered: Indeed, acids, bases, and salts can interact with charged surfaces and influence polymorph selection. In pure water, surfaces with a net charge density are not stable, but they can be passivated by a double layer, given sufficient ionic strength. For glycine, nucleation of the polar gamma-polymorph is favored by small amounts of salt. One of our recent papers models the effects of salt on glycine nucleation using Debye–Hückel theory, and also using alchemical transformations[1]. The situation with pH is perhaps more complicated, because different surfaces will probably have different isoelectric points.

1. N. Duff, Y. R. Dahal, J. D. Schmidt and B. Peters, Salting out the polar polymorph: Analysis by alchemical solvent transformation, *J. Chem. Phys.*, 2014, **140**, 014501.

Allan Myerson asked: Those of us with children have made the "diet cola geyser" by adding the candy Mentos® rapidly to a large bottle of diet cola. The smooth sugar coating on the Mentos® roughens in the diet cola and provide heterogeneous nucleation sites for the carbon dioxide. We thus can clearly accelerate nucleation due to heterogeneous effects, and virtually all work to demonstrate this can be taken back to the heterogeneous nucleation of the surface. A different question is, can you chemically affect nucleation by affecting molecular ordering? The only example I am aware of is the use of nucleators (also called clarifying agents) in polymers to induce nucleation. A number of companies manufacture nucleators; some are not crystalline but are added on a molecular basis. Could this be an example of a molecular accelerator?

Baron Peters responded: Thank you, Prof. Myerson, for helping to delineate our proposal from more established epitaxial and heterogeneous nucleation strategies. Thank you also for the tip about clarifiers in polymer nucleation. Please refer to my earlier response to Prof. Veesler for additional examples of dissolved nucleants.

Hugo Meekes remarked: You use a scalar for the surface free energy, and you suggest that for real crystals one could do contact angle measurements to see the effect of additives on surface energy. In reality the surface free energy is a tensor, due to the anisotropy of the crystal structure. As an example, our earlier work on the morphology of anthracene[1] shows that one would never observe a crystal without a one single dislocation on the basal face of the crystals at moderate supersaturation, while at very high supersaturation crystals with a thickness of only a few monolayers are found, without such a dislocation. Still the crystals usually found, that is with a screw dislocation, in a contact angle measurement would show an very low surface energy corresponding to the extremely anisotropic morphology of the dislocation free crystals. Could you generalize your model to include such an anisotropic surface energy?

1. H. M. Cuppen, W. S. Graswinckel and H. Meekes, Screw Dislocations on Polycenes: A Requirement for Crystallization, *Cryst. Growth Des.*, 2004, **4**, 1351.

Baron Peters answered: Thank you for pointing me to this interesting work. In our model, the interfacial free energy is averaged over all facets of the equilibrium shape, and we did not consider the effects of screw dislocations on the time for growth to observable sizes. The fraction of nuclei with such dislocations could indeed be important for quantitative comparisons to experimental nucleation rates. It may be possible to model the free energy of defect formation at the scale of nuclei, and to combine that with the effects of additives on interfacial free energy as described in our model.

Jim De Yoreo commented: The results are reminiscent of thin film epitaxy experiments in which introduction of a surfactant lowers the interfacial energy for nucleation of the next layer. These surfactants act as the molecular agents in your model.

Martin Ward communicated: This is an excellent paper, and really got me to think about the dynamics of what may be going on when these surfactant-type additives are present in the system. The lowering of the free energy barrier to nucleation by reduction of the nucleus–solution interfacial tension is clear. However, I can see two routes to nucleation in this scheme: (i) the lowering of the interfacial tension allows the nucleus to grow until it reaches critical size, and/or (ii) depending on the magnitude of the free energy change, inclusion of additive at the surface of the nucleus is sufficient to promote sub-critical nuclei to become super-critical, and so then able to nucleate. The difference between these two scenarios is subtle, but it may be the case that the dynamics involved in the nucleation step dictate which scenario is favoured.

I would like to find out if your current model would, perhaps, favour one of these routes to nucleation over the other?

Baron Peters communicated in reply: Thank you, Professor Ward, for this excellent question. Our current model follows classical nucleation theory in assuming a nucleus size coordinate. Additionally, we have assumed a mean field additive coverage (from the Langmuir isotherm) at each nucleus size. Because of these assumptions, the present model cannot distinguish between processes (i) and (ii) as you describe. Geoffrey Poon and I are currently developing a generalized model of the free energy, dynamics, and steady-state fluxes on the two-dimensional landscape of nucleus size and additive coverage. The two-dimensional model may provide more insight into the sequence of events along typical nucleation pathways.

Jim De Yoreo opened a general discussion of the paper by Franca Jones: Do the particles come together in an oriented fashion, or are they misaligned? Are they coming together as particles or through atomic transfer?

Franca Jones answered: From the timed TEM images, we know that the barium sulfate forms roughly spherical aggregates of ~50nm in size, that are themselves composed of smaller particles of about 5nm. At the one hour mark these aggregates are disordered, according to SAED. However, as time evolves (5 h), the appearance of the aggregates appears unaltered, but the SAED shows that the aggregate is now developing crystallinity or order. No obvious signs of dissolution

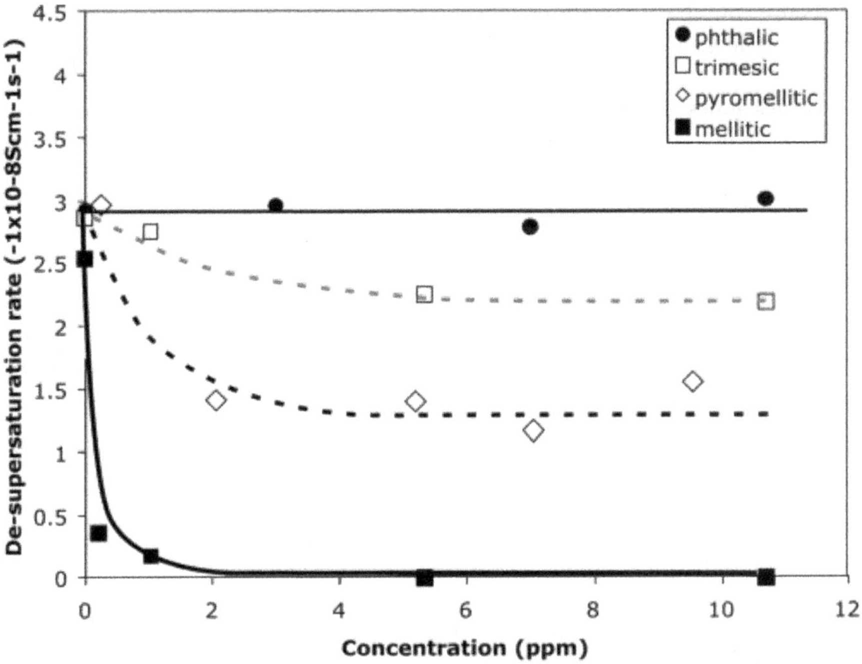

Fig. 2 De-supersaturation rates versus organic acid concentration for di-, tri-, tetra-, and hexabenzoic acid. Lines are drawn to aid reader only.

are observable, and therefore we believe that the aggregates are ordering themselves over time by realigning themselves in an oriented/ordered manner. Thus, we believe that this step is an oriented attachment process (nonclassical process). Having said this, there is a ripening process at longer periods (≥1 days) that does involve a dissolution-recrystallization step, which is clearly observable as thinning or hollowing out of the interior of the aggregates in the TEM images; this process is associated with the oriented mesocrystal fusing together. This ripening process, then, would be a growth unit-by-growth unit transfer process (classical process).

Gerard Coquerel asked: What happens if you do not use this hexa-carboxylic acid, but another benzoic acid derivative decorated with a different number of carboxylic functions?

Franca Jones responded: When other benzoic acid derivatives are used, standard single crystals are obtained, albeit with modified morphologies. This has been observed previously.[1]

1. S. R. Freeman, F. Jones, M. I. Ogden, A. Oliviera and W. R. Richmond, Effect of Benzoic Acids on Barite and Calcite Precipitation, *Cryst. Growth Des.*, 2006, **6**, 2579.

Gerard Coquerel asked: With reference to the previous question, what happen if you increase the number of tricarboxylic acid in your medium so that there is the same number of carboxylic functions? (*i.e.* double the concentration of tricarboxylic acid compared to the concentration of hexacarboxylic acid)?

Franca Jones responded: As mentioned previously, we have investigated the effect of many of the different acids at varying concentrations.[1] If 2ppm of the hexabenzoic acid is present and you wish to compare the equivalent concentration of acid groups present, you would need to compare to 7ppm trimesic acid. Doing this comparison, we find that the degree of inhibition is still greater for the hexabenzoic acid and still no mesocrystal formation occurs.

I would like to clarify, the presence of 10ppm trimesic acid does not induce mesocrystal formation of barium sulfate while as little as 2ppm mellitic acid can induce mesocrystal formation. Also the de-supersaturation curves for the acids can be seen in Fig. 2.

1. S. R. Freeman, F. Jones, M. I. Ogden, A. Oliviera and W. R. Richmond, Effect of Benzoic Acids on Barite and Calcite Precipitation, *Cryst. Growth Des.*, 2006, **6**, 2579

Richard Sear asked: Do you know what effect, if any, mellitic acid has on the nucleation of barium sulphate? Do you have experimental data on this?

Franca Jones replied: This is a difficult question to answer. Sparingly soluble salts such as barium sulfate, at supersaturations equivalent to those used in this study, have high nucleation rates that are difficult to measure properly. We use a turbidimetric method that compares the control situation to that with mellitic acid, however, as with any method investigating nucleation these methods are often only detecting particles when they reach a certain size dictated by the detection method. Using this turbidimetric method, it appeared that mellitic acid inhibited the nucleation rate of barium sulfate significantly. However, when we measured the time evolution of the particle size using a Malvern Nano ZS, which is a dynamic light scattering method, the particle counts (which should be related to the nucleation rate) appeared to be similar whether or not mellitic acid was present. Thus, it is unclear at this point whether mellitic acid impacts homogenous nucleation. Using AFM measurements we can say, however, that on the (001) face of barium sulfate, the presence of mellitic acid promotes 2D nucleation. We measure this by looking at the number of new 2D nuclei formed over many scans, for a fixed scan area, and then take the average. Because we are at supersaturations that should be below the homogenous nucleation regime, all new surface nuclei should be due to 2D nucleation. In the case of barium sulfate grown in the presence of mellitic acid at the higher concentrations, the presence of new nuclei was so great it was difficult to measure them all, but the mellitic acid certainly promoted the formation of 2D nuclei.

Celso Aparecido Bertran communicated: Figure 6 of the paper shows that the zeta potential of the barium sulphate particles that formed with a higher mellitate (20ppm) concentration are less negative than the particles formed at a lower concentration (5ppm). However, if the zeta potential results from the adsorption of the mellitate ion, a negative ion, on the particle surface, the expected behavior should be the opposite. Why didn't the expected behavior happen?

Franca Jones communicated in reply: This is an excellent question; however, the answer is yet to be conclusively determined. We are still investigating this effect, but we know the following: that the zeta potential alters with mellitic acid concentration, and therefore, the effect is related to the mellitic acid and not to barium or sulfate. One hypothesis is that at low concentrations the fully de-protonated mellitate ion is preferentially adsorbed, while as the concentration of mellitic acid is increased, the protonated versions of the acid are adsorbed in increasing amounts. Another possibility is that as more mellitic acid is adsorbed, the manner in which it is adsorbed leads to fewer "exposed" carboxylate groups. The data collected thus far cannot differentiate between these possibilities, but an obvious possible means of supporting one or other hypothesis is to undertake FTIR of the solids at a high (say 20 or 50ppm) mellitic acid concentration and see what the mellitic bands looks like, and whether they are consistent with the fully de-protonated or somewhat protonated forms of the acid.

Kevin Back opened a general discussion of the papers by Marco Mazzotti: Returning to Prof. Kevin Roberts' point about industry's interest in one day being able to control primary nucleation to the extent that we can design the right polymorph and particle size for our active pharmaceutical ingredient crystallisation, I would like to ask your opinion on measuring nucleation rates. If I was designing a non-seeded cooling crystallisation for my process, and was interested in measuring nucleation parameters to help me understand and control my process, how should I go about measuring them? During this meeting, we have heard a number of ways to obtain these parameters: you have produced two models with wide applicability based on polythermal data from Prof. Joop ter Horst's group, Prof. Joop ter Horst has also earlier described an isothermal method, Prof. Ake Rasmuson described a driving force method which is also isothermal but with each supersaturation measured at a different temperature, and Dimitrios Toroz and Diana Camacho have described another polythermal method. Which of these methods is best suited to obtain the nucleation parameters for my industrial crystallisation, and what else should I measure?

Marco Mazzotti answered: This question deserves an answer at two different levels. Firstly, with the model that we propose, both polythermal and isothermal experiments can be analysed and interpreted effectively. Of course, the former provide more information than the latter. Secondly, it remains to be seen whether microscopic experiments, such as those carried out in mL-scale crystallisers, are sufficient to characterise nucleation rates applicable in larger, industrial crystallisers. This is indeed the ultimate goal of our research program, of which this contribution is part.

Thomas Vetter asked: In your paper, you present a model that relaxes some of the more stringent assumptions typically employed when analyzing induction time distributions that result from stochastic nucleation. Specifically, the assumption that only the first nucleation event behaves in a stochastic manner is relaxed, and you allow for all subsequent nucleation events to behave stochastically as well. I think such a model is a step in the right direction, but should we not also account for dispersion in the growth rate, which is quite often observed? Surely this would affect the results of such a stochastic model as well?

Marco Mazzotti responded: Growth rate dispersion (GRD) has been observed in certain systems,[1] but it is very difficult to characterise and to explain in general. It might be due to crystal properties that are distributed within the population of particles, *e.g.* the lattice strain or the surface characteristics . It does not seem to be likely that GRD plays an important role in the physics of the early stages of nucleation, as all crystals are formed in the same way and in the same environment. If, however, experimental evidence indicated that GRD was also important here, then we should extend our stochastic model, as well as the mathematical tools that we utilise

1. For example: D. Ochsenbein, S. Schorsch, F. Salvatori, T. Vetter, M. Morari and M. Mazzotti, Modeling the facet growth rate dispersion of β l-glutamic acid—Combining single crystal experiments with nD particle size distribution data, *Chem. Eng. Sci.*, 2015, DOI:10.1016/j.ces.2015.02.026.

(522:[522]522) **Richard Sear** asked: Your Fig. 1 shows the fraction of samples that have crystallised, as a function of the supercooling in temperature, and you propose models for data of this type. Data of this type are very common in studies of the nucleation of ice in supercooled water droplets, and they have been studied quantitatively since the 1940s.[1] Do you think that we (the community studying crystallisation from solution), could learn from this body of literature?

For example, they have simple (two parameter) models called singular models[1,2,3] that are widely and successfully used. They could be used when we want to keep the number of parameters down to a minimum. These differ from your model in that a particular droplet freezes deterministically, *i.e.*, at a set temperature, although due to randomness in the impurities that are assumed to be driving nucleation in the droplets, this set temperature varies from droplet to droplet.

There are also models which, like your model, include randomness when nucleation occurs in a droplet, but there are still differences. For example, some models used for ice nucleation include both randomness in the impurities as well as randomness when nucleation occurs on a given impurity. See Herbert *et al.*[4] and references therein for recent work, and the textbook of Pruppacher and Klett[3] for an introduction to this literature.

1. J. Levine, *Natl. Advis. Comm. Aeronaut., Tech. Notes*, 1950, 2234
2. R. P. Sear, Generalisation of Levine's prediction for the distribution of freezing temperatures of droplets: a general singular model for ice nucleation, *Atmos. Chem. Phys.*, 2013, **13**, 7215
3. H. R. Pruppacher and J. D. Klett, *Microphysics of Clouds and Precipitation*, Reidel Publishing, Dordrecht, 1978
4. R. J. Herbert, B. J. Murray, T. F. Whale, S. J. Dobbie, and J. D. Atkinson, Representing time-dependent freezing behaviour in immersion mode ice nucleation, *Atmos. Chem. Phys.*, 2014, **14**, 8501

Marco Mazzotti answered: In the literature cited in this question, the observation that the nucleation temperatures of ice in water droplets are distributed is attributed to the following fact: the properties of the active sites (*e.g.* dust), where heterogeneous nucleation occurs in the different droplets, are themselves distributed. In our work, and in the literature that we use to get data, experimental conditions are identical in every repetition of the same experiment, but nucleation times are distributed because of the intrinsic stochasticity of nucleation, which is considered to be a rare event that can be described as a generalised

Poisson process. Although the mathematical treatment is similar, and the analogy drawn by the question is intriguing, we believe that the physics behind the two phenomena is different enough to call for caution in comparing them.

Simon Black remarked: Most of what we have heard about nucleation in this meeting so far supports the view that seeded crystallisations are more robust than un-seeded ones, particularly for scale-out and scale-up.

Joop Ter Horst remarked: In your model you are only considering primary nucleation and growth. It is shown that secondary nucleation plays an important role in these experiments[1,2], eventually causing the detection of the presence of the suspension and herewith the presence of the primary nuclei. Should secondary nucleation be considered in the model to get realistic results comparable to the experimental data?

1. S. S. Kadam, H. J. M. Kramer and J. H. ter Horst, Combination of a Single Primary Nucleation Event and Secondary Nucleation in Crystallization Processes, *Cryst. Growth Des.*, 2011, **11**, 1271–1277.
2. S. A. Kulkarni, H. Meekes and J. H. ter Horst, Polymorphism Control through a Single Nucleation Event, *Cryst. Growth Des.*, 2014, **14**, 1493–1499.)

Marco Mazzotti responded: In this work we show that it is possible to describe, with good accuracy, the distribution of detection times observed in experiments by combining a stochastic phenomenon, *i.e.* primary nucleation, and a deterministic one, *i.e.* crystal growth. We have used rigorous temperature- and supersaturation-dependent models for the rates of both mechanisms, and we have shown that the two sets of model parameters are correlated. If and when secondary nucleation plays an important role, our model will be able to account for it as a second deterministic mechanism, besides growth, that leads to crystal detection. Of course, the degree of correlation among the model parameters will be, in this case, even more acute.

Jan Sefcik said: Thank you for an interesting contribution; in order to understand nucleation better we need to know how to properly measure nucleation rates. There are three components to this; the first is a well defined experimental system, where we know what the conditions (concentration, temperature, shear rate, surfaces *etc.*) are at the point where nucleation actually happens. The second is the understanding of quantities we measure when attempting to detect the presence of crystals in our system, and the third is how to link what we're measuring to nucleation kinetics. Regarding the first component: it seems that all models available in literature, including the one presented in this contribution, assume that the temperature is the same throughout the vessel, even under cooling conditions, and that the flow field is the same everywhere. However, by definition, under cooling conditions the temperature on the wall of a vessel will be lower than in the bulk, so the supersaturation at the wall is higher than in the bulk. Therefore, it is most likely that the nucleation happens near or at the wall, where the flow field will also be very different than in the bulk, especially near an impeller. Therefore, in order to use these models properly, one needs to do isothermal measurements or ensure that the supersaturation variation across the vessel is negligible. This should also be taken into account when scaling up.

Kevin Roberts remarked: On Dr Price's point about seeding, I feel that whilst this approach is quite common, it is still an admission of failure, as it would be best to crystallise the sample within a "right first time" approach to achieve the particle size and structures required. An alternative approach to seeding would be to develop a better engineering model for scaling-up primary nucleation building on the work described by John Mullin in his book.[1] Such models have been found to scale-up with Reynold's number remarkably well over the 0.5-20 litre reactor size range for vessels with the same hydrodynamic ratio and impeller design.[2,3]

Also, the industry uses a pragmatic approach of milling crystals, but intellectually I'm not comfortable with this method; milling changes the solid-form properties (polymorphic form, morphology, surface energy, crystalinity etc.) of crystals and in doing so it can undo the "good work" of a well optimised crystallisation process crystal structure. Thus it's not the ideal approach to use.

1. J. W. Mullin, *Crystallisation*, Butterworth-Heinmann, Oxford, UK, 3rd edn, 1961.
2. K. Liang, G. White, D. Wilkinson, L. J. Ford, K. J. Roberts and W. M. L. Wood, Examination of the process scale dependence of L-glutamic acid batch crystallised from supersaturated aqueous solution in relation to reactor hydrodynamics, *Ind. Eng. Chem. Res.*, 2004, **43**, 1227–1234.
3. K. Liang, G. White, D. Wilkinson, L. J. Ford, K. J. Roberts, W. M. L. Wood, An examination into the effect of stirrer material and agitation rate on the nucleation of L-glutamic acid batch crystallised from slow-cooled supersaturated solutions, *Cryst. Growth Des.*, 2004, **4**, 1039–1044

Marco Mazzotti addressed Baron Peters: I believe that it is perfectly legitimate to assume that an additive present in a supersaturated solution adsorbs at the interface between subcritical clusters and solution, thus modifying the surface tension of the cluster hence of the forming nucleus. The amount adsorbed will be controlled by a dynamic equilibrium, as adsorption is typically very fast for the molecules considered in this contribution. In adsorption this is well known, and the spreading pressure - as the surface tension is called - depends on the additive concentration in solution, and it is affected by it even when such a concentration is very low and the adsorption isotherm reduces to a linear relationship, *i.e.* the Langmuir isotherm reduces to the Henry's law.

Baron Peters answered: Thank you Professor Mazzotti for these encouraging remarks. We are currently investigating our assumption of rapid adsorption up to a mean field coverage.

Peter Vekilov commented: In your models, you assume equilibrium between critical and subcritical clusters and the monomers in the solution. Firstly, this assumption seems unrealistic: the system is far from equilibrium, since nucleation typically occurs in highly supersaturated solutions. Secondly, it seems unnecessary: the steady state problem of nucleation as a transition over a barrier in the space of cluster dimensions was solved in 1942 by Zeldovich (although the English translation of the paper is from 1943)[1]; this work is often cited in chemical kinetics texts as a new paradigm in kinetics modeling.[2] Is there a compelling reason for the equilibrium assumption?

1. Y. B. Zel'dovich, *Acta Physicochim. URSS*, 1943, **18**, 1–22.
2. P. L. Houston, *Chemical Kinetics and Reaction Dynamics*, McGraw-Hill Higher Education, New York, 2001.

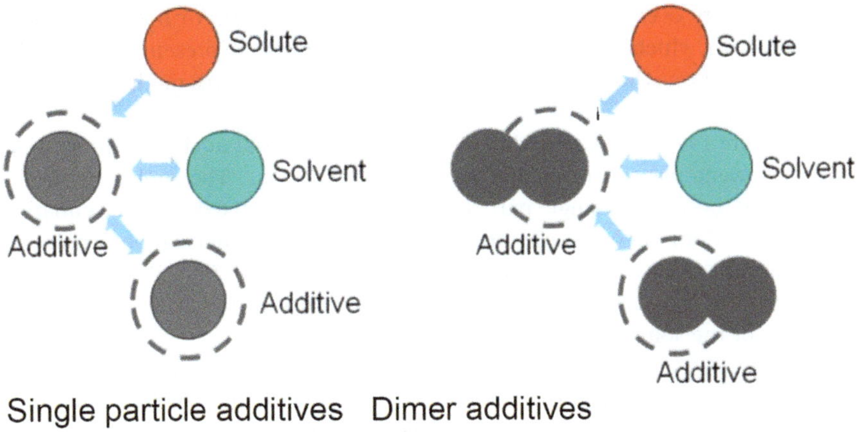

Fig. 3 The three types of molecules: solute, solvent and additive. Single particle and dimer additives were considered. Changes in characteristics of the additive molecules included variation in affinity parameter between the molecules (the light blue arrow), and the size of the additive molecules (shown by the broken circumference outlines).

Baron Peters replied: We started from the non-equilibrium steady-state framework of classical nucleation theory (CNT). Equilibrium properties appear in the CNT rate expression because the Zeldovich factor allows us to write the non-equilibrium steady-state flux in terms of the equilibrium free energy barrier. Specifically,

$$
\begin{aligned}
J_{CNT} &= -D(n^{\dagger})\left.\frac{\partial \rho_{SS}}{\partial n}\right|_{\dagger} \\
&= \rho_1 D(n^{\dagger}) Z \exp\left[-\Delta F(n^{\dagger})/k_b T\right]
\end{aligned}
\tag{1}
$$

where $\Delta F(n^{\dagger}) = F(n^{\dagger}) - F(1)$. The details of this transformation and a discussion of its relationship to theories of chemical kinetics can be found in ref. 1. We have not assumed equilibrium, but also we did not account for the effects of additives on the mobility $D(n^{\dagger})$ or on the Zeldovich factor Z in our theoretical model. We included the effects of additives on both barriers and prefactors (including the Zeldovich factor) in our simulations.

1. V. Agarwal and B. Peters, Solute Precipitate Nucleation: A Review of Theory and Simulation Advances, *Adv. Chem. Phys.*, 2014, **155**, 97–153. See in particular, pgs. 106–108.

Marco Mazzotti added: The assumption we made is that nucleation is given as a function of temperature (T) and supersaturation (S) by the expression in Equation (12). That is consistent with both the classical nucleation theory and its extension as made by Zeldovich. Hence we do not need any equilibrium assumption about clusters.

Jamshed Anwar presented some slides on simulating crystal nucleation from a solvent using molecular modelling: Whilst we have a good understanding of how additives promote or inhibit crystal growth, our understanding of how additives affect nucleation has been very much lacking. A number of studies on nucleation inhibitors carry over ideas of molecular recognition related to how molecules inhibit growth, which are probably not appropriate. I wish to present earlier work by us[1] that informs the present discussion on the how additives modulate crystal nucleation.

In the late 1990s we showed that we can simulate crystal nucleation processes from a solvent using molecular dynamics simulations.[2] The molecules were represented by simple, single-particle models that interacted with a Lennard-Jones interaction. These simple models are appropriate as nucleation processes of interest are generic being observed in a wide class of materials. We monitored the onset and extent of nucleation using the peaks of the pair distribution function for the solute particles.

We then investigated how third-party (auxiliary) molecules would modulate the observed nucleation in the simulations by following the kinetics using the pair distribution function of the solute particles. The systems contained three types of molecules: solute, solvent, and additive. The characteristics of the additive molecules were systematically changed and their effect on nucleation monitored. The additives were represented as single particles or as dimers. For the single particles the particle size was varied, as was the affinity of the additive for the solute, for the solvent and for itself. For the case of the dimer additive, one of the particles had characteristics of the solvent, while the affinity of the second particle for the solute, the solvent and for itself was varied (see Fig. 3). One of the dimer molecules was effectively a solute and a solvent particle stuck together – a perfect surfactant.

The results yielded design rules which were intuitive. For inhibition of nucleation, firstly the additive must have a strong affinity for the solute; a lack of affinity would abandon the additive to remain in the solvent making it ineffective. The second condition is that the additive (on being incorporated into the solute clusters) is able to disrupt the emerging lattice. It may do this by steric effects or strong energetic interactions. For promotion of nucleation, the most effective additives had surfactant-type properties in that they minimised the solute-solvent interfacial interaction.

1. J. Anwar, P. K. Boateng, R. Tamaki and S. Odedra, Mode of action and design rules for additives that modulate crystal nucleation, *Angew. Chem., Int. Ed.*, 2009, **48**, 1596–1600
2. J. Anwar and P. K. Boateng, Computer simulation of crystallization from solution, *J. Am. Chem. Soc.*, **120**, 9600–9604.

Helmut Cölfen answered: There exist quantitative experimental studies on the effect of additives on the nucleation and growth of $CaCO_3$,[1-4] which might be useful to compare with the theoretical predictions.

1. D. Gebauer, H. Cölfen, A. Verch and M. Antonietti, The Multiple Roles of Additives in $CaCO_3$ Crystallization: A Quantitative Case Study, *Adv. Mater.*, 2009, **21**, 435–439.
2. D. Gebauer, A. Verch, H. G. Börner and H. Cölfen, Influence of Selected Artificial Peptides on Calcium Carbonate Precipitation - A Quantitative Study, *Cryst. Growth Des.*, 2009, **9**, 2398–2403.
3. A. Verch, D. Gebauer, M. Antonietti and H. Cölfen, How to control the scaling of $CaCO_3$: a "fingerprinting technique" to classify additives, *Phys. Chem. Chem. Phys.*, 2011, **13**, 16811–16820.

Joop Ter Horst addressed Jamshed Anwar: You say that a small concentration of additives (ppm) can inhibit the nucleation rate. In cases of small additive concentrations, a large part of the volume is unaffected by the additives. In this part of the volume, crystal nucleation can take place unhindered. Therefore, the overall nucleation rate would not be affected by the additive. You can, however, expect an inhibiting effect on the growth rate in cases of small concentrations of additive.

Jamshed Anwar responded: I do not believe that I mentioned ppm levels, but your general point is sound. The simulations indicate that effective inhibitor molecules will end up in the solute molecule populations. In the case of inhibitor molecules, once incorporated within the solute populations or clusters, the individual molecules can perturb the emerging lattice over considerable distances, and one would expect that low concentrations would be effective.

For the promotion-type additives, particularly surfactants, they would end up at the interfaces, and we expect that higher concentrations would be required. The simulations do, however, indicate that it is not necessary to have full coverage of the interface by such molecules.

Joop Ter Horst asked: The picture in your slide shows a lot of additive molecules around a cluster; what is the concentration of the additive? It must be high. If it is high, does this then change the thermodynamics as well?

Jamshed Anwar responded: This was the surfactant-type additive, and the concentration was circa 5% (solute basis). The chemical potential of the system does indeed change, as does the effective interfacial free energy (which of course is by design). The pertinent point is that a lowering of the chemical potential (as might be expected for a surfactant-type additive) should decrease the probability for nucleation, but these additives actually enhance the nucleation. So it appears that the reduction in the interfacial free energy dominates.

Jim De Yoreo returned to the discussion of the paper by Baron Peters: One important detail that should be considered is the potential competition in timescales for impurity and solute adsorption. Some years ago we looked at the effect of different additives in a number of crystal systems. We found that when surface exposure times were short compared to the characteristic adsorption time of the impurity, they had no effect. When the exposure times became longer than that characteristic time, the impurities suddenly had big effects. For simple ionic additives, the time scale for adsorption was seconds, but when larger, organic molecules such as polypeptides were used, the time scale was much longer (10–100x). Therefore, whether or not an impurity has an impact on growth depends on the nature of the impurity, and the terrace lifetime (growth rate). I think the same will be true with nucleation; the effects will depend on the time scale of the adsorption process relative to the time scale of monomer attachment.

Baron Peters responded: The presentation by Prof. Davey illustrated an important link between growth rates and attachment frequencies in the nucleation prefactor. Certainly, one expects strong analogies between the effects of additives on 2D and 3D nucleation barriers. However, the effects of additives on spiral growth at low supersaturations may not translate to nucleation, except perhaps that additives might modulate kink energies in a similar way that they modulate interfacial free energies. As noted in my response to Prof. Vekilov, our model assumes a mean field additive coverage (from the Langmuir isotherm) at each nucleus size. In reality, both the nucleus size and the number of adsorbed additives are dynamically fluctuating variables. The relative attachment frequencies for solutes and additives will depend on the relative concentrations, diffusivities, and perhaps on the chemical steps of attachment. Geoffrey Poon and I are currently developing a model of the free energy, dynamics, and steady-state fluxes on the two-dimensional landscape of nucleus size and additive coverage. The new model should provide an answer to your question.

Jessica Lovelock asked: As a more general question: how widely applicable is the model you have put forward? Your model makes a number of simplifying assumptions - for instance, that available adsorption sites are small, discretized, and non-interacting. This may not hold for, say, a system with large organic molecules/proteins, or one with strongly charged additives. In an overall sense, what sorts of restrictions would you expect to have to keep in mind when trying to generalize your model to other systems?

Baron Peters replied: Indeed, we assumed uniform sites on all facets, non-interacting adsorbates, and monomeric additives that bind *via* a single interaction site. Many of these assumptions will not be valid, making the model primarily useful for qualitative predictions. We hope the model is just accurate enough to identify the quantities which determine whether an additive is an effective low dosage nucleant.

Alasdair Mackenzie said to Joop Ter Horst: Although you accept that an impurity particle could promote nucleation, you express doubt at Prof. Anwar's suggestion that an impurity on the scale of one percent or ppm could inhibit nucleation, because it would not likely meet in solution. However, given that stable prenucleation clusters can be hundreds of nanometers in diameter, could it not be that the molecule could enter the diffuse cluster and stabilise it with respect to crystal formation? Furthermore, it need only enter the clusters which are close to critical size, and inhibit nucleation for long enough that the solute has gone back into free solution and the clusters are no longer critical. In this way, the likelihood of inhibiting nucleation could increase for an impurity which makes up a small amount of the solution.

Joop Ter Horst replied: Low concentration impurities will only affect a negligible part of the supersaturated volume. In the unaffected volumes, nucleation can occur unaffected. So in order to influence the measured nucleation rate in the entire volume, the impurities should significantly speed up the formation of nuclei. This can be done if the impurity acts as a heterogeneous particle or surface.

There may conceivably be indirect scenarios, however, that could lead to an inhibition of nucleation rate. The impurity might, for instance, deactivate heterogeneous particles through adsorption on the particle surface, so that formation of a nucleus on this heterogeneous particle is no longer possible.

Kevin Roberts commented: It may be worthwhile to state that restricting or controlling nucleation can be just as important as promoting nucleation. An example of the former might be when you are making highly concentrated (supersaturated) solutions for *trans*-dermal patches, and you don't want the solutions to crystallise.[1] Another example could be when you have fuels operating in cold weather conditions, where you don't want the less soluble components in the fuel to crystallise and cause fuel starvation due to crystals blocking the fuel filters.[2]

1. See: http://drugtopics.modernmedicine.com/drug-topics/news/clinical/clinical-pharmacology/neupro-patches-recalled-due-crystallization?page=full, (accessed 30/4/15).
2. A. R. Gerson, K. J. Roberts and J. N. Sherwood, An instrument for the examination of nucleation from solution and its application to the study of precipitation from diesel fuels and solutions of n-alkanes, *Powder Technol.*, 1991, **65**, 243–249.

Jessica Lovelock remarked to Baron Peters: Fundamentally, your model is based on the differences in interfacial free energy both before and after the additives are adsorbed. You describe only cases which are used to accelerate nucleation, but if the sign were flipped, could it not also be applicable to systems where additives inhibited nucleation instead? The model would not, of course, be directly applicable - if only the sign were changed, this would imply unfavourable binding between the additive and the growing nucleus, which would mean that the additive would not adsorb at all. Would it, however, be possible to modify the model to take into account a more complex system - one in which binding between the additive and the particle it adsorbs onto would be favourable, but which, after adsorption, made other particle interactions less favourable?

Baron Peters replied: Thank you for the question. Additives might also influence aggregation of nuclei or other processes, but I have not thought about these.

Wenhao Sun presented some slides on metastable aragonite $CaCO_3$ nucleation in seawater (see ESI†): Many people have been asking questions on how additives might serve as nucleation inhibitors. I would like to present some work on how Mg^{2+} serves as a nucleation inhibitor of the stable calcite phase in the calcium carbonate system. In marine environments, Mg^{2+} serves as a nucleaton inhibitor of calcite, the stable phase of calcium carbonate. In high Mg^{2+} concentrations, the metastable phase, aragonite, forms in its stead. There are two ways that Mg^{2+} could affect the calcite nucleation barrier: it could increase its solubility, reducing supersaturation at a given ion activity product, or it could increase its surface energy. I show here that calcite incorporates Mg^{2+} spontaneously, even though the free-energy of mixing is positive. This spontaneous incorporation is enabled by the free-energy of solution–mineral exchange. When calcite incorporates Mg^{2+}, both the surface energy and solubility of calcite increases. However, because the nucleation barrier scales as the surface energy cubed, while it is only reduced by $(\ln S)^2$, the surface energy increase inhibits nucleation far more than the increase in solubility. We map the influence of surface energy and solubility on the competitive nucleation rates between calcite and aragonite. We show that in seawater Mg : Ca ratios, aragonite becomes kinetically preferred. For more details, see our published work.[1]

1. W. Sun *et al.*, Nucleation of Metastable Aragonite CaCO3 in Seawater, *Proc. Natl. Acad. Sci. U. S. A.*, 2015, **112**, 3199–3204.

Stéphane Veesler addressed Wenhao Sun: I do not think you are describing what we call an additive effect in crystal growth. Usually additives are added at low concentrations (< 1%).

Wenhao Sun replied: Mg^{2+} might not be considered an additive in a traditional sense; you might also call it a spectator ion, or a solution impurity. In my opinion, these are just semantics. The take-away message is that, under the same thermodynamic conditions (temperature and pressure), altering the solution chemistry can bias the formation of metastable polymorphs, by inhibition of the stable phase.

Jessica Lovelock said: As a note on the cross-applicability between your work and that of Baron Peters: yours incorporates magnesium directly into the bulk, whereas Baron's explicitly only considers surface interactions. Both models are certainly interesting and informative, but not directly comparable.

Wenhao Sun replied: It should be mentioned that Baron Peters' model could not be used to identify additives that inhibit nucleation - as the Langmuir adsorption mechanism of nucleation acceleration described requires that additive adsorption be thermodynamically preferable - *e.g.* its adsorption energy must be negative (thus reducing the surface energy). An additive that raises the surface energy would not adsorb in the first place. The Mg-calcite system is quite interesting in the sense that it is "thermodynamically preferable" for calcite to incorporate Mg into the bulk, which has a side-effect of increasing the surface energy, thus inhibiting calcite nucleation. I would argue that the two works are comparable, in the sense that the underlying arguments are based on altering the surface energy in the nucleation barrier - $\Delta Gc \sim y^3/(-RT\ln\sigma)^2$. We both identify the importance of the surface energy on the nucleation barrier, and his work suggests a mechanism that reduces surface energy, whereas our work suggests a mechanism that increases surface energy, which accelerates or retards nucleation accordingly.

Allan Myerson remarked: Antifreeze protein isolated from a winter fish, the flounder, works by specifically absorbing on the basal plane of ice (based on the spacings of the histadines in the protein), thus changing the shape from long needles to very small particles. These are examples of shape and growth modifiers rather than nucleation inhibitors.

David T. Wu addressed Baron Peters: In your model, favorable adsorption leads to a reduction in the free energy barrier to nucleation and hence acceleration of the rate. Would there be a modification of the model that would accommodate inhibition of nucleation with favorable adsorption? For instance, what if additives have a preference for clusters of a particular size (*e.g.*, forming micelles), that lower the free energy formation of this cluster, but raise the barrier to nucleation?

Baron Peters replied: Thank you, Prof. Wu, for this interesting question. Our model only considers additives that bind to the surfaces of homogeneous nuclei, and that process should only accelerate homogeneous nucleation. However, Prof. Sear has noted that the same adsorption effects might inhibit heterogeneous nucleation. Please refer to his question and my response for details.

It may be possible to design or identify additives that favor a particular size, creating (as you've suggested) a sort of stable micelle. Understanding that situation would require a more detailed model of effects like additive-additive interactions on the surface.

Jim De Yoreo commented: The results remind me of amelogenin, which is responsible for the formation of tooth enamel. *In vitro* experiments have shown that amelogenin assembles into a dodecamer, within which amorphous calcium phosphate forms. These mineral-containing dodecamers then assemble into chains, but the mineral can not change or grow because it is capped by the dodecamers.

Gerard Coquerel said: There are small transparent animals called tardigrades that live in the Alps at very low temperatures, down to −40 degrees or below. 30% percent of their mass is just trehalose; it is used as a cryoprotectant to prevent the crystallization of ice.

Baron Peters responded: Creatures which use trehalose to preserve themselves through deep-freeze conditions are fascinating. I have learned from Prof. Juan de Pablo that trehalose can also preserve creatures (like brine shrimp) through dessication. I suspect that in both cases, the trehalose prevents crystallization by creating a glassy matrix within the organism.

Ken Lewtas commented: There have been presentations and discussions about the possibility of designing tailor-made additives which can specifically inhibit (or, alternatively, encourage) crystallisation. There have also been questions asking if it is possible to achieve this effect. It might be interesting to draw the attendees' attention to the fact that such additives have been modelled, designed, synthesised and commercialised many decades ago. Designed Crystal Nucleators and Growth Inhibitors have been successfully used in almost every diesel fuel around the world for over thirty years, during winter conditions. They modify the crystallisation of waxes that precipitate from such fuels and oils, thus preventing them from blocking fuel lines and filters which would otherwise cause vehicle failure. Molecular lattice recognition (matching) aids the binding of the additive. The additive also has a second moiety which, by design, either prevents or encourages further crystal growth. Most of the work is consequently decades old, and much of it has been commercial rather than academic. References that might be of interest include those covering the design and fitting of a growth-inhibiting additive to the crystal,[1] the effect of polymer design on crystal growth[2] and the early academic computational modelling of such systems.[3]

1. K. Lewtas, *Eur. Pat.*, EP0261959A2, WO8802393A, 1986.
2. D. H. M. Beiny *et al.*, Crystallization of n-dotriacontane from hydrocarbon solution with polymeric additives, *J. Cryst. Growth*, 1990, **102**, 801–806

3. G. Clydesdale *et al.*, Modelling the morphology of molecular crystals in the presence of blocking tailor-made additives, *J. Cryst. Growth*, 1994, **141**, 443–450

Haihua Pan returned to a discussion of the paper by Baron Peters: In equation 9, $\ln((J(x)/J)$ is proportional to $K\ln K$. When $K > 1$, an additive promotes nucleation, which is the situation you have discussed in this paper. What about $K < 1$? In eq. 9, it predicts the inhibition of additions on nucleation. It is weird to me that one additive that dislikes the nuclei or crystallites will inhibit nucleation. Could you discuss this situation?

Baron Peters answered: Thank you, Prof. Pan, for this excellent question. The model does predict a counterintuitive inhibition of nucleation for $K < 1$. Note that $K < 1$ means that the surface of the nucleus preferentially interacts with solvents as opposed to additives. When $K < 1$, growing a nucleus in the presence of additives, *i.e.* with non-zero x , requires some extra work to "exclude" additives from the first solvation shell around the nucleus. The inhibition effect makes sense within the assumptions of the model, but there are two dubious aspects of this prediction. Firstly, note that the inhibition effect is very weak compared to the predicted promotion effects. For $K < 1$(promoting additives) $K\ln K$ can easily become a very large number, but for $K < 1$(inhibiting additives) the quantity $K\ln K$ never exceeds 0.37 in magnitude. The quantity $xK\ln K$ would be even smaller if, as assumed, $x \ll 1$. Thus, the predicted inhibition effects are extremely small, and likely undetectable if they do exist. Secondly, one could further question their existence. If additives bind to the surfaces of nuclei, they may similarly interact with free solutes in solution. Interactions between solutes and additives in solution are not accounted for in our theoretical analysis, but they do interact in the simulations. Note that the simulations do not show an inhibition.

Sophie Janbon said: Additives in API crystallisation processes are not a common practice in the pharmaceutical industry. However some impurities, even at a low level, may interact with existing clusters in supersaturated states, and those interactions could potentially accelerate the formation of a crystalline phase, or otherwise inhibit it.[1,2] The paper presented by Baron Peters is insightful in the sense that it highlights a potential mechanism to why an additive may actually accelerate the nucleation.

1. J. Bauer, S. Spanton, J. Henry, J. Quick, W. Dziki, W. Porter and J. Morris, Ritanovir: An Extraordinary Example of Conformational Polymorphism, *Pharm. Res.*, 2001, **18**, 859–866.
2. C.-H. Gu, K. Chaterjee, V. Young Jr and D. J. W. Grant, Stabilization of a metastable polymorph of Sulfamerazine by structurally related additives, *J. Cryst. Growth*, 2002, **235**, 471–481.

Baron Peters responded: Thank you for the encouraging remarks and for these references.

Faraday Discussions

PAPER

Do solid-to-solid polymorphic transitions in DL-norleucine proceed through nucleation?

Joost A. van den Ende,[a] Mireille M. H. Smets,[a] Daniël T. de Jong,[a] Sander J. T. Brugman,[a] Bernd Ensing,[b] Paul T. Tinnemans,[a] Hugo Meekes[a] and Herma M. Cuppen*[a]

Received 10th November 2014, Accepted 4th December 2014

DOI: 10.1039/c4fd00214h

DL-Norleucine is a molecular crystal exhibiting two enantiotropic phase transitions. The high temperature $\alpha \leftrightarrow \gamma$ transition has been shown to proceed through nucleation and growth [Mnyukh *et al.*, *J. Phys. Chem. Solids*, 1975, **36**, 127]. We focus on the low temperature $\beta \leftrightarrow \alpha$ transition in a combined computational and experimental study. The temperature dependence of the structural and energetic properties of both polymorphic forms is nearly identical. Molecular dynamics simulations and nudged elastic band calculations of the transition process itself, suggest that the transition is governed by cooperative movements of bilayers over relatively large energy barriers.

1 Introduction

The occurrence of polymorphism in molecular crystals, *i.e.*, the ability of a compound to crystallise in different crystal structures, can have important implications in different areas of industry ranging from the food industry[1] to the production of dyes and pharmaceutical products.[2] Since the crystal structure determines several properties such as dissolution rate and solubility of the manufactured product, control over the obtained polymorph and its stability is desired. Transformations between polymorphic forms can be divided in two categories: solvent mediated[3] and solid-to-solid transformations. Examples of the latter category are transformations induced by temperature,[4] gas,[5] large pressures,[6] or grinding.[7]

Most strategies to inhibit or induce solid-to-solid transformations are empirical in nature. In the case of transformations in which the parent and daughter phase have similar structures, there is an ongoing debate whether the

[a]*Institute for Molecules and Materials, Radboud University Nijmegen, Heyendaalseweg 135, 6525 AJ Nijmegen, The Netherlands. E-mail: h.cuppen@science.ru.nl*

[b]*Van't Hoff Institute for Molecular Sciences, University of Amsterdam, Science Park 904, 1098 XH Amsterdam, The Netherlands*

transformation mechanism involves a cooperative movement of molecules or occurs through nucleation and growth.[8–10]

DL-Norleucine (2-aminohexanoic acid) is an example of a molecular crystal that exhibits interesting enantiotropic polymorphism, *i.e.*, polymorphism in which the relative thermodynamic stability of the forms changes as a function of temperature in a reversible way. The different polymorphic forms all consist of tightly packed molecular bilayers which are strongly connected by hydrogen bonds between the amino and acid groups. These molecular bilayers are mutually van der Waals bonded by unbranched hydrophobic side chains.

Mathieson[11] predicted polymorphism to exist for DL-norleucine (DL-NLE) based on X-ray diffraction measurements. More recent determinations of the polymorphic forms β[12] (stable at low temperatures), α[13] (stable at room temperature), and γ[14] (stable at high temperature) confirmed and extended his ideas. The β and α forms consist of highly similar structures in which the molecular conformations are nearly identical.[15] Fig. 1 shows the structure of both polymorphs projected along the *b* and *a* axes. The molecules are coloured according to their handedness: *R*-norleucine is yellow and *S*-norleucine is blue. The difference between the forms is in the relative position of the bilayers.

The temperature of the α ↔ γ transition is well defined. The α and the γ polymorphic forms differ both in the molecular conformation and in the relative position of the bilayers. The relative position of the bilayers is similar for the γ and the β polymorphic forms, their differences are dominated by a change in molecular conformation. The α ↔ γ transition is accompanied by cracking of the crystals due to a discontinuity in the density at the transition.[9,14] In contrast, the transition from the β to the α form is reported to be 'extraordinarily variable' both in behaviour and in transition temperature.[14] It is still an open question which processes govern the onset of the low temperature transition and whether this solid-to-solid transition can be described by a nucleation-and-growth mechanism or not.

A distinct difference between polymorphic transitions occurring through a nucleation-and-growth mechanism and those occurring through a displacive

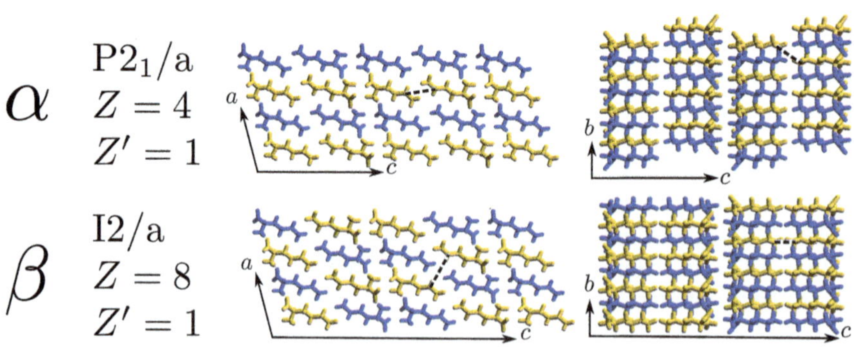

Fig. 1 An overview of the structures of the α and β polymorphic form of DL-norleucine. The molecules are coloured according to their chirality: *R*-molecules in yellow and *S*-molecules in blue. The dotted lines depict the distance parameters (DP1 and DP2) which are averaged in-plane distances between molecules of the same chirality in opposing bilayers.

mechanism is the effect of defects on the occurrence of the transition.[9] When the transitions proceed through a nucleation-and-growth mechanism, they can be induced by defects that can act as centres of nucleation. Contrarily, for transitions with a displacive character, a defect acts as an obstacle for the transition and has the capability to hinder it. The nucleation-and-growth characteristics of the $\alpha \leftrightarrow \gamma$ transition have convincingly been proven on the basis of these arguments by Mnyukh *et al.*[9]

Recently, some of us have reported that both the lattice parameters and the potential energies of the α and β polymorphic form behave identically as a function of temperature in a Molecular Dynamics (MD) simulations study.[16] We conclude that the difference in volume between the experimentally obtained structures[12,13] is a consequence of thermal expansion and not of differences between the polymorphic forms themselves. This continuous behaviour of the crystal properties, suggests the possibility of a transition without a clear nucleation centre which is in accordance with earlier modelling studies on this transition.[17–20] Furthermore, the recent MD simulations showed partial phase transitions from the β to the α polymorphic form, which result in an intermediate structure.[16]

This discussion tries to pin down the mechanism that governs the poorly understood $\beta \leftrightarrow \alpha$ transition by means of a further computational and experimental characterisation of the polymorphic forms and the solid-to-solid polymorphic transition. We use single crystal X-ray diffraction to verify computational results concerning the lattice parameters of the two polymorphic forms. From differential scanning calorimetry results and simulations we deduce energetic properties of the polymorphic forms. To study the mechanism of the transition we use MD simulations with different sizes of the simulation cell. Finally, nudged elastic band calculations are used to probe the energy landscape of the transition.

The outline of the discussion is as follows. Section 2 introduces the properties of the polymorphic forms α and β. In Section 3 we describe our insights concerning the transition mechanism between the polymorphic forms. The main findings and conclusions of these two sections are written in Section 4, followed by the used methodology in Section 5.

2 Polymorphic properties

2.1 Structural properties

Discontinuous behaviour of lattice vectors, angles and/or volume of a crystal as a function of temperature is a sign of a first order phase transition,[8,10] which typically occurs through a nucleation-and-growth mechanism. In this discussion we combine experimental (single crystal X-ray diffraction) and modelling (MD simulations) results of the temperature dependence of the lattice parameters of β and α DL-NLE. The use of MD allows for the study of a polymorphic form even in its metastable regime where experimentally this form might be inaccessible because of its limited lifetime.

Along the lines of our earlier work,[16] we have determined the values of the lattice parameters through MD simulations for both the α and β polymorphic forms at a number of temperatures between 120 and 350 K. Fig. 2 shows the behaviour of all three lattice vectors, the monoclinic angle and the volume as a function of temperature. For the α polymorphic form we have doubled the unit

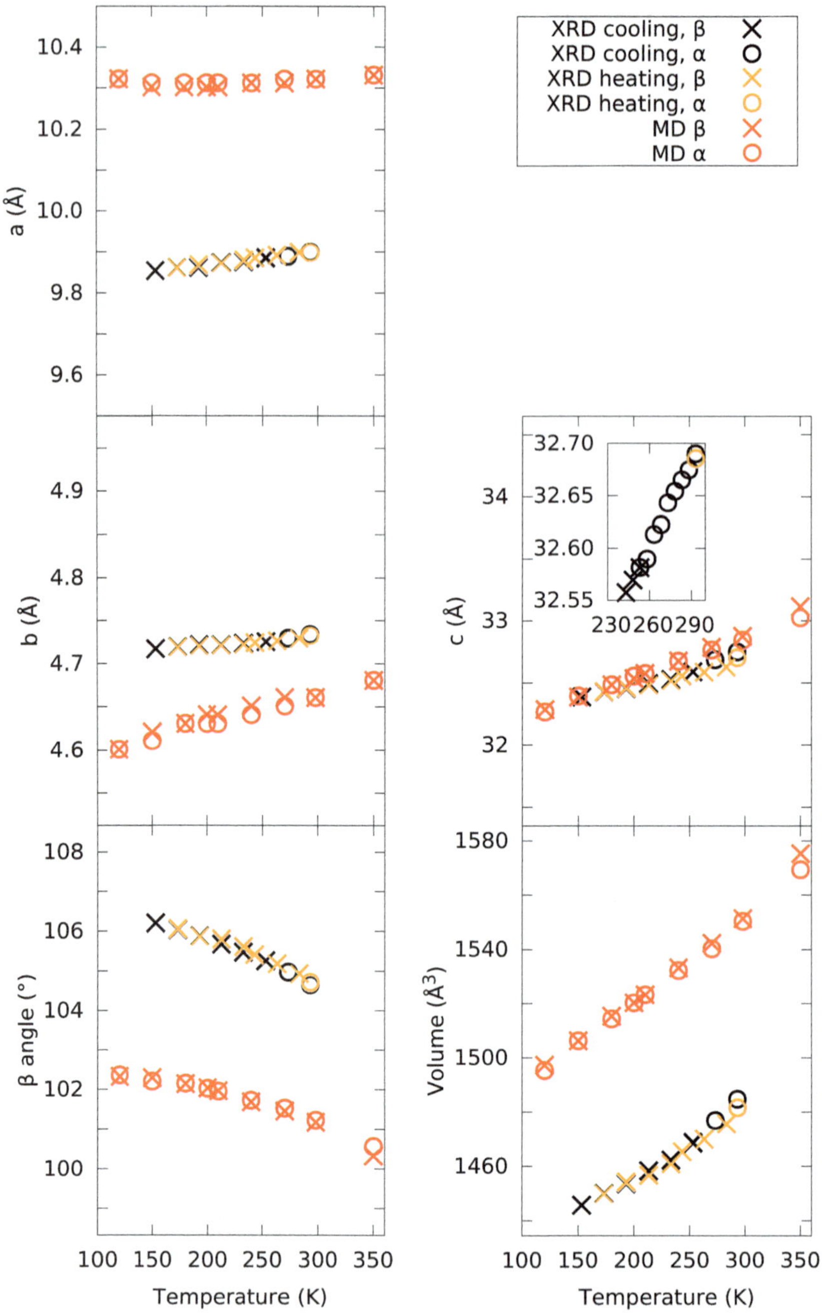

Fig. 2 The lattice parameters of DL-norleucine as a function of temperature as determined using SCXRD and MD simulations. The inset in the *c*-axis panel shows a second measurement of the same single crystal with additional points measured in the temperature range of the transition.

cell in the c direction to have the same number of molecules in the unit cell as β and to make the forms easily comparable. As one can see, the cell parameters behave identically for the α and β polymorphic form. This implies that all these five properties and their derivatives with respect to temperature, depend on the temperature in a continuous way, thereby allowing the possibility of a transition governed by a displacive mechanism. Notably, the thermal expansion is almost only affecting the c-axis, which is perpendicular to the bilayers, and the β-angle.

These results can be compared with the experimental lattice parameters of DL-NLE determined using single crystal X-ray diffraction (SCXRD). The results of such a measurement on a good quality single crystal with a clear α → β transition is shown in Fig. 2. The α phase unit cell was found above 253 K and the β unit cell was found upon cooling for temperatures between 253 K and 113 K. When the crystal was heated again, the β phase was retained until room temperature. However, after leaving the crystal at room temperature for about 60 hours the crystal had spontaneously transformed into the α polymorph.

From Fig. 2 it is clear that the lattice parameters and the thermal expansion of α and β are very similar to the results of the MD simulations, apart from an offset. No clear discontinuity is observed in the volume of the unit cell nor in the cell parameters. The inset of the panel for the c-axis shows the results of another measurement of the same single crystal with smaller increments in the temperature region of the transition. One can see that only for one specific temperature (253 K) both polymorphic forms were assigned to the obtained diffraction pattern. For this particular crystal, the transition was quite fast and a clear transition temperature could be identified. For other crystals this was not always the case. Typically, there is a temperature regime where diffraction peaks of both phases are present, which indicates coexistence of the α and β phase. This temperature regime can be quite large (even up to 100 K). Subsequent measurements of the same crystal resulted in similar lattice parameters as a function of temperature, but the amount of disorder in the layer stacking upon cooling and the temperature region of the phase transition varied from crystal to crystal. In some measurements extra peaks were visible that indicate disorder in the layer stacking or a super cell in the c-direction perpendicular to the layers. In other cases the reflections showed severe streaking. Besides, in some crystals no signs of a phase transition were visible down to 173 K. In summary, the behaviour of the crystals of DL-norleucine is extremely variable and the α ↔ β transition remains difficult to characterise.

The large similarity between the experimental and simulation results, both quantitatively and qualitatively, is reassuring. The fact that in both cases no discontinuities are observed and that the MD reproduces the anisotropy of the thermal expansion tells us that the MD simulations and the applied force field are accurate and this gives confidence that we can use the atomistic information from the simulations to learn more about the mechanism of the transformation.

2.2 Energetic properties

To assess whether the relative experimental stability of the polymorphic forms could be represented by our computational description of the system (see Section 5.4.1), we performed a 0 K geometry optimization. The minimal energies of the three polymorphic forms are: -93.5 kJ mol^{-1} for β, -92.3 kJ mol^{-1} for α and

−84.6 kJ mol^{-1} for γ. The order β < α < γ agrees with the experimental stability order. The energy difference between the β and α polymorphic forms (1.2 kJ mol^{-1}) is lower than the available thermal energy at room temperature (2.5 kJ mol^{-1}). The difference in energy between the α and γ polymorphic form is clearly higher at 7.7 kJ mol^{-1}. Particularly, the small energy difference between the β and α polymorphic form in combination with the large hysteresis in transition temperature, suggests the presence of a relatively large energy barrier for this transition.

Besides the geometry optimization, we have studied the enthalpy of the polymorphic forms as a function of temperature by means of MD simulations. Fig. 3 shows the dependence of the enthalpy on temperature. Again, these values are identical for the β and α polymorphic forms at all temperatures. Combined with the information depicted in Fig. 2, we conclude that the difference in relative orientation of the molecular bilayers with respect to each other is not a source for an observable enthalpy difference between the polymorphic forms. Moreover, this implies that transitions from β to α or *vice versa*, can occur without a net discontinuity in the enthalpy. An energy barrier between the two phases might still exist.

The heat capacity at constant pressure (C_p) can be obtained from the temperature dependence of the enthalpy. Not surprisingly, these are similar for both polymorphic forms: $C_p(\alpha) = C_p(\beta) = 5.6 \times 10^2$ J mol^{-1} K^{-1}. From this, one can obtain the difference in entropy over the temperature range:

$$S(T_f) - S(T_i) = \int_{T_i}^{T_f} \frac{C_p}{T} \mathrm{d}T \approx C_p \ln\frac{T_f}{T_i} = 6.0 \times 10^2 \text{ J mol}^{-1} \text{ K}^{-1},$$

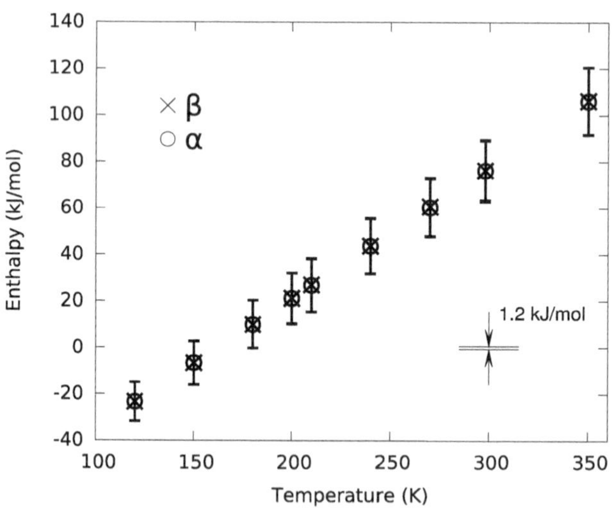

Fig. 3 The enthalpy of the β and α polymorphic forms as obtained from the MD simulations show identical behaviour in their dependence on temperature. The error bars denote the standard deviation of the enthalpy during the simulation, they overlap for both polymorphic forms because the enthalpies show identical behaviour. Crosses denote β and circles denote α. The inset shows the 1.2 kJ mol^{-1} lattice energy difference between the polymorphic forms.

for both polymorphic forms, in which $T_i = 120$ K and $T_f = 350$ K. Since, the reversible polymorphic transition occurs in this temperature range, there must be a finite difference in absolute entropy between the two polymorphic forms at 120 K. An estimate of this difference obtained from a calculation of the crossing of the Gibbs free energies, is:

$$S(\alpha, T_i) - S(\beta, T_i) = (H(\alpha) - H(\beta))/T_{trans} \approx 4 \text{ J mol}^{-1} \text{ K}^{-1},$$

in which T_{trans} is approximated by 300 K. Indeed, the transition is driven by a small entropic difference and moreover, this difference has a very small temperature dependence.

The energetic resemblance of the polymorphic forms is confirmed by differential scanning calorimetry (DSC) results. Certain single crystals show several small exothermic peaks between 254 K and 242 K during cooling (Fig. 4), which suggest a very small enthalpy change upon transforming from the α to the β polymorph. For these crystals no peak is observed during the heating process. However, the occurrence of the small peaks upon subsequent cooling indicates that the backward transition must have taken place. The peaks occur typically at slightly different temperatures and have different intensities (~0.02 kJ mol^{-1}). Several other crystals show no peak during the cooling and subsequent heating of the crystal below 383 K. In powdered samples of similar weight, which intrinsically contain more defects than single crystals, no enthalpy change is observed. Moreover, after the single crystal undergoes the α → γ transformation, no peaks were observed in the DSC measurements around 250 K, most probably due to cracking of the crystal, which results in more defects. These are arguments against a nucleation-and-growth mechanism for the α ↔ β transition, since in such a mechanism the probability of transitions increases with the amount of defects. On the other hand, the α ↔ γ transition of DL-NLE is clearly observed at

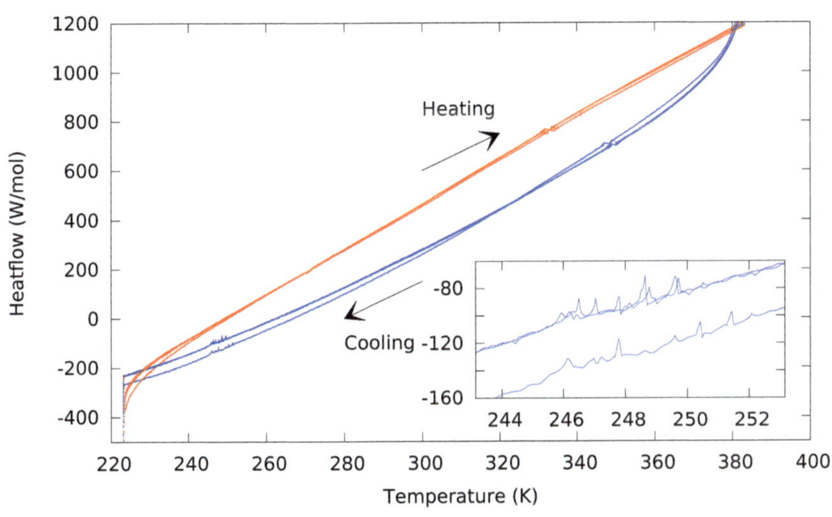

Fig. 4 Three DSC measurements of a single crystal (0.1 mg) between 383 K and 223 K with a heating rate of 2 K min^{-1}. Several small peaks indicating the α → β transition are visible during cooling between 252 K and 244 K (see inset). During heating no peak is observed.

391 K during heating as a strong endothermic peak with a small hysteresis (~2 K) both in powders and single crystals. It has an enthalpy of transition of 4.8 ± 0.2 kJ mol^{-1} in a powdered sample (which is identical to the enthalpy reported in ref. 21) and 4.6 ± 0.4 kJ mol^{-1} in a single crystal, compared to ~0.1 kJ mol^{-1} for the $\alpha \rightarrow \beta$ transition of a single crystal.

3 Transition mechanism

The β and α polymorphic forms differ from each other by a shift of each second bilayer over $a/2$ and $b/2$, as is depicted in Fig. 1. In earlier work we have exploited this difference by introducing two two-dimensional average distance parameters: one in the *bc* plane (DP1) and one in the *ac* plane (DP2). Only distances between molecules of the same chirality are included. For interfaces between the bilayers at the alkane end which have α character, DP1 is roughly 3.5 (high) and DP2 4.5 (low) at 350 K. Interfaces with β character typically have DP1 and DP2 values of 3 (low) and 6.5 (high), respectively. The distance parameters are schematically depicted in Fig. 1. In ref. 16 we were able to simulate shifts of individual interfaces from the β (low–high) to an intermediate (high–high) form. Shifts in the perpendicular directions towards (low–low) were never observed (see Fig. 8), suggesting an anisotropy in the energy barriers related to this sliding. To extend our insight in the transition mechanism, we present simulations monitored at a higher sampling rate, simulations with different sizes of the simulation cell, and results obtained with the nudged elastic band method[22,23] to probe the different barriers.

3.1 Thermally induced partial phase transitions using MD

One of the observations made in ref. 16 is that the interfaces between bilayers change rather independently of each other. In most of the cases, shifts of bilayers only affected one interface and the extra stress due to a structural mismatch was counteracted by a change in the angle of the simulation cell. Shifts of one bilayer could also result in a change of the local rearrangements of the interfaces on both sides, in which case no volume or angle change of the simulation cell was observed. Since the interfaces can change their character in terms of DP1 and DP2 rather independently, a change of one interface in the simulation cell – which typically has four interfaces – can be seen as a first step in the partial phase transition from β to the intermediate (high–high) form.

Fig. 5 shows the change in DP1 and DP2 parameters for such a partial phase transition. In this case, two interfaces are affected, which gives no effective volume change. In previous simulations, the sampling rate was too low to observe intermediate values of DP1. Here we have increased the sampling rate by a factor of ten, which allows us to actually track the transition. Since DP1 and DP2 are parameters depending on an average distance parameter, an intermediate value is not very descriptive. This is because an intermediate value can be the result of a nucleus of molecules which made the full shift and a remaining group of molecules which still needs to follow, or it could be that all molecules are half way, suggesting a cooperative mechanism without a clear centre of nucleation.

Fig. 6 shows a frame of the MD simulation at 52.95 ps, which is halfway through the partial phase transition shown in Fig. 5. As one can see, there is no

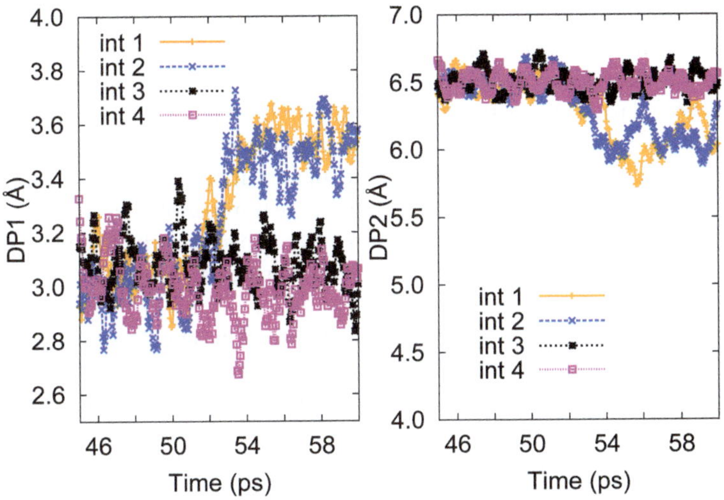

Fig. 5 The time evolution of the two distance parameters during a simulation of the β-polymorphic form at 350 K for the four interfaces of the simulation cell. The changes in DP1 show the partial phase transition along *b* in which two interfaces are affected. Due to a high sampling frequency of the trajectories, intermediate values for DP1 can be distinguished. A snapshot of the system at intermediate DP1 is shown in Fig. 6 for 52.95 ps.

clear centre of nucleation of the transition. All molecules of the most right bilayer move together along the two interfaces in a cooperative manner. This indicates the possible displacive character of the mechanism, at least at the studied length scale of approximately 24 Å.

3.2 Simulation cell size dependence

To study the influence of the size of the simulation cell on the occurrence of the partial phase transitions, we have performed MD simulations with 14 different simulation cell sizes. For a full treatment of the methodology, we refer to Section 5.4.3. The dimensions are changed along all three directions of the crystal,

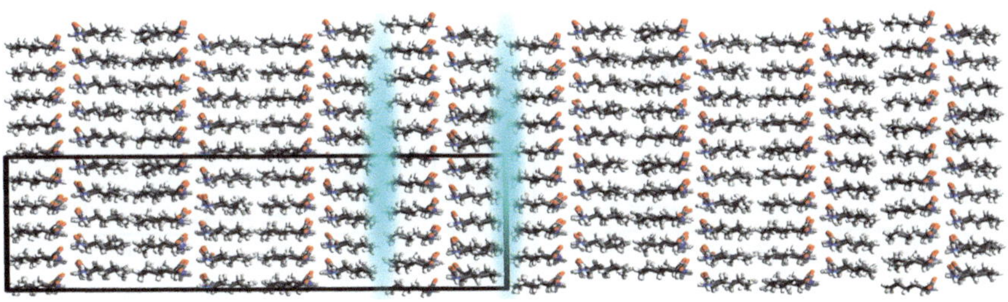

Fig. 6 A snapshot of the *bc*-plane of the MD-simulation halfway (52.95 ps) the partial phase transition that is depicted in Fig. 5. There is no clear centre of nucleation of the partial phase transition, since all the molecules in the interfaces (highlighted in cyan) on both sides of the up moving most right bilayer of the simulation cell are halfway between β and α-character. Shown are the *R*-molecules in the simulation cell projected along *a*. The simulation cell is replicated in the *c*-direction and in the *b*-direction.

leading to a change in the size of the bilayers themselves or in the number of bilayers. For all cases periodic boundary conditions are applied and hence the observed size effects are not due to edge effects because the interactions continue over the boundaries of the simulation box. The inverse size dependence of the number of observed partial phase transitions for the *a*-direction (Fig. 7a) and the *b*-direction (Fig. 7b) is clearly visible. Again, all partial phase transitions occurred along *b*. When the cell size is enlarged along *c*, the number of bilayers grows. This explains the growing number of shifts in Fig. 7c. However, when the transitions are normalised to the number of bilayers, the inverse size dependence is retained (Fig. 7d).

A distinct difference between transitions governed by a nucleation-and-growth mechanism and those having a displacive character, is the dependence on the size of the crystal.[9] The rate of nucleation will increase with size, since there are more possible sites to nucleate, while cooperative motion is harder when more molecules are involved in this cooperative movement. Therefore, the diminishing number of partial phase transitions as a function of the size of the simulation cell could point towards a displacive character of the transition. An alternative explanation for the inverse size dependence of the occurrence of the partial phase transitions would be the ratio between the size of the simulation cell and the critical nucleus of the transition in a nucleation-and-growth mechanism. If the simulation cell is too small to sustain the full critical nucleus size, a partial phase

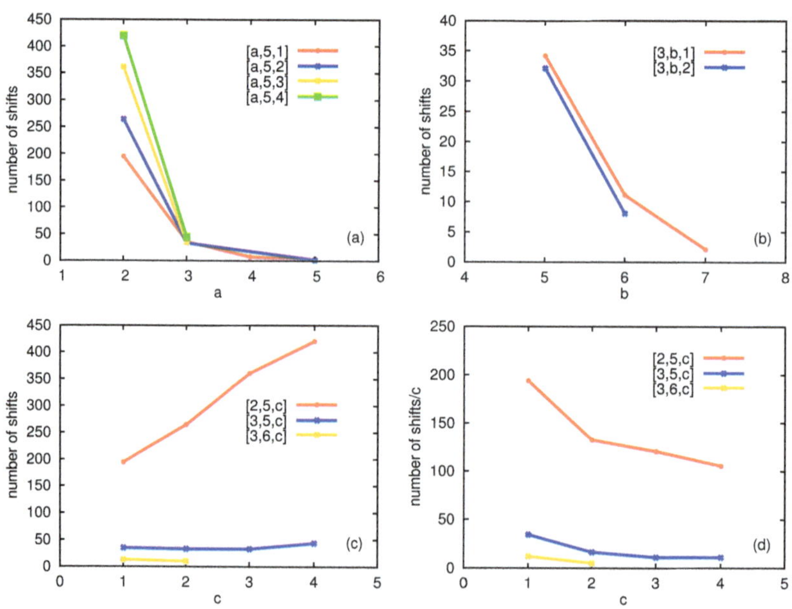

Fig. 7 The number of partial phase transitions in the MD simulations are plotted for different sizes as a function of the number of lattice parameters *a*, *b*, and *c* in the simulation cell (panels a–c). In panel d the number of partial phase transitions is normalized to the number of interfaces through a division by *c* and plotted against *c*. An increase of *a* or *b* (panels a and b) leads to a decrease of the occurrence of partial phase transitions. The same decrease takes place as a function of *c* after the normalization (panel d). These results suggest a displacive character of the partial phase transition.

transition resulting from a fully grown nucleus would be severely hindered. However, the gathering of a few molecules in the process to reach the critical nucleus should be observable, which we never did in our simulations.

3.3 Energy barriers of bilayer shifts

The full scheme of possible transitions for interfaces to go from β character (low–high) to α character (high–low) is depicted in Fig. 8: a direct mechanism in which both shifts occur simultaneously, and two mechanisms in which the transitions occur in a two step fashion with (high–high) and (low–low) as the two possible intermediate forms. By conventional molecular dynamics simulations, we have only been able to simulate directly the transition from (low–high) to (high–high), *i.e.*, a slide along *b*. This section presents Nudged Elastic Band calculations (NEB) which give the forward and backward energy barriers for all five processes to check whether the slide along *b* is indeed the most likely transition pathway. For details on the exact usage, we refer to Section 5.4.4.

NEB provides several evaluations of the system along the minimum energy path as well as the transition state structure, given an initial and a final structure. The initial structure in this case is the β form. Because of the implementation of NEB in LAMMPS,[24] it is only possible to perform these calculations within the NVT ensemble, and hence no volume changes during the process can be accounted for. For this reason, we have chosen to look at transitions involving the shift of one bilayer, affecting two interfaces instead of a shift of one interface which is accompanied by rearrangements and a volume change of the simulation cell. This is very similar to the process presented in Section 3.1 and hence the results on the (low–high) to (high–high) transition for both cases are directly comparable. The

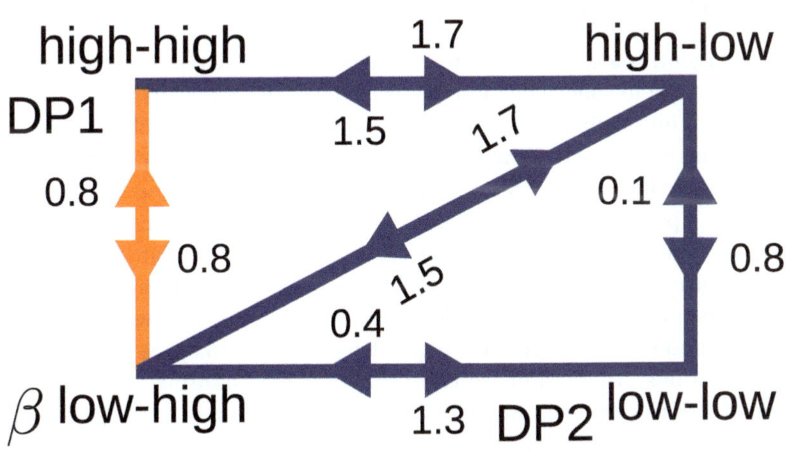

Fig. 8 A schematic overview of the shifts along *b* (DP1) and *a* (DP2) and their energy barriers in kJ mol^{-1} DL-norleucine molecules obtained from nudged elastic band calculations. Starting from the β-polymorphic form (low–high) a bilayer has been shifted first along *b* and then along *a* or *vice versa* or shifted along *b* and *a* simultaneously. The end state is (high–low) in which 2 out of the 4 interfaces possess α-character. The transitions from (low–high) to (high–high), which could be probed directly in MD simulations, are depicted in orange. This process has indeed the lowest energy barrier. The most likely path from (low–high) to (high–low) goes *via* (low–low).

resulting barriers for all back- and forward transitions are calculated by dividing the total energy barrier by the number of molecules in one bilayer and are summarised in Fig. 8. One can see that starting from β the (low–high) to (high–high) transition has indeed the lowest barrier (0.8 kJ mol^{-1}), compared to the other possible end states, (low–low, 1.3 kJ mol^{-1}) and (high–low, 1.7 kJ mol^{-1}). This explains why the shift towards the (high–high) state is the only one which we could probe with MD simulations (denoted in orange in Fig. 8).

The use of a fixed simulation cell is a severe limitation in determining the precise values of the energy barriers involved in the shifting of the bilayers. Therefore, the absolute values of the barriers should not be taken too strictly. This could be the reason for the difference in the value of the barrier to (high–high) when compared to the 3.0 kJ mol^{-1} obtained with a transition path sampling study.[20] However, the relative sizes of the energy barriers related to the different shifts can be deduced from these results and can point at the most likely transition.

Contrary to our earlier suggestion based on the observed partial phase transitions that a transition from β to α would proceed with a shift along *b* followed with a shift along *a*,[16] the most likely mechanism for the transition proceeds in the reverse order. The two-independent-shifts picture still holds, but the (low–low) conformation is the most likely intermediate structure, since the process involving the (low–low) conformation has the lowest maximum energy barrier (1.3 kJ mol^{-1}) along all three possible pathways which end in the (high–low) conformation. The second step from (low–low) to (high–low) has a much smaller barrier of 0.1 kJ mol^{-1}. Please note that the state (high–low) has α-character for 2 out of the 4 simulated interfaces of bilayers, which means that it is not completely the α polymorphic form.

4 Conclusions

This discussion is a contribution to the debate about the possible mechanisms governing solid-to-solid polymorphic transitions in molecular crystals. The possibility or impossibility of cooperative motion to be such a mechanism instead of nucleation and growth of the new phase, is at the heart of this debate. We have studied both computationally and experimentally the low temperature enantiotropic β ↔ α polymorphic transition of the amino acid DL-norleucine. We observe an identical behaviour in the temperature dependence of the lattice parameters for both polymorphs, both with X-ray diffraction measurements and in molecular dynamics simulations. The same identical behaviour of the two polymorphic forms is found for the temperature dependence of the enthalpy as obtained from MD simulations. These similarities in enthalpy are confirmed by differential scanning calorimetry measurements that show only very small peaks upon cooling and no peaks upon heating up to the α ↔ γ transition temperature. This identical behaviour in the properties of the polymorphic forms might point to a concerted or cooperative mechanism.

The congruence between experimental and computational results both in energetic and structural properties, shows the relevance of MD simulations used as a computational microscope to study the transitions on a molecular level. In MD simulations at a high sampling frequency of the trajectories, we did not observe a clear centre of nucleation when zooming in at a partial phase transition along the *b*-axis. There was also no sign of a centre of nucleation in our simulations with different sizes of the simulation cell. On the basis of nudged elastic band calculations we obtained the most likely mechanism of the transition, which is first sliding along the *a*-axis followed by a shift along the *b*-axis.

In summary, the answer to the question of the title of this paper is two-fold. For the high-temperature transition $\alpha \leftrightarrow \gamma$, the answer is definitely "yes" as is proven by Mnyukh *et al.*[9] This discussion studies the other, $\beta \leftrightarrow \alpha$, polymorphic transition of the compound at lower temperatures. On the basis of our findings we conclude that it is very likely that the polymorphic transition proceeds through a cooperative mechanism with an energy barrier instead of through nucleation and growth.

5 Methodology

5.1 Materials

DL-Norleucine (98% pure) was purchased from Alfa Aesar and recrystallised by vapour diffusion from an ethanol/water mixture to form single crystals.

5.2 Differential scanning calorimetry (DSC)

A Mettler Toledo DSC822^e calorimeter in combination with a Julabo FT900 immersion cooler, a TSO 801RO Sample Robot and STAR^e software 11.0 was used for differential scanning calorimetry measurements. Powder samples and single crystals of DL-norleucine have been investigated with this method using heating and cooling rates of 2 to 10 K min^{-1} in the temperature range of 223 to 423 K. Samples of a few milligrams were sealed in an aluminium pan (40 μL) and the heat flow was measured in comparison to an empty reference pan. The calorimeter was calibrated with the melting points of indium ($T_{on} = 429.5$ K and $\Delta H = -28.13$ J g^{-1}) and zinc ($T_{on} = 692.85$ K and $\Delta H = -104.77$ J g^{-1}), both supplied by Mettler Toledo.

5.3 Single crystal X-ray diffraction (SCXRD)

A DL-norleucine single crystal of (0.6 × 0.2 × 0.1 mm) was annealed at 383 K for at least 15 min to ensure only the α polymorph was present at the start of the measurements. The crystal was mounted on a goniometer head. Scans for unit cell determination were collected on a Nonius KappaCCD diffractometer in ϕ and ω scan mode using Mo-Kα radiation and a graphite monochromator. The scans were measured at various temperatures during cooling between room temperature and 113 K with a cooling rate of 5 K min^{-1}. In addition, at each temperature one scan was collected to observe the changes in the direction perpendicular to the layers of the crystal in more detail during the transition of the α to the β form of DL-norleucine. The unit cell was determined from the scans using the Nonius EvalCCD program suite software[25] and either the α or the β form, or both were chosen. All α unit cells were converted from standard setting $P2_1/c$ to a $P2_1/a$ unit cell with a doubling in the direction perpendicular to the bilayers (*c*-direction). The β unit cells were converted from a $C2/c$ to an $I2/a$ setting, so that the orientation of the molecules w.r.t. the cell axes of both α and β were comparable.

5.4 Computational settings

5.4.1 Force field. The same computational settings for the force field and charges as in ref. 16 are used. In summary, this means the amber force field[26] in combination with AM1-BCC charges.[27,28] For the non-bonded interactions a cutoff of 10 Å and an Ewald summation with a precision factor of 10^{-6} were used.

5.4.2 Structures and optimization. LAMMPS[24] was used to obtain the minimal energies of the different polymorphic forms. The experimentally determined structures (DLNLUA01, [13] DLNLUA02,[12] and DLNLUA05[14]) were minimised with a convergence of 3.5×10^{-2} kJ mol^{-1} Å^{-1} through an approximated second order method (hftn). To make the cell parameters of the two polymorphic forms easily comparable the same conversion of the settings of the unit cells as described in Section 5.3 was used.

5.4.3 Molecular Dynamics. For a general overview of the issues involved when using MD simulations for studies of molecular crystals we refer to Nemkevich *et al.*[29] All MD simulations have been performed with DL_POLY_4.05 (ref. 30) with input files generated with the help of DL_FIELD[31] and GDIS.[32] The anisotropic isothermal–isobaric (NPT) ensemble[33] has been applied with a barostat and thermostat parameter of 0.4 and 0.04 ps, respectively, to simulate at a constant pressure of 1 atm. An integration time step of 0.5 fs was used. Statistics and trajectories have been recorded every 0.05 and 0.5 ps, respectively. After a three step equilibration process (NVT @ 10 K, NVT @ 120 K (β) or 298 K (α), NPT @ same temperature), 500 ps simulations between 120 and 350 K were performed. All the settings are in accordance with ref. 16.

For the study of the thermal dependence of the polymorphic properties, the simulation cell consisted of $3 \times 5 \times 2$ unit cells containing 240 molecules. To study the transition mechanism, 20 different trajectories of 500 ps of the β polymorphic form were simulated at 350 K. This was done for 14 different sizes of the simulation cell: $2 \times 5 \times c$ and $3 \times 5 \times c$ with c ranging from 1 to 4, $3 \times 6 \times 1$, $3 \times 6 \times 2$, $3 \times 7 \times 1$, $4 \times 5 \times 1$, $5 \times 5 \times 1$, and $5 \times 5 \times 2$ unit cells. For the systems $2 \times 5 \times c$ the cutoff radius had to be decreased to 9.8 Å. The occurrence of partial phase transitions was followed with the help of two specifically designed distance parameters, which probe the shifts along the planes and are explained in ref. 16.

5.4.4 Nudged elastic band. The nudged elastic band calculations[22,23,34] are performed in LAMMPS.[24] The 14 replicas are connected with a spring constant of 1.4 kJ mol^{-1} Å^{-1}. The time step used in the damped dynamics for the minimisation of the replicas was 0.5 fs. The starting structure (denoted as (low–high) in Fig. 8) was a minimised frame of a trajectory without a partial phase transition of an MD simulation of the β polymorphic form at 350 K. From there the other initial structures were obtained by shifting over $b/2$, $a/2$ and $b/2 + a/2$. In total 12 atoms were connected between the replicas. The atoms (C_α, C_δ, and C_ε) were part of four molecules. These molecules formed two pairs which were located in two interfaces of bilayers. The replicas were minimised with a convergence of the force of 4.5×10^{-2} kJ mol^{-1} Å^{-1}. The obtained total energy barrier was divided by 60, the number of molecules in one bilayer.

Acknowledgements

H. M. C and M. M. H. S are grateful for support from the VIDI research program 700.10.427, which is financed by The Netherlands Organisation for Scientific Research (NWO). H. M. C, J. A. v. d. E, and M. M. H. S acknowledge the European Research Council (ERC-2010-StG, Grant Agreement no. 259510-KISMOL) for financial support. We thank René de Gelder for stimulating discussions.

References

1 K. Roth, *Chem. Unserer Zeit*, 2005, **39**, 416–428.
2 J. Bernstein, *Polymorphism in Molecular Crystals*, Oxford Science Publications, Oxford, 2002.
3 M. Aucamp, N. Stieger, N. Barnard and W. Liebenberg, *Int. J. Pharm.*, 2013, **449**, 18–27.
4 A. Grunenberg, D. Bougeard and B. Schrader, *Thermochim. Acta*, 1984, **77**, 59–66.
5 J. Tian, S. J. Dalgarno and J. L. Atwood, *J. Am. Chem. Soc.*, 2011, **133**, 1399–1404.
6 A. Dawson, D. R. Allan, S. A. Belmonte, S. J. Clark, W. I. F. David, P. A. McGregor, S. Parsons, C. R. Pulham and L. Sawyer, *Cryst. Growth Des.*, 2005, **5**, 1415–1427.
7 B. D. Altheimer, S. Pagola, M. Zeller and M. A. Mehta, *Cryst. Growth Des.*, 2013, **13**, 3447–3453.
8 F. H. Herbstein, *Acta Crystallogr., Sect. B: Struct. Sci.*, 2006, **62**, 341–383.
9 Y. Mnyukh, N. Panfilova, N. Petropavlov and N. Uchvatova, *J. Phys. Chem. Solids*, 1975, **36**, 127–144.
10 J. D. Dunitz, *Pure Appl. Chem.*, 1991, **63**, 177–185.
11 A. M. Mathieson, *Acta Crystallogr.*, 1953, **6**, 399–403.
12 B. Dalhus and C. H. Görbitz, *Acta Crystallogr., Sect. C: Cryst. Struct. Commun.*, 1996, **52**, 1761–1764.
13 M. M. Harding, B. M. Kariuki, L. Williams and J. Anwar, *Acta Crystallogr., Sect. B: Struct. Sci.*, 1995, **51**, 1059–1062.
14 S. J. Coles, T. Gelbrich, U. J. Griesser, M. B. Hursthouse, M. Pitak and T. Threlfall, *Cryst. Growth Des.*, 2009, **9**, 4610–4612.
15 U. Mukhopadhyay and I. Bernal, *Mendeleev Commun.*, 2004, **14**, 270–276.
16 J. A. van den Ende and H. M. Cuppen, *Cryst. Growth Des.*, 2014, **14**, 3343–3351.
17 S. C. Tuble, J. Anwar and J. D. Gale, *J. Am. Chem. Soc.*, 2004, **126**, 396–405.
18 J. Anwar, S. C. Tuble and J. Kendrick, *J. Am. Chem. Soc.*, 2007, **129**, 2542–2547.
19 D. Zahn and J. Anwar, *Chem.–Eur. J.*, 2011, **17**, 11186–11192.
20 D. Zahn and J. Anwar, *RSC Adv.*, 2013, **3**, 12810–12815.
21 M. Matsumoto and K. S. Kunihisa, *Chem. Lett.*, 1984, **13**, 1279–1282.
22 G. Henkelman and H. Jónsson, *J. Chem. Phys.*, 2000, **113**, 9978–9985.
23 G. Henkelman, B. P. Uberuaga and H. Jónsson, *J. Chem. Phys.*, 2000, **113**, 9901–9904.
24 S. Plimpton, *J. Comput. Phys.*, 1995, **117**, 1–19.
25 A. J. M. Duisenberg, L. M. J. Kroon-Batenburg and A. M. M. Schreurs, *J. Appl. Crystallogr.*, 2003, **36**, 220–229.
26 W. D. Cornell, P. Cieplak, C. I. Bayly, I. R. Gould, K. M. Merz, D. M. Ferguson, D. C. Spellmeyer, T. Fox, J. W. Caldwell and P. A. Kollman, *J. Am. Chem. Soc.*, 1995, **117**, 5179–5197.
27 A. Jakalian, B. L. Bush, D. B. Jack and C. I. Bayly, *J. Comput. Chem.*, 2000, **21**, 132–146.
28 A. Jakalian, D. B. Jack and C. I. Bayly, *J. Comput. Chem.*, 2002, **23**, 1623–1641.
29 A. Nemkevich, H.-B. Burgi, M. A. Spackman and B. Corry, *Phys. Chem. Chem. Phys.*, 2010, **12**, 14916–14929.
30 I. T. Todorov, W. Smith, K. Trachenko and M. T. Dove, *J. Mater. Chem.*, 2006, **16**, 1911–1918.

31 C. W. Yong, *DL_FIELD version 2.2*, http://www.ccp5.ac.uk, STFC Daresbury Laboratory, 2012.

32 S. Fleming and A. Rohl, *Z. Kristallogr.*, 2005, **220**, 580–584.

33 G. J. Martyna, M. E. Tuckerman, D. J. Tobias and M. L. Klein, *Mol. Phys.*, 1996, **87**, 1117–1157.

34 A. Nakano, *Comput. Phys. Commun.*, 2008, **178**, 280–289.

Faraday Discussions

PAPER

Zeolite synthesis in hydrated silicate ionic liquids

Leen van Tendeloo,[a] Mohamed Haouas,[b] Johan A. Martens,[a] C. E. A. Kirschhock,[a] Eric Breynaert[a] and Francis Taulelle*[ab]

Received 25th November 2014, Accepted 2nd December 2014
DOI: 10.1039/c4fd00234b

Hydrated alkali silicate ionic liquids (HSIL) were prepared by hydrolysis of tetraethoxysilane (TEOS) in alkali hydroxide–water mixtures, inducing coacervation and phase separation. The resulting optically clear, homogenous silicate ionic liquid offers exceptional potential for monitoring zeolite crystallisation. This enhanced synthesis route provides access to analysis of speciation, mechanistic details of zeolite formation, and brings organic-template-free zeolite synthesis by design within reach.

1. Introduction

Over the past half century the industrial interest in zeolite materials for application in catalysis,[1] molecular separation purification,[2] and detergents[3] has stimulated intensive investigation of zeolite synthesis. Despite this intense research, some basic scientific questions persist[4] and zeolite formation remains a source for some of the most resistant conundrums in crystallisation science: how come so many hypothetical networks can be predicted while so few frameworks are actually obtained?[5,6] Why are so few data available about zeolite solubility?[7] Why does crystallisation of natural zeolite topologies from homogeneous liquids fail in the laboratory?[8] As these questions remain, despite of the dozens of zeolite frameworks successfully obtained through what is known as the hydrothermal gel method, a paradigm shift is essential to unravel zeolite formation on the molecular scale. In this paper we propose zeolite synthesis from homogeneous hydrated silicate ionic liquids.

Hydrolysis of TEOS in aqueous hydroxide solution can induce coacervation, resulting in the separation of a dense liquid silicate phase and a water–ethanol phase. The dense phase is a hydrated silicate ionic liquid (HSIL); *i.e.* the number of water molecules in this phase does not exceed the coordination number of the alkaline cation (4–8).[9] Doping such HSILs with traces of aluminate, followed by

[a]*Centrum voor Oppervlaktechemie en Katalyse, KU Leuven, Kasteelpark Arenberg 23, bus 2461, 3001, Heverlee, Belgium. E-mail: francis.taulelle@uvsq.fr; francis.taulelle@biw.kuleuven.be; Fax: +3216321998*
[b]*Institut Lavoisier de Versailles, University of Versailles Saint-Quentin en Yvelines, 45 Avenue des Etats-Unis, 78035 Versailles Cedex, France. Fax: +33139254476*

controlled heating, yields a number of homo-ionic zeolites with natural zeolite topologies. The outcome of these syntheses depends on choice of the alkaline cations, the Si/OH ratio, the H_2O/MOH ratio and temperature. It was observed presence of aluminate is essential to obtain zeolite phases and commonly the yield of the product approaches 100% yield in aluminium. In contrast with the hydrothermal gel method, the optically transparent HSILs offer the exceptional potential for monitoring zeolite crystallisation with unprecedented accuracy.

Inspired by natural zeolite formation, the pioneers of zeolite synthesis departed from purely inorganic systems. Breck and Flanigen for example thoroughly studied zeolite crystallisation in the Na_2O–SiO_2–Al_2O_3–H_2O phase diagram.[10] Crystallisation of these traditional mixtures, readily forming complex gels and strongly depending on structure directing effects of inorganic cations, has generated most of the industrially exploited framework types of the synthetic zeolite collection.[6–9] These syntheses typically yield zeolites with high cation exchange capacity that are of practical importance in many fields, and particularly in pollution abatement, CO_2 sequestration, and catalysis for renewables.[11]

The introduction of organic structure directing agents (SDAs) allowed the synthesis of high silica zeolites[12] and led to the discovery of a score of new framework types. Despite of their success, these templates are costly, often involve complicated synthesis, and in most cases cannot be recovered after zeolite formation but need to be sacrificed in calcination. Similar to traditional inorganic zeolite synthesis mixtures, most of these compositions still form complex gels that impair complete characterization of zeolite precursors and study of the mechanism of zeolite formation.

Some compositions however yielded clear liquid ('sol'-type) precursors, thereby allowing improved characterization of the precursor with NMR, mass spectrometry and photon scattering techniques such as SAXS and DLS.[13] This has been demonstrated for the synthesis of silicalite-1 from TEOS-tetrapropylammonium precursors that historically were called 'clear solutions',[14] but actually are clear sols containing 1–2 nm objects, *i.e.*, assembly of silicate oligomers and templates, and wherein the zeolite formation is initiated. Owing to the clear appearance of these suspensions a combination of different characterization techniques led to considerable insight into the formation process.[14,15] Today such knowledge is still absent for purely inorganic syntheses where the commonly encountered gel phase prevents thorough characterization. In a few cases, colloidal suspensions ('sol') consisting of gel particles with uniform and very small size are generated by mixing aluminium and silicon sources.[16,17] Ng *et al.* for example devised a Na-based organic template free synthesis of Na-EMT, a zeolite synthesis originally involving the use of 18-crown-6 ether template.[17,18] This achievement is in full accordance with the predicted phase diagram.[19] However, also in these cases NMR characterization suffers from line broadening as crystallisation occurs within the particles. To allow full exploitation of high resolution NMR for mechanistic studies on template free, alkaline zeolite synthesis, clear precursor solutions are essential.

This manuscript presents the concept of homogeneous liquids of simple preparation, being room temperature hydrated molten silicate salts, which readily crystallise into zeolites. While other clear zeolite precursors using Na or K have been reported earlier, the creation of such mixtures required heating and/or the introduction of other elements (anions).[20] The current method is based on hydrated alkali silicate ionic liquids (HSILs) which significantly differs from ionothermal syntheses,[21–24] the ionic liquid anions being the silicates themselves. The inorganic ionic liquids are essentially solvent free, exclusively containing water molecules locked into the primary hydration shell of cations and oligomeric silicate units.

2. Experimental details

Mixing TEOS (Acros Organics, 98%) : KOH (AnalaR Normapur, 86%) : H_2O (MilliQ) with ratio 1 : 1 : 20 or $Al(^iOPr)_3$ (Acros Organics, 98^+%) : KOH : H_2O with ratio 0.1 : 1 : 3, under vigorous stirring, initially yields a homogeneous emulsion. Upon complete hydrolysis these alkoxide emulsions spontaneously undergo phase separation into two liquid systems.

Gravimetry revealed the composition of the lower, dense silicate phase as SiO_2 : KOH : H_2O in a 1 : 1 : 6 molar ratio, while the upper phase was identified as a pure ethanol–water mixture resulting from TEOS hydrolysis. The composition of both phases was analysed with ^1H, ^{29}Si, and ^{39}K NMR. After removal of the upper, ethanol–water phase, the $SiO_2/H_2O/KOH$ ratio of the silicate phase can be adjusted by addition of KOH–H_2O mixture, while keeping a clear ionic liquid.

By dissolution of $Al(OH)_3$ in KOH–H_2O mixture or by exploiting the phase separation in the $Al(^iOPr)_3$: KOH : H_2O system, addition of small volumes of aluminate homogeneous liquid with final composition 0.1 : 1 : 2.85 allowed the introduction of up to 5% of T-atoms as Al without provoking gel formation. The resulting homogeneous liquids can be used to crystallise zeolites as demonstrated by formation of MER zeolite, obtained by autoclaving such a precursor with composition: $0.5SiO_2$: 1KOH : $8H_2O$: $0.013Al_2O_3$ in a temperature range from 60 to 170 °C.

The HSIL precursors were investigated by DLS on a ALV/CGS-3 instrument (ALV, Langen, Germany). 60 s measurements were performed at scattering angles from 30 to 150 and a wavelength of 632.8 nm. SAXS was measured at room temperature on a SAXSess MC^2 instrument (Anton Paar) with line-collimated CuKα radiation.

SEM images of the zeolite products were made on a Nova NanoSEM 450.

XRD patterns were recorded on a STOE STADI MP diffractometer with a focusing Ge(111) monochromator (CuKα_1 radiation) in Debye-Scherrer geometry with a linear position sensitive detector (PSD) (6 2°θ window) with a step width of 0.5° and internal PSD resolution 0.01°.

3. Results and discussion

TEOS hydrolysis is considered a reproducible route to molecular, monomeric silicic acid and is therefore preferred for mechanistic studies of zeolite formation. Despite this advantage, ethanol released during hydrolysis tends to decrease the dielectric constant of the liquid, favouring sol formation and precipitation.[25] In zeolite synthesis mixtures, the essential introduction of aluminate amplifies this effect by its strong tendency towards flocculation and gelation.

Zeolites from ionic liquid precursors

In this study TEOS-based silicate systems with composition $SiO_2 : KOH : H_2O : EtOH$ in $n : 1 : 20 : 4n$ ratio were observed to spontaneously de-mix upon full TEOS hydrolysis for some values of *n*.[9] This phase separation into two clear liquids exclusively occurs in the range from *ca.* 0.75 : 1 : 20 : 3 to 1.25 : 1 : 20 : 5 and strictly depends on the amount of water per Si. Upon dilution of the system with water, the phase separation disappears, returning the system to a transparent homogeneous solution. TEOS concentrations outside the indicated range result in the formation of gels or apparently single-phase liquids (Table 1). In contrast to the true ionic liquids obtained after successful coacervation, the apparent single-phase systems with ethanol contain nano-aggregates and are better described as colloidal suspensions or 'sols'.[26] Similar coacervation phenomena have been described for aqueous solutions combining sodium silicate and salts of small tetra-alkylammonium ions, or ethanol.[27,28]

To demonstrate effective zeolite crystallisation from the HSILs, precursors with different water and aluminium content were prepared by the addition of $KOH–H_2O$ mixture and/or an aluminate mixture of composition $0.1Al : 1 KOH : 2.85H_2O$. In the evaluated systems, the SiO_2/KOH ratio was fixed at 0.5. This resulted in the crystallisation of phase pure zeolites even for very low $H_2O : KOH$ ratio (≥ 4), much below a fully hydrated K ion, with a typical coordination number of 8.

Phase pure MER type zeolite was obtained from an HSIL precursor with composition $0.5SiO_2 : 1KOH : 8H_2O : 0.013Al_2O_3$ (Fig. 1). The HSIL precursor was clear to the eye and according to DLS and SAXS did not contain any object larger than 1 nm radius. The absence of peak broadening in ^{29}Si NMR spectra further confirmed the absence nano-aggregates, or nanoparticles in the precursor liquid at room temperature, while detailed spectral analysis revealed these precursors exclusively contained small oligomers.[29]

In contrast with the low content of aluminium in the above precursor, MER type zeolites are typically obtained from gels with the Si/Al ratio 2.5,[30] resulting in a ratio around 1.8 in the final framework. In this work, MER with a typical Si/Al ratio of around 2, was formed from precursors with Si/Al ratios as high as 36. As no congruent liquid to crystal conditions are used the yield must be calculated on the element of lowest concentration. It reaches 100% related to aluminium content. MER crystal morphology was observed to be dependent on the Al content of the precursor liquid and on the crystallisation temperature. While low aluminium (and water) content precursor compositions yield very large (up to 150 μm) isolated single crystals of high quality, other conditions resulted in lath-shaped crystals and bundles of these.[43]

Table 1 Chemical composition leading to de-mixing into two clear solutions

TEOS	Molar composition KOH	H_2O	Outcome
0.25	1	20	Solution, clear single phase
0.50	1	20	
0.75	1	20	Coacervation, 2 clear liquids
1.00	1	20	
1.25	1	20	
1.50	1	10	Coacervation, gel + liquid

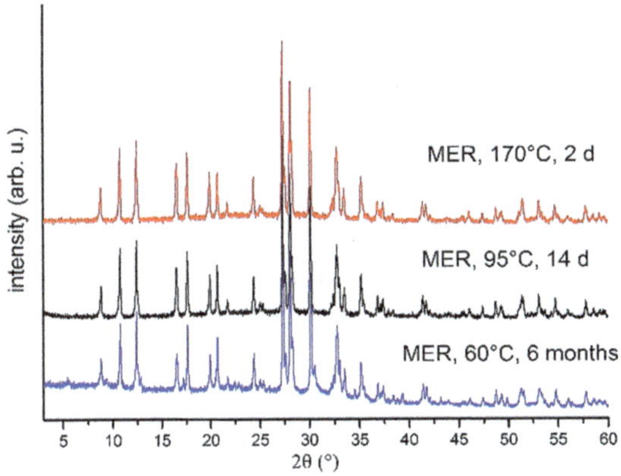

Fig. 1 XRD patterns of MER type zeolite recovered after hydrothermal treatment of potassium aluminosilicate ionic liquid in the system $0.5SiO_2 : 1KOH : 8H_2O : 0.013Al_2O_3$ at 170 °C (2 d), 95 °C (14 d), and 60 °C (6 m). Small reflections belonging to CHA (*e.g.* at 12.85, 17.20 and 30.55° 2θ) were present in the 60 °C sample.

MER zeolite formed at temperatures from 60 °C to 170 °C (Fig. 2). At 60 °C, traces of CHA were detected by XRD and SEM. In more diluted conditions, ($0.8SiO_2 : 1KOH : 92H_2O : 0.013Al_2O_3$) combined with higher Si/OH ratio, LTL formed at 85 °C, consistent with the synthesis conditions reported for classical gel systems.[31–33] In accordance with the phase diagram established by Lillerud *et al.*,[33] phase pure CHA could not yet be obtained because the maximum Si/KOH ratio for a clear precursor is not high enough.

The principle of preparing HSILs in the TEOS–KOH–H_2O system can also be exploited with other alkali cations, such as Na. Also in this case, a coacervation process takes place, but Na was observed to have stronger affinity for gel

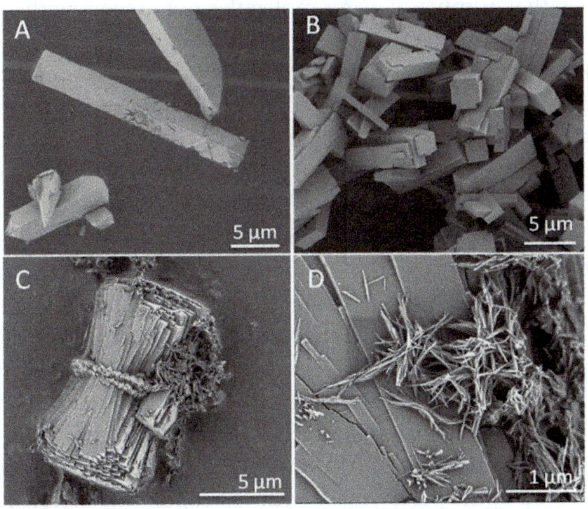

Fig. 2 SEM images of the products crystallised at 170 °C for 16 h (A), at 90 °C for 30 days (B) and at 60 °C for 6 months (C and D) from the precursor with composition $0.5SiO_2 : 1KOH : 8H_2O : 0.013Al_2O_3$.

formation at lower Si/OH ratios. Here again, two separate phases were obtained at $0.75SiO_2 : 1NaOH : 20H_2O$. Instead of dissolving as in the KOH system, $Al(^iOPr)_3$ precipitated as $Al(OH)_3$ in the NaOH–H_2O mixture. Therefore a mixture with adjusted composition of $0.1Al(OH)_3 : 1.3NaOH : 6H_2O$ was chosen as the Al source. A Na based precursor with final composition $0.5SiO_2 : 1NaOH : 8H_2O : 0.013Al_2O_3$ yielded phase pure GIS at 90 °C after 3 days. Combination of Na- and K-based HSILs yielded pure GIS for $x = Na/(Na + K) = 1$, pure PHI for $x = 0.75$ to 0.50, and pure MER for $x = 0.25$ to 0.00. The same zeolite types were encountered by Donahoe *et al.* in the $Na_2O : K_2O : Al_2O_3 : SiO_2$ gel systems.[20]

Mechanism of formation of zeolites

A recent review by Cundy and Cox[34] lists an impressive number of contributions proposing mechanisms of formation of zeolites. Despite of the wealth of literature on this topic a comprehensive description can yet not be reached due to a lack of information about the molecular steps of the lattice construction. With the advent of HSILs, alternative routes to elucidate those elementary steps are within reach. In the next paragraphs, the steps of zeolite formation derived from previous contributions and from these novel silicate ionic liquid precursors will be enumerated in relation to the reaction types indicated in Table 2.

Silicic acid formation and deprotonation. The monomeric silicon source TEOS slowly hydrolyses into silicic acid in a water–alkali hydroxide mixture. TEOS does not mix with the water phase spontaneously, so energetic stirring is indicated (reaction 1 of Table 2). In the highly alkaline solution (about pH 13), silicic acid readily deprotonates (reaction 2 of Table 2). This deprotonation is approximately limited to one proton per $Si(OH)_4$ producing a silicate anion ion-paired to an alkaline cation, $Si(OH)_3O^- \cdots M^+$. When prepared with 20 water molecules per MOH, and 0.5 to 1.2 TEOS, this liquid mixture separates into a water–ethanol upper phase and a denser hyposolvated alkali–silicate phase. The dense phase is an ionic liquid with a number of water molecules not exceeding the coordination number of the alkaline cation (4–8).[4]

Limited condensation by oxolation. With a progressively increasing Si/OH ratio due to hydrolysis, small oligomers develop. The main reaction involved is Si–O–Si formation by an oxolation reaction, (reaction 3 and 4 of Table 2). The average connectivity, the number of bridging O, varies progressively from 0 in the monomer to about 3.0 for a Si/OH ratio of 3. Those oligomers interconvert slowly, which can be detected by 2D EXSY NMR experiments (reaction 5 of Table 2).[35] The oligomers formed have a progressively increasing nuclearity, *i.e.*, an increasing number of silicon atoms per oligomer. At the maximum oligomer connectivity of *ca.* 3.0 most species are still below a nuclearity of about twelve. A small fraction of the population however exhibits nuclearity as high as 31 as detected by mass spectrometry.[36] This condensation stage had already been identified as early as at the beginning of zeolite syntheses studies in works by Barrer.[37,38]

Si/OH ratio, the titration of OH^- by $Si(OH)_4$. Titration of hydroxides by silicic acid can be observed by the pH decrease with corresponding increase of the Si/OH ratio. In non-separated hydro-alcoholic systems nano-aggregation occurs when reaching a Si/OH ratio of one, as detected by DLS and SAXS. In hyposolvated mixtures (with $H_2O/MOH \leq 20$) de-mixing occurs, resulting in a dense silicate

Table 2 Reactions for the different stages of TEOS hydrolysis into silicic acid, its deprotonation, successive condensation and dehydration stages. The H_2O subscript indicates solvated species

Reactions		Main characteristic	Parameter observable
1	$Si(OEt)_4$ pure + $4H_2O$ = $Si(OH)_4 + 4EtOH$	Hydrolysis	Solubility
2	$Si(OH)_4 + MOH_{H2O} = Si(OH)_3O^-M^+ + H_2O$	Deprotonation, acidity	Titration
3	$(OH)_3Si(OH)_{H2O} + {}_{H2O}(OH)Si(OH)_2O^- = (OH)_3Si–O–Si(OH)_2O^- + H_2O$	Dimerization condensation	Oligomers formation, increase of $\langle n \rangle$
4	$nSi(OH)_3O^-M_{H2O}^+ = Si_nO_p(OH)_{3n-2p}O^{n-}(M^+)_n + pH_2O$	Oligomerization condensation	Increase of $\langle n \rangle$ Si/OH < 1
5	$Si_nO_p(OH)_{3n-2p}O^{n-} + Si_qO_r(OH)_{3q-2r}O^{q-} = Si_jO_k(OH)_{3j-2k}O^{j-} + Si_lO_m(OH)_{3l-2m}O^{l-}$	Exchange	Interconversion $\langle n \rangle$ invariant
6	$n[Si(OH)_3O^-M^+,xH_2O] = [Si(OH)_3O^-M^+, pH_2O]_n + n(x-p)H_2O$	Aggregation, dehydration coacervation or de-mixing	$\langle n \rangle$ invariant after Si/MOH = 1 or 0.5
7	$nSi(OH)_3O^-M_{H2O}^+ = Si_nO_p(OH)_{3n-2p}O^{n-}(M^+)_n + pH_2O$	Oligomerization condensation within nanoaggregates, and water exclusion/ dehydration	$\langle n \rangle$ increases Si/OH > 1
8	$Si(OH)_3O^-M_{H2O}^+ + HOSi(OH)_3 = Si_2O(OH)_5O^- + OH^-M_{H2O}^+$	Condensation, MOH release	$\langle n \rangle$ invariant by dismutation with T if $\langle n \rangle > 3.0$
9	$Si(OH)_3O^-M^+ + M^{+-}OSi(OH)_3 + H_2O = Si_2O(OH)_5O^- + 2\,OH^-M_{H2O}^+$	Condensation, M_2O release $O^{2-}M^+_2$ and reaction with H_2O	$\langle n \rangle > 3$

ionic liquid with composition xSi(OH)$_4$: 1MOH : yH$_2$O and a y value of about 6 for potassium. This silicate ionic liquid producing phase separation occurs for x values between 0.5 and 1. The resulting ionic liquids exhibit "dressed" cations, coordinated by water molecules and/or silicate species, thereby representing ion pairs of silicate anions with fully or partially hydrated cations. The free energy of formation arises from the Coulombic energy between ions, with an additional stabilization by coordinative hydration of the cation. Additional water would progressively lead to aqueous solutions, where the lattice energy would be dominated by solvation of both ions, rather than by Coulombic interaction as the ions are progressively separated by dilution. The latter case is conventionally described by the Debye–Hückel solvent model. The former case resembling a molten alkali-silicate hydrate has been called hydrated or solvated ionic liquids as coined by C. A. Angell since 1965.[39]

Nano-aggregation, or coacervation, or de-mixing. Ionic liquid formation. Increasing concentrations of silicate salts go hand in hand with an increasing degree of silicate oligomerization, corresponding to H_2O release (reaction 3 of Table 2, oxolation), MOH release (reaction 8 of Table 2), or $M_2O/2MOH$ release (reaction 9 of Table 2). With increasing oligomerisation, the dielectric constant of the medium changes, initially fairly gradually, but at Si/OH $\cong$ 1.0 for ethanol containing mixtures or at Si/OH $\cong$ 0.5 in spontaneously de-mixing systems, quite abruptly. The macroscopic effect of the resulting coacervation phenomenon depends on the solubility of the hydroxide cation M^+ in ionic liquid compared to the ethanol–water phase of the source solution. Depending on this solubility it leads to formation of suspended nanoaggregates[26] (high solubility) or to separation into a dense liquid silicate and a water–ethanol phase (low solubility).[43] This results in the formation of nano-aggregates in the former case and in complete phase separation in the latter. Both processes are related, and have been designated with a variety of names such as coacervation, liquid–liquid de-mixing and even ion-pairing, with or without a potential spinodal decomposition when slow conformation rearrangements of complex organic cations occur. Though a general theory describing these liquid–liquid de-mixing processes is lacking, their main characteristic is that the inorganic phases de-mix on macroscopic or nanoscale, consequently leading to phase separation or the formation of nano-aggregates. In both cases, the denser phase contains the inorganic species in significantly hyposolvated state compared to the initial mixture.

This dehydration step is analogous to a salting out process, or more adequately described, an ethanol "washing out" process.[44] Most noteworthy, this process does not directly affect the connectivity of the oligomer distribution, thereby indicating that the dominant aspect of this stage is a free energy discontinuity resulting from dehydration and Coulombic lattice energy development (reaction 7 of Table 2).

Condensation by hydroxide release, pores formation. Provision of additional energy to the nano-aggregate precursors by heating liquid precursors with a Si/OH ratio of 3.0 corresponding to an average connectivity $\langle n \rangle$ of 3, initiates a second stage of aggregation with particles increasing in size from nano-aggregates of about 2–3 nm, to 6 to 100 nm, before crystallisation. This period of induction, lasting about 24 h at 95 °C for a clear silicalite precursor sol, exhibits no change in $\langle n \rangle$ though the size of particles grows considerably. At this stage no Bragg crystallinity is detected by *in situ* X-ray diffraction.

Until a Si/OH ratio of 1.0, silicic acid is deprotonated only once, which allows it to reach an average connectivity $\langle n \rangle$ of 3 by oxolation only. At increasing Si/OH ratios this situation changes. For example at Si/OH = 3.0, and when $\langle n \rangle$ has already reached 3 by further addition of silicic acid, so that each silicon in average is connected to 3 neighbours, 1/3 of the silicon atoms are deprotonated, presenting one SiO^- group while 2/3 still exhibit a SiOH function. These are the groups left available for further condensation, as described by reaction 8 of Table 2. Since this reaction forms a Si–O–Si bond by releasing an OH^- group, the latter can attack a Si–O–Si bond and open it. Consequently $\langle n \rangle$ does not vary globally during this stage. Instead, two phenomena take place: super-aggregation of higher connectivity silicon atoms in a network and release of lesser connectivity silicon inside nano-aggregates. During this induction period, Si–O–Si bond breaking and bond formation allow for the initiation of network formation. The network connects but does not yet reach the periodic organization of the crystal. It is formally analogous to the correction scheme of supramolecular crystal formation. The dominant process is driven by the free energy evolution of the lattice by the pore formation enclosing hydrated cations or template molecules, which is essentially an entropic contribution.

Heating a HSIL at a Si/OH ratio of 0.5 leads to a similar situation. Heating decreases the dielectric constant of the medium, and density fluctuations take place. The process might be faster as compared to the nano-aggregate condensation-super aggregation of sols. The presence of OH^- groups, with a $\langle n \rangle$ of *ca.* 1.5 allows an overlap of oxolation and MOH release condensations, which are therefore no longer fully separated reaction steps. The limited number of water molecules favours formation of clathrate types or clusters around the cations. Si–O–Si bonds formation at this stage requires heat, due to a higher activation energy of the water/OH assisted bond breaking and bond reformation, in which the water clusters around cations have to contribute their coordination water.

Crystallisation, condensation by hydroxide/oxide release. Finally at crystallisation a dramatic change occurs, hallmarked by an entropy driven free energy evolution. Organisation of the almost connected network fragments into a periodically ordered crystal representing a first order phase transition. At this stage reactions 8 and 9 of Table 2 take place, two reactions generating a massive release of OH^- and an increase of pH, which almost returns to its initial value of the MOH water mixture before introduction of TEOS. The phase obtained at this stage is a dense phase with pores filled by water or cation-cluster clathrate-like structures.

Though this mechanism of formation has only been established in a few studies of the formation of MFI, BEA, MER, CHA, LTL, GIS, or PHI, it overlaps consistently with most of the observations or assumptions described in Table 1 of Cundy and Cox (2005).[34] The nature of the chemical origin of activation energy for Si–O–Si bond breaking and reformation is related to the different water-assisted mechanisms of condensation controlling its formation at each stage. They depend on the cation type, the water clathrate structure, the composition of the ionic liquid, and the concerted reaction involving the reagents with Si–O–Si.

The limitation of zeolites to a severely restricted number of real topologies compared to the innumerable list of hypothetical ones can also be explained by the small number of units, cages and rings, exhibiting enough stability to be found within zeolite networks, thereby leading to a limited set of crystals formed with minimal lattice energy.[5]

The templating effect of organic as well as alkaline cations of zeolites can be revisited using this new paradigm. As Barrer indicated,[41] the chemical potential of the hydrated, clathrate-like structure of cations has to be taken into account for calculating the chemical potential of the crystal. Since zeolites form as dense crystals with hydrated cations or organic templates filling the cavity system, their nanoporous character appears only at the post-crystallisation stage when pores are emptied. It will be essential to re-determine the crystal structures of zeolites filled with cations[41] and water to derive rules of stabilization of the large cages as suggested by the work of Blatov and Proserpio.[42]

The global scheme depicting the description of the zeolite formation discussed above is provided in Fig. 3.

Data still missing

This paper demonstrates a new practical route for the synthesis of natural zeolites from silicate ionic liquids. In contrast with traditional zeolite synthesis, zeolites can now be crystallised from a hydrated silicate ionic liquid. In case of excess aluminium, gels phases are obtained which coexist with HSIL. The excess aluminium used in gel syntheses serves several purposes: as an aluminium source, as a nucleation surface for heterogeneous nucleation and as a diffusion slowing agent.

Silicate ionic liquid synthesis of zeolites will enable the study of all molecular aspects of the syntheses, in conditions very comparable to gel phases, with very low water content. This is a requirement to assign the activation energy and free energy of each stage of the synthesis to a definite molecular process. There are however missing data still to be acquired. Among them, one can cite the densities

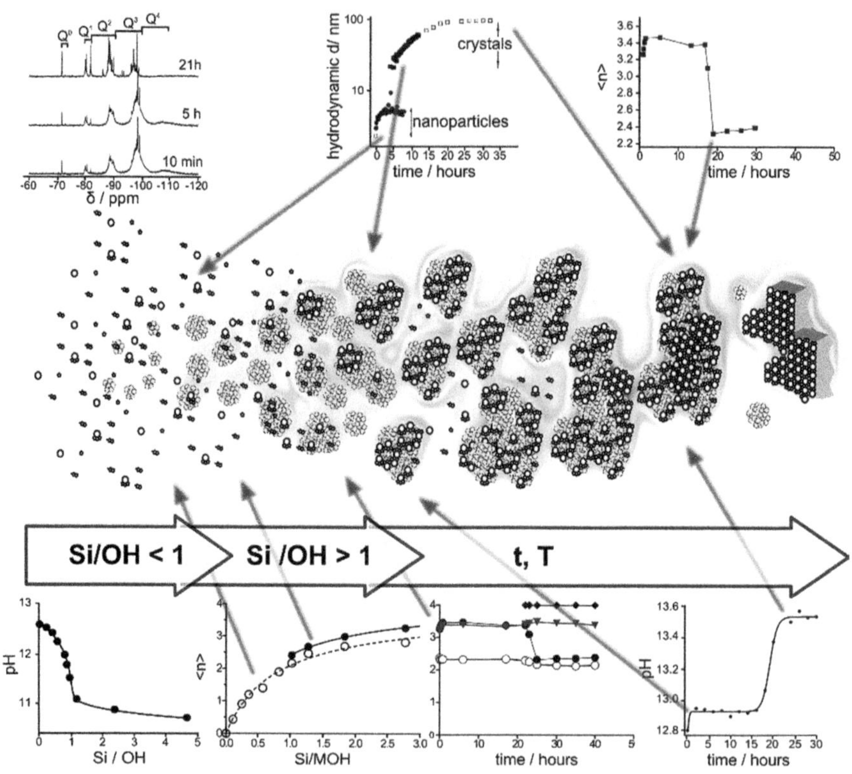

Fig. 3 Mechanism of homogeneous liquid with nano-aggregation or ionic liquid formation from room temperature precursor formation to hydrothermal heating of precursor into zeolite crystal. Precursor state varies with Si/OH. Before Si/OH = 1, neither aggregation nor de-mixing occurs. For Si/OH > 1.0 nano-aggregation occurs for ethanol containing sols. De-mixing with ethanol exclusion and hypohydrated ionic liquid starts at Si/OH of 0.5. When heating, an induction phase starts, and ends at crystallisation. NMR, pH, Dynamic Light Scattering and $\langle n \rangle$ determined from NMR are plotted in insets.

of the HSIL, their viscosities and conductivities. Solubility measurements in HSIL would be directly accessible, even in non congruent crystallisation conditions by using the analytical power of *in situ* NMR for all nuclei. Solubility of a direct or retrograde character will therefore be accessible and formulated unambiguously. This represents an essential step to allow formalising of the crystallisation of zeolites. In addition to direct solubility measurements, also interconversion of zeolites could lead to critical assessment of the proper understanding of zeolite formation. Indeed, the interconversion also is determined by crystallisation kinetics and thermodynamics which depend on the activation energy of the slowest crystallisation step and the solubility of the zeolites involved.

4. Conclusions

In this contribution a new formalism to zeolite formation has been proposed. It was necessary to identify that the phase from which zeolites nucleate, and at a later stage crystallise, is in a state of low hydration, a hydrated silicate ionic liquid (HSIL). The latter cannot be confused with standard water solutions, as the amount of water does not exceed the water molecules in the coordination sphere of the silicate salt cation. Their study will lead to understanding of any type of zeolite precursors, clear solutions, sols or gels. Additionally, depending on the presence of an excess of solid reagent like the aluminium source, heterogeneous nucleation can take over. This will allow also to evaluate the change of homogeneous to heterogeneous nucleation. The reacting medium from which zeolites form in a traditional gel synthesis is probably an interstitial ionic liquid formed by salting out the excess water. The HSIL route will allow application of most methods of molecular-level characterisation. This can lead to unification of the different zeolite synthesis routes.

With the actual results presented in this study, it seems that zeolite crystallisation from hydrated alkaline silicate ionic liquids does not represent any exception from classical crystallisation mechanisms. In contradiction to common claims in the literature depicting zeolite formation as a special case, zeolite crystallisation, like classical crystallisation, is governed by well-defined physicochemical rules. This is where the paradigm shifts lies. However, compared to classical crystallisations zeolite crystallisation proceeds in slow motion. By studying the HSIL precursors, one can establish and quantify a wealth of parameters for molecular processes related to nucleation and growth that are relevant not only for zeolites, but also can shed light on crystallisation phenomena in general. With the identification of the complexity of all stages of zeolite formation, the proposal of a grand scheme for crystallisation is within reach, covering different interconnected models of different grain sizes form coarse to fine. This is yet to come.

Acknowledgements

M.H. and F.T acknowledge support by the CNRS and the Region Ile de France. L.V.T. and E.B. acknowledge a fellowship as respectively Aspirant and Postdoctoraal onderzoeker van FWO-Vlaanderen. F.T. acknowledges KU Leuven for an appointment as visiting professor. This work was supported by long-term structural funding by the Flemish Government (Methusalem), Belgian IAP-PAI networking and BelSpo in the frame of Prodex.

Notes and references

1 B. Yilmaz and U. Müller, *Top. Catal.*, 2009, **52**, 888–895.
2 E. J. García, J. Pérez-Pellitero, G. D. Pirngruber, C. Jallut, M. Palomino, F. Rey and S. Valencia, *Ind. Eng. Chem. Res.*, 2014, **53**, 9860–9874.
3 A. De Lucas, M. A. Uguina, I. Covian and L. Rodriguez, *Ind. Eng. Chem. Res.*, 1992, **31**, 2134–2140.
4 J. D. Rimer, M. Kumar, R. Li, a. I. Lupulescu and M. D. Oleksiak, *Catal. Sci. Technol.*, 2014, **4**, 3762–3771.
5 V. A. Blatov, G. D. Ilyushin and D. M. Proserpio, *Chem. Mater.*, 2013, **25**, 412–424.
6 R. M. Barrer, *J. Chem. Soc.*, 1948, 127–132.
7 R. M. Barrer, *Zeolites*, 1981, **1**, 130–140.
8 R. M. Barrer, L. Hinds and E. A. White, *J. Chem. Soc.*, 1953, 1466–1475.
9 R. M. Barrer and C. Marcilly, *J. Chem. Soc. A*, 1970, 2735–2745.
10 E. M. Flanigen and W. D. Breck, in *137th Meeting of the ACS, Division of Inorganic Chemistry*, Abstracts p. 33-M, Cleveland, OH, April 1960.
11 T. D. Pham, Q. L. Liu and R. F. Lobo, *Langmuir*, 2013, **29**, 832–839.
12 E. M. Flanigen, J. M. Bennett, R. W. Grose, J. P. Cohen, R. L. Patton, R. L. Kirchner and J. V. Smith, *Nature*, 1978, **271**, 512–516.
13 N. Pienack and W. Bensch, *Angew. Chem., Int. Ed.*, 2011, **50**, 2014–2034.
14 A. Aerts, C. E. A. Kirschhock and J. A. Martens, *Chem. Soc. Rev.*, 2010, **39**, 4626–4642.
15 L. Tosheva, B. Mihailova, L. H. Wee, B. Gasharova, K. Garbev and A. M. Doyle, *Angew. Chem., Int. Ed.*, 2008, **47**, 8650–8653.
16 S. Mintova, N. H. Olson and T. Bein, *Angew. Chem., Int. Ed.*, 1999, **38**, 3201–3204.
17 E. P. Ng, D. Chateigner, T. Bein, V. Valtchev and S. Mintova, *Science*, 2012, **335**, 70–73.
18 E. P. Ng, J. M. Goupil, A. Vincente, C. Fernandez, R. Retoux, V. Valtchev and S. Mintova, *Chem. Mater.*, 2012, **24**, 4758–4765.
19 M. Maldonado, M. D. Oleksiak, S. Chinta and J. D. Rimer, *J. Am. Chem. Soc.*, 2013, **135**, 2641–2652.
20 R. J. Donahoe, J. G. Liou and S. Guldman, *Clays Clay Miner.*, 1984, **32**, 433–443.
21 P. J. Byrne, D. S. Wragg, J. E. Warren and R. E. Morris, *Dalton Trans.*, 2009, 795–799.
22 E. R. Parnham, E. A. Drylie, P. S. Wheatley, A. M. Z. Slawin and R. E. Morris, *Angew. Chem., Int. Ed.*, 2006, **45**, 4962–4966.
23 M. F. Tang, Y. H. Liu, P. C. Chang, Y. C. Liao, H. M. Kao and K. H. Lii, *Dalton Trans.*, 2007, 4523–4528.
24 Y. S. Wang, Y. P. Xu, Z. J. Tian and L. W. Lin, *Chin. J. Catal.*, 2012, **33**, 39–50.
25 S. G. Thoma and T. M. Nenoff, *Microporous Mesoporous Mater.*, 2000, **34**, 301–306.
26 D. P. Petry, M. Haouas, S. C. C. Wong, A. Aerts, C. E. A. Kirschhock, J. A. Martens, S. J. Gaskell, M. W. Anderson and F. Taulelle, *J. Phys. Chem. C*, 2009, **113**, 20827–20836.

27 J. C. J. van der Donck and H. N. Stein, *Langmuir*, 1993, **9**, 2270–2275.

28 N. E. Prikhid'ko and V. S. Molchanov, *Kolloidn. Zh.*, 1951, **13**, 450.

29 C. T. G. Knight, R. J. Balec and S. D. Kinrade, *Angew. Chem., Int. Ed.*, 2007, **46**, 8148–8152.

30 In IZA Synthesis Commission, *Verified Synthesis of Zeolitic Materials*, http://www.iza-online.org.

31 A. Bieniok, K. Bornholdt, U. Brendel and W. H. Baur, *J. Mater. Chem.*, 1996, **6**, 271–275.

32 J. C. Quirin, L. T. Yuen and S. I. Zones, *J. Mater. Chem.*, 1997, **7**, 2489–2494.

33 B. M. Skofteland, O. H. Ellestad and K. P. Lillerud, *Microporous Mesoporous Mater.*, 2001, **43**, 61–71.

34 C. S. Cundy and P. A. Cox, *Microporous Mesoporous Mater.*, 2005, **82**, 1–78.

35 C. T. G. Knight, R. J. Kirkpatrick and E. Oldfield, *J. Magn. Reson.*, 1988, **78**, 31–40.

36 D. P. Petry, S. C. C. Wong, M. Haouas, F. Taulelle, S. J. Gaskell and M. W. Anderson, in *Studies in Surface Science and Catalysis*, ed. P. M. Antoine Gédéon and B. Florence, Elsevier, 2008, vol. 174, Part B, pp. 941–944.

37 R. M. Barrer, J. W. Baynham, F. W. Bultitude and W. M. Meier, *J. Chem. Soc.*, 1959, 195–195.

38 R. M. Barrer, *Chem. Br.*, 1966, 380–394.

39 C. T. Moynihan, C. R. Smalley, C. A. Angell and E. J. Sare, *J. Phys. Chem.*, 1969, **73**, 2287–2293.

40 R. M. Barrer, *Hydrothermal Chemistry of Zeolites*, Academic Press, 1982.

41 B. M. Skofteland, O. H. Ellestad and K. P. Lillerud, *Microporous Mesoporous Mater.*, 2001, **43**, 61–71.

42 N. A. Anurova, V. A. Blatov, G. D. Ilyushin and D. M. Proserpio, *J. Phys. Chem. C*, 2010, **114**, 10160–10170.

43 M. Haouas, L. Lakiss, C. Martineau, J. El Fallah, V. Valtchev and F. Taulelle, *Microporous Mesoporous Mater.*, 2014, **198**, 35–44.

44 L. I. N. Tomé, J. F. B. Pereira, R. D. Rogers, M. G. Freire, J. R. B. Gomes and J. A. P. Coutinho, *Phys. Chem. Chem. Phys.*, 2014, **16**, 2271.

Faraday Discussions

PAPER

Amorphous calcium phosphate phase-mediated crystal nucleation kinetics and pathway†

Shuqin Jiang,[ab] Haihua Pan,*[a] Yan Chen,[b] Xurong Xu[b] and Ruikang Tang[ab]

Received 7th November 2014, Accepted 28th January 2015
DOI: 10.1039/c4fd00212a

Generally, a solution nucleation model is used to study biomineralization kinetics. However, we found that the amorphous calcium phosphate (ACP)-mediated hydroxyapatite (HAP) nucleation in simulated body fluids (SBF) had a different profile from the linear relationship between $\ln J$ and $\ln^{-2} S$ (J, nucleation rate; S, supersaturation). This behaviour was alternatively explained by a developed heterogeneous nucleation theory, which indicated that HAP was nucleated at the ACP–solution interface *via* a polymorph transformation. Based upon this new model, we demonstrated experimentally that the embedded polymer molecules inside ACP were inert on HAP nucleation kinetics; rather, the polymers adsorbed on ACP surface could inhibit HAP nucleation from ACP. It further confirmed the heterogeneous nucleation pathway of HAP on the precursor phase. The present study provides an in-depth understanding of HAP formation for ACP-mediated crystallization.

Introduction

The formation of biominerals is believed to be controlled by the principle of crystallization. During typical crystallization processes, nucleation is the first step which determines crystal phase, size and orientation.[1] Among the inorganic crystal materials, hydroxyapatite (HAP, $Ca_5(PO_4)_3OH$) has attracted much attention because it is the main component of bones and teeth.[2] As a sparingly soluble mineral phase, HAP is actually highly supersaturated in simulated body fluids (SBF)[3] but its direct nucleation from solution is well inhibited *in vivo*. It has been shown that amorphous calcium phosphate (ACP) phase is the intermediate phase for HAP crystallization in both hard tissue formation (*in vivo*) and biomimetic

[a]Centre for Biomaterials & Biopathways, and Department of Chemistry, Zhejiang University, Hangzhou, Zhejiang 310027, China. E-mail: rtang@zju.edu.cn
[b]Qiushi Academy for Advanced Studies, Zhejiang University, Hangzhou, Zhejiang 310027, China. E-mail: panhh@zju.edu.cn

† Electronic supplementary information (ESI) available: Solution compositions, figures and tables. See DOI: 10.1039/c4fd00212a

mineralization (*in vitro*).[4] Recently, the ACP-mediated HAP crystallization rather than direct HAP crystallization has been accepted as the biomineralization pathway in living organisms.[4c] This non-classical pathway exhibits different behaviour from the conventional understanding for solution crystallization. For example, in a previous work we observed an unexpected faster crystallization of HAP at lower pH (means lower supersaturation) in ACP-mediated HAP crystallization.[5] These nucleation data cannot be explained by the established linear relationship between $\ln J_N$ (J_N for nucleation rate) and $\ln^{-2} S$ (S for supersaturation) (J–S relationship), in which lower supersaturation should result in slower nucleation. It is interesting to learn whether this abnormal profile is valid in a broad spectrum of calcium phosphate solutions that are close to physiological fluid. In this work, we examined the nucleation kinetics over a variety of pH, supersaturation and temperature conditions in the absence and presence of polymers. The experimental results suggested that HAP was always nucleated at the ACP–solution interface *via* a polymorph transformation. It was also demonstrated that the solution species were greatly altered in the presence of ACP precursor phase and, thereby, the direct use of supersaturation respective to HAP for the kinetic discussion became inappropriate. We suggested a developed heterogeneous nucleation model, which highlighted the ACP–solution interaction in the kinetic control of HAP nucleation.

Experiments

Materials

The following materials were used: polyacrylate (PAA, M_w = 2.1 kDa, Sigma-Aldrich, USA), poly-aspartate (p-Asp, M_w = 27 kDa, Alamanda Polymers, USA) and poly-glutamate (p-Glu, M_w = 2–15 kDa, GL Biochem Ltd, Shanghai, China). All other chemicals were analytical grade and purchased from Aladdin Reagent (Shanghai, China) unless specially described. The solutions were filtered through 0.22 μm Millipore membranes prior to use.

HAP crystallization

Calcium solutions (Ca) contained $CaCl_2$ and $MgCl_2$. Phosphate solutions (P) contained Na_2HPO_4, Na_2SO_4, NaCl, KCl, *N*-2-hydroxyethylpiperazine-*N*′-2-ethane sulfonic acid (HEPES, GenomBioMed Technology Inc., Hangzhou, China). SBF solutions were prepared by a rapid mixing of equal volumes of the Ca and P solutions with designated concentrations, and solution compositions were close to body fluids (Table S1, ESI†). After mixing, the pH curve of the reaction solution was monitored by a PHSJ-3F pH meter with an E-201-C composite electrode (Leici Instrument, Shanghai, China). The induction time, t_i, for HAP nucleation was determined by the abrupt change in pH curves as described in a previous work.[6] Since the ACP were formed immediately (within few seconds) after the mixing, the induction times for ACP nucleation were not investigated in the current work. Each experiment was repeated at least four times and the relative standard deviations for induction time were around 8%, which were marked as the error bars in the plots. The typical reproducibility of pH curves are shown in Fig. S1, ESI.†

Polymer additives

PAA, p-Asp, and p-Glu were used as prototypes for non-collagenous proteins (NCPs) in biomineralization. There were two protocols for the introduction of polymer additives (stock solution: 10 mg mL^{-1}) into the mineralization reaction: (i) pre-mixing: 20 μL was added into 5 mL of P solution prior to mixing with 5 mL Ca (as explained later, in this protocol the polymer molecules were mainly embedded inside the resultant ACP). (ii) post-mixing: 20 μL polymer stock solution was added into the ACP suspension after the mixing of 5 mL P and 5 mL Ca for 10 min (as explained later, the polymer molecules were mainly absorbed on the ACP surface). All the experiments were conducted at 37.0 ± 0.5 °C, pH = 7.4 ± 0.02, and the final concentration of the polymer additive was 20 μg mL^{-1} (solution compositions are listed in Table S2, ESI†).

Characterizations

During the reaction, 10 mL suspension was withdrawn and centrifuged at 10 000 g for 2 min (Luxian, GL-22M, China), and washed with ethanol. The obtained mineral phase was characterized by FITR (Shimadzu Iraffinity-1) and X-ray diffraction (XRD, Rigaku D/MAX-2550pc, Cu Kα radiation $\lambda = 1.54$ Å). For TEM/SAED (Hitachi HT-7700 or JEOL 2100F) examination, the specimens were obtained by paddling a copper grid in suspension, and washing this with water and ethanol, before drying by lamp light. The zeta potentials of precipitate were determined using a Zetasizer Nano S (Malvern).

The calcium and phosphate concentrations in the solution were measured using ICP-AES (Thermo Fisher iCAP 6300). For this purpose, aliquots were taken from the supernatant solution periodically, filtered with a syringe filter with a pore size of 25 nm to remove the ACP precipitates, and then immediately diluted with 5% hydrochloric acid.

Solution chemistry calculation

The supersaturations of the solutions were calculated using VMINTEQ 3.0.[7a] Davies' approximation of the Debye–Hückel equation was used for calculating the activity coefficients (parameter $b = 0.3$ is applied). The solubility product, K_{sp}, of HAP ($pK_{sp} = 58.33$, MINTEQ database: NIST 46.7), and ACP ($pK_{sp} = 25.5$ (ref. 7b and c)) were used in the calculation.

J–S relationship

In classical nucleation kinetics, the nucleation rate J_N is given as[8]

$$J_N = K_1 \exp\left[\frac{-16\pi\gamma^3\Omega^2}{3(k_B T)^3(\ln S)^2}\right] = K_1 \exp\left[\frac{K_2}{(\ln S)^2}\right] \tag{1}$$

where K_1 is a pre-exponential factor; k_B is the Boltzmann constant; T is the absolute temperature; γ is the interfacial energy between the crystals and the mother phase; Ω is the volume of the growth units and S is the supersaturation. Eqn (1) shows the curve of $\ln J_N$ *vs.* $\ln^{-2} S$ should exhibit a linear relationship.

The supersaturations with respect to HAP and ACP are written as S_{HAP} and S_{ACP} respectively, which are given by,

$$S_{\text{HAP}} = \frac{\{\text{Ca}^{2+}\}^5\{\text{PO}_4{}^{3-}\}^3\{\text{OH}^-\}}{K_{\text{sp,HAP}}} \tag{2}$$

$$S_{\text{ACP}} = \frac{\{\text{Ca}^{2+}\}^3\{\text{PO}_4{}^{3-}\}^2}{K_{\text{sp,ACP}}} \tag{3}$$

where {X}, is the activity of species X and $K_{\text{sp,X}}$ is the solubility product of X. To measure the nucleation rate, one of the most common methods is to measure the induction time of nucleation at different supersaturations. In practice, J_N is also defined as,[8]

$$J_N = 1/(t_i V) \tag{4}$$

in which V is the volume of the system and t_i is the induction time.

Results and discussion

Nucleation kinetics

Generally, crystal nucleation from solutions should obey the linear J–S relationship between $\ln J_N$ and $\ln^{-2} S$. However, such a relationship was not found in the current study of HAP nucleation in SBF. There were two different liner regions in the plot of $\ln J_N$–$\ln^{-2} S$ in the current study (Fig. 1a). At first glance, this kind of bimodal nucleation profile had been reported as the transition from homogeneous nucleation to heterogeneous nucleation[9] or from structurally matched nucleation to structurally mismatched heterogeneous nucleation.[10] However, in both cases, a smaller slope (low nucleation rate) should be expected in the lower supersaturation region, which was in disagreement with the observed data in our experiments (Fig. 1b). This unexpected phenomenon could be observed over a broad spectrum of calcium phosphate solutions at various temperatures (Fig. 2a) and pH (Fig. 2b) in our studies. The previously established J–S relationship could not explain this HAP nucleation kinetic profile in SBF.

ACP-mediated crystallization pathway

The crystallization of HAP in SBF is monitored by the solution pH drop (Fig. 3a). Typically, the pH curve can be divided into three different stages: I, induction

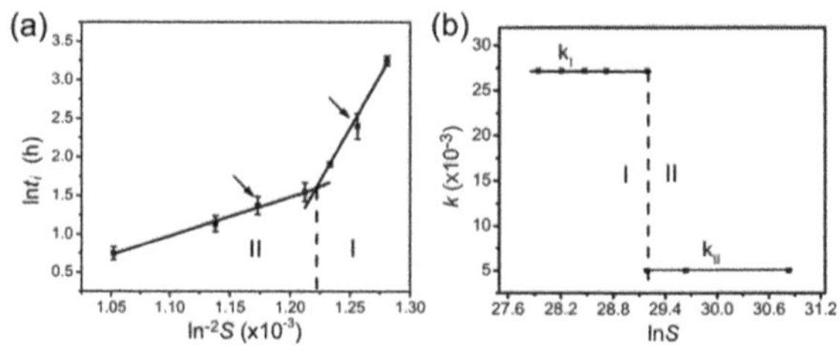

Fig. 1 (a) Plot of $\ln t_i$ *vs.* $\ln^{-2} S$ for HAP nucleation at 25 °C, pH = 7.4. The nucleation kinetics can be divided into two regions, named I and II respectively. Solutions and precipitation used for further characterization were indicated by arrows. (b) The slope is higher than expected for low supersaturation.

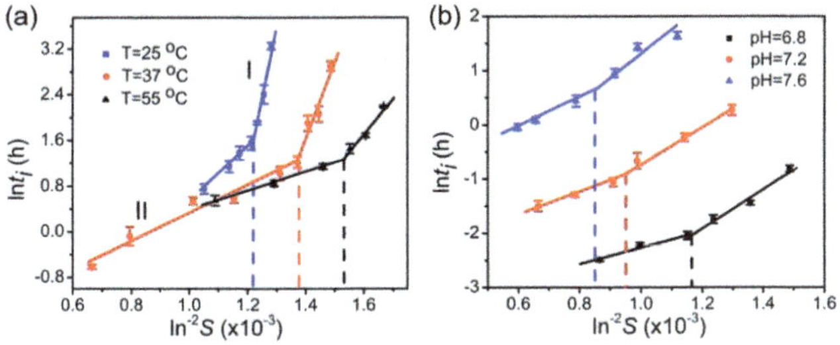

Fig. 2 Plot of ln t_i *vs.* $\ln^{-2} S$ for ACP-mediated HAP nucleation at different temperatures (a) and pH (b).

period; II, HAP crystallization stage; III, post crystallization stage.[5,6] The mineral phase at each stage was *ex situ* characterized by FTIR (Fig. 3b), XRD (Fig. 3c) and TEM (Fig. 3d–f). In terms of the conventional understanding, the supersaturated solution is in a metastable state and there is no solid precipitation in the induction period (stage I). Actually, we noted that the precursor solid phase, ACP, was already precipitated in this stage. The formation and growth or ripening of ACP particles was observed (*cf.* Fig. 3d and S2 in ESI†) at large scale. The broad FTIR absorption bands around 1055 cm^{-1} (phosphate ν_3 vibrations) and 570 cm^{-1} (phosphate ν_4 bending) indicated the amorphous state of calcium phosphate. The

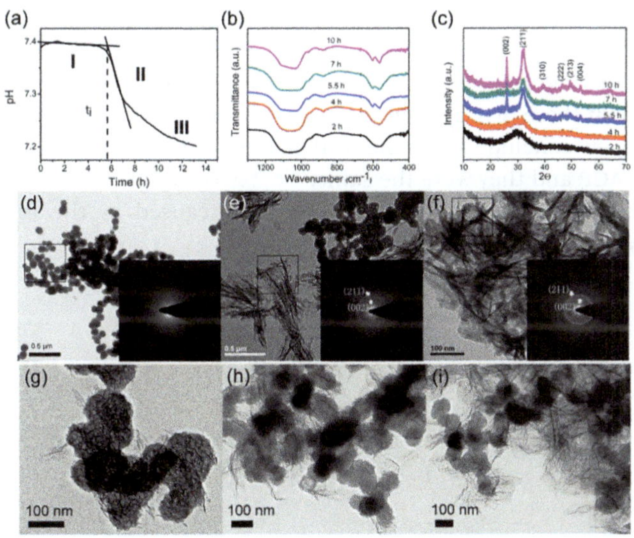

Fig. 3 Crystal nucleation kinetics and crystallization pathway. (a) The representative pH curve of calcium phosphate solutions (ln $S = 28.47$, pH = 7.4, $T = 25$ °C, see Table S1† for details). The crystallization process could be divided into three stages. I, induction period; II, crystallization stage; III, post crystallization stage. The FTIR spectra (b) and XRD patterns (c) of precipitations at stage I, II and III respectively. At 2 h and 4 h, we see ACP; at 5.5 h and 7 h, we find weakly crystallized HAP; at 10 h, we observe crystallized HAP. (d–f) TEM images and SAED patterns (inset) show the typical morphologies and phases of (d) stage I (ACP), (e) II (ACP, HAP) and (f) III (HAP), respectively. (g–i) TEM images of mineral at the induction time (5.5 h).

ACP phase was also confirmed by the broad XRD diffraction peak at about 30° (2θ). The splitting of phosphate ν_3 vibration and ν_4 bending absorption bands in FTIR (Fig. 3b), and the appearance of sharp XRD diffraction peaks (Fig. 3c) indicated the formation HAP at stage II. Under TEM, the newly formed platelet-like crystallites were seen mainly on the existing ACP aggregates (Fig. 3e). This followed the heterogeneous nucleation behaviour of the crystallites on the precursors (Fig. 3f). The electron diffraction patterns showed that the crystallites were HAP (Fig. 3e inset). For the specimens collected at the phase transition period (at the induction time point in the pH curve, *cf.* Fig. 3a), outgrowths of platelet-like nuclei on the ACP surface were frequently observed (Fig. 3g–i). These evidences suggested the scenario of an ACP-mediated crystallization pathway: (1) a primary formation of ACP; (2) polymorphic transformation to HAP at ACP–solution interface.

Accordingly, the nucleation kinetics data could be divided into two regions (I, II in Fig. 1b). Previously, Eanes *et al.* suggested two types of ACP, which were called discoidal-ACP and spheroidal-ACP.[7d] Therefore, it might be reasonable to suggest that the two kinds nucleation behaviours were attributed to the formation of different types of ACP in the experimental system. However, no significant difference between the resulting ACP in these two regions was detected using FTIR (Fig. 4a), XRD (Fig. 4b), and TEM (Fig. 4c–f). In addition, the ICP-AES study showed that the Ca/P ratios of the resultant ACP were 1.54 ± 0.05 and 1.54 ± 0.04 in the regions of I and II (*cf.* Fig. 1a), respectively. We also examined their solubility products, pK_{sp}, which were 25.87 ± 0.04 and 25.96 ± 0.08 for the ACP samples obtained from the regions I and II, respectively (see Tables S3 and S4 in ESI†). Moreover, the zeta potentials of these ACPs were also exactly the same (*cf.* Fig. S3†). In fact, all applied characterization methods could not distinguish one ACP from the other for these regions. As the morphology of these ACP particles was always spherule (Fig. 4c and e) without any change, we concluded that the ACP phases formed at region I and II were spheroidal-ACP and they were the same.[7b,7d,11] Accordingly, the different nucleation behaviours for region I and II were not caused by the formation of different types of ACP.

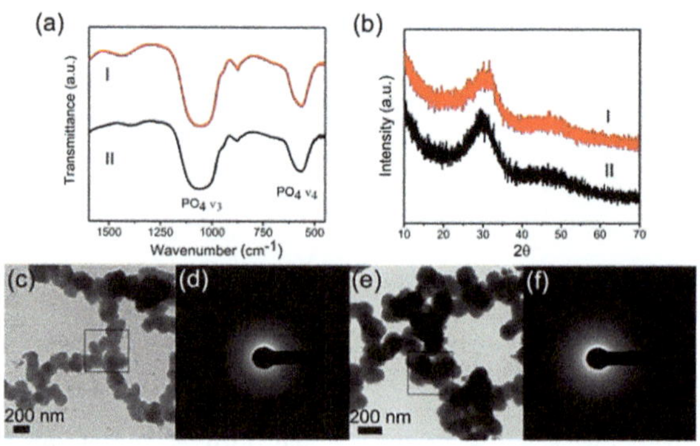

Fig. 4 Characterizations of ACP in region I and II (see Fig. 1) (a) FTIR spectra, (b) XRD patterns, (c and e) TEM images and (d and f) SAED patterns. There was no obvious difference between these ACPs.

ACP-mediated HAP nucleation kinetics

Since the HAP was precipitated *via* an ACP-mediated pathway, the direct nucleation of HAP from solutions is not valid in the kinetic discussion. Previously, we have developed a heterogeneous nucleation model for ACP-mediated HAP nucleation, and the nucleation rate of HAP, $J_{N,hetero}$, is given as,[5]

$$J_{\mathrm{N,hetero}} = KC_{\mathrm{ACP}}\{\mathrm{Ca}\}_{\mathrm{eff}}^{2} \tag{5}$$

$$K = K_1 \exp\left(\frac{-\Delta G^{*}_{\mathrm{react}}}{k_{\mathrm{B}}T}\right)\exp\left(\frac{-\Delta G^{*}_{\mathrm{nucl,hetero}}}{k_{\mathrm{B}}T}\right)f \tag{6}$$

$$\ln J_{\mathrm{N,hetero}} = \ln C_{\mathrm{ACP}}\{\mathrm{Ca}\}_{\mathrm{eff}}^{2} - \left[\frac{\Delta G^{*}_{\mathrm{react}} + \Delta G^{*}_{\mathrm{nucl,hetero}}}{k_{\mathrm{B}}T} - \ln K_2\right] \tag{7}$$

$$K_2 = K_1 f \tag{8}$$

where K_1 is the pre-exponential factor; ΔG^{*}_{react} the chemical reaction barrier for the incorporation of free calcium into HAP nuclei; $\Delta G^{*}_{nucl,hetero}$, the nucleation barrier for heterogeneous nucleation of HAP from ACP; f, the converting factor from surface area to the amount of ACP (*i.e.* $A_{particles} = fC_{ACP}$). C_{ACP}, the amount of ACP formed; and $\{Ca\}_{eff}$ is the effective activity of calcium ions (after the precipitation of ACP). Due to the precipitation of ACP, all solutions become saturated with ACP in spite of their different supersaturations at the original solution states. Therefore, K could be treated as a constant in the ACP-mediated HAP crystallization. Thus, there should be a linear relationship between $\ln J_{N,hetero}$ and $\ln(\{Ca\}_{eff}^{2}C_{ACP})$ (*cf.* eqn (7); the slope was fixed as 1.0 according to eqn (5); C_{ACP} was obtained by a solution chemistry calculation, see Tables S1 and S5 in the ESI† for the details) and it was named as the *J*–*C* relationship. We found that this nucleation rule could be used to fit the nucleation kinetics well for a wide spectrum of calcium phosphate solutions even at different temperatures (Fig. 5a) and pH (Fig. 5b).

Eqn (7) shows that the nucleation barriers ($\Delta G^{*}_{react} + \Delta G^{*}_{nucl,hetero}$) are correlated to the intercept of the *J*–*C* relationship. By measuring the intercepts of *J*–*C* curves at different temperatures, the values of $\Delta G^{*}_{react} + \Delta G^{*}_{nucl,hetero}$ could be estimated. In our experiments, the energy of $\Delta G^{*}_{react} + \Delta G^{*}_{nucl,hetero}$ was about

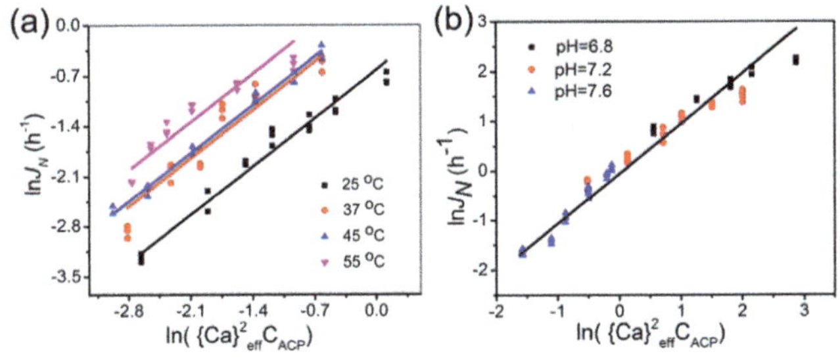

Fig. 5 Nucleation rate $\ln J_N$ (h^{-1}) *vs.* $\ln(\{Ca\}_{eff}^{2}C_{ACP})$ for calcium phosphate solutions at different (a) temperature and (b) pH. All kinetics data obey a *J*–*C* relationship.

13.6 k_BT (*cf.* Fig. 6), which was much smaller than that of direct HAP nucleation from the solution, which was >150 k_BT.[10b,12] It followed that the nucleation barrier of the ACP-mediated HAP crystallization pathway was ten-fold lower than its direct nucleation from solutions. Therefore, the former pathway might be more preferred for HAP formation in SBF solutions.

Free calcium ions and HAP nucleation

In the ACP-mediated formation pathway, the polymorphic transformation from ACP to HAP is the key to generate HAP. Stoichiometrically, the Ca/P ratio of the ACP precursor (1.5) is less than that of the HAP crystallites (1.67). So, during the polymorphic transformation of ACP to HAP, additional calcium ions should be required from the solution as detailed in the following chemical equation:

$$3Ca_3(PO_4)_2 + Ca^{2+} + 2H_2O \rightarrow 2Ca_5(PO_4)_3OH + 2H^+ \quad (9)$$

According to eqn (7), the nucleation rate of HAP from the ACP precursor should be dependent on the effective activity of calcium and the amount of ACP. With the increasing of the initial solution supersaturation, more ACP would be precipitated (*cf.* Table S5†) in the system but the change of the effective activity of calcium was negligible (*cf.* Table S5†) because all solutions became saturated with ACP. Thus it followed that the effective activity of calcium was not sensitive to the change of initial (apparent) supersaturation (*cf.* Table S5†) when pH and temperature were fixed. However, we found that the effective calcium activity was related to the solution pH (Table S6†). In the solution, $\{PO_4\}$ was extremely sensitive to pH due to the association and disassociation balance, {Ca} could be also in turn affected by pH due to the constant of the activity product of $\{Ca\}^3\{PO_4\}^2$ in the system. In order to exclude the effect from different ACP amounts, we examined the HAP nucleation kinetics in the solutions with a fixed amount of ACP (C_{ACP} = 1.648 mM). By slightly decreasing the solution pH from 7.6 to 6.8 (Table S6†), HAP always nucleated faster at higher levels of $\{Ca\}_{eff}$ (Fig. 7). This supported the idea that the reaction of free calcium ions in solution with ACP was a key factor to control the heterogeneous nucleation of HAP.

Polymer additives and nucleation kinetics

In biomineralization and biomimetic mineralization, organic additives play paramount roles in the regulation of crystallization. The inhibitory effect of PAA

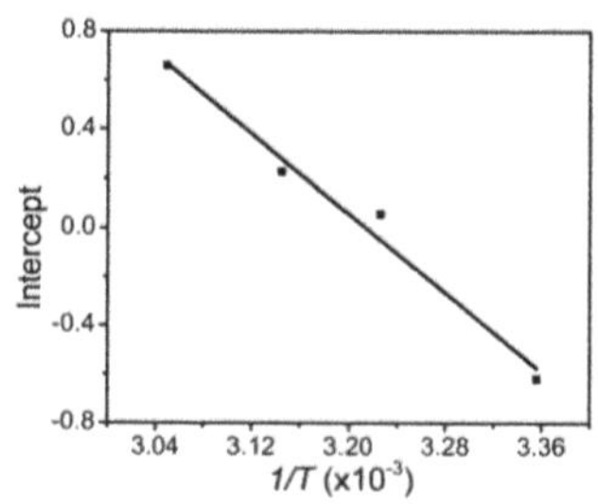

Fig. 6 Plot of intercepts of the *J*–*C* relationship *vs.* 1/*T*. The slope of this plot is the nucleation barrier (*cf.* eqn (7)).

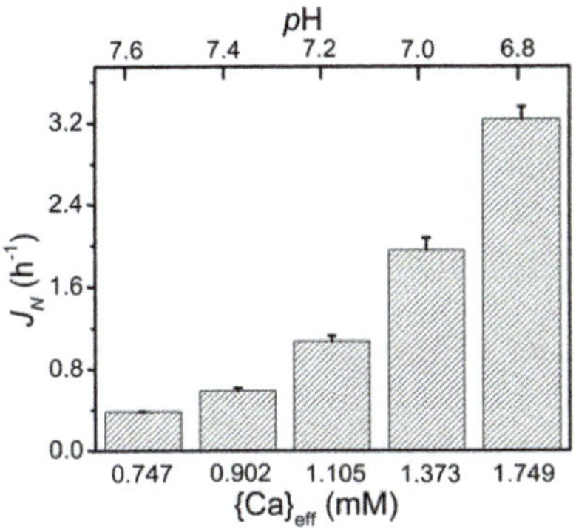

Fig. 7 Plot of J_N (h^{-1}) *vs.* $\{Ca\}_{eff}$ (mM) at 37 °C with fixed C_{ACP} (1.648 mM, relative error is less then ± 0.8%), and $\{Ca\}_{eff}$ was sensitive to the changing of pH. The nucleation rate was increased with the increasing of $\{Ca\}_{eff}$.

on crystallization has been widely reported.[13] However, in this work we found that the inhibition is highly dependant upon the protocols for the introduction of PAA. When a small amount of PAA (20 μg mL^{-1}) was introduced before the formation of ACP (the pre-mixing protocol), there was no obvious inhibiting effect for HAP nucleation (Fig. 8a). But, if the addition of PAA occurred after the formation of ACP (post-mixing protocol), the induction time of HAP nucleation was significantly prolonged. It demonstrated the strong inhibition effect of the later addition of PAA. Similar profiles were also found in the case of p-Asp and p-Glu (Fig. S4†). It was interesting to note that TEM observations did not show any obvious change in the morphology and size of ACP by using the different protocols (*cf.* Fig. S5†). As the compositions for both protocols were exactly the same, $\{Ca\}_{eff}^{2}C_{ACP}$ should be the same as well (the effect of 20 μg mL^{-1} PAA on the free calcium concentration reduction was negligible). By using eqn (5) and the

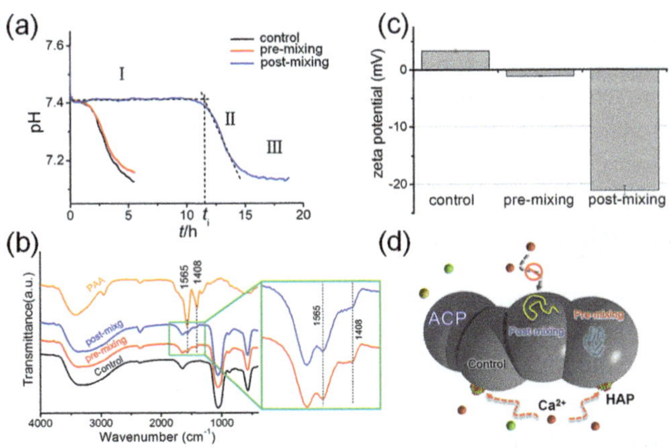

Fig. 8 (a) Representative pH curves of HAP crystallization in the presence (pre-mixing and post-mixing) and absence (control) of PAA. (b) FTIR spectra and (c) zeta potentials of precipitates in different protocols at 10 min. (d) The scheme of ACP-mediated HAP crystallization.

ACP-mediated crystallization, the factor K should be affected by PAA. This implies that the chemical processe for the interaction of free calcium ions with the ACP surface would be affected. We assumed that in the pre-mixing protocol, PAA molecules were mainly embedded inside ACP; in post-mixing protocols, PAA were adsorbed on the ACP surface (shown in the scheme in Fig. 8d). Thus, the ACP–solution interface was almost the same as that of the control experiment (without PAA) using the pre-mixing protocol. Therefore, the introduction of the PAA additive could not affect the HAP nucleation rate on ACP. But for the post-mixing protocol, the adsorbed PAA might affect the direct reaction between the free calcium and ACP, and subsequently inhibit HAP nucleation. This phenomenon would provide an alternative novel pathway to control HAP nucleation in addition to ACP amount or free calcium concentration.

The further characterizations of ACP particles revealed the possible locations of PAA. Firstly, FTIR spectra indicated the presence of PAA for ACP particles obtained by both protocols. The appearance of additional bands around 1565 and 1408 cm^{-1} (for the anti-symmetrical and symmetrical vibrations of the carboxyl group for PAA) supported the co-precipitation of PAA with ACP (Fig. 8b). Secondly, a distinct zeta potential value has been observed for ACP particles by using the different protocols (Fig. 8c). The zeta potential of ACP particles in pre-mixing protocol (-1.12 ± 0.08 mV; mean ± s.d., $n = 5$) was close to that of the control (3.34 ± 0.22 mV, $n = 5$). Thus it follows that the PAA molecules were almost embedded inside ACP particles. In contrast, in the case of post-mixing, the zeta potential of ACP became highly negative (-11.70 ± 0.39 mV, $n = 5$), implying the adsorption of the negatively charged PAA on the ACP surface.

Conclusions

It has been demonstrated that the well-established J–S relationship for solution nucleation cannot be used to match the ACP-mediated HAP nucleation kinetics. In the presence of the ACP precursor phase, HAP is heterogeneously nucleated at the ACP–solution interface *via* a polymorph transformation. Accordingly, the ACP phase and free calcium ions (Fig. 8d) are the keys to control the nucleation kinetics of HAP from ACP. By using polymers, we further confirm that the protocols of mineralization experiments can also affect HAP nucleation in the ACP-mediated crystallization. This work provides an alternative kinetic understanding of HAP formation under physiological conditions and highlights the phase transition pathway by using the developed heterogeneous nucleation model.

Acknowledgements

We thank James J. De Yoreo for helpful discussions. This work was supported by the Fundamental Research Funds for the Central Universities, National Natural Science Foundation of China (91127003).

Notes and references

1 F. C. Meldrum and H. Cölfen, *Chem. Rev.*, 2008, **108**, 4332.
2 (*a*) E. Beniash, *Wiley Interdiscip. Rev.: Nanomed. Nanobiotechnol.*, 2011, **3**, 47; (*b*) R. Z. Legeros, *Adv. Dent. Res.*, 1988, **2**, 164.

3 (*a*) A. Oyane, H. M. Kim, T. Furuya, T. Koukubo, T. Miyazaki and T. Nakamura, *J. Biomed. Mater. Res., Part A*, 2003, **65**, 188; (*b*) T. Kokubo and H. Takadama, *Biomaterials*, 2006, **27**, 2907.

4 (*a*) J. D. Termine and A. S. Posner, *Calcif. Tissue Res.*, 1967, **1**, 8; (*b*) A. L. Boskey, *J. Dent. Res.*, 1997, **76**, 1433; (*c*) J. Mahamid, B. Aichmayer, E. Shimoni, *et al.*, *Proc. Natl. Acad. Sci. U. S. A.*, 2010, **107**, 6316; (*d*) L. B. Gower, *Chem. Rev.*, 2008, **108**, 4551.

5 S. Jiang, Y. Chen and H. Pan, *Phys. Chem. Chem. Phys.*, 2013, **15**, 12530.

6 (*a*) Y. Chen, W. Gu, H. Pan, S. Jiang and R. Tang, *CrystEngComm*, 2014, **16**, 1864; (*b*) P. B. Y. Ofir, R. G. Lippman, N. Garti and H. Füredi-Milhofer, *Cryst. Growth Des.*, 2003, **4**, 177; (*c*) X. Yang, B. Xie, L. Wang, Y. Qin, Z. J. Henneman and G. H. Nancollas, *CrystEngComm*, 2011, **13**, 1153; (*d*) B. Xie, T. J. Halter, B. M. Borah and G. H. Nancollas, *Cryst. Growth Des.*, 2014, **14**, 1659.

7 (*a*) J. P. Gustafsson, *Visual MINTEQ, Version 3.0*, Stockholm, available from http://www.lwr.kth.se./English/OurSoftware/vminteq, 2010; (*b*) M. R. Christoffersen, J. Christoffersen and W. Kibalczyc, *J. Cryst. Growth*, 1990, **106**, 349; (*c*) S. V. Dorozhkin and M. Epple, *Angew. Chem., Int. Ed.*, 2002, **41**, 3130; (*d*) E. D. Eanes, J. D. Termine and M. U. Nylen, *Calcif. Tissue Int.*, 1973, **12**, 143.

8 X. Y. Liu and K. Sato, in *Advances in Crystal Growth Research*, Elsevier, Amsterdam, 2001, p. 42.

9 (*a*) O. Söhnel and J. M. Mullin, *J. Cryst. Growth*, 1978, **44**, 377; (*b*) J. M. Mullin and H. M. Ang, *Faraday Discuss. Chem. Soc.*, 1976, **61**, 141.

10 (*a*) X. Y. Liu and S. W. Lim, *J. Am. Chem. Soc.*, 2002, **125**, 888; (*b*) H. Jiang, X. Y. Liu, G. Zhang and Y. Li, *J. Biol. Chem.*, 2005, **280**, 42061.

11 (*a*) C. Combes and C. Rey, *Acta Biomater.*, 2010, **6**, 3362; (*b*) S. V. Dorozhkin, *Acta Biomater.*, 2010, **6**, 4457.

12 W. J. E. M. Habraken, J. Tao, L. J. Brylka, *et al.*, *Nat. Commun.*, 2013, **4**, 1507.

13 (*a*) S. Schweizer and A. Taubert, *Macromol. Biosci.*, 2007, **7**, 1085; (*b*) Z. Amjad, *Langmuir*, 1989, **5**, 1222; (*c*) A. Bigi, E. Boanini, G. Cojazzi, G. Falini and S. Panzavolta, *Cryst. Growth Des.*, 2001, **1**, 239; (*d*) V. Ball, M. Michel, F. Boulmedais, J. Hemmerle, Y. Hailel, P. Schaaf and J. C. Voegel, *Cryst. Growth Des.*, 2006, **6**, 327.

Faraday Discussions

PAPER

Nucleation rate analysis of methane hydrate from molecular dynamics simulations

Daisuke Yuhara,[a] Brian C. Barnes,†[b] Donguk Suh,[a] Brandon C. Knott,[c] Gregg T. Beckham,[c] Kenji Yasuoka,[a] David T. Wu[bd] and Amadeu K. Sum*[b]

Received 11th November 2014, Accepted 6th January 2015
DOI: 10.1039/c4fd00219a

Clathrate hydrates are solid crystalline structures most commonly formed from solutions that have nucleated to form a mixed solid composed of water and gas. Understanding the mechanism of clathrate hydrate nucleation is essential to grasp the fundamental chemistry of these complex structures and their applications. Molecular dynamics (MD) simulation is an ideal method to study nucleation at the molecular level because the size of the critical nucleus and formation rate occur on the nano scale. Various analysis methods for nucleation have been developed through MD to analyze nucleation. In particular, the mean first-passage time (MFPT) and survival probability (SP) methods have proven to be effective in procuring the nucleation rate and critical nucleus size for monatomic systems. This study assesses the MFPT and SP methods, previously used for monatomic systems, when applied to analyzing clathrate hydrate nucleation. Because clathrate hydrate nucleation is relatively difficult to observe in MD simulations (due to its high free energy barrier), these methods have yet to be applied to clathrate hydrate systems. In this study, we have analyzed the nucleation rate and critical nucleus size of methane hydrate using MFPT and SP methods from data generated by MD simulations at 255 K and 50 MPa. MFPT was modified for clathrate hydrate from the original version by adding the maximum likelihood estimate and growth effect term. The nucleation rates calculated by MFPT and SP methods are within 5%, and the critical nucleus size estimated by the MFPT method was 50% higher, than values obtained through other more rigorous but computationally expensive estimates. These methods can also be extended to the analysis of other clathrate hydrates.

[a]*Department of Mechanical Engineering, Keio University, Yokohama, Japan*

[b]*Center for Hydrate Research, Chemical & Biological Engineering Department, Colorado School of Mines, Golden, Colorado, USA. E-mail: asum@mines.edu*

[c]*National Renewable Energy Laboratory, Golden, Colorado, USA*

[d]*Chemistry Department, Colorado School of Mines, Golden, Colorado, USA*

† Present address: Weapons and Materials Research Directorate, U.S. Army Research Laboratory, Aberdeen Proving Ground, Maryland, USA.

1 Introduction

Clathrate hydrates are ice-like structures in which guest molecules are trapped inside water cages connected by a hydrogen-bonded network.[1] The formation of clathrate hydrates of natural gases, also called gas hydrates, is a serious problem in the flow assurance of oil/gas flow lines. Inhibiting and mitigating hydrate formation in flow lines are crucial for the safety and the reduction of operating costs of maintaining flow lines.[1,2] Gas hydrates are also abundant in the seafloor and have attracted attention as a potential energy resource.[3] Depressurization of these hydrate deposits is projected to be an efficient method to produce natural gas from the hydrates in the sea.[4] To depressurize the hydrate reservoir, it is important to induce dissociation to release the gas from the hydrates, which will consequently generate thermodynamically favorable conditions for hydrate reformation. Therefore, understanding hydrate formation is required to develop efficient energy production strategies. At the most fundamental level, the mechanism of hydrate formation must be understood, since the incipient hydrate crystallization phenomenon can be controlled in flow lines during the production of oil/gas and for the gathering of gas from hydrate reservoirs. Moreover, a thorough understanding of the hydrate formation process will improve the efficiency of other challenging hydrate applications such as gas transport,[2,3] hydrogen storage,[5] and capture/sequestration of carbon dioxide.[6]

Nucleation is the first stage in hydrate formation, which is generally an activated process where small clusters of the new phase are formed from a supersaturated phase. A free energy barrier exists between the two phases, and small clusters which exceed the critical size located at the peak of the barrier become the nucleus of the new phase.[7] Hydrate growth can be observed experimentally,[8] however, hydrate nucleation cannot be observed, as it is a molecular level process that involves nuclei size and time scales that are on the nano scale. Molecular dynamics (MD) simulation has proven to be an invaluable tool to observe hydrate nucleation.[9–11] There have been numerous studies on hydrate nucleation,[12–14] growth,[15–17] and stability[18–20] by simulations. Previous studies have thoroughly examined the formation mechanisms of various hydrate structures.[21–26] Order parameters that characterize the hydrate structures formed in the nucleation process have also been developed.[27–31] Nonetheless, the analysis of hydrate nucleation is still in the early stages, and there are remarkably scarce reports in which the nucleation rate and the critical nucleus size are discussed.[32,33]

Nucleation is a stochastic process and commonly considered to be a "rare event" in molecular simulations.[34] Direct molecular simulations typically require long calculation times (100s nanoseconds to microseconds) to observe hydrate nucleation. But more importantly, the critical nucleus size of hydrate formation, which is essential in calculating the nucleation rate, cannot be found beforehand. Although there are many difficulties in the simulations of hydrate nucleation, Walsh *et al.* succeeded in observing hydrate nucleation in microsecond simulations.[12] Recently, Barnes *et al.* performed a number of similar simulations using high-performance computing and analyzed the nucleation rate and critical nucleus size.[35]

Nucleation rate analyses using MD simulations have actually been well documented for monatomic systems forming droplets or bubbles through homogeneous and heterogeneous nucleation.[36–45] This study is initiated from the assumption that the various methods to analyze the nucleation rate and critical nucleus size may be applicable for other nucleation processes, and therefore we have used these methods to analyze hydrate nucleation.[46–49] In this study, we analyzed the simulation results from Barnes *et al.*[35] and calculated the nucleation rate and critical nucleus size of the methane hydrates by implementing methods originally applied to analyze vapor-to-liquid nucleation. This work introduces a methodology for analyzing methane hydrate nucleation but the findings are easily applicable to other complex clathrate hydrate structures.

2 Methods

2.1 Simulation details

Barnes *et al.* performed 200 MD simulations of methane hydrate nucleation at $T=$ 255 K, P = 50 MPa.[35] Simulation cells included 2944 water and 512 methane molecules with a cylindrical water/methane interface. The initial configuration was created by melting 64 unit cells of structure I hydrate at T = 550 K.[12] Water and methane models were TIP4P/Ice[50] and the unified atom model,[51] respectively, and the Lorentz–Berthelot combining rules were used to calculate water–methane interactions. GROMACS 4.5 and 4.6 were used to perform the simulations, using the Verlet leapfrog algorithm for time integration.[52] The isobaric–isothermal ensemble was applied, where the pressure was controlled by the Parrinello–Rahman barostat[53] with a time constant of 4 ps, and the temperature was controlled by the Nosé–Hoover thermostat[54,55] with a time constant of 2 ps. The SETTLE algorithm was used to constrain the bond lengths and angles of water molecules.[56] A time step of 2 fs was used, with short-range interactions truncated at 1 nm, and the long-range electrostatic interactions were calculated by the particle mesh Ewald algorithm[57,58] with a Fourier spacing of 0.12 nm. Over 90% of the simulations were performed for a minimum of 3 μs.

2.2 Analysis of nucleation

An activated process typified by nucleation is the formation of small embryos of a new phase from an existing metastable phase by overcoming a free energy barrier.[7] The rate of which the critical-sized embryos are formed during the nucleation process is the nucleation rate. The nucleation process is considered to be a diffusion process over a barrier in an internal space, which can be described by the Fokker–Planck equation in terms of a variable X that is an internal coordinate or degree of freedom,

$$\frac{\partial \rho}{\partial t} = \frac{\partial}{\partial X}\left(D(X,t)\frac{\partial \rho}{\partial X} + A(X,t)\rho\right), \tag{1}$$

where ρ is the density in the internal space, and D and A are the diffusion and drift coefficients in this space, respectively. In the nucleation process, eqn (1) can be expressed as a function of a variable n, which is the number of molecules constituting a cluster, instead of X. In this case, ρ indicates the cluster size distribution, and eqn (1) can be expressed in the form of a continuity equation in the internal space,

$$\frac{\partial}{\partial t}\rho(n,t) = -\frac{\partial}{\partial n}j(n,t), \tag{2}$$

where $\rho(n,t)$ is the number density of clusters containing n monomers at time t and $j(n,t)$ is the formation rate of size n clusters in the system. At steady state, $\partial\rho(n,t)/\partial t = 0$ and $j(n,t)$ is constant, so the nucleation rate at steady state can be defined as $J = j/V$, where V is the volume of the system. The nucleation rate can also be described by means of integrating eqn (2) with respect to n, $\partial N(n_t,t)/\partial t = j(n_t,t)\cdot N(n_t,t)$ is the total number of clusters larger than a threshold size of n_t. The nucleation rate obtained by this expression is

$$J(n_t,t) = \frac{1}{V}\frac{\partial N(n_t,t)}{\partial t}. \tag{3}$$

In this equation, the nucleation rate is described by a time derivative of the number of clusters greater than n_t per unit volume. This rate should be constant at steady state and independent of n_t, so n_t must be greater than the critical size. In this work, the nucleation rate was evaluated by this definition based on eqn (3).

2.2.1 Definition of clusters. The definition of cluster size is essential in the analysis of nucleation. Recent hydrate nucleation studies indicate that amorphous structures are initially formed and these may anneal and crystallize.[59,60] Nucleation and growth of hydrates are usually characterized by order parameters that distinguish the phase of water molecules, structure of cages, and coordinates of guest molecules.[27–31,61] However, it is difficult to identify clusters that have formed initially using these order parameters due to the complex molecular geometries. Most recently, Barnes *et al.* developed an order parameter, called the Mutually Coordinated Guest (MCG) order parameter, that identifies guest molecules separated by water clusters consisting of five or six-member rings.[62] This order parameter (OP) can estimate the cluster size of methane hydrates sufficiently, so the MCG-1 OP from the MCG algorithm was used as the cluster size n in this work.

2.2.2 Mean first-passage time method (MFPT). This method is suitable to analyze nucleation in MD simulations. The method makes the simplifying assumption that the free energy barrier is high, and only the height and curvature of the barrier control the nucleation kinetics, and that growth is much faster than nucleation.[48,49] MFPT does not require large systems for the simulation, but rather demands numerous replications of small nucleating systems to analyze the statistics of the phenomenon. Full reaction coordinate analysis of methane hydrate nucleation requires enormous calculation time, so MFPT is advantageous in analyzing this event because it only uses the trajectories of direct MD simulations of nucleating systems.

The mean first-passage time in the case of nucleation is defined as the mean time $\tau(n)$ that the largest cluster in each system requires to reach or exceed a threshold size n for the first time. If the free energy barrier is high enough, MFPT as a function of the cluster size n can be described by a specific sigmoidal curve

$$\tau(n) = \frac{\tau_J}{2}\left[1 + \mathrm{erf}\left(Z\sqrt{\pi}(n - n^*)\right)\right], \tag{4}$$

where τ_J is the nucleation time, erf(x) is the error function, and Z is the Zeldovich factor. A fitting of the simulation results to eqn (4) directly yields the nucleation time τ_J, critical nucleus size n^*, and the Zeldovich factor. In the case of a high free energy barrier, the MFPT curve has a clear plateau at the end of the sigmoidal shape, which indicates the nucleation time τ_J. The critical nucleus size is considered to be the size at time $\tau_J/2$ in the MFPT curve because the probability of the transition at the top of the barrier is 50%. The nucleation rate J is calculated from the nucleation time τ_J as

$$J = \frac{1}{V\tau_J}. \tag{5}$$

2.2.3 Survival probability. The formation of a large enough post-critical cluster in the presence of a high free energy barrier is a random event, which follows the Poisson distribution

$$P_k(t) = \frac{(t/\tau_n)^k \mathrm{e}^{-t/\tau_n}}{k!}, \tag{6}$$

where $P_k(t)$ is the probability that k clusters larger than n appear at time t, and τ_n is the average time when these clusters appear. In the case of $k = 0$, eqn (6) shows the survival probability (SP) $P_0(t)$,[8,47,49] *i.e.*, the probability that there are no clusters larger than n after a time t in the system, to be

$$P_0(t) = \mathrm{e}^{-t/\tau_n}. \tag{7}$$

On the other hand, the nucleation probability $P_{\mathrm{nuc}}(t)$ is given by

$$P_{\mathrm{nuc}}(t) = \frac{N_{\mathrm{nuc}}}{N_{\mathrm{all}}}, \tag{8}$$

where N_{nuc} is the number of systems in which nucleation is observed and N_{all} is the number of all simulation systems. From eqn (7) and (8), the SP $P_{\mathrm{surv}}(t)$ becomes

$$P_{\mathrm{surv}}(t) = \exp\left(-\frac{t - t_0}{\tau_J}\right) = 1 - P_{\mathrm{nuc}}(t), \tag{9}$$

where t_0 is the fastest time a cluster takes to reach the threshold size n among all simulations. If the free energy barrier is high, these probabilities do not depend on the threshold size of cluster n. A fitting of the simulation results to eqn (9) yields the nucleation time τ_J, and the nucleation rate can be obtained with eqn (5).

3 Results and discussion

Nucleation and growth of methane hydrate was observed in 46 out of the 200 replications of the MD simulation trajectories. We analyze the nucleation rate from this set using the MFPT and SP, and the critical nucleus size is also calculated from MFPT.

3.1 Mean first-passage time

The MFPT of each cluster size n (up to $n = 400$, corresponding to near complete solidification of the system) was calculated and plotted in Fig. 1. The statistics for small n are better than those for large n. This is mainly due to the observation

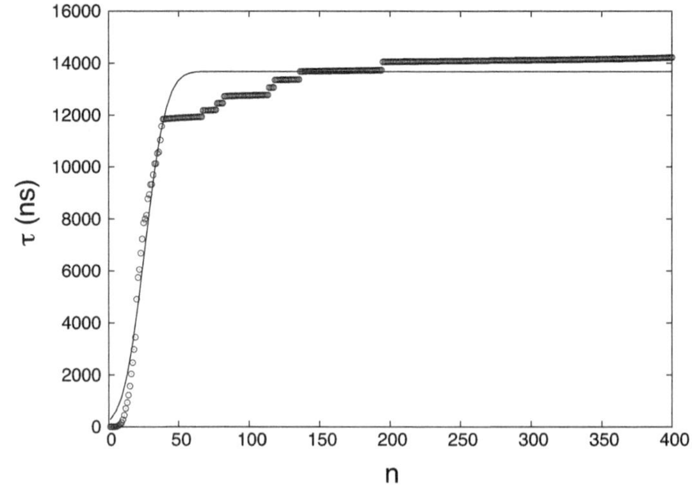

Fig. 1 The MFPT curve obtained from 200 nucleation trajectories. τ is the MFPT and n is the threshold size of a cluster. The symbols are mean first-passage times calculated from the results of the simulations and the solid line is a fitting of the MFPT data.

time for small clusters being relatively short (even though this "short" time consumed massive computational resources) compared to those needed to fully observe complete nucleation. The ratio of nucleated to non-nucleated systems most likely influences the calculation of MFPT, so τ is calculated by a supplemental equation, which is based on the maximum likelihood estimate,[32,63]

$$\tau = \frac{\sum_{i=1}^{N_R} \tau_i + \sum_{k=1}^{N_{N_R}} \tau_k}{N_R}, \tag{10}$$

where N_R (= 46) is the number of reacted (clusters reaching or exceeding a particular size n) trajectories and N_{N_R} (= 154) is the number of remaining trajectories. τ_i is the nucleation time, and τ_k is the total simulation time for non-nucleating trajectories. The nucleation time τ_J and critical nucleus size n^* are obtained by fitting the data to eqn (4). The nucleation rate J is calculated by plugging the obtained values into eqn (5) with $V = 84.9\ \text{nm}^3$ (estimated by considering the volume of the aqueous phase in a non-nucleating trajectory – volume of bulk methane phase subtracted from total volume). The nucleation rate from Fig. 1 is $J = 8.61 \times 10^{23}\ \text{cm}^{-3}\ \text{s}^{-1}$, and the critical nucleus size is $n^* = 25.9$.

The fitting of the MFPT curve from eqn (4) is poor in the range of $40 \leq n \leq 120$. Under the thermodynamic conditions considered, the free energy barrier is relatively high, so it is difficult to observe nucleation (only about a quarter of replications nucleated). In systems where nucleation solely occurs, the plateau is expected to appear right after the sigmoidal curve.[48] Compared to nucleation, the time scale of growth is commonly much shorter, but in our case, there seems to be an overlap of the time scales in the nucleation and growth processes, which can affect the MFPT results. Therefore, a modification to the MFPT curve is applied as suggested by Yi *et al.* by adding an additional term to eqn (4) to account for finite growth rates of post-critical clusters, with[64]

$$\tau(n) = 0.5\tau_J\left[1 + \mathrm{erf}\left(Z\sqrt{\pi}(n - n^*)\right)\right] + G^{-1}(n - n^*)H(n - n^*), \tag{11}$$

where G is the growth rate and $H(x)$ is the Heaviside function. The Heaviside function becomes effective when a cluster size exceeds the critical nucleus size, *i.e.*, when growth occurs after nucleation. The smooth approximation of the Heaviside function can also be presented by an error function, transforming eqn (11) into

$$\tau(n) = 0.5\tau_J\left[1 + \mathrm{erf}\left(Z\sqrt{\pi}(n - n^*)\right)\right] + 0.5G^{-1}(n - n^*)[1 + \mathrm{erf}(C(n - n^*))], \tag{12}$$

where C is required to be a large positive number. Fig. 2 shows the fitting given by eqn (12). The plateau in Fig. 1 becomes a positive slope that incorporates the growth contribution that is absent in Fig. 1. The newly calculated nucleation rate is $J = 9.43 \times 10^{23}$ cm^{-3} s^{-1}, and the critical nucleus size becomes $n^* = 23.8$. The variation in the nucleation rate is greater than that of the critical nucleus size without growth.

To analyze how the number of replications influences the results when using the MFPT method, a sensitivity analysis is conducted. Subsets of replications are randomly taken from the total of 200 and the nucleation times are estimated by the maximum likelihood method. The MFPT curve is generated from the averages of the subsets. Table 1 shows the averaged results of MFPT for different numbers of subsets. The results show that the differences in both the nucleation rate and critical nucleus are within 20%, which is an insignificant variation for typical nucleation studies.[46,48,49] Despite the variation, a greater number of replications are expected to produce more accurate results.

As shown in Table 1, from all 200 replications, the nucleation rate and critical nucleus size are estimated as 9.43×10^{23} cm^{-3} s^{-1} and $n^* = 23.8$, respectively. Nucleation in the monatomic systems can be observed relatively quickly, so the

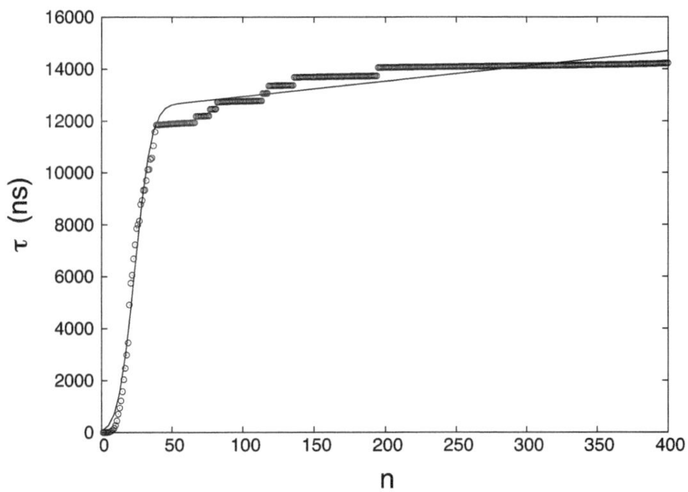

Fig. 2 The MFPT curve considering nucleation and growth simultaneously. The symbols are mean first-passage times calculated from the results of the simulations and the solid line is a fitting of MFPT data points.

Table 1 The MFPT results calculated from the various number of simulations. N_{all} is the number of total simulation, J is the nucleation rate, and n^* is the critical nucleus size

N_{all}	20	50	100	150	200
J (10^{23} cm^{-3} s^{-1})	8.19	10.28	10.53	8.49	9.43
n^*	21.8	24.9	23.6	23.1	23.8

MFPT curve has a clear plateau right after the sigmoidal shape.[48,49] On the other hand, hydrate nucleation required a long calculation time due to a high free energy barrier, so nucleation may not be observed for many systems within the limited calculation time. Furthermore, hydrate formation is a complex process and growth just after the nucleation process tend to occur at the same time scale of nucleation. Therefore, MFPT for methane hydrate was modified from its original function using the maximum likelihood estimate by adding the growth term. The nucleation rate result is within 5% of that obtained by Barnes *et al.*,[35] though the critical nucleus size is larger by 50%. The critical nucleus size obtained by Barnes *et al.* was estimated from a p_b histogram test and this seems to have a higher variance.[35,65–67] However, MFPT is a function of the nucleus size n only, so there are limitations in fully capturing the complexities of hydrate nucleation. The source of the discrepancy in the critical nucleus size also may be from the aforementioned comparable nucleation and growth time scales as well as the diffusion kinetics depending on the non-classical shape of the free energy curve.[35] Although further studies are required to clarify the discrepancy in the nucleus size, the agreement in the nucleation rate confirms the effectiveness of the MFPT analysis method for a qualitative estimate without further free energy computations.

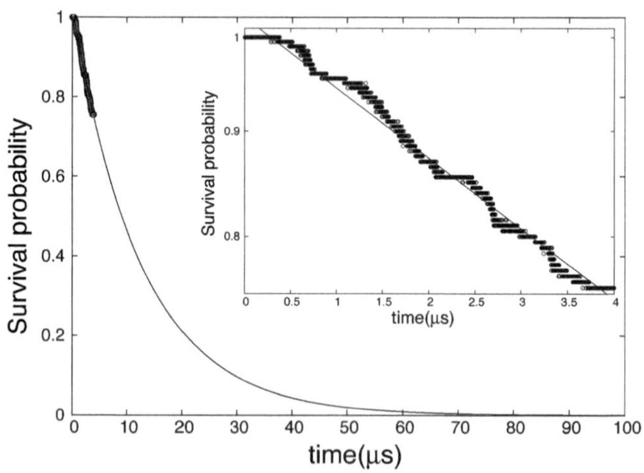

Fig. 3 Survival probability in the case of threshold $n_t = 40$. Points are calculated from the results of the simulations and the solid line is the fitting of the data points. The inset is a single logarithmic graph in the range of SP calculated from the results of the simulations ($0.76 \le P_{surv}(t) \le 1.00$).

3.2 Survival probability

The survival probability is calculated by eqn (9) and the results are shown in Fig. 3. Unlike completely nucleated systems ($n_t = 400$) that had 46 out of 200 occurrences, $n_t = 40$ had 49, and the range of the survival probability is within $0.76 \leq P_{surv}(t) \leq 1.00$. The systems that nucleated are found to grow and not dissociate. If the threshold size is increased to $n_t = 80$, the number of nuclei converges to 46. In other words, three replications did not have a nucleus that reached 80 in size within the simulation time. The variation in the threshold size is analyzed and is found to have little effect on the results of SP. The fitting line in Fig. 3 illustrates SP at a longer time scale, and presents the asymptotic tendency at infinite time. Based on Fig. 3, one can see that if the simulations of this study are performed to around 70 μs, the SP will become close to zero, meaning nearly all systems will have likely nucleated. The inset in Fig. 3 portrays a single logarithmic transformation from the original SP plot, and the nucleation rate is calculated from the slope of this graph. The nucleation rate is $J = 9.31 \times 10^{23}$ cm^{-3} s^{-1}, which is close to that obtained by MFPT.

SP is commonly independent of the threshold cluster size in the case that the free energy barrier is high. We verified this by changing the threshold cluster size, and Table 2 contains the results. The difference among the values is around 5%, which is remarkable considering the deviation in typical nucleation studies. The nucleation rate will slightly decrease with an increasing threshold size because the number of nucleated systems decreases. The rate at $n_t = 20$, however, was the smallest. The reason for this is from the fact that 20 is around the critical nucleus size obtained by Barnes *et al.* and MFPT, hence the inconsistency.[49] Therefore, when using the SP method, the threshold size should be set by an adequately larger number than the critical nucleus size ($n_t \geq 40$ was sufficient in this study). On the other hand, if the threshold size is too large, the analysis range may spill over to the growth stage rather than staying purely in the nucleation stage, which is the primary target.

The nucleation rate calculated by Walsh *et al.* is from the maximum likelihood estimate.[31] They performed the six replications at $T = 250$ K, $P = 50$ MPa and the induction times were verified by analyzing the evolution of the global F_4[61] order parameter as well as the appearance of cages larger than seven through the MD trajectories. The nucleation rate from Walsh *et al.* is around $J_{sim} = 5.00 \times 10^{24}$ cm^{-3} s^{-1}. Their results are one order of magnitude greater than our results from MFPT and SP. This difference may be attributed from the small number of samples and the definition of induction time. On the other hand, the nucleation rate calculated by Barnes *et al.* is from the maximum likelihood estimate based on prior knowledge of the critical nucleus size.[32,35] The nucleation rate from Barnes *et al.* is $J_{sim} = 9.07 \times 10^{23}$ cm^{-3} s^{-1}, which is very close to our results from MFPT and SP. The biggest advantage of this study compared to previous studies is that

Table 2 The SP of the various threshold cluster sizes. n_t is the threshold size, N_{nuc} is the number of nucleated systems, and J is the nucleation rate

n_t	10	20	40	60	80	100
N_{nuc}	54	49	49	47	46	46
J (10^{23} cm^{-3} s^{-1})	9.52	8.98	9.31	9.25	9.10	9.10

the MFPT and SP methods can generate the nucleation rate directly from the MD trajectories.

4 Conclusion

The nucleation of methane hydrate is a phenomenon that is complicated to analyze at the molecular level due to a high free energy barrier. Recently, this process has been observed by molecular dynamics simulations using high performance computing. In this study, we analyzed the nucleation rate and critical nucleus size of methane hydrate using MFPT and SP, and assessed the applicability of these methods for methane hydrate nucleation analysis. In this study, MFPT was modified from its original function using the maximum likelihood estimate and adding the growth term. The nucleation rates obtained by MFPT and SP are in good agreement (within 20%) and these results are also close to the rate Barnes *et al.* estimated using direct calculations based on the maximum likelihood estimate. MFPT and SP are convenient methods to calculate the methane hydrate nucleation rate since they only require simulation trajectories. The critical nucleus size was also calculated by MFPT, which was found to be larger than that of Barnes *et al.* based on the p_b histogram test. This difference is likely to come from the non-classical shape of the free energy surface and the methane hydrate system having comparable nucleation and growth time scales, which will influence the critical nucleus size generated from MFPT. However, a modification for growth for MFPT slightly improves the fit of the simple fitting function. Though the MFPT and SP methods are only applied for the estimation of the nucleation rate and critical nucleus size of methane hydrate, the methods can easily be extended to the analysis of other clathrate hydrates.

Acknowledgements

This work was supported (in part) by MEXT Grant-in-Aid for the "Program for Leading Graduate Schools". This project was partially funded by the U.S. National Science Foundation (CHE-1125235). B.C.K. and G.T.B. thank the NREL Directors Fellowship Program for Funding. High-performance computing resources were provided by the National Renewable Energy Laboratory, Golden Energy Computing Organization (Colorado School of Mines), and Sandia National Laboratories.

References

1 E. D. Sloan Jr and C. Koh, *Clathrate Hydrates of Natural Gases*, CRC press, Boca Raton, 2007.
2 E. D. Sloan, *Nature*, 2003, **426**, 353–363.
3 R. Boswell, *Science*, 2009, **325**, 957–958.
4 M. R. Walsh, S. H. Hancock, S. J. Wilson, S. L. Patil, G. J. Moridis, R. Boswell, T. S. Collett, C. A. Koh and E. D. Sloan, *Energy Econ.*, 2009, **31**, 815–823.
5 L. J. Florusse, C. J. Peters, J. Schoonman, K. C. Hester, C. A. Koh, S. F. Dec, K. N. Marsh and E. D. Sloan, *Science*, 2004, **306**, 469–471.
6 Y. Park, D.-Y. Kim, J.-W. Lee, D.-G. Huh, K.-P. Park, J. Lee and H. Lee, *Proc. Natl. Acad. Sci. U. S. A.*, 2006, **103**, 12690–12694.

7 D. Reguera, J. Rubı and A. Pérez-Madrid, *Phys. A*, 1998, **259**, 10–23.
8 R. Ohmura, M. Ogawa, K. Yasuoka and Y. H. Mori, *J. Phys. Chem. B*, 2003, **107**, 5289–5293.
9 C. P. Ribeiro Jr and P. L. Lage, *Chem. Eng. Sci.*, 2008, **63**, 2007–2034.
10 J. A. Ripmeester and S. Alavi, *ChemPhysChem*, 2010, **11**, 978–980.
11 B. C. Barnes and A. K. Sum, *Curr. Opin. Chem. Eng.*, 2013, **2**, 1–7.
12 M. R. Walsh, C. A. Koh, E. D. Sloan, A. K. Sum and D. T. Wu, *Science*, 2009, **326**, 1095–1098.
13 S. Sarupria and P. G. Debenedetti, *J. Phys. Chem. Lett.*, 2012, **3**, 2942–2947.
14 N. J. English, M. Lauricella and S. Meloni, *J. Chem. Phys.*, 2014, **140**, 204714.
15 S. Liang and P. G. Kusalik, *Chem. Phys. Lett.*, 2010, **494**, 123–133.
16 D. Yuhara, M. Hiratsuka, D. Takaiwa and K. Yasuoka, *Mol. Simul.*, 2014, 1–5.
17 D. Bai, B. Liu, G. Chen, X. Zhang and W. Wang, *AIChE J.*, 2013, **59**, 2621–2629.
18 N. J. English and G. M. Phelan, *J. Chem. Phys.*, 2009, **131**, 074704.
19 S. A. Bagherzadeh, P. Englezos, S. Alavi and J. A. Ripmeester, *J. Phys. Chem. B*, 2012, **116**, 3188–3197.
20 M. Conde and C. Vega, *J. Chem. Phys.*, 2013, **138**, 056101.
21 M. Matsumoto, *J. Phys. Chem. Lett.*, 2010, **1**, 1552–1556.
22 L. C. Jacobson and V. Molinero, *J. Am. Chem. Soc.*, 2011, **133**, 6458–6463.
23 A. H. Nguyen, L. C. Jacobson and V. Molinero, *J. Phys. Chem. C*, 2012, **116**, 19828–19838.
24 F. Jimenez-Angeles and A. Firoozabadi, *J. Phys. Chem. C*, 2014, **118**, 11310–11318.
25 F. Jimenez-Angeles and A. Firoozabadi, *J. Phys. Chem. C*, 2014, **118**, 26041–26048.
26 G. S. Smirnov and V. V. Stegailov, *J. Chem. Phys.*, 2012, **136**, 044523.
27 R. W. Hawtin, D. Quigley and P. M. Rodger, *Phys. Chem. Chem. Phys.*, 2008, **10**, 4853–4864.
28 M. R. Walsh, J. D. Rainey, P. G. Lafond, D.-H. Park, G. T. Beckham, M. D. Jones, K.-H. Lee, C. A. Koh, E. D. Sloan, D. T. Wu, *et al.*, *Phys. Chem. Chem. Phys.*, 2011, **13**, 19951–19959.
29 L. C. Jacobson, W. Hujo and V. Molinero, *J. Phys. Chem. B*, 2009, **113**, 10298–10307.
30 M. Matsumoto, A. Baba and I. Ohmine, *J. Chem. Phys.*, 2007, **127**, 134504.
31 S. N. Chakraborty, E. M. Grzelak, B. C. Barnes, D. T. Wu and A. K. Sum, *J. Phys. Chem. C*, 2012, **116**, 20040–20046.
32 M. R. Walsh, G. T. Beckham, C. A. Koh, E. D. Sloan, D. T. Wu and A. K. Sum, *J. Phys. Chem. C*, 2011, **115**, 21241–21248.
33 B. C. Knott, V. Molinero, M. F. Doherty and B. Peters, *J. Am. Chem. Soc.*, 2012, **134**, 19544–19547.
34 D. Frenkel and B. Smit, *Understanding Molecular Simulation*, 2nd edn, Academic, San Diego, 2002.
35 B. C. Barnes, B. C. Knott, G. T. Beckham, D. T. Wu and A. K. Sum, *J. Phys. Chem. B*, 2014, **118**, 13236–13243.
36 K. Yasuoka and M. Matsumoto, *J. Chem. Phys.*, 1998, **109**, 8463–8470.
37 T. Kinjo, K. Ohguchi, K. Yasuoka and M. Matsumoto, *Comput. Mater. Sci.*, 1999, **14**, 138–141.
38 P. R. Ten Wolde and D. Frenkel, *J. Chem. Phys.*, 1998, **109**, 9901–9918.
39 K. Yasuoka and X. C. Zeng, *J. Chem. Phys.*, 2007, **126**, 124320.

40 H. Matsubara, T. Koishi, T. Ebisuzaki and K. Yasuoka, *J. Chem. Phys.*, 2007, **127**, 214507.
41 B. R. Novak, E. J. Maginn and M. J. McCready, *Phys. Rev. B: Condens. Matter Mater. Phys.*, 2007, **75**, 085413.
42 J. Wedekind, D. Reguera and R. Strey, *J. Chem. Phys.*, 2006, **125**, 214505.
43 J. Wedekind and D. Reguera, *J. Phys. Chem. B*, 2008, **112**, 11060–11063.
44 D. Suh and K. Yasuoka, *J. Phys. Chem. B*, 2011, **115**, 10631–10645.
45 D. Suh and K. Yasuoka, *J. Phys. Chem. B*, 2012, **116**, 14637–14649.
46 K. Yasuoka and M. Matsumoto, *J. Chem. Phys.*, 1998, **109**, 8451.
47 M. Sekine, K. Yasuoka, T. Kinjo and M. Matsumoto, *Fluid Dynam. Res.*, 2008, **40**, 597–605.
48 J. Wedekind, R. Strey and D. Reguera, *J. Chem. Phys.*, 2007, **126**, 134103.
49 G. Chkonia, J. Wölk, R. Strey, J. Wedekind and D. Reguera, *J. Chem. Phys.*, 2009, **130**, 064505.
50 J. Abascal, E. Sanz, R. G. Fernández and C. Vega, *J. Chem. Phys.*, 2005, **122**, 234511.
51 W. L. Jorgensen, J. D. Madura and C. J. Swenson, *J. Am. Chem. Soc.*, 1984, **106**, 6638–6646.
52 B. Hess, C. Kutzner, D. Van Der Spoel and E. Lindahl, *J. Chem. Theory Comput.*, 2008, **4**, 435–447.
53 M. Parrinello and A. Rahman, *J. Appl. Phys.*, 1981, **52**, 7182–7190.
54 S. Nosé, *J. Chem. Phys.*, 1984, **81**, 511.
55 W. Hoover, *Phys. Rev. A*, 1985, **31**, 1695.
56 S. Miyamoto and P. A. Kollman, *J. Comput. Chem.*, 1992, **13**, 952–962.
57 T. Darden, D. York and L. Pedersen, *J. Chem. Phys.*, 1993, **98**, 10089.
58 U. Essmann, L. Perera, M. Berkowitz, T. Darden, H. Lee and L. Pedersen, *J. Chem. Phys.*, 1995, **103**, 8577.
59 L. C. Jacobson, W. Hujo and V. Molinero, *J. Phys. Chem. B*, 2010, **114**, 13796–13807.
60 L. C. Jacobson, W. Hujo and V. Molinero, *J. Am. Chem. Soc.*, 2010, **132**, 11806–11811.
61 C. Moon, R. Hawtin and P. M. Rodger, *Faraday Discuss.*, 2007, **136**, 367–382.
62 B. C. Barnes, G. T. Beckham, D. T. Wu and A. K. Sum, *J. Chem. Phys.*, 2014, **140**, 164506.
63 D. D. Dunlop and A. C. Tamhane, *Statistics and Data Analysis: From Elementary to Intermediate*, Prentice Hall, 2000.
64 P. Yi, C. R. Locker and G. C. Rutledge, *Macromolecules*, 2013, **46**, 4723–4733.
65 R. Du, V. S. Pande, A. Y. Grosberg, T. Tanaka and E. S. Shakhnovich, *J. Chem. Phys.*, 1998, **108**, 334–350.
66 P. L. Geissler, C. Dellago and D. Chandler, *J. Phys. Chem. B*, 1999, **103**, 3706–3710.
67 P. G. Bolhuis, C. Dellago and D. Chandler, *Proc. Natl. Acad. Sci. U. S. A.*, 2000, **97**, 5877–5882.

Faraday Discussions

PAPER

Formation of new polymorphs without any nucleation step. Desolvation of the rimonabant monohydrate: directional crystallisation concomitant to smooth dehydration†

Baptiste Fours, Yohann Cartigny, Samuel Petit and Gérard Coquerel*

Received 17th November 2014, Accepted 7th January 2015
DOI: 10.1039/c4fd00222a

Rimonabant monohydrate can be dehydrated at 100 °C or above with complete loss of structural information; in this case the amorphous material can lead to nucleation and crystal growth. The water molecules can also be removed by a smooth process below T_g (78 °C) of the anhydrous phase. In that latter process there is a structural filiation between the mother phase and the daughter phase. The solvent molecules escape from the mother structure by using a network of specific channels; the new non-solvated material undergoes a relaxation process similar to a directional crystallization. By this soft mode of desolvation inside a material which has a very limited mobility, the nucleation of a non-solvated material can be avoided. The structural information contained in the mother phase is not used as a template for crystal growth but it is more a progressive rearrangement of the new desolvated material towards the nearest well in energy. Thus, a metastable new polymorph of a non-solvated component can be obtained by: (i) the crystallization of the component as a solvate and (ii) a smooth desolvation at $T < T_g$. Other parameters liable to interfere with that transmission of structural information are discussed.

Introduction

Crystal polymorphism of molecular compounds is usually assumed to be determined by the nucleation step,[1] as soundly deduced in the 90's from puzzling phenomena such as 'disappearing polymorphs'[2] and 'concomitant polymorphs'.[3] However the fundamental mechanisms at a molecular level of nucleation from

Normandie Université, EA 3233 SMS, Crystal Genesis Unit, University of Rouen, F-76821 Mont Saint Aignan, France. E-mail: gerard.coquerel@univ-rouen.fr; Fax: +33 23552 2927; Tel: +33 23552 2927

† Electronic supplementary information (ESI) available: CIF file for the crystal structure of rimonabant monohydrate; DSC curve showing the glass transition for the amorphous form of rimonabant; DVS results of rimonabant forms I and II at 20 °C. CCDC 1033743. For ESI and crystallographic data in CIF or other electronic format see DOI: 10.1039/c4fd00222a

solution remain insufficiently understood and constitute a subject of controversial discussions,[4] mainly related to the difficulty of exploring the size range (and possibly time scale) of nucleation with suitable experimental techniques.[5]

In order to get new insights into nucleation mechanisms, it is therefore of high interest to consider situations in which spontaneous nucleation can be either triggered or inhibited by imposing suitable environmental conditions. In that respect, desolvation processes might constitute an interesting situation since their associated mechanism(s) can involve (or not) a nucleation step after a more or less cooperative departure of solvent molecules from the initial crystal packing.

Almost twenty years ago, a so-called 'unified model for the dehydration of molecular crystals' was proposed, in which four criteria were identified as decisive parameters for cooperative or destructive mechanisms.[6] These views have been reinforced by other consistent classifications based on strictly structural[7] or phenomenological approaches.[8]

In the present paper, it is suggested that a thorough analysis of desolvation behaviours as well as the use of suitable criteria may constitute valuable methodological tools for the study of systems in which crystallisation might occur either without nucleation (by means of a cooperative mechanism involving a structural filiation) or with nucleation (and growth) when operating conditions do not meet pre-defined criteria. This approach, here applied to a monohydrated form of the pharmaceutical ingredient rimonabant (Fig. 1) is a first step toward a putative better control of the nucleation step in the frame of solid–solid transformations, and might be used as a rational strategy for the design of new polymorphs.

Rimonanbant was an API (Active Pharmaceutical Ingredient) developed by Sanofi-Aventis for alcohol and drug withdrawal. After some years this active pharmaceutical ingredient has been itself withdrawn from market. Like many pharmaceutical substances, rimonabant exhibits polymorphism. Two modifications (form I, $T_m = 156.3$ °C and form II, $T_m = 156.8$ °C) have been identified prior to this work.[9] These two polymorphs are very close in energy whatever the temperature so that they appeared almost as isoenergetic varieties. After melting and quenching a glass is obtained with a T_g at 78 °C (mid-point transition – see ESI†). Moreover, like other pharmaceuticals (*e.g.* olanzapine[10]), rimonanbant shows a great ability to form solvates. The following list gives an idea of these

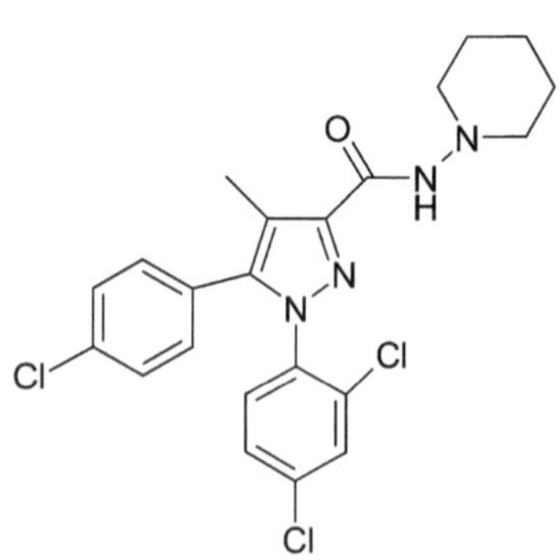

Fig. 1 Developed formula of rimonabant: *N*-piperido-5-(4-chlorophenyl)-1-(2,4-dichlorophenyl)-4-methylpyrazole-3-carboxamide.

solvent molecules able to become a partner of crystallisation: water; methanol, ethanol, 2-methoxyethanol, isopropanol, 1-methyl butanol, DMF, NMP, DMSO, methylcyclohexane (2 polymorphs), dimethylacetamide, 1,4 dioxane. This is by no mean an exhaustive list and only ketones have resisted to give any solvate. This study is focused on the monohydrate only.[11]

Experimental section

Crystalline samples of rimonabant were provided by the Sanofi-Aventis Company, with a chemical purity higher than 99%. The monohydrated form was produced by dissolving 8 g of rimonabant in 40 g of acetone at room temperature. Then, 10 g of water was added dropwise and the monohydrate crystallized out spontaneously. Temperature can be dropped to 5 °C to improve the yield (6.5 g after filtration) and drying to a fixed composition in water 3.7% w/w. Single crystals with a needle-like shape were obtained by slow evaporation at RT of the same acetone–water solution.

X-ray powder diffraction (XRPD)

XRPD analyses were performed using a D8 diffractometer (Bruker, Germany) equipped with a modified goniometer of reverse-geometry ($-\theta/-\theta$)[12] and a LynxEye® detector (Bruker, Germany). Using Cu Kα ($\lambda = 1.54059$ Å) with a tube voltage and amperage set at 40 kV and 40 mA respectively, XRPD analyses were performed with a step of 0.04° (2θ), and a 4 s per step counting time from 3 to 30° (2θ).

Single crystal X-ray diffraction (SC-XRD)

The selected single crystal was stuck on a glass fibre and mounted on the full three-circle goniometer of a Bruker SMART APEX diffractometer equipped with a CCD area detector and operating with Mo $K_{\alpha 1}$ ($\lambda = 0.71071$ Å) as incident beam. The SMART software[13] was used to determine the cell parameters and the orientation matrix. Intensities were integrated and corrected for Lorentz polarization and absorption effects using SAINT software.[13] The structure was solved by direct methods using the SHELX-97 suite of programs,[14] and anisotropic displacement parameters were refined for all non-hydrogen atoms. All hydrogen atoms were located by Fourier-difference synthesis and fixed geometrically according to their environment with a pre-defined isotropic thermal factor.

Thermal analysis (TGA and DSC)

Thermogravimetric-Differential Scanning Calorimetry (TG-DSC) measurements were carried out by using Netzsch STA 449C Jupiter apparatus (Selb, Germany). The purge gas was helium (flow = 40 mL min^{-1}) and the reference material was an empty covered aluminium pan. Enthalpy and temperature calibrations were performed with biphenyl and indium. The samples (5 mg ± 0.05 mg) were weighed in covered pierced aluminium pans, and then placed in the analyser. Analyses were performed in the temperature range 20–150 °C using a 5 K min^{-1} heating rate. After acquisition, the PROTEUS software was used for data processing (v. 4.8.4)

Dynamic vapour sorption (DVS)

Moisture sorption isotherms at 20 °C of the monohydrate and the amorphous form were obtained by using a DVS-1 automated water sorption analyser (Surface Measurement Systems, Alperton, U.K.). About 2 mg of solid were placed in the analyser for each experiment. Mass variations were recorded while relative humidity (R.H.) was successively decreased or increased from 90% to 0% R.H. by adapted steps of R.H. The automated analyser was allowed to proceed with the following step as soon as the mass variation of the sample was less than $5 \times 10^{-4}\%\ min^{-1}$.

Optical microscopy analysis of single crystals

Crystal morphology was observed by optical microscopy using a Hirox high resolution numerical microscope which enables a magnification up to ×7000. The evolution of the observed crystals as a function of temperature was monitored with a CCD camera coupled to the microscope and connected to a computer. The cooling or heating rates 5 K min^{-1} were controlled with a THMS 600 hot-stage setup (Linkham).

Results and discussion

After preparation of the monohydrated form of rimonabant from an acetone–water solution, single crystals of sufficient quality and size could be obtained and the crystal structure was analysed at room temperature (Table 1, structure

Table 1 Crystal data and refinement parameters for rimonabant monohydrate

Formula	$C_{22}H_{21}Cl_3N_4O \cdot H_2O$
Molecular weight (g mol^{-1})	481.79
CSD deposition number	CCDC 1033743
T (K)	296
Wavelength (Å)	0.71073
Crystal dimension (mm)	2 × 0.2 × 0.2
Crystal morphology	Colourless needle
System	Triclinic
Space group	$P\bar{1}$
a (Å) α (°)	7.4245(5) 96.898(1)
b (Å) β (°)	13.2237(8) 96.177(1)
c (Å) γ (°)	24.718(2) 90.664(1)
Z, Z'	4, 2
V (Å^3)	2394.5(3)
d_{calc} (g cm^{-3})	1.336
μ (mm^{-1})	0.408
F_{000}	1000
θ range (°)	1.55 to 26.43
h, k, l ranges	−9/9, −16/15, −30/30
Nb of collected/unique reflections	19 375/9674
Nb of reflections ($I > 2\sigma I$)/parameters	4953/585
Residual electronic density (e Å^{-3})	+0.233/−0.225
R_1, wR_2 ($I > 2\sigma I$)	0.0453, 0.0950
R_1, wR_2 (all data)	0.0928, 0.1072
Goodness of fit	0.829

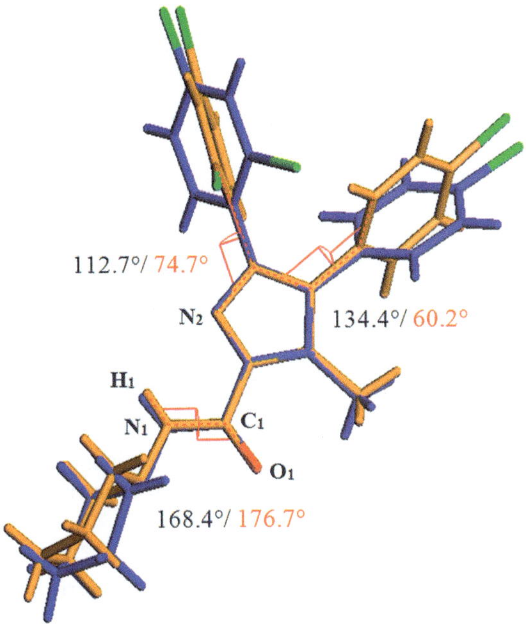

Fig. 2 Comparison between the conformations of the two independent molecules in the crystal structure of rimonabant monohydrate.

deposited in the CSD as CCDC 1033743†). It could be established by X-ray diffraction that the main axis of the elongated single crystals corresponds to the crystallographic *a* axis. The asymmetric unit in the structure contains two molecules with different conformations, as illustrated in Fig. 2. They mainly differ by the orientation of the 4-chlorophenyl and 2,4-dichlorophenyl substituents with reference to the pyrazole ring. Fig. 3 depicts the packing of the monohydrate, with two independent water molecules connected to the independent rimonabant molecules (labelled A and B) *via* their amide groups. When considering only the H-bonds as intermolecular interactions (Table 2), this pattern, as schematized in Fig. 4, induces the existence of Periodic Bond Chains (PBCs) running along the *b* axis, which is in apparent contradiction with the main direction of crystal

Table 2 Geometric parameters of hydrogen bonds in the structure of rimonabant monohydrate (see atom labelling in Fig. 2)

Hydrogen bond	(D···A) (Å)	(D–H···A) (°)	(H···A) (Å)
$N_{1A}–H_{1A}\cdots O_{wA}$	2.89	177.4	2.05
$N_{1B}–H_{1B}\cdots O_{wB}$	2.86	172.7	2.02
$O_{wA}–H_{wA}\cdots O_{1B}$	2.76	167.4	2.01
$O_{wB}–H_{wB}\cdots O_{1A}$	2.65	169.3	1.82
$O_{wB}–H_{wB}\cdots N_{2B}$	2.95	—	—
$O_{wA}–H_{2A}\cdots N_{2A}$	3.11	—	—

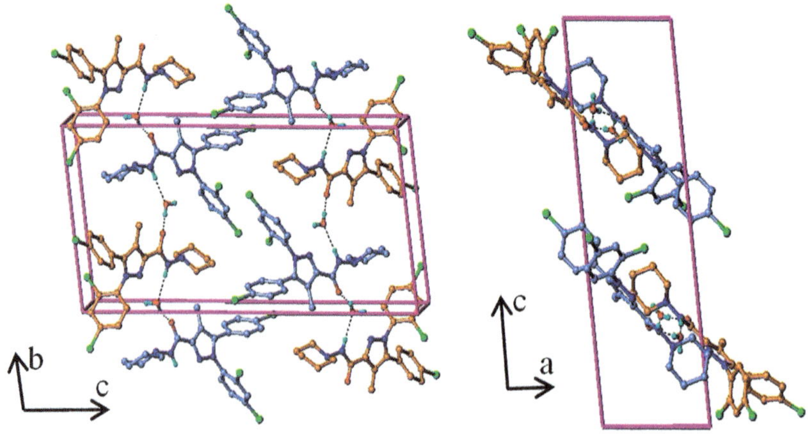

Fig. 3 Crystal packing and H-bond network involving the two independent molecules (A: orange; B: blue) in the monohydrated structure of rimonabant.

development. One can see from Fig. 3 that this stoichiometric hydrate constitutes a channel hydrate according to the definition of Morris.[7]

The thermal behavior of the monohydrate is displayed in Fig. 5. The dehydration starts at *ca.* 40 °C, and the mass loss is completed by 90 °C when heating is operated at 5 K min^{-1}. The mass loss fits well with the expected value for a monohydrate *i.e.* 3.77% w/w. Close to the end of the dehydration, another large endothermic phenomenon takes place. TR-XRPD with similar experimental conditions shows above *circa* 100 °C a flat XRPD pattern leading to the hypothesis that the second phenomenon is actually the melting of the new anhydrous material.

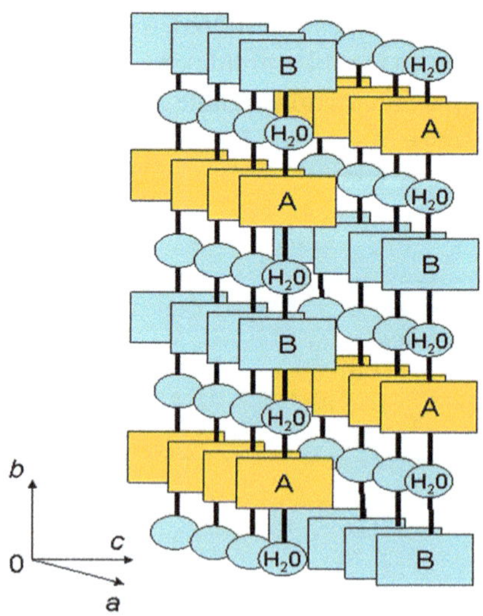

Fig. 4 Schematic representation of rimonabant monohydrate as a stoichiometric channel hydrate.

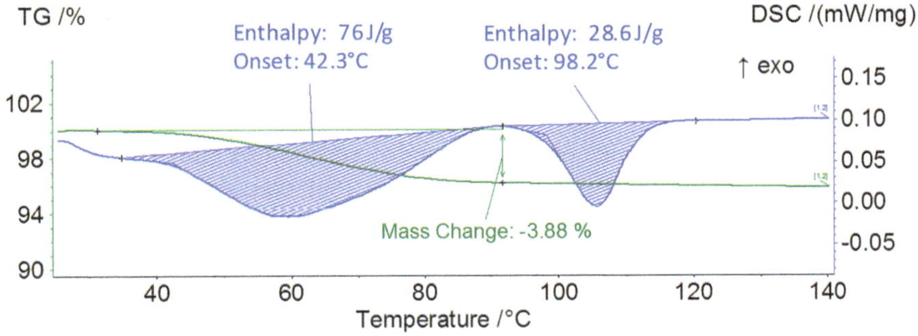

Fig. 5 TGA (green curve) DSC (blue curve) of rimonabant monohydrate (heating rate 5 K min^{-1} under a constant flow of helium).

Softer desolvation conditions were also tested such as: exposure to P_2O_5 in a closed chamber at room temperature (RT), constant flow of dry nitrogen at RT, isotherm at 60 °C (*i.e.*, 18 °C below T_g). Fig. 6 displays at various magnifications the initial state and the final state of the dehydration under soft conditions: isotherm at 60 °C under normal pressure. The initial single crystal is transparent; it shows the morphology of the particle which is elongated along the *a* axis. After a while the crystal becomes translucent. At greater magnifications (×320 and then ×2100) it appears that the initial needle has been segmented, more or less

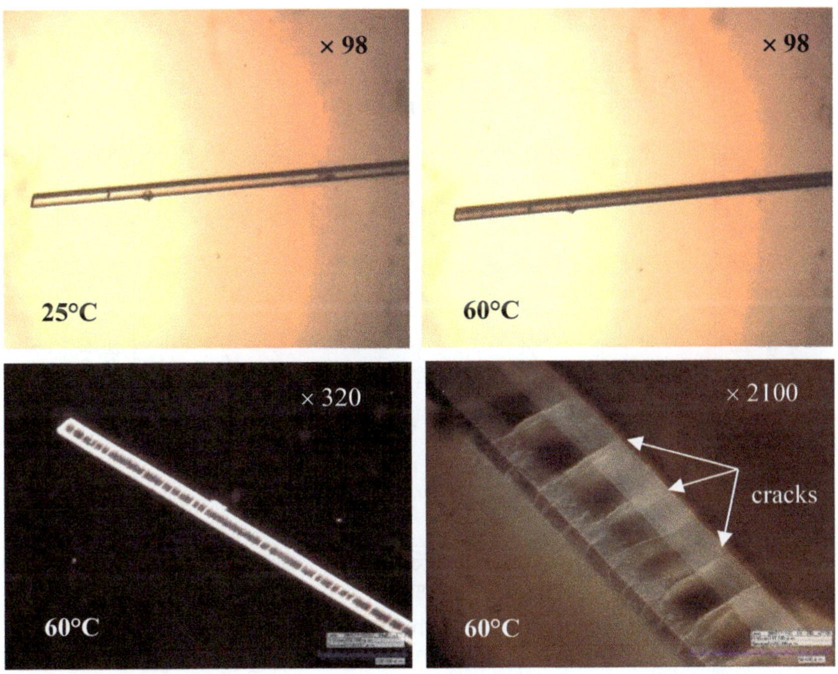

Fig. 6 Optical microscopy photographs showing a single crystal of rimonabant monohydrate undergoing dehydration under smooth conditions (upper part). The macroscopic consequences of dehydration at 60 °C are shown at higher magnification (lower part).

periodically, into small regions of some micrometers long. The channels containing the water molecules being parallel to the main axis of the needle, it is not surprising with this set of kinetic parameters that water molecules could find a short cut.

Fig. 7 shows several cycles of sorption and desorption of moisture at 20 °C (DVS studies) from 0% to 90% Relative Humidity (RH). Starting from the monohydrate at 90% RH, it shows that this phase is stable at 20 °C down to 4% RH. On sorption, a sharp take-off appears at 14% RH demonstrating the reversibility of the dehydration and rehydration when operated under smooth conditions. Even though the mass *versus* RH graph shows two abrupt events with hysteresis, the time-resolved corresponding representation shows that actually, for a small mass

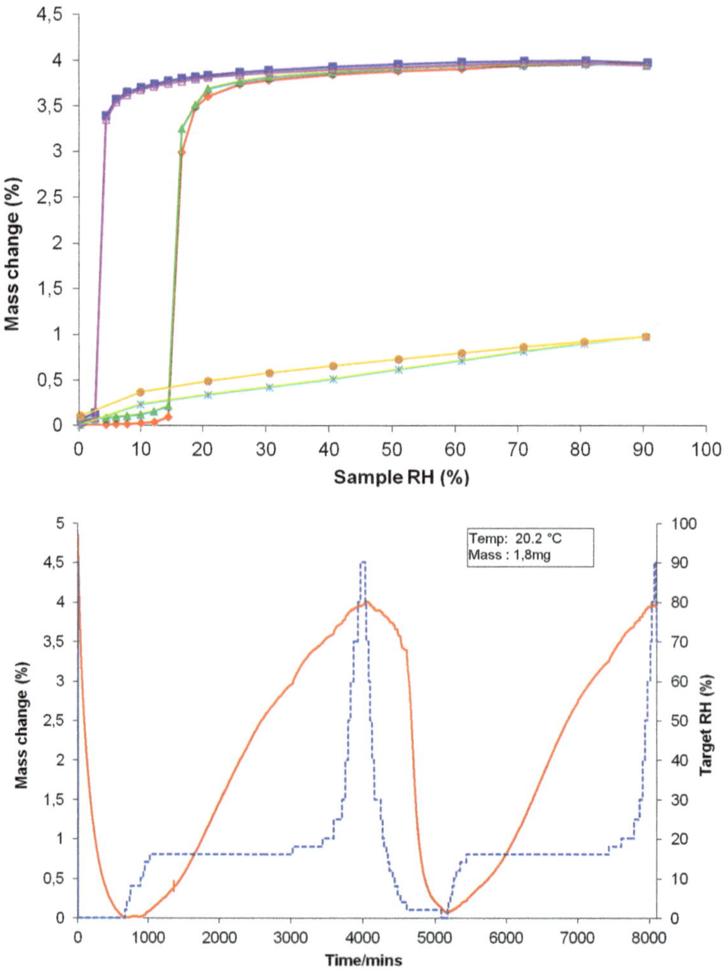

Fig. 7 DVS studies at 20 °C of rimonabant showing the monohydrate–anhydrous equilibria (blue; pink, dark green and red isotherms). In the upper part, the green and orange curves illustrate the absence of evolution (limited to a weak adsorption) for the amorphous form under the same conditions. The lower part shows the time-resolved DVS evolution for successive dehydrations and rehydrations (red: mass variations, dashed blue line: RH variations).

(1.8 mg), the desorption and sorption of water molecules required several thousands of minutes for completion when operated smoothly under a constant flow of nitrogen with a precise RH. By contrast, if these cycles are performed on the amorphous API obtained after a dehydration up to 100 °C and cooling down to RT, no stepped evolution is obtained even after more than 2 × 2000 minutes per cycle. In a more general way, we never observed the spontaneous nucleation of the monohydrate by sorption of humidity on the following starting materials: amorphous state, form I and form II, even after inoculation of the monohydrate at 20 °C (see ESI† for DVS analyses of forms I and II). Rimonabant shows a poor hydrophilicity; there is thus a weak plasticizing effect and this can, at least partly, explain the difficulty of crystallizing the monohydrate by means of amorphous–vapor interactions.

Fig. 8 shows the XRPD patterns of the monohydrate and the anhydrous phase obtained by desolvation at room temperature *via* desorption of water in a closed vial with P_2O_5. Even if the crystallinity of the dehydrated phase has dropped, the filiation of structural information is rather obvious. Thus, it can be concluded that smooth desolvation of this channel hydrate below T_g leads to a crystallized phase with a clear structural filiation with the mother phase, and the situation looks like a case of 'isomorphic desolvate' similar to those studied by Stephenson *et al.*[15] Fig. 9 shows the superimposition of the monohydrate behavior and that of the anhydrous phase resulting from the soft dehydration below T_g. The TGA profile of the anhydrous form is indeed flat and the endothermic phenomenon at 98 ± 0.2 °C (onset temperature) is the metastable fusion of what has to be defined as a new polymorph H (to recall that it comes from the desolvation of the monohydrate). The melting temperature of form H is thus 60 °C below that of forms I (156.3 °C) and II (156.8 °C).

From this experimental study we can conclude that the dehydration of the rimonabant monohydrate depends on the environmental conditions applied. If the dehydration is performed above 100 °C a complete destruction of the initial phase occurs and the system will suffer from a complete amnesia about the former structural order. A reappearance of structural order will need to proceed through a nucleation and growth process. By contrast, if a smooth dehydration is performed below T_g (78 °C) there is a transmission of structural information

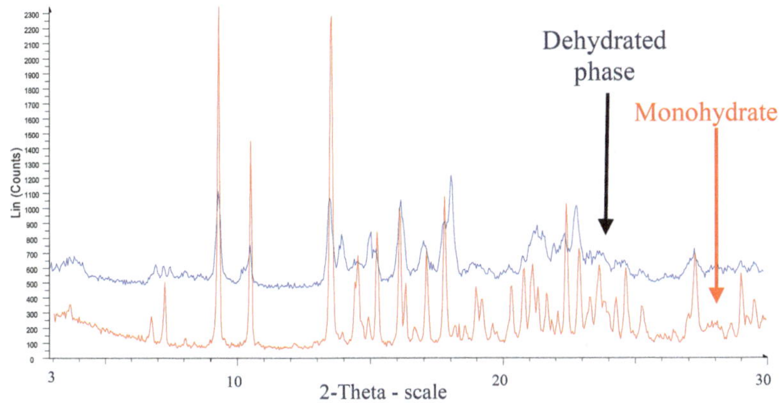

Fig. 8 XRPD patterns of the monohydrate and smoothly dehydrated forms of rimonabant.

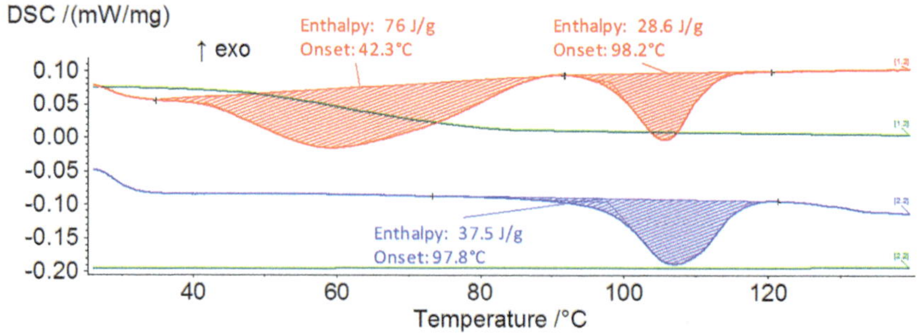

Fig. 9 Comparison of the TGA and DSC curves for the monohydrated (upper) and smoothly dehydrated (lower) forms of rimonabant (heating rate 5 K min^{-1}).

between the mother phase (monohydrate crystal lattice) and the new metastable polymorph H so no nucleation is needed. A schematic representation of such behaviour is proposed in Fig. 10, and illustrates the possibility of identifying new polymorphic forms, as recently shown in the case of paroxetine HCl.[16]

Concerning rimonabant, the channel hydrate has probably some degrees of elasticity to ensure the release of water molecules. Nevertheless, these channels are aligned along the long axis of the needle making the pathways for a release in the gas phase too long a voyage. The defects perpendicular to the *a* axis can be used as short cuts. It is remarkable that under the environmental conditions used for the desolvation, the needle is segmented by periodic cracks and in between these large defects there are other small cracks which are more or less periodic as well. It is likely that the free migration path in the crystal lattice depends on the driving force for desolvation.

Extended discussion

Many hydrates lead to a schematic binary system displayed in Fig. 11a with a so-called non-congruent fusion. The hydrate is stable up to T_p, the temperature of the peritectic transition:[17]

$$\langle \text{Hydrate} \rangle \Leftrightarrow \langle \text{anhydrous A} \rangle + \text{doubly saturated solution.}$$

Even if it sounds weird, it is then perfectly possible to desolvate a hydrate in aqueous solution if the temperature of the system is adjusted above T_p.

The schematic binary system displayed in Fig. 11b represents the corresponding metastable congruent fusion at T_F AH_2O and starting from that point a metastable liquidus (in red dashed line) and solidus (in blue dashed line) converge to the metastable melting point of A_H at T_F A_H. Dehydration or re-hydration corresponds to the crossing of the monophasic domain $\langle A(H_2O)_x \rangle$ with $0 \leq x \leq 1$ *i.e.* a non-stoichiometric hydrate (*i.e.* continuum in composition) whose crystal lattice features *versus* *x* show little changes if any. This situation should be distinguished from a physical mixture of hydrated and anhydrous forms, that may also exist as inter-grown domains. In the case of a dynamic solid solution envisaged here, the access to the putative polymorph A_H implies other

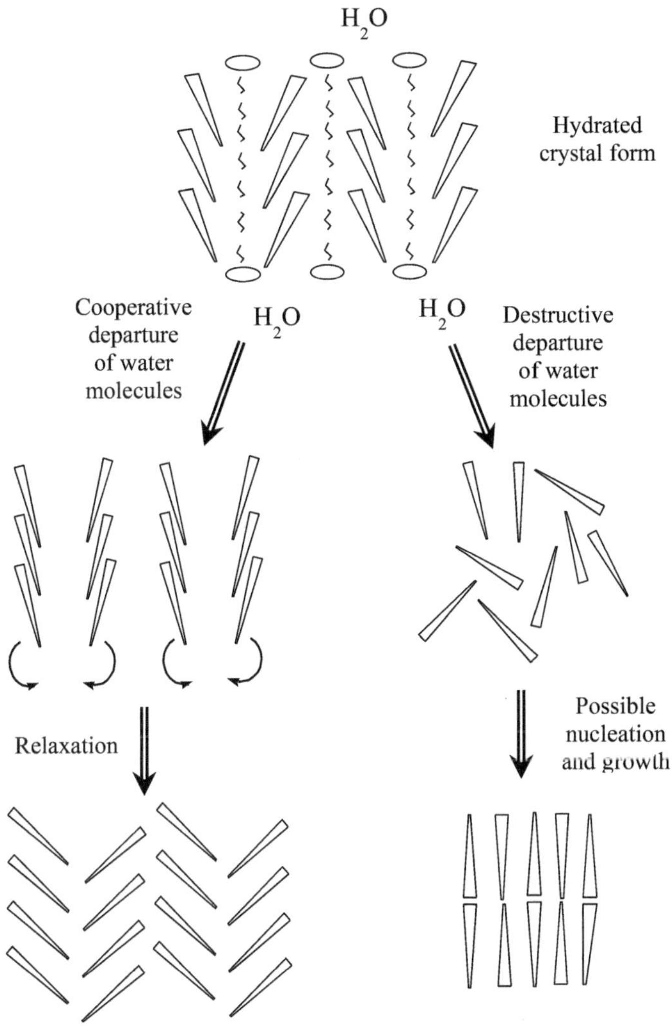

Fig. 10 Schematic representation showing the possibility to produce different crystal forms and possibly amorphous material by means of distinct dehydration mechanisms from a channel hydrate.

requirements need to be fulfilled. The glass transition is correlated with molecular mobility and thus a safe way to access to A_H is to complete the desolvation below T_g so that the new anhydrous material is 'frozen' with some limited abilities to re-arrange from the purely isomorphous desolvate to the local lowest energy crystal lattice. It should however be highlighted that water activity may decrease the temperature of glass transition through a plasticizing effect, so the residence time of water in the sample may affect the dehydration behaviour under smooth conditions.[18]

The access to A_H is thus conditioned by a set of 4 temperatures: temperature of fusion of the stable form (no other metastable form than A_H is considered for clarity), the temperature of fusion of the metastable form A_H, the temperature of

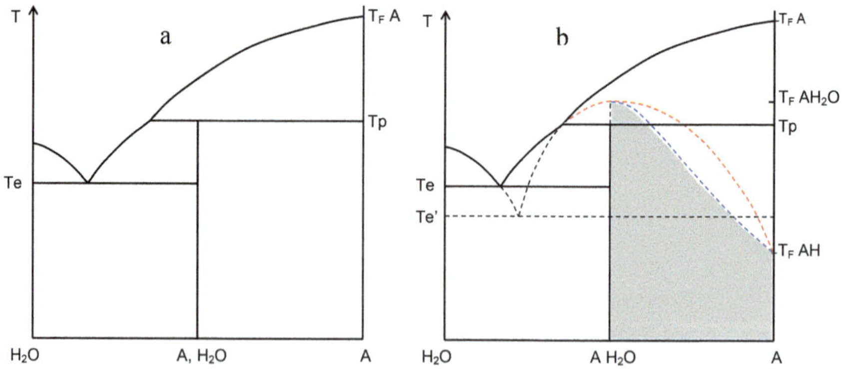

Fig. 11 (a) Stable equilibria involving a stoichiometric monohydrate with its reversible decomposition at the peritectic temperature T_p (non-congruent fusion); (b) represents, in addition to the stable equilibria, the metastable equilibria including: the congruent fusion at T_F AH_2O, the grey zone represents the domain of the non-stoichiometric phase spanning from pure A – up to T_F AH – to AH_2O. This monophasic domain could be labelled $\langle A(H_2O)_x \rangle$ with $0 \leq x \leq 1$.

the glass transition T_g and the temperature of desolvation. Some combinations only are discussed below:

$$T_F\ \mathrm{A} > T_F\ \mathrm{A_H} > T_g > T_{\mathrm{desolvation}}$$

This case has been treated in this experimental study. If the crystals are not excessively large and thus the residence time of the solvent molecules out of their crystallographic sites is short,[18] there is a good probability that a new desolvated phase appears. This anhydrous form can be obtained only by a smooth release of the volatile partner of crystallization. Therefore, an exhaustive exploration of the landscape of a molecule (survey of its unary system) needs to go through investigations of several binary systems with volatile solvents with a view to crystallize solvates. Smooth desolvations at a temperature below T_g and the temperature of fusion of this desolvated phases lead to new crystal packings impossible to be obtained by simple adjustments of T and P in the unary system.

$$T_F\ \mathrm{A} > T_{\mathrm{desolvation}} > T_F\ \mathrm{A_H} > T_g$$

This is also a case exemplified in this study. Because of the 'high' temperature of desolvation, the dehydrated material has lost the structural information contained in the solvated phase. The material is ready for nucleation and crystal growth of the stable anhydrous form.[19]

$$T_F\ \mathrm{A} > T_g > T_F\ \mathrm{A_H} > T_{\mathrm{desolvation}}$$

In that interesting case the temperature of fusion of the desolvated phase is clearly below the T_g of the anhydrous phase. If the release of solvent molecules is smooth and fast enough, the resulting phase will still contain long-range order.

But in that particular case the amorphous solid is more stable than this flimsy crystallized phase. On ageing the crystallized phase will collapse progressively. This phenomenon called superheating has been observed with alpha trehalose.[20] It is even possible to speed up the desolvation at a temperature above T_F A_H but still below T_g for a limited period of time (*i.e.* $T_F\,A > T_g > T_{desolvation} > T_F\,A_H$) and to observe this superheated crystallized phase.

Conclusion

The experimental part of this paper gives evidence that rimonabant monohydrate can be desolvated with a complete loss of structural information of the former crystallized phase (*e.g.* by heating up to 100 °C). The subsequent evolution must go through a nucleation and crystal growth process.

The monohydrate can also be submitted to a smooth desolvation process below T_g of the anhydrous phase which leads to a new polymorph A_H exhibiting a clear structural filiation with the mother hydrated phase. The relaxation of the solid after departure of the water molecules is ensured by directional cooperative movements; there is no need for nucleation and growth of this metastable phase whose melting point is 60 °C below that of the two, almost iso-energetic, polymorphs I and II.

Extension of these data to a desolvation mechanism is proposed by considering the following 4 temperatures: temperature of fusion of the stable form T_F A, the temperature of fusion of the metastable form A_H, the temperature of the glass transition T_g and the temperature of desolvation. A prediction of the outcome of a desolvation is possible by considering the order of T_F A, T_g, T_F A_H and $T_{desolvation}$.

Of course other parameters such as: crystallinity of the initial phase; crystal structure; crystal size distribution of the solvate; mode of desolvation; *etc.* have an impact on the course of the desolvation but the order of these 4 temperatures gives a guide to predict the likelihood of the desolvation mechanism.

One of the general consequences of this scheme is that a thorough investigation of the solid phases of a pure component (unary system) should include the search for solvates with volatile solvents. If some of them exist they could be desolvated at a temperature below T_g and T_F $A_{desolvate}$ (*e.g.* $T_F\,A > T_F\,A_{desolvate} > T_g > T_{desolvation}$) to produce metastable phases when compared to other polymorphs obtained from a solution or from the molten state. Their crystallographic features have some filiations with the corresponding solvates. These phases are not accessible directly in the unary system. Therefore, the experimenter needs to explore binary systems before going back to the pure component with the original packing(s).

Acknowledgements

Sanofi is thanked for its support to this study.

References

1 J. Bernstein, Polymorphism in Molecular Crystals, *IUCr Monographs on Crystallography*, Oxford University Press, Oxford, 2002, vol. 14; T. L. Threlfall, *Analyst*, 1995, **120**, 2435.

2 J. D. Dunitz and J. Bernstein, *Acc. Chem. Res.*, 1995, **28**, 193.
3 J. Bernstein, R. J. Davey and J.-O. Henck, *Angew. Chem., Int. Ed.*, 1999, **38**, 3440.
4 A. Gavezzotti, *Chem.–Eur. J.*, 1999, **5**, 567; D. Erdemir, A. Y. Lee and A. S. Myerson, *Acc. Chem. Res.*, 2009, **42**, 621; R. J. Davey, S. L. M. Schroeder and J. H. ter Horst, *Angew. Chem., Int. Ed.*, 2013, **52**, 2.
5 P. G. Vekilov, *Cryst. Growth Des.*, 2004, **4**, 671.
6 S. Petit and G. Coquerel, *Chem. Mater.*, 1996, **8**, 2247.
7 K. R. Morris, Structural aspects of hydrates and solvates, in *Polymorphism in Pharmaceutical Solids*, ed. H. G. Brittain, Marcel Dekker, New York, 1999, vol. 95, p. 125.
8 A. K. Galwey, *Thermochim. Acta*, 2000, **355**, 181.
9 B. J. Venhuis, M. V. Vredenbregt, N. Kaun, J. K. Maurin, Z. Fijałek and D. de Kaste, *J. Pharm. Biomed. Anal.*, 2011, **54**, 21.
10 C. Cavallari, B. Pérez-Artacho Santos and A. Fini, *J. Pharm. Sci.*, 2013, **102**, 4046.
11 G. Coquerel, H. Duplaa, B. Fours, O. Monnier and P. Ochsenbein, Monohydrate of rimonabant, preparation method thereof and pharmaceutical compositions containing same, Patent WO2007/090949, 2007
12 G. Coquerel, M. Sanselme and A. Lafontaine, Method and measuring scattering of X-rays, its applications and implementation device, Patent WO2012/136921 A1, 2012.
13 Bruker AXS Inc., *Bruker SMART (5.059) and SAINT (6.01)*, Madison, Wisconin, 1997.
14 G. M. Sheldrick, *SHELXS97 and SHELXL97*, Institüt für Anorganische Chemie der Universität, Gottingen, Germany, 1997.
15 G. A. Stephenson, E. G. Groleau, R. L. Kleemann, W. Xu and D. R. Rigsbee, *J. Pharm. Sci.*, 1998, **87**, 536.
16 M. F. Pina, M. Zhao, J. F. Pinto, J. J. Sousa, C. S. Frampton, V. R. Diaz, O. Suleiman, L. Fábián and D. Q. M. Craig, *Cryst. Growth Des.*, 2014, **14**, 3774.
17 G. Coquerel, *Chem. Soc. Rev.*, 2014, **43**, 2286.
18 Y. Amharar, S. Petit, M. Sanselme, Y. Cartigny, M.-N. Petit and G. Coquerel, *Cryst. Growth Des.*, 2011, **11**, 2453.
19 M. Descamps and E. Dudognon, *J. Pharm. Sci.*, 2014, **103**, 2615.
20 J. F. Willart, A. Hedoux, Y. Guinet, F. Danede, L. Paccou, F. Capet and M. Descamps, *J. Phys. Chem. B*, 2006, **110**, 11040.

Faraday Discussions

PAPER

Localizing and inducing primary nucleation

Zoubida Hammadi,[a] Romain Grossier,[a] Shuheng Zhang,[a] Aziza Ikni,[b] Nadine Candoni,[a] Roger Morin[a] and Stéphane Veesler*[a]

Received 22nd December 2014, Accepted 27th January 2015

DOI: 10.1039/c4fd00274a

Do the differing properties of materials influence their nucleation mechanisms? We present different experimental approaches to study and control nucleation, and shed light on some of the factors affecting the nucleation process.

Introduction

With the control of material properties being directly related to their methods of production, understanding the mechanisms governing these methods is vital. With crystallized materials, such as minerals, pharmaceuticals, proteins, biominerals and nanomaterials, the most important properties of crystal morphology, habit, size distribution and phases are controlled by the nucleation step. Nucleation was clearly defined by D. Kashchiev as "the process of random generation of those nanoscopically small formations of the new phase that have the ability for irreversible overgrowth to macroscopic sizes".[1] The size of the critical cluster and the nucleation rate are determined by the classical nucleation theory (CNT) derived from the Gibbs treatment of droplet nucleation.[2] Interestingly, this simple model fits experimental results very well but is many orders of magnitude out in predicting nucleation frequencies, as shown with protein crystallization.[3] It is noteworthy that even experimental data on the same system, in the same conditions, can have discrepancies of more than 10 orders of magnitude,[3–7] as illustrated in Fig. 1.

Moreover, in several studies on protein crystallization using solution scattering techniques, ultracentrifugation–sedimentation and dialysis kinetics, data were interpreted as indicating the formation of prenucleation aggregates, clusters or oligomeric species, ranging between monomer and crystal.[8–19] However, other authors[20–26] did not observe such aggregates and offered the following explanation: "the failure to account for direct and indirect protein interactions in the solutions results in unrealistic aggregation scenarios".[23] Note that the stable entity in solution is not necessarily a monomer; it can be an oligomer, for

[a]CINaM-CNRS, Aix-Marseille Université, Campus de Luminy, F-13288 Marseille, France. E-mail: veesler@cinam.univ-mrs.fr

[b]SPMS-CNRS, UMR 8580, Grande Voie des Vignes, F-92295 Châtenay-Malabry, France

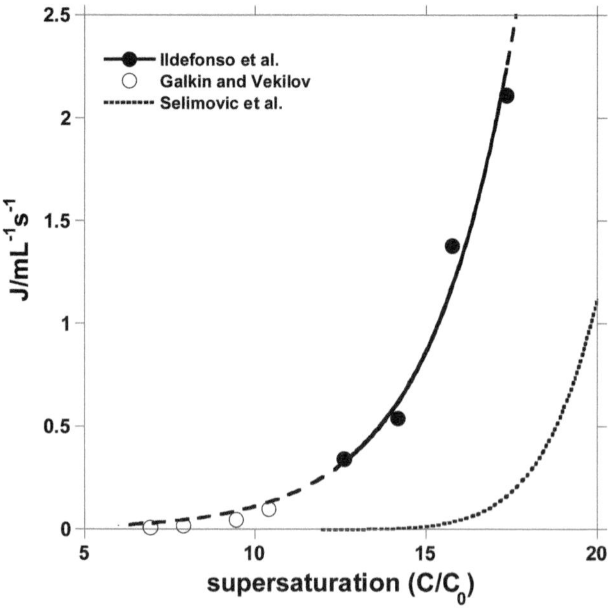

Fig. 1 Primary nucleation rate of lysozyme *vs.* supersaturation; (●) data at 20 °C-$C_0 = 3.2$ mg mL^{-1} (NaCl = 0.7 M and pH = 4.5) from Ildefonso *et al.* in microfluidics,[44] (○) data, at 12.6 °C-C_0 = 1.6 mg mL^{-1} (NaCl = 0.7 M and pH = 4.5), from Galkin and Vekilov in microbatches in oil[7] and (…) data at 12 °C-C_0 = 5 mg mL^{-1} (NaCl = 0.6 M and pH = 4.5) from Selimovic *et al.* in microfluidics.[4]

biological or physicochemical reasons.[27] Finally, simulation papers published in the same period proposed a 2-step process for protein crystal nucleation[28–30] using the presence of a metastable LLPS (liquid–liquid phase separation) to reduce the nucleation free-energy barrier. The first step could be a local densification (liquid-like dense phase) and the second step could be the nucleation of the solid inside this dense phase. There has even been a third nucleation pathway proposed, linked to biomineralization, consisting of nucleation *via* stable prenucleation clusters.[31]

To summarize, in the CNT only one order parameter, density, describes the transition between the two phases, whereas in the 2-step theory two order parameters, density and structure, are used. For reviews see ref. 32 and 33. This appears to us as an attempt to find a universally applicable explanation for experimental discrepancies. Firstly, the limiting step, densification or structuration, will depend on the solute, as suggested by Knezic *et al.*[34] For example, macromolecules take longer to rearrange themselves into crystalline nuclei than smaller molecules. Secondly, the fact that the location, nature and lifetime of the dense phase and/or the critical nucleus are unknown (at the moment) explains difficulties encountered in experimental observations and interpretations. In this paper, however, we explore nucleation from a purely practical point of view, without any *a priori* assumption on the nucleation mechanism. We describe the two experimental approaches taken here: indirect experiments, measuring nucleation frequency by counting crystals or by measuring induction time, and direct experiments, observing the critical nucleus. Kinetic measurements will provide us with data to be tested by classical and 2-step nucleation models.[35] Meanwhile, direct observation will yield information on the size and structure of the critical nucleus. The objective is to shed light on nucleation mechanisms in solution.

Experimental

I. Microfluidic set-up (~nL)

Experiments were performed in PDMS (poly(dimethylsiloxane)) and PEEK (polyether ether ketone) and Teflon devices based on HPLC techniques; T-shaped junctions as described previously[36,37] or cross-shaped (Zhang *et al.* submitted) ("plug-factory") were used to form the droplets with volumes in the nanoliter range, without the addition of surfactants. We used FMS oil (Hampton Research) for aqueous solvents. This oil shows no or very low miscibility with water and good wettability with PDMS and Teflon. The droplet concentration and composition were varied using a programmable multi-channel syringe pump (neMESYS) and by controlling the relative and absolute flow rates of the different solutions. (Zhang *et al.* submitted) The tubing containing the droplets was placed in a thermostatted tubing-holder and incubated to obtain crystallization. Droplets were observed using an XYZ-motorized camera.

II. Production of microdroplets with microinjectors (~pL to fL)

We used a simply-constructed and easy-to-use fluidic device that generates arrayed aqueous phase microdroplets in oil, with volumes ranging from nanoliter to femtoliter, without the addition of surfactant.[38] The device enables the entire volume range to be attained in the course of one experiment. All experiments were performed on a hydrophobic coverslip, at 20 °C in a thermostatted room. The coverslip was covered with approximately 100 µL of inert DMS oil (Hampton Research HR3-419, refractive index = 1.390). The micrometer-sized droplets of water solution were generated on the coverslip by a microinjector (Femtojet Eppendorf) with a micropipette of 0.5 µm internal diameter (Femtotip Eppendorf). Two home-made micromanipulators (MS30 Mechonics) consisting of 3 miniature translation stages allowed displacement of the injector (capillary holder) and the tip in X, Y and Z, with a displacement of 18 mm in the 3 directions by steps of 16 nm. Here, we made the sharp tips from tungsten (W) wires (125 µm diameter); for a detailed description see ref. 39–41.

III. Localized DC electric field

The experimental set-up was composed of a crystallization cell with two electrodes, at least one of which had a sharp W-tip, as previously described.[41] Two home-made micromanipulators allowing displacement of the electrodes in X, Y and Z were added.

IV. Protein solutions

Hen-egg white lysozyme (14 600 Da, pI = 11.2) was purchased from Sigma (batch 057K7013 L 2879) and used without further purification. The purity of lysozyme was checked by molecular sieving. Proper amounts of lysozyme and NaCl were dissolved in pure water (ELGA UHQ reverse osmosis system) to obtain the necessary stock solutions. The different solutions were buffered with 80 mM acetic acid, adjusted to pH = 4.5 with NaOH (1 M) and filtered through 0.22 μm Millipore filters. The pH was checked with a pH meter (Schott Instrument, Prolab 1000) equipped with a pH microelectrode. Lysozyme concentrations were checked by optical density measurements (Shumadzu, UV-1800) using an extinction coefficient of 2.64 mL cm^{-1} mg^{-1} at 280 nm for lysozyme.

V. Gel solutions

A stock solution of 1% agarose type V from Sigma (gel point 42 °C) was prepared using the method developed by Robert and Lefaucheux.[42] Gel lysozyme solutions were prepared by mixing at 42 °C the solutions of agarose gel, sodium chloride and protein, yielding a final agarose concentration of 1% w/v. The mixture gelified in a few minutes once the temperature was decreased.

VI. Mineral solutions

NaCl (R.P. Normapur, analytical reagent) solutions were prepared by dissolution of the proper amount of powder in pure water.

VII. Oils

Dodecanemethylpentasiloxane (DMS) oil (Hampton Research HR2-593, refractive index = 1.390 at 20 °C) and paraffin oil (Hampton Research HR3-421, refractive index = 1.467) were used as continuous phases.

Results and discussions

I. Indirect experiments

The indirect measurement of nucleation is obtained through nucleation frequencies or induction times. Because nucleation has a stochastic nature, it is important to perform a large number of experiments, in order to obtain reliable data. Here nucleation was performed in small volumes, in order to reduce the number of crystals and to render their observation easier. Moreover, this has the advantage of decreasing the quantity of molecules while speeding up heat and mass transfer.[43] In addition, the supersaturation range experimentally accessible is increased for kinetic[44] and thermodynamic[45] reasons. One kinetic limitation in small volumes arises from the nucleation frequency (J/number of nucleus by unit time and volume): the smaller the volume, the longer the induction time and the wider the metastable zone, thus requiring greater supersaturation for nucleation.[46,47]

I.1. Small volumes of μL to nL range, the kinetic effect. Our experiments used microfluidics technologies. Microfluidics has proved an efficient way of measuring nucleation frequency, by measuring the probability of crystal presence in the droplets as a function of time[48–50] and by counting the number of crystal nuclei per droplet.[4,44] Fig. 1 summarizes data from different studies obtained using the double pulse technique.[6,51,52] This technique allows direct determination of the steady-state rate of primary nucleation, separating crystal nucleation from the growth process, by counting nucleated crystals instead of estimating the induction times. The discrepancies between the data presented in Fig. 1 chiefly stem from the experimental method applied, where the supersaturation during the quench is not constant, according to the authors.[4] Note the agreement between data obtained in microbatches[7] (~1 μL) and microfluidics[44] (~100 nL). When the volume is diminished, the range of experimentally measurable J is increased. One of the limitations usually involved in measuring J is the supersaturation range over which the experiment can be performed. When β is too low, nucleation frequency is low and nucleation is sensitive to local heterogeneity. When β is too high, J is difficult to measure because it is too rapid. This is why in practice, we are only able to measure the nucleation frequency in the vicinity of the metastable zone limit.

I.2. Small volumes of pL range, the thermodynamic effect. Emulsion-based methods such as microfluidics have advantages; they permit the production, storage and observation of a large number of microdroplets of controlled chemical composition, for instance. But microfluidics has also drawbacks: microdroplet size is controlled by channel size and the microdroplets are not accessible (hardware limitation). Microdroplet generation with micropipettes or microinjectors render microdroplets accessible and makes the size range easy to control. However, its drawback is that a single microdroplet is produced.[43,53,54] Consequently, we recently presented a fluidic device[38] that generates arrayed aqueous phase microdroplets under oil (sessile geometry). This set-up combines the advantage of channelled microfluidic techniques, generating thousands of droplets (Fig. 2), with the advantage of micropipette techniques, giving control over size and microdroplet accessibility.

As pointed out by Bempah and Hileman,[55] "One major problem associated with the use of the droplet technique in nucleation studies is the creation of sufficient supersaturation within each droplet to ensure crystallization". In other words, the smaller the volume, the longer the induction time and the wider the metastable zone. Hence a thermodynamic limitation appears because, contrary to CNT which supposes an infinite reservoir of molecules, supersaturation is no longer constant during the nucleation process (aggregation of molecules) but is decreasing.[45] In these small-volume systems, unexpected high-supersaturated metastable solutions are observed. In practice, there is an influence of the volume on nucleation from a picoliter range down.[45] This is in agreement with experimental results showing effects under nanoscopic confinement.[56–60]

We therefore used the droplet contraction method to generate sufficient supersaturation, *i.e.* diffusion of water from the microdroplet into the oil. The experiment presented in Fig. 3 and 4 is isothermal (20 °C) and the volume is decreasing linearly[61] with time and can be monitored throughout the experiment.

Fig. 2 Array of droplets (~720) of NaCl solutions generated through the layer of inert DMS oil.

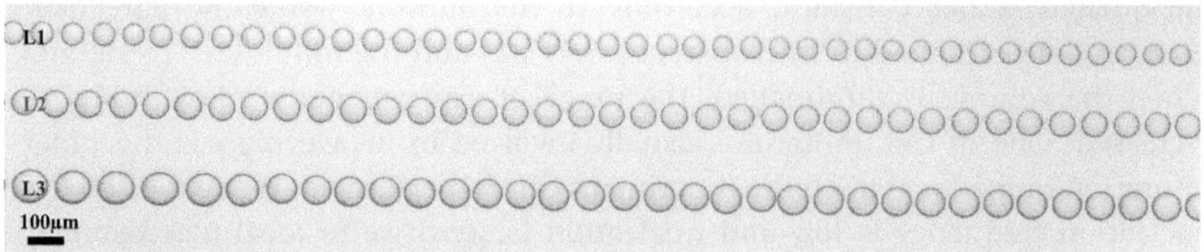

Fig. 3 Array of droplets of NaCl solutions at $\beta = 0.1$ generated through the layer of inert DMS oil. Microdroplet volumes are 104, 172 and 284 pL for rows L1, L2 and L3 respectively, 104 pL assuming a spherical shape for the droplet with a contact angle of 130° as measured previously.[61]

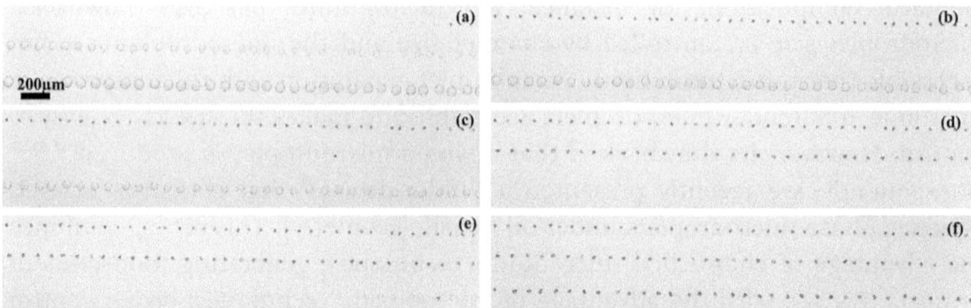

Fig. 4 Time sequence showing nucleation of NaCl crystals (a) at $t = 1106$ s, (b) $t = 1305$ s and (c) $t = 1390$ s, (d) at $t = 1563$ s, (e) $t = 1680$ s and (f) $t = 1995$ s. In (a), (c) and (e) the first microdroplet in L1, L2 and L3 has crystallized and in (b), (d) and (f) all microdroplets in L1, L2 and L3 have crystallized. All micrographs are at the same magnification. Rows L1, L2 and L3 are from Fig. 3.

In the first stage (Fig. 3), 3 microdroplet rows containing NaCl solution at $\beta = 0.1$ are generated, and supersaturation β is defined as the ratio of the NaCl concentration in solution *versus* the solubility of NaCl, 6.15 M at 20 °C in water.[62] Droplets slowly evaporate until critical supersaturation is reached. The critical supersaturation for nucleation is the maximum supersaturation that a solution can withstand without nucleating a new phase.[47] The nucleation times of the microdroplets from the experiment (Fig. 3 and 4) are plotted in Fig. 5, where $P(t)$ represents the normalized fraction of nucleated microdroplets. Due to the fast growth rate, the time required for the newly-formed nuclei to grow to a detectable size is negligible with regards to the induction time (see Fig. 6). Thus, the time when a detectable crystal is observed can be considered the induction time. In the experiment shown in Fig. 3 and 4, we observe that the smaller the volume, the faster the evaporation rate and the faster the nucleation (Fig. 5).

II. Direct experiments

According to Davey *et al.*,[63] "The biggest challenge is to identify the structure of low-concentration, nanosized dynamic clusters of molecules". Thus, with the development of new experimental tools such as *in situ* electron microscopy,[64] cryogenic transmission electron microscopy,[65,66] fluctuation transmission electron microscopy,[67] AFM,[68] laser scanning confocal microscopy,[69] laser confocal

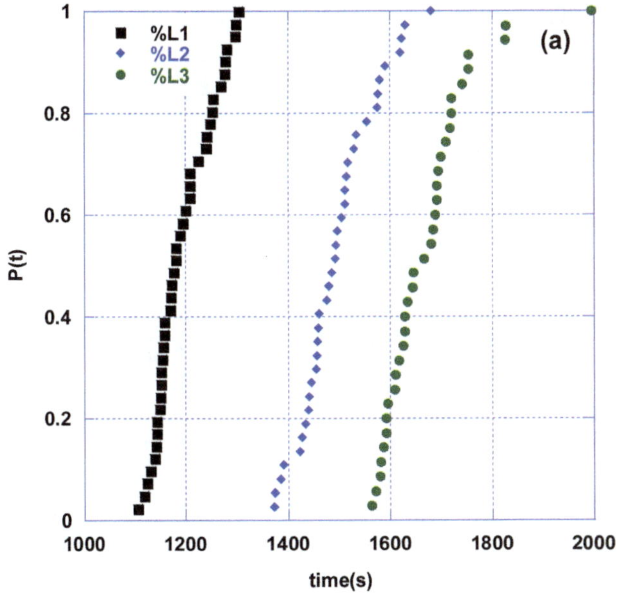

Fig. 5 Nucleation probability $P(t)$ *vs.* time. $P(t)$ represent the normalized fraction of nucleated microdroplets. Rows L1, L2 and L3 are from Fig. 3.

differential interference contrast microscopy[70] or even classical microscopy for colloidal nucleation,[71] an increasing focus is placed on how to catch the critical cluster. The challenge is that nucleation is stochastic: during an experiment, we do not know where and when nucleation will occur. All we can do in our different laboratories, for the moment, is to perform many experiments and hope that the law of large numbers will allow us to observe critical clusters. Whether or not this experimental observation is representative of nucleation is still an open question (see the controversy over the existence of prenucleation aggregates).

II.1. Confinement by volume and external field. Here, we address the unpredictability of the spatial and temporal location of the critical nucleus. In the experiment presented in Fig. 3 and 4, confinement by volume allows us to have spatial control at the picoliter scale. With this set-up, a resolution better than 1 μm would be difficult to obtain, which is still far from the expected size of a critical cluster.

Thus, we propose confinement coupled to an external field, in order to control the location of the nucleation event. The implications of an external field for crystal growth in solution were highlighted by Voss,[72] Oxtoby[73] and Revalor.[52] Two effects on the supersaturated solutions are expected: molecular orientation and density fluctuation. The principle of the experiment is that first solute microdroplets are generated in oil with the microinjector, and the microdroplets then concentrate contraction to generate supersaturation. Secondly, when the microdroplets completely disappear a sharp tip is used to induce a nucleation event, by touching the supersaturated metastable microdroplet.

It was previously shown that any disturbance, for instance a crystal touching dense droplets[74] or mechanical contact with a nanotip,[75] triggers nucleation. In

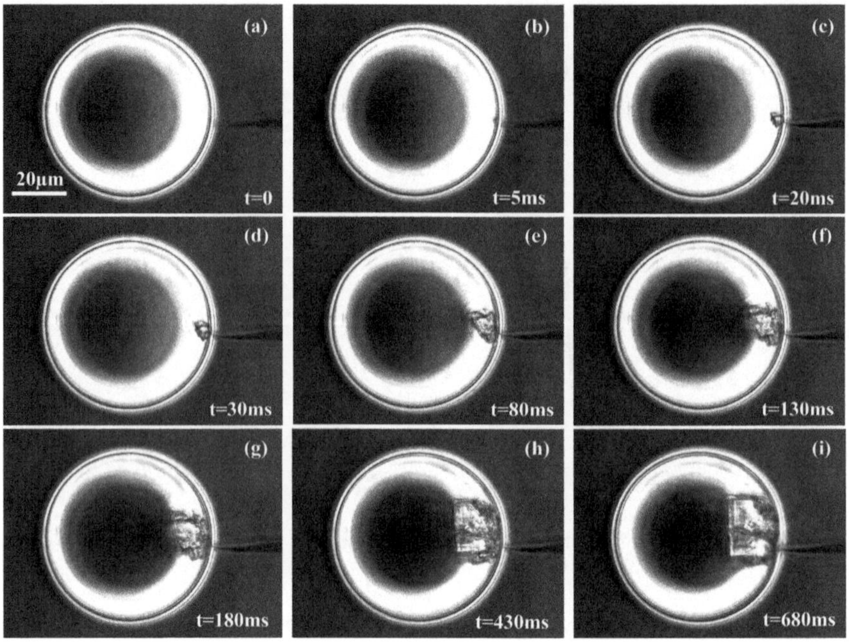

Fig. 6 Panels representing a time sequence showing nucleation induced by a sharp tip and growth of NaCl in paraffin oil. (a) Initial condition: $\beta > 1.24$, (a) droplet size 60 μm (95 pL assuming a spherical shape for the droplet with a contact angle of 120° at the time of nucleation as measured previously[61]). All micrographs (phase contrast mode) are at the same magnification.

the experiment presented in Fig. 6, we induced a structural transformation *via* mechanical contact at precisely determined points (at 16 nm accuracy) and times, using an sCMOS camera at 200 frames per s (Andeor Neo). The time between tip contact and observation of nucleation was shorter than 5 ms (Fig. 6a and b) and the mean crystal growth rate was greater than 200 μm s^{-1} during the first 20 ms (Fig. 6a–c). The nucleated crystal was rough, and transition to a faceted crystal was observed in less than 1 s, *i.e.* the transition between the nucleation form and the equilibrium form. In practice, because the tip position is controlled with micromanipulators, the position of the critical nucleus can be determined with an accuracy of 16 nm.

This method has some drawbacks, however. First, it is a trial and error method: the droplet is repeatedly tapped with the tip during the generation of supersaturation by water diffusion in order to launch nucleation. Second, it is less successful with protein, probably because the crystallization medium is not binary but ternary – that is to say, composed of a solvent, a solute and crystallization agents. During droplet contraction, both protein and crystallization agents concentrate and reach supersaturation. Then, for kinetic reasons,[76] salt crystallizes first and/or polymer induces an LLPS, often before protein nucleation. For these reasons, we use another experimental procedure to generate supersaturation for protein crystallization experiments, as described in the next section of the paper.

II.2. Confinement by fluxes. In this section, we describe how we use an external localized DC electric field to control the location of the nucleation event.

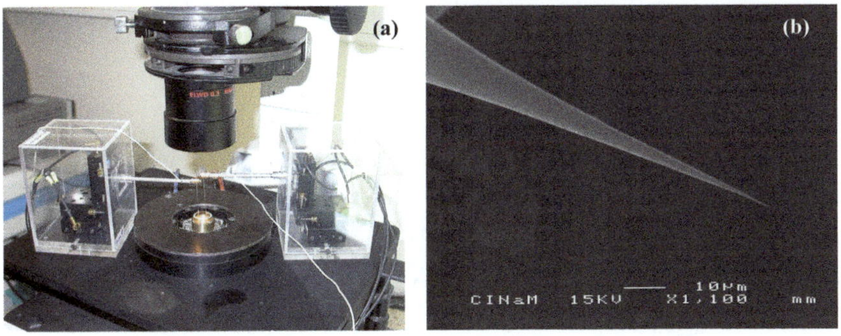

Fig. 7 (a) Temperature-controlled crystallization cell with the 2 micromanipulators and (b) SEM image of a W-tip.

Two electrodes were placed in a supersaturated metastable lysozyme solution (Fig. 7a), one of the electrodes being sharp (Fig. 7b). Because of the nanometer size of the tip, large electric fields[77] and large field gradients are encountered near the tip at low DC voltage. This geometry also induces a high current density inside the solution, close to the region of high curvature. Here, supersaturation was increased by a flux and an accumulation of molecules at a precise point in the vicinity of the tip apex.[41] This accumulation created inside the crystallization cell a concentration gradient, that in solution is counteracted by convection. Therefore, the experiments were performed in gel in order to eliminate convection and keep the confinement in the vicinity of the electrode until the concentration reached the critical supersaturation.

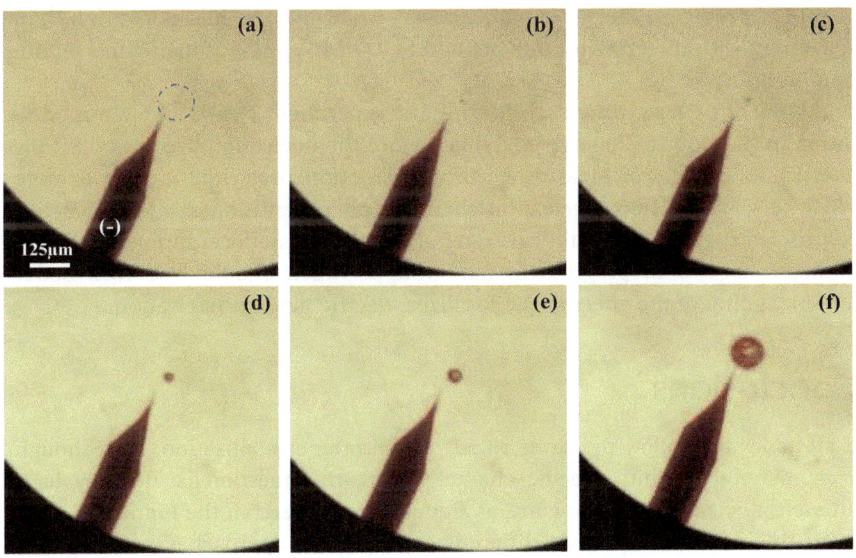

Fig. 8 Panels representing a time sequence showing nucleation and growth of lysozyme 20 mg mL^{-1} NaCl 0.7 M in agarose gel 1% (initial conditions: 0.5 V and 0.35 μA) at (a) t = 600 s, (b) t = 1200 s, (c) t = 1800 s, (d) t = 7200 s, (e) t = 10 800 s and (f) t = 30 000 s (the (-) indicates the anode).

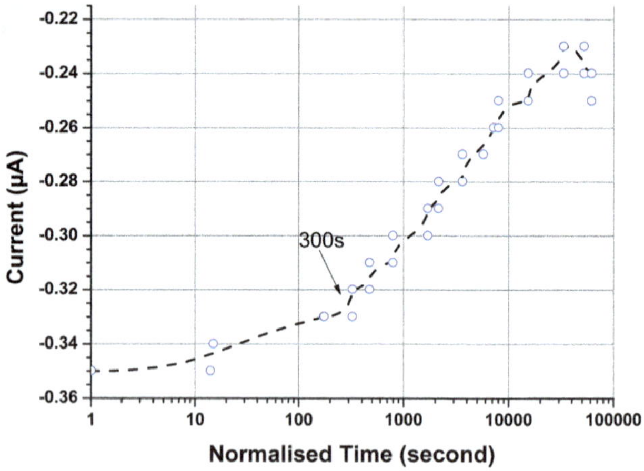

Fig. 9 Evolution of the current over time during the experiment, semi-log scale. Line is a guide for the eye.

The sequence presented in Fig. 8 shows that the use of gel as a crystallization medium clearly enables better control of the location of nucleation, and the nucleus appears in the vicinity of the cathode at the tip apex where the electric field is the strongest. As a result:

(1) The nucleation time in presence of the DC electric field is shorter than 600 s (instead of $t > 24$ h in the absence of an electric field); the nucleus is depicted by a circle in Fig. 8a.

(2) The particle is rough, due to high local supersaturation encountered during nucleation and growth.

(3) The average crystal growth rate (between 600 and 10 800 s) is 15 μm h^{-1}, in agreement with the growth rate obtained by Durbin *et al.*[78] at high supersaturation for lysozyme.

(4) The current *vs.* time curve during the experiment clearly shows a modification in the slope at 300 s (Fig. 9), just before the nucleation becomes optically observable in Fig. 8a. Measuring current variation over time is thus a more sensitive way to detect nucleation than optical observation suggesting better control is expected. A theoretical study of the field nano-localization with alternating voltage is in progress in our group,[79] which should lead to a better understanding of the effect of the localized electric field on nucleation.

Conclusions

If all molecules follow the same rules[80] concerning crystallization, even though each material exhibits specific characteristics, the question is: do they have different nucleation mechanisms, as frequently proposed in the literature?

In this paper, we present different experimental approaches to study and control nucleation. The use of small volumes makes it possible to measure nucleation kinetics. We also show that coupling volume reduction with use of an external field makes it possible to obtain spatial and temporal control over nucleation for small molecules. Application to macromolecules raises some difficulties; however, the use of an electric field to create localized field and fluxes shows promising results, leading to the spatial control of nucleation.

These experiments also shed light on some of the factors influencing the nucleation process.

Acknowledgements

We thank N. Ferté for protein characterization and fruitful discussions. We thank M. Sweetko for English revision.

References

1 D. Kashchiev, *Nucleation: basic theory with applications*, Butterworth-Heinemann, Oxford, 2000.
2 J. Gibbs, *The Collected Works, Vol. 1: Thermodynamics*, Yale University Press, 1948.
3 N. M. Dixit, A. M. Kulkarni and C. F. Zukoski, *Colloids Surf., A*, 2001, **190**, 47–60.
4 S. Selimovic, Y. Jia and S. Fraden, *Cryst. Growth Des.*, 2009, **9**, 1806–1810.
5 M. Ildefonso, N. Candoni and S. Veesler, *Cryst. Growth Des.*, 2013, **13**, 2107–2110.
6 O. Galkin and P. G. Vekilov, *J. Phys. Chem. B*, 1999, **103**, 10965–10971.
7 O. Galkin and P. G. Vekilov, *J. Am. Chem. Soc.*, 2000, **122**, 156–163.
8 T. Azuma, K. Tsukamoto and I. Sunagawa, *J. Cryst. Growth*, 1989, **98**, 371–376.
9 Y. Georgalis, A. Zouni and W. Saenger, *J. Cryst. Growth*, 1992, **118**, 360–364.
10 Y. Georgalis, A. Zouni, W. Eberstein and W. Saenger, *J. Cryst. Growth*, 1993, **126**, 245–260.
11 Y. Georgalis and W. Saenger, *Adv. Colloid Interface Sci.*, 1993, **46**, 165–183.
12 F. Boue, F. Lefaucheux, M. C. Robert and I. Rosenman, *J. Cryst. Growth*, 1993, **133**, 246–254.
13 A. J. Malkin and A. Mcpherson, *Acta Crystallogr., Sect. D: Biol. Crystallogr.*, 1994, **50**, 385–395.
14 N. Niimura, Y. Minezaki, M. Ataka and T. Katsura, *J. Cryst. Growth*, 1994, **137**, 671–675.
15 N. Niimura, Y. Minezaki, M. Ataka and T. Katsura, *J. Cryst. Growth*, 1995, **154**, 136–144.
16 N. Niimura, M. Ataka, Y. Minezaki and T. Katsura, *Phys. B*, 1995, **213 et 214**, 745–747.
17 Y. Minezaki, N. Niimura, M. Ataka and T. Katsura, *Biophys. Chem.*, 1996, **58**, 355–363.
18 N. Niimura, Y. Minezaki, I. Tanaka, S. Fujiwara and M. Ataka, *J. Cryst. Growth*, 1999, **200**, 265–270.
19 A. Stradner, H. Sedgwick, F. Cardinaux, W. C. K. Poon, S. U. Egelhaaf and P. Schurtenberger, *Nature*, 2004, **432**, 492–495.
20 J. B. Bishop, W. J. Fredericks, S. B. Howard and T. Sawada, *J. Cryst. Growth*, 1992, **122**, 41–49.
21 S. Veesler, S. Marcq, S. Lafont, J. P. Astier and R. Boistelle, *Acta Crystallogr., Sect. D: Biol. Crystallogr.*, 1994, **50**, 355.
22 M. Muschol and F. Rosenberger, *J. Cryst. Growth*, 1996, **167**, 738–747.

23 F. Rosenberger, P. G. Vekilov, M. Muschol and B. R. Thomas, *J. Cryst. Growth*, 1996, **168**, 1–27.
24 S. Finet, F. Bonnete, J. Frouin, K. Provost and A. Tardieu, *Eur. Biophys. J.*, 1998, **27**, 263–271.
25 M. Budayova-Spano, F. Bonnete, J. P. Astier and S. Veesler, *J. Cryst. Growth*, 2002, **235**, 547–554.
26 A. Shukla, E. Mylonas, E. Di Cola, S. Finet, P. Timmins, T. Narayanan and D. I. Svergun, *Proc. Natl. Acad. Sci. U. S. A.*, 2008, **105**, 5075–5080.
27 F. Bonneté, N. Ferté, J. P. Astier and S. Veesler, *J. Phys. IV*, 2004, **118**, 3–13.
28 H. N. W. Lekkerkerker, *Phys. A*, 1997, **244**, 227–237.
29 P. R. Ten Wolde and D. Frenkel, *Science*, 1997, **277**, 1975–1978.
30 V. J. Anderson and H. N. W. Lekkerkerker, *Nature*, 2002, **416**, 811–815.
31 D. Gebauer and H. Cölfen, *Nano Today*, 2011, **6**, 564–584.
32 P. G. Vekilov, *Cryst. Growth Des.*, 2004, **4**, 671–685.
33 D. Erdemir, A. Y. Lee and A. S. Myerson, *Acc. Chem. Res.*, 2009, **42**, 621–629.
34 D. Knezic, J. Zaccaro and A. S. Myerson, *J. Phys. Chem. B*, 2004, **108**, 10672–10677.
35 R. P. Sear, *CrystEngComm*, 2014, **16**, 6506–6522.
36 R. D. Dombrowski, J. D. Litster, N. J. Wagner and Y. He, *Chem. Eng. Sci.*, 2007, **62**, 4802–4810.
37 M. Ildefonso, N. Candoni and S. Veesler, *Org. Process Res. Dev.*, 2012, **16**, 556–560.
38 R. Grossier, Z. Hammadi, R. Morin, A. Magnaldo and S. Veesler, *Appl. Phys. Lett.*, 2011, **98**, 091916.
39 E. W. Muller and T. T. Tsong, *Field ion microscopy: principles and applications*, American Elsevier Publishing Company, New York, 1969.
40 Z. Hammadi, M. Gauch and R. Morin, *J. Vac. Sci. Technol., B: Microelectron. Nanometer Struct.-Process., Meas., Phenom.*, 1999, **17**, 1390–1394.
41 Z. Hammadi, J. P. Astier, R. Morin and S. Veesler, *Cryst. Growth Des.*, 2007, **7**, 1476–1482.
42 M. C. Robert and F. Lefaucheux, *J. Cryst. Growth*, 1988, **90**, 358–367.
43 S. Lee and J. Wiener, *J. Chem. Educ.*, 2010, **88**, 151–157.
44 M. Ildefonso, N. Candoni and S. Veesler, *Cryst. Growth Des.*, 2011, **11**, 1527–1530.
45 R. Grossier and S. Veesler, *Cryst. Growth Des.*, 2009, **9**, 1917–1922.
46 Z. Kozisek, K. Sato, S. Ueno and P. Demo, *J. Chem. Phys.*, 2011, **134**, 094508–094509.
47 D. Kashchiev, *J. Chem. Phys.*, 2011, **134**, 196102.
48 P. Laval, J.-B. Salmon and M. Joanicot, *J. Cryst. Growth*, 2007, **303**, 622–628.
49 L. Goh, K. Chen, V. Bhamidi, G. He, N. C. S. Kee, P. J. A. Kenis, C. F. Zukoski and R. D. Braatz, *Cryst. Growth Des.*, 2010, **10**, 2515–2521.
50 S. V. Akella, A. Mowitz, M. Heymann and S. Fraden, *Cryst. Growth Des.*, 2014, **14**, 4487–4509.
51 D. Tsekova, S. Dimitrova and C. N. Nanev, *J. Cryst. Growth*, 1999, **196**, 226–233.
52 E. Revalor, Z. Hammadi, J. P. Astier, R. Grossier, E. Garcia, C. Hoff, K. Furuta, T. Okutsu, R. Morin and S. Veesler, *J. Cryst. Growth*, 2010, **312**, 939–946.
53 P. B. Duncan and D. Needham, *Langmuir*, 2006, **22**, 4190–4197.
54 K. Allain, R. Bebawee and S. Lee, *Cryst. Growth Des.*, 2009, **9**, 3183–3190.
55 O. A. Bempah and O. E. Hileman Jr, *Can. J. Chem.*, 1973, **51**, 3435–3442.

56 J.-M. Ha, J. H. Wolf, M. A. Hillmyer and M. D. Ward, *J. Am. Chem. Soc.*, 2004, **126**, 3382–3383.
57 M. Beiner, G. T. Rengarajan, S. Pankaj, D. Enke and M. Steinhart, *Nano Lett.*, 2007, **7**, 1381–1385.
58 K. Kim, I. S. Lee, A. Centrone, T. A. Hatton and A. S. Myerson, *J. Am. Chem. Soc.*, 2009, **131**, 18212–18213.
59 C. J. Stephens, S. F. Ladden, F. C. Meldrum and H. K. Christenson, *Adv. Funct. Mater.*, 2010, **20**, 2108–2115.
60 R. Grossier, A. Magnaldo and S. Veesler, *J. Cryst. Growth*, 2010, **312**, 487–489.
61 I. Rodríguez-Ruiz, Z. Hammadi, R. Grossier, J. Gómez-Morales and S. Veesler, *Langmuir*, 2013, **29**, 12628–12632.
62 H. Langer and H. Offermann, *J. Cryst. Growth*, 1982, **60**, 389–392.
63 R. J. Davey, S. L. M. Schroeder and J. H. Ter Horst, *Angew. Chem. Int. Ed.*, 2013, **52**, 2166–2179.
64 Y. Kimura, H. Niinomi, K. Tsukamoto and J. M. García-Ruiz, *J. Am. Chem. Soc.*, 2014, **136**, 1762–1765.
65 E. M. Pouget, P. H. H. Bomans, J. A. C. M. Goos, P. M. Frederik, G. De With and N. A. J. M. Sommerdijk, *Science*, 2009, **323**, 1455–1458.
66 J. Baumgartner, A. Dey, P. H. H. Bomans, C. Le Coadou, P. Fratzl, N. A. J. M. Sommerdijk and D. Faivre, *Nat. Mater.*, 2013, **12**, 310–314.
67 B.-S. Lee, G. W. Burr, R. M. Shelby, S. Raoux, C. T. Rettner, S. N. Bogle, K. Darmawikarta, S. G. Bishop and J. R. Abelson, *Science*, 2009, **326**, 980–984.
68 S. T. Yau and P. G. Vekilov, *Nature*, 2000, **406**, 494–497.
69 U. Gasser, E. R. Weeks, A. Schofield, P. N. Pusey and D. A. Weitz, *Science*, 2001, **292**, 258–262.
70 M. Sleutel and A. E. S. Van Driessche, *Proc. Natl. Acad. Sci. U. S. A.*, 2014, **111**, E546–E553.
71 K.-Q. Zhang and X. Y. Liu, *Nature*, 2004, **429**, 739–743.
72 D. Voss, *Science*, 1996, **274**, 1325.
73 D. W. Oxtoby, *Nature*, 2002, **420**, 277–278.
74 D. Vivares, E. W. Kalera and A. M. Lenhoff, *Acta Crystallogr., Sect. D: Biol. Crystallogr.*, 2005, **61**, 819–825.
75 R. Grossier, Z. Hammadi, R. Morin and S. Veesler, *Phys. Rev. Lett.*, 2011, **107**, 025504.
76 A. A. Chernov, *Physics Reports I.M.Lifshitz and Condensed Matter Theory*, 1997, vol. 288, pp. 61–75.
77 R. Gomer, *Field emission and field ionization*, Harvard University press, Cambridge, 1961.
78 S. D. Durbin and G. Feher, *J. Cryst. Growth*, 1986, **76**, 583–592.
79 Z. Hammadi, R. Morin and J. Olives, *Appl. Phys. Lett.*, 2013, **103**, 223106.
80 A. A. Chernov, *J. Mater. Sci.: Mater. Electron.*, 2001, **12**, 437–449.

Faraday Discussions

DISCUSSIONS

Nucleation in complex multi-component and multi-phase systems: general discussion

Sarah Price, Bart Rimez, Wenhao Sun, Baron Peters, Hugo Christenson, Colan Hughes, Changquan Calvin Sun, Stéphane Veesler, Haihua Pan, Clement Brandel, Beatrice Biscans, Hugo Meekes, Ian Rosbottom, Wieslaw J. Roth, Linda Seton, Francis Taulelle, Simon Black, Terence Threlfall, Peter Vekilov, Sendhil Poornachary, Jürg Diemand, Dimitrios Toroz, Matteo Salvalaglio, Pratchaya Tipduangta, Jan Sefcik, Samuel Booth, Ake Rasmuson, Sophie Janbon, Joop Ter Horst, Elena Simone, Robert Hammond, Celso Aparecido Bertran, Thomas Vetter, Richard Sear, Jim De Yoreo, Kenneth Harris, Radoljub Ristic, Anne Kavanagh, Kevin Roberts, Eric Breynaert, Allan Myerson, Gerard Coquerel, David Wu, Helmut Cölfen, Herma Cuppen, Mireille Smets and David T. Wu

DOI: 10.1039/c5fd90039e

Kenneth Harris opened a general discussion of the paper by Herma Cuppen: It is well established from solid-state NMR and other techniques that the NH_3^+ group in crystalline amino acids undergoes rapid rotation about the C–NH_3^+ bond. Is the phase transition between the β and α phases of DL-norleucine associated with any significant discontinuity in the rate of this motion, and/or the temperature dependence of the rate of this motion? Furthermore, is there any evidence for disorder (dynamic or static) of the alkyl chain of the norleucine molecules in the α and β phases, and if so, does the nature of this disorder change significantly at the phase transition?

Herma Cuppen answered: Rotation barriers about the C–NH_3^+ bond depend on the motive of the hydrogen bonding network in the crystalline phases. For DL-norleucine this motive stays intact during the transition, and I therefore do not expect a large difference for the two phases. The changes occur where the aliphatic chains interact. In a previous study,[1] we determined the rotation barrier of the methyl group by means of molecular dynamics simulations for both phases. We found them to be identical, and to be in agreement with experimental values for DL-norvaline[2] and DL-norleucine.[3] We did not find any correlation between internal movement of the molecules and the onset of the transition.

1. J. A. van den Ende and H. M. Cuppen, *Cryst. Growth Des.*, 2014, **14**, 3343.

2. H. J. Svec and D. D. Clyde, *J. Chem. Eng. Data*, 1965, **10**, 151–152.
3. P. Ren, D. Reichert, Q. He, L. Zhang and H. Tang, *J. Phys. Chem. B*, 2011, **115**, 2814–2823.

Terence Threlfall remarked: The site of the transition is not the amino acid end of the molecule - that is firmly fixed as hydrogen bonded dimers - but the hydrocarbon end. The hydrocarbon ends slip over each other, twisting as they do so. We have previously described this motion as like that of a bicycle being pedalled.

Kenneth Harris responded: I think you need to be careful to define what you mean when saying that "the amino acid end of the molecule...is firmly fixed as hydrogen bonded dimers". It is very well established from solid-state NMR studies that, in crystalline amino acids, the NH_3^+ group in the (zwitterionic) amino acid head-group typically undergoes a 3-site 120° jump motion about the C–NH_3^+ bond. Examples may be found in ref. 1 and 2 below, as well as in several other papers. Given the fact that X-ray diffraction provides a time-averaged and space-averaged representation of the crystal structure, and given that the time-average over a 3-site 120° jump motion of an NH_3^+ group is indistinguishable from a static NH_3^+ group (as a consequence of the symmetry of the jump process matching the local symmetry of the site), an NH_3^+ group undergoing a 3-site 120° jump motion would still look to be completely "fixed" and static in the time-averaged representation given by X-ray diffraction, while solid-state NMR techniques would reveal very clearly that the 3-site 120° jump motion occurs. Indeed, as in the examples given in ref. 1 and 2, small changes in hydrogen-bond geometry can lead to significant changes in the rate of the 3-site 120° jump motion of the NH_3^+ group. So, even if the region involving the amino acid head-groups in the α and β phases of DL-norleucine is geometrically very similar in the time-averaged crystal structures determined from X-ray diffraction, the rate of dynamics of the NH_3^+ group could still be significantly different in the α and β phases. I think this issue merits future investigation in the case of DL-norleucine, rather than being dismissed as not relevant.

1. S. J. Kitchin, S. Ahn and K. D. M. Harris, *J. Phys. Chem. A*, 2002, **106**, 7228–7234.
2. S. J. Kitchin, G. Tutoveanu, M. R. Steele, E. L. Porter and K. D. M. Harris, *J. Phys. Chem. B*, 2005, **109**, 22808–22813.

Terence Threlfall replied: I agree that it is likely that there will be rotation of the $NH3^+$ group in this case. Indeed there is evidence from the X-ray structural studies for it.[1] However there is no indication that the rotation results in any loosening of the amine–carbonyl conjunction. As there is by contrast clear evidence of hydrocarbon chain slippage and twisting resulting in the transformation and a plausible mechanism for it, it seems unneccesary to invoke any participation of the amino-acid end of the molecules.

1. C. H. Goerbitz, *Acta Crystallogr.*, 2104, **E780**, 341–343.

Terence Threlfall said to Herma Cuppen: You have described the mechanism of the transition as cooperative, but I agree with Mynukh that a concurrent cooperative mechanism must be energetically unfeasible. So are you proposing a sequential one? This might be a wave motion and might be regarded as a phonon mode and hence observable.

Herma Cuppen answered: I do indeed believe that the cooperative motion occurs only at a microscopic scale, and not over the full length of the crystal. This would result in a propagating front, and less in a wave. The transition is of first order, but because of the strong anisotropy of the crystal structure it occurs, in the light of classical nucleation theory, through this special mechanism.

Linda Seton addressed Herma Cuppen: When looking at DSC of the physical crystals, your paper states that the β to α transition is detected in some crystals and not in others, and assert that it has taken place even when not detected. Can you be sure the transition has taken place if there are no peaks in the DSC? Did you undertake any additional analysis to verify that the transition had occurred? Why do you think the transition was detectable in some crystals and not in others?

Herma Cuppen responded: Concerning your first question: upon cooling the sample for a second time, the peaks reoccur in the DSC traces. This means that a transition had taken place. Furthermore, our XRD results show that the transition indeed falls in this temperature range. Since the transition is expected to occur in a layer-wise fashion, some crystals transform with many layers simultaneously, leading to a detectable signal, whereas others transform over a larger temperature range.

Bart Rimez asked: What influence would a faster heating and cooling rate in DSC have on your results?

I would suggest you have not yet fully explored all possibilities to actually see the reverse change back to the previous lattice whilst heating. To start with, you should try to get a much more horizontal baseline with no curvature at all. This can be done by measuring your sample with empty aluminum pans on the sample and reference side. This way, you could subtract this instrument baseline from your actual measurement and get much straighter lines. Tthis will make it easier to actually see the transition. You mentioned that you couldn't observe the phase change when you were cooling at higher rates, and that's fair enough, but what about cooling at the same rate and then heating it up later at a faster rate, so that you would increase any possible signal to be able to actually see it? If it changes, a huge supercooling will provoke the change and your signals weren't that flat during heating (so there could be something hidden inside) - as you will of course know, anything that is formed with any sort of latent heat should have a resulting heat signal, unless it falls beneath the detection limits of the system. However, that's not true for your case, I think.

Additionally, you could also try to measure the difference in specific heat before and after the phase change. For this you would need to do the subtraction of a measurement taken with empty pans, and then calculate the step change, if any. This would give you a result in J g^{-1} K^{-1} which you can easily compare back again to the proposed 4 J mol^{-1} K^{-1} that is in the paper.

Finally, there are much more sensitive calorimeters available, where you put your crystal straight onto a sensor chip, meaning that the chances of you measuring the actual heat is much higher. If you are interested in doing that, we should get in contact, as I am working with people who specialise in doing this for polymer analysis (where solid–solid transitions are quite well-known, I might add). Good luck!

Mireille Smets answered: Thank you for the suggestions. Please be aware that the *y*-axis of Fig. 4 is extremely zoomed in, in order to visualize the very small enthalpy change, and the unit is in W mol^{-1}. This exaggerates the curvature of the baseline. We did measure our sample compared to an empty reference pan, but an extra measurement with two empty pans could perhaps improve the baseline. However, we do not believe this will significantly affect the measured enthalpy of transition. We performed new experiments with different heating rates during heating to enhance the heating signal as you suggested, but so far we have observed no significant effect.

Kevin Roberts commented: I wonder if you have considered using a temperature gradient cell for studying the phase transformation process in this system?. For example, if you use a hot stage set-up with a micro-temperature gradient, it would allow you to characterise whether you might have a nucleation event. Such a set-up would also allow you to have control over the movement of the transformation or nucleation front. Once you have established a temperature gradient across the crystal, you can adjust the temperature to obtain a phase boundary. Thereafter you can very gently change the temperature and move the phase boundary back and forth through the sample. With an appropriate microscopy technique, it should be feasible to characterise the structures observed. If the cell can be set up vertically, then gravity would provide a natural gradient.[1,2]

1. R. A. Becker, R. Docherty, A. El Korashy, H.-D. Jennissen, H. Klapper and K. J. Roberts, An examination of the 140°C phase transition in triammonium hydrogen sulphate $(NH_4)_3H(SO_4)_2$ using synchrotron radiation white-beam topography, *Phase Transitions*, 1992, **39**, 171–183.
2. H. L. Bhat, H. Klapper and K. J. Roberts, An X-ray topographic study of the paraelectric to ferroelectric phase transformation in nearly perfect single crystals of ammonium sulphate, *J. Appl. Crystallogr.*, 1995, **28**, 168–188.

Herma Cuppen replied: This is an interesting suggestion. However, since the transition temperature in DL-norleucine shows such a large hysteresis of several tens of degrees, I am not sure whether it would be possible to freeze in the transition for this particular system.

Kevin Roberts asked: Could you advise regarding the nature of the relationship between the space groups of the two phases? It would be interesting to know if the two structures were in a group/sub-group inter-relationship. In such a case, then, might one anticipate a solid-to-solid displacive and/or enantiotropic transformation between the two phases?

Herma Cuppen responded: The α polymorphic form crystallizes in $P2_1/c$, which is indeed a subgroup of $C2/c$, the space group of the β structure. However, since the lattice points change as well, this relationship does not appear to be relevant for this transition.

Wenhao Sun remarked: I find solid–solid transformations very interesting, as we usually observe these phase transformations to occur *via* a solution-mediated dissolution–reprecipitation route. I noticed that in your paper, you said that the metastable polymorph lasts for about 60 h before transforming to the stable phase, and you also mapped out the energy barrier to this transformation. I wonder, do you have any formulation for relating transformation energy barrier to lifetimes?

Herma Cuppen answered: In principle, this could give us some information on the kinetic prefactor in classical nucleation theory. However, the exact mechanism and even the size of the barrier for the transformation - since it scales with the number of molecules involved - is not yet understood fully enough to make this correlation.

Wenhao Sun said: You said that this barrier is in energy mol^{-1}, and since it is a cooperative transformation, the barrier applies to the whole crystal and therefore scales with the size of the crystal. Would you expect smaller crystals to exhibit lower polymorph lifetimes, given a solid–solid transformation?

Herma Cuppen replied: These barriers are indeed expressed in terms of energy per mole molecules in the moving interface. We have not performed an extensive study on how this scales with system size. We observe this cooperative motion at least on the molecular scale of several unit cell lengths. I do not expect this cooperative motion to extend through the full crystal, since, as you point out, the barrier will simply become too high. What I would expect is that the phase transition propagates through the crystal by a moving front.

Dimitrios Toroz asked: I was wondering if there is any change in the energy difference and also in the angles between the corners and edges of the crystalline forms, moving from one polymorph to the other, when these calculations are assessed in a wide range of crystal sizes (from small to large), even if the simulations performed in constant volume?

Herma Cuppen replied: The molecular dynamics simulations are performed in the NPT ensembles, and hence the volume is not constant. Here we do not see any significant change in volume and energy going from one phase to the other, as can be seen in Figures 2 and 3. Figures 2 and 4 show the experimental confirmation of this picture.

Jim De Yoreo opened a general discussion of the paper by Francis Taulelle: This was a very interesting talk and paper. I couldn't quite tell if you think that the nanoaggregates transform or if they simply supply the ions for nucleation; can you explain which process you think is happening?

Francis Taulelle replied: Actually, nanoaggregates are slowly adding to their initial pure aggregation state some additional interoligomeric species bonds. With their lower hydration state, they have quite a long life compared to the oligomers interconversion. After slightly increasing their average silicon connectivity, the charge density of such nanoaggregates is decreased, favoring a secondary aggregation between them. These phenomena have been monitored by DLS, SAXS and NMR.[1]

But as, during the induction period, the average connectivity does not evolve, one must deduce that some nano-aggregates transform in a much higher connection state and mostly undergo a second aggregation process, while some are undergoing a lessening of their connectivity. This sort of dismutation between nanoaggregates, leading to highly connected amorphous nanoparticles and weakly connected nano-aggregates, is revealed by the evolution of the nanoaggregates/nanoparticles in figure 10.b of ref. 2. This redissolution of amorphous nanoparticles has been studied in detail in ref. 3 and in figure 10 of ref. 2. Dissolution of nano-aggregates proceeds by decreasing their connectivity without changing the global size of the particles, and then all at the sudden, the nanoparticles return to oligomers and disappear. One may therefore describe the process as a two-step aggregation; firstly, a super aggregation of highly connected nanoparticles with a simultaneous decrease in the connectivity of the remaining nanoaggregates due to OH release from the highly connected nanoparticles, followed secondly by crystallization from the large amorphous super-aggregates and dissolution of the nano-aggregates. At this stage, the oligomers resulting from nanoaggregate dissolution participate in crystallization in the growth regime, not in the nucleation part. To me, this is just Oswald ripening in very slow motion. Oswald ripening in its first expression is oversimplifying the sequence of events in order to describe it with some sort of thermodynamic description, bypassing the intermediate processes and the complexity of the molecular events.

1. A. Aerts, M. Haouas, T. P. Caremans, L. R. A. Follens, T. S. V. Erp, F. Taulelle, J. Vermant, J. A. Martens and C. E. A. Kirschhock, Investigation of the Mechanism of Colloidal Silicalite-1 Crystallization by Using DLS, SAXS, and Si-29 NMR Spectroscopy, *Chem. - Eur. J.*, 2010, **16**, 2764–2774.
2. A. Aerts, C. E. A. Kirschhock and J. A. Martens, Methods for *in situ* spectroscopic probing of the synthesis of a zeolite, *Chem. Soc. Rev.* 2010, **39**, 4626–4642.
3. T. P. Caremans, B. Loppinet, L. R. A. Follens, T. S. V. Erp, J. Vermant, B. Goderis, C. E. A. Kirschhock, J. A. Martens and A. Aerts, Investigation of Nanoparticles Occurring in the Colloidal Silicalite-1 Zeolite Crystallization Process Using Dissolution Experiments, *Chem. Mater.*, 2010, **22**, 3619–3629.

Jim De Yoreo commented: It seems like you end up with an unlikely structure - I would expect a more dense state. Do you understand why this low density structure is obtained?

Francis Taulelle answered: Actually, when I state that zeolites are nanoporous structures, this is a simplification of language. Zeolites become nanoporous only after crystallization and an "activation" process, usually calcination, that empties them without breaking the silicate or aluminosilicate network. When zeolites crystallize, they do as a dense phase. After activation the empty pores are filled with cations surrounded by water molecules, or a clathrate type of organization. Their interaction with the framework is weaker than the covalent aluminosilicate network, but is essential to crystallization.

Beatrice Biscans asked: According to your hypothesis, nanoaggregates don't change state for a long time, and the system is stable for NMR measurements.

Have you checked this assumption? Are the measurements reproducible? Can you explain why these naoaggregates are stable when forming; and can you correlate your data to the DLS and SAXS measurements and comment on the results?

Francis Taulelle replied: The state of nanoaggregates has been studied by NMR before.[1] Experiments are extremely reproducible, and many PhD students have reproduced them at KU Leuven.

The same methdology has been applied to other zeolitic systems.[2-4]

The nanoaggregates forms at a Si : OH ratio of 1. At this value, titration of all OH^- has been reached and the pH drops, as well as the dielectric constant of the medium. Therefore ion-pairing is preferred, leading to a critical aggregation concentration. As a low solubility of tetrapropylammonium exists in the hydro-alcoholic liquid, nanoaggregates are defined by a narrow size distribution, of about 2 nm, at a Si : OH ratio of 1. As soon as the critical aggregation constant has been reached, any further addition of a silicon source leads to a progressive integration of oligomers into the nanoaggregates, as their population growth is made at the expense of the oligomer population. When a Si : OH ratio of 2.5 to 2.8 is reached, most of the oligomer population has been converted into nano-aggregates. Crystallization can therefore proceed, though at a very slow pace at room temperature (about 3 years to get a reasonable fraction of crystals) compared to about 25 h at 95°C.

DLS and SAXS have been systematically used to correlate with NMR connectivity evolution. Such a correlation was first described in ref. 5. After formation of the nanoaggregates, if the system is heated, the size of the nanoaggregates increases in steps up to 4–6 nm after about 5h of heating, and from 5 to 100 nm between 5h and 25 h, transforming the nanoaggregates into amorphous connected pre-crystalline nanoparticles. At about 25 h, this amorphous phase starts transforming into a periodic crystal, as indicated by Bragg peaks; the nucleation is starting from inside the large amorphous nanoparticles.

This applies to ionic liquids as well, except that with these ionic liquids the reduced amount of water leads to an onset of nanoaggregation at a Si : OH ratio of about 0.5. Furthermore, with temperature increase the dielectric constant of the medium decreases, leading to a more long-lived state of aggregation and promoting further successful condensation of the proto-crystalline network, before the onset of crystallization.

1. D. P. Petry, M. Haouas, S. C. C. Wong, A. Aerts, C. E. A. Kirschhock, J. A. Martens, S. J. Gaskell, M. W. Anderson and F. Taulelle, Connectivity Analysis of the Clear Sol Precursor of Silicalite: Are Nanoparticles Aggregated Oligomers or Silica Particles? *J. Phys. Chem. C*, 2009, **113**, 20827–20836.
2. N. Hould, M. Haouas, V. Nikolakis, F. Taulelle and R. Lobo, Mechanisms of Quick Zeolite Beta Crystallization, *Chem. Mater.*, 2012, **24**, 3621–3632.
3. E. A. Eilertsen, M. Haouas, A. B. Pinar, N. D. Hould, R. F. Lobo, K. P. Lillerud and F. Taulelle, NMR and SAXS Analysis of Connectivity of Aluminum and Silicon Atoms in the Clear Sol Precursor of SSZ-13 Zeolite, *Chem. Mater.*, 2012, **24**, 571–578.
4. M. Castro, M. Haouas, F. Taulelle, I. Lim, E. Breynaert, G. Brabants, C. E. A. Kirschhock and W. Schmidt, Multidiagnostic analysis of silicate speciation in clear solutions/sols for zeolite synthesis, *Microporous Mesoporous Mater.*, 2014, **189**, 158–162.
5. A. Aerts, M. Haouas, T. P. Caremans, L. R. A. Follens, T. S. V. Erp, F. Taulelle, J. Vermant, J. A. Martens and C. E. A Kirschhock, Investigation of the Mechanism of Colloidal Silicalite-1 Crystallization by Using DLS, SAXS, and Si-29 NMR Spectroscopy, *Chem. - Eur. J.*, 2010, **16**, 2764–2774.

Beatrice Biscans asked: The mechanism of formation of the zeolites is decribed with nanoaggregates precursors. Is it possible to obtain experimental evidence of these nanoaggregates? Which are the scales reachable by the different techniques (DLS, SAXS): individual nanoparticles, nanoaggregates, or aggregates of smaller nanoaggregates? Is there a fractal dimension in such systems?

Francis Taulelle answered: The experimental evidence is provided by the evolution of the connectivity of silicon. A fully connected network reaches 4 for such crystals. From a topological point of view, a crystal realization is an ensemble of nodes connected by links. When all the links are realized, the net of the crystal is reached. However, periodicity may not yet be present. After some evolution, the periodicity can be reached. The critical number for following the evolution of crystallization is to measured the number of links realized. This has previously been done by NMR.[1] NMR allows us to perform such counting of realized links for every sort of interaction, leading to measurable proximities; for inorganic systems like zeolites, for biological entities like peptides of protein, for small organic molecules with hydrogen bonding and for salts with electrostatic interaction. In each situation, measurement of proximities is made possible by solid state NMR. This is the critical chemical information about the network development. DLS and SAXS provide information about the size and shape. Scales from DLS and SAXS are described in ref. 1. Both network advancement, and size and shape, are necessary to qualify the crystallization process. In the case of sol-gel silicate formation, its dimensionality is described by a fractal dimension. In the case of silicates in hyper-basic conditions, the objects are following standard geometrical dimensions for oligomerization and crystallization. However, nanoaggregation is still not described as a fractal dimension, though the network realization fraction (the fraction of total links needed to form the crystal) can be analyzed by a percolation theory and with fractal dimensions. From nanoaggregation to super-aggregation of nanoparticles, the beads in the necklace are usually following 1 dimension, but at crystallization this evolves from 1 to 3 dimensions. Again, the fraction of the maximal number of links to form the crystal is the critical number to monitor. Currently, this is the original contribution of our work: measuring such a network advancement by their fraction of links realized.

1. D. P. Petry, M. Haouas, S. C. C. Wong, A. Aerts, C. E. A. Kirschhock, J. A. Martens, S. J. Gaskell, M. W. Anderson and F. Taulelle, Connectivity Analysis of the Clear Sol Precursor of Silicalite: Are Nanoparticles Aggregated Oligomers or Silica Particles? *J. Phys. Chem. C*, 2009, **113**, 20827–20836.

Colan Hughes remarked: Your solution-state NMR data clearly tells you a lot about the initial stages of the crystallization. Have you considered using *in situ* solid-state NMR to look at the later stages of the process, as larger particles form?

Francis Taulelle responded: As I am myself a solid-state NMR spectroscopist, the answer is yes. Focussing on the early stages of nucleation puts the emphasis on NMR liquid-state-like methods, while the last part of nucleation and the beginning of growth focuses more on solid-state NMR-like methods. For zeolites, solid-state methods are exquisitely useful, especially if you use cross-polarization schemes between protons and silicon, reducing the frequently encountered burden of very long relaxation times for silicon-29 when measuring it directly.

Kevin Roberts commented: Do you think that there might be an opportunity to use X-ray absorption spectroscopy in the soft X-ray region to probe the local structure, and in particular the chemistry associated with the oxygen environment? It's possible to assemble a radical distribution around individual atoms, and my thought is that one could use techniques such as these to determine, for example, the oxidation state.

Francis Taulelle answered: Radial distribution functions around atoms can be probed by X-ray methods. And they might help to contribute to the elucidation of the local order. As an NMR spectroscopist myself, before anything else I would use ^{29}Si-, ^{27}Al-, ^{17}O- and ^{1}H NMR. They would provide very valuable information to probe local order by their chemical shifts around individual atoms. Additionally, the magnetic dipolar interaction between nuclei lead, by methods like REDOR, to a progressive build-up of magnetization transfer onto the observed nucleus from the distant nuclei; this produces selective radial distribution functions that can be correlated to those obtained by diffraction methods. They are powerful tools to investigate the local order, and the distribution of heteroatoms in zeolites.

Eric Breynaert remarked: During the discussion, the question has been raised whether Si K-edge NEXAFS could contribute to resolving the early stages of zeolite nucleation. Concerning this topic, it is important to discriminate between different systems in which zeolite nucleation can occur. In silicate ionic liquids, the speciation of silicon is accessible with ^{29}Si MR, while DLS, SAXS and NMR also provide information on pre-nucleation stages and aggregated clusters.[1] For larger colloidal species or gel systems, the information that can be obtained from NMR, DLS and SAXs is much more limited. In such cases, Si K-edge NEXAFS could contribute to the characterisation of the system, although more detailed information can be expected from High Energy X-ray Scattering (HEXS) in combination with Pair Distance distribution Function analysis (PDF). While the information derived from NEXAFs is limited to the local chemical environment of Si up to a few angstroms, the length scale probed by HEXS/PDF could reach up to 15Å. In combination with ^{29}Si NMR (both liquid state and MAS-NMR), this could provide more insight into the evolution of the silicon speciation in gel systems and systems containing larger colloidal species, which are less easily accessible for multidiagniostic analysis with SAXS, DLS and liquid state NMR.[1]

1. Castro *et al.*, Multidiagnostic analysis of silicate speciation in clear solutions/sols for zeolite synthesis, *Microporous Mesoporous Mater.*, 2014, **189**, 158–162.

Jim De Yoreo opened a general discussion of the papers by Haihua Pan: I'm not sure I understand how you are expressing nucleation rate in terms of calcium and amorphous calcium phosphate (ACP) concentration. Wouldn't you need to consider the total surface area of ACP instead of the mass? Why does it depend on the square of the calcium concentration?

Haihua Pan responded: For heterogenous nucleation, nucleation rate is proportional to the square of monomer concentration. In our system, the transformation of ACP to HAP needs one additional calcium:

$3Ca_3(PO_4)_2 + Ca^{2+} + 2H_2O \rightleftarrows 2Ca_5(PO_4)_3(OH) + 2H^+$.

For ACP particles with a similar size, the surface area of ACP is proportional to the mass of ACP. In this case, we can use mass or molar concentration of ACP. Otherwise, the total surface area of ACP should be applied.

Richard Sear asked: As I understand your findings, you see hydroxyapatite (HAP) crystals apparently nucleate at the surface of ACP particles, and grow out into the surrounding solution, not into the ACP particles themselves. Is this correct? If so, do you know why? Are the dynamics of the ions in ACP very slow, *i.e.*, is it glassy which thus inhibits crystallisation inside the ACP particles?

Haihua Pan responded: Yes, it is. In our understanding, additional calcium is needed to transfer from ACP to HAP:

$$3Ca_3(PO_4)_2 + Ca^{2+} + 2H_2O \rightleftarrows 2Ca_5(PO_4)_3(OH) + 2H^+.$$

Free calcium ions are available in solution, so HAP is readily formed on the ACP surface. We do have direct evidence of the dynamics of the ions in ACP. Since ACP could be in a form of a partially merged peanut or a dumbbell morphology, it is safe to say that ACP is either a solid (glassy) phase or a highly viscous gel. The diffusion of ions in ACP could be extremely slow compared with that in solution.

Richard Sear asked: Can you measure the viscosity and/or elasticity of ACP? Or measure diffusion constants for calcium or phosphate ions in ACP?

Haihua Pan replied: That would be an interesting topic. Currently we have not measured these properties.

Beatrice Biscans commented: In the ACP–HAP–polymer system, supersaturation of the solutions is a key parameter. In the process, supersaturation is mixing dependent, and evolves (so the supersaturation is not constant). Can you explain your supersaturation calculations, and how the activity coefficients were calculated? What is the range of ionic force values? Can you reproduce the pH evolution with this calculation?

Haihua Pan responded: This is a good question. The supersaturations of the solutions were calculated using VMINTEQ 3.0. Davies' approximation of the Debye–Huckel equation was used for calculating the activity coefficients (the parameter $b = 0.3$ is applied). The solubility products, pKsp, of HAP (pKsp = 58.33, MINTEQ database: NIST 46.7), and ACP (pKsp = 25.5, taken from ref. 1 and 2, which is close to our measurements of 25.9, *cf.* Table S3 and S4 in our paper, DOI: 10.1039/c4fd00212a) were used in the calculation. The ionic strength was about 0.15 M in our solutions. Quantitatively, we have checked that the pH dropped as the formation of ACP occurred, and it dropped further as the precipitation of HAP occurred. But we cannot reproduce the pH evolution by calculation, as we are missing some details: the adsorption of species on the ACP surface, the fraction of prenucleation clusters with respect to the free ions, the solubility of poorly crystallized HAP and its chemical compositions. We agree that it would be wonderful if one could reproduce the pH evolution using solution chemistry calculations by the first principle (without too many assumptions or over-fitting).

1. M. R. Christoffersen, J. Christoffersen and W. Kibalczyc, *J. Cryst. Growth*, 1990, **106**, 349.
2. S. V. Dorozhkin and M. Epple, *Angew. Chem., Int. Ed.*, 2002, **41**, 3130.

Richard Sear said: The solubility of a thermodynamic phase coexisting with another should only depend on thermodynamic parameters such as temperature, pH, *etc.*, not on how the phase is prepared (for example, from starting concentrations). Do you observe that ACP behaves like a thermodynamic phase in this sense? Or do ACP's properties depend on how it is prepared?

Haihua Pan answered: This is a really good question. We cannot fully answer these questions. In our limited measurements, the solubility of ACP is almost the same at different supersaturations (see Table S3 and S4 in our paper). The size and aggregation state of ACP depends on how it is prepared (*e.g.* adding one solution drop-by-drop into another while stirring, fast injecting, mixing by a Y-tube or mixing tube, *etc.*). If the solubility of ACP is size-dependent, then the solubility of ACP depends on how it is prepared (which is why in nucleation kinetics studies, the protocols should be kept internally consistent.) As for whether ACP is a thermodynamic phase, this is a complicated question. Whether it is a single phase or a spectrum of phases is still a question. For the amorphous calcium carbonate (ACC) system, proto-vaterite and proto-aragonite have been reported. The evolution from an oligomer or a chain-like ionic cluster to condensed amorphous spheres has been observed for both ACC and ACP. In this regard, we prefer to say that ACP is a spectrum of phases. But eventually, they will evolve into a certain kind of amorphous phase before crystal phase nucleation, as indicated by the solubility and the compositions of these amorphous phases (see Table S3, S4 in our paper). This question is still open, as we only have limited known properties. The density, compressibility, viscosity, mechanical and optical properties are largely unknown.

Jim De Yoreo commented: In our experiments, if one uses concentrations slightly below the solubility limit of ACP, as expected, ACP doesn't form at all. Instead the solution forms HAP first. However, in all of our experiments performed under conditions that give ACP first, it transforms to OCP before forming HAP. You observe ACP go directly to HAP. Why are your results different?

Haihua Pan responded: In our observations, whether ACP will transform to OCP first or directly to HAP depends on the solution conditions (additives, pH, temperature, Ca : P ratio, etc). In our recipes (Table S1) and protocols, it directly transforms to HAP. We found that at some lower pH conditions (say, a pH of less than 6.2), OCP formed first before HAP. This would be another interesting topic.

Celso Aparecido Bertran communicated: The pre-mixing and post-mixing experiments must be more detailed. Why is the time between the mixture of the ionic solutions and the addition of the 20μL of the polymer solution in the post-mixing experiments?

My question is; Figure 8a in your paper represents the pH variation with the time for the pre-mixing and the post-mixing experiments. Could the time between the mixing of the ionic solutions and the addition of the polymer solution in the post-mixing experiment change the time at which the pH goes down in the post-mixing experiment?

Haihua Pan communicated in reply: Thank you for your question. In post-mixing experiments, after the mixing of calcium and phosphate solutions, we waited for 10 mins before adding 20 μL of PAA solution into the suspension. This was to make sure that ACP was formed, but that it was still far before the nucleation of HAP (the induction time for the control experiment is about 2 h). The pH curve was still monitored during the mixing of PAA with the suspensions, and the mixing process was finished within seconds. So, it will not affect the determination of the induction time for post-mixing experiments (note that the induction time for HAP nucleation was more than 10 h in this protocol).

Terence Threlfall returned to the discussion of the paper by Herma Cuppen: Can I clarify the problems with the DSC of D,L-norleucine. The high temperature transition at 117 °C is easy to see. The other transition is of very low enthalpy (typically 0.6 J g^{-1}) and is very variable in both energy and temperature, pointing to defect mediation. We have done the DSC at different rates.[1] At 300 degrees per minute, rather than becoming more intense, it vanishes completely because the ramp exceeds the kinetics of the process. Even at lower rates, sometime it can be seen and sometimes not, as it is at the limit of instrumental sensitivity. Typically it occurs in two stages, one at about −10 °C and one about −20 °C, at around 10 degrees per minute cooling. So we have done variable temperature single crystal diffraction, at +10 °C, between −10 and −20 °C, and below −20 °C, and we observe a P16, a P32, a C16 and a C32 cell (P = primitive, C = centred, 16 and 32 refer to unit cell lengths). These correspond to the 4 corners of the box shown by Dr Cuppen. So theory matches experiment, but the experiments are very difficult, and we have never been able to solve the intermediate structures or understand why they arise.

1. Coles *et al. Cryst. Growth Des,* 2009, **9**, 4610–4612

Mireille Smets replied: Our experience with DSC measurements of the α ↔ β transition in DL-norleucine is that the transition is indeed extremely variable, and the enthalpy change is often in the order of the detection limit. However, at low heating rates of 2 K min^{-1}, the transition during cooling is sometimes visible and reproducible in single crystals. We do not observe two distinct peaks at −10 and −20 °C, but rather a few small peaks at around −20 °C, indicating the layer-wise transition. In SCXRD we also observed a "P32" or "C16" cell during some measurements, but we believe that this is due to imperfect translational symmetry in the direction perpendicular to the layers of the transforming crystals.

Haihua Pan remarked: In your work, it tells us that there is no nucleation for a solid-to-solid polymorphic transformation. But I have noticed a hysteric loop for the transition from one to another. There should be a barrier between these two phases. Could you tell us how to differentiate the chemical reaction barrier from the nucleation barrier in this work? Both barriers seem to lead to a hysteric loop during phase transformation.

Herma Cuppen answered: In our contribution, we conclude that the transition indeed occurs with a barrier, but that the mechanism is *via* a cooperative motion, and not through a standard nucleation and growth mechanism that would start from one or a few molecules. In Fig. 8 of our paper, we consider the different possible kinetic pathways. We expect, however, that the transitions in both directions proceed through the same route.

Wenhao Sun said: Is it possible to do lattice vibration calculations for these crystals? If so, would it be possible to calculate the phonon spectra for each phase, as well as the phonon spectra at the transition point/structure? Perhaps this would allow us to quantify and predict lifetimes. If you can calculate the phonon spectra, it would give you an idea of how much the crystal vibrates towards the transition point as a function of temperature. Then you could estimate the probability of a thermal fluctuation bringing the energy of your system beyond that transition point, as a function of time. This would give a predicted estimate on the lifetime of polymorphs.

Herma Cuppen replied: Phonon spectra can be easily determined for the stable phases. For the transition point this will be more challenging, but this is an interesting suggestion.

Francis Taulelle commented: If, as some colleagues have mentioned, dielectric spectroscopy or phonon modes could be used to investigate the differences between DL-norleucine polymorphs, then using solid state NMR with T1ρ measurement of protons, indirectly detected by ^{13}C-NMR when using cross-polarization magic angle spinning NMR experiments, could be an alternative method that could be much easier or accessible.

Herma Cuppen replied: This is indeed an interesting suggestion. In fact, a manuscript dealing with such measurements is in preparation.

Kenneth Harris addressed Herma Cuppen: Following up the question raised earlier concerning the possible role of phonons in driving the phase transition between the β and α phases: in this regard, it would be interesting to explore any possible differences in the phase transition behaviour between a sample of DL-norleucine with natural isotopic abundances (as studied in your work), and a sample of fully deuterated DL-norleucine.

Herma Cuppen responded: This is an interesting suggestion.

Jan Sefcik returned to a discussion of the paper by Francis Taulelle: Would it be correct to say that zeolites are essentially covalent crystals? Could you think of zeolite formation as a polymerisation process? If solubility of zeolites increases with temperature, why do you need to heat to induce nucleation, and if you have to heat, is this really an activated reaction/polymerisation process? If there is an analogy of zeolites with molecular crystals, how should solubility of zeolites then be defined? Why does it seem to be difficult to determine this? Regarding zeolite solubility, see ref. 1.

1. J. Sefcik and A. V. McCormick, *Microporous Mater.* 1997, **10**, 173–179.

Francis Taulelle answered: Zeolites are definitely covalent crystals. And the analogy with polymerization holds as a first approximation of crystal formation. However, Si-*O*-Si bonds are not all having and equal energy, this is the point where the analogy with polymerization processes breaks down. This applies also to the so-called "coordination polymers". Chemically identical bonds are formed but as soon as the crystalline network appears they do not have the same bond energy, due mainly to the topological and anisotropic character of the crystal network. The work of M. Anderson *et al.* at the University of Manchester[1] has proven that when dissolving zeolites, units leaving the crystal are not randomly distributed but are a limited number in shape and nuclearity. Furthermore, such an identification of the units in zeolites has been attempted by topological crystallography.[2]

These findings are quite different from the concept of secondary building units introduced much earlier in zeolite description. The belief that one could find the crystal sub-units inside the solution has disappeared. I made such a statement in ref. 3 and 4. This is the underlying reason why such units cannot be recognized in solution. The oligomers are interconverting in solution, so as soon as the detachment of units occurs, these oligomers, stable as such within the crystal network, are no longer stable with the same topology under solvation state. Therefore, they interconvert readily and change their stable forms. Conversely, at inclusion they need to de-solvate and connect to the network; in doing so they change their conformation, or isomerize to match the network. In doing this, they connect into a topological network, that may be completely or almost completely connected, but does not yet exhibit the periodic order of the crystal. This phenomenon is identical to the concept of two-step nucleation. Solubility is an essential issue for the understanding of zeolite formation. I apologize for not having immediately recognized one of the co-authors of the reference indicated. I myself quote this paper as a example of what needs to be done in zeolite chemistry. In the case of the direct synthesis of K-ABW using hydrated silicate ionic liquid, a phase that has previously never been directly synthesized but formed through the ion exchange of Li-ABW with K, we directly measured the solubility of the zeolite by measuring all components in solution by NMR of each nucleus, K, Si and Al.

If zeolites with a 1 : 1 Si : Al ratio, solubility can be rigorously defined, using Sefcik and McCormick notation of the solubility product. For a ratio different from 1 : 1, two cases occur, with zoning and with random distribution of Al respecting the statistical distribution indicated in Dempsey's rule. In the first case, solubility can still be defined considering the crystal as a solid solution of a Si : Al of 1 : 1 ratio and a purely siliceous zeolite. In the case of Dempsey's rule, solubility remains definable as a macroscopic measurement. However, in terms of the kinetic evolution of dissolution, the situation will be different, because the lability and stability of Si–O and Al–O bonds are different; Al–O are more stable than Si–O, but Al–O are more labile than Si–O. Concerning the solubility as a function of temperature, this is an open matter. Very few solubilities of zeolites have been determined, among them the ones from Sefcik and McCormick, and even fewer as a function of temperature. Quartz still remains a reference for silicate, but is probably not representative of zeolites. Therefore, are zeolites, as commonly admitted, exhibiting an increase of solubility with temperature, or like alumino-phosphates, with many isotypic zeolitic structures, or carbonates, do they show a reverse solubility? Or is it a variation about direct and reverse solubility? What is for sure is the following. If a zeolite crystal forms by network connection, Si-*O*-Si connectivity will reach approximately 4 at crystallization, and different modes of Si-*O*-Si, with processes implying water-assisted and temperature-assisted reactions, are taking place; depending on the solubility and bond formation activation needed, one must have an optimal temperature for forming each zeolite. This field is still open, but the syntheses from HSIL directly open the possibility of measuring the solubility product in *in situ* conditions by NMR.

1. R. Brent, P. Cubillas, S. M. Stevens, K. E. Jelfs, A. Umemura, J. T. Gebbie, B. Slater, O. Terasaki, M. A. Holden and M. W. Anderson, Unstitching the Nanoscopic Mystery of Zeolite Crystal Formation,*J. Am. Chem. Soc.*, 2010, **132**, 3858–13868.
2. V. A. Blatov, G. D. Ilyushin and D. M. Proserpio, The Zeolite Conundrum: Why Are There so Many Hypothetical Zeolites and so Few Observed? A Possible Answer from the Zeolite-Type Frameworks Perceived As Packings of Tiles, *Chem. Mater.*, 2013, **25**, 412–424.
3. F. Taulelle, Crystallogenesis of microporous metallophosphates. *Curr. Opin. Solid State Mater. Sci.*, 2001, **5**, 397–405.
4. F. Taulelle, M. Pruski, J. Amoureux, D. Lang, A. Bailly, C. Huguenard, M. Haouas, C. Gerardin, T. Loiseau and G. Ferey, Isomerization of the prenucleation building unit during crystallization of ALPO(4)-CJ2: An MQMAS, CP-MQMAS, and HETCOR NMR study, *J. Am. Chem. Soc.*, 1999, **121**, 12148–12153.

Wieslaw Roth addressed Jan Sefcik, Francis Taulelle and Eric Breynaert: Dr Sefcik mentioned the solubility of zeolite A, which has the simplest composition we know, *i.e.* Si/Al = 1. Most zeolites have higher Si/Al and exhibit Al zoning with different concentrations in various regions, for example inside or at the external surface. The solubility is reduced near Al centers, and thus their non-uniform distribution means that there is no well-defined single value of solubility. This is now exploited in synthesis of zeolites with hierarchical pore structure by desilication, *i.e.* limited dissolution of siliceous regions, leaving behind less soluble Al-richer parts.

Even if a defined solubility existed and we knew it, the modeling at this point would be extremely difficult (although it should be pursued). Dr Sefcik and Prof. Taulelle suggested that silicalite might be suitable for modeling studies.

Silicalite nucleation and crystal growth have been already studied quite extensively, including population balance methodology[1]. So far little attention has been paid to these results, and there has been no sustained effort to develop a comprehensive understanding of zeolite formation, which is obscure. One of the recent reviews of zeolite nucleation[2] suggests some agreement between some experimental results and the theory of zeolite nucleation, but overall there is almost total separation between theories of zeolite formation, especially quantitative approaches and practical synthesis. The latter is quite advanced and highly valuable.

1. D. Creaser, *Stud. Surf. Sci. Catal.*, 1999, **125**, 117.
2. B. Subotic *et al.*, *Ordered Porous Solids*, ed. V. Valtchev, S. Mintova and M. Tspatsis, Elsevier, 2009, pp 127–186.

Francis Taulelle answered: Dr Wieslaw Roth, you commented on the solubility differences beyond zeolites of a Si : Al ratio of 1. For those zeolites with a Si : Al ratio of 1 : 1, solubility can be well defined. For Si : Al beyond 1, zoning may occur and solubility may differ from place to place, depending on the local Si : Al ratio.

We should be quite careful with the definition of solubility. By dissolution, one compares the free energy of components making a crystal - the lattice free energy - to the free energy of the solvated state of components in solution. Reference states are different, the pure crystal phase and the infinitely diluted solvated species.

When studying zeolite with a Si : Al ratio of 1 : 1, all those terms are well defined. However, for a Si : Al ratio higher than 1, several cases are at stake; the statistical distribution of Al follows Dempsey's rule, while the zoning of Al infringes Dempsey's rule as Dr Roth mentions. Usually no situation that infringes Lowenstein's rule is considered. Dissolution under hydroxide hyper-basic media usually leads to de-silication of zeolites, but does not de-aluminate them. In acidic dissolution, or steaming, aluminum dissolves but silicon does not. Two phenomena are needed to understand such variations:

1. The kinetics of Al–O and Si–O bond breaking and formation.
2. The relative stability of extracted entities: monomers or oligomers of Al or Si, as well as other amorphous forms like amorphous oxyhydroxyaluminates, amorphous silica, and alumino-silicates, crystalline or not.

If zoning is related to a thermodynamically stable situation, it can be viewed as solid-solid solutions, and can therefore be studied as such. If zoning is dominated by kinetics, the zeolite formed is metastable, and a solubility measurement is still possible; it is a macroscopic measurement, but may be difficult to accurately reproduce due to kinetics of dissolution. Apparent solubility can be measured with accurately defined protocols. Properties that are controlled by morphology are related to kinetics of growth; there is faster Al integration due to a lower activation energy of formation of 4 connected Al T sites, and a higher activation energy for 4 connected Si T sites. But simultaneously, there is higher stability for Al–O bonds compared to Si–O bonds.

The rest of the solubility discussion has already been covered in my previous response.

Jan Sefcik responded: It is also possible to address the solubility of zeolites with Si : Al ratios higher than one using similar thermodynamic models as those used to describe the solubility of zeolite A.[1] This was illustrated for zeolite X where equilibrium crystallisation diagrams were calculated in sodium aluminosilicate aqueous systems.[2]

1. J. Sefcik and A. V. McCormick, *Microporous Mater.*, 1997, **10**, 173–179.
2. J. Sefcik and A. V. McCormick, *Chem. Eng. Sci.*, 1999, **54**, 3513–3519.

Eric Breynaert replied: The discrepancy between practical zeolite synthesis and theories on zeolite nucleation originates from the fact that zeolites are metastable phases traditionally synthesized by hydrothermal treatment of a reactive hydrogel, prepared from sources of aluminium and silicon with organic and/or inorganic cations as structure directing agents (SDA). The synthesis outcome depends on chemical composition, ripening, alkalinity, temperature and reaction time. Ageing of the gel below the temperature of the actual hydrothermal process is often needed to prepare the desired zeolite phase. A multitude of parameters have an impact on nucleation and growth kinetics, phase purity, final chemical composition, crystal size and morphology.

The transformation of traditional gel precursors into crystalline product is difficult to access experimentally. Therefore, the study of zeolite nucleation processes remains a topic of substantial academic interest, even after several decades of zeolite research.[1–3] Significant scientific progress has been made in the understanding of zeolite formation from precursor clear solutions with organic SDA, a system which contributed significantly to the elucidation of the molecular aspects of this type of zeolite assembly process.[4–7] Compared to gels, solutions are much more accessible to study early molecular events that initiate zeolite nucleation and crystal growth. These optically transparent solutions are typically prepared by hydrolysis of a monomeric silicon source such as tetraethylorthosilicate (TEOS) in the presence of tetrapropylammonium hydroxide (TPAOH). Self-condensation of the in situ generated silicic acid generates oligomeric silicates with finite nuclearity.[8,9] In clear solutions with a Si : TPAOH molar ratio exceeding unity, silica oligomerizes and assembles into nano aggregates.[10–12] These suspended primary nanoparticles measuring 1–2 nm have been characterized using ^{29}Si NMR and scattering techniques such as SAXS and DLS.[8] While each characterization technique yields fragmented information, only a combination of the different techniques provided detailed insight into the transformation process of the primary nanoparticles, which over time adopt a silicate connectivity resembling the final zeolite.[13,14] Recently, this work was extended to precursor clear solutions containing different organic SDAs by integrating SAXS, DLS and NMR in a unified characterisation scheme.[15]

These advances in understanding of zeolite formation with organic SDA have not yet been matched by an improved molecular understanding of zeolite nucleation in purely inorganic syntheses, a discrepancy which is related to the ready formation of gels in traditional inorganic synthesis mixtures. Interestingly, colloidal suspensions of nanoparticles with uniform and very small sizes have been prepared in concentrated solution by combining aluminium and silicon sources via advanced mixing schemes.[16,17] Ng *et al.*, for example, devised a Na-based organic free synthesis of Na–EMT, a zeolite synthesis originally involving the use of a 18-crown-6 ether template.[17,18] This achievement was in full accordance with the predicted phase diagram.[19] However, in these cases NMR characterization also suffered from line broadening as crystallisation occurred within the solid phase.

This setback has very recently been overcome by the discovery of hydrated silicate ionic liquids (HSILs), which are highly concentrated, optically transparent solutions containing hypohydrated alkali and alkaline earth metal silicates from which zeolites can nucleate and grow. In contrast to the hydrothermal gel method, the optically transparent HSILs offer exceptional potential for monitoring zeolite nucleation and crystallisation with SAXS, NMR and DLS.

1. C. S. Cundy and P. A. Cox, The hydrothermal synthesis of zeolites: history and development from the earliest days to the present time, *Chem. Rev.*, 2003, **103**, 663–702.
2. C. S. Cundy and P. A. Cox, The hydrothermal synthesis of zeolites: Precursors, intermediates and reaction mechanism, *Microporous Mesoporous Mater.*, 200, **82**, 1–78.
3. J. D. Epping and B. F. Chmelka, Nucleation and growth of zeolites and inorganic mesoporous solids: Molecular insights from magnetic resonance spectroscopy, *Curr. Opin. Colloid Interface Sci.*, 2006, **11**, 81–117.

4. A. Aerts, L. R. A. Follens, E. Biermans, S. Bals, G. Van Tendeloo, B. Loppinet, C. E. A. Kirschhock and J. A. Martens, Modelling of synchrotron SAXS patterns of silicalite-1 zeolite during crystallization, *Phys. Chem. Chem. Phys.*, 2011. **13**, 4318–4325.
5. T. Verstraelen, B. M. Szyja, D. Lesthaeghe, R. Declerck, V. Speybroeck, M. Waroquier, A. P. J. Jansen, A. Aerts, L. R. A. Follens, J. A. Martens, C. E. A. Kirschhock and R. A. Santen, Multi-level Modeling of Silica–Template Interactions During Initial Stages of Zeolite Synthesis, *Top. Catal.*, 2009, **52**, 1261–1271.
6. T. P. Caremans, B. Loppinet, L. R. A. Follens, T. S. van Erp, J. Vermant, B. Goderis, C. E. A. Kirschhock, J. A. Martens and A. Aerts, Investigation of Nanoparticles Occurring in the Colloidal Silicalite-1 Zeolite Crystallization Process Using Dissolution Experiments, *Chem. Mater.*, 2010, **22**, 3619–3629.
7. A. Aerts, M. Haouas, T. P. Caremans, L. R. a Follens, T. S. Van Erp, F. Taulelle, J. Vermant, J. A. Martens and C. E. A. Kirschhock, Investigation of the mechanism of colloidal silicalite-1 crystallization by using DLS, SAXS, and 29Si NMR spectroscopy, *Chem. - Eur. J.*, 2010, **16**, 2764–2774.
8. L. R. A. Follens, A. Aerts, M. Haouas, T. P. Caremans, B. Loppinet, B. Goderis, J. Vermant, F. Taulelle, J. A. Martens and C. E. A. Kirschhock, Characterization of nanoparticles in diluted clear solutions for Silicalite-1 zeolite synthesis using liquid 29Si NMR, SAXS and DLS, *Phys. Chem. Chem. Phys.*, 2008, **10**, 5574–5583.
9. S. A. Pelster, W. Schrader and F. Schüth, Monitoring temporal evolution of silicate species during hydrolysis and condensation of silicates using mass spectrometry, *J. Am. Chem. Soc.*, 2006, **128**, 4310–4317.
10. E. A. Eilertsen, M. Haouas, A. B. Pinar, N. D. Hould, R. F. Lobo, K. P. Lillerud and F. Taulelle, NMR and SAXS analysis of connectivity of aluminum and silicon atoms in the clear sol precursor of SSZ-13 zeolite, *Chem. Mater.*, 2012, **24**, 71–578.
11. N. D. Hould and R. F. Lobo, Nanoparticle precursors and phase selectivity in hydrothermal synthesis of zeolite, *Chem. Mater.*, 2008, **20**, 5807–5815.
12. A. Rivas-Cardona, M. Chovanetz and D. F. Shantz, A systematic investigation of silicalite-1 precursor mixtures with varying degrees of dilution, *Microporous Mesoporous Mater.*, 2012, **155**, 56–64.
13. A. Aerts, C. E. A. Kirschhock and J. A. Martens, Methods for in situ spectroscopic probing of the synthesis of a zeolite, *Chem. Soc. Rev.*, 2010, **39**, 4626–4642.
14. L. Tosheva, B. Mihailova, L. H. Wee, B. Gasharova, K. Garbev and A. M. Doyle, Indirect observation of structured incipient zeolite nanoparticles in clear precursor solutions, *Angew. Chemie - Int. Ed.*, 2008, **47**, 8650–8653.
15. M. Castro, M. Haouas, F. Taulelle, I. Lim, E. Breynaert, G. Brabants, C. E. A. Kirschhock and W. Schmidt, Multidiagnostic analysis of silicate speciation in clear solutions/sols for zeolite synthesis, *Microporous Mesoporous Mater.*, 2014, **189**, 158–162.
16. S. Mintova, N. H. Olson and T. Bein, Electron microscopy reveals the nucleation mechanism of zeolite Y from precursor colloids, *Angew. Chemie - Int. Ed.*, 1999, **38**, 3201–3204.
17. E.-P. Ng, D. Chateigner, T. Bein, V. Valtchev and S. Mintova, Capturing Ultrasmall EMT Zeolite from Template-Free Systems, *Science*, 2012, **335**70–73.
18. E. P. Ng, J. M. Goupil, A. Vicente, C. Fernandez, R. Retoux, V. Valtchev and S. Mintova, Nucleation and crystal growth features of EMT-type zeolite synthesized from an organic-template-free system, *Chem. Mater.*, 2012, **24**, 4758–4765.
19. M. Maldonado, M. D. Oleksiak, S. Chinta and J. D. Rimer, Controlling crystal polymorphism in organic-free synthesis of na-zeolites, *J. Am. Chem. Soc.*, 2013, **135**, 2641–2652.

Kenneth Harris remarked: Thank you for your interesting paper. What are the prospects for applying your experimental strategy to gain fundamental information on the role of organic templates (structure directing agents) in zeolite synthesis? For example, in the case of templated zeolite synthesis, can you elucidate insights on specific interactions between organic template molecules and zeolite building units in the aggregates formed in the early stages of the crystallization process?

Francis Taulelle responded: For organic structure directing agents, there are four states of possible environments. 1. They can charge compensate oligomers as fully solvated or as ion-paired. 2. For the nanoaggregates, they can be ion-paired with internal ion-pairing, the most desolvated case, or, with more than one shared layered of water, externally ion-paired. 3. When nanoaggregates increase their average silicon connectivity, they at the same time expel water and MOH, where M^+ is the OSDA. Some nanoaggregates decrease their global charge, increasing their nanoparticle size, whereas some other nanoaggregates decrease their average connectivity, increasing their global charge and decreasing their average size. This is a form of dismutation. The OSDA to Si ratio must decrease in the first population and increase in the second. 4. When the high silicon connectivity nanoparticles start crystallizing, they must reach the crystalline OSDA : Si ratio.

There are several NMR tools that can be used to separate the different states and try to quantify the proximity of template from silicon containing objects. Firstly, ^1H relaxation of the OSDA must depend on the state in which it is located, so a TOSY (Time of relaxation ordered spectroscopy, or dispersion of relaxation in ref. 1) should separate the different states nicely.

Secondly, DOSY (diffusion ordered spectroscopy) allows us to separate the different diffusing OSDA, and differentiate them by class of sizes. Thirdly, CPMAS between ^1H and ^{29}Si allows us to demonstrate proximity. Actually, CPMAS works even at small scales; an example is TMS (tetramethylsilane), a non-viscous small molecule for which CPMAS works well. So OSDA, like TMA, TEA, TPA and TBA (tretra methyl, ethyl, propyl and butyl ammoniums), must work for such purposes. M. Davis has used CPMAS to demonstrate the proximity of OSDA nanoparticles.

TOSY, unlike DOSY, is quantitative, and the fraction of each state can be determined as well as their amount of water, exhibiting as well different diffusion states related to each environment.

So the NMR toolbox is rich and will allow us to quantitatively address most aspects of OSDA interaction with oligomers, nanoaggregates or nanoparticles, even of crystals. For each aspect, a variant of the experiments optimized for each type of information to be extracted can be used.

1. V. Gilard, S. Trefi, S. Balayssac, M. A. Delsuc, T. Gostan, M. Malet-Martino, R. Martino, Y. Prigent and F. Taulelle, DOSY NMR for Drug Analysis, pp. 269–289, in *NMR Spectroscopy in Pharmaceutical Analysis*, ed. U. Holzgrabe, I. Waver and B. Diehl, 2008.

Haihua Pan commented: You are an expert on silicate. Could you help us on these findings from our experiments? We found that if we put in a tiny amount of sodium silicate during the formation of ACP, we got smaller ACP; but if the concentration of sodium silicate was increased to a certain level, we got much bigger ACP. Can you tell us what would be the possible reason for this, in your experience?

Francis Taulelle answered: I cannot say much about this effect. What I would say, as a general statement, is that the size of the ACP is most probably governed by the prior nanoaggregation or ion-pairing with lower solvation of the carbonates. Therefore, depending on the observed size, we may infer what the dielectric screening of the nanoaggregates by the liquid medium would be. If silicate concentration was higher, what would be the pH compared to the lower amount of sodium silicate? What would be the oligomerization level of the silicate species? These things must be examined before attempting any explanation.

Anne Kavanagh presented some slides on the benefits of seeded crystallisation: Many of the papers presented at this meeting have stated the importance of the nucleation step in controlling the particle properties achieved at the end of the crystallisation step. However, in some systems, a seeded and controlled crystallisation can achieve control over the final particle size, and deliver robust and reproducible crystallisation processes.

Key features of such processes are the addition of the seed within the metastable zone, usually close to the solubility curve, and remaining within the metastable zone (to avoid spontaneous nucleation) during the course of the crystallisation. Fig. 1 shows how particle size has been controlled in a seeded, cooling crystallization of a pharmaceutical compound, using micronised seed particles at different seed loadings (further details can be found in ref. 1), and includes an unseeded crystallization for comparison.

It can be seen in the figure that seeded crystallisations have produced particles quite different from unseeded crystallisations. The amount of seed added is expressed as a fraction of the total drug present, and it can be seen that the final particle size varies inversely with the amount of seed added, as would be expected if crystal growth is only occurring on the seed crystals. Replicates of the seeded experiments showed that the final particle size was reproducible, whereas it was not reproducible for unseeded crystallisations.

1. G. Steele, P. Morgan, Z. Huang, L. Griffiths, S. Gracin, S. Black, F. Muller, C. MacLeod, S. J. Lawton and A. Kavanagh, Manufacturing drug substance to meet formulation needs, AlChE Annual Meeting, Conference Proceedings, Salt Lake City, UT, United States, Nov. 7–12, 2010, a204.

Stéphane Veesler asked: Are you sure that in the case of 20% of seeds, you are still getting secondary nucleation?

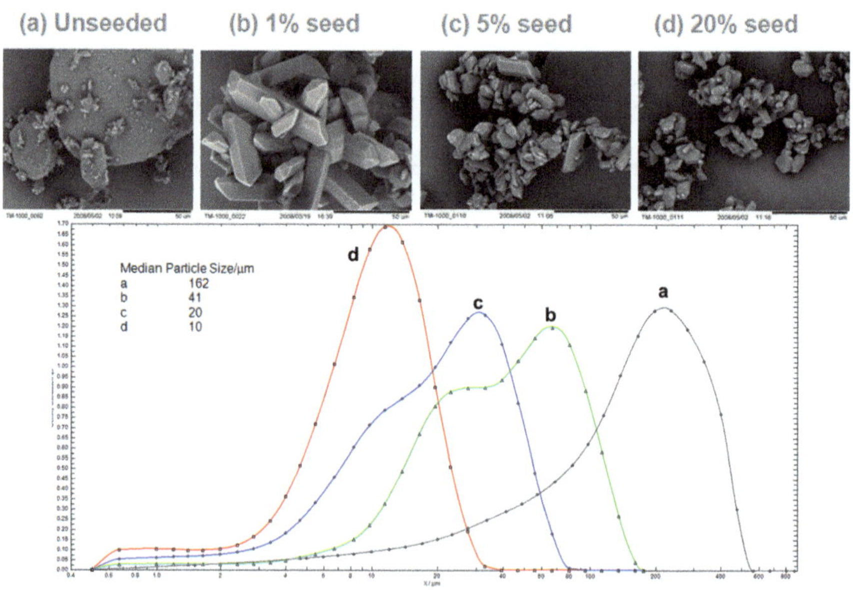

Fig. 1 SEM and particle size analysis showing the effects of seed loading on the size of particles obtained from a cooling crystallisation.

Anne Kavanagh responded: The aim of the seeded crystallisations was to avoid secondary or spontaneous nucleation and to ensure that crystal growth only occurred on the surfaces of the seed crystals. We used a Lasentec FBRM probe to monitor the behaviour of the seed, and to check for spontaneous nucleation. The data showed that the seed crystals grew when they were introduced to the supersaturated solution, and spontaneous nucleation was not observed. The absence of secondary nucleation is further indicated by the fact that the crystallisations were reproducible, and the final particle size was controlled by the amount of seed added.

Terence Threlfall remarked: As to the possibility of obtaining crystals by quench cooling, crystals of D,L-norvaline normally form as small flakey plates, often too thin or multiply laminated, so it is unlikely that crystallising from highly supersaturated solutions would produce crystals that could be examined by single crystal diffraction. Furthermore, the close similarity would make it likely that any polymorph would nucleate any other. Finally, the temperature dependence of the polymorphic form would ensure that only the form stable at that temperature could be recovered.

I should have stated D,L-norleucine rather than D,L-norvaline. However D,L-norleucine, D,L-norvaline and many other straight-chain amino acid crystals behave similarly, in respect of their crystallisation characteristics and in respect of their transitions.

Peter Vekilov opened a general discussion of the paper by Amadeu Sum†: Your order parameter counts the number of carbon atom pairs at a certain distance. It seems that it does not distinguish between liquid and ordered solid new phase domains. How is your simulation relevant to the nucleation of gas hydrate crystals?

David T. Wu responded: From direct observation of brute force molecular dynamics simulation of nucleation, we have seen that the early structures distinguished from those in liquid involve adjacent methane molecules coordinated by groups (typically pentagonal faces) of water, starting from a single pair of water-coordinated methane molecules. We have called this a "Mutually Coordinated Guest" structure. These early structures are not necessarily subsets of a perfect solid crystal lattice. They can be amorphous, and often form cages, but these early cages can be different from the cages found in the solid crystal. Thus, this order parameter detects the evolution of structure distinct from liquid, but before it shows crystal-like order. It does detect "solid"-like order, if we broaden the concept of solid from crystal order to include fluctuating amorphous clusters (that are not liquid-like in the sense that the timescales for rearrangement are much longer than those in a liquid droplet). In short, we have found that the relevant order in the first stages of nucleation are described by an order parameter capturing structures intermediate between liquid and the final solid crystal. I would add that the order parameter does nonetheless distinguish (have different values for) the liquid and the ordered solid.

† Amadeu Sum's paper was presented by David T. Wu, Colorado School of Mines, Golden, Colorado, USA.

Robert Hammond commented: Could you make some general observations about the comparative merits of using metadynamics, or an approach such as yours (mean free passage time MFPT and survival probability) when performing molecular dynamics simulations to interrogate nucleation processes?

For your system the molecules do not have conformational flexibility; do you think that your general approach would work well for treating larger molecules with conformational degrees of freedom?

David T. Wu answered: The MFPT approach of Regeura that we test in this paper, as well as the more rigorous reaction coordinate analysis we performed in ref. 35 of the paper, directly measure kinetic rates. These methods should work for flexible molecules as well.

Joop Ter Horst asked: You determine the mean passage time in your simulations. This is an efficient method at higher supersaturations. For lower supersaturations, you could fill your simulation box with a larger cluster from such simulation. Then you can do a number of simulations at a certain supersaturation and see whether the cluster decays or grows out. This would be following a previously outlined procedure.[1] This allows the study of cluster behavior at lower, more realistic supersaturations. You can determine the growth probability when doing more simulation runs under the same conditions. This growth probability, then, is related to nucleus size and Zeldovich factor, similarly to that stated in the mean passage time equation in your article.

1. J. H. ter Horst and D. Kashchiev, Determination of the nucleus size from the growth probability of clusters, *J. Chem. Phys.*, 2003, **119**, 2241–2246.

David T. Wu replied: This is a good observation, and very close to what we did as part of our reaction coordinate analysis (specifically, the transition probabilities as a function of cluster size, the estimation of an effective cluster size diffusion constant, and the Equilibrium Path Sampling analysis for the free energy surface) presented in our earlier publication,[1] ref. 35 in our paper; that provided the trajectories for the mean passage time analysis of this work. I would highlight one extra challenge in our multicomponent system: a good order parameter, with which one would choose the test clusters to insert, is not necessarily obvious from the start. Our reaction coordinate analysis and validation allowed us to define the relevant multicomponent clusters with specific solvent/solute ordering to be used. (Another distinction: in our study we do not insert the clusters, but harvest them from locally equilibrated simulations.)

1. B. C. Barnes, B. C. Knott, G. T. Beckham, D. T. Wu and A. K. Sum, *J. Phys. Chem. B*, 2014, **118**, 13236–13243.

Matteo Salvalaglio asked: I would like to add a further consideration on the use of metadynamics *vs.* unbiased approaches to simulate nucleation in multicomponent liquid systems.

In small, confined, systems such as those simulated with molecular dynamics approaches, the number of molecules is, by construction, constrained. As a consequence, the free energy profile associated to nucleation can assume a significantly different shape with respect to its counterpart in a macroscopic system maintained at constant chemical potential. The distortion of the free energy profile associated to nucleation is due to the coupling between the cluster size and the chemical potential of the mother phase in very small volumes, and is extensively discussed in ref. 1 and 2. This phenomenon is similar to the effect of confinement observed in small confined droplets, as discussed in Ref. 3.

Depending on both intrinsic properties of the solute (solubility and surface tension) and simulation setup (volume and concentration) the free energy profile may become a monotonically increasing function of the nucleus size, without any barrier associated to the nucleation event, even if the initial state of the system is supersaturated (See Fig. 1 of our paper, DOI: 10.1039/c4fd00235k). In such cases, enhanced sampling, and in particular the methods based on the introduction of a bias potential such as metadynamics, are essential to sampling a wide range of cluster sizes. In these cases, unbiased MD simulations like those used to compute the MFPT will sample only the supersaturated initial state, without displaying any transition. That being said, it is also worth highlighting that metadynamics simulations like those described in our paper are aimed at providing a qualitative mechanistic description of the nucleation mechanism, and do not provide a direct estimate of the nucleation rate.

1. M. Salvalaglio, C. Perego, F. Giberti, M. Mazzotti, and M. Parrinello, *Proc. Natl. Acad. Sci.*, 2015, **112**, E6–E14.
2. D. Reguera, R. K. Bowles, Y. Djikaev and H. Reiss, 2013, *J. Chem. Phys.*, **118**, 340–353.
3. R. Grossier and S. Veesler, *Cryst. Growth Des.*, 2009, **9**, 1917–1922.

David T. Wu answered: Yes, this is a good point that the finite size of the system can affect the free energy curve due to the changing chemical potential of the mother phase. For small enough systems, as you point out, the barrier can be lost altogether. For our study, the system is large enough that we do have a barrier (as measured by free energy techniques - Equilibrium Path Sampling in this work), consistent with the observations in our unbiased simulations of a well-established pre-equilibrium before nucleation occurs stochastically and spontaneously. Moreover, for large enough systems that the formation of the critical nucleus has not significantly depleted the mother phase, and thus not significantly changed the chemical potential, these finite-size corrections to the free energy curve in the region that controls nucleation rate will be small. In our case, the critical nucleus corresponds to about 3% of the mother phase, and so we expect only a small correction here. This being said, the free energy curve for large clusters will need a growing correction; for these situations, the methods you have referred to in your question, and that are included in your paper, can be applied to correct for these finite size effects. The question of comparing metadynamics to unbiased approaches was also raised earlier by Robert Hammond, which relates to my response here.

Jürg Diemand said: The mean first passage time (MFPT) and also the survival probability (SP) methods do not really measure the nucleation rate, but mostly the lag time (or induction time) between the start of the simulation and the first nucleation event. Therefore, the rates derived from them can be orders of magnitudes smaller than the true steady state nucleation rate that one would observe in a large volume simulation of the same system after the lag time has passed.

We do not agree that these methods have proven to be effective in monoatomic systems: The plot in Fig. 2 (replicated from Fig. 9 in ref. 1) shows an example of a large-scale Lennard-Jones nucleation simulation. The long lag time of these runs would indicate a very low nucleation if the MFPT or the SP method were used. However, the large scale simulation shows clearly that the true steady state nucleation rate established after the lag time is much larger; *e.g.* for run T3n14 it is a factor of 40 larger the MFPT estimate. A similar point was made recently by others.[1,2]

1. J. Diemand, R. Angélil, K. K. Tanaka, and H. Tanaka, *J. Chem. Phys.*, 2013, **139**, 074309.
2. V. A. Shneidman, *J. Chem. Phys.*, 2014, **141**, 051101.
3. A. V. Mokshin and B. N. Galimzyanov, *J. Chem. Phys.*, 2014, **140**, 024104.

David T. Wu answered: This is a good point that one needs to keep in mind, that the lag time includes not only the characteristic nucleation time (inverse nucleation rate) but also the relaxation time from quench to the establishment of a steady state. Whether this is a significant distinction depends on the relative magnitudes of the relaxation time to the characteristic nucleation time. In general, this will depend on the degree of supersaturation, *e.g.* for low supersaturation, the nucleation time can be very long compared to the relaxation time.

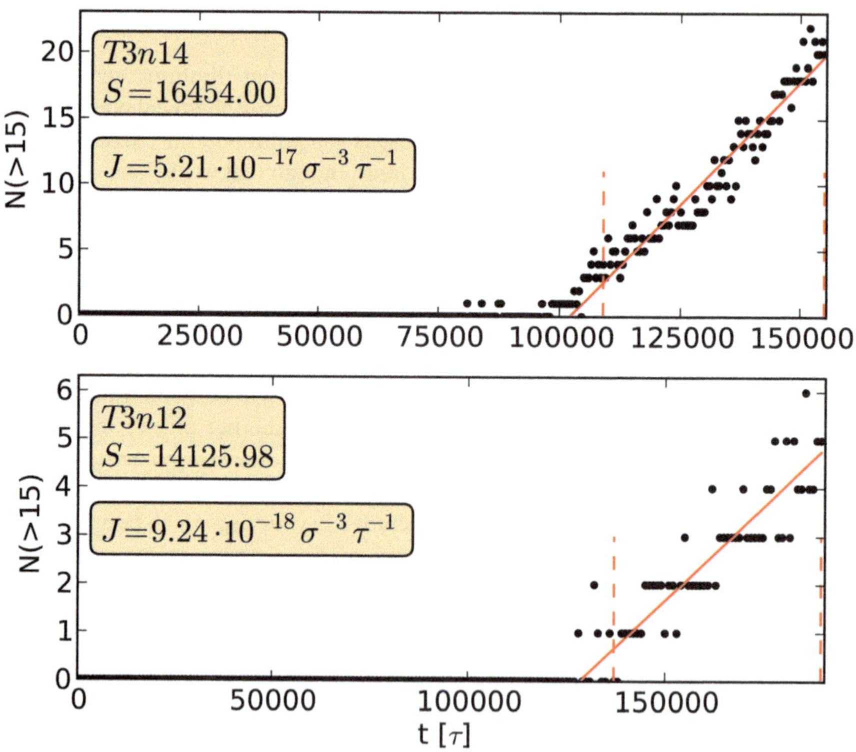

Fig. 2 Number of liquid-like clusters with more than 15 atoms forming in two billion atom Lennard-Jones molecular dynamics simulations of vapor-to-liquid nucleation (replicated from ref. 1). Note that the time from the start of the run until the formation of the first clusters ("lag time") is many times larger than the steady-state nucleation timescale measured later on.

This is a complementary observation to the other limitations of MFPT that are brought out in our paper, such as in the asymmetry of the free energy surface (deviating from CNT) and in the fact that the critical nucleus size doesn't always correspond to the top of the barrier.

As with the threshold methods that focus on the steady-state nucleation rate by effectively subtracting out a fit constant time offset (the relaxation time), one could in principle do something similar with the MFPT methods, with a time shift fit to account for the relaxation time. Our measured MFPT traces don't appear to correspond to large vertical shifts, and so suggest that the relaxation time is not overwhelming the nucleation time in our systems.

Baron Peters remarked: There are two ways of doing mean first passage time (MFPT) calculations. You can run short trajectories from the top of the free energy barrier, or you can run long trajectories to directly estimate the MFPT at each size using the error function + growth formula, as done in your paper. But your trajectories, using the latter approach, are about 1000 times longer. Plus, the movies of these trajectories have a spherical bubble for a gas reservoir, and therefore there is a large Laplace pressure that varies as the bubble is consumed by the growing hydrate nucleus. I don't see the advantages of the approach you used. Why are you doing it this way?

David T. Wu responded: To establish the correct top of the free energy barrier, and thus a statistically validated good reaction coordinate, was the subject of our recent prior work.[1]

This step by itself is a very extensive calculation, taking roughly double the total amount of CPU time than needed for the long trajectories. If we had guessed a reaction coordinate and assumed it was a good one, then indeed short trajectories run from the presumed top of the barrier would be much shorter. However, given we already had the long trajectories as an output from our earlier work, we set out to test the long trajectory MFPT method proposed by Reguera,[2,3] as this required no new simulation, only analysis.

On the issue of the varying Laplace pressure that varies with bubble size: this is a good point, and a compromise that was made in this study for computational speed. The first point to note is that if the nucleus is small (which it is), the amount of methane consumed is not very large, and hence the bubble size and Laplace pressure have not changed much during the period of nucleation that determines the rate. Nonetheless, the direct way to eliminate this problem altogether is to use a flat interface between vapor and liquid; however, to establish a gas reservoir in a layer geometry that is stable against closing into a cylinder or sphere requires a much larger volume of gas, and therefore a much larger simulation system. Given that the total computation time for our foundational work[1] is stretching current computational resource limits, we decided to perform the calculation with a curved interface (it is actually cylindrical, not spherical). We have previously reported the effects of interface curvature on the nucleation rate, as a result of different Laplace pressures, including simulations of these large flat interface systems (but without reaction coordinate analysis).[4]

1. B. C. Barnes, B. C. Knott, G. T. Beckham, D. T. Wu and A. K. Sum, *J. Phys. Chem. B*, 2014, **118**, 13236–13243.

2. J. Wedekind, R. Strey and D. Reguera, *J. Chem. Phys.*, 2007, **126**, 134103.
3. G. Chkonia, J.Wölk, R. Strey, J. Wedekind and D. Reguera, *J. Chem. Phys.*, 2009, **130**, 064505.
4. M. R. Walsh, G. T. Beckham, C. A. Koh, E. D. Sloan, D. T. Wu and A. K. Sum, *J. Phys. Chem. C*, 2011, **115**, 21241–21248.

Radoljub Ristic asked: Which order parameters did you use in your simulation and why?

David T. Wu responded: We used an order parameter, the Mutually Coordinated Guest (MCG) order parameter (OP), which we had identified and statistically validated as a good reaction coordinate in a recent paper.[1] The MCG order parameter is a two-component OP that tracks the connectivity of clusters of methane guest molecules separated by water clusters. Two methane molecules are considered to be "connected" (part of the same cluster) if they obey certain geometric constraints (*e.g.*, distance cutoffs and the presence of at least five water molecules in a defined intervening region). The MCG cluster size, then, is the number of guests in a connected cluster, and the MCG order parameter used in this study is the size of the largest MCG cluster. We chose this two-component order parameter based on observations that the initial steps of nucleation involved only pairs of methane molecules connected in the MCG sense above, and that the early steps observed involved the recruitment of methane molecules to a growing cluster, with the concomitant recruitment of intervening hydrogen-bonded water molecules. This network of hydrogen-bonded water molecules connecting guests eventually becomes the cage framework for hydrates, and hence is characteristic of the crystal state. However, it also captures non-crystalline order, such as amorphous solid networks (MCG-clusters). Thus, this OP captured the observed stages of nucleation involving both guest and solvent species, and was validated by basin commitment analyses to be a good reaction coordinate.

1. B. C. Barnes, B. C. Knott, G. T. Beckham, D. T. Wu and A. K. Sum, *J. Phys. Chem. B*, 2014, **118**, 13236–13243.

Haihua Pan commented: The classical nucleation theory presented a molecular model of chain reactions for nucleation, which is still a black box to us. But MD simulation is useful to see what is inside this black box. Have you directly compared MD simulation results with the molecular models of CNT? And what can we learn from your MD simulations on the details of nucleation process?

David T. Wu responded: Indeed, this was a major focus of our earlier work.[1]

We identified and validated a "good" reaction coordinate, *i.e.*, we established that the essential idea of a chain of reactions of clusters of growing "size" (defined in a specific way, with geometric criteria involving both solute and solvent) does provide a free-energetic description of the progress towards the critical nucleus and subsequent growth. Furthermore, this free energy surface deviates quantitatively from the simple, standard CNT expression based on volume/surface considerations. One consequence is that the critical nucleus size (the tipping point for further growth as compared to shrinkage) does not correspond to the peak in the free energy, as assumed by CNT. In our system, the critical nucleus size is found to be 1/3 larger than would have been predicted by using the free energy peak.

1. B. C. Barnes, B. C. Knott, G. T. Beckham, D. T. Wu and A. K. Sum, *J. Phys. Chem. B*, 2014, **118**, 13236–13243.

Colan Hughes opened a general discussion of the paper by Gerard Coquerel: In your presentation, you have shown a DSC plot (not included in your paper) of the heating of the glass formed after dehydration. In this plot, you indicated a glass transition which actually looks more like an endotherm, rather than a simple change in the heat capacity. If it is a glass transition, then it is a super-heated glass transition, implying that the glass transition should occur at a lower temperature. What are the implications of this super heating of the glass transition?

Gerard Coquerel responded: In my opinion, this is a simple glass transition with a rather classical overshot. It is worth noting that upon annealing, let us say, 5°C below the T_g for 24 h, the overshot will increase further. This is a classical relaxation of the glass, and confirms that it is indeed a glass transition.

Colan Hughes commented: You said that you observe relaxation of the glass if you hold it below the glass transition for some time, resulting in a more endothermic character to the glass transition in the DSC. Does this relaxation also generate cracks, similar to those you observe during the dehydration process?

Gerard Coquerel replied: Very good question! It did not occur to us to make this correlation.

Changquan Calvin Sun asked: What is the origin of the cracks? Were they caused by mechanical strain due to different unit cell parameters between the hydrate and the anhydrate crystals?

Gerard Coquerel answered: We think that it is a combination of mechanical strain during the release of water, and the initial defects which existed in the former hydrate crystal.

Changquan Calvin Sun said: Cracking after dehydration is quite common. Is the anhydrate crystal structure available? Comparing the unit cells of the anhydrate and hydrate may give us some insights on the origin of the cracks.

Gerard Coquerel answered: Cracking during/after dehydration is indeed quite common. Unfortunately, with the set of experimental conditions that we used in this study, we found that dehydration induces a lot of noise even if we performed the dehydration at 20°C under relatively smooth conditions. The quality of the XRPD pattern is thus not good enough even for a refinement of the crystal structure, and so is not good enough for a determination of the unit cell of the anhydrate.

Changquan Calvin Sun remarked: We have one example of a single crystal-to-single crystal transformation from a hydrate to an anhydrate. It may be possible to obtain single crystals of anhydrous rimonabant by using gentler conditions than those tried. We have found that controlling the relative humidity of the air is very important. To get high quality anhydrate crystals, you need to be patient when removing water.

Gerard Coquerel replied: I fully agree with your remark.

Clement Brandel said: XRPD patterns of the solvated form, and the new polymorphic form recrystallizing from the new anhydrous material (NAM), are indeed very similar. However, how far can we consider this as a indication that the molecular mechanism of the growth of the new polymorphic form does not rely on nucleation, but on a cooperative rearangement of the NAM? Is it possible to solve the crystal structure of the new polymorphic form from a smoothly desolvated single crystal of the solvate?

Gerard Coquerel answered: As you said, the striking analogy between the two XRPD patterns has led us to think that a cooperative rearrangement of the NAM is probable.

Due to the rather poor crystallinity we did not dare to resolve the crystal structure of this highly metastable phase.

Pratchaya Tipduangta commented: From the DSC thermogram, I would like to suggest using MTDSC in order to separate a relaxation from the glass transition of the drug. The reverse heat flow would give you a clear transition of the drug.

Gerard Coquerel responded: Indeed, modulated DSC would be preferable here.

Sarah Price asked: You have a nice argument that from suitable phase diagrams, you can form a desolvated solvate by smooth desolvation below the glass transition temperature. Such desolvated solvates could, as in this case, be highly metastable. However, if they are long-lived, they could have practical advantages such as higher solubility. Such desolvated solvates are a nightmare for predicting such polymorphs, though there is some indication that this is possible.[1,2] How critical are the structural features of the monohydrate to your ability to produce the new polymorph?

1. A. J. Cruz-Cabeza, G. M. Day and W. Jones, Predicting Inclusion Behaviour and Framework Structures in Organic Crystals, *Chem. - Eur. J.*, 2009, **15**, 13033–13040.
2. R. M. Bhardwaj, L. S. Price, S. L. Price, S. M. Reutzel-Edens, G. J. Miller, I. D. H. Oswald, B. Johnston and A. J. Florence, Exploring the Experimental and Computed Crystal Energy Landscape of Olanzapine, *Cryst. Growth Des.*, 2013, **13**, 1602–1617.

Gerard Coquerel answered: Thank you, Sally, for this pertinent comment and references.

In my opinion, channel hydrates could be best, because the water molecules are likely to be confined in a tubular space to avoid too much disturbance in the rest of the structure. Thus, the probability of a good transmission of structural information from the mother phase (*i.e.* the hydrate) to the daughter phase (the anhydrous phase) is maximized.

Sarah Price commented: You describe this as a stoichiometric channel hydrate, which is an unusual combination of two terms normally associated with different dehydration behaviours. Usually, channel hydrates are non-stoichiometric, with varying amounts of water loosely bound in the channels, and the structure is maintained during dehydration. If you perform a void space analysis on the monohydrate structure, would you see a connected set of voids, sterically allowing for the water or other small molecules to move through the channels? In your case you have a stoichiometric hydrate, with the water molecules clearly hydrogen bonded to the API molecule and not to each other, and yet you manage to remove the water without the framework structure changing much, and it appears not to be travelling down the channel but instead getting out through the sides. Is there anything in the structure perpendicular to the channel that suggests this could provide an escape route?

Gerard Coquerel answered: I'd like to comment a little bit on this classification of a "stoichoimetric channel hydrate". I agree with you that in many papers "channel hydrate" means a non-stoichiometric hydrate where water molecules are loosely bound to the solute. Here, it is clearly a stoichiometric hydrate when you look at the sharp water uptake and release of water molecule on sorption and desorption respectively, but the water molecules are located in a channel.

The crystal structure shows us that the water molecules could move by discrete jumps from one crystallographic site to another along "channels" parallel to the A axis, which corresponds to the main axis of particle development. When performing a thermal dehydration below the glass transition (77°C), let us say 60°C, the water molecules in the middle of the particle should make a long voyage inside the channels before release in the gas phase. The defects in the crystals help in finding short cuts for a faster release of the water molecules. The more-or-less periodic location of cracks along the needle after dehydration may have some connections with the temperature of dehydration. We can assume that a lower temperature of dehydration (we know that it is possible to dehydrate at 20°C in a dry atmosphere) could lead to a distance between those cracks longer than that obtained at 60°C.

Kenneth Harris remarked: I would like to comment on the prospects for structure determination of the dehydrated form of rimonabant from powder XRD data. The powder XRD pattern of the dehydrated form shown in Figure 8 of your paper (DOI: 10.1039/c4fd00222a) has rather broad lines and significant peak overlap, and therefore indexing and structure solution using this powder XRD data may be challenging. However, as you comment in your paper, it is clear that there is significant similarity between the powder XRD patterns of the monohydrate and dehydrated forms, suggesting that the structures of these forms may share several common features. Under these circumstances, it may be possible to determine the structure of the dehydrated form simply by using the structure of the monohydrate (but with water molecules omitted) as the initial structural model for a Rietveld refinement, thus bypassing the indexing and structure solution stages of the structure determination process. This type of strategy was used recently in structure determination of form II of L-phenylalanine from powder XRD data.[1]

1. P. A. Williams, C. E. Hughes, A. B. M. Buanz, S. Gaisford and K. D. M. Harris, *J. Phys. Chem. C*, 2013, **117**, 12136–12145.

Gerard Coquerel answered: I fully agree with you. In order to maximise the chance of success, it would be better to dehydrate under the smoothest possible environmental conditions.

My concern with this approach is how to avoid new dehydrated material with different crystallinity in the same preparation (different structural purity). It might be difficult to ensure a similar stress for every single particle.

Francis Taulelle commented: It seems that molecular compounds may behave like zeolites. I would like to mention that Irbesartan A, the structure of which is incompletely published but has been recently redetermined (unpublished results), exhibits hexagonal channels with a water molecule at the center of the channel. The water molecule can leave the structure without a loss of the crystalline lattice. My question is, do you think water may template the molecular structure as it does in zeolites, even hydrophobic ones?

Gerard Coquerel responded: The water molecules (two independent water molecules, because $Z' = 2$) act as a relay between N–H and 0=C, which are not directly linked each other. In other words, they create the following bridge:

N–H···OHH···0=C

Thus, I would not say that the water molecules are just space filler here, but are more of a connector between the polar parts of the rimonabant molecules.

Hugo Meekes asked: Relating to the question of whether this is a nucleation-driven transformation: did you also observe the generation of the cracks? Do they start from one side of the needles or simultaneously all over the crystal, assuming no temperature gradient across the crystals?

Gerard Coquerel replied: Very good question, Hugo. Unfortunately, we did not have any monitoring of the desolvation in real time.

Thomas Vetter opened a general discussion of the paper by Stéphane Veesler: When you use the tip of the needle to induce nucleation of a crystal in a droplet, how often does the crystal stick to the tip? Is it controllable in any way whether the crystal sticks to the needle tip or not?

Stéphane Veesler responded: It depends on the way you approach the droplet with the tip:[1]

Using the "perpendicular approach", nucleation is launched, and the crystal stays in the droplet, when you withdraw the tip.[2] Using the "tangential approach", with increasing tip penetration, also launches instantaneous nucleation, but here the crystal remains attached to the tip during growth.[3]

1. R. Grossier, Z. Hammadi, R. Morin and S. Veesler, Predictive Nucleation of Crystals in Small Volumes and Its Consequences, *Phys. Rev. Lett.*, 2011, **107**, 025504.

http://www.cinam.univ-mrs.fr/pro_perso/veesler/video/video%20KNO3%20 induction.wmv

http://www.cinam.univ-mrs.fr/pro_perso/veesler/video/LC13676_video_fig4.wmv

Baron Peters said: I have a related question about the electrode tip. Some of the bubbles stick to the tip and some leave immediately. Is it possible that the bubbles which stick were formed heterogeneously and those which immediately leave the tip were formed homogeneously?

Stéphane Veesler answered: The electrochemical reaction rate (related to the bubble nucleation rate) is proportional to the rate of charge transfer through the electrode–electrolyte interface. This interface is known to be a few water layers thick. We consider that nucleation occurs at this interface.

Concerning the larger bubbles: they stick to the electrode and coalesce after their nucleation.

Hugo Christenson remarked: Yours is a very interesting paper, and the images are amazing. In the paper you refer to a "disturbance" that triggers nucleation when a sharp tip touches a supersaturated droplet of NaCl solution. Do you not simply mean that the presence of a new surface - that of the tip - induces heterogeneous nucleation?

Stéphane Veesler replied: I assume you are referring to this sentence: "It was previously shown that any disturbance, for instance a crystal touching dense droplets or mechanical contact with a nanotip, triggers nucleation." My answer is yes.

Hugo Christenson commented: I agree with Stéphane Veesler that significant deviations from bulk values of the surface tension could only occur for droplets of nanometer size. And although there would be an excess pressure inside these droplets, of the order of a few kPa, such pressures would have no effect on the precipitation of crystals from liquid solutions.

Stéphane Veesler responded: Another effect related to the reduction of volume, which is the increase of pressure in the droplet, has been dismissed in our experiments. Indeed, we can apply the Young–Laplace equation, $\Delta P = 2\gamma/R$, which describes the equilibrium pressure difference sustained across the interface between two static fluids. Here P is the excess of pressure inside the droplet due to its curvature, γ is the surface tension of the solution in contact with the oil, and R is the droplet radius. Using this, we can observe that for a representative droplet of our system, with $R = 25$ μm, and considering a γ value of 72 mN m^{-1} (water droplet in contact with air), ΔP is 0.057 atm. This extra pressure is not a significant amount to be considered in our nucleation experiments. Only for submicrometer droplets (R < 1 μm) would the pressure be exerting an important effect.[1]

1 I. Rodríguez-Ruiz, S. Veesler, J. Gómez-Morales, J. M. Delgado-López, O. Grauby, Z. Hammadi, N. Candoni and J. M. García-Ruiz, Transient Calcium Carbonate Hexahydrate (Ikaite) Nucleated and Stabilized in Confined Nano- and Picovolumes, *Cryst. Growth Des.*, 2014, **14**, 792–802.

Baron Peters addressed Stéphane Veesler: Can you describe the amount of oil above the droplets? That could have a significant effect on the rate at which their water escapes into the oil. Also, can the droplets see each other laterally? I'm not worried about the droplets touching, or having some energetic interactions. My concern is that, as water diffuses out of the droplet, it saturates the surrounding oil. If the droplets are close enough to each other then they mutually saturate their environments, and that may cause their internal supersaturation to grow more slowly than would otherwise be expected.

Stéphane Veesler answered: Firstly, in our experiments, water microdroplets in the pL (10-100 pL) range are formed in an oil medium with a volume of ~100 µL (6 orders of magnitude larger than that of the microdroplets). Thus, we can consider the oil as an infinite medium.

Secondly, our preliminary analysis has shown no effect during water diffusion.[1]

1. I. Rodríguez-Ruiz, Z. Hammadi, R. Grossier, J. Gómez-Morales and S. Veesler, Monitoring Picoliter Sessile Microdroplet Dynamics Shows That Size Does Not Matter, *Langmuir*, 2013, **29**, 12628–12632. Also unpublished results on a larger number of droplets.

Ake Rasmuson asked: do you generate droplets and then stop the flow and wait for nucleation, or do you have a continuous flow?

Stéphane Veesler replied: Droplets are generated through the droplet-factory. Once the tube is filled the flow is stopped, the set-up is stored in a thermostatted sample holder and droplets are observed.

Ake Rasmuson said: Looking at the outcome from different droplets, I can see that the results appear fairly random. Have you analysed whether the order of the droplets has an effect, and if there is bias or cross-contamination between droplets?

Stéphane Veesler answered: In the microfluidic experiments with micro-injectors (~pL to fL), we never observed contamination. Droplets are static; there are sessile droplets.

In the microfluidic experiments in chips or tubes (~nL), to the best of our observations, no cross-contamination was observed. What did we observe sometimes, at high supersaturation, was precipitation as soon as the solutions are mixed.

This is a potential problem with microfluidic methods, because precipitates tend to adhere to the walls of the channels and perturb or even obstruct the flow. Consequently, there is a risk of clogging the channel, one of the major drawbacks in microfluidics.

Thomas Vetter commented: When measuring induction time distributions and then correlating nucleation rates from this data using microfluidics, do you think that the nucleation rate changes when the droplets are kept in motion, rather than in a stagnant environment? Professor Davey has drawn my attention to a recent study[1] which indicates that this could indeed be the case. Even in the absence of breakage/agglomeration/secondary nucleation, would these findings not present quite a challenge when transferring nucleation kinetics obtained on a miniature scale to a manufacturing scale?

1. Rossi *et al.*, *Cryst. Growth Des.*, 2015, **15**, 1784–1791.

Stéphane Veesler responded: In fact, I believe that you have answered the first part of your question already, as this was tested in the paper you mention. It is probably the case. It is like comparing stagnant medium to stirred medium.

Ake Rasmuson said: How long do you typically wait before nucleation occurs - is it hours or weeks? Is there a temperature control during this time?

Stéphane Veesler answered: In the example presented here - protein crystallization - induction times are between minutes and hours. The whole experiment lasts 24 h.

Sendhil Poornachary commented: In the microdroplet (nL to pL sample volumes) nucleation experiments, how do you sample crystal formation *in situ*? Is the size of the droplet typically much smaller than the crystal?

Stéphane Veesler replied: Experiments are performed under an optical microscope. Thus, time evolution of droplets is observed, and we know the time when a detectable crystal is observed in each droplet. In our experimental configuration, due to the fast growth rate (>200μm s^{-1}), the time required for the newly-formed nuclei to grow to a detectable size is negligible with regard to the induction time. Thus, the time when a detectable crystal is observed is really the induction time. The size of the crystal is smaller than the droplet diameter, and is controlled by the number of solute molecules in the droplet at the beginning of the experiment (see theoretical development in ref 1.)

1. R. Grossier and S. Veesler, Reaching one single and stable critical cluster through finite sized systems, *Cryst. Growth Des.*, 2009, **9**, 917–1922.

Sendhil Poornachary said: In the microdroplet nucleation experiments, do you observe crystals nucleating at the contact line between the solution droplet and oil phase? If so, how can we relate the nucleation kinetics data obtained from these experiments to crystal nucleation rates in bulk solution?

Stéphane Veesler responded: This was also mentioned by Thomas Vetter. The location of nucleation can be at: (1) the solution–oil interface, (2) the solution–coverslip interface, (3) the contact line solution–oil–coverslip or (4) in the bulk. What we know from previous measurements using microfluidics is that the chemical nature of the solution–oil interface influences nucleation rate.[1]

Experiments presented here concern heterogeneous nucleation, but our belief is that most of the experiments presented in the literature (except maybe the experiments in Levitated Droplets from the Myerson group) are heterogeneous nucleation. Thus, extrapolation to larger crystallizers is always difficult.

1. M. Ildefonso, N. Candoni and S. Veesler, Heterogeneous nucleation in droplet-based nucleation measurements, *Cryst. Growth Des.*, 2013, **13**, 2107–2110).

Samuel Booth communicated: I was interested in your potential induced nucleation system. With this work, have you tried varying the applied potential in order to see how this might cause a change in the nucleation time that you observe? Also, with your system would you be able to apply a 3-electrode system, in order to better define the potential applied?

Stéphane Veesler communicated in reply: For the first part of your question: yes, we tested different applied voltages greater than 0 V and smaller than the current required for Faradaic reaction (electrolysis) - see Fig 10 in ref. 1. From this, we obtained different results with increasing V values, from no nucleation to nucleation of crystals, concomitant nucleation of crystals and liquid–liquid phase separation (LLPS), and finally only LLPS.

For the second part of your question: as specified in the first part of my answer, we are in experimental conditions in which no reaction occurs. The important effect here is the electric field (E) which for a tip is: $E = V/5r$, with V being the potential applied and r the curvature radius of the tip (in the nm range - see the supplementary material of ref. 2). Thus, control and reproducibility is assured by the control of the current at the beginning of the experiment, in the μA range.

1. Z. Hammadi and S. Veesler, New approaches on crystallization under electric fields, *Prog. Biophys. Mol. Biol.*, 2009, **101**, 38–44.
2. R. Grossier, Z. Hammadi, R. Morin and S. Veesler, Predictive Nucleation of Crystals in Small Volumes and Its Consequences, *Phys. Rev. Lett.*, 2011, **107**, 025504.

Ian Rosbottom returned to the discussion of the paper by Gerard Coquerel: Do you think that by doing AFM you could see any crack periodicity that is not observable by eye, and see just how periodic the cracks are, and also whether the size of, or distance between, the cracks correspond to any lattice vectors?

Gerard Coquerel responded: My gut feeling is that indeed AFM could show smaller cracks with periodicity. I believe that the size of those cracks and the more-or-less periodic distance between them will depend on the external conditions applied to the solvated phase, the size of the crystals and the crystallinity of the solvate. Under harsh conditions, the distance should be small. Conversely, under very smooth desolvation conditions the distance between those cracks can be much larger.

Changquan Calvin Sun asked: You mentioned in your paper that the dehydrated crystal is less stable than the amorphous phase. This doesn't make sense to me. Could you explain why?

Gerard Coquerel replied: This unusual fact has been observed just for trehalose (not for rimonanbant). For trehalose, the group of Professor Descamps has observed that: $TFA > T_g > TF\ AH > T_{desolvation}$, leading to a superheating of the newly formed anhydrous phase (see the extended discussion in our paper, DOI: 10.1039/c4fd00222a). In this case, like in the vast majority of the others, the crystallized anhydrous phase is more stable than the amorphous solid obtained by desolvation, up to 97°C (greater than the melting point of the newly dehydrated material).

Robert Hammond said: Do you think there could be an analogy between the pathway to forming a new polymorph in the case you have described, and the observation of polymorphs that are apparently stable but have unexpectedly low density, such as the trigonal polymorph of carbamazepine[1] (space group $R\bar{3}$)? For the carbamazepine polymorph it has been speculated[2] that some residual solvent or water could remain within contiguous voids, parallel to the crystallographic c-axis, and more recently experimental evidence has been presented supporting this.[3]

1. M. M. J. Lowes, M. R. Caira, A. P. Lotter, J. G. van der Watt, *J. Pharm. Sci.*, 1987, **76**, 766–752.
2. R. K. Harris, P. Y. Ghi, H. Puschmann, D. C. Apperley, U. J. Griesser, R. B. Hammond, C. Ma, K. J. Roberts, G. J. Pearce, J. R. Yates and C. J. Pickard, *Org. Process Res. Dev.*, 2005, **9**, 902–910.
3. F. P. A. Fabbiani, L. T. Byrne, J. J. McKinnon, M. A. Spackman, *CrystEngComm*, 2007, **9**, 728–731.

Gerard Coquerel responded: When there are solid solutions (by means of insertion in the case of carbamazepine ($R\bar{3}$), but possibly also by substitution) there is a wide range of possibilities, including the effect that you mention.

Ref. 1 details some rationale between polymorphism and solid solution. Thank you for your references.

1. G. Coquerel, Thermodynamic predictions of physical properties - Prediction of solid solutions in molecular solutes exhibiting polymorphism, *Chem. Eng. Technol.*, 2006, **29**, 182–186.

Ian Rosbottom remarked: You say that there are two conformations in the asymmetric unit. Are these close to the conformational mininum? Is water inducing a conformational change in the molecule?

Gerard Coquerel replied: Indeed, there are two distinct conformations in the monohydrate (SG = $P\bar{1}$; $Z' = 2$). A simple calculation (using molecular mechanics) has shown that they are pretty close in energy.

I would say that it is the crystal packing which imposes these two different conformations.

Ian Rosbottom asked: Have you considered removing the water from the structure and then optimising the structure, possibly using a dispersion-corrected *ab initio* approach such as the M06 functional, and then predicting the PXRD pattern? You could compare how close that predicted PXRD is to the experimental one you get.

Gerard Coquerel replied: That's indeed a good idea. We will try also to improve the quality of the XRPD pattern by a very smooth desolvation process.

Simon Black asked: In the crystal structure of the hydrate, amide groups are connected by the water molecules. When the water molecules leave, could the amide groups form hydrogen bonds directly with each other? Could this be detected by spectroscopy?

Gerard Coquerel replied: It is likely that the N–H and O=C groups would be directly connected after the departure of the water molecules. Nevertheless, this probably means that there is a substantial change in conformation of the molecules, so that the two groups could be brought together.

In the two known anhydrous forms, there are indeed classical N–H···O=C bonds.

Clement Brandel commented: In the introduction of your paper, you mention other solvates of rimonabant. How do they behave upon desolvation, as compared to the hydrate investigated in your paper?

Gerard Coquerel replied: Basically, every solvent different from ketone can give a solvate with rimonabant (like olanzapine). We did not make any detailed study with these solvates in comparison with the monohydrate.

Sophie Janbon communicated: Does the expression "channel hydrate", as mentioned in this paper, refer to the physico-chemical stability, *i.e.* the tendency to lose water easily from the channel structures, or a feature in the crystal structure, or potentially both aspects?

Gerard Coquerel communicated in reply: The semantic aspect of channel hydrate is an interesting issue! I would rather go for both aspects: structural aspect and behaviour.

Sophie Janbon communicated: CCDC has developed new tools using prior knowledge from the CSD - one example, Full Interaction Maps (FIMs), was presented by Roger Davey in his introductory talk. Some of these tools would be good to use on rimonabant monohydrate to enhance the current understanding on this form, especially the one related specifically to hydrates.

Elena Simone communicated: Did you try to produce metastable phases of different compounds, other than rimonabant, using smooth dehydration? Would it be possible? And which characteristic should a compound have to be potentially able to produce a new metastable phase through smooth dehydration?

Gerard Coquerel communicated in reply: Rimonabant is far from a unique example. If, for instance, you dehydrate the disaccharide trehalose, a similar behaviour can be observed. Under smooth conditions of dehydration, and below T_g, a metastable anhydrous phase can be obtained.

Richard Sear opened a general discussion of the paper by Allan S. Myerson: Thank you for an interesting survey talk. I was particularly interested in the part in which you gave three examples of data for the fraction of a set of nominally identical droplets that has not crystallised, $F(t)$, as a function of time. The data sets are for lysozyme, glycine and ammonium sulphate. In each case, the data was fitted with a Weibull function (AKA a stretched/compressed exponential), *i.e.*, $F(t)$

$= \exp[-(t/\tau)^{\beta}]$. The best-fit values of β were less than one (lysozyme), greater than one (glycine), and very close to one (ammonium sulphate).

If nucleation occurs in one step, at a rate that is not changing with time, and if this rate is the same in all droplets, then $F(t)$ is an exponential (*i.e.*, $\beta = 1$). For ammonium sulphate, this is what you found. Roughly speaking, a two-step (or multi-step) process, or any process that makes the effective rate increase with time, will give $F(t)$ curves that are best fit[1] by $\beta > 1$. The exponent beta was in this range for glycine. And if the rate varies from one droplet to another, due to some droplets having "better" impurities than others, this results in an effective rate that decreases with time[2]; then $\beta < 1$. The exponent was in that range for lysozyme. Thus, although there are limits to what $F(t)$ curves on their own can tell us, this suggests that in your lysozyme experiments there is considerable droplet-to-droplet variability in the nucleation rate, as there may also be in, for example, your work on aspirin and some other molecules with Ying Diao, Bernhardt Trout and others.[3,4] Whereas for glycine, the fact that $\beta > 1$, suggests that nucleation may be *via* two steps, or perhaps that some relatively slow chemical process is acting to gradually increase the effective nucleation rate. This is rather speculative, but does show that data such as yours do provide evidence for and against possible mechanisms. Perhaps in the future, using data for $F(t)$ for varying supersaturation, volume, *etc.*, we will see how parameters such as β and t vary with these variables, and so further advance our understanding of heterogeneous nucleation.

1. R. P. Sear, *CrystEngComm*, 2014, **16**, 6506.
2. F. Proschan, *Technometrics*, 1963, **5**, 375.
3. Y. Diao, A. S. Myerson, T. A. Hatton and B. L. Trout, *Langmuir*, 2011, **27**, 5324.
4. Y. Diao, M. E. Helgeson, Z. A. Siam, P. S. Doyle, A. S. Myerson, T. A. Hatton and B. L. Trout, *Crys. Growth Des.*, 2012, **12**, 508.

Ake Rasmuson asked: Are we prepared to remove the question about whether clusters exist or not from the 1988 Gordon conference list of questions? Do we have sufficient experimental and computational evidence to state that yes, clusters do exist?

Helmut Cölfen answered: We would answer this question with a clear "yes". At least for the system $CaCO_3$, there exists conclusive theoretical and experimental evidence that clusters exist, as summarized in Ref. 1.

1. D. Gebauer, M. Kellermeier, J. D. Gale, L. Bergstrom and H. Cölfen, *Chem. Soc. Rev.*, 2014, **43**, 2348–2371.

Sarah Price addressed Allan Myerson: Is there any tension between your optimism that simulations can provide insights into what is happening during two-step nucleation, and the assumptions needed to define the simulation? These include specifying how many of which molecules are present, the starting configuration, and thermodynamic conditions at the minimum. There are usually other decisions, such as choice of collective variables, order parameters or other rare events methods that may restrict what behaviour can be seen in the simulations.

Allan Myerson replied: Certainly it is possible to bias simulations based on choices such as the number of molecules, starting configuration and thermodynamic conditions at the minimum, as well as the level of supersaturation and other issues. I am still optimistic about potential progress because of the continuing advances in computing power, and thus the ability to do more complex calculations with more molecules for longer times.

Baron Peters remarked: At points in the meeting, the single term "two-step nucleation" was used in discussions of very different processes. I have an *Adv. Chem. Phys.* review where I try to classify the different processes according to the stability of the intermediate phase and the height of the first and second barriers. Firstly, there is a "ripening regime two-step", where the dense liquid-like intermediates are stable so that they can ripen until crystals form within them. I believe this is the mechanism that Prof. Vekilov presented. There are two cases where the intermediate phase is metastable with respect to solution: one that is "droplet nucleation limited" and one that is "crystallization limited". The crystallization limited case is that of ten Wolde and Frenkel. I am unsure whether the droplet limited case can actually occur. I also discuss a fourth mechanism, resembling two-step nucleation, that I called "concentration fluctuation gating". I think this is the situation that Prof. Sefcik has described. Clusters form and redissolve slowly throughout the solution, but only the largest clusters have enough material to grow a stable nucleus before redissolving. I should note that during this meeting, Prof. Sear noted that he has reviewed the effects of impurities on the stability of the intermediate phase, which would clearly impact the classifications above. Distinguishing between all of these different cases is a first step towards determining rate laws and experimentally testable signatures of each one.

Wenhao Sun commented: Thank you, Prof. Myerson, for your wonderful conclusion, as well as summarizing the major questions in nucleation. I'd like to bring another perspective to this community. I come from a materials design group that focuses on energy materials, primarily Li-ion batteries and solar cell materials. A major question that we ask is: how does metastability affect the viability of a material technology? In our field, we aim to synthesize the structure with the most desirable properties (battery performance, solar cell voltage and absorption, *etc.*). There is rarely any correlation between desirable properties and stability - the material with the most desirable property can just as likely be a (particular) metastable phase as the stable phase. In the case that we desire a metastable phase, we are interested in the following questions: how can we direct materials synthesis toward this particular metastable phase, instead of the stable phase? Can it resist transformation to the stable phase under operation conditions? If it cannot, what is the lifetime of this metastable polymorph? On the other hand, sometimes we want the most stable phase of a material. In this case, we ask: why do metastable phases form before the stable phase? Is there a synthesis pathway/synthesis condition that circumvents the formation of the metastable polymorphs that interfere with the synthesis of the stable phase? Ultimately, the questions that we ask are the same as presented here: how can we control synthesis pathways *via* processing conditions? What are the governing mechanisms of structure-selection during nucleation? What are the lifetimes of transition polymorphs? But we ask these questions with other goals in mind - to optimize and utilize functional materials. This is a perspective that I did not see represented at the meeting.

Haihua Pan responded: We might not get to the answers to some of the specific or general questions that you, Prof. Myerson or I have asked. So many good questions have arisen during this Faraday Discussion. Actually, most of them have not been fully or satisfactory answered in this Discussion. But I think it is a good opportunity to ask questions about classical nucleation theory (CNT), and to rethink the physical chemistry of nucleation as Gibbs had done a hundred years ago. There are so many *meta*/stable species and phases (including prenucleation clusters, oligomers, and liquid phases) that could be formed in a supersaturated solution before the starting of the nucleation of the final stable phase. As Prof. Davey commented at the beginning of this Faraday Discussion, Dunning had stated the importance of chemical kinetics barriers for the selection of nucleation pathways ("maybe it has nothing to do with the stability in thermodynamics"). We need to think as Dunning did, but take recent new findings (prenucleation clusters, oligomers, liquid phases) into mind. We need to stand on the shoes of solutions: where are we now - what is my supersaturation now? How many species have we gotten - and where to go tomorrow - from what will I transfer and to what, and what are their kinetics barriers?

This reminds me of Prof. Myerson's comments in concluding remarks, "... thermodynamics is always true, but not necessary useful. And in our case you need to understand clearly kinetics!" I would say - great minds think alike!

Bart Rimez communicated: I have some personal thoughts about the remarks you made on some kinetic investigations you mentioned during your talk. I have read some of your publications that addressed the kinetics of different nucleating systems.[1] I was wondering: suppose that the factor *A* that you introduce to explain the deviation from Turnbull's analysis would actually express the existence/non-existence of a one-, two- or multi-step nucleation? That is, *A* going to one for assemblies that organise so fast they can be regarded as actual one-step nucleation; whereas the more you deviate from a value of one for *a*, going toward zero, the longer a sort of intermediate state or states must take place before actual nucleation is effected? In that case, indeed, this expression could be regarded as a more universal way to look at/determine nucleation rates; rather than discussing whether or not it is one- or multi-step, we could this question behind us within the crystallization community. Do you have any comments on this? To be honest, I am quite new to the crystallization community, so I was a bit stunned by the fact that these issues have not yet been fully addressed (as you make apparent when comparing hot topics between 1988 and now).

1. For example, D. Knezic, J. Zaccaro and A. S. Myerson, *J. Phys. Chem. B*, 2004, 108, 10672–10677.

Allan Myerson communicated in reply: The simple model equation you discuss from our work was simply used to demonstrate that a stretched exponential was a better fit than classical nucleation theory, and that differences could be seen. The equation itself was not derived to have any physical significance. The data, however, do demonstrate that the postulated second step will vary significantly based on the type of system being studied.

Hugo Christenson communicated: It would appear that almost all of us agree that nucleation is almost always heterogeneous, and typically occurs on container walls or particulate impurities of unknown character and source. Yet many of the contributions in this meeting have implicitly treated nucleation as if it were homogeneous. The structure of the preferentially nucleating polymorph is related to the configuration of some precursor species in solution, without considering what effect the nucleant surface might have on this. Do you have any comment on this?

Allan Myerson communicated in reply: I agree that virtually all nucleation is heterogeneous, and that the type of surface and impurity can influence both the rate of nucleation and the polymorph nucleated. This has been demonstrated experimentally by a number of investigators, using various surfaces to template nucleation. That being said, it is also true that even if such surfaces are present, solvent composition can influence nucleation through its influence on both molecular conformation and the formation of dimers and higher order entities. Thus, these studies are certainly worthwhile and help elucidate part of the nucleation process.

Faraday Discussions

PAPER

Concluding remarks

Allan S. Myerson*

Received 27th April 2015, Accepted 1st May 2015
DOI: 10.1039/c5fd00042d

Crystallization from solution is a crucial process used in the manufacture of a wide variety of materials. The first step in the crystallization process is the birth of a new crystalline phase, which is known as nucleation. Nucleation plays a key role in determining the results of any crystallization process with respect to the size, shape and crystal form obtained. Classical nucleation theory does not adequately explain the crystal nucleation process. Work described in the literature and at this Faraday Discussion describe more complex nucleation mechanisms which are generally known as two-step nucleation models. In addition, as most nucleation is influenced by dust, dirt and container surfaces, the importance of heterogeneous nucleation and the use of templates to accelerate nucleation and influence crystal form are promising methods for the study and control of nucleation. It is also clear from this Faraday Discussion that interest in this topic has grown, and new and novel experimental and modeling approaches are being used for the study of crystal nucleation from solution.

A. Introduction and motivation

Crystallization from solution is used in the manufacture of a wide variety of materials in the chemical, pharmaceutical and food industries. The goal of the crystallization process is the production of a material with the desired purity and process yield while obtaining crystals of the desired size, shape, and crystal form. In a crystallization process, a solution containing the species to be crystallized undergoes a change of state, so as to form a supersaturated solution in which crystals will nucleate and then grow. Nucleation plays a decisive role in determining the crystal form and size distribution. Thus, improved control of crystallization cannot be achieved without understanding the fundamentals of nucleation. However, our understanding of the nucleation process is still developing, and no accurate description of the process exists.

All theories and descriptions of the nucleation of crystals postulate the existence of molecule clusters (or concentration fluctuations). In 1988 I gave a lecture at the Gordon Conference on Crystal Growth (Colby Sawyer College, New Hampshire) which was entitled "Structure of Supersaturated Solutions". At the

Massachusetts Institute of Technology - Chemical Engineering, 77 Massachusetts Avenue 66-568, Cambridge, Massachusetts 02139, USA. E-mail: myerson@mit.edu

beginning of the lecture I asked a series of questions about pre-nucleation clusters, which are listed below:

1. Do clusters exist?

2. What is their size and size distribution?

3. What is the structure of the cluster—lattice arrangement, amorphous or something intermediate?

4. Are the clusters solvated?

5. How do the clusters become crystalline nuclei?

While there is a fairly general consensus that clusters exist (question 1), the presentations and discussions at this conference indicate that we are still asking questions 2–5, and that these issues are under study using both experimental and computational methods.

B. Nucleation theory: classical and two-step models

Classical nucleation theory (CNT) is the most widely used theory to describe the nucleation process. CNT was derived for the condensation of a vapour but is applied to the nucleation of crystalline solids from solution.

The thermodynamic description of this process was developed by Gibbs, who defined the free energy change required for cluster formation (ΔG) as the sum of the free energy change for the phase transformation (ΔG_v) and the free energy change for the formation of a surface (ΔG_s). The first of these describes the spontaneous tendency of a supersaturated solution to change from a solution to the more stable solid state, resulting in a negative value for ΔG_v. The introduction of a solid/liquid interface increases the free energy by an amount proportional to the surface area of the cluster. The growth of clusters is thus a competition between a decrease in ΔG_v, which favours growth, and an increase in ΔG_s, which favours dissolution. This results in the classic free energy diagram for nucleation and the concept of the critical size, as shown in Fig. 1.

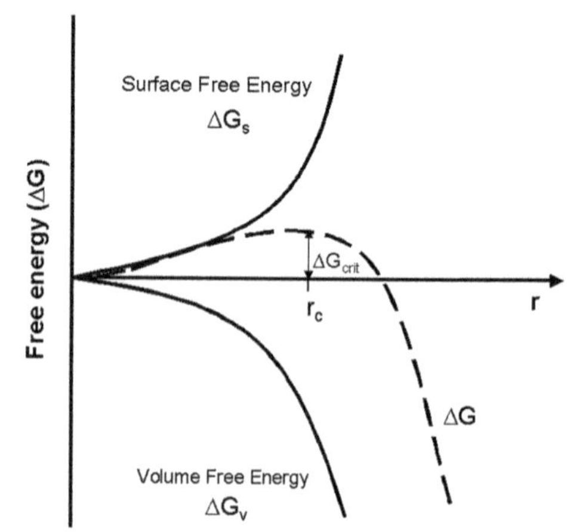

Fig. 1 Free energy diagram for nucleation.

CNT has a number of major assumptions,[1] but I would like to focus on the assumption which relates to the crystalline state. CNT models clusters as spherical droplets having uniform interior densities and sharp interfaces, with the density of the droplet independent of size and equal to the macroscopic density of the bulk condensed phase. For crystallization from solution, this assumption requires that the molecular arrangement in a cluster is identical to that of the crystal produced.

Given the wide range of molecules which form crystal nuclei from solution (ionic species, organic molecular crystals, proteins), which all form lattice structures with varying degrees of symmetry, this assumption at best cannot describe nucleation of all crystalline materials, and at worst is incorrect in virtually all cases involving the formation of crystalline nuclei from solution.

The flaws of CNT when applied to the nucleation of crystals from solution has led to theories which postulate more complex routes to nucleation. These routes are generally referred to as two-step nucleation, but actually encompass a number of potential mechanisms.[2,3] A general description of these models involves a clustering step in which the molecules are not in the lattice structure of the final crystalline solid, followed by a rearrangement step into an ordered crystalline structure, as shown in Fig. 2. The organization step is proposed as the rate-determining step. This is consistent with the observation that nucleation from solution takes longer as the complexity of molecules increases, since it would be more difficult for more complex molecules to arrange themselves in the appropriate lattice structures due to their high degree of conformational flexibility. In addition, the existence of isolated site and channel hydrates whose structures are stabilized by solvent molecules requires the active involvement of solvent molecules in the nucleation process, further complicating the mechanism.

In discussing the work that has been done relating to the two-step nucleation mechanism, it is convenient to divide the studies into three groups based on the types of compounds studied. These group are organic molecular crystals, inorganic (ionic) crystals and protein crystals.

Experimental evidence of a two-step nucleation mechanism for organic molecular crystals was provided through a phenomenon known as non-photochemical laser induced nucleation (NPLIN).[4,5] Supersaturated solutions of small organic molecules exposed to the laser nucleate much faster compared with control solutions. In addition, polymorphs of glycine can be preferentially nucleated by changing the polarization state of the laser. Linear and circularly polarized light appears to influence the alignment of the molecular building blocks of the alpha and gamma polymorphs of glycine. NPLIN appears to work

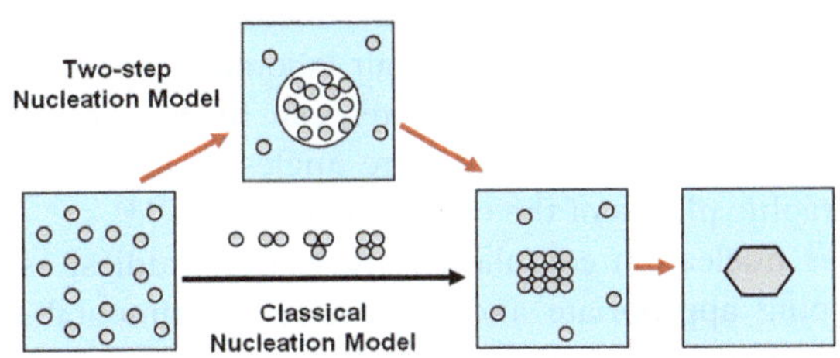

Fig. 2 Classical *vs.* Two-Step Nucleation Models.

through a rearrangement of clusters which are not yet in the correct lattice structure, thus increasing the rate of the second step in the two-step process.

Studies of protein nucleation indicate that the first step in the nucleation process is the formation of dense protein liquid droplets. Crystalline nuclei then form within these droplets. This has been demonstrated for a variety of protein systems.[6,7] The size of these dense liquid protein clusters varies from ten to several hundred nanometers and thus they are experimentally observable. They have low volume fractions (0.1%) and extended lifetimes.

While a variety of inorganic systems have been examined experimentally, calcium carbonate has been extensively studied because of its importance in biological systems.[8,9] Novel experimental studies using Cryo-Transmission Electron Microscopy (Cryo-TEM) and Scanning Electron Microscopy (SEM)[8,9] have reported liquid-like precursors, which have been described as liquid-like and amorphous. Further experimental studies have provided evidence for stable $CaCO_3$ clusters with diameters as large as 2 nm. The authors suggest that these large clusters coalesce and then rearrange to form nuclei.

The studies discussed here, as well as the work presented at this meeting, demonstrate the complexity of the nucleation process, as well as the significant progress made in attempting to elucidate nucleation mechanisms.

C. Heterogeneous nucleation

Heterogeneous nucleation refers to nucleation which is influenced by the presence of dust, dirt or container surfaces. Generally these decrease the nucleation induction time at a given supersaturation. Virtually all experiments which involve studies of nucleation and crystallization involve heterogeneous nucleation (except for methods such as levitated droplets and isolated droplets in microfluidic devices).

Surfaces can influence nucleation through epitaxy, where the molecules in a cluster near the surface are ordered by the surface into their lattice structure. One way this can be done is through the use of a crystalline substrate, which provides a lattice match to the crystallizing species.[10] This has been shown to both reduce nucleation induction time and allow control of the nucleated polymorph. Functional group chemistry can also be used to reduce nucleation induction time and control polymorphism. Self Assembled Monolayers (SAMs) can be used to alter the crystallographic plane nucleated on a surface, as well as to influence the polymorph obtained.[11] Their ability to functionalize a surface with the desired properties, and their highly ordered structure, make self-assembled monolayers attractive templates for nucleation. Amorphous polymer surfaces can also be used as heterogeneous nucleating agents, and can reduce induction time and influence the polymorph obtained. Recent work in our laboratory has demonstrated that polymers imprinted with angular nanopores can further enhance nucleation, through angular matching of the nanopore angles to the angles between the major crystallographic planes of the crystalline material.[12,13]

Heterogeneous nucleation complicates nucleation studies, as care must be taken in developing appropriate and repeatable experimental techniques. In addition, the stochastic nature of the nucleation process requires that a statistically significant number of experiments be performed.

D. Computational studies

In addition to experiments, computational studies can aid in our understanding of the crystal nucleation process. The difficulty in applying molecular simulations to the nucleation process lies in the large time scale and the large number of molecules involved, making the simulations very long and computationally expensive. In addition, models used must be accurate enough to model real systems. A number of groups are attempting to address the problem of time scale by developing novel sampling methods.[14–16] Computational methods can generally be categorized as biased or unbiased. Biased methods use functions which are assumed to be governing parameters of the nucleation process. The difficulty of biased methods is knowing whether the functions used are correct. Unbiased methods sample a system without imposing collective variables or limiting their motion; however, this limits the complexity of the problem that can be studied. Improved methods and faster computing should continue to aid in the development of computational methods for studying crystal nucleation.

E. Conclusion

The study of crystalline nucleation has increased dramatically in recent years. In addition, new experimental methods and improved instrumentation has greatly improved our ability to probe this complex phenomenon. Computational methods are also improving through new methods and faster computers, and can add to our understanding of the nucleation process. The quality of the work presented at this meeting, including excellent posters and robust discussions, make me optimistic about future advances in our understanding of this complex problem. I would like to thank the Scientific Committee and the RSC for their excellent work in organizing and running this meeting.

References

1 D. Erdemir, A. Y. Lee and A. S. Myerson, *Acc. Chem. Res.*, 2009, **42**, 621.
2 P. G. Vekilov, *Cryst. Growth Des.*, 2010, **10**, 5007.
3 R. P. Sear, *Int. Mater. Rev.*, 2012, **57**, 328.
4 B. Garetz, J. Aber, N. Young and A. S. Myerson, *Phys. Rev. Lett.*, 1996, **77**, 3475.
5 B. Garetz, J. Matic and A. S. Myerson, *Phys. Rev. Lett.*, 2002, **89**, 175501.
6 O. Galkin and P. G. Vekilov, *Proc. Natl. Acad. Sci. U. S. A.*, 2000, **97**, 6277.
7 P. G. Vekilov, *Cryst. Growth Des.*, 2004, **4**, 671.
8 D. Gebauer, A. Völkel and H. Cölfen, *Science*, 2008, **322**, 1819.
9 E. Pouget, P. Bomans, J. Goos, P. Frederik, G. de With and N. A. Sommerdijk, *Science*, 2009, **323**, 1455.
10 C. Mitchell, L. Yu and M. D. Ward, *J. Am. Chem. Soc.*, 2001, **123**, 10830.
11 A. Singh, I. Lee and A. S. Myerson, *CrystEngComm*, 2011, **13**, 24.
12 Y. Diao, T. Harada, A. S. Myerson, T. A. Hatton and B. L. Trout, *Nat. Mater.*, 2011, **10**, 867.
13 V. Lopez-Mejias, A. S. Myerson and B. L. Trout, *Cryst. Growth Des.*, 2013, **13**, 3835.
14 A. F. Wallace, *et al.*, *Science*, 2013, **341**, 885.
15 E. E. Santiso and B. L. Trout, *J. Phys. Chem. B*, 2011, **134**, 064109.
16 F. Giberti, G. A. Tribello and M. Parrinello, *J. Chem. Theory Comput.*, 2013, **9**, 2526.

Poster titles

Printing nanoparticle patterns in a polymer: shedding from a ferrofluid droplet, **A. R. Abdel Fattah, S. Ghosh and I. K. Puri**, *Mechanical Engineering Departement, McMaster University, Canada*

Following nanoparticle self-assembly in levitating colloidal droplets by time-resolved SAXS, **M. Agthe, T. Plivelic, A. Labrador, L. Bergström and G. Salazar-Alvarez**, *MMK - Stockholm University, Sweden*

Sophisticated nanopatterned metal networks for transparent flexible electrodes and optoelectronic applications, **E. M. Akinoglu and M. Giersig**, *Freie Universität Berlin, Department of Physics, Germany*

Deformable hydrogen capping layer influences self-assembly of NC superlattices, **M. Boles and D. Talapin**, *University of Chicago, USA*

Hybrid nanoparticle multilayers: development, synthesis and applications, **W. Brullot and T. Verbiest**, *KU Leuven, Belgium*

Synthesis of nanostructured ZnO using precursor-insensitive water-in-oil micro-emulsion, **A. Bumajdad and M. Madkour**, *Kuwait University, Faculty of Science, Kuwait*

Luminescent properties of dysprosium-activated hydroxyapatite nanoparticles, **D. Cervantes, S. Castro, I. Ponce and G. Hirata**, *Universidad Autónoma de Baja California, Mexico*

Multiscale modeling of material self-assembly controlled by external fields, **H. Chan and P. Král**, *University of Illinois at Chicago, USA*

Protein-mediated assembly of plasmonic nanoparticles into modular 3D-clusters, **R. Höller, C. Kuttner, M. Dulle, T. A. F. König, S. Förster, A. Fery and M. Chanana**, *Physical Chemistry II, University of Bayreuth, Germany*

pH-responsive nano-polymeric-micelles based on 6-O-dodecyl-chitosan carbamate for gene delivery, **H. Chen, S. Cui, Y. Zhao, S. Zhang and X. Peng**, *Dalian Nationalities University, China*

Preparation of stable suspensions of magnetic nanoparticle clusters of controlled size by competitive stabilizer desorption method to a liquid interface, **F. El Haddassi, E. K. Fox, J. M. Hierrezuelo and D. Brougham**, *Dublin City University, Ireland*

Structural origins of pink light emission in ultra-small gold nanoparticles, **O. Ersoy, Y. Zhang, A. Karatutlu, N. Tarakina and A. Sapelkin**, *Queen Mary University of London, UK*

Probing the surface of nanocrystals with electrochemistry *in situ*, **I. Fedin and D. Talapin**, *University of Chicago, USA*

Intact flying nanoparticles for gas-phase investigations, **A. Gallego, U. Sezer, M. Arndt and M. Mayor**, *University of Basel, Switzerland*

An anode buffer layer with size-controlled Ag nanoparticles for polymer solar cells with improved efficiencies, **D. Ganeshan, S.-C. Chen, Z. Yin and Q. Zheng**, *Fujian Institute of Research on the Structure of Matter, CAS, China*

Synthesis of novel bio-element geocomposites and their biochemical characterization, **J. Wolak, K. Ghosh, J. Smith, K. Ekstam, A. Wanekaya, N. Monteiro-Riviere, P. Chandran, C. Sorensen, S. Bossmann, S. Aryal, J. Comer, G. Glaspell and R. DeLong**, *Missouri State University, Chemistry, USA*

Formulation and preliminary evaluation of delivery vehicles for the boron neutron capture therapy of cancer, **T. O. B. Olusanya, T. Stich, S. C. Higgins, R. E. Lloyd, G. J. Pilkington, D. G. Fatouros, G. Calabrese, J. R. Smith and J. Tsibouklis**, *University of Portsmouth, UK*

Design of carbon dots, carbon dot clusters and their applications, **J. Gu, D. Hu, X. Jia and K. Xi**, *Nanjing University, China*

Digestive ripening of gold nanorods in aqueous solution, **P. Guimera Coll and C. Sorensen**, *Kansas State University, USA*

Synthesis and assembly of titania nanocubes, **H. Huesmann, P. Schneider and W. Tremel**, *Johannes Gutenberg University Mainz, Germany*

Self-assembly of cyclodextrin-functionalized particles, **D. John, J. A. Krings, B. J. Ravoo and A. Böker**, *Fraunhofer - Institut für Angewandte Polymerforschung, Germany*

Tunable plasmonic lattices, **K. Volk, T. Honold and M. Karg**, *University of Bayreuth, Germany*

Emulsion-assisted assembly of nanoparticles into 3D clusters, **T. Kister, M. Mravlak, T. Schilling and T. Kraus**, *INM - Leibniz Institute for New Materials, Germany*

Photoresponsive self-assembly of anisotropic silica nanoparticles, **J. Krings, D. John, A. Böker and B. J. Ravoo**, *University Muenster, Germany*

Influence of line tension on spherical nanoparticles at liquid-vapor surfaces, **S. McBride, H. S. Wi and B. Law**, *Kansas State University, USA*

Surface patterning through spontaneous disintegration of monolayer protected nanoclusters, **Z. Li, S. Mertens, H. Seyyed Amirhossein and T. Wandlowski**, *Ball State University, USA*

Directed assembly of nanoparticles for deterministic construction of plasmonic heterostructures, **S. Biswas, J. W. Jarrett, K. L. Knappenberger Jr., R. Vaia and P. F. Nealey**, *University of Chicago, USA*

Synthesis of highly monodisperse alkanethiolate protected silver nanoparticles by modified aging process, **Y. Liu and Y. J. Tong**, *Georgetown University, USA*

Solution and substrate assembly of anisotropic gold nanoparticles, **E. Malachosky and P. Guyot-Sionnest**, *University of Chicago, USA*

Synthesising colloidal molecules with externally tuneable interaction sites, **L. K. Månsson, J. N. Immink, A. M. Mihut, J. J. Crassous and P. Schurtenberger**, *Division of Physical Chemistry, Lund University, Sweden*

Quantum dots/chitosan-tripeptide nanoconjugates for cell biolabelling and bio-imaging, **H. Mansur, S. Carvalho and A. Mansur**, *Federal University of Minas Gerais- UFMG, Brazil*

Biopolymer-quantum dots fluorescent immunoconjugates: a novel strategy for cancer theranostics, **H. Mansur and A. Mansur**, *Federal University of Minas Gerais-UFMG, Brazil*

Synchrotron based X-ray absorption studies on the effects of the structure of metal precursor species in nanoparticle formation, **A. Mantalidi, G. Sankar, T. Hyde, E. C. Corbos and P. R. Ellis**, *University College London, UK*

Microwave-assisted synthesis of colloidal semiconductor nanocrystals, **R. Siramdas, M. Yazdanparast, B. Kebede, A. Pfeifer and E. McLaurin**, *Kansas State University, USA*

Insights into mechanisms of capillary assembly, **S. Ni, J. Leeman, A. Omeradzic, M. Thielmann, H. J. Herrmann, H. Wolf and L. Isa**, *ETH Zurich, Switzerland*

Growth of anisotropic gold nanoparticles in photorheological fluid, **R. Pallares, S. Hon Limc, N. Thanh and S. Xiaodic**, *University College London (UCL), UK*

Contact angle and adsorption energies of nanoparticles at the air-liquid interface determined by neutron reflectivity and molecular dynamics, **J. Reguera, E. Ponomarev, T. Geue, F. Bresme, F. Stellacci and M. Moglianetti**, *CIC biomaGUNE, Spain*

Controlled living nanowire growth – precise control of the morphology and optical properties of AgAuAg bimetallic nanowires, **L. Scarabelli, M. Mayer, K. March, T. Altantzis, M. Tebbe, F. García de Abajo, M. Kociak, S. Bals, A. Fery and L. Liz-Marzan**, *CIC biomaGUNE, Spain*

Synthesis of water soluble N-heterocyclic carbene functionalized Pd- and Au-NPs and their application in catalysis, **K. Schaepe, A. Ferry, P. Tegeder, C. Richter, F. Glorius and B. Ravoo**, *Westfälische Wilhelms-Universität Münster, Germany*

Electrochemistry studies of HgTe quantum dot films, **G. Shen and P. Guyot-Sionnest**, *University of Chicago, USA*

When core/shell transforms to nanodumbbell, **S.-G. Kwon, G. Krylova, S. Chattopadhyay, T. Shibata, E. Bunel, V. Prakapenka, B. Lee and E. Shevchenko**, *Argonne National Laboratory, USA*

Microwave synthesis of different size indium phosphide (InP) nanocrystals, **R. Siramdas and E. McLaurin,** *Kansas State University, USA*

Understanding the synthesis of CdSe/CdS dot-in-rod nanocrystals, **J. Souza Junior, P. Cunningham, L. Varanda and D. Talapin,** *University of Sao Paulo, Brazil*

In situ study on the growth and transformation of colloidal noble metal nanocrystals, **S. Peng, Z. Li, Q. Liu, Y. Ren, J. Almer, J. Okasinski and Y. Sun,** *Argonne National Laboratory, USA*

Microspherical hybrid particles with inorganic nanoparticle-assembled shell layer using supercritical carbon dioxides, **M. Takafuji, H. Yamamoto, M. A. Alam, Y. Kuwahara and H. Ihara,** *Kumamoto University, Japan*

Silver-overgrowth of gold nanorods to Au@Ag cuboids: from non-invasive monitoring of growth kinetics to tailored internal mirror charges, **M. Tebbe, C. Kuttner, M. Mayer, M. Männel, P. Probst, T. König and A. Fery,** *University of Bayreuth, Germany*

Synthesis and self-assembly of novel amphiphilic silver nanoparticles capped with polyene antibiotic amphotericin B, **K. Tutaj, R. Szlazak, J. Starzyk, W. Grudzinski, R. Luchowski and W. Gruszecki,** *Maria Curie-Sklodowska University, Poland*

Synthesis of hierarchical metal dendrimers by sequential seeded growth, **R. Weiner and S. Skrabalak,** *Indiana University, USA*

Synthesis and self-assembly of iron oxide nanoparticles, **K. Wichmann, E. Schechtel, M. Panthöfer, B. Barton, V. Ksenofontov and W. Tremel,** *Institute for Inorganic Chemistry and Analytical Chemistry, Germany*

Ligand-mediated interactions between nanoscale surfaces depend sensitively and non-linearly on temperature, facet dimensions, and ligand coverage, **A. Widmer-Cooper and P. Geissler,** *University of Sydney, Australia*

Mechanistic studies of direct dissolution process of silver, **A. Yapa, H. Wang, E. McLaurin, X.-M. Lin, C. Sorensen and S. Bossmann,** *Kansas State University, USA*

Colloidal nanocrystals with inorganic halide, pseudohalide, and halometallate ligands, **H. Zhang, J. Jang, W. Liu, M. Kurley and D. Talapin,** *University of Chicago, USA*

Water based synthesis of gold nanoparticles, **J. Changstrom, P. Guimera-Coll, J. Powell, S. Bossmann, X.-M. Lin, E. McLaurin and C. Sorensen,** *Kansas State University, USA*

Photocurrent from intraband transitions in quantum dots, **Z. Deng and P. Guyot-Sionnest,** *University of Chicago/Department of Chemistry, James Frank Institute, USA*

Temperature-dependent solubility of gold nanoparticle suspensions, **E. Herman, K. Bayliff, J. Powell and C. Sorensen,** *Kansas State University, USA*

Novel gemini cationic lipids with carbamate groups for gene delivery, **Y. Zhao, S. Cui, H. Chen and S. Zhang,** *Dalian Nationalities University, China*

Synthesis, self-assembly and disassembly of coated metal nanoparticles, **M. Martin, A. Allen and J. Havsky,** *Khalifa University, UAE*

Aerosol lithography assisted fabrication of nanoscale silver dots and rings, **J.-H. Byeon,** *Yeungnam University, South Korea*

Self-assembled protein nanotubes triggering gelation, **O. Tarhan, B. Hamaker and O. Campanella,** *Purdue University, USA*

Visible light utilization via micro/mesoporous tungstate titanate composite, **S. Sajjad, S. Ahmed Khan Leghari and J. Zhang,** *International Islamic University, Pakistan*

The Skinner Prize for the best poster was jointly awarded to Daniela John, of Fraunhofer Institut für Angewandte Polymerforschung, Germany, for her poster on Self-assembly of cyclodextrin-functionalized particles, and Igor Fedin, of University of Chicago, USA, for his poster on Probing the surface of nanocrystals with electrochemistry *in situ*.

List of participants

Miss Maria Adobes Vidal, *University of Warwick, United Kingdom*
Dr Jamshed Anwar, *Lancaster University, United Kingdom*
Mrs Raquel Arribas Bueno, *SSPC, Ireland*
Dr Peter Aspin, *GlaxoSmithKline, United Kingdom*
Dr Kevin Back, *Pfizer, United Kingdom*
Dr Angela Bejarano-Villafuerte, *University of Leeds, United Kingdom*
Professor Dr Celso Aparecido Bertran, *State University of Campinas, Brazil*
Professor Dr Beatrice Biscans, *CNRS, France*
Mr James Black, *University of Manchester, United Kingdom*
Dr Simon Black, *AstraZeneca, United Kingdom*
Mr Raitis Bobrovs, *University of Latvia, Latvia*
Mr Samuel Booth, *University of Manchester, United Kingdom*
Dr Clement Brandel, *GSK, United Kingdom*
Dr Eric Breynaert, *KULeuven - Center for Surface Chemistry and Catalysis, Belgium*
Miss Diana Milena Camacho Corzo, *University of Leeds, United Kingdom*
Prof Dr Nadine Candoni, *Centre Interdisciplinaire de Nanoscience de Marseille, France*
Miss Sin Yuen Chang, *The University of Manchester, United Kingdom*
Dr Hugo Christenson, *University of Leeds, United Kingdom*
Professor Dr Gerard Coquerel, *University of Rouen, France*
Dr Aurora J. Cruz-Cabeza, *Hoffmann-La Roche, Switzerland*
Dr Herma Cuppen, *Radboud University Nijmegen, Netherlands*
Professor Dr Helmut Cölfen, *University of Konstanz, Germany*
Professor Roger Davey, *University of Manchester, United Kingdom*
Dr Jim De Yoreo, *Pacific Northwest National Laboratory, USA*
Professor Dr Jürg Diemand, *University of Zurich, Switzerland*
Dr Michael Dietzsch, *BASF Construction Solutions GmbH, Germany*
Dr Sven Dorosz, *University of Luxemburg, Luxembourg*
Ms Petra Dral, *University of Twente, Netherlands*
Dr Laszlo Fabian, *Univeristy of East Anglia, United Kingdom*
Miss Carol Forsyth, *University of Strathclyde, United Kingdom*
Dr Colin Freeman, *University of Sheffield, United Kingdom*
Dr Katharina Fucke, *Durham University, United Kingdom*
Mr Adrian Gainar, *The University of Manchester, United Kingdom*
Dr Denis Gebauer, *University of Konstanz, Germany*
Mr Edward Gent, *University of Birmingham, United Kingdom*
Dr Martí Gich, *Institut de Ciència de Materials de Barcelona-Consejo Superior de Investigaciones Científicas, Spain*
Dr Sara Gilks, *The University of Manchester, United Kingdom*
Dr Romain Grossier, *Centre Interdisciplinaire de Nanoscience de Marseille, France*
Dr Robert Hammond, *University of Leeds, United Kingdom*
Dr Alan Hare, *MRSC, without affiliation (retired), United Kingdom*
Professor Kenneth Harris, *Cardiff University, Wales*

Dr Dritan Hasa, *University of Cambridge*, *United Kingdom*
Dr Mark Holden, *University of Leeds*, *United Kingdom*
Dr Colan Hughes, *Cardiff University*, *United Kingdom*
Professor Dr Jürg Hulliger, *University of Bern*, *Switzerland*
Mr Joshua Humphreys, *University of Nottingham*, *United Kingdom*
Mr Riccardo Innocenti Malini, *The University of Sheffield*, *United Kingdom*
Dr Sophie Janbon, *AstraZeneca*, *United Kingdom*
Professor William Jones, *University of Cambridge*, *United Kingdom*
Dr Franca Jones, *Curtin University*, *Australia*
Mr Mark Joswiak, *University of California, Santa Barbara*, *USA*
Dr Anne Kavanagh, *AstraZeneca*, *United Kingdom*
Dr Matthias Kellermeier, *BASF SE*, *Germany*
Dr Dikshitkumar Khamar, *University of Limerick*, *Ireland*
Dr Yi-Yeoun Kim, *School of Chemistry, University of Leeds*, *United Kingdom*
Dr Alexander Kulak, *School of Chemistry, University of Leeds, United Kingdom*
Dr Xiaojun Lai, *University of Leeds*, *United Kingdom*
Miss Sarah Latham, *Royal Society of Chemistry*, *United Kingdom*
Professor Ken Lewtas, *Lewtas Science & Technologies*, *United Kingdom*
Mr Yuandi Li, *Royal Society of Chemistry*, *United Kingdom*
Dr Maxim Likhatskiy, *ICCT SB RAS*, *Russia*
Miss Jessica Lovelock, *University of Toronto*, *Canada*
Miss Kirsty Lunnon, *Royal Society of Chemistry*, *United Kingdom*
Mr Alasdair Mackenzie, *The University of Edinburgh*, *United Kingdom*
Miss Faduma Maddar, *University of Warwick*, *United Kingdom*
Professor Dr Dominique Maes, *Vrije Universiteit Brussel*, *Belgium*
Mr Giovanni Maria Maggioni, *ETH Zurich*, *Switzerland*
Mr Pablo Martin Soladana, *University of Leeds*, *United Kingdom*
Miss Susannah May, *Royal Society of Chemistry*, *United Kingdom*
Professor Dr Marco Mazzotti, *ETH Zurich*, *Switzerland*
Dr Hugo Meekes, *Radboud University Nijmegen*, *Netherlands*
Dr Paul Meenan, *Pfizer*, *USA*
Professor Fiona Meldrum, *University of Leeds*, *United Kingdom*
Professor Allan Myerson, *Massachusetts Institute of Technology*, *USA*
Professor Dr Igor Neelov, *ITMO University & Russian Academy of Sciences*, *Russian Federation*
Mr Michael H. Nielsen, *UC Berkeley / Lawrence Berkeley National Lab*, *USA*
Dr Fredrik Nordstrom, *AbbVie*, *USA*
Miss Roisin O'Connell, *University of Leeds*, *United Kingdom*
Professor Haihua Pan, *Zhejiang University*, *China*
Professor Baron Peters, *UCSB Chemical Engineering*, *USA*
Prof Dr Wolfgang Peukert, *University Erlangen*, *Germany*
Dr Sendhil Poornachary, *A*STAR Institute of Chemical and Engineering Sciences*, *Singapore*
Dr Flavia Pop, *University of Nottingham*, *United Kingdom*
Professor P Popelier, *MIB, University of Manchester*, *United Kingdom*
Professor Sarah Price, *UCL*, *United Kingdom*
Dr Marc Radu, *Max Planck Institute for Polymer Research*, *Germany*
Professor Ake Rasmuson, *University of Limerick*, *Ireland*
Dr Bart Rimez, *Université Libre de Bruxelles*, *Belgium*
Dr Radoljub Ristic, *University of Sheffield*, *United Kingdom*
Professor Kevin Roberts, *University of Leeds*, *UK*
Mr Ian Rosbottom, *University of Leeds*, *United Kingdom*
Dr Wieslaw Roth, *Jagiellonian University in Kraków*, *Poland*

Dr Matteo Salvalaglio, *ETH Zurich, Switzerland*
Dr Erik Santiso, *North Carolina State University, USA*
Ms Andrea Sauter, *Institut für Angewandte Physik, Universität Tübingen, Germany*
Mrs Johanna Scheck, *University of Konstanz, Germany*
Professor Sven Schroeder, *University of Leeds, United Kingdom*
Dr Andrew Scott, *University of Leeds, United Kingdom*
Dr Richard Sear, *University of Surrey, United Kingdom*
Dr Colin Seaton, *University of Bradford, United Kingdom*
Prof Jan Sefcik, *University of Strathclyde, United Kingdom*
Dr Linda Seton, *LJMU, UK, United Kingdom*
Dr Igor Shvab, *Newcastle University, United Kingdom*
Miss Elena Simone, *Loughborough University, United Kingdom*
Ms Mireille Smets, *Radboud University Nijmegen, Netherlands*
Professor Dr Nico Sommerdijk, *Eindhoven University of Technology, Netherlands*
Dr Joanna Stevens, *The University of Manchester, United Kingdom*
Miss Rachel Sullivan, *University of Manchester, United Kingdom*
Dr Wenhao Sun, *Massachusetts Institute of Technology, USA*
Professor Dr Changquan Calvin Sun, *University of Minnesota, USA*
Miss Xue Tang, *University of Leeds, United Kingdom*
Professor Dr Francis Taulelle, *Katholieke Universiteit Leuven, Belgium*
Professor Joop Ter Horst, *University of Strathclyde, United Kingdom*
Miss Danielle Thomas, *University of Leeds, United Kingdom*
Ms Alpana Thorat, *IIT Gandhinagar, India*
Dr Terence Threlfall, *University of Southampton, United Kingdom*
Mr Pratchaya Tipduangta, *school of pharmacy university of East Anglia, United Kingdom*
Dr Dimitrios Toroz, *University of Leeds, United Kingdom*
Mr Thomas Turner, *University of Leeds, United Kingdom*
Mr Joost van den Ende, *Radboud University, Nijmegen Institute for Molecules and Materials, Netherlands*
Dr Stéphane Veesler, *CINaM-CNRS Aix-Marseille Université, France*
Professor Peter Vekilov, *University of Houston, USA*
Dr Thomas Vetter, *University of Manchester, United Kingdom*
Ms Cornelia Völkle, *Universität Konstanz, Germany*
Dr Martin Ward, *University of Edinburgh, United Kingdom*
Dr Helen Wheatcroft, *AstraZeneca, United Kingdom*
Mr Robert Willacy, *GlaxoSmithKline, United Kingdom*
Mr Stefan Wolf, *Universtity of Konstanz, Germany*
Dr Rachel Wood, *Royal Society of Chemistry, United Kingdom*
Mr Grahame Woollam, *University of Manchester, United Kingdom*
Professor Chi-Man Wu, *City University of Hong Kong, Hong Kong*
Professor David Wu, *Colorado School of Mines, USA*
Dr Jacek Zeglinski, *University of Limerick, Ireland*
Dr Fajun Zhang, *University of Tuebingen, Germany*
Miss Shuheng Zhang, *Centre Interdisciplinaire de Nanoscience de Marseille, France*
Miss Boyang Zou, *University of Leeds, United Kingdom*